RANDOM MATRICES

Third Edition

This is volume 142 in the PURE AND APPLIED MATHEMATICS series
Founding Editors: Paul A. Smith and Samuel Eilenberg

RANDOM MATRICES

Third Edition

Madan Lal Mehta

Saclay, Gif-sur-Yvette, France

Amsterdam Boston Heidelberg London New York Oxford
Paris San Diego San Francisco Singapore Sydney Tokyo

ELSEVIER B.V.
Sara Burgerhartstraat 25
P.O. Box 211, 1000 AE Amsterdam,
The Netherlands

ELSEVIER Inc.
525 B Street, Suite 1900
San Diego, CA 92101-4495, USA

ELSEVIER Ltd
The Boulevard, Langford Lane
Kidlington, Oxford OX5 1GB, UK

ELSEVIER Ltd
84 Theobalds Road
London WC1X 8RR UK

First edition 2004

Library of Congress Cataloging in Publication Data
A catalog record is available from the Library of Congress.

British Library Cataloguing in Publication Data
A catalogue record is available from the British Library.

Series ISSN 0079-8169
ISBN-13: 978-0-12-088409-4
ISBN-10: 0-12-088409-7

Printed in the Netherlands.

CONTENTS

PREFACE TO THE THIRD EDITION

In the last decade following the publication of the second edition of this book the subject of random matrices found applications in many new fields of knowledge. In heterogeneous conductors (mesoscopic systems) where the passage of electric current may be studied by transfer matrices, quantum chromo dynamics characterized by some Dirac operator, quantum gravity modeled by some random triangulation of surfaces, traffic and communication networks, zeta function and L-series in number theory, even stock movements in financial markets, wherever imprecise matrices occurred, people dreamed of random matrices.

Some new analytical results were also added to the random matrix theory. The noteworthy of them being, the awareness that certain Fredholm determinants satisfy second order nonlinear differential equations, power series expansion of spacing functions, a compact expression (one single determinant) of the general correlation function for the case of hermitian matrices coupled in a chain, probability densities of random determinants, and relation to random permutations. Consequently, a revision of this book was felt necessary, though in the mean time four new books (Girko, 1990; Effetof, 1997; Katz and Sarnak, 1999; Deift, 2000), two long review articles (di Francesco et al., 1995; Guhr et al., 1998) and a special issue of J. Phys. A (2003) have appeared. The subject matter of them is either complimentary or disjoint. Apart from them the introductory article by C.E. Porter in his 1965 collection of reprints remains instructive even today.

In this new edition most chapters remain almost unaltered though some of them change places. Chapter 5 is new explaining the basic tricks of the trade, how to deal with integrals containing the product of differences $\prod |x_i - x_j|$ raised to the power 1, 2 or 4. Old Chapters 5 to 11 shift by one place to become Chapters 6 to 12, while Chapter 12 becomes 18. In Chapter 15 two new sections dealing with real random matrices and the probability density of determinants are added. Chapters 20 to 27 are new. Among the appendices some have changed places or were regrouped, while 16, 37, 38 and 42 to 54 are new. One major and some minor errors have been corrected. It is really surprising how such a major error could have creeped in and escaped detection by so many experts reading it. (Cf. lines 2, 3 after Eq. (1.8.15) and line 6 after Eq. (1.8.16); $h(d)$ is not the number of different quadratic forms as presented, but is the number of different primitive inequivalent quadratic forms.) Not to hinder the fluidity of reading the original source of the material presented is rarely indicated in the text. This is done in the "notes" at the end.

While preparing this new edition I remembered the comment of B. Suderland that from the presentation point of view he preferred the first edition rather than the second. As usual, I had free access to the published and unpublished works of my teachers, colleagues and friends F.J. Dyson, M. Gaudin, H. Widom, C.A. Tracy, A.M. Odlyzko, B. Poonen, H.S. Wilf, A. Edelman, B. Dietz, S. Ghosh, B. Eynard, R.A. Askey and many others. G. Mahoux kindly wrote Appendix A.16. M. Gingold helped me in locating some references and L. Bervas taught me how to use a computer to incorporate a figure as a .ps file in the TEX files of the text. G. Cicuta, O. Bohigas, B. Dietz, M. Gaudin, S. Ghosh, P.B. Kahn, G. Mahoux, J.-M. Normand, N.C. Snaith, P. Sarnak, H. Widom and R. Conte read portions of the manuscript and made critical comments thus helping me to avoid errors, inaccuracies and even some blunders. O. Bohigas kindly supplied me with a list of minor errors of references in the figures of Chapter 16. It is my pleasant duty to thank all of them. However, the responsibility of any remaining errors is entirely mine. Hopefully this new edition is free of serious errors and it is self-contained to be accessible to any diligent reader.

February, 2004
Saclay, France

Madan Lal MEHTA

PREFACE TO THE SECOND EDITION

The contemporary textbooks on classical or quantum mechanics deal with systems governed by differential equations which are simple enough to be solved in closed terms (or eventually perturbatively). Hence the entire past and future of such systems can be deduced from a knowledge of their present state (initial conditions). Moreover, these solutions are stable in the sense that small changes in the initial conditions result in small changes in their time evolution. Such systems are called integrable. Physicists and mathematicians now realize that most of the systems in nature are not integrable. The forces and interactions are so complicated that either we can not write the corresponding differential equation, or when we can, the whole situation is unstable; a small change in the initial conditions produces a large difference in the final outcome. They are called chaotic. The relation of chaotic to integrable systems is something like that of transcendental to rational numbers.

For chaotic systems it is meaningless to calculate the future evolution starting from an exactly given present state, because a small error or change at the beginning will make the whole computation useless. One should rather try to determine the statistical properties of such systems.

The theory of random matrices makes the hypothesis that the characteristic energies of chaotic systems behave locally as if they were the eigenvalues of a matrix with randomly distributed elements. Random matrices were first encountered in mathematical statistics by Hsu, Wishart and others in the 1930s, but an intensive study of their properties in connection with nuclear physics began only with the work of Wigner in the 1950s. In 1965 C.E. Porter edited a reprint volume of all important papers on the subject, with a critical and detailed introduction which even today is very instructive. The first edition of the present book appeared in 1967. During the last two decades many new results have been discovered, and a larger number of physicists and mathematicians got interested in the subject owing to various possible applications. Consequently it was felt that this book has to be revised even though a nice review article by Brody et al. has appeared in the mean time (*Rev. Mod. Phys.*, 1981).

Among the important new results one notes the theory of matrices with quaternion elements which serves to compute some multiple integrals, the evaluation of n-point spacing probabilities, the derivation of the asymptotic behaviour of nearest neighbor spacings, the computation of a few hundred millions of zeros of the Riemann zeta function and the analysis of their statistical properties, the rediscovery of Selberg's 1944 paper giving rise to hundreds of recent publications, the use of the diffusion equation to evaluate an integral over the unitary group thus allowing the analysis of non-invariant Gaussian ensembles and the numerical investigation of various systems with deterministic chaos.

After a brief survey of the symmetry requirements the Gaussian ensembles of random Hermitian matrices are introduced in Chapter 2. In Chapter 3 the joint probability density of the eigenvalues of such matrices is derived. In Chapter 5 we give a detailed treatment of the simplest of the matrix ensembles, the Gaussian unitary one, deriving the n-point correlation functions and the n-point spacing probabilities. Here we explain how the Fredholm theory of integral equations can be used to derive the limits of large determinants. In Chapter 6 we study the Gaussian orthogonal ensemble which in most cases is appropriate for applications but is mathematically more complicated. Here we introduce matrices with quaternion elements and their determinants as well as the method of integration over alternate variables. The short Chapter 8 introduces a Brownian motion model of Gaussian Hermitian matrices. Chapters 9, 10 and 11 deal with ensembles of unitary random matrices, the mathematical methods being the same as in Chapters 5 and 6. In Chapter 12 we derive the asymptotic series for the nearest neighbor spacing probability. In Chapter 14 we study a non-invariant Gaussian Hermitian ensemble, deriving its n-point correlation and cluster functions; it is a good example of the use of mathematical tools developed in Chapters 5 and 6. Chapter 16 describes a number of statistical quantities useful for the analysis of experimental data. Chapter 17 gives a detailed account of Selberg's integral and of its consequences. Other chapters deal with questions or ensembles less important either for applications or for the mathematical methods used. Numerous appendices treat secondary mathematical questions, list power series expansions and numerical tables of various functions useful in applications.

The methods explained in Chapters 5 and 6 are basic, they are necessary to understand most of the material presented here. However, Chapter 17 is independent. Chapter 12 is the most difficult one, since it uses results from the asymptotic analysis of differential equations, Toeplitz determinants and the inverse scattering theory, for which in spite of a few nice references we are unaware of a royal road. The rest of the material is self-contained and hopefully quite accessible to any diligent reader with modest mathematical background.

Contrary to the general tendency these days, this book contains no exercises.

October, 1990 M.L. MEHTA
Saclay, France

PREFACE TO THE FIRST EDITION

Though random matrices were first encountered in mathematical statistics by Hsu, Wishart, and others, intensive study of their properties in connection with nuclear physics began with the work of Wigner in the 1950s. Much material has accumulated since then, and it was felt that it should be collected. A reprint volume to satisfy this need had been edited by C.E. Porter with a critical introduction (see References); nevertheless, the feeling was that a book containing a coherent treatment of the subject would be welcome.

We make the assumption that the local statistical behavior of the energy levels of a sufficiently complicated system is simulated by that of the eigenvalues of a random matrix. Chapter 1 is a rapid survey of our understanding of nuclear spectra from this point of view. The discussion is rather general, in sharp contrast to the precise problems treated in the rest of the book. In Chapter 2 an analysis of the usual symmetries that quantum system might possess is carried out, and the joint probability density function for the various matrix elements of the Hamiltonian is derived as a consequence. The transition from matrix elements to eigenvalues is made in Chapter 3, and the standard arguments of classical statistical mechanics are applied in Chapter 4 to derive the eigenvalue density. An unproven conjecture is also stated. In Chapter 5 the method of integration over alternate variables is presented, and an application of the Fredholm theory of integral equations is made to the problem of eigenvalue spacings. The methods developed in Chapter 5 are basic to an understanding of most of the remaining chapters. Chapter 6 deals with the correlations and spacings for less useful cases. A Brownian motion model is described in Chapter 7. Chapters 8 to 11 treat circular ensembles; Chapters 8 to 10 repeat calculations analogous to those of Chapter 4 to 7. The integration method discussed in Chapter 11 originated with Wigner and is being published here for the first time. The theory of non-Hermitian random matrices, though not applicable to any physical problems, is a fascinating subject and must be studied for its own sake. In this direction an impressive effort by Ginibre is described in Chapter 12. For the Gaussian ensembles the level density in regions where it is very low is discussed in Chapter 13. The investigations of Chapter 16 and Appendices A.29 and A.30 were recently carried out in collaboration with Professor Wigner at Princeton University. Chapters 14, 15, and 17 treat a number of other topics. Most of the material in the appendices is either well known or was published elsewhere and is collected here for ready reference. It was surprisingly difficult to obtain the proof contained in Appendix A.21, while Appendices A.29, A.30 and A.31 are new.

October, 1967 M.L. MEHTA
Saclay, France

1

INTRODUCTION

Random matrices first appeared in mathematical statistics in the 1930s but did not attract much attention at the time. In the theory of random matrices one is concerned with the following question. Consider a large matrix whose elements are random variables with given probability laws. Then what can one say about the probabilities of a few of its eigenvalues or of a few of its eigenvectors? This question is of pertinence for the understanding of the statistical behaviour of slow neutron resonances in nuclear physics, where it was proposed in the 1950s and intensively studied by the physicists. Later the question gained importance in other areas of physics and mathematics, such as the characterization of chaotic systems, elastodynamic properties of structural materials, conductivity in disordered metals, the distribution of the values of the Riemann zeta function on the critical line, enumeration of permutations having certain particularities, counting of certain knots and links, quantum gravity, quantum chromo dynamics, string theory, and others (cf. *J. Phys. A* **36** (2003), special issue: random matrices). The reasons of this pertinence are not yet clear. The impression is that some sort of a law of large numbers is in the back ground. In this chapter we will try to give reasons why one should study random matrices.

1.1 Random Matrices in Nuclear Physics

Figure 1.1 shows a typical graph of slow neutron resonances. There one sees various peaks with different widths and heights located at various places. Do they have any definite statistical pattern? The locations of the peaks are called nuclear energy levels, their widths the neutron widths and their heights are called the transition strengths.

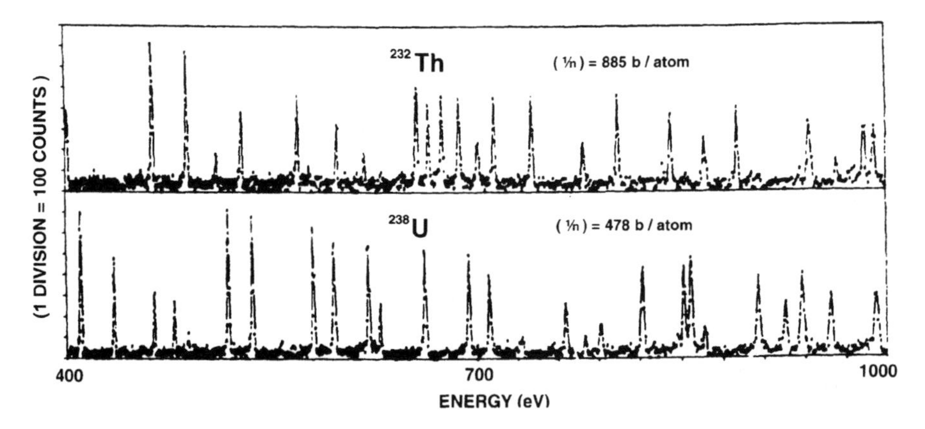

Figure 1.1. Slow neutron resonance cross-sections on thorium 232 and uranium 238 nuclei. Reprinted with permission from The American Physical Society, Rahn et al., Neutron resonance spectroscopy, X, *Phys. Rev. C* 6, 1854–1869 (1972).

The experimental nuclear physicists have collected vast amounts of data concerning the excitation spectra of various nuclei such as shown on Figure 1.1 (Garg et al., 1964, where a detailed description of the experimental work on thorium and uranium energy levels is given; (Rosen et al., 1960; Camarda et al., 1973; Liou et al., 1972b). The ground state and low lying excited states have been impressively explained in terms of an independent particle model where the nucleons are supposed to move freely in an average potential well (Mayer and Jensen, 1955; Kisslinger and Sorenson, 1960). As the excitation energy increases, more and more nucleons are thrown out of the main body of the nucleus, and the approximation of replacing their complicated interactions with an average potential becomes more and more inaccurate. At still higher excitations the nuclear states are so dense and the intermixing is so strong that it is a hopeless task to try to explain the individual states; but when the complications increase beyond a certain point the situation becomes hopeful again, for we are no longer required to explain the characteristics of every individual state but only their average properties, which is much simpler.

The statistical behaviour of the various energy levels is of prime importance in the study of nuclear reactions. In fact, nuclear reactions may be put into two major classes—fast and slow. In the first case a typical reaction time is of the order of the time taken by the incident nucleon to pass through the nucleus. The wavelength of the incident nucleon is much smaller than the nuclear dimensions, and the time it spends inside the nucleus is so short that it interacts with only a few nucleons inside the nucleus. A typical example is the head-on collision with one nucleon in which the incident nucleon hits and ejects a nucleon, thus giving it almost all its momentum and energy. Consequently in such cases the coherence and the interference effects between incoming and outgoing nucleons is strong.

Another extreme is provided by the slow reactions in which the typical reaction times are two or three orders of magnitude larger. The incident nucleon is trapped and all its energy and momentum are quickly distributed among the various constituents of the target nucleus. It takes a long time before enough energy is again concentrated on a single nucleon to eject it. The compound nucleus lives long enough to forget the manner of its formation, and the subsequent decay is therefore independent of the way in which it was formed.

In the slow reactions, unless the energy of the incident neutron is very sharply defined, a large number of neighboring energy levels of the compound nucleus are involved, hence the importance of an investigation of their average properties, such as the distribution of neutron and radiation widths, level spacings, and fission widths. It is natural that such phenomena, which result from complicated many body interactions, will give rise to statistical theories. We shall concentrate mainly on the average properties of nuclear levels such as level spacings.

According to quantum mechanics, the energy levels of a system are supposed to be described by the eigenvalues of a Hermitian operator H, called the Hamiltonian. The energy level scheme of a system consists in general of a continuum and a certain, perhaps a large, number of discrete levels. The Hamiltonian of the system should have the same eigenvalue structure and therefore must operate in an infinite-dimensional Hilbert space. To avoid the difficulty of working with an infinite-dimensional Hilbert space, we make approximations amounting to a truncation keeping only the part of the Hilbert space that is relevant to the problem at hand and either forgetting about the rest or taking its effect in an approximate manner on the part considered. Because we are interested in the discrete part of the energy level schemes of various quantum systems, we approximate the true Hilbert space by one having a finite, though large, number of dimensions. Choosing a basis in this space, we represent our Hamiltonians by finite dimensional matrices. If we can solve the eigenvalue equation

$$H\Psi_i = E_i\Psi_i,$$

we shall get all the eigenvalues and eigenfunctions of the system, and any physical information can then be deduced, in principle, from this knowledge. In the case of the nucleus, however, there are two difficulties. First, we do not know the Hamiltonian and, second, even if we did, it would be far too complicated to attempt to solve the corresponding equation.

Therefore from the very beginning we shall be making statistical hypotheses on H, compatible with the general symmetry properties. Choosing a complete set of functions as basis, we represent the Hamiltonian operators H as matrices. The elements of these matrices are random variables whose distributions are restricted only by the general symmetry properties we might impose on the ensemble of operators. And the problem is to get information on the behaviour of its eigenvalues. "The statistical theory will not predict the detailed level sequence of any one nucleus, but it will describe the general appearance and the degree of irregularity of the level structure that is expected to occur in any nucleus which is too complicated to be understood in detail" (Dyson, 1962a).

In classical statistical mechanics a system may be in any of the many possible states, but one does not ask in which particular state a system is. Here we shall renounce knowledge of the nature of the system itself. "We picture a complex nucleus as a black box in which a large number of particles are interacting according to unknown laws. As in orthodox statistical mechanics we shall consider an ensemble of Hamiltonians, each of which could describe a different nucleus. There is a strong logical expectation, though no rigorous mathematical proof, that an ensemble average will correctly describe the behaviour of one particular system which is under observation. The expectation is strong, because the system might be one of a huge variety of systems, and very few of them will deviate much from a properly chosen ensemble average. On the other hand, our assumption that the ensemble average correctly describes a particular system, say the U^{239} nucleus, is not compelling. In fact, if this particular nucleus turns out to be far removed from the ensemble average, it will show that the U^{239} Hamiltonian possesses specific properties of which we are not aware. This, then will prompt one to try to discover the nature and the origin of these properties" (Dyson, 1962b).

Wigner was the first to propose in this connection the hypothesis alluded to, namely that the local statistical behaviour of levels in a simple sequence is identical with the eigenvalues of a random matrix. A simple sequence is one whose levels all have the same spin, parity, and other strictly conserved quantities, if any, which result from the symmetry of the system. The corresponding symmetry requirements are to be imposed on the random matrix. There being no other restriction on the matrix, its elements are taken to be random with, say, a Gaussian distribution. Porter and Rosenzweig (1960a) were the early workers in the field who analyzed the nuclear experimental data made available by Harvey and Hughes (1958), Rosen et al. (1960) and the atomic data compiled by Moore (1949, 1958). They found that the occurrence of two levels close to each other in a simple sequence is a rare event. They also used the computer to generate and diagonalize a large number of random matrices. This Monte Carlo analysis indicated the correctness of Wigner's hypothesis. In fact it indicated more; the density and the spacing distribution of eigenvalues of real symmetric matrices are independent of many details of the distribution of individual matrix elements. From a group theoretical analysis Dyson found that an irreducible ensemble of matrices, invariant under a symmetry group G, necessarily belongs to one of three classes, named by him orthogonal, unitary and symplectic. We shall not go into these elegant group theoretical arguments but shall devote enough space to the study of the circular ensembles introduced by Dyson. As we will see, Gaussian ensembles are equivalent to the circular ensembles for large orders. In other words, when the order of the matrices goes to infinity, the limiting correlations extending to any finite number of eigenvalues are identical in the two cases. The spacing distribution, which depends on an infinite number of correlation functions, is also identical in the two cases. It is remarkable that standard thermodynamics can be applied to obtain certain results which otherwise require long and difficult analysis to derive. A theory of Brownian motion of matrix elements has also been created by Dyson (1962b) thus rederiving a few known results.

Various numerical Monte Carlo studies indicate, as Porter and Rosenzweig (1960a) noted earlier, that a few level correlations of the eigenvalues depend only on the overall symmetry requirements that a matrix should satisfy and they are independent of all other details of the distribution of individual matrix elements. The matrix has to be Hermitian to have real eigenvalues, the diagonal elements should have the same distribution and the off-diagonal elements should be distributed symmetrically about the zero mean and the same mean square deviation for all independent parts entering in their definition. What is then decisive is whether the matrix is symmetric or self-dual or something else or none of these. In the limit of large orders other details are not seen. Similarly, in the circular ensembles, the matrices are taken to be unitary to have the eigenvalues on the circumference of the unit circle, what counts then is whether they are symmetric or self-dual or none of these. Other details are washed out in the limit of large matrices.

This independence is expected; but apart from the impressive numerical evidence, some heuristic arguments of Wigner and the equivalence of Gaussian and circular ensembles, no rigorous derivation of this fact has yet been found. Its generality seems something like that of the central limit theorem.

1.2 Random Matrices in Other Branches of Knowledge

The physical properties of metals depend characteristically on their excitation spectra. In bulk metal at high temperatures the electronic energy levels lie very near to one another and are broad enough to overlap and form a continuous spectrum. As the sample gets smaller, this spectrum becomes discrete, and as the temperature decreases the widths of the individual levels decrease. If the metallic particles are minute enough and at low enough temperatures, the spacings of the electronic energy levels may eventually become larger than the other energies, such as the level widths and the thermal energy, kT. Under such conditions the thermal and the electromagnetic properties of the fine metallic particles may deviate considerably from those of the bulk metal. This circumstance has already been noted by Fröhlich (1937) and proposed by him as a test of the quantum mechanics. Because it is difficult to control the shapes of such small particles while they are being experimentally produced, the electronic energy levels are seen to be random and the theory for the eigenvalues of the random matrices may be useful in their study.

In the mathematical literature the Riemann zeta function is quite important. It is suspected that all its non-real zeros lie on a line parallel to the imaginary axis; it is also suspected (Montgomery, 1974) that the local fluctuation properties of these zeros on this line are identical to those of the eigenvalues of matrices from a unitary ensemble.

Random matrices are also encountered in other branches of physics. For example, glass may be considered as a collection of random nets, that is, a collection of particles with random masses exerting random mutual forces, and it is of interest to determine the distribution of frequencies of such nets (Dyson, 1953). The most studied model of glass is the so called random Ising model or spin glass. On each site of a 2- or 3-dimensional regular lattice a spin variable σ_i is supposed to take values $+1$ or -1, each with a probability equal to $1/2$. The interaction between neighboring spins σ_i and σ_j is $J_{ij}\sigma_i\sigma_j$, and that between any other pair of spins is zero. If $J_{ij} = J$ is fixed, we have the Ising model, whose partition function was first calculated by Onsager (cf. McCoy and Wu, 1973). If J_{ij} is a random variable, with a symmetric distribution around zero mean, we have the random Ising model, the calculation of the partition function of which is still an open problem.

A problem much studied during the 1980s is that of characterizing a chaotic system. Classically, a system is chaotic if small differences in the initial conditions result in large differences in the final outcome. A polygonal billiard table with incommensurate sides and having a circular hole inside is such an example; two billiard balls starting on nearby paths have diverging trajectories. According to their increasing chaoticity, systems are termed classically as ergodic, mixing, a K-system or a Bernoulli shift. Quantum mechanically, one may consider, for example, a free particle confined to a finite part of the space (billiard table). Its possible energies are discrete; they are the eigenvalues of the Laplace operator in the specified finite space. Given the sequence of these discrete energy values, what can one say about its chaoticity; whether it is ergodic or mixing or ...; i.e. is there any correspondence with the classical notions? A huge amount of numerical evidence tells us that these energies behave as if they were the eigenvalues of a random matrix taken from the Gaussian orthogonal ensemble (GOE). Finer recognition of the chaoticity has not yet been possible.

Weaver (1989) found that the ultrasonic resonance frequencies of structural materials, such as aluminium blocks, do behave as the eigenvalues of a matrix from the GOE. This is expected of the vibrations of complex structures at frequencies well above the frequencies of their lowest modes. Thus random matrix theory may be of relevance in the non-destructive evaluation of materials, architectural acoustics and the decay of ultrasound in heterogeneous materials.

Le Caër (1989) considered the distribution of trees in the Scandinavian forests and found that their positions look like the eigenvalues of a random complex matrix while the distribution of capital cities of districts of mainland France (Le Caër and Delannay, 1993) looks statistically different.

The series of nuclear energy levels, or any sequence of random numbers for that matter, can be thought to have two distinct kinds of statistical properties, which may be called global and local. A global property varies slowly, its changes being appreciable only on a large interval. The average number of levels per unit of energy or the mean level density, for example, is a global property; it changes appreciably only on intervals containing thousands of levels. Locally it may be treated as a constant. A local property on the other hand fluctuates from level to level. The distance between two successive levels, for example, is a local property. Moreover, global and local properties of a complex system seem to be quite disconnected. Two systems having completely different global properties may have identical local fluctuation properties, and inversely two systems having the same global properties may have different local fluctuation properties.

The random matrix models studied here will have quite different global properties, none of them corresponding exactly to the nuclear energy levels or to the sequence of zeros on the critical line of the Riemann zeta function. One may even choose the global properties at will (Balian, 1968)! However, the nice and astonishing thing about them is that their local fluctuation properties are always the same and determined only by the over-all symmetries of the system. From the extended numerical experience (cf. Appendix A.1) one might state the kind of central limit theorem, referred to earlier, as follows.

Conjecture 1.2.1. Let H be an $N \times N$ real symmetric matrix, its off-diagonal elements H_{ij}, for $i < j$, being independent identically distributed (i.i.d.) random variables with mean zero and variance $\sigma > 0$, i.e. $\langle H_{ij} \rangle = 0$, and $\langle H_{ij}^2 \rangle = \sigma^2 \neq 0$. Then in the limit of large N the statistical properties of n eigenvalues of H become independent of the probability density of the H_{ij}, i.e. when $N \to \infty$, the joint probability density (j.p.d.) of arbitrarily chosen n eigenvalues of H tends, for every finite n, with probability one, to the n-point correlation function of the Gaussian orthogonal ensemble studied in Chapter 7.

Note that the off-diagonal elements H_{ij} for $i > j$ are determined from symmetry, and the diagonal elements H_{ii} may have any distributions.

For Hermitian matrices we suspect the following

Conjecture 1.2.2. Let H be an $N \times N$ Hermitian matrix with complex numbers as elements. Let the real parts of H_{ij} for $i < j$ be i.i.d. random variables with mean zero and variance $\sigma_1 > 0$, while let the imaginary parts of H_{ij} for $i < j$ be i.i.d. random variables with mean zero and variance $\sigma_2 > 0$. Then as the order $N \to \infty$, the j.p.d. of n arbitrarily chosen eigenvalues of H, for every finite n, tends with probability one to the n-point correlation function of the Gaussian unitary ensemble studied in Chapter 6.

Note that, as for real symmetric matrices, the elements H_{ij} for $i > j$ are determined from Hermiticity and the distributions of the diagonal elements H_{ii} seem irrelevant.

A similar result is suspected for self-dual Hermitian quaternion matrices; all finite correlations being identical to those for the Gaussian symplectic ensemble studied in Chapters 8 and 11.

In other words, the local statistical properties of a few eigenvalues of a large random matrix seem to be independent of the distributions of individual matrix elements. What matters is, whether the matrix is real symmetric, or self-dual (quaternion) Hermitian or only Hermitian. The rest does not seem to matter. And this seems to be true even under less restrictive conditions; for example, the probability law for different matrix elements (their real and imaginary parts) may be different.

For level density, i.e. the case $n = 1$, we have some arguments to say that it follows Wigner's "semi-circle" law (cf. Chapter 4). Except for this special case we have only numerical evidence in favor of the above conjectures.

Among the Hermitian matrices, the case of the Gaussian distributions of matrix elements is the one treated analytically by Hsu, Selberg, Wigner, Mehta, Gaudin, Dyson, Rosenzweig, Bronk, Ginibre, Pandey, des Cloizeaux, and others. Circular ensembles of unitary matrices have similarly been studied. We will describe these developments in great detail in the following pages.

1.3 A Summary of Statistical Facts about Nuclear Energy Levels

1.3.1 Level Density. As the excitation energy increases, the nuclear energy levels occur on the average at smaller and smaller intervals. In other words, level density increases with the excitation energy. The first question we might ask is how fast does this level density increase for a particular nucleus and what is the distribution of these levels with respect to spin and parity? This is an old problem treated by Bethe (1937). Even a simple model in which the nucleus is taken as a degenerate Fermi gas with equidistant single-particle levels gives an adequate result. It amounts to determining the number of partitions $\lambda(n)$ of a positive integer n into smaller positive integers $\nu_1, \nu_2, \ldots$

$$n = \nu_1 + \nu_2 + \cdots + \nu_\ell, \qquad \nu_1 \geqslant \nu_2 \geqslant \cdots \geqslant \nu_\ell > 0.$$

For large n this number, according to the Hardy–Ramanujan formula, is given by

$$\lambda(n) \sim \exp\left[(\theta\pi^2 n/3)^{1/2}\right],$$

where θ is equal to 1 or 2 according to whether the ν_i are all different or whether some of them are allowed to be equal. With a slight modification due to later work (Lang and Lecouteur, 1954; Cameron, 1956). Bethe's result gives the level density as

$$\rho(E, j, \pi) \propto (2j+1)(E-\Delta)^{-5/4}\exp\left[-j(j+1)/2\sigma^2\right]\exp\left[2a(E-\Delta)^{1/2}\right],$$

where E is the excitation energy, j is the spin and π is the parity. The dependence of the parameters σ, a and Δ on the neutron and proton numbers is complicated and only imperfectly understood. However, for any particular nucleus a few measurements will suffice to determine them all; the formula will then remain valid for a wide range of energy that contains thousands and even millions of levels.

1.3.2 Distribution of Neutron Widths. An excited level may decay in many ways; for example, by neutron ejection or by giving out a quantum of radiation. These processes are characterized by the corresponding decay widths of the levels. The neutron reduced widths $\Gamma_n^0 = \Gamma_n / E^{1/2}$, in which Γ_n is the neutron width and E is the excitation energy of the level, show large fluctuations from level to level. From an analysis of the available data Scott (1954) and later Porter and Thomas (1956) concluded that they had a χ^2-distribution with $\nu = 1$ degree of freedom:

$$\begin{aligned} P(x) &= (\nu/2)\left[\Gamma(\nu/2)\right]^{-1}(\nu x/2)^{(\nu/2)-1}e^{-\nu x/2} \\ &= (2\pi x)^{-1/2}e^{-x/2}, \\ x &= \Gamma_n^0/\bar{\Gamma}_n^0, \end{aligned}$$

where $\bar{\Gamma}_n^0$ is the average of Γ_n^0 and $P(x)$ is the probability that a certain reduced width will lie in an interval dx around the value x. This indicates a Gaussian distribution for the reduced width amplitude

$$\left(\frac{2}{\pi}\right)^{1/2} \exp\left[-\frac{1}{2}(\sqrt{x})^2\right] d(\sqrt{x})$$

expected from the theory. In fact, the reduced width amplitude is proportional to the integral of the product of the compound nucleus wave function and the wave function in the neutron-decay channel over the channel surface. If the contributions from the various parts of the channel surface are supposed to be random and mutually independent, their sum will have a Gaussian distribution with zero mean.

1.3.3 Radiation and Fission Widths. The total radiation width is almost a constant for particular spin states of a particular nucleus. The total radiation width is the sum of partial radiation widths

$$\Gamma = \sum_{i=1}^{m} \Gamma_i.$$

If we assume that each of these $\Gamma_i/\bar{\Gamma}_i$ ($\bar{\Gamma}_i$ denoting the average of Γ_i), has a χ^2-distribution with one degree of freedom like the neutron widths and all the $\bar{\Gamma}_i$ are the same, then $\Gamma/\bar{\Gamma}$ ($\bar{\Gamma}$ being the average of Γ), will have a χ^2-distribution with m degrees of freedom. For (moderately) large m, this is a narrow distribution. This conclusion remains valid even when the $\bar{\Gamma}_i$ are not all equal.

It is difficult to measure the partial radiation widths.

Little is known about the fission-width distributions. Some known fission widths of U^{235} have been analyzed (Bohr, 1956) and a χ^2-distribution with 2 to 3 degrees of freedom has been found to give a satisfactory fit.

From now on we shall no longer consider neutron, radiation, or fission widths.

1.3.4 Level Spacings. Let us regard the level density as a function of the excitation energy as known and consider an interval of energy δE centered at E. This interval is much smaller compared to E, whereas it is large enough to contain many levels; that is,

$$E \gg \delta E \gg D,$$

where D is the mean distance between neighboring levels. How are the levels distributed in this interval? On Figure 1.2 a few examples of level series are shown. In all these cases the level density is taken to be the same, i.e. the scale in each case is chosen so that the average distance between the neighboring levels is unity. It is evident that these different level sequences do not look similar. There are many coincident pairs or sometimes even triples of levels as well as large gaps when the levels have no correlations, i.e. the Poisson distribution; whereas the zeros of the Riemann zeta function are more or less equally spaced. The case of prime numbers, the slow neutron resonance levels of erbium 166 nucleus and the possible energy values of a free particle confined to a billiard table of a specified shape are far from either regularly spaced uniform series or the completely random Poisson series with no correlations.

Although the level density varies strongly from nucleus to nucleus, the fluctuations in the precise positions of the levels seem not to depend on the nucleus and not even on the excitation energy. As the density of the levels is nearly constant in this interval, we might think that they occur at random positions without regard to one another, the only condition being that their density be a given constant. However, as we see on Figure 1.2(c), such is not the case. It is true that nuclear levels with different spin and parity or atomic levels with different sets of good quantum numbers seem to have no influence on each other. However, levels with the same set of good quantum numbers show a large degree of regularity. For instance, they rarely occur close together.

A more detailed discussion of the experimental data regarding the above quantities as well as the strength functions may be found in the review article by Brody et al. (1981).

1.4 Definition of a Suitable Function for the Study of Level Correlations

To study the statistical properties of the sequence of eigenvalues one defines suitable functions such as level spacings or correlation and cluster functions, or some suitable quantity called a statistic such as the Δ-, Q-, F- or the Λ-statistic of the literature. We will consider a few of them in due course. For the spacings let $E_1, E_2, \ldots$ be the positions of the successive levels in the interval δE $(E_1 \leqslant E_2 \leqslant \cdots)$ and let $S_1, S_2, \ldots$ be their distances apart, $S_i = E_{i+1} - E_i$. The average value of S_i is the mean spacing D. We define the relative spacing $s_i = S_i/D$. The probability density function $p(s)$ is defined by the condition that $p(s)\,ds$ is the probability that any s_i will have a value between s and $s + ds$.

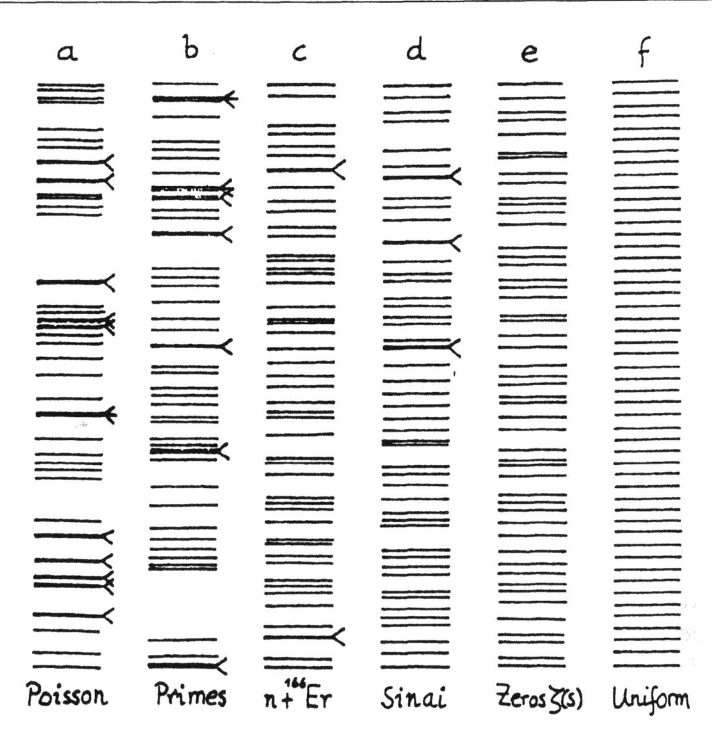

Figure 1.2. Some typical level sequences. From Bohigas and Giannoni (1984). (a) Random levels with no correlations, Poisson series. (b) Sequence of prime numbers. (c) Slow neutron resonance levels of the erbium 166 nucleus. (d) Possible energy levels of a particle free to move inside the area bounded by 1/8 of a square and a circular arc whose center is the mid point of the square; i.e. the area specified by the inequalities, $y \geqslant 0$, $x \geqslant y$, $x \leqslant 1$, and $x^2 + y^2 \geqslant r$. (Sinai's billiard table.) (e) The zeros of the Riemann zeta function on the line $\operatorname{Re} z = 1/2$. (f) A sequence of equally spaced levels (Bohigas and Giannoni, 1984).

For the simple case in which the positions of the energy levels are not correlated the probability that any E_i will fall between E and $E + dE$ is independent of E and is simply $\rho\, dE$, where $\rho = D^{-1}$ is the average number of levels in a unit interval of energy. Let us determine the probability of a spacing S; that is, given a level at E, what is the probability of having no level in the interval $(E, E + S)$ and one level in the interval $(E + S, E + S + dS)$. For this we divide the interval S into m equal parts.

$E \quad E + S/m \quad E + 2S/m \quad \ldots \quad E + (m-1)S/m \quad E + S \quad E + S + dS$

Because the levels are independent, the probability of having no level in $(E, E + S)$ is the product of the probabilities of having no level in any of these m parts. If m is large, so that S/m is small, we can write this as $(1-\rho S/m)^m$, and in the limit $m \to \infty$,

$$\lim_{m\to\infty}\left(1-\rho\frac{S}{m}\right)^m = e^{-\rho S}.$$

Moreover, the probability of having a level in dS at $E+S$ is $\rho\, dS$. Therefore, given a level at E, the probability that there is no level in $(E, E+S)$ and one level in dS at $E+S$ is

$$e^{-\rho S}\rho\, dS,$$

or in terms of the variable $s = S/D = \rho S$

$$p(s)\, ds = e^{-s}\, ds. \tag{1.4.1}$$

This is known as the Poisson distribution or the spacing rule for random levels.

That (1.4.1) is not correct for nuclear levels of the same spin and parity or for atomic levels of the same parity and orbital and spin angular momenta is clearly seen by a comparison with the empirical evidence (Figures 1.3 and 1.4). It is not true either for the eigenvalues of a matrix from any of the Gaussian ensembles, as we will see.

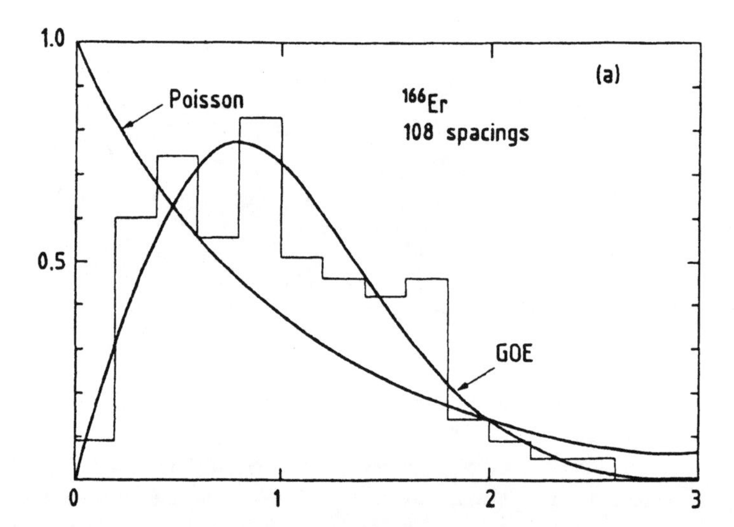

Figure 1.3. The probability density for the nearest neighbor spacings in slow neutron resonance levels of erbium 166 nucleus. The histogram shows the first 108 levels observed. The solid curves correspond to the Poisson distribution, i.e. no correlations at all, and that for the eigenvalues of a real symmetric random matrix taken from the Gaussian orthogonal ensemble (GOE). Reprinted with permission from The American Physical Society, Liou et al., Neutron resonance spectroscopy data, *Phys. Rev. C* 5 (1972) 974–1001.

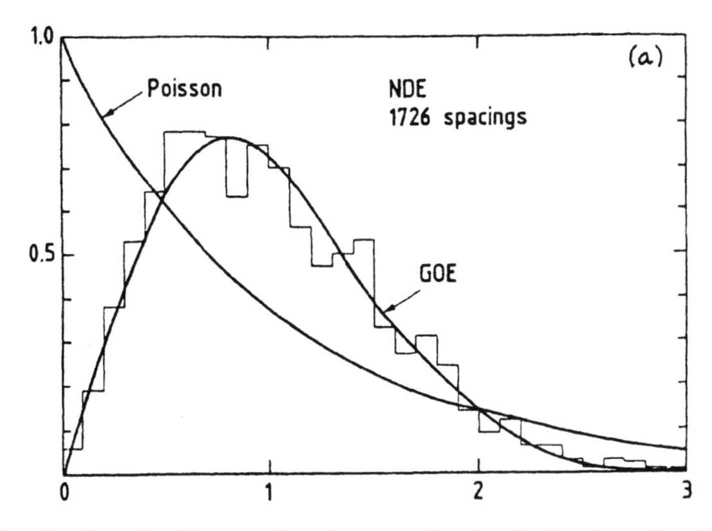

Figure 1.4. Level spacing histogram for a large set of nuclear levels, often referred to as nuclear data ensemble. The data considered consists of 1407 resonance levels belonging to 30 sequences of 27 different nuclei: (i) slow neutron resonances of Cd(110, 112, 114), Sm(152, 154), Gd(154, 156, 158, 160), Dy(160, 162, 164), Er(166, 168, 170), Yb(172, 174, 176), W(182, 184, 186), Th(232) and U(238); (1146 levels); (ii) proton resonances of Ca(44) ($J = 1/2+$), Ca(44) ($J = 1/2-$), and Ti(48) ($J = 1/2+$); (157 levels); and (iii) (n, γ)-reaction data on Hf(177) ($J = 3$), Hf(177) ($J = 4$), Hf(179) ($J = 4$), and Hf(179) ($J = 5$); (104 levels). The data chosen in each sequence is believed to be complete (no missing levels) and pure (the same angular momentum and parity). For each of the 30 sequences the average quantities (e.g. the mean spacing, spacing/mean spacing, number variance μ_2, etc., see Chapter 16) are computed separately and their aggregate is taken weighted according to the size of each sequence. The solid curves correspond to the Poisson distribution, i.e. no correlations at all, and that for the eigenvalues of a real symmetric random matrix taken from the Gaussian orthogonal ensemble (GOE). Reprinted with permission from Kluwer Academic Publishers, Bohigas O., Haq R.U. and Pandey A., Fluctuation properties of nuclear energy levels and widths, comparison of theory with experiment, in: *Nuclear Data for Science and Technology*, Bökhoff K.H. (Ed.), 809–814 (1983).

1.5 Wigner Surmise

When the experimental situation was not yet conclusive, Wigner proposed the following rules for spacing distributions:

(1) In the sequence of levels with the same spin and parity, called a simple sequence, the probability density function for a spacing is given by

$$p_W(s) = \frac{\pi s}{2} \exp\left(-\frac{\pi}{4} s^2\right), \quad s = \frac{S}{D}. \tag{1.5.1}$$

(2) Levels with different spin and parity are not correlated. The function $p(s)$ for a mixed sequence may be obtained by randomly superimposing the constituent simple sequences (cf. Appendix A.2).

Two simple arguments give rise to Rule 1. As pointed out by Wigner (1951) and by Landau and Smorodinski (1955), it is reasonable to expect that, given a level at E, the probability that another level will lie around $E + S$ is proportional to S for small S. Now if we extrapolate this to all S and, in addition, assume that the probabilities in various intervals of length S/m obtained by dividing S into m equal parts are mutually independent, we arrive at

$$p(S/D)\,dS = \lim_{m\to\infty} \prod_{r=0}^{m-1} \left(1 - \frac{Sr}{m}\frac{S}{m}a\right) aS\,dS = aSe^{-aS^2/2}\,dS. \tag{1.5.2}$$

The constant a can be determined by the condition that the average value of $s = S/D$ is unity:

$$\int_0^\infty sp(s)\,ds = 1. \tag{1.5.3}$$

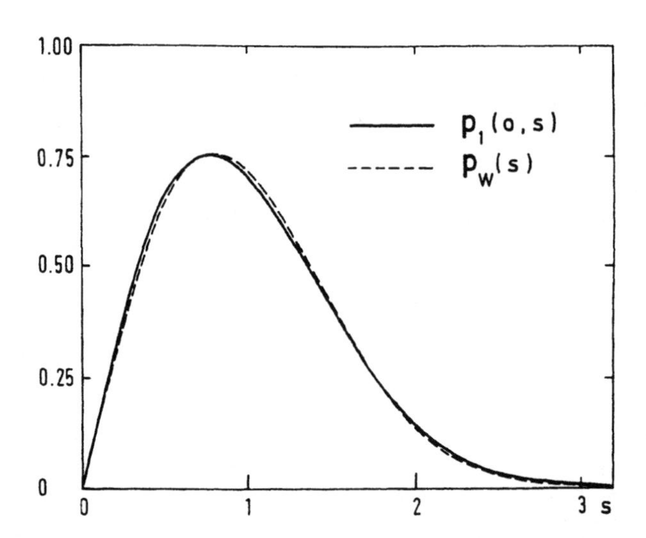

Figure 1.5. The probability density $p_1(0; s)$ of the nearest neighbor spacings for the eigenvalues of a random real symmetric matrix from the Gaussian orthogonal ensemble, Eq. (7.4.18) and the Wigner surmise $p_W(s)$; Eq. (1.5.1). Reprinted with permission from Elsevier Science Publishers, Gaudin M., Sur la loi limite de l'espacements des valeurs propres d'une matrice aléatoire, *Nucl. Phys.* 25, 447–458 (1961).

Let us, at this point, define the n-point correlation function $R_n(E_1, \ldots, E_n)$ so that $R_n\, dE_1\, dE_2 \cdots dE_n$ is the probability of finding a level in each of the intervals $(E_1, E_1 + dE_1), \ldots, (E_n, E_n + dE_n)$ all other levels being unobserved. The two simple arguments of Wigner given in the derivation of Rule 1 are equivalent to the following. The two-point correlation function $R_2(E_1, E_2)$ is linear in the variable $|E_1 - E_2|$, and three and higher order correlation functions are negligibly small.

We shall see in Chapter 7 that both arguments are inaccurate, whereas Rule 1 is very near the correct result (cf. Figure 1.5). It is surprising that the two errors compensate so nearly each other.

1.6 Electromagnetic Properties of Small Metallic Particles

Consider small metallic particles at low temperatures. The number of electrons in a volume V is $n \approx 4\pi p_0^3 V/3h^3$, where p_0 is the Fermi momentum and h is Planck's constant. The energy of an excitation near the Fermi surface is $E_0 \approx p_0^2/2m^*$, where m^* is the effective mass of the electron. The level density at zero excitation is therefore $\sigma = dn/dE_0 \approx 4\pi p_0 V m^*/h^3$, and the average level spacing is the inverse of this quantity $D \approx \sigma^{-1}$. For a given temperature we can easily estimate the size of the metallic particles for which $D \gg kT$, where k is Boltzmann's constant and T is the temperature in degrees Kelvin. For example, the number of electrons in a metallic particle of size 10^{-6}–10^{-7} cm may be as low as a few hundred and, at $T \approx 10$ K, $D \approx 1$ eV, whereas $kT \approx 10^{-3}$ eV. It is possible to produce particles of this size experimentally and then to sort them out according to their size (e.g., by centrifuging and sampling at a certain radial distance). Thus we have a large number of metallic particles, each of which has a different shape and therefore a different set of electronic energy levels but the same average level spacing, for the volumes are equal. It would be desirable if we could separate (e.g., by applying a non uniform magnetic field) particles containing an odd number of conduction electrons from those containing an even number. The energy-level schemes for these two types of particles have very different properties.

Given the position of the electron energies, we can calculate the partition function in the presence of a magnetic field and then use thermodynamic relations to derive various properties such as electronic specific heat and spin paramagnetism. Fröhlich (1937) assumed that the energies were equally spaced and naturally obtained the result that all physical quantities decrease exponentially at low temperatures as $\exp(-D/kT)$ for $1 \ll D/kT$. Kubo (1969) repeated the calculation with the assumption that the energies were random without correlations and that their spacings therefore follow a Poisson law. He arrived at a linear law for the specific heat $\sim kT/D$. The constants are different for particles containing an odd number of electrons from those containing an even number. For spin paramagnetism even the dependence on temperature is different for the two sets of particles. Instead of Fröhlich's equal spacing rule or Kubo's Poisson law, it would perhaps be better to suppose with Gorkov and Eliashberg (1965), that these energies behave as the eigenvalues of a random matrix. This point of view may be justified as follows. The energies are the eigenvalues of a fixed Hamiltonian with random boundary conditions. We may incorporate these boundary conditions into the Hamiltonian by the use of fictitious potentials. The energies are thus neither equally spaced, nor follow the Poisson law, but they behave as the eigenvalues of a random matrix taken from a suitable ensemble. In contrast to nuclear spectra, we have the possibility of realizing in practice all three ensembles considered in various sections of this book. They apply in particular when (a) the number of electrons (in each of the metallic particles) is even and there is no external magnetic field, (b) the number of electrons (in each of the metallic particles) is odd and there is no external magnetic field, (c) there is an external magnetic field much greater than D/μ, where μ is the magnetic moment of the electron.

As to which of the three assumptions is correct should be decided by the experimental evidence. Unfortunately, such experiments are difficult to perform neatly and no clear conclusion is yet available. See the discussion in Brody et al. (1981).

1.7 Analysis of Experimental Nuclear Levels

The enormous amount of available nuclear data was analyzed in the 1980s by French and coworkers, specially in view of deriving an upper bound for the time reversal violating part of the nuclear forces (cf. Haq et al. (1982), Bohigas et al. (1983, 1985)). Actually, if nuclear forces are time reversal invariant, then the nuclear energy levels should behave as the eigenvalues of a random real symmetric matrix; if they are not, then they should behave as those of a random Hermitian matrix. The level sequences in the two cases have different properties; for example, the level spacing curves are quite distinct. Figures 1.6 and 1.7 indicate that nuclear and atomic levels behave as the eigenvalues of a random real symmetric matrix and the admixture of a Hermitian anti-symmetric part, if any, is small. Figure 1.4 shows the level spacing histogram for a really big sample of available nuclear data from which French et al. (1985) could deduce an upper bound for this small admixture. Figure 1.8 shows how the probability density of the nearest neighbor spacings of atomic levels changes as one goes from one long period to another in the periodic table of elements. For further details on the analysis of nuclear levels and other sequences of levels, see Chapter 16.

1.8 The Zeros of The Riemann Zeta Function

The zeta function of Riemann is defined for $\mathrm{Re}\, z > 1$ by

$$\zeta(z) = \sum_{n=1}^{\infty} n^{-z} = \prod_{p} (1 - p^{-z})^{-1}, \tag{1.8.1}$$

and for other values of z by its analytical continuation. The product in Eq. (1.8.1) above is taken over all primes p, and is known as the Euler product formula.

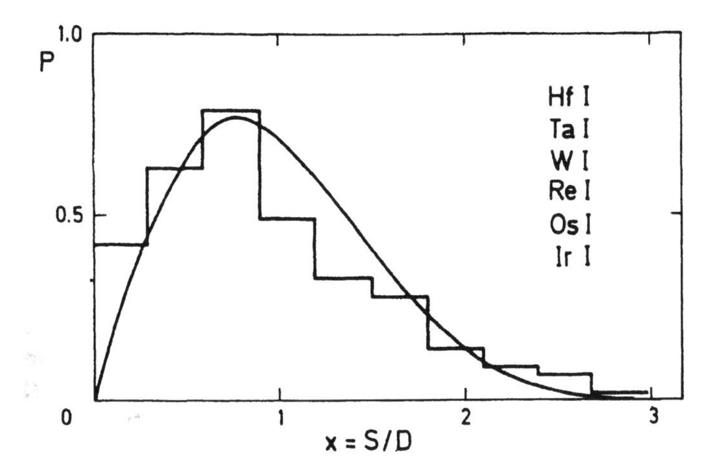

Figure 1.6. Plot of the density of nearest neighbor spacings between odd parity atomic levels of a group of elements in the region of osmium. The levels in each element were separated according to angular momentum, and separate histograms were constructed for each level series, and then combined. The elements and the number of contributed spacings are HfI, 74; TaI, 180; WI, 262; ReI, 165; OsI, 145; IrI, 131 which lead to a total of 957 spacings. The solid curve corresponds to the Wigner surmise, Eq. (1.5.1). Reprinted with permission from Annales Academiae Scientiarum Fennicae, Porter C.E. and Rosenzweig N., Statistical properties of atomic and nuclear spectra, *Annale Academiae Scientiarum Fennicae, Serie A VI, Physica* 44, 1–66 (1960).

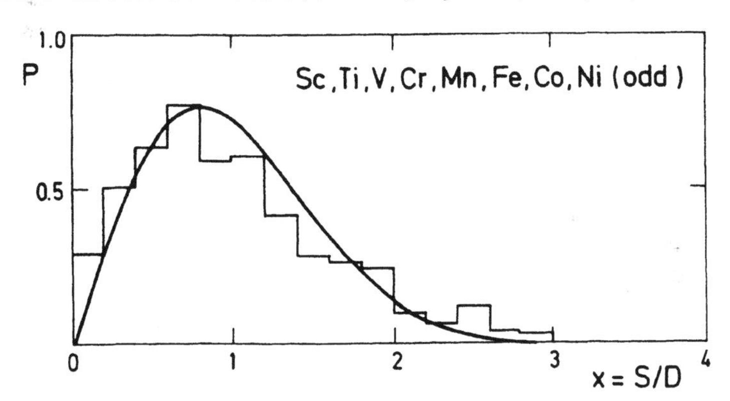

Figure 1.7. Plot of the density of nearest neighbor spacings between odd parity levels of the first period corresponding to a separation of the levels into sequences each of which is labeled by definite values of S, L and J. Comparison with the Wigner surmise (solid curve) shows a good fit if the (approximate) symmetries resulting from (almost) complete absence of spin-orbit forces are taken into consideration. Reprinted with permission from Annales Academiae Scientiarum Fennicae, Porter C.E. and Rosenzweig N., Statistical properties of atomic and nuclear spectra, *Annale Academiae Scientiarum Fennicae, Serie A VI, Physica* 44, 1–66 (1960).

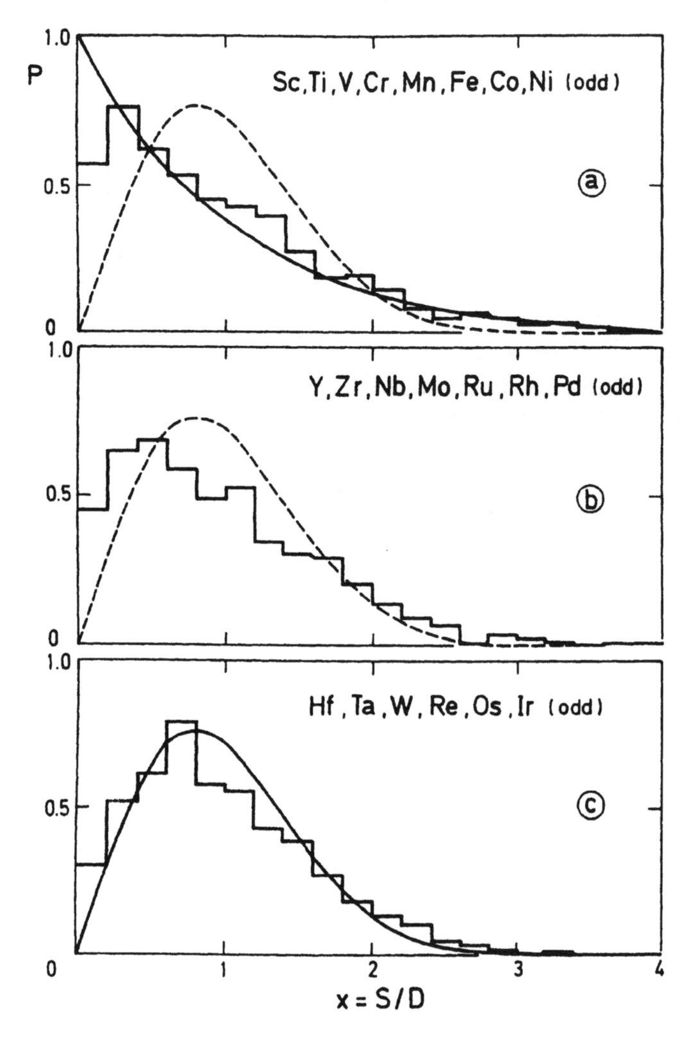

Figure 1.8. Empirical density of nearest neighbor spacings between odd parity levels of elements in the first, second and third long periods (histograms a, b and c respectively). To obtain these figures separate histograms were constructed for the J sequences of each element and then the results were combined. Comparison with Poisson (or exponential) distribution and Wigner surmise (also shown) indicates that curves go from Poisson to Wigner curves as one goes from the first to the second and then finally to the third period. This variation can be understood in terms of the corresponding increase in strength of the spin dependent forces. Reprinted with permission from Annales Academiae Scientiarum Fennicae, Porter C.E. and Rosenzweig N., Statistical properties of atomic and nuclear spectra, *Annale Academiae Scientiarum Fennicae, Serie A VI, Physica* 44, 1–66 (1960).

It is well-known that $\zeta(z)$ is zero for $z = -2n$, $n = 1, 2, \ldots$; these are called the "trivial zeros". All the other zeros of $\zeta(z)$ lie in the strip $0 < \operatorname{Re} z < 1$, and are symmetrically situated with respect to the critical line $\operatorname{Re} z = 1/2$. It is conjectured that they actually lie on the critical line $\operatorname{Re} z = 1/2$ (Riemann hypothesis, 1876). Since $\zeta(z^*) = \zeta^*(z)$, if z is a zero of $\zeta(z)$, then z^* is also a zero of it. Assuming the truth of the Riemann hypothesis (RH), let $z = 1/2 \pm i\gamma_n$, γ_n real and positive, be the "non-trivial" zeros of $\zeta(z)$. How the γ_n are distributed on the real line?

These questions have their importance in number theory. Many, as yet unsuccessful, attempts have been made to prove or disprove the RH. With the advent of electronic computers of ever faster speeds and larger memories, a series of efforts have also been made to verify the truth of the RH numerically and to evaluate some statistical properties of the γ_n.

Until the 1970s few persons were interested in the distribution of the γ_n. The main reason seems to be the feeling that if one cannot prove (or disprove!) the γ_n to be all real (RH), then there is no point in asking even harder questions. Montgomery (1973) took the lead in investigating questions about the distribution of the γ_n. It was known (Titchmarsh, 1951, Chapter 10) that the number $N(T)$ of the γ_n with $0 < \gamma_n \leqslant T$ is

$$N(T) = \frac{T}{2\pi} \ln\left(\frac{T}{2\pi}\right) - \frac{T}{2\pi} + S(T) + \frac{7}{8} + \mathrm{O}(T^{-1}), \tag{1.8.2}$$

as $T \to \infty$, where

$$S(t) = \frac{1}{\pi} \arg \zeta\left(\frac{1}{2} + it\right). \tag{1.8.3}$$

Maximum order of $S(T)$ remains unknown. Probably,

$$S(T) = \mathrm{O}\left(\left(\frac{\ln T}{\ln \ln T}\right)^{1/2}\right), \tag{1.8.4}$$

or more likely (Montgomery, private communication)

$$S(T) = \mathrm{O}\left((\ln T \ln \ln T)^{1/2}\right). \tag{1.8.4$'$}$$

Assuming the RH Montgomery studied $D(\alpha, \beta)$, the number of pairs γ, γ' such that $\zeta(1/2 + i\gamma) = \zeta(1/2 + i\gamma') = 0$, $0 < \gamma \leqslant T$, $0 < \gamma' \leqslant T$, and $2\pi\alpha/(\ln T) \leqslant \gamma - \gamma' \leqslant 2\pi\beta/(\ln T)$. Or taking the Fourier transforms, it amounts to evaluate the function

$$F(\alpha) = \frac{2\pi}{T \ln T} \sum_{0 < \gamma, \gamma' \leqslant T} T^{i\alpha(\gamma - \gamma')} \frac{4}{4 + (\gamma - \gamma')^2} \tag{1.8.5}$$

for real α. Since F is symmetric in γ, γ', it is real and even in α. Montgomery (1973) showed that if RH is true then $F(\alpha)$ is nearly non-negative, $F(\alpha) \geqslant -\varepsilon$ uniformly in α

for $T > T_0(\varepsilon)$, and that

$$F(\alpha) = (1 + o(1))T^{-2\alpha} \ln T + \alpha + o(1) \tag{1.8.6}$$

uniformly for $0 \leqslant \alpha \leqslant 1$. For $\alpha > 1$, the behaviour changes. He also gave heuristic arguments to suggest that for $\alpha \geqslant 1$,

$$F(\alpha) = 1 + o(1). \tag{1.8.7}$$

And from this conjecture he deduced that

$$\frac{2\pi}{T \ln T} D(\alpha, \beta) \approx \int_\alpha^\beta \left(1 - \left(\frac{\sin r}{r}\right)^2 + \delta(\alpha, \beta)\right) dr \tag{1.8.8}$$

for any real α, β with $\alpha < \beta$. Here $\delta(\alpha, \beta) = 1$ if $\alpha < 0 < \beta$, and $\delta(\alpha, \beta) = 0$ otherwise. This $\delta(\alpha, \beta)$ is there because for $\alpha < 0 < \beta$, $D(\alpha, \beta)$ includes terms with $\gamma = \gamma'$.

Equation (1.8.8) says that the two-point correlation function of the zeros of the zeta function $\zeta(z)$ on the critical line is

$$R_2(r) = 1 - (\sin(\pi r)/(\pi r))^2. \tag{1.8.9}$$

As we will see in Chapter 6.2, this is precisely the two-point correlation function of the eigenvalues of a random Hermitian matrix taken from the Gaussian unitary ensemble (GUE) (or that of a random unitary matrix taken from the circular unitary ensemble, cf. Chapter 11.1). This is consistent with the view (quoted to be conjectured by Polya and Hilbert) that the zeros of $\zeta(z)$ are related to the eigenvalues of a Hermitian operator. As the two-point correlation function seems to be the same for the zeros of $\zeta(z)$ and the eigenvalues of a matrix from the GUE, it is natural to think that other statistical properties also coincide. With a view to prove (or disprove!) this expectation, Odlyzko (1987) computed a large number of zeros $1/2 + i\gamma_n$ of $\zeta(z)$ with great accuracy. Taking γ_n to be positive and ordered, i.e. $0 < \gamma_1 \leqslant \gamma_2 \leqslant \cdots$ with $\gamma_1 = 14.134\ldots$, $\gamma_2 = 21.022\ldots$ etc. he computed sets of 10^5 zeros γ_n with a precision $\pm 10^{-8}$, and $N + 1 \leqslant n \leqslant N + 10^5$, for $N = 0$, $N = 10^6$, $N = 10^{12}$, $N = 10^{18}$ and $N = 10^{20}$. (This computation gives, for example, $\gamma_n = 15202440115920747268.6290299\ldots$ for $n = 10^{20}$, cf. Odlyzko, 1989 or Cipra, 1989.) This huge numerical undertaking was possible only with the largest available supercomputer CRAY X-MP and fast efficient algorithms devised for the purpose. The numerical evidence so collected is shown on Figures 1.9 to 1.14.

Note that as one moves away from the real axis, the fit improves for both the spacing as well as the two point correlation function. The convergence is very slow; the numerical curves become indistinguishable from the GUE curves only far away from the real axis, i.e. when $N \geqslant 10^{12}$. In contrast, for Hermitian matrices or for the unitary

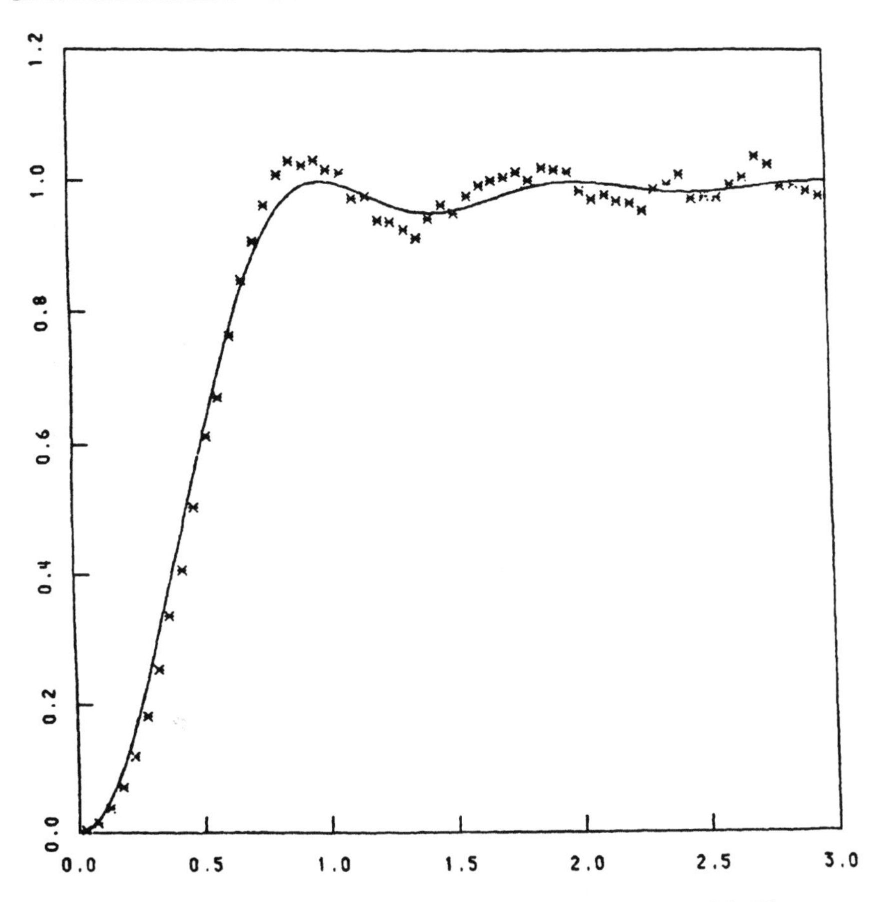

Figure 1.9. Two point correlation function for the zeros $0.5 \pm i\gamma_n$, γ_n real, of the Riemann zeta function; $1 < n < 10^5$. The solid curve is Montgomery's conjecture, Eq. (1.8.9). Reprinted from "On the distribution of spacings between zeros of the zeta function," A.M. Odlyzko, *Mathematics of Computation* pages 273–308 (1987), by permission of The American Mathematical Society.

matrices the limit is practically reached for matrix orders 100×100 and 20×20 respectively. This can be understood by the fact that the density of the zeros of the zeta function around $N \approx 10^{12}$ is comparable to the eigenvalue density for the Hermitian or the unitary matrices of orders 100 or 20.

One may compare the Riemann zeta function on the critical line to the characteristic function of a random unitary matrix. In other words, consider the variations of the Riemann zeta function as one moves along the critical line $\mathrm{Re}\, z = 1/2$ far away from the real axis as well as the variations of the characteristic function of an $N \times N$ random

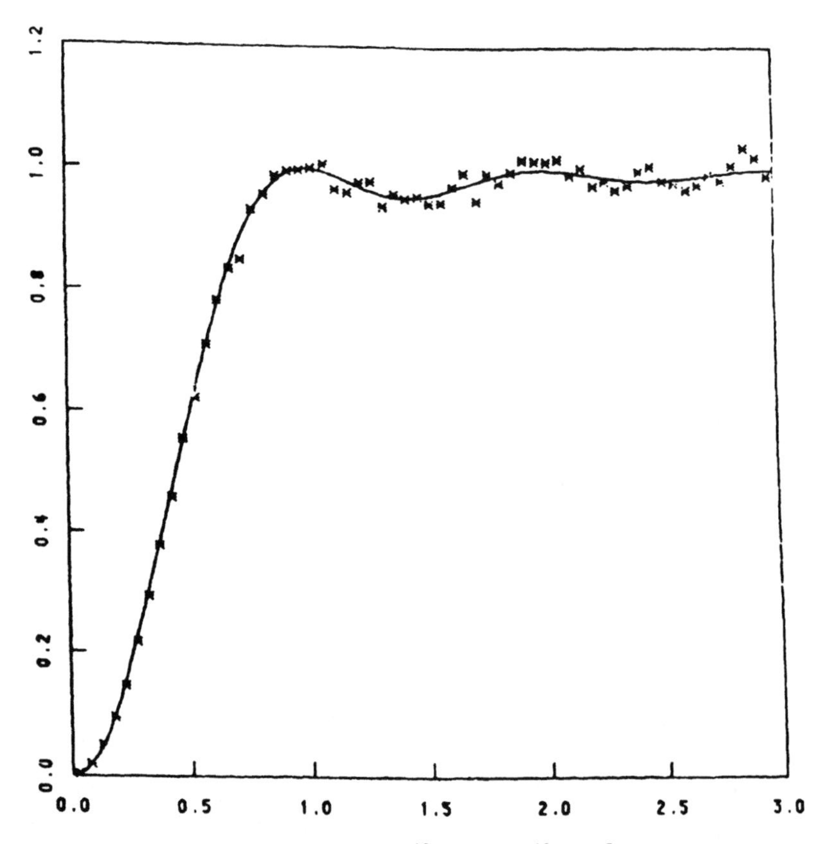

Figure 1.10. The same as Figure 1.9 with $10^{12} < n < 10^{12} + 10^5$; Note that the curves fit much better. Reprinted from "On the distribution of spacings between zeros of the zeta function," A.M. Odlyzko, *Mathematics of Computation* pages 273–308 (1987), by permission of The American Mathematical Society.

unitary matrix,

$$f(x) = \zeta\left(\frac{1}{2} + it\right), \quad t = T + \frac{x}{2\pi}\ln\left(\frac{T}{2\pi}\right), \tag{1.8.10}$$

$$g(x) = \det[e^{i\theta} \cdot \mathbf{1} - U], \quad \theta = x \cdot \frac{N}{2\pi}, \tag{1.8.11}$$

then in the limit of large T and N empirically the zeros of $f(x)$ behave as if they were the zeros of $g(x)$; they have the same m-point correlation functions for $m = 2, 3, 4, \ldots$; they have the same spacing functions $p(m, s)$ for $m = 0, 1, \ldots$ (cf. Chapters 6 and 16).

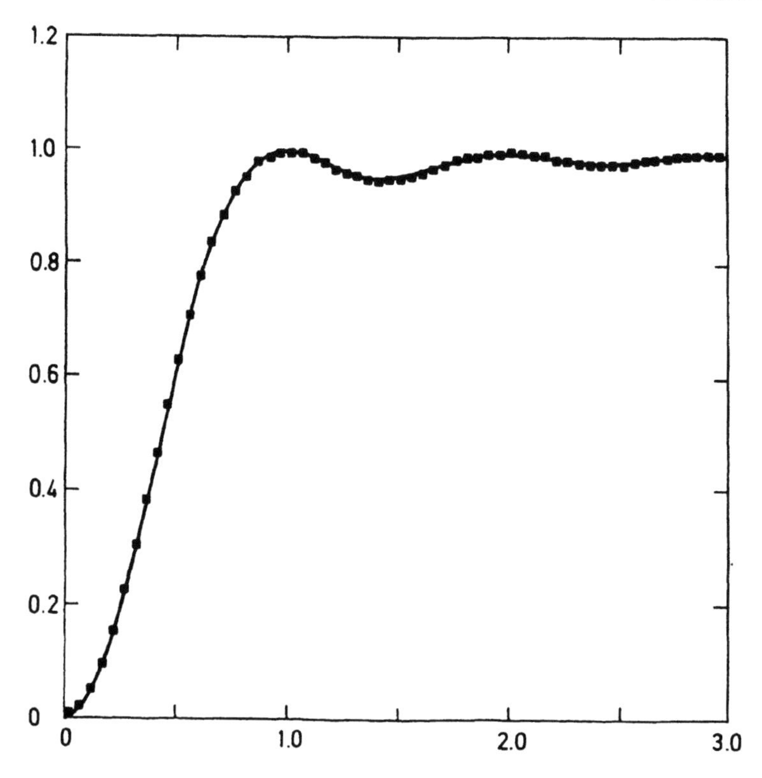

Figure 1.11. The same as Figure 1.9, but for 79 million zeros around $n \approx 10^{20}$. From Odlyzko (1989). Copyright © 1989 American Telephone and Telegraph Company, reprinted with permission.

Quantities other than the zeros of these functions have been considered and found to match in all the known cases. For example, Odlyzko computed the density of values of the real and imaginary parts of the logarithm of $\zeta(1/2 + it)$ for large values of t and noticed that their convergence to the asymptotic limit, known to be Gaussian, is quite slow. Keating and Snaith (2000a) computed the density of values of the real and imaginary parts of the logarithm of $\zeta(1/2 + it)$ for t around the 10^{20}th zero of the zeta function from the data of Odlyzko and compared it with the probability density of the real and imaginary parts of the logarithm of the characteristic function of a random 42×42 unitary matrix (since 10^{20}th zero of the zeta function is at $t \approx 1.52 \times 10^{19}$ and $\ln[1.52 \times 10^{19}/(2\pi)] \approx 42$). The two curves are indistinguishable and visibly differ from their asymptotic limits, known to be Gaussian. See Figures 1.15 and 1.16 (taken from Keating and Snaith (2000a), *Comm. Math. Phys.* **214**, 57–89).

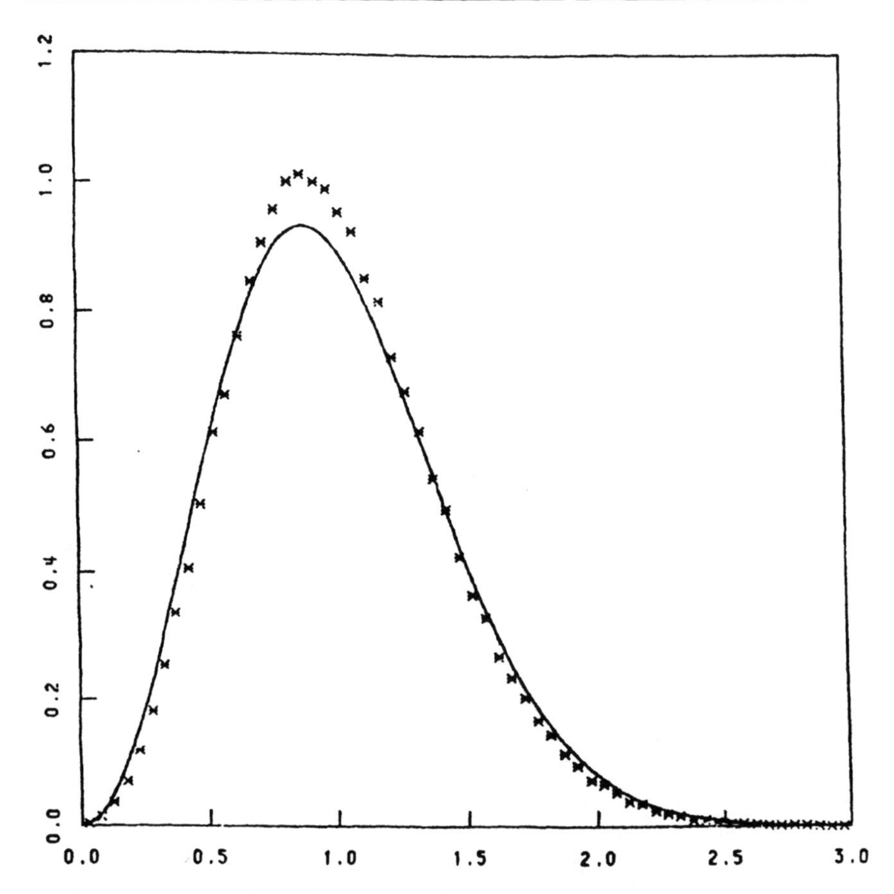

Figure 1.12. Plot of the density of normalized spacings for the zeros $0.5 \pm i\gamma_n$, γ_n real, of the Riemann zeta function on the critical line. $1 < n < 10^5$. The solid curve is the spacing probability density for the Gaussian unitary ensemble, Eq. (6.4.32). From Odlyzko (1987). Reprinted from "On the distribution of spacings between zeros of the zeta function," *Mathematics of Computation* (1987), pages 273–308, by permission of The American Mathematical Society.

Also the distribution of the zeros of $d\zeta(1/2 + it)/dt$ and those of the derivative of the characteristic function of a random unitary matrix has been investigated and found empirically to be the same when centered and normalized properly (Mezzadri, 2003).

It is hard to imagine the zeros of the zeta function as the eigenvalues of some unitary or Hermitian operator. It is even harder to imagine the zeta function on the critical line as the characteristic function of a unitary operator.

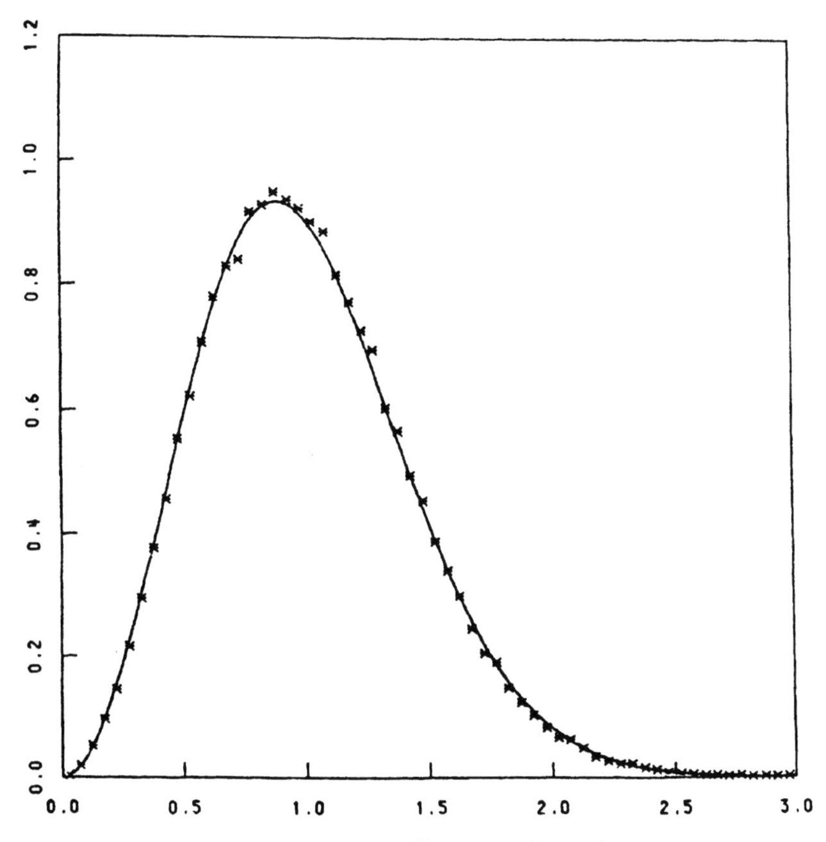

Figure 1.13. The same as Figure 1.12 with $10^{12} < n < 10^{12} + 10^5$. Note the improvement in the fit. From Odlyzko (1987). Reprinted from "On the distribution of spacings between zeros of the zeta function," *Mathematics of Computation* (1987), pages 273–308, by permission of The American Mathematical Society.

A generalization of the Riemann zeta function is the function $\zeta(z, a)$ defined for $\operatorname{Re} z > 1$, by

$$\zeta(z, a) = \sum_{n=0}^{\infty} (n + a)^{-z}, \quad 0 < a \leqslant 1, \tag{1.8.12}$$

and by its analytical continuation for other values of z. For $a = 1/2$ and $a = 1$, one has

$$\zeta(z, 1/2) = (2^z - 1)\zeta(z), \quad \zeta(z, 1) = \zeta(z), \tag{1.8.13}$$

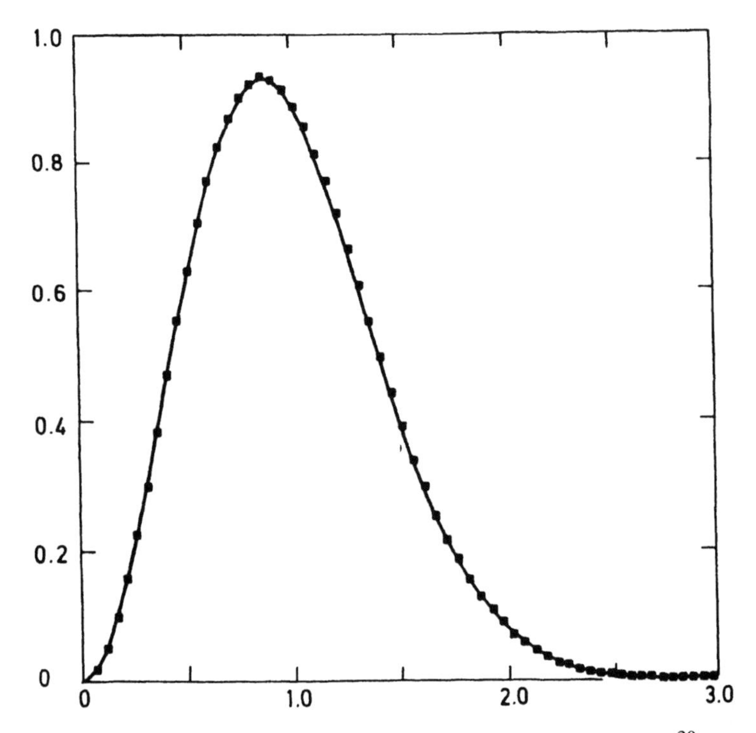

Figure 1.14. The same as Figure 1.12 but for the 79 million zeros around the 10^{20}th zero. From Odlyzko (1989). Copyright © 1989, American Telephone and Telegraph Company, reprinted with permission.

so there is nothing more about their zeros. For rational values of a other than $1/2$ or 1 or for transcendental values of a, it is known that $\zeta(z, a)$ has an infinity of zeros with $\text{Re}\, z > 1$. For irrational algebraic values of a it is not known whether there are any zeros with $\text{Re}\, z > 1$ (Davenport and Heilbronn, 1936).

A quadratic form (in two variables) $Q(x, y) = ax^2 + bxy + cy^2$, a, b, c integers, is positive definite if $a > 0$, $c > 0$ and the discriminant $d = b^2 - 4ac < 0$. It is primitive, if a, b, c have no common factor other than 1. Let the integers $\alpha, \beta, \gamma, \delta$ be such that $\alpha\delta - \beta\gamma = \pm 1$. When x and y vary over all the integers, the set of values taken by the quadratic forms $Q(x, y)$ and

$$Q'(x, y) := Q(\alpha x + \beta y, \gamma x + \delta y) = a'x^2 + b'xy + c'y^2$$

are identical; the two forms have the same discriminant d, they are said to be equivalent. The number $h(d)$ of inequivalent primitive positive definite quadratic forms with a given discriminant d is finite and is called the class function. (See Appendix A.53.)

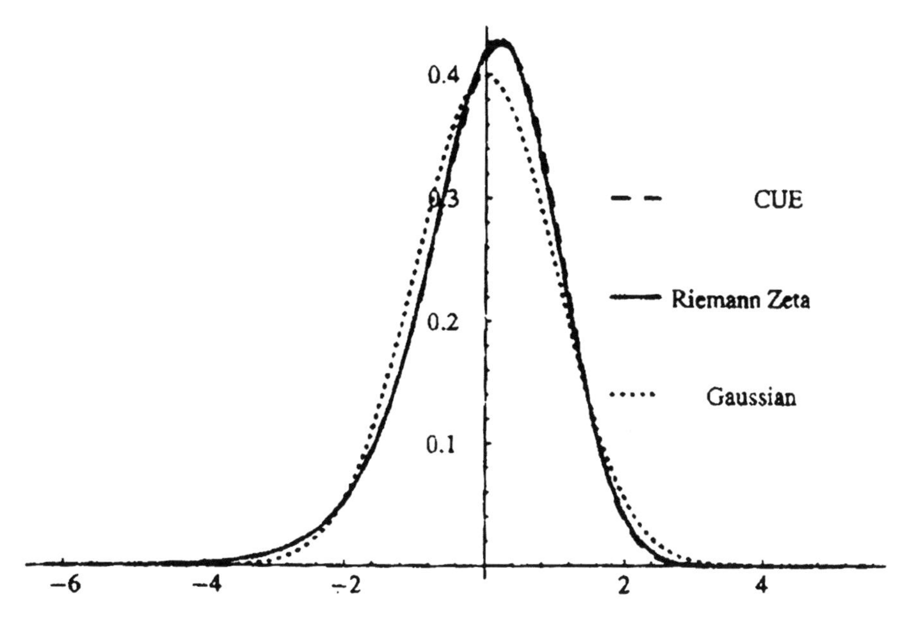

Figure 1.15. The density of values of $\log|f(x)|$ and the probability density of $\log|g(x)|$ for $T =$ the 10^{20}th zero of the Riemann zeta function and $N = 42$ for matrices from the CUE; Eqs. (1.8.10), (1.8.11). The asymptotic limit, the Gaussian, is also shown. From Keating and Snaith (2000a), reprinted with permission.

For any positive definite quadratic form

$$Q(x, y) = ax^2 + bxy + cy^2, \quad a, b, c \text{ integers}, \tag{1.8.14}$$

the Epstein zeta function is defined for $\operatorname{Re} z > 1$ by

$$\zeta(z, Q) = \sum_{\substack{-\infty<m,n<\infty \\ m,n\neq 0,0}} Q(m, n)^{-z}, \tag{1.8.15}$$

and for other values of z by its analytical continuation. This Epstein zeta function satisfies the functional equation

$$\left(\frac{2\pi}{\sqrt{-d}}\right)^{-z} \Gamma(z)\zeta(z, Q) = \left(\frac{2\pi}{\sqrt{-d}}\right)^{z-1} \Gamma(1-z)\zeta(1-z, Q), \tag{1.8.16}$$

where d is the discriminant of Q, $d = b^2 - 4ac < 0$. But in general it has no Euler product formula. Let $h(d)$ be the class function. If $h(d) = 1$, then the RH seems to be

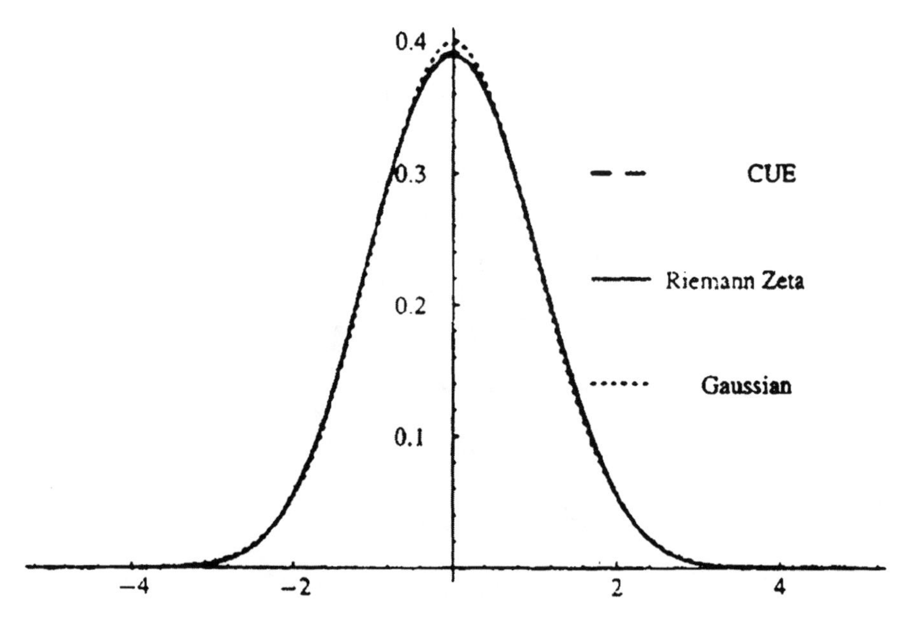

Figure 1.16. The density of values of the phase of $f(x)$ and the probability density of the phase of $g(x)$ for $T =$ the 10^{20}th zero of the Riemann zeta function and $N = 42$ for matrices from the CUE; Eqs. (1.8.10), (1.8.11). The asymptotic limit, the Gaussian, is also shown. From Keating and Snaith (2000a), reprinted with permission.

true, i.e. all the non-trivial zeros of $\zeta(z, Q)$ seem to lie on the critical line $\text{Re}\, z = 1/2$. If $h(d) > 1$, then it is known that $\zeta(z, Q)$ has an infinity of zeros with $\text{Re}\, z > 1$ (Potter and Titchmarsh, 1935).

In case $h(d) > 1$, it is better to consider the sum

$$\zeta_d(z) = \sum_{r=1}^{h(d)} \zeta(z, Q_r), \tag{1.8.17}$$

where the inequivalent primitive positive definite quadratic forms $Q_1, Q_2, \ldots, Q_r$ all have the same discriminant d. This function is proportional to a Dedekind zeta function, satisfies a functional equation and has an Euler product formula. It has no zeros with $\text{Re}\, z > 1$, and it is believed that all its non-trivial zeros lie on the critical line $\text{Re}\, z = 1/2$ (Potter and Titchmarsh, 1935).

For example, $h(-4) = 1$, has the unique primitive positive definite quadratic form up to equivalence $Q_1 = x^2 + y^2$, while the nonequivalent primitive positive definite quadratic forms $Q_2 = x^2 + 5y^2$ and $Q_3 = 3x^2 + 2xy + 2y^2$ both have $d = -20$, $h(-20) = 2$ (the quadratic form $3x^2 + 4xy + 3y^2$ taken by Potter and Titchmarsh is equivalent to our Q_3). Thus it is known that each of the functions

$$\zeta_2(z) = \sum_{\substack{-\infty<m,n<\infty \\ m,n\neq 0,0}} (m^2+5n^2)^{-z} \quad \text{and} \quad \zeta_3(z) = \sum_{\substack{-\infty<m,n<\infty \\ m,n\neq 0,0}} (3m^2+2mn+2n^2)^{-z}$$

has an infinity of zeros in the half plane $\operatorname{Re} z > 1$, while their sum $\zeta_2 + \zeta_3$ and the function

$$\zeta_1(z) = \sum_{\substack{-\infty<m,n<\infty \\ m,n\neq 0,0}} (m^2+n^2)^{-z}$$

are never zero in the same half plane. Moreover, it is believed that all the non-trivial zeros of $\zeta_1(z)$ and of $\zeta_2(z)+\zeta_3(z)$ lie on the critical line $\operatorname{Re} z = 1/2$.

What about their distribution? And in general, what about the distribution of the values of $\zeta_1(1/2+it)$ and of $\zeta_2(1/2+it)+\zeta_3(1/2+it)$ for real t? Do they have anything in common with the distribution of the values of the characteristic function of a random unitary matrix?

As another generalization one may consider for example,

$$Z_k(z) = \sum_{n_1,\ldots,n_k} (n_1^2+\cdots+n_k^2)^{-z}, \tag{1.8.18}$$

where the sum is taken over all integers $n_1,\ldots,n_k$, positive, zero and negative, except when $n_1 = \cdots = n_k = 0$. The zeta function, Eq. (1.8.18), is defined with respect to a hypercubic lattice in k dimensions. Equation (1.8.1) is the special case $k = 1$ of Eq. (1.8.18). Instead of the hypercubic lattice one can take any other lattice and define a zeta function corresponding to this lattice. Or one could consider the Dirichlet series

$$L(z) = \sum_{n\in\mathcal{N}^r} P(n)^{-z}, \tag{1.8.19}$$

where $P(x)$ is a polynomial in r variables $x_1,\ldots,x_r$ with real non-negative coefficients, the sum is taken over all positive integers $n_1,\ldots,n_r$ except for the singular points of $P(x)$, if any. Thus if $P(x) = x$, we have $L(z) \approx \zeta(z)$; if $P(x) = a+bx$, a, b real and > 0, then we have the Hurwitz zeta function

$$\zeta_{a,b}(z) = \sum_{n=0}^{\infty} (a+bn)^{-z}, \quad \operatorname{Re} z > 1. \tag{1.8.20}$$

Such Dirichlet series or general zeta functions, though do not have a functional equation in general, they have many important properties similar to those of the Riemann or Hurwitz zeta functions. For example, $\Gamma(z)\zeta_{a,b}(z)$ (with Hurwitz function $\zeta_{a,b}(z)$,

Eq. (1.8.20), and the gamma function $\Gamma(z)$), has simple poles at $z = 1, 0, -1, -2, \ldots$ with residues rational in (a, b). The "non-trivial" zeros of the hypercubic lattice zeta function $Z_k(z)$, Eq. (1.8.18), for example, for $k = 2, 4, 8, \ldots$, may be expected to be on the critical line $\text{Re}\, z = 1/2$. What about their distribution?

Little is known about the zeros of such general zeta functions or of the Dirichlet L-series, even empirically. It will be interesting to see what is the distribution of their (non-trivial) zeros and if they have any thing to do with the GUE results.

An eventual proof of the RH for the zeta function and its generalizations is important for at least two reasons; (i) it has remained a challenge for such a long time, (ii) it will imply a significant improvement in the estimation of various arithmetic functions, such as $\pi(x)$, the number of primes $\leqslant x$. However, one needs some further knowledge of the distribution of the γ_n to answer other questions like how the largest gap between two consecutive primes $\leqslant x$ increases with x. Or can one approximate real numbers by rational numbers whose numerator and denominator are both primes; more specifically, is it true that for every irrational number θ there are infinitely many prime numbers p, q such that $|\theta - p/q| < q^{-2+\varepsilon}$.

Katz and Sarnak (1999) found some particular L-series whose zeros would mimic not the eigenvalues of a random unitary matrix (circular unitary ensemble) but those of a random real orthogonal matrix or those of a random symplectic matrix, not related to the circular orthogonal ensemble (COE) or the circular symplectic ensemble (CSE). According to knowledgeable persons no number theoretic functions have yet been found that show COE or CSE statistics.

The fluctuation properties of the nuclear energy levels or that of the levels of a chaotic system are quite different, they behave, in the absence of a strong magnetic field, as the eigenvalues of a matrix from the orthogonal ensemble, cf. Chapter 7.

1.9 Things Worth Consideration, But Not Treated in This Book

Another much studied problem is that of percolation, or a random assembly of metals and insulators, or that of normal and super-conductors. Consider again a 2- or 3-dimensional lattice. Let each bond of the lattice be open with probability p and closed with probability $1 - p$. Or p and $1 - p$ may be considered as the respective probabilities of the bond being conducting or insulating or it being a normal and a super-conductor. The question is what is the probability that one can pass from one end to the other of a large lattice? Or how conducting or super-conducting a large lattice is? It is clear that if $p = 0$, no bond is open, and the probability of passage is zero. What is not so clear is that it remains zero for small positive values of p. Only when p increases and passes beyond a certain critical value p_c, this probability attains a non-zero value. It is known from numerical studies that the probability of passage is proportional to $(p - p_c)^\alpha$ for $p \geqslant p_c$, where the constant α, known as the critical index, is independent of the lattice and depends only on the dimension. The mathematical problem here can be characterized in its simplest form as follows. Given two fixed matrices A and B of the same order, let the matrix M_i for $i = 1, 2, \ldots, N$ be equal to A with probability p and equal to B with probability $1 - p$. How does the product $M = M_1 M_2 \ldots M_N$ behave when N is large? For example, what is the distribution of its eigenvalues? Of course, instead of just two matrices A and B and probabilities p and $1 - p$, we can take a certain number of them with their respective probabilities, but the problem is quite difficult even with two. No good analytical solution is known.

The actual problem of percolation or conduction is more complicated. Numerical simulation is performed on a ribbon of finite width, ≈ 10, for two dimensions (see Figure 1.17), or on a rod of finite cross-section, $\approx 5 \times 5$, for three dimensions. If all the resistances of the lattice lines are given, then we can explicitly compute the currents flowing through the open ends $B_1, B_2, \ldots, B_n$, when an electric field is applied between A and these ends. Adding one more layer in the length of the ribbon, the currents in the ends $B_1, \ldots, B_n$ change and this change can be characterized by an $n \times n$ matrix T, called the transfer matrix. This matrix T gives the new currents in terms of the old ones; it depends on the resistances of the lattice lines we have just added. As these resistances are random, say c with probability p and ∞ with probability $1 - p$, the matrix T is random; its matrix elements are constructed from the added random resistances according to known laws. For a length L of the ribbon, the currents in the ends $B_1, \ldots, B_n$ are determined by the product of L matrices $T_1 T_2 \ldots T_L$. Computing the spectrum of the product of L matrices for L large, $\approx 10^6$, and studying how it depends on L and on the order n, one can get a fair estimate for the critical index α.

Chaotic systems are simulated by a particle free to move inside a certain 2-dimensional domain. If the plane is Euclidean, then the shape of the domain is chosen so that the Laplace equation in the domain with Dirichlet boundary conditions has no analytic solution. This usually happens when the classical motion of the particle in the domain with elastic bouncing at the boundary has no, or almost no, periodic orbits. If the plane has a negative curvature, the boundary is usually taken to be polygonal with properly chosen angles. Sometimes one takes two quartic coupled oscillators with the potential

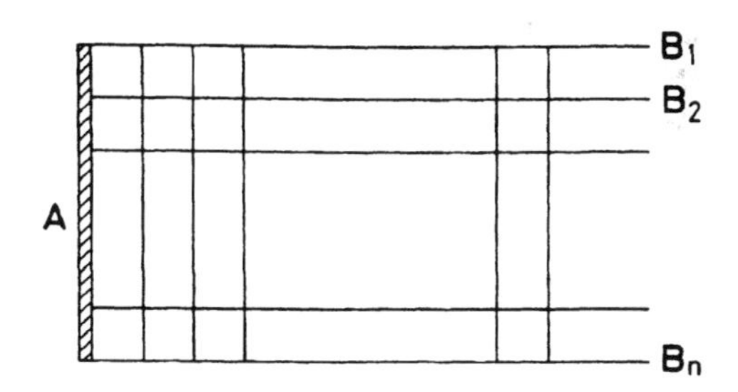

Figure 1.17. A model for the numerical simulation of conduction through a random assembly of conductors and insulators. At each step a new layer of randomly chosen conductors and insulators is added to the right and the subsequent change in the currents through the open ends $B_1, \ldots, B_n$ is studied.

energy $x^4 + y^4 + \lambda x^2 y^2$ as a model of a chaotic system. Why the spectrum of all these systems has to obey the predictions of the random matrix theory is not at all clear.

Still another curious instance has been reported (Balazs and Voros, 1989). Consider the matrix of the finite Fourier transform with matrix elements $A_{jk}(n) = n^{-1/2}\omega_n^{jk}$, $\omega_n = \exp(2\pi i/n)$. This $A(n)$ is unitary for every $n \geqslant 1$. Its eigenvalues are either ± 1 or $\pm i$. Let $B(n) = A^{-1}(2n)[A(n)\dot{+}A(n)]$ be the $2n \times 2n$ matrix, the product of the inverse of $A(2n)$ and the direct sum of $A(n)$ and $A(n)$. This $B(n)$ is also unitary, so that its eigenvalues lie on the unit circle. The matrix $B(n)$ is in some sense the quantum analog of the baker's transformation.

Now when n is very large, the eigenvalues of $B(n)$ and the eigenvalues of a random matrix taken from the circular unitary ensemble (see Chapters 10 and 11) have almost the same statistical properties. Instead of the above $B(n)$, we could have considered $B(n) = A^{-1}(3n)[A(n)\dot{+}A(2n)]$ or $B(n) = A^{-1}(3n)[A(n)\dot{+}A(n)\dot{+}A(n)]$, and the result would have been the same. Why this should be so, is not known.

Fluctuations in the electrical conductivity of heterogeneous metal junctions in a magnetic field and in general the statistical properties of transmission through a random medium is related to the theory of random matrices. In this connection see the articles by Al'tshuler, Mailly, Mello, Muttalib, Pichard, Zano and coworkers.

The hydrogen atom in an intense magnetic field has a spectrum characteristic of matrices from the Gaussian unitary ensemble, see Delande and Gay (1986).

In the last decade random matrix models have found applications in subjects as divers as quantum gravity, string theories, quantum chromo dynamics, counting of certain knots and links, counting of tree graphs (i.e. graphs having no closed paths) having special structures, and others. We will not deal with them either for lack of competence. Fortunately good review articles dealing with the vast literature on these topics are available. See for example, di Francesco et al. (1995), 2-D gravity and random matrix models, *Phys. Rep.* **254**, 1–133; Guhr et al. (1998), Random matrix theories in quantum physics: common concepts, *Phys. Rep.* **299**, 189–425; Katz and Sarnak (1999), *Random Matrices, Frobenius Eigenvalues and Monodromy*, Amer. Math. Soc., Providence, RI and many references therein.

2

GAUSSIAN ENSEMBLES. THE JOINT PROBABILITY DENSITY FUNCTION FOR THE MATRIX ELEMENTS

After examining the consequences of time-reversal invariance, we introduce Gaussian ensembles as a mathematical idealization. They are implied if we make the hypothesis of maximum statistical independence allowed under the symmetry constraints.

2.1 Preliminaries

In the mathematical model our systems are characterized by their Hamiltonians, which in turn are represented by Hermitian matrices. Let us look into the structure of these matrices. The low-lying energy levels (eigenvalues) are far apart and each may be described by a different set of quantum numbers. As we go to higher excitations, the levels draw closer, and because of their mutual interference most of the approximate quantum numbers lose their usefulness, for they are no longer exact. At still higher excitations the interference is so great that some quantum numbers may become entirely meaningless. However, there may be certain exact integrals of motion, such as total spin or parity, and the quantum numbers corresponding to them are conserved whatever the excitation may be. If the basis functions are chosen to be the eigenfunctions of these conserved quantities, all Hamiltonian matrices of the ensemble will reduce to the form of diagonal

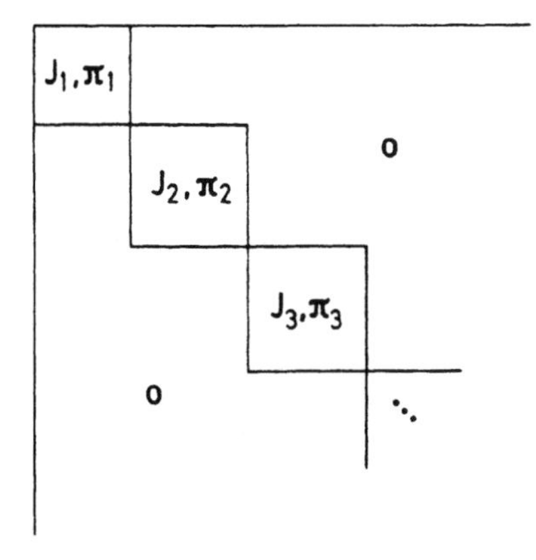

Figure 2.1. Block diagonal structure of a Hamiltonian matrix. Each diagonal block corresponds to a set of exact symmetries or to a set of exactly conserved quantum numbers. The matrix elements connecting any two diagonal blocks are zero, whereas those inside each diagonal block are random.

blocks. One block will correspond uniquely to each set of exact quantum numbers. The matrix elements lying outside these blocks will all be zero, and levels belonging to two different blocks will be statistically uncorrelated. As to the levels corresponding to the same block, the interactions are so complex that any regularity resulting from partial diagonalization will be washed out. (See Figure 2.1.)

We shall assume that such a basis has already been chosen and restrict our attention to one of the diagonal blocks, an $N \times N$ Hermitian matrix in which N is a large but fixed positive integer. Because nuclear spectra contain at least hundreds of levels with the same spin and parity, we are interested in (the limit of) very large N.

With these preliminaries, the matrix elements may be supposed to be random variables and allowed the maximum statistical independence permitted under symmetry requirements. To specify precisely the correlations among various matrix elements we need a careful analysis of the consequences of time-reversal invariance.

2.2 Time-Reversal Invariance

We begin by recapitulating the basic notions of time-reversal invariance. From physical considerations, the time-reversal operator is required to be antiunitary (Wigner, 1959) and can be expressed, as any other antiunitary operator, in the form

$$T = KC, \tag{2.2.1}$$

where K is a fixed unitary operator and the operator C takes the complex conjugate of the expression following it. Thus a state under time reversal transforms to

$$\psi^R = T\psi = K\psi^*, \tag{2.2.2}$$

ψ^* being the complex conjugate of ψ. From the condition

$$(\Phi, A\psi) = (\psi^R, A^R\Phi^R),$$

for all pairs of states ψ, Φ, and (2.2.2), we deduce that under time reversal an operator A transforms to

$$A^R = KA^TK^{-1}, \tag{2.2.3}$$

where A^T is the transpose of A. A is said to be self-dual if $A^R = A$. A physical system is invariant under time reversal if its Hamiltonian is self-dual, that is, if

$$H^R = H. \tag{2.2.4}$$

When the representation of the states is transformed by a unitary transformation, $\psi \to U\psi$, T transforms according to

$$T \to UTU^{-1} = UTU^\dagger, \tag{2.2.5}$$

or K transforms according to

$$K \to UKU^T. \tag{2.2.6}$$

Because operating twice with T should leave the physical system unchanged, we have

$$T^2 = \alpha \cdot 1, \quad |\alpha| = 1, \tag{2.2.7}$$

where 1 is the unit operator; or

$$T^2 = KCKC = KK^*CC = KK^* = \alpha \cdot 1. \tag{2.2.8}$$

But K is unitary:

$$K^*K^T = 1.$$

From these two equations we get

$$K = \alpha K^T = \alpha(\alpha K^T)^T = \alpha^2 K.$$

Therefore

$$\alpha^2 = 1 \quad \text{or} \quad \alpha = \pm 1, \tag{2.2.9}$$

so that the unitary matrix K is either symmetric or antisymmetric. In other words, either

$$KK^* = 1, \tag{2.2.10}$$

or

$$KK^* = -1. \tag{2.2.11}$$

These alternatives correspond, respectively, to an integral or a half-odd integral total angular momentum of the system measured in units of $\hbar$ (Wigner, 1959), for the total angular momentum operator $\mathbf{J} = (J_1, J_2, J_3)$ must transform as

$$J_\ell^R = -J_\ell, \quad \ell = 1, 2, 3. \tag{2.2.12}$$

For brevity we call the two possibilities the even-spin and odd-spin case, respectively.

2.3 Gaussian Orthogonal Ensemble

Suppose now that the even-spin case holds and (2.2.10) is valid. Then a unitary operator U will exist such that (cf. Appendix A.3)

$$K = UU^T. \tag{2.3.1}$$

By (2.2.6) a transformation $\psi \to U^{-1}\psi$ performed on the states ψ brings K to unity. Thus in the even-spin case the representation of states can always be chosen so that

$$K = 1. \tag{2.3.2}$$

After one such representation is found, further transformations $\psi \to R\psi$ are allowed only with R a real orthogonal matrix so that (2.3.2) remains valid. The consequence of (2.3.2) is that self-dual matrices are symmetric. In the even spin case every system invariant under time reversal will be associated with a real symmetric matrix H if the representation of states is suitably chosen. For even-spin systems with time-reversal invariance the Gaussian orthogonal ensemble E_{1G}, defined below, is therefore appropriate.

Definition 2.3.1. The Gaussian orthogonal ensemble E_{1G} is defined in the space T_{1G} of real symmetric matrices by two requirements:

(1) The ensemble is invariant under every transformation

$$H \to W^T H W \tag{2.3.3}$$

of T_{1G} into itself, where W is any real orthogonal matrix.

(2) The various elements H_{kj}, $k \leqslant j$, are statistically independent.

These requirements, expressed in the form of equations, read as follows:

(1) The probability $P(H)\,dH$ that a system of E_{1G} will belong to the volume element $dH = \prod_{k\leqslant j} dH_{kj}$ is invariant under real orthogonal transformations:

$$P(H')\,dH' = P(H)\,dH, \tag{2.3.4}$$

where

$$H' = W^T H W, \tag{2.3.5}$$

and

$$W^T W = W W^T = 1. \tag{2.3.6}$$

(2) This probability density function $P(H)$ is a product of functions, each of which depends on a single variable:

$$P(H) = \prod_{k\leqslant j} f_{kj}(H_{kj}). \tag{2.3.7}$$

Suppose, next, that we are dealing with a system invariant under space rotations. The spin may now be even or odd. The Hamiltonian matrix H which represents the system commutes with every component of **J**. If we use the standard representation of the J matrices with J_1 and J_3 real and J_2 pure imaginary, (2.2.12) may be satisfied by the usual choice (Wigner, 1959)

$$K = e^{i\pi J_2} \tag{2.3.8}$$

for K. With this choice of K, H and K commute and H^R reduces to H^T. Thus a rotation-invariant system is represented by a real symmetric matrix H, and once again the ensemble E_{1G} is appropriate.

2.4 Gaussian Symplectic Ensemble

In this section we discuss a system to which E_{1G} does not apply, a system with odd-spin, invariant under time reversal, but having no rotational symmetry. In this case (2.2.11) holds, K cannot be diagonalized by any transformation of the form (2.2.6), and there is no integral of the motion by which the double-valuedness of the time-reversal operation can be trivially eliminated.

Every antisymmetric unitary operator can be reduced by a transformation (2.2.6) to the standard canonical form (cf. Appendix A.3)

$$Z = \begin{bmatrix} 0 & +1 & 0 & 0 & \cdots \\ -1 & 0 & 0 & 0 & \cdots \\ 0 & 0 & 0 & +1 & \cdots \\ 0 & 0 & -1 & 0 & \cdots \\ \cdots & \cdots & \cdots & \cdots & \cdots \\ \cdots & \cdots & \cdots & \cdots & \cdots \end{bmatrix} \equiv \begin{bmatrix} 0 & +1 \\ -1 & 0 \end{bmatrix} \dotplus \begin{bmatrix} 0 & +1 \\ -1 & 0 \end{bmatrix} \dotplus \cdots, \tag{2.4.1}$$

which consists of (2×2) blocks

$$\begin{bmatrix} 0 & +1 \\ -1 & 0 \end{bmatrix}$$

along the leading diagonal; all other elements of Z are zero. We assume that the representation of states is chosen so that K is reduced to this form. The number of rows and columns of all matrices must now be even, for otherwise K would be singular in contradiction to (2.2.11). It is convenient to denote the order of the matrices by $2N$ instead of N. After one such representation is chosen, for which $K = Z$, further transformations $\psi \to B\psi$ are allowed, only with B a unitary $(2N \times 2N)$ matrix for which

$$Z = BZB^T. \tag{2.4.2}$$

Such matrices B form precisely the N-dimensional symplectic group (Weyl, 1946), usually denoted by $\mathrm{Sp}(N)$.

It is well known (Chevalley, 1946; Dieudonné, 1955) that the algebra of the symplectic group can be expressed most conveniently in terms of quaternions. We therefore introduce the standard quaternion notation for (2×2) matrices,

$$e_1 = \begin{bmatrix} i & 0 \\ 0 & -i \end{bmatrix}, \quad e_2 = \begin{bmatrix} 0 & 1 \\ -1 & 0 \end{bmatrix}, \quad e_3 = \begin{bmatrix} 0 & i \\ i & 0 \end{bmatrix}, \tag{2.4.3}$$

with the usual multiplication table

$$e_1^2 = e_2^2 = e_3^2 = -1, \tag{2.4.4}$$

$$e_1e_2 = -e_2e_1 = e_3, \qquad e_2e_3 = -e_3e_2 = e_1, \qquad e_3e_1 = -e_1e_3 = e_2. \tag{2.4.5}$$

Note that in (2.4.3), as well as throughout the rest of this book, i is the ordinary imaginary unit and not a quaternion unit. The matrices e_1, e_2, and e_3, together with the (2×2) unit matrix

$$1 = \begin{bmatrix} 1 & 0 \\ 0 & 1 \end{bmatrix},$$

form a complete set, and any (2×2) matrix with complex elements can be expressed linearly in terms of them with complex coefficients:

$$\begin{bmatrix} a & b \\ c & d \end{bmatrix} = \frac{1}{2}(a+d)1 - \frac{i}{2}(a-d)e_1 + \frac{1}{2}(b-c)e_2 - \frac{i}{2}(b+c)e_3. \tag{2.4.6}$$

All the $(2N \times 2N)$ matrices will be considered as cut into N^2 blocks of (2×2) and each (2×2) block expressed in terms of quaternions. In general, a $(2N \times 2N)$ matrix with complex elements thus becomes an $(N \times N)$ matrix with complex quaternion elements. In particular the matrix Z is now

$$Z = e_2 I, \tag{2.4.7}$$

where I is the $(N \times N)$ unit matrix. It can be verified that the rules of matrix multiplication are not changed by this partitioning.

Let us add some definitions. We call a quaternion "real" if it is of the form

$$q = q^{(0)} + \mathbf{q} \cdot \mathbf{e} \equiv q^{(0)} + q^{(1)}e_1 + q^{(2)}e_2 + q^{(3)}e_3, \tag{2.4.8}$$

with real coefficients $q^{(0)}$, $q^{(1)}$, $q^{(2)}$, and $q^{(3)}$. Thus a real quaternion does not correspond to a (2×2) matrix with real elements. Any complex quaternion has a "conjugate quaternion"

$$\bar{q} = q^{(0)} - \mathbf{q} \cdot \mathbf{e}, \tag{2.4.9}$$

which is distinct from its "complex conjugate"

$$q^* = q^{(0)*} + \mathbf{q}^* \cdot \mathbf{e}. \tag{2.4.10}$$

A quaternion with $q^* = q$ is real; one with $q^* = -q$ is pure imaginary; and one with $\bar{q} = q$ is a scalar. By applying both types of conjugation together, we obtain the "Hermitian conjugate"

$$q^\dagger = \bar{q}^* = q^{(0)*} - \mathbf{q}^* \cdot \mathbf{e}. \tag{2.4.11}$$

A quaternion with $q^{\dagger} = q$ is Hermitian and corresponds to the ordinary notion of a (2×2) Hermitian matrix; one with $q^{\dagger} = -q$ is anti-Hermitian. The conjugate (Hermitian conjugate) of a product of quaternions is the product of their conjugates (Hermitian conjugates) taken in the reverse order:

$$\overline{q_1 q_2 \dots q_n} = \bar{q}_n \dots \bar{q}_2 \bar{q}_1, \tag{2.4.12}$$

$$(q_1 q_2 \dots q_n)^{\dagger} = q_n^{\dagger} \dots q_2^{\dagger} q_1^{\dagger}. \tag{2.4.13}$$

Now consider a general $(2N \times 2N)$ matrix A which is to be written as an $(N \times N)$ matrix Q with quaternion elements q_{kj}; $k, j = 1, 2, \dots, N$. The standard matrix operations on A are then reflected in Q in the following way:

Transposition

$$(Q^T)_{kj} = -e_2 \bar{q}_{jk} e_2. \tag{2.4.14}$$

Hermitian conjugation

$$(Q^{\dagger})_{kj} = q_{jk}^{\dagger}. \tag{2.4.15}$$

Time reversal

$$(Q^R)_{kj} = e_2 (Q^T)_{kj} e_2^{-1} = \bar{q}_{jk}. \tag{2.4.16}$$

The matrix Q^R is called the "dual" of Q. A "self-dual" matrix is one with $Q^R = Q$. That is if $q_{jk} = \left[\begin{smallmatrix} a_{jk} & b_{jk} \\ c_{jk} & d_{jk} \end{smallmatrix}\right]$, then $Q = [q_{jk}]$ is self-dual if

$$a_{jk} = d_{kj}, \qquad b_{jk} = -b_{kj} \quad \text{and} \quad c_{jk} = -c_{kj}. \tag{2.4.17}$$

The usefulness of quaternion algebra is a consequence of the simplicity of (2.4.15) and (2.4.16). In particular, it is noteworthy that the time-reversal operator K does not appear explicitly in (2.4.16) as it did in (2.2.3). By (2.4.15) and (2.4.16) the condition

$$Q^R = Q^{\dagger} \tag{2.4.18}$$

is necessary and sufficient for the elements of Q to be real quaternions. When (2.4.18) holds, we call Q "quaternion real".

A unitary matrix B that satisfies (2.4.2) is automatically quaternion real. In fact, it satisfies the conditions

$$B^R = B^{\dagger} = B^{-1}, \tag{2.4.19}$$

which define the symplectic group. The matrices H which represent the energy operators of physical systems are Hermitian as well as self-dual:

$$H^R = H, \qquad H^\dagger = H, \tag{2.4.20}$$

hence are also quaternion real. From (2.4.15) and (2.4.16) we see that the quaternion elements of a self-dual Hermitian matrix must satisfy

$$q_{jk}^\dagger = \bar{q}_{jk} = q_{kj}, \tag{2.4.21}$$

or $q_{jk}^{(0)}$ must form a real symmetric matrix, whereas $q_{jk}^{(1)}$, $q_{jk}^{(2)}$, and $q_{jk}^{(3)}$ must form real antisymmetric matrices. Thus the number of real independent parameters that define a $(2N \times 2N)$ self-dual Hermitian matrix is

$$\frac{1}{2}N(N+1) + \frac{1}{2}N(N-1)\cdot 3 = N(2N-1).$$

From this notational excursion, let us come back to the point. Systems having odd-spin, invariance under time-reversal, but no rotational symmetry, must be represented by self-dual, Hermitian Hamiltonians. Therefore the Gaussian symplectic ensemble, as defined below, should be appropriate for their description.

Definition 2.4.1. The Gaussian symplectic ensemble E_{4G} is defined in the space T_{4G} of self-dual Hermitian matrices by the following properties:

(1) The ensemble is invariant under every automorphism

$$H \to W^R H W \tag{2.4.22}$$

of T_{4G} into itself, where W is any symplectic matrix.

(2) Various linearly independent components of H are also statistically independent.

These requirements put in the form of equations read as follows:

(1) The probability $P(H)\,dH$ that a system E_{4G} will belong to the volume element

$$dH = \prod_{k \leqslant j} dH_{kj}^{(0)} \prod_{\lambda=1}^{3} \prod_{k<j} dH_{kj}^{(\lambda)} \tag{2.4.23}$$

is invariant under symplectic transformations; that is,

$$P(H')\,dH' = P(H)\,dH, \tag{2.4.24}$$

if

$$H' = W^R H W, \tag{2.4.25}$$

where

$$W^R W = 1 \quad \text{or} \quad W Z W^T = Z. \tag{2.4.26}$$

(2) The probability density function $P(H)$ is a product of functions each of which depends on a single variable:

$$P(H) = \prod_{k \leqslant j} f_{kj}^{(0)}\left(H_{kj}^{(0)}\right) \prod_{\lambda=1}^{3} \prod_{k<j} f_{kj}^{(\lambda)}\left(H_{kj}^{(\lambda)}\right). \tag{2.4.27}$$

2.5 Gaussian Unitary Ensemble

Mathematically a much simpler ensemble is the Gaussian unitary ensemble E_{2G} which applies to systems without invariance under time reversal. Such systems are easily created in principle by putting an ordinary atom or nucleus, for example, into an externally generated magnetic field. The external field is not affected by the time-reversal operation. However, for the unitary ensemble to be applicable, the splitting of levels by the magnetic field must be at least as large as the average level spacing in the absence of the magnetic field. The magnetic field must, in fact, be so strong that it will completely "mix up" the level structure that would exist in zero field; for otherwise our random hypothesis cannot be justified. This state of affairs could never occur in nuclear physics. In atomic or molecular physics a practical application of the unitary ensemble may perhaps be possible.

A system without time-reversal invariance has a Hamiltonian that may be an arbitrary Hermitian matrix not restricted to be real or self-dual. Thus we are led to the following definition.

Definition 2.5.1. The Gaussian unitary ensemble E_{2G} is defined in the space T_{2G} of Hermitian matrices by the following properties:

(1) The probability $P(H)\,dH$ that a system of E_{2G} will belong to the volume element

$$dH = \prod_{k \leqslant j} dH_{kj}^{(0)} \prod_{k<j} dH_{kj}^{(1)}, \tag{2.5.1}$$

where $H_{kj}^{(0)}$ and $H_{kj}^{(1)}$ are real and imaginary parts of H_{kj}, is invariant under every automorphism

$$H \to U^{-1} H U \tag{2.5.2}$$

of T_{2G} into itself, where U is any unitary matrix.

(2) Various linearly independent components of H are also statistically independent.

In mathematical language these requirements are

(1)

$$P(H')\,dH' = P(H)\,dH, \tag{2.5.3}$$

if

$$H' = U^{-1} H U, \tag{2.5.4}$$

where U is any unitary matrix.

(2) $P(H)$ is a product of functions, each of which depends on a single variable:

$$P(H) = \prod_{k \leqslant j} f_{kj}^{(0)}\left(H_{kj}^{(0)}\right) \prod_{k<j} f_{kj}^{(1)}\left(H_{kj}^{(1)}\right). \tag{2.5.5}$$

2.6 Joint Probability Density Function for the Matrix Elements

We now come to the question of the extent to which we are still free to specify the joint probability density function $P(H)$. It will be seen that the two postulates of invariance and statistical independence elaborated above fix uniquely the functional form of $P(H)$.

The postulate of invariance restricts $P(H)$ to depend only on a finite number of traces of the powers of H. We state this fact as a lemma (Weyl, 1946).

Lemma 2.6.1. *All the invariants of an $(N \times N)$ matrix H under nonsingular similarity transformations A,*

$$H \to H' = A H A^{-1},$$

can be expressed in terms of the traces of the first N powers of H.

Actually the trace of the jth power of H is the sum of the jth powers of its eigenvalues λ_k, $k = 1, 2, \ldots, N$, of H,

$$\operatorname{tr} H^j = \sum_{k=1}^{N} \lambda_k^j \equiv p_j, \quad \text{say},$$

and it is a well-known fact that any symmetric function of the λ_k can be expressed in terms of the first N of the p_j; see, for example, Macdonald (1979) or Mehta (1989MT).

The postulate of statistical independence excludes everything except the traces of the first two powers, and these, too, may occur only in an exponential. To see this we will need the following lemma.

Lemma 2.6.2. *If three continuous and differentiable functions $f_k(x)$, $k = 1, 2, 3$, satisfy the equation*

$$f_1(xy) = f_2(x) + f_3(y), \tag{2.6.1}$$

then they are necessarily of the form $a \ln x + b_k$ ($k = 1, 2, 3$), with $b_1 = b_2 + b_3$.

Proof. Differentiating (2.6.1) with respect to x, we have

$$f_1'(xy) = \frac{1}{y} f_2'(x),$$

which, on integration with respect to y, gives

$$\frac{1}{x} f_1(xy) = f_2'(x) \ln y + \frac{1}{x} g(x), \tag{2.6.2}$$

where $g(x)$ is still arbitrary. Substituting $f_1(xy)$ from (2.6.2) into (2.6.1),

$$x f_2'(x) \ln y + g(x) - f_2(x) = f_3(y). \tag{2.6.3}$$

Therefore the left-hand side of (2.6.3) must be independent of x; this is possible only if

$$x f_2'(x) = a \quad \text{and} \quad g(x) - f_2(x) = b_3,$$

that is, only if

$$f_2(x) = a \ln x + b_2 = g(x) - b_3,$$

where a, b_2 and b_3 are arbitrary constants.

Now (2.6.3) gives

$$f_3(y) = a \ln y + b_3,$$

and finally (2.6.1) gives

$$f_1(xy) = a \ln(xy) + (b_2 + b_3).$$

□

Let us now examine the consequences of the statistical independence of the various components of H. Consider the particular transformation

$$H = U^{-1} H' U, \tag{2.6.4}$$

where

$$U = \begin{bmatrix} \cos\theta & \sin\theta & 0 & \dots & 0 \\ -\sin\theta & \cos\theta & 0 & \dots & 0 \\ 0 & 0 & 1 & \dots & 0 \\ \dots & \dots & \dots & \dots & \dots \\ 0 & 0 & 0 & \dots & 1 \end{bmatrix}, \tag{2.6.5}$$

or, in quaternion notation (provided N is even),

$$U = \begin{bmatrix} \cos\theta + e_2 \sin\theta & 0 & \dots & 0 & \dots & 0 \\ 0 & 1 & \dots & 0 & \dots & 0 \\ \dots & \dots & \dots & \dots & \dots & \dots \\ 0 & 0 & \dots & 0 & \dots & 1 \end{bmatrix}. \tag{2.6.6}$$

This U is, at the same time, orthogonal, symplectic, and unitary.

Differentiation of (2.6.4) with respect to θ gives

$$\frac{\partial H}{\partial \theta} = \frac{\partial U^T}{\partial \theta} H' U + U^T H' \frac{\partial U}{\partial \theta} = \frac{\partial U^T}{\partial \theta} U H + H U^T \frac{\partial U}{\partial \theta}, \tag{2.6.7}$$

and by substituting for U, U^T, $\partial U/\partial\theta$ and $\partial U^T/\partial\theta$ from (2.6.5) or (2.6.6) we get

$$\frac{\partial H}{\partial \theta} = AH + HA^T, \tag{2.6.8}$$

where

$$A = \frac{\partial U^T}{\partial \theta} U = \begin{bmatrix} 0 & -1 & 0 & 0 \\ 1 & 0 & 0 & 0 \\ 0 & 0 & 0 & 0 \\ \dots & \dots & \dots & \dots \\ 0 & 0 & 0 & 0 \end{bmatrix}, \tag{2.6.9}$$

or, in quaternion notation, A is diagonal.

$$A = \begin{bmatrix} -e_2 & 0 & \dots & 0 \\ 0 & 0 & \dots & 0 \\ \dots & \dots & \dots & \dots \\ 0 & 0 & \dots & 0 \end{bmatrix}. \tag{2.6.10}$$

If the probability density function

$$P(H)=\prod_{(\alpha)}\prod_{j\leqslant k} f_{kj}^{(\alpha)}\left(H_{kj}^{(\alpha)}\right) \tag{2.6.11}$$

is invariant under the transformation U, its derivative with respect to θ must vanish; that is

$$\sum \frac{1}{f_{kj}^{(\alpha)}} \frac{\partial f_{kj}^{(\alpha)}}{\partial H_{kj}^{(\alpha)}} \frac{\partial H_{kj}^{(\alpha)}}{\partial \theta}=0. \tag{2.6.12}$$

Let us write this equation explicitly, say, for the unitary case. Equations (2.6.8) and (2.6.12) give

$$\begin{aligned}
&\left[\left(-\frac{1}{f_{11}^{(0)}} \frac{\partial f_{11}^{(0)}}{\partial H_{11}^{(0)}}+\frac{1}{f_{22}^{(0)}} \frac{\partial f_{22}^{(0)}}{\partial H_{22}^{(0)}}\right)\left(2 H_{12}^{(0)}\right)+\frac{1}{f_{12}^{(0)}} \frac{\partial f_{12}^{(0)}}{\partial H_{12}^{(0)}}\left(H_{11}^{(0)}-H_{22}^{(0)}\right)\right] \\
&+\sum_{k=3}^{N}\left(-\frac{1}{f_{1k}^{(0)}} \frac{\partial f_{1k}^{(0)}}{\partial H_{1k}^{(0)}} H_{2k}^{(0)}+\frac{1}{f_{2k}^{(0)}} \frac{\partial f_{1k}^{(0)}}{\partial H_{2k}^{(0)}} H_{1k}^{(0)}\right) \\
&+\sum_{k=3}^{N}\left(-\frac{1}{f_{1k}^{(1)}} \frac{\partial f_{1k}^{(1)}}{\partial H_{1k}^{(1)}} H_{2k}^{(1)}+\frac{1}{f_{2k}^{(1)}} \frac{\partial f_{2k}^{(1)}}{\partial H_{2k}^{(1)}} H_{1k}^{(1)}\right)=0.
\end{aligned} \tag{2.6.13}$$

The braces at the left-hand side of this equation depend on mutually exclusive sets of variables and their sum is zero. Therefore each must be a constant; for example,

$$-\frac{H_{2k}^{(0)}}{f_{1k}^{(0)}} \frac{\partial f_{1k}^{(0)}}{\partial H_{1k}^{(0)}}+\frac{H_{1k}^{(0)}}{f_{2k}^{(0)}} \frac{\partial f_{2k}^{(0)}}{\partial H_{2k}^{(0)}}=C_{k}^{(0)}. \tag{2.6.14}$$

On dividing both side of (2.6.14) by $H_{1k}^{(0)} H_{2k}^{(0)}$ and applying the Lemma 2.6.2, we conclude that the constant $C_{k}^{(0)}$ must be zero, that is,

$$\frac{1}{H_{1k}^{(0)}} \frac{1}{f_{1k}^{(0)}} \frac{\partial f_{1k}^{(0)}}{\partial H_{1k}^{(0)}}=\frac{1}{H_{2k}^{(0)}} \frac{1}{f_{2k}^{(0)}} \frac{\partial f_{2k}^{(0)}}{\partial H_{2k}^{(0)}}=\text{constant}=-2a, \quad \text{say}, \tag{2.6.15}$$

which on integration gives

$$f_{1k}^{(0)}\left(H_{1k}^{(0)}\right)=\exp\left[-a\left(H_{1k}^{(0)}\right)^{2}\right]. \tag{2.6.16}$$

In the other two cases we also derive a similar equation. Now because the off-diagonal elements come on as squares in the exponential and all invariants are expressible in terms of the traces of powers of H, the function $P(H)$ is an exponential that contains traces of at most the second power of H.

Because $P(H)$ is required to be invariant under more general transformations than we have here considered, one might think that the form of $P(H)$ is further restricted. This, however, is not so, for

$$\begin{aligned} P(H) &= \exp(-a \operatorname{tr} H^2 + b \operatorname{tr} H + c) \\ &= e^c \prod_j \exp(b H_{jj}^{(0)}) \prod_{k \leqslant j} \exp[-a(H_{kj}^{(0)})^2] \prod_\lambda \prod_{k<j} \exp[-a(H_{kj}^{(\lambda)})^2] \end{aligned} \tag{2.6.17}$$

is already a product of functions, each of which depends on a separate variable. Moreover, because we require $P(H)$ to be normalizable and real, a must be real and positive and b and c must be real.

Therefore we have proved the following theorem (Porter and Rosenzweig, 1960a).

Theorem 2.6.3. *In all the above three cases the form of $P(H)$ is automatically restricted to*

$$P(H) = \exp(-a \operatorname{tr} H^2 + b \operatorname{tr} H + c), \tag{2.6.18}$$

where a is real and positive and b and c are real.

In the foregoing discussion we have emphasized the postulate of statistical independence of various components of H even at the risk of frequent repetitions. This statistical independence is important in restricting $P(H)$ to the simple form (2.6.18), and hence makes the subsequent analytical work tractable. However, it lacks a clear physical motivation and therefore looks somewhat artificial.

The main objection to the assumption of statistical independence, leading to (2.6.18), is that all values of $H_{kj}^{(\lambda)}$ are not equally weighted and therefore do not correspond to all "interactions" being "equally probable". By a formal change Dyson (1962a, I) has defined his "circular ensembles", which are esthetically more satisfactory and equally easy to work with. We shall come to them in Chapters 9 to 11. They give equivalent results as we will see in Chapter 11. On the other hand, Rosenzweig (1963), has emphasized the "fixed strength" ensemble briefly considered in Chapter 27. Others (Leff, 1963; Fox and Kahn, 1964) have arbitrarily tried the so-called "generalized" ensembles related to classical orthogonal polynomials other than the Hermite polynomials. We will study them in Chapter 19.

If we keep only the first requirement that $P(H)$ is invariant under $H \to UHU^{-1}$, then $P(H)$ may be any function of the traces of powers of H. People have studied in particular the case when $P(H) \propto \exp(-\operatorname{tr} V(H))$, where $V(x)$ is a polynomial, preferably of even degree with the coefficient of the highest power positive. This case is briefly mentioned in Section 19.4.

2.7 Gaussian Ensemble of Hermitian Matrices With Unequal Real and Imaginary Parts

The ensembles so far considered were characterized by two requirements: (i) the probability $P(H)\,dH$ that a system belongs to the volume element dH is such that $P(H)$ is invariant under $H \to U^{-1}HU$, where U is any matrix which is either real orthogonal, symplectic or unitary according to the symmetry of the system; and (ii) various linearly independent components of H are also statistically independent.

If for our system the time reversal invariance is only weakly violated, then the appropriate ensemble will be almost an orthogonal or symplectic ensemble slightly mixed with the unitary ensemble. Keeping the hypothesis (ii) that various linearly independent parts of H are also statistically independent, we should now take

$$P(H) \propto \exp\bigl(-\operatorname{tr}\bigl(H_1^2/c_1 + H_2^2/c_2\bigr)\bigr), \tag{2.7.1}$$

where $H = H_1 + H_2$, H_1 and H_2 are Hermitian, H_1 is symmetric (self-dual) and H_2 is anti-symmetric (anti-self-dual). If $c_2 = 0$, then $H_2 = 0$ with probability 1, and we have the orthogonal (symplectic) ensemble; if $c_2 = c_1$, then we have the unitary ensemble. For a small violation of the time reversal invariance, $c_2 \ll c_1$. Since it does not increase the mathematical difficulties and the analytical solution is as elegant, we will treat in Chapter 14 the general case where c_1 and c_2 are arbitrary real numbers.

Note that under real orthogonal transformations the traces of powers of H_1 and H_2 (i.e. of real and imaginary parts of H) are invariant and so is the probability density $P(H)$ of Eq. (2.7.1). However, under unitary transformations the real and imaginary parts of H mix up and the above $P(H)$ is no longer invariant unless $c_1 = c_2$.

2.8 Anti-Symmetric Hermitian Matrices

Though physically not relevant, the mathematical analysis of a Gaussian ensemble of anti-symmetric (or that of anti-self-dual quaternion) Hermitian matrices is equally elegant. As above, the probability will be taken as

$$P(H)\,dH, \quad dH = \prod_{j<k} dH_{jk},$$

and

$$P(H) \propto \exp\bigl(-a \operatorname{tr} H^2\bigr).$$

Summary of Chapter 2

The probability density $P(H)$ of a random matrix H is proportional to $\exp(-a \operatorname{tr} H^2 + b \operatorname{tr} H + c)$ with certain constants a, b, c in the following three cases:

(1) If H is a Hermitian symmetric random matrix, its elements H_{jk} with $j \geqslant k$ are statistically independent, and $P(H)$ is invariant under all real orthogonal transformations of H. The resulting ensemble is named as Gaussian orthogonal.
(2) If H is a Hermitian random matrix, its diagonal elements H_{jj} and the real and imaginary parts of its off-diagonal elements H_{jk} for $j > k$ are statistically independent, and $P(H)$ is invariant under all unitary transformations of H. The resulting ensemble is named as Gaussian unitary.
(3) If H is a Hermitian self-dual random matrix, its diagonal elements H_{jj} and the four quaternionic components of its off-diagonal elements H_{jk} with $j > k$ are statistically independent, and $P(H)$ is invariant under all symplectic transformations of H. The resulting ensemble is named as Gaussian symplectic.

Moreover,

(4) For a Hermitian anti-symmetric random matrix H, it is not unreasonable to take the elements H_{jk} with $j > k$ as Gaussian variables with the same variance.
(5) Similarly, for a Hermitian random matrix H, with $P(H)$ not invariant under unitary transformations of H, it is not unreasonable to take its symmetric and anti-symmetric parts to have the probability densities prescribed under cases (1) and (4) above.

Invariance of $P(H)$ under orthogonal, unitary or symplectic transformations of H is required by physical considerations and depend on whether the system described by the Hamiltonian H has or does not have certain symmetries like time reversal or rotational symmetry. The statistical independence of the various real parameters entering H is assumed for simplicity.

3

GAUSSIAN ENSEMBLES. THE JOINT PROBABILITY DENSITY FUNCTION FOR THE EIGENVALUES

In this chapter we will derive the joint probability density function for the eigenvalues of H implied by the Gaussian densities for its matrix elements. Finally an argument based on information theory is given. This argument rationalizes any of the Gaussian probability densities for the matrix elements; it even allows to define an ensemble having a preassigned eigenvalue density.

3.1 Orthogonal Ensemble

The joint probability density function (abbreviated j.p.d.f. later in the chapter) for the eigenvalues $\theta_1, \theta_2, \ldots, \theta_N$ can be obtained from (2.6.18), by expressing the various components of H in terms of the N eigenvalues θ_j and other mutually independent variables, p_μ, say, which together with the θ_j form a complete set. In an $(N \times N)$ real symmetric matrix the number of independent real parameters which determine all H_{kj} is $N(N+1)/2$. We may take these as H_{kj} with $k \leqslant j$. The number of extra parameters p_μ needed is therefore

$$\ell = \frac{1}{2}N(N+1) - N = \frac{1}{2}N(N-1). \tag{3.1.1}$$

Because

$$\operatorname{tr} H^2 = \sum_1^N \theta_j^2, \qquad \operatorname{tr} H = \sum_1^N \theta_j, \tag{3.1.2}$$

the probability density that the N roots and $N(N-1)/2$ parameters will occur around the values $\theta_1, \ldots, \theta_N$ and $p_1, p_2, \ldots, p_\ell$ is, according to (2.6.18)

$$P(\theta_1, \ldots, \theta_N; p_1, \ldots, p_\ell) = \exp\left(-a \sum_1^N \theta_j^2 + b \sum_1^N \theta_j + c\right) J(\theta, p), \tag{3.1.3}$$

where J is the Jacobian

$$J(\theta, p) = \left| \frac{\partial(H_{11}, H_{12}, \ldots, H_{NN})}{\partial(\theta_1, \ldots, \theta_N, p_1, \ldots, p_\ell)} \right|. \tag{3.1.4}$$

Hence the j.p.d.f. of the eigenvalues θ_j can be obtained by integrating (3.1.3) over the parameters $p_1, \ldots, p_\ell$. It is usually possible to choose these parameters so that the Jacobian (3.1.4) becomes a product of a function f of the θ_j and a function g of the p_μ. If this is the case, the integration provides the required j.p.d.f. as a product of the exponential in (3.1.3), the function f of the θ_j and a constant. The constant can then be absorbed in c in the exponential.

To define the parameters p_μ (Wigner, 1965a) we recall that any real symmetric matrix H can be diagonalized by a real orthogonal matrix (cf. Appendix A.3):

$$H = U \Theta U^{-1} \tag{3.1.5}$$

$$= U \Theta U^T, \tag{3.1.5$'$}$$

where Θ is the diagonal matrix with diagonal elements $\theta_1, \theta_2, \ldots, \theta_N$ arranged in some order, say, $\theta_1 \leqslant \theta_2 \leqslant \cdots \leqslant \theta_N$, and U is a real orthogonal matrix

$$UU^T = U^T U = 1, \tag{3.1.6}$$

whose columns are the normalized eigenvectors of H. These eigenvectors are, or may be chosen to be, mutually orthogonal. To define U completely we must in some way fix the phases of the eigenvectors, for instance by requiring that the first nonvanishing component be positive. Thus U depends on $N(N-1)/2$ real parameters and may be chosen to be U_{kj}, $k > j$. If H has multiple eigenvalues, further conditions are needed to fix U completely. It is not necessary to specify them, for they apply only in regions of lower dimensionality which are irrelevant to the probability density function. At any rate enough appropriate conditions are imposed on U so that it is uniquely characterized by the $N(N-1)/2$ parameters p_μ. Once this is done, the matrix H, which completely determines the Θ and the U subject to the preceding conditions, also determines the θ_j and the p_μ uniquely. Conversely, the θ_j and p_μ completely determine the U and Θ, and hence by (3.1.5) all the matrix elements of H.

Differentiating (3.1.6), we get

$$\frac{\partial U^T}{\partial p_\mu} U + U^T \frac{\partial U}{\partial p_\mu} = 0, \tag{3.1.7}$$

and because the two terms in (3.1.7) are the Hermitian conjugates of each other,

$$S^{(\mu)} = U^T \frac{\partial U}{\partial p_\mu} = -\frac{\partial U^T}{\partial p_\mu} U \tag{3.1.8}$$

is an antisymmetric matrix.

Also from (3.1.5) we have

$$\frac{\partial H}{\partial p_\mu} = \frac{\partial U}{\partial p_\mu} \Theta U^T + U \Theta \frac{\partial U^T}{\partial p_\mu}. \tag{3.1.9}$$

On multiplying (3.1.9) by U^T on the left and by U on the right, we get

$$U^T \frac{\partial H}{\partial p_\mu} U = S^{(\mu)} \Theta - \Theta S^{(\mu)}. \tag{3.1.10}$$

In terms of its components, (3.1.10) reads

$$\sum_{j,k} \frac{\partial H_{jk}}{\partial p_\mu} U_{j\alpha} U_{k\beta} = S^{(\mu)}_{\alpha\beta} (\theta_\beta - \theta_\alpha). \tag{3.1.11}$$

In a similar way, by differentiating (3.1.5) with respect to θ_γ,

$$\sum_{j,k} \frac{\partial H_{jk}}{\partial \theta_\gamma} U_{j\alpha} U_{k\beta} = \frac{\partial \Theta_{\alpha\beta}}{\partial \theta_\gamma} = \delta_{\alpha\beta}\delta_{\alpha\gamma}. \tag{3.1.12}$$

The matrix of the Jacobian in (3.1.4) can be written in partitioned form as

$$[J(\theta, p)] = \begin{bmatrix} \dfrac{\partial H_{jj}}{\partial \theta_\gamma} & \dfrac{\partial H_{jk}}{\partial \theta_\gamma} \\ \dfrac{\partial H_{jj}}{\partial p_\mu} & \dfrac{\partial H_{jk}}{\partial p_\mu} \end{bmatrix}. \tag{3.1.13}$$

The two columns in (3.1.13) correspond to N and $N(N-1)/2$ actual columns; $1 \leqslant j < k \leqslant N$. The two rows in (3.1.13) correspond again to N and $N(N-1)/2$ actual rows: $\gamma = 1, 2, \ldots, N$; $\mu = 1, 2, \ldots, N(N-1)/2$. If we multiply the $[J]$ in (3.1.13) on the right by the $N(N+1)/2 \times N(N+1)/2$ matrix written in partitioned form as

$$[V] = \begin{bmatrix} (U_{j\alpha} U_{j\beta}) \\ (2U_{j\alpha} U_{k\beta}) \end{bmatrix}, \tag{3.1.14}$$

in which the two rows correspond to N and $N(N-1)/2$ actual rows, $1 \leqslant j < k \leqslant N$, and the column corresponds to $N(N+1)/2$ actual columns, $1 \leqslant \alpha \leqslant \beta \leqslant N$, we get by using (3.1.11) and (3.1.12)

$$[J][V] = \begin{bmatrix} \delta_{\alpha\beta}\delta_{\alpha\gamma} \\ S^{(\mu)}_{\alpha\beta}(\theta_\beta - \theta_\alpha) \end{bmatrix}. \tag{3.1.15}$$

The two rows on the right-hand side correspond to N and $N(N-1)/2$ actual rows and the column corresponds to $N(N+1)/2$ actual columns. Taking the determinant on both sides of (3.1.15), we have

$$J(\theta, p) \det V = \prod_{\alpha<\beta} (\theta_\beta - \theta_\alpha) \det \begin{bmatrix} \delta_{\alpha\beta}\delta_{\alpha\gamma} \\ S^{\mu}_{\alpha\beta} \end{bmatrix}$$

or

$$J(\theta, p) = \prod_{\alpha<\beta} |\theta_\beta - \theta_\alpha| f(p), \tag{3.1.16}$$

where $f(p)$ is independent of the θ_j and depends only on the parameters p_μ.

By inserting this result in (3.1.3) and integrating over the variables p_μ we get the j.p.d.f. for the eigenvalues of the matrices of an orthogonal ensemble

$$P(\theta_1, \ldots, \theta_N) = \exp\left[-\sum_1^N \left(a\theta_j^2 - b\theta_j - c\right) \right] \prod_{j<k} |\theta_k - \theta_j|, \tag{3.1.17}$$

where c is some new constant. Moreover, if we shift the origin of the θ to $b/2a$ and change the energy scale everywhere by a constant factor $\sqrt{2a}$, we may replace θ_j with $(1/\sqrt{2a})x_j + b/2a$. By this formal change (3.1.17) takes the simpler form

$$P_{N1}(x_1, \ldots, x_N) = C_{N1} \exp\left(-\frac{1}{2} \sum_1^N x_j^2 \right) \prod_{j<k} |x_j - x_k|, \tag{3.1.18}$$

where C_{N1} is a constant. (Subscript 1 is to remind of the power of the product of differences.)

3.2 Symplectic Ensemble

As the analysis is almost identical in all three invariant cases, we have presented the details for one particular ensemble, the orthogonal one. Here and in Sections 3.3 and 3.4 we indicate briefly the modifications necessary to arrive at the required j.p.d.f. in the other cases.

Corresponding to the result that a real symmetric matrix can be diagonalized by a real orthogonal matrix, we have the following:

Theorem 3.2.1. *Given a quaternion-real, self-dual matrix H, there exists a symplectic matrix U such that*

$$H = U\Theta U^{-1} = U\Theta U^{R}, \tag{3.2.1}$$

where Θ is diagonal, real, and scalar (cf. Appendix A.3).

The fact that Θ is scalar means that it consists of N blocks of the form

$$\begin{bmatrix} \theta_j & 0 \\ 0 & \theta_j \end{bmatrix} \tag{3.2.2}$$

along the main diagonal. Thus the eigenvalues of H consist of N equal pairs. The Hamiltonian of any system which is invariant under time reversal, which has odd spin, and no rotational symmetry satisfies the conditions of Theorem 3.2.1. All energy levels of such a system will be doubly degenerate. This is Kramer's degeneracy (Kramer, 1930), and Theorem 3.2.1 shows how it appears naturally in the quaternion language.

Apart from the N eigenvalues θ_j, the number of real independent parameters p_μ needed to characterize an $N \times N$ quaternion-real, self-dual matrix H is

$$\ell = 4 \cdot \frac{1}{2} N(N-1) = 2N(N-1). \tag{3.2.3}$$

Equations (3.1.2) and (3.1.3) are replaced, respectively, by

$$\operatorname{tr} H^2 = 2 \sum_{j=1}^{N} \theta_j^2, \qquad \operatorname{tr} H = 2 \sum_{j=1}^{N} \theta_j, \tag{3.2.4}$$

and

$$P(\theta_1, \ldots, \theta_N; p_1, \ldots, p_\ell) = \exp\left[-\sum_{j=1}^{N}\left(2a\theta_j^2 - 2b\theta_j - c\right)\right] J(\theta, p), \quad (3.2.5)$$

where $J(\theta, p)$ is now given by

$$J(\theta, p) = \left|\frac{\partial(H_{11}^{(0)}, \ldots, H_{NN}^{(0)}, H_{12}^{(0)}, \ldots, H_{12}^{(3)}, \ldots, H_{N-1,N}^{(0)}, \ldots, H_{N-1,N}^{(3)})}{\partial(\theta_1, \ldots, \theta_N, p_1, \ldots, p_{2N(N-1)})}\right|. \quad (3.2.6)$$

Equation (3.1.5) is replaced by (3.2.1); (3.1.6)–(3.1.10) are valid if U^T is replaced by U^R. Note that these equations are now in the quaternion language, and we need to separate the four quaternion parts of modified (3.2.1). For this we let

$$H_{jk} = H_{jk}^{(0)} + H_{jk}^{(1)} e_1 + H_{jk}^{(2)} e_2 + H_{jk}^{(3)} e_3, \quad (3.2.7)$$

$$S_{\alpha\beta}^{(\mu)} = S_{\alpha\beta}^{(0\mu)} + S_{\alpha\beta}^{(1\mu)} e_1 + S_{\alpha\beta}^{(2\mu)} e_2 + S_{\alpha\beta}^{(3\mu)} e_3, \quad (3.2.8)$$

and write (3.1.10) and the equation corresponding to (3.1.12) in the form of partitioned matrices:

$$\begin{bmatrix} \dfrac{\partial H_{jj}^{(0)}}{\partial \theta_\gamma} & \dfrac{\partial H_{jk}^{(0)}}{\partial \theta_\gamma} & \dfrac{\partial H_{jk}^{(1)}}{\partial \theta_\gamma} & \cdots & \dfrac{\partial H_{jk}^{(3)}}{\partial \theta_\gamma} \\[2ex] \dfrac{\partial H_{jj}^{(0)}}{\partial p_\mu} & \dfrac{\partial H_{jk}^{(0)}}{\partial p_\mu} & \dfrac{\partial H_{jk}^{(1)}}{\partial p_\mu} & \cdots & \dfrac{\partial H_{jk}^{(3)}}{\partial p_\mu} \end{bmatrix} \begin{bmatrix} v & w \\ A^{(0)} & B^{(0)} \\ \cdots & \cdots \\ A^{(3)} & B^{(3)} \end{bmatrix}$$
$$= \begin{bmatrix} \rho_{\gamma,\alpha} & \sigma_{\gamma,\alpha\beta}^{(0)} & \cdots & \sigma_{\gamma,\alpha\beta}^{(3)} \\ \varepsilon_\alpha^{(\mu)} & S_{\alpha\beta}^{(0\mu)}(\theta_\beta - \theta_\alpha) & \cdots & S_{\alpha\beta}^{(3\mu)}(\theta_\beta - \theta_\alpha) \end{bmatrix},$$
$$1 \leqslant j < k \leqslant N, \quad 1 \leqslant \alpha < \beta \leqslant N, \quad 1 \leqslant \gamma \leqslant N,$$
$$1 \leqslant \mu \leqslant 2N(N-1), \quad (3.2.9)$$

where the matrices $\partial H_{jj}^{(0)}/\partial\theta_\gamma$, v, and ρ are $N \times N$, the matrices $\partial H_{jk}^{(\lambda)}/\partial\theta_\gamma$ and $\sigma_{\gamma,\alpha\beta}^{(\lambda)}$, with $\lambda = 0, 1, 2, 3$, are $N \times N(N-1)/2$, the $A^{(\lambda)}$ are all $N(N-1)/2 \times N$, the $\partial H_{jj}^{(0)}/\partial p_\mu$ and the $\varepsilon_\alpha^{(\mu)}$ are $2N(N-1) \times N$, the w is $N \times 2N(N-1)$, the $\partial H_{jk}^{(\lambda)}/\partial p_\mu$ and the $S_{\alpha\beta}^{(\lambda\mu)}$ are $2N(N-1) \times N(N-1)/2$ and the matrices $B^{(\lambda)}$ are $N(N-1)/2 \times 2N(N-1)$. The matrices ρ and the σ appear as we separate the result of differentiation of (3.2.1) with respect to θ_γ into quaternion components. Because Θ is diagonal and scalar, the $\sigma^{(\lambda)}$ are all zero matrices. Moreover, the matrix ρ does not depend on θ_γ, for Θ depends linearly on the θ_γ. The computation of the matrices $v, w, A^{(\lambda)}$, and $B^{(\lambda)}$ is straightforward, but we do not require them. All we need is to note that they are formed of the various components of U, and hence do not depend on θ_γ.

Now we take the determinant on both sides of (3.2.9). The determinant of the first matrix on the left is the Jacobian (3.2.6). Because the $\sigma^{(\lambda)}$ are all zero, the determinant of the right-hand side breaks into a product of two determinants:

$$\det[\rho_{\gamma.\alpha}]\det\left[S_{\alpha\beta}^{(\lambda\mu)}(\theta_\beta-\theta_\alpha)\right], \tag{3.2.10}$$

the first one being independent of the θ_γ, whereas the second is

$$\prod_{\alpha<\beta}(\theta_\beta-\theta_\alpha)^4\det\left[S_{\alpha\beta}^{(\lambda\mu)}\right]. \tag{3.2.11}$$

Thus

$$J(\theta,p)=\prod_{\alpha<\beta}(\theta_\beta-\theta_\alpha)^4 f(p), \tag{3.2.12}$$

which corresponds to (3.1.16).

By inserting (3.2.12) into (3.2.5) and integrating over the parameters, we obtain the j.p.d.f.

$$P(\theta_1,\ldots,\theta_N)=\exp\left(-2a\sum_{j=1}^N\theta_j^2+2b\sum_{j=1}^N\theta_j+c\right)\prod_{j<k}(\theta_j-\theta_k)^4. \tag{3.2.13}$$

As before, we may shift the origin to make $b=0$ and change the scale of energy to make $a=1$. Thus the j.p.d.f. for the eigenvalues of matrices in the symplectic ensemble in its simple form is

$$P_{N4}(x_1,\ldots,x_N)=C_{N4}\exp\left(-2\sum_{j=1}^N x_j^2\right)\prod_{j<k}(x_j-x_k)^4, \tag{3.2.14}$$

where C_{N4} is a constant. (Subscript 4 to remind again of the power of the product of differences.)

3.3 Unitary Ensemble

In addition to the real eigenvalues, the number of real independent parameters p_μ needed to specify an arbitrary Hermitian matrix H completely is $N(N-1)$. Equations

(3.1.2) and (3.1.3) remain unchanged, but (3.1.4) is replaced by

$$J(\theta, p) = \frac{\partial(H_{11}^{(0)}, \ldots, H_{NN}^{(0)}, H_{12}^{(0)}, H_{12}^{(1)}, \ldots, H_{N-1,N}^{(0)}, H_{N-1,N}^{(1)})}{\partial(\theta_1, \ldots, \theta_N, p_1, \ldots, p_{N(N-1)})}, \tag{3.3.1}$$

where $H_{jk}^{(0)}$ and $H_{jk}^{(1)}$ are the real and imaginary parts of H_{jk}. Equations (3.1.5) to (3.1.10) are valid if U^T replaced by $U^\dagger$. Instead of (3.1.11) and (3.1.12) we now have

$$\sum_{j,k} \frac{\partial H_{jk}}{\partial p_\mu} U_{j\alpha}^* U_{k\beta} = S_{\alpha\beta}^{(\mu)}(\theta_\beta - \theta_\alpha), \tag{3.3.2}$$

$$\sum_{j,k} \frac{\partial H_{jk}}{\partial p_\gamma} U_{j\alpha}^* U_{k\beta} = \frac{\partial \Theta_{\alpha\beta}}{\partial \theta_\gamma} = \delta_{\alpha\beta}\delta_{\alpha\gamma}. \tag{3.3.3}$$

By separating the real and imaginary parts we may write these equations in partitioned matrix notation as

$$\begin{bmatrix} \dfrac{\partial H_{jj}^{(0)}}{\partial \theta_\gamma} & \dfrac{\partial H_{jk}^{(0)}}{\partial \theta_\gamma} & \dfrac{\partial H_{jk}^{(1)}}{\partial \theta_\gamma} \\ \dfrac{\partial H_{jj}^{(0)}}{\partial p_\mu} & \dfrac{\partial H_{jk}^{(0)}}{\partial p_\mu} & \dfrac{\partial H_{jk}^{(1)}}{\partial p_\mu} \end{bmatrix} \begin{bmatrix} v & w \\ A^{(0)} & B^{(0)} \\ A^{(1)} & B^{(1)} \end{bmatrix}$$
$$= \begin{bmatrix} \rho_{\gamma,\alpha} & \sigma_{\gamma,\alpha\beta}^{(0)} & \sigma_{\gamma,\alpha\beta}^{(1)} \\ \varepsilon_\alpha^{(\mu)} & S_{\alpha\beta}^{(0\mu)}(\theta_\beta - \theta_\alpha) & S_{\alpha\beta}^{(1\mu)}(\theta_\beta - \theta_\alpha) \end{bmatrix},$$
$$1 \leqslant j < k \leqslant N, \quad 1 \leqslant \alpha < \beta \leqslant N,$$
$$1 \leqslant \mu \leqslant N(N-1), \quad 1 \leqslant \gamma \leqslant N, \tag{3.3.4}$$

where $S_{\alpha\beta}^{(0\mu)}$ and $S_{\alpha\beta}^{(1\mu)}$ are the real and imaginary parts of $S_{\alpha\beta}^{(\mu)}$. The matrices $\partial H_{jj}^{(0)}/\partial\theta_\gamma$, v and ρ are $N \times N$; the $\partial H_{jk}^{(\lambda)}/\partial\theta_\gamma$ and the $\sigma_{\gamma,\alpha\beta}^{(\lambda)}$ are $N \times N(N-1)/2$; the $A^{(\lambda)}$ are $N(N-1)/2 \times N$; the $\partial H_{jk}^{(\lambda)}/\partial p_\mu$ and $S_{\alpha\beta}^{(\lambda\mu)}$ are $N(N-1) \times N(N-1)/2$; the $B^{(\lambda)}$ are $N(N-1)/2 \times N(N-1)$; the $\partial H_{jj}^{(0)}/\partial p_\mu$ and the $\varepsilon_\alpha^{(\mu)}$ are $N(N-1) \times N$; and the matrix w is $N \times N(N-1)$. To compute $v, w, A^{(\lambda)}, \rho, \varepsilon, \sigma^{(\lambda)}$, etc., is again straightforward, but we do not need them explicitly. What we want to emphasize is that they are either constructed from the components of U or arise from the differentiation of Θ with respect to θ_j and consequently are all independent of the eigenvalues θ_j. Similarly, $S^{(\mu)}$ is independent of θ_j. One more bit of information we need is that $\sigma^{(0)}$ and $\sigma^{(1)}$ are zero matrices, which can be verified.

Thus by taking the determinants on both sides of (3.3.4) and removing the factors $(\theta_\beta - \theta_\alpha)$ we have

$$J(\theta, p) = \prod_{\alpha<\beta} (\theta_\beta - \theta_\alpha)^2 f(p), \tag{3.3.5}$$

where $f(p)$ is some function of the p_μ.

By inserting (3.3.5) into (3.1.3) and integrating over the parameters p_μ we get the j.p.d.f. for the eigenvalues of matrices in the unitary ensemble

$$P(\theta_1, \ldots, \theta_N) = \exp\left(-a\sum_1^N \theta_j^2 + b\sum_1^N \theta_j + c\right) \prod_{j<k} (\theta_j - \theta_k)^2, \tag{3.3.6}$$

and, as before, by a proper choice of the origin and the scale of energy we have

$$P_{N2}(x_1, \ldots, x_N) = C_{N2} \exp\left(-\sum_1^N x_j^2\right) \prod_{j<k} (x_j - x_k)^2. \tag{3.3.7}$$

We record (3.1.18), (3.2.14) and (3.3.7) as a theorem.

Theorem 3.3.1. *The joint probability density function for the eigenvalues of matrices from a Gaussian orthogonal, Gaussian symplectic or Gaussian unitary ensemble is given by*

$$P_{N\beta}(x_1, \ldots, x_N) = C_{N\beta} \exp\left(-\frac{1}{2}\beta\sum_1^N x_j^2\right) \prod_{j<k} |x_j - x_k|^\beta, \tag{3.3.8}$$

where $\beta = 1$ if the ensemble is orthogonal, $\beta = 4$ if it is symplectic, and $\beta = 2$ if it is unitary. The constant $C_{N\beta}$ is chosen in such a way that the $P_{N\beta}$ is normalized to unity:

$$\int_{-\infty}^{\infty} \cdots \int_{-\infty}^{\infty} P_{N\beta}(x_1, \ldots, x_N)\, dx_1 \cdots dx_N = 1. \tag{3.3.9}$$

According to Selberg the normalization constant $C_{N\beta}$ is given by (see Chapter 17),

$$C_{N\beta}^{-1} = (2\pi)^{N/2} \beta^{-N/2-\beta N(N-1)/4} [\Gamma(1+\beta/2)]^{-N} \prod_{j=1}^N \Gamma(1+\beta j/2). \tag{3.3.10}$$

For the physically interesting cases $\beta = 1, 2$ and 4 we will recalculate this value in a different way later (see Sections 6.2, 7.2 and 8.2).

It is possible to understand the different powers β that appear in Eq. (3.3.8) by a simple mathematical argument based on counting dimensions. The dimension of the space T_{1G} is $N(N+1)/2$, whereas the dimension of the subspace T'_{1G}, composed of the matrices in T_{1G} with two equal eigenvalues, is $N(N+1)/2-2$. Because of the single restriction, the equality of two eigenvalues, the dimension should normally have decreased by one; as it is decreased by two, it indicates a factor in Eq. (3.3.8) linear in $(x_j - x_k)$. Similarly, when $\beta = 2$, the dimension of T_{2G} is N^2, whereas that of T'_{2G} is $N^2 - 3$. When $\beta = 4$, the dimension of T_{4G} is $N(2N-1)$, whereas that of T'_{4G} is $N(2N-1)-5$ (see Appendix A.4).

3.4 Ensemble of Anti-Symmetric Hermitian Matrices

The eigenvalues and eigenvectors of anti-symmetric Hermitian matrices come in pairs; if θ is an eigenvalue with the eigenvector V_θ, then $-\theta$ is an eigenvalue with the eigenvector V_θ^*. The vectors V_θ and V_θ^* can be normalized, and if $\theta \neq 0$ they are orthogonal. Thus if $V_\theta = \xi^{(\theta)} + i\eta^{(\theta)}$, $\xi^{(\theta)}$ and $\eta^{(\theta)}$ are real, then

$$\left(V_\theta, V_\theta^*\right) = \textstyle\sum_j \left(\xi_j^{(\theta)} + i\eta_j^{(\theta)}\right)^2 = 0, \tag{3.4.1}$$

$$\left(V_\theta, V_\theta\right) = \textstyle\sum_j \left|\xi_j^{(\theta)} + i\eta_j^{(\theta)}\right|^2 = 2. \tag{3.4.2}$$

These two equations are equivalent to

$$\sum_j \left(\xi_j^{(\theta)}\right)^2 = \sum_j \left(\eta_j^{(\theta)}\right)^2 = 1, \qquad \sum_j \xi_j^{(\theta)}\eta_j^{(\theta)} = 0. \tag{3.4.3}$$

Therefore if $\theta \neq 0$, the real and imaginary parts of V_θ have the same length and are orthogonal to each other. Moreover, the real and imaginary parts of V_θ are each orthogonal to the real and imaginary parts of V_λ if $\theta \neq \lambda$. If the matrices are of odd order, $2N+1$, then zero is an additional eigenvalue with an essentially real eigenvector V_0 orthogonal to all the ξ and η. As before, it is not necessary to consider the case in which two or more eigenvalues coincide.

With this information on the matrix diagonalizing H, we can derive the j.p.d.f. for the eigenvalues. Denoting the positive eigenvalues of H by $\theta_1, \ldots, \theta_n$, one has

$$\operatorname{tr} H^2 = \sum_{j=1}^{n} 2\theta_j^2.$$

The Jacobian will again be a product of the differences of all pairs of eigenvalues and of a function independent of them. Thus the j.p.d.f. for the eigenvalues will be proportional to

$$\prod_{1\leqslant j<k\leqslant n} (\theta_j^2-\theta_k^2)^2 \exp\left(-2\sum_{j=1}^{n}\theta_j^2\right), \tag{3.4.4}$$

if the order N of H is even, $N=2n$, and

$$\prod_{j=1}^{n}\theta_j^2 \prod_{1\leqslant j<k\leqslant n} (\theta_j^2-\theta_k^2)^2 \exp\left(-2\sum_{j=1}^{n}\theta_j^2\right), \tag{3.4.5}$$

if $N=2n+1$ is odd. The constants of normalization will be calculated in Chapter 13.

3.5 Gaussian Ensemble of Hermitian Matrices With Unequal Real and Imaginary Parts

To derive the j.p.d.f. for the eigenvalues is a little tricky in this case. This is so because the matrix element probability densities depend on the eigenvalues and the angular variables characterizing the eigenvectors; and one has to integrate over these angular variables. When either $c_1=0$ or $c_2=0$ or $c_1=c_2$, the matrix element probability densities depend only on the eigenvalues. Also the Jacobian separates into a product of two functions, one involving only the eigenvalues and the other only the eigenvectors; therefore the integral over the eigenvectors, giving only a constant need not be calculated. For arbitrary c_1 and c_2 this simplification is not there. In view of these difficulties we will come back to this question in Chapter 14, when we are better prepared with quaternion determinants (Section 5.1), with the method of integration over alternate variables (Section 5.5), and the integral over the unitary group

$$\begin{aligned}&\int \exp\left(\operatorname{tr}(A-U^{\dagger}BU)^2\right)dU\\&\quad=\text{const}\cdot\det\left[\exp(a_j-b_k)^2\right]\cdot\prod_{j<k}\left[(a_j-a_k)(b_j-b_k)\right]^{-1},\end{aligned} \tag{3.5.1}$$

valid for arbitrary Hermitian matrices A and B having eigenvalues $a_1,\ldots,a_n$ and $b_1,\ldots,b_n$ respectively (see Appendix A.5).

3.6 Random Matrices and Information Theory

A reasonable specification of the probability density $P(H)$ for the random matrix H can be supplied from a different point of view, which is more satisfactory in some ways. Let us define the amount of information $\mathcal{I}(P(H))$ carried by the probability density $P(H)$. For discrete events $1,\ldots,m$, with probabilities $p_1,\ldots,p_m$, additivity and continuity of the information fixes it uniquely, apart from a constant multiplicative factor, to be (Shannon, 1948; Khinchin, 1957)

$$\mathcal{I} = -\sum_j p_j \ln p_j. \tag{3.6.1}$$

For continuous variables entering H it is reasonable to write

$$\mathcal{I}(P(H)) = -\int dH\, P(H) \ln P(H), \tag{3.6.2}$$

where dH is given by Eq. (2.5.1). One may now adopt the point of view that H should be as random as possible compatible with the constraints it must satisfy. In other words, among all possible probabilities $P(H)$ of a matrix H constrained to satisfy some given properties, we must choose the one which minimizes the information $\mathcal{I}(P(H))$; $P(H)$ must not carry more information than what is required by the constraints. The constraints may, for example, be the fixed expectation values k_i of some functions $f_i(H)$,

$$\langle f_i(H)\rangle \equiv \int dH\, P(H) f_i(H) = k_i. \tag{3.6.3}$$

To minimize the information (3.6.2) subject to (3.6.3) one may use Lagrange multipliers. This gives us for an arbitrary variation $\delta P(H)$ of $P(H)$,

$$\int dH\, \delta P(H)\left(1 + \ln P(H) - \sum_i \lambda_i f_i(H)\right) = 0 \tag{3.6.4}$$

or

$$P(H) \propto \exp\left(\sum_i \lambda_i f_i(H)\right), \tag{3.6.5}$$

and the Lagrange multipliers λ_i are then determined from (3.6.3) and (3.6.5).

For example, requiring H Hermitian, $H = H_1 + iH_2$, H_1, H_2 real, and

$$\langle \operatorname{tr} H_1^2\rangle = k_1, \qquad \langle \operatorname{tr} H_2^2\rangle = k_2,$$

will give us Eq. (2.7.1) for $P(H)$.

Another example is to require the level density to be a given function $\sigma(x)$

$$\langle \operatorname{tr} \delta(H - x)\rangle = \sigma(x). \tag{3.6.6}$$

The Dirac delta function $\delta(H - x)$ is defined through the diagonalization of H, $\delta(H - x) = U^\dagger \delta(\theta - x) U$, if $H = U^\dagger \theta U$, U unitary, θ (or $\delta(\theta - x)$) diagonal real with diagonal elements θ_i (or $\delta(\theta_i - x)$). This gives then

$$P(H) \propto \exp\left(\int dx \, \lambda(x) \operatorname{tr} \delta(H - x) \right) = \exp(\operatorname{tr} \lambda(H)) = \det[\mu(H)]; \quad (3.6.7)$$

the Lagrange multipliers $\lambda(x) \equiv \ln \mu(x)$ are then determined by Eq. (3.6.6). Thus one may determine a $P(H)$ giving a preassigned level density. For more details see Balian (1968).

It is important to note that if $P(H)$ depends only on the traces of powers of H, then the joint probability density of the eigenvalues will contain the factor $\prod |\theta_j - \theta_k|^\beta$, $\beta = 1, 2$ or 4; coming from the Jacobian of the elements of H with respect to its eigenvalues. And the local fluctuation properties are mostly governed by this factor.

Summary of Chapter 3

For the random Hermitian matrices H considered in Chapter 2, the joint probability density of its real eigenvalues $x_1, x_2, \ldots, x_N$ is derived. It turns out to be proportional to

$$\prod_{1 \leqslant j < k \leqslant N} |x_j - x_k|^\beta \exp\left(-\beta \sum_{j=1}^{N} x_j^2/2 \right), \quad (3.3.8)$$

where $\beta = 1, 2$ or 4 according as the ensemble is Gaussian orthogonal, Gaussian unitary or Gaussian symplectic.

For the Gaussian ensemble of Hermitian anti-symmetric random matrices the joint probability density of the eigenvalues $\pm x_1, \ldots, \pm x_m$ with $x_j \geqslant 0$, is proportional to

$$(x_1 \cdots x_m)^\gamma \prod_{1 \leqslant j < k \leqslant m} \left(x_j^2 - x_k^2\right)^2 \exp\left(- \sum_{j=1}^{m} x_j^2 \right),$$

where $\gamma = 0$, if the order N of the matrices is even, $N = 2m$, and $\gamma = 2$, if this order is odd, $N = 2m + 1$. In the $N = 2m + 1$, odd case, zero is always an eigenvalue.

The non-invariant ensemble of Hermitian random matrices with unequal real and imaginary parts being more complicated is taken up later in Chapter 14.

An additional justification of our choices of the ensembles is supplied from an argument of the information theory.

4

GAUSSIAN ENSEMBLES. LEVEL DENSITY

In this short chapter we reproduce a statistical mechanical argument of Wigner to "derive" the level density for the Gaussian ensembles. The joint probability density of the eigenvalues is written as the Boltzmann factor for a gas of charged particles interacting via a two dimensional Coulomb force. The equilibrium density of this Coulomb gas is such as to make the potential energy a minimum, and this density is identified with the level density of the corresponding Gaussian ensembles. That the eigenvalue density so deduced is correct, will be seen later when in Chapters 6 to 8 and 14 we will compute it for each of the four Gaussian ensembles for any finite $N \times N$ matrices and take the limit as $N \to \infty$.

Another argument, again essentially due to Wigner, is given to show that the same level density, a "semi-circle", holds for random Hermitian matrices with elements having an average value zero and a common mean square value.

4.1 The Partition Function

Consider a gas of N point charges with positions $x_1, x_2, \ldots, x_N$ free to move on the infinite straight line $-\infty < x < \infty$. Suppose that the potential energy of the gas is given by

$$W = \frac{1}{2}\sum_i x_i^2 - \sum_{i<j} \ln |x_i - x_j|. \tag{4.1.1}$$

The first term in W represents a harmonic potential which attracts each charge independently towards the point $x = 0$; the second term represents an electrostatic repulsion between each pair of charges. The logarithmic function comes in if we assume the universe to be two-dimensional. Let this charged gas be in thermodynamical equilibrium at a temperature T, so that the probability density of the positions of the N charges is given by

$$P(x_1, \ldots, x_N) = C \exp(-W/kT), \tag{4.1.2}$$

where k is the Boltzmann constant. We immediately recognize that (4.1.2) is identical to (3.3.8) provided β is related to the temperature by

$$\beta = (kT)^{-1}. \tag{4.1.3}$$

This system of point charges in thermodynamical equilibrium is called the Coulomb gas model, corresponding to the Gaussian ensembles.

Following Dyson (1962a, 1962b, 1962c), we can define various expressions that relate to our energy-level series in complete analogy with the classical notions of entropy, specific heat, and the like. These expressions, when computed from the observed experimental data and compared with the theoretical predictions, provide a nice method of checking the theory.

In classical mechanics the joint probability density in the velocity space is a product of exponentials

$$\prod_j \exp(-C_j v_j^2)$$

with constant C_j, and its contribution to the thermodynamic quantities of the model are easily calculated. We simply discard these trivial terms. The nontrivial contributions arise from the partition function

$$\psi_N(\beta) = \int_{-\infty}^{\infty} \cdots \int_{-\infty}^{\infty} e^{-\beta W}\, dx_1 \cdots dx_N \tag{4.1.4}$$

and its derivatives with respect to β. Therefore it is important to have an analytical expression for $\psi_N(\beta)$. Fortunately, this can be deduced from an integral evaluated by Selberg (see Chapter 17).

Theorem 4.1.1. *For any positive integer N and real or complex β we have, identically,*

$$\psi_N(\beta) = (2\pi)^{N/2} \beta^{-N/2-\beta N(N-1)/4} (\Gamma(1+\beta/2))^{-N} \prod_{j=1}^{N} \Gamma(1+\beta j/2). \tag{4.1.5}$$

Let us note the fact that the energy W given by (4.1.1) is bounded from below. More precisely,

$$W \geqslant W_0 = \frac{1}{4}N(N-1)(1+\ln 2) - \frac{1}{2}\sum_{j=1}^{N} j \ln j, \tag{4.1.6}$$

and this minimum is attained when the positions of the charges coincide with the zeros of the Hermite polynomial $H_N(x)$ (cf. Appendix A.6).

Once the partition function is known, other thermodynamic quantities such as free energy, entropy, and specific heat can be calculated by elementary differentiation. Because all the known properties are identical to those of the circular ensembles, studied at length in Chapter 12 we do not insist on this point here.

4.2 The Asymptotic Formula for the Level Density. Gaussian Ensembles

Since the expression (3.3.8) for $P(x_1, \ldots, x_N)$, the probability that the eigenvalues will lie in unit intervals around $x_1, x_2, \ldots, x_N$, is valid for all values of x_i, the density of levels

$$\sigma_N(x) = N \int_{-\infty}^{\infty} \cdots \int_{-\infty}^{\infty} P(x, x_2, \ldots, x_N)\, dx_2 \cdots dx_N \tag{4.2.1}$$

can be calculated for any N by actual integration (Mehta and Gaudin, 1960). The details of this tedious calculation are not given here, since an expression for $\sigma_N(x)$, derived by a different method, appears in Chapters 6, 7 and 8.

However, if one is interested in the limit of large N, as we certainly are, these complications can be avoided by assuming that the corresponding Coulomb gas is a classical fluid with a continuous macroscopic density. More precisely, this amounts to the following two assumptions:

(1) The potential energy W given by (4.1.1) can be approximated by the functional

$$W(\sigma) = \frac{1}{2}\int_{-\infty}^{\infty} dx\, x^2 \sigma(x) - \frac{1}{2}\int_{-\infty}^{\infty} dx\, dy\, \sigma(x)\sigma(y) \ln|x-y|. \tag{4.2.2}$$

(2) The level density $\sigma(x)$ will be such as to minimize the expression (4.2.2), consistent with the requirements

$$\int_{-\infty}^{\infty} dx\, \sigma(x) = N, \tag{4.2.3}$$

and

$$\sigma(x) \geqslant 0. \tag{4.2.4}$$

The first integral in (4.2.2) reproduces the first sum in (4.1.1) accurately in the limit of large N. The same is not true of the second integral, for it neglects the two-level correlations, which may be expected to extend over a few neighboring levels; however, because the total number of levels is large their effect may be expected to be small. The factor $1/2$ in the second term of (4.2.2) comes from the condition $i < j$ in (4.1.1).

The problem of finding the stationary points of the functional $W(\sigma)$, (4.2.2), with the restriction (4.2.3) leads us to the integral equation

$$-\frac{1}{2}x^2 + \int_{-\infty}^{\infty} dy\, \sigma(y) \ln|x-y| = C, \tag{4.2.5}$$

where C is a Lagrange constant. Actually (4.2.5) has to hold only for those values of x for which $\sigma(x) > 0$. One cannot add a negative increment to $\sigma(x)$ where $\sigma(x) = 0$, and therefore the functional differentiation is not valid; hence (4.2.5) cannot be derived for such values of x. It is not difficult to solve (4.2.5) (Mushkelishvili, 1953). This will not be done here, but the solution will be given and then verified.

Differentiation of (4.2.5) with respect to x eliminates C. Before carrying it out, we must replace the integral with

$$\lim_{\varepsilon\to 0}\left(\int_{-\infty}^{x-\varepsilon} dy + \int_{x+\varepsilon}^{\infty} dy\right)\sigma(y) \ln|x-y|. \tag{4.2.6}$$

When (4.2.6) is differentiated with respect to x, the terms arising from the differentiation of the limits drop out and only the derivative of $\ln|x-y|$ remains. The integral becomes a principal value integral and (4.2.5) becomes

$$P\int_{-\infty}^{\infty} \frac{\sigma(y)}{x-y}\, dy = x. \tag{4.2.7}$$

Conversely, if (4.2.7) is satisfied by some $\sigma(y)$ and this σ is an even function, then it will satisfy (4.2.5) also. We try

$$\sigma(y) = \begin{cases} C\left(A^2 - y^2\right)^{1/2}, & |y| < A, \\ 0, & |y| > A. \end{cases} \tag{4.2.8}$$

Elementary integration gives

$$\begin{aligned} \int \frac{(A^2-y^2)^{1/2}}{x-y}\, dy &= x \sin^{-1}\left(\frac{y}{A}\right) - \left(A^2-y^2\right)^{1/2} \\ &+ \left(A^2-x^2\right)^{1/2} \ln\left(\frac{A(x-y) - x(A^2-y^2)^{1/2} - y(A^2-x^2)^{1/2}}{A(x-y) - x(A^2-y^2)^{1/2} + y(A^2-x^2)^{1/2}}\right). \end{aligned} \tag{4.2.9}$$

Taking the principal value of (4.2.9) between the limits $(-A, A)$, we find that only the first term gives a nonzero contribution, which is πx. Hence (4.2.7) gives

$$C = 1/\pi, \tag{4.2.10}$$

and (4.2.3) gives

$$\frac{1}{\pi}\frac{\pi}{2}A^2 = N. \tag{4.2.11}$$

Thus

$$\sigma(x) = \begin{cases} \dfrac{1}{\pi}\left(2N - x^2\right)^{1/2}, & |x| < (2N)^{1/2}, \\ 0, & |x| > (2N)^{1/2}. \end{cases} \tag{4.2.12}$$

This is the so-called "semicircle law" first derived by Wigner.

Actually the two-level correlation function can be calculated (cf. Sections 6.2, 7.2 and 8.2) and the above intuitive arguments put to test. Instead, we shall derive an exact expression for the level-density valid for any N. The limit $N \to \infty$ can then be taken (cf. Appendix A.9) to obtain the "semicircle law".

We have noted in Section 4.1 that without any approximation whatever the energy W attains its minimum value when the points $x_1, x_2, \ldots, x_N$ are the zeros of the Nth order Hermite polynomial. The postulate of classical statistical mechanics then implies that in the limit of very large N the level density is the same as the density of zeros of the Nth order Hermite polynomial. This later problem has been investigated by many authors, and we may conveniently refer to the relevant mathematical literature (Szegö, 1959).

4.3 The Asymptotic Formula for the Level Density. Other Ensembles

Numerical evidence shows, as we said in Chapter 1, that the local statistical properties of the Gaussian ensembles are shared by a much wider class of matrices. In particular even the eigenvalue density, which is a global property, follows the "semi-circle law". Wigner (1955) first considered bordered matrices, i.e., real symmetric matrices H with elements

$$H_{jk} = \begin{cases} \pm h, & \text{if } |j-k| \leqslant m, \\ 0, & \text{if } |j-k| > m. \end{cases} \tag{4.3.1}$$

Except for the symmetry of H, the signs of H_{jk} are random. He then calculates the moments of the level density and derives an integral equation for it. The calculations are long. The final result is that in the limit of $h^2/m \to 0$ and the order of the matrices is infinite, the eigenvalue density is a "semi-circle". Here we present still another heuristic argument in its support, again essentially due to Wigner.

Consider a matrix H with elements H_{ij} all having an average value zero and a mean square value V^2. Let the order N be large enough so that the density of its eigenvalues may be taken to be a continuous function. Let this function be $\sigma(\varepsilon, V^2)$, so that the number of eigenvalues lying between ε and $\varepsilon + \delta\varepsilon$ is given by $\sigma(\varepsilon, V^2)\, d\varepsilon$. If we change the matrix elements by small quantities δH_{ij} such that the δH_{ij} themselves all have the average value zero and a mean square value v^2, the change in a particular eigenvalue at ε_i can be calculated by the second order perturbation theory

$$Z(\varepsilon, V^2) = \delta H_{ii} + \sum_{j \neq i} \frac{|\delta H_{ij}|^2}{\varepsilon_i - \varepsilon_j} + \cdots. \tag{4.3.2}$$

The δH_{ii} do not produce, on the average, any change in ε_i. The eigenvalues ε_j which lie nearest to ε_i give the largest contribution to (4.3.2) with an absolute value $v^2/\bar{s}$ where $\bar{s}$ is the mean spacing at ε_i. But as there are eigenvalues on both sides of ε_i, the two contributions arising from the two nearest eigenvalues nearly cancel out, leaving quantities of a higher order in v^2. The sum in (4.3.2) can therefore be approximated by

$$Z(\varepsilon, V^2) \approx v^2 \int \frac{\sigma(\varepsilon', V^2)}{\varepsilon - \varepsilon'} d\varepsilon', \tag{4.3.3}$$

where the integral in (4.3.3) is a principal value integral and

$$V^2 = \langle |H_{ij}|^2 \rangle, \qquad v^2 = \langle |\delta H_{ij}|^2 \rangle. \tag{4.3.4}$$

The ensemble averages being indicated by $\langle\ \rangle$. Let us calculate the change in the number of eigenvalues lying in an interval $(\varepsilon, \varepsilon + \delta\varepsilon)$. This can be done in two ways; one gives, as is obvious from the way of writing,

$$\sigma(\varepsilon, V^2) Z(\varepsilon, V^2) - \sigma(\varepsilon + \delta\varepsilon, V^2) Z(\varepsilon + \delta\varepsilon, V^2) \approx -\frac{\partial(\sigma Z)}{\partial \varepsilon} \delta\varepsilon, \tag{4.3.5}$$

while the other gives in a similar way

$$v^2 \frac{\partial \sigma}{\partial V^2}. \tag{4.3.6}$$

If all the matrix elements H_{ij} are multiplied by a constant c, the values ε_i are also multiplied by c, while V^2 is multiplied by c^2. Hence,

$$\sigma(c\varepsilon, c^2 V^2) c\, d\varepsilon = \sigma(\varepsilon, V^2)\, d\varepsilon. \tag{4.3.7}$$

Setting $cV = 1$ the last equation gives

$$\sigma\left(\varepsilon, V^2\right) = \frac{1}{V}\sigma(\varepsilon/V, 1),$$

which could have been inferred by dimensional arguments. Putting

$$Z\left(\varepsilon, V^2\right) = \frac{v^2}{V}Z_1(\varepsilon/V), \qquad \sigma\left(\varepsilon, V^2\right) = \frac{1}{V}\sigma_1(\varepsilon/V), \tag{4.3.8}$$

in (4.3.3), (4.3.5) and (4.3.6), we obtain

$$\frac{\partial(Z_1\sigma_1)}{\partial x} = \frac{1}{2}\frac{\partial(x\sigma_1)}{\partial x}, \qquad x = \varepsilon/V, \tag{4.3.9}$$

$$Z_1(x) = P\int \frac{\sigma_1(x')}{x - x'}\,dx'. \tag{4.3.10}$$

When $x = 0$, by symmetry requirement $Z_1 = 0$; therefore (4.3.9) gives, on integration

$$Z_1(x) = x/2. \tag{4.3.11}$$

Finally we have the boundary condition

$$\int \sigma\left(\varepsilon, V^2\right) d\varepsilon = \int \sigma_1(x)\,dx = N. \tag{4.3.12}$$

Equations (4.3.10), (4.3.11) and (4.3.12) together are equivalent to the integral equation (4.2.7) together with (4.2.3). The solution, as there, is the semi-circle law (4.2.12):

$$\sigma\left(\varepsilon, V^2\right) = \begin{cases} \dfrac{1}{2\pi V^2}\left(2NV^2 - \varepsilon^2\right)^{1/2}, & \varepsilon^2 < 2NV^2, \\ 0, & \varepsilon^2 > 2NV^2. \end{cases} \tag{4.3.13}$$

Olson and Uppulury (1972) and later Wigner extended these considerations to include a still wider class of matrices to have the "semi-circle law" as their eigenvalue density.

For the two level correlation function or the spacing distribution no such argument has yet been found.

Summary of Chapter 4

As a consequence of Selberg's integral one has the partition function

$$\psi_N(\beta) = \int_{-\infty}^{\infty}\cdots\int_{-\infty}^{\infty} e^{-\beta W}\,dx_1\cdots dx_N \tag{4.1.4}$$

$$= (2\pi)^{N/2}\beta^{-N/2-\beta N(N-1)/4}(\Gamma(1+\beta/2))^{-N}\prod_{j=1}^{N}\Gamma(1\beta j/2), \qquad (4.1.5)$$

where

$$W = \frac{1}{2}\sum_{j=1}^{N} x_j^2 - \sum_{1\leqslant j<k\leqslant N} \ln|x_j - x_k|. \qquad (4.1.1)$$

For a large class of random matrices the asymptotic density of eigenvalues is the "semi-circle"

$$\sigma(x) = \begin{cases} \frac{1}{\pi}(2N - x^2)^{1/2}, & |x| < (2N)^{1/2}, \\ 0, & |x| > (2N)^{1/2}. \end{cases} \qquad (4.2.12)$$

5

ORTHOGONAL, SKEW-ORTHOGONAL AND BI-ORTHOGONAL POLYNOMIALS

We will often encounter integrals of the form

$$\int \left[\prod_{k=1}^{N} w(x_k)\right]\left[\prod_{1 \leqslant i < j \leqslant N} |x_i - x_j|^{\beta}\right] dx_{m+1} \cdots dx_N, \tag{5.0.1}$$

for the three values $\beta = 1$, 2 and 4, and integers m and N with $0 \leqslant m < N$. The case $\beta = 2$ is the oldest known and the easiest one; for its evaluation one needs to know the properties of determinants and of orthogonal polynomials. For the other two cases $\beta = 4$ and $\beta = 1$, one needs to know the properties of Pfaffians, of anti-symmetric scalar products and the associated skew-orthogonal polynomials. The treatment of these two cases can be made parallel to the more familiar case $\beta = 2$ if one introduces quaternions and one particular definition of the determinant of matrices with (non-commuting) quaternion elements. By this transcription of a Pfaffian as a quaternion determinant one gets an explicit compact form of these integrals in an elegant way. However, to understand how these quaternion determinants are constructed one has to go through the laborious procedure of the usual Pfaffian calculations.

Integrals of the form

$$\int \prod_{j=1}^{N} w(x_j, y_j) \prod_{1 \leqslant i < j \leqslant N} (x_i - x_j)(y_i - y_j)\, dx_{m_1+1} \cdots dx_N\, dy_{m_2+1} \cdots dy_N, \quad (5.0.2)$$

for $0 \leqslant m_1, m_2 < N$, can also be written in a compact form using non-local scalar products and bi-orthogonal polynomials.

5.1 Quaternions, Pfaffians, Determinants

Let $B = [b_{ij}]$ be an $N \times N$ anti-symmetric matrix (elements b_{ij} complex numbers, $b_{ij} = -b_{ji}$), with $N = 2s$ or $N = 2s + 1$, $s > 0$. The Pfaffian of B is defined by

$$\operatorname{pf} B = (2^s s!)^{-1} \sum_P \sigma(P) b_{i_1 i_2} b_{i_3 i_4} \cdots b_{i_{2s-1} i_{2s}}, \quad (5.1.1)$$

where the summation is taken over all $N!$ permutations P of the N indices $\{1, 2, \ldots, N\}$ and $\sigma(P)$ is the sign of P. The factor $(2^s s!)^{-1}$ is there so that each distinct term is counted only once and one can omit it provided the sum is taken over all permutations P with the restrictions $i_1 < i_2, i_3 < i_4, \ldots, i_{2s-1} < i_{2s}$, $i_1 < i_3 < \cdots < i_{2s-1}$. The square of the Pfaffian of B is equal to the determinant of B if its order N is even, and is equal to the determinant of a matrix obtained from B by adding a column of ones and a row of minus ones at the end, $b_{i,N+1} = -b_{N+1,i} = 1$, $i = 1, \ldots, N$, $b_{N+1,N+1} = 0$, so as to make it an anti-symmetric matrix of even order $N + 1$. See e.g. Mehta, *Matrix Theory* (Mehta, 1989MT), Section 2.3.

The evaluation of the integral (5.0.1) where $w(x)$ is a positive weight function with all its moments finite, and $\beta = 1, 2$ or 4, depends on a few theorems given below, after recalling some elementary facts about quaternions.

A quaternion a has the form (cf. Section 2.4)

$$a = a_0 \mathbf{1} + a \cdot \mathbf{e} := a_0 \mathbf{1} + a_1 \mathbf{e}_1 + a_2 \mathbf{e}_2 + a_3 \mathbf{e}_3, \quad (5.1.2)$$

where a_0, a_1, a_2, a_3 are real or complex numbers, the four units $\mathbf{1}, \mathbf{e}_1, \mathbf{e}_2, \mathbf{e}_3$ satisfy the multiplication rules

$$\mathbf{1e}_i = \mathbf{e}_i \mathbf{1} = \mathbf{e}_i, \quad \mathbf{e}_i \mathbf{e}_j = -\delta_{ij} \mathbf{1} + \varepsilon_{ijk} \mathbf{e}_k, \quad i, j, k = 1, 2, 3, \quad (5.1.3)$$

where δ_{ij}, the Kronecker symbol, is equal to 1 or 0 according as $i = j$ or $i \neq j$, ε_{ijk} is the completely antisymmetric tensor with $\varepsilon_{123} = +1$, and multiplication is associative. The scalar part of a is a_0. The dual of a is $\bar{a} = a_0 \mathbf{1} - a \cdot \mathbf{e}$.

Any quaternion can be represented by a 2×2 matrix with complex elements; for example

$$\mathbf{1} \to \begin{bmatrix} 1 & 0 \\ 0 & 1 \end{bmatrix}, \quad \mathbf{e}_1 \to \begin{bmatrix} \sqrt{-1} & 0 \\ 0 & -\sqrt{-1} \end{bmatrix},$$
$$\mathbf{e}_2 \to \begin{bmatrix} 0 & 1 \\ -1 & 0 \end{bmatrix}, \quad \mathbf{e}_3 \to \begin{bmatrix} 0 & \sqrt{-1} \\ \sqrt{-1} & 0 \end{bmatrix}. \tag{5.1.4}$$

Conversely, any 2×2 matrix with complex elements can be represented by a quaternion.

Quaternion matrices form the subject of a chapter in Mehta (1989MT), and so we will be brief in discussing them here. The dual of the quaternion matrix $A = [a_{ij}]$ is the quaternion matrix $\overline{A} = [\bar{a}_{ji}]$. A self-dual quaternion matrix is such that $a_{ij} = \bar{a}_{ji}$. In any $N \times N$ matrix $A = [a_{ij}]_N$ with quaternion elements a_{ij}, if we replace a_{ij} by its 2×2 matrix representation, we get a $2N \times 2N$ matrix with complex elements, which we denote $\Theta[A]$. When the a_{ij} are complex, the multiplication is commutative, and there is no ambiguity about the determinant of the $N \times N$ matrix $[a_{ij}]_N$ defined in the usual way. However, when the elements a_{ij} are quaternions, the multiplication is no longer commutative, and there is no unique definition of a determinant. We will adopt the following definition (Moore, 1935; Dyson, 1970; Mehta, 1989MT, Chapter 8)

$$\det[a_{ij}] = \sum_P \sigma(P) \prod_{\text{cycles}} [a_{rs} a_{st} \cdots a_{ur}]^{(0)}, \tag{5.1.5}$$

where the permutation P consists of the exclusive cycles $(r \to s \to t \to \cdots \to u \to r)$, $\sigma(P)$ is its sign, and the sum is taken over all $N!$ permutations. The superscript $^{(0)}$ on each cycle means that we take the scalar part of the product. One might think that the definition (5.1.5) is incomplete for two reasons. The product of elements in a particular cycle depends on the index by which one begins the cycle and the value of the whole product on the right-hand side depends on the order in which the different cyclic factors are written. Both these ambiguities are removed by taking the scalar parts.

Remark 5.1.1. If A is self-dual, we do not have to take the scalar part in Eq. (5.1.5) provided the same ordering of the ℓ cyclic factors in (5.1.5) be used for the permutation P and for other permutations obtained from P by reversing one or more of its cycles. To make this point clearer, let

$$P = C_1 C_2 \cdots C_\ell \tag{5.1.6}$$

where $C_1, C_2, \ldots$ are cycles operating on different sets of indices. Let one of these cycles be

$$C := (j_1 j_2 \ldots j_r) := \begin{pmatrix} j_1 j_2 \ldots j_r \\ j_2 j_3 \ldots j_1 \end{pmatrix}. \tag{5.1.7}$$

The reverse or inverse cycle is then

$$C^{-1} = (j_r \ldots j_2 j_1) := \begin{pmatrix} j_1 \ldots j_3 j_2 \\ j_r \ldots j_2 j_1 \end{pmatrix}. \tag{5.1.8}$$

Permutations derived from P by reversing one or more of its cycles include for example

$$P_1 = C_1^{-1} C_2 C_3 \cdots C_\ell, \quad P_2 = C_1 C_2^{-1} C_3 \cdots C_\ell, \quad P_3 = C_1^{-1} C_2^{-1} C_3 \cdots C_\ell, \tag{5.1.9}$$

etc. As the conjugate of a product is the product of conjugates taken in the reverse order

$$\overline{a_{j_1 j_2} a_{j_2 j_3} \cdots a_{j_r j_1}} = \overline{a_{j_r j_1}} \cdots \overline{a_{j_2 j_3}}\ \overline{a_{j_1 j_2}} = a_{j_1 j_r} \cdots a_{j_3 j_2} a_{j_2 j_1}, \tag{5.1.10}$$

the factors corresponding to a cycle C and the reverse cycle C^{-1} are conjugates to each other, and their sum is already scalar. In Eq. (5.1.5) the sum is taken over all permutations, and therefore if the order of the ℓ cyclic factors is kept the same for $P, P_1, P_2, P_3, \ldots$, then it amounts to taking the scalar parts. For self-dual A we rewrite Eq. (5.1.5) without scalar parts

$$\det A = \sum_P (-1)^{N-\ell} \prod_1^\ell (a_{rs} a_{st} \cdots a_{ur}). \tag{5.1.11}$$

Let $\overline{A}$ be the dual of the quaternion matrix A and let Z be the anti-symmetric matrix consisting of the 2×2 blocks

$$\begin{bmatrix} 0 & 1 \\ -1 & 0 \end{bmatrix}$$

along the main diagonal and zeros elsewhere; i.e. Z is the direct sum of 2×2 matrices

$$Z = \begin{bmatrix} 0 & 1 \\ -1 & 0 \end{bmatrix} \dotplus \begin{bmatrix} 0 & 1 \\ -1 & 0 \end{bmatrix} \dotplus \cdots. \tag{5.1.12}$$

Then it can be verified that

$$\Theta[\overline{A}] = Z \{\Theta[A]\}^T Z^{-1}. \tag{5.1.13}$$

Consequently, A is self-dual if and only if $Z\,\Theta[A]$ is anti-symmetric.

Theorem 5.1.2 (Dyson, 1970). *Let $A = [a_{ij}]_N$ be a $N \times N$ self-dual quaternion matrix, and $\Theta[A]$ be its representation by a $2N \times 2N$ complex matrix. Then*

$$\operatorname{pf} Z\Theta[A] = \det A. \tag{5.1.14}$$

Here, pf denotes the Pfaffian. Hence for quaternion self-dual matrices A one may take Eq. (5.1.14) as the definition of its determinant. When the elements of A are quaternion scalars not containing the units **e**, i.e. ordinary complex numbers, then this definition coincides with the ordinary determinant.

For a proof see e.g. Mehta (1989MT), Section 8.9.4.

A first consequence of the Theorem 5.1.2 is the following formula

$$\det \Theta[A] = (\det A)^2. \tag{5.1.15}$$

A second consequence is:

Corollary 5.1.3. *Let $A = [a_{ij}]_N$ be a $N \times N$ quaternion matrix, $\overline{A} = [\bar{a}_{ji}]_N$ be its dual, and $\Theta[A]$ be as defined above. Then*

$$\det \Theta[A] = \det \Theta[\overline{A}] = \det(A\overline{A}). \tag{5.1.16}$$

Note that the determinants on the left-hand side of Eqs. (5.1.15) and (5.1.16) are of $2N \times 2N$ matrices with complex elements, while those on the right-hand sides of Eqs. (5.1.14)–(5.1.16) are of $N \times N$ matrices with quaternion elements.

Proof. Let A be a $N \times N$ quaternion matrix, and $\overline{A}$ be its dual. The $2N \times 2N$ complex matrix representations $\Theta[A]$ and $\Theta[\overline{A}]$ of A and $\overline{A}$ are related by Eq. (5.1.13). Thus

$$\det \Theta[A] = \det \Theta[\overline{A}], \tag{5.1.17}$$

and

$$(\det \Theta[A])^2 = \det \Theta[A]\Theta[\overline{A}] = \det \Theta[A\overline{A}] \tag{5.1.18}$$

$$= [\det(A\overline{A})]^2. \tag{5.1.19}$$

To write the last equality, we have used Theorem 5.1.2, taking advantage of the self-duality of the matrix $A\overline{A}$. Then

$$\det \Theta[A] = \pm \det(A\overline{A}). \tag{5.1.20}$$

Next, we remark that both sides of this equality are continuous functions of the elements of $\Theta[A]$, which excludes one of the two signs $\pm$. It suffices now to take a diagonal matrix A to exclude the minus sign, and this ends the proof. □

Theorem 5.1.4 (Dyson, 1970). *Let $K(x, y)$ be a function with real, complex or quaternion values, such that*

$$\overline{K}(x, y) = K(y, x), \tag{5.1.21}$$

where $\overline{K} = K$ if K is real, $\overline{K}$ is the complex conjugate of K if it is complex, and $\overline{K}$ is the dual of K if it is quaternion. Assume that

$$\int K(x, y)K(y, z)\, dy = K(x, z) + \lambda K(x, z) - K(x, z)\lambda, \tag{5.1.22}$$

or symbolically

$$K * K = K + \lambda K - K\lambda \tag{5.1.22$'$}$$

with λ a constant quaternion. Let $[K(x_i, x_j)]_N$ denote the $N \times N$ matrix with its (i, j) element equal to $K(x_i, x_j)$. Then

$$\int \det[K(x_i, x_j)]_N\, dx_N = (c - N + 1)\det[K(x_i, x_j)]_{N-1}, \tag{5.1.23}$$

where

$$c = \int K(x, x)\, dx. \tag{5.1.24}$$

Note that when $K(x, y)$ is real or complex, λ vanishes. Conditions (5.1.21) and (5.1.22) then mean that the linear operator defined by the kernel $K(x, y)$ is a projector, and the constant c is its trace (hence a nonnegative integer).

Proof. From the definition of the determinant,

$$\det[K(x_i, x_j)]_N = \sum_P \sigma(P) \prod_1^{\ell} (K(\xi, \eta)K(\eta, \zeta) \cdots K(\theta, \xi)), \tag{5.1.25}$$

where the permutation P consists of ℓ cycles of the form $(\xi \to \eta \to \zeta \to \cdots \to \theta \to \xi)$. Now the index N occurs somewhere. There are two possibilities: (i) it forms a cycle by itself, and $K(x_N, x_N)$ gives on integration the scalar constant c; the remaining factor is by definition $\det[K(x_i, x_j)]_{N-1}$; (ii) it occurs in a longer cycle and integration on x_N reduces the length of this cycle by one. This can happen in $(N - 1)$ ways since the index N can be inserted between any two indices of the cyclic sequence $1, 2, \ldots, N - 1$. Also the resulting permutation over $N - 1$ indices has an opposite sign. The presence of $\lambda K(x, z) - K(x, z)\lambda$ in Eq. (5.1.22) does not matter since the terms containing λ sum to zero (see Mehta (1989MT), Appendix A.12 for details). The remaining expression is by definition $\det[K(x_i, x_j)]_{N-1}$, and the contribution from this possibility is $-(N - 1)\det[K(x_i, x_j)]_{N-1}$. Adding the two contributions we get the result.

The evaluation of the integral (5.0.1) depends on expressing its integrand $[\prod_{k=1}^{N} w(x_k)]|\Delta_N(x)|^\beta$, with $\Delta_1(x) := 1$ and

$$\Delta_N(x) := \prod_{1 \leqslant j < i \leqslant N} (x_i - x_j), \quad N > 1, \tag{5.1.26}$$

as $\det[K_\beta(x_i, x_j)]_N$ where $K_\beta(x, y)$ satisfies Eqs. (5.1.21) and (5.1.22). The simplest case is $\beta = 2$, the only one where no quaternions are needed. The next difficult case is $\beta = 4$, where the constant quaternion λ of Theorem 5.1.4 vanishes. Next in difficulty is the case $\beta = 1$ with an even number of variables and the most difficult one from the mathematical point of view is the case $\beta = 1$ and an odd number of variables.

In Sections 5.3 to 5.5 below we describe a method of systematically constructing a kernel $K_\beta(x, y)$ for the three values 1, 2 and 4 of β satisfying the requirements:

(i) $\overline{K}_\beta(x, y) = K_\beta(y, x)$;
(ii) $\det[K_\beta(x_i, x_j)]_N = \text{const} \cdot |\Delta_N(x)|^\beta \prod_{i=1}^{N} w(x_i)$;
(iii) $K * K = K + \lambda K - K\lambda$ with λ a constant quaternion.

However, given a weight function $w(x)$, such a $K_\beta(x, y)$ is not unique and any kernel $K_\beta(x, y)$ satisfying the above three requirements will be as good. □

5.2 Average Value of $\prod_{j=1}^{N} f(x_j)$; Orthogonal and Skew-Orthogonal Polynomials

For the three values 1, 2 and 4 of β let us consider the multiple integral

$$\mathcal{I}_\beta(f) := \int \prod_{j=1}^{N} f(x_j) w(x_j) |\Delta_N(x)|^\beta \, dx_1 \cdots dx_N. \tag{5.2.1}$$

The "average value" of $f(\mathbf{x}) := \prod_{j=1}^{N} f(x_j)$ will then be

$$\langle f(\mathbf{x}) \rangle_\beta := \mathcal{I}_\beta(f) / \mathcal{I}_\beta(1). \tag{5.2.2}$$

Here the positive weight $w(x)$ has all its moments finite $\int x^j w(x)\, dx < \infty$, for $j = 0, 1, 2, \ldots$ and $f(x)$ is any function provided that the integral converges.

Evaluation of the integral (5.2.1) is basic, since by making various choices of the function $f(x)$ it will allow us to calculate many quantities of interest such as correlation functions, spacing functions, probability density of the determinants, average value of characteristic polynomials, etc.

5.3 Case $\beta = 2$; Orthogonal Polynomials

We write $\Delta_N(x)$, the product of differences, Eq. (5.1.26), as an $N \times N$ determinant. Let $P_i(x)$, $i = 0, 1, \ldots, N-1$, be a set of N linearly independent polynomials each of degree less than N; i.e. let

$$P_i(x) = \sum_{j=0}^{N-1} a_{i,j} x^j \tag{5.3.1}$$

with the $N \times N$ matrix $[a_{i,j}]$, $i, j = 0, 1, \ldots, N-1$, non-singular. Then since the determinant of the product is the product of the determinants, one has

$$\begin{aligned} \det[P_i(x_j)]_{i=0,\ldots,N-1;\, j=1,\ldots,N} &= \det[a_{i,j}]_{i,j=0,\ldots,N-1} \det[x_j^{i-1}]_{i,j=1,\ldots,N} \\ &= \det[a_{i,j}]_{i,j=0,\ldots,N-1} \cdot \Delta_N(x), \end{aligned} \tag{5.3.2}$$

or

$$\begin{aligned} \Delta_N(x) &= \det[x_j^{i-1}]_{i,j=1,\ldots,N} = \det[C_{i-1}(x_j)]_{i,j=1,\ldots,N} \\ &= \prod_{i=0}^{N-1} c_i \det\left[\frac{1}{c_{i-1}} C_{i-1}(x_j)\right]_{i,j=1,\ldots,N}, \end{aligned} \tag{5.3.3}$$

where $C_i(x)$, $i = 0, 1, \ldots, N-1$, are any set of N linearly independent polynomials each of degree less than N, suitably normalized to have equality in the above equation and c_i are any non-zero constants. We will usually take $C_i(x)$ as a monic polynomial of precise degree i. Recall that a polynomial is called monic when the coefficient of the highest power, i.e. x^i, is one. In Eq. (5.3.3) the first equality is the Vandermonde determinant while the second one is obtained by adding to the ith row suitable multiples of other rows above it. This amounts to taking $a_{i,j} = 0$ for $i < j$ and $a_{i,i} = 1$ in Eq. (5.3.2). Then

$$\prod_{i=1}^{N} w(x_i)\Delta_N^2(x) = \prod_{i=1}^{N} w(x_i) \det[C_{i-1}(x_j)] \det[\overline{C}_{i-1}(x_j)] \tag{5.3.4}$$

$$= \prod_{i=1}^{N} w(x_i) \prod_{i=0}^{N-1} c_i \det\left[\sum_{k=0}^{N-1} \frac{1}{c_k} \overline{C}_k(x_i) C_k(x_j)\right]_{i,j=1,\ldots,N}. \tag{5.3.5}$$

In the above, $C_i(x)$ and $\overline{C}_i(x)$ are any monic polynomials and c_i are any non-zero constants. Notice that in Eq. (5.3.5) each variable occurs in one row and one column.

Expanding the determinants in (5.3.4) one has

$$\Delta_N^2(x) = \sum_{(i)} \sum_{(j)} \sigma(i)\sigma(j) C_{i_1}(x_1) \ldots C_{i_N}(x_N) \overline{C}_{j_1}(x_1) \ldots \overline{C}_{j_N}(x_N), \tag{5.3.6}$$

where $\sigma(i)$ is the sign of the permutation $(i) = \binom{0 \;\ldots\; N-1}{i_1 \;\ldots\; i_N}$, the sum (i) is over all the $N!$ permutations (i), and similarly for the permutations (j). Hence

$$\begin{aligned} \mathcal{I}_2(f) &= \prod_{i=0}^{N-1} c_i \sum_{(i)} \sum_{(j)} \sigma(i)\sigma(j) \prod_{k=1}^{N} c_{i_k}^{-1} \int dx_k\, C_{i_k}(x_k) \overline{C}_{j_k}(x_k) w(x_k) f(x_k) \\ &= \prod_{i=0}^{N-1} c_i \sum_{(i)} \sum_{(j)} \sigma(i)\sigma(j) \Psi_{i_1, j_1}(f) \cdots \Psi_{i_N, j_N}(f) \\ &= \left(\prod_{i=0}^{N-1} c_i \right) N! \det[\Psi_{i,j}(f)]_{i,j=0,\ldots,N-1}, \end{aligned} \tag{5.3.7}$$

where

$$\Psi_{i,j}(f) := c_i^{-1} \int C_i(x) \overline{C}_j(x) w(x) f(x)\, dx. \tag{5.3.8}$$

We will use Eqs. (5.3.7), (5.3.8) quite often with various choices of f to calculate the correlation functions, spacing functions, probability density of the determinants, mean value of the characteristic polynomial, etc.

As we said earlier, $C_i(x)$ and $\overline{C}_i(x)$ are arbitrary monic polynomials of order i and c_i are arbitrary non-zero constants. We can choose them as we like. For example, if we choose them to be orthogonal with the weight $w(x)$, $C_i(x) = \overline{C}_i(x)$,

$$\int C_i(x) C_j(x) w(x)\, dx = c_i \delta_{ij}, \tag{5.3.9}$$

and take $f(x) = 1$, then we get the "normalization constant" or the "partition function" $A_2 := \mathcal{I}_2(1)$

$$A_2 = \int \prod_{j=1}^{N} w(x_j) \Delta_N^2(x)\, dx_1 \cdots dx_N = N! \prod_{j=0}^{N-1} c_j. \tag{5.3.10}$$

If we choose them orthogonal as above, and take $f(x) = 1 + a(x)$, then we get

$$\left\langle \prod_{i=1}^{N} (1 + a(x_i)) \right\rangle_2 = A_2^{-1} \det[\Psi_{i,j}(1 + a)]_{i,j=0,\ldots,N-1}$$

$$= \det[\delta_{i,j} + \Psi_{i,j}(a)]_{i,j=0,\dots,N-1}, \tag{5.3.11}$$

$$\Psi_{i,j}(a) = c_i^{-1} \int w(x) C_i(x) C_j(x) a(x)\, dx. \tag{5.3.12}$$

Differentiating Eq. (5.3.11) m times functionally with respect to $a(x)$ and then setting $a(x) = 0$ we will get the m-point "correlation function"

$$X_m^{(2)}(x_1, \dots, x_m) := A_2^{-1} \frac{N!}{(N-m)!} \int |\Delta_N(x)|^2 \left[\prod_{k=1}^{N} w(x_k) \right] dx_{m+1} \cdots dx_N. \tag{5.3.13}$$

Expanding Eq. (5.3.11) in powers of $a(x)$ (see e.g. Mehta (1989MT), Section 2.5.3) the term containing m times $a(x)$ is the sum of all diagonal $m \times m$ sub-determinants of the $N \times N$ matrix $[\Psi_{i,j}(a)]$ of (5.3.12), i.e.

$$\left\langle \prod_{i=1}^{N} (1 + a(x_i)) \right\rangle_2 = 1 + \sum_{m=1}^{N} \sum_{0 \leqslant i_1 < i_2 < \cdots < i_m \leqslant N-1} \det[\Psi_{i_j, i_k}(a)]_{j,k=1,\dots,m}. \tag{5.3.14}$$

Interchanging any two indices amounts to interchanging simultaneously two rows and two columns in the $m \times m$ determinant above. Also if two indices are equal, the determinant vanishes. So the indices can be summed independently giving

$$\left\langle \prod_{i=1}^{N} (1 + a(x_i)) \right\rangle_2 = 1 + \sum_{m=1}^{N} \frac{1}{m!} \sum_{i_1=0}^{N-1} \cdots \sum_{i_m=0}^{N-1} \det[\Psi_{i_j, i_k}(a)]_{j,k=1,\dots,m}. \tag{5.3.15}$$

Now

$$\det[\Psi_{i_j, i_k}(a)]_m = \int \prod_{k=1}^{m} [dx_k\, a(x_k) w(x_k) c_{i_k}^{-1}] \det[C_{i_j}(x_k) C_{i_k}(x_k)]_{j,k=1,\dots,m} \tag{5.3.16}$$

$$= \int \prod_{k=1}^{m} [dx_k\, a(x_k) w(x_k) c_{i_k}^{-1} C_{i_k}(x_k)] \det[C_{i_j}(x_k)]_{j,k=1,\dots,m}$$

$$= \int \prod_{k=1}^{m} [dx_k\, a(x_k) w(x_k) c_{i_k}^{-1}] \det[C_{i_j}(x_j) C_{i_j}(x_k)]_{j,k=1,\dots,m}, \tag{5.3.17}$$

so that

$$\left\langle \prod_{i=1}^{N} (1 + a(x_i)) \right\rangle_2 = 1 + \sum_{m=1}^{N} \frac{1}{m!} \int \prod_{k=1}^{m} [dx_k \, a(x_k) w(x_k)]$$

$$\times \det \left[\sum_{i=0}^{N-1} \frac{1}{c_i} C_i(x_j) C_i(x_k) \right]_{j,k=1,\ldots,m} . \qquad (5.3.18)$$

The m-point correlation function is therefore

$$X_m^{(2)}(x_1, \ldots, x_m) = \prod_{k=1}^{m} w(x_k) \det \left[\sum_{i=0}^{N-1} \frac{1}{c_i} C_i(x_j) C_i(x_k) \right]_{j,k=1,\ldots,m}$$

$$= \det[K_2(x_j, x_k)]_{j,k=1,\ldots,m}, \qquad (5.3.19)$$

where $K_2(x, y)$ can be taken as

$$K_2(x, y) = w(x) \sum_{i=0}^{N-1} \frac{1}{c_i} C_i(x) C_i(y), \qquad (5.3.20)$$

or more symmetrically

$$K_2(x, y) = \sqrt{w(x) w(y)} \sum_{i=0}^{N-1} \frac{1}{c_i} C_i(x) C_i(y). \qquad (5.3.21)$$

In the above two examples we chose the polynomials $C_i(x)$ to be orthogonal with the weight $w(x)$. However, depending on the function $f(x)$ other choices of the polynomials $C_i(x)$ and $\overline{C}_i(x)$ may be more convenient. The only requirement is that each of the N polynomials $C_0(x), \ldots, C_{N-1}(x)$ be of degree less than N and that they be linearly independent. Similarly for the polynomials $\overline{C}_i(x)$. For example, if $w(-x) = w(x)$, the domain of integration is symmetric about the origin and $f(-x) = \pm f(x)$, then it is better to choose $C_i(-x) = (-1)^i C_i(x)$, say $C_i(x) = \overline{C}_i(x) = x^i$, so that $\Psi_{i,j}(f)$ in Eq. (5.3.8) satisfies $\Psi_{i,j}(f) = \pm(-1)^{i+j} \Psi_{i,j}(f)$, alternate elements in the determinant (5.3.7) are zero and its computation may be simpler.

5.4 Case $\beta = 4$; Skew-Orthogonal Polynomials of Quaternion Type

We now take the next difficult case, $\beta = 4$. Write $\Delta_N^4(x)$ as a confluent alternant (see e.g. Mehta (1989MT), Section 7.1.2)

$$\prod_{1 \leqslant i < j \leqslant N} (x_j - x_i)^4 = \det[x_i^j \quad j x_i^{j-1}]_{i=1,\dots,N;\, j=0,\dots,2N-1}$$

$$= \det[Q_j(x_i) \quad Q'_j(x_i)]_{i=1,\dots,N;\; j=0,\dots,2N-1}, \tag{5.4.1}$$

where $Q_i(x), i = 0, 1, \dots, 2N-1$, are any set of $2N$ linearly independent polynomials each of degree less than $2N$, suitably normalized to have equality in the above equation; and a prime denotes the derivative. We will usually take $Q_i(x)$ as a monic polynomial of precise degree i. As each variable occurs in two columns, we expand the determinant according to pairs of columns in the Laplace manner (see e.g. Mehta (1989MT), Section 2.5.1)

$$\Delta_N^4(x) = \sum_{(i)} \sigma(i) \det\begin{bmatrix} Q_{i_1}(x_1) & Q'_{i_1}(x_1) \\ Q_{i_2}(x_1) & Q'_{i_2}(x_1) \end{bmatrix} \cdots \det\begin{bmatrix} Q_{i_{2N-1}}(x_N) & Q'_{i_{2N-1}}(x_N) \\ Q_{i_{2N}}(x_N) & Q'_{i_{2N}}(x_N) \end{bmatrix}, \tag{5.4.2}$$

where $\sigma(i)$ is the sign of the permutation $(i) = \binom{0 \; \dots \; 2N-1}{i_1 \; \dots \; i_{2N}}$, the sum (i) is over all of those $(2N)!$ permutations (i) which satisfy the restrictions $i_1 < i_2, \dots, i_{2N-1} < i_{2N}$. Notice that each variable occurs in only one 2×2 sub-determinant.

Multiplying with $\prod_{j=1}^N f(x_j) w(x_j)$ and integrating over all the x_j, we make the following observations.

(i) The integral is the sum of products of N factors

$$\alpha_{i,j} = \int f(x) w(x) (Q_i(x) Q'_j(x) - Q'_i(x) Q_j(x))\, dx, \tag{5.4.3}$$

with all the indices distinct, they are the indices 0 to $2N-1$ taken in some order.

(ii) The sign of the factor $\alpha_{i_1,i_2} \cdots \alpha_{i_{2N-1},i_{2N}}$ is $\sigma(i)$, the sign of the permutation (i).

(iii) The sum is taken over all those of the $(2N)!$ permutations which satisfy the restrictions $i_1 < i_2, \dots, i_{2N-1} < i_{2N}$.

So recalling the definition of the Pfaffian the result is

$$\mathcal{I}_4(f) = N! \,\mathrm{pf}[\alpha_{i,j}]_{i,j=0,\dots,2N-1}$$

$$= N! \prod_{i=0}^{N-1} q_i \,\mathrm{pf}\left[\frac{1}{\sqrt{q_{[i/2]} q_{[j/2]}}} \alpha_{i,j}\right]_{i,j=0,\dots,2N-1}, \tag{5.4.4}$$

with any positive constants q_i. Here $[x]$ is the integer part of x.

Recall that $Q_i(x)$ are arbitrary monic polynomials and q_i are arbitrary positive constants. We can choose them as we like. If we choose them skew-orthogonal of "quaternion" type with the weight $w(x)$,

$$\int w(x)(Q_{2i}(x)Q'_{2j}(x) - Q'_{2i}(x)Q_{2j}(x))\,dx = 0,$$

$$\int w(x)(Q_{2i+1}(x)Q'_{2j+1}(x) - Q'_{2i+1}(x)Q_{2j+1}(x))\,dx = 0,$$

$$\int w(x)(Q_{2i}(x)Q'_{2j+1}(x) - Q'_{2i}(x)Q_{2j+1}(x))\,dx = q_i\delta_{i,j}, \tag{5.4.5}$$

then taking $f(x) = 1$ in Eq. (5.2.1) we get the "normalization constant" or the "partition function"

$$A_4 = \mathcal{I}_4(1) = N! \prod_{i=0}^{N-1} q_i. \tag{5.4.6}$$

Choosing $Q_i(x)$ and q_i as above and taking $f(x) = 1 + a(x)$, we get

$$\left\langle \prod_{i=1}^{N}(1 + a(x_i)) \right\rangle_4 := \mathcal{I}_4(f)/\mathcal{I}_4(1) = \mathrm{pf}[Z_{i,j} + \Psi_{i,j}(a)]_{i,j=0,1,\ldots,2N-1}$$

$$= \mathrm{pf}\begin{bmatrix} \Psi_{2i,2j}(a) & \delta_{i,j} + \Psi_{2i,2j+1}(a) \\ -\delta_{i,j} + \Psi_{2i+1,2j}(a) & \Psi_{2i+1,2j+1}(a) \end{bmatrix}_{i,j=0,1,\ldots,N-1},$$

$$\Psi_{i,j}(a) := \frac{1}{\sqrt{q_{[i/2]}q_{[j/2]}}} \int a(x)w(x)(Q_i(x)Q'_j(x) - Q'_i(x)Q_j(x))\,dx. \tag{5.4.7}$$

To get the correlation functions one can expand the Pfaffian in powers of $a(x)$ and pick up the corresponding term. The term containing m times $a(x)$ is the sum of all the quasi-diagonal $m \times m$ sub-Pfaffians $\Psi_{i,j}(a)$ in the expansion

$$\left\langle \prod_{i=1}^{N}(1 + a(x_i)) \right\rangle_4$$

$$= 1 + \sum_{m=1}^{N} \sum_{0 \leqslant i_1 < \cdots < i_m \leqslant N-1} \int \left[\prod_{k=1}^{m} dx_k\, a(x_k)w(x_k)\right] \mathrm{pf}[\alpha_{i_j,i_k}(x_k)]_{j,k=1,\ldots,m},$$

$$\alpha_{j,k}(x) = \frac{1}{\sqrt{q_j q_k}} \begin{bmatrix} \Psi_{2j,2k}(x) & \Psi_{2j,2k+1}(x) \\ \Psi_{2j+1,2k}(x) & \Psi_{2j+1,2k+1}(x) \end{bmatrix}. \tag{5.4.8}$$

As in the case $\beta = 2$, one can sum independently over the indices i_j, $j = 1, \ldots, m$, from 0 to $N-1$, change roles of the indices denoting rows or columns and the variables, and take the summations inside the Pfaffian. However, as we will see in Section 5.7 below, it is much easier to get the same result introducing quaternions, treating the Pfaffians as quaternion determinants and using Theorems 5.1.2 and 5.1.4. So we will not pursue this line any further.

Equation (5.4.4) is valid for any choice of the polynomials $Q_i(x)$ and positive constants q_i. Depending on the function $f(x)$, choices other than skew-orthogonal ones for $Q_i(x)$ may be more convenient. The only requirement is that the order of each $Q_i(x)$ be less than $2N$ and the $2N$ polynomials $Q_0(x), \ldots, Q_{2N-1}(x)$ be linearly independent. For example, if $w(-x) = w(x)$, the domain of integration is symmetric about the origin and $f(-x) = \pm f(x)$, then it is better to choose $Q_i(-x) = (-1)^i Q_i(x)$, say $Q_i(x) = x^i$, so that alternate elements in the Pfaffian (5.4.4) are zero and its computation may be simpler.

5.5 Case $\beta = 1$; Skew-Orthogonal Polynomials of Real Type

When $\beta = 1$, the absolute value sign of $\Delta_N(x)$ in the integrand is very inconvenient. As the integrand is completely symmetric in all the N variables, we can order the variables, for example, $x_1 \leqslant x_2 \leqslant \cdots \leqslant x_N$, and multiply the result by $N!$.

$$\mathcal{I}_1(f) = \int_a^b \prod_{j=1}^N f(x_j) w(x_j) |\Delta_N(x)| \, dx_1 \cdots dx_N$$

$$= N! \int_{a \leqslant x_1 \leqslant x_2 \leqslant \cdots \leqslant x_N \leqslant b} \prod_{j=1}^N f(x_j) w(x_j) \Delta_N(x) \, dx_1 \cdots dx_N. \tag{5.5.1}$$

Now write $\Delta_N(x)$ as an $N \times N$ determinant

$$\prod_{j=1}^N f(x_j) w(x_j) \Delta_N(x) = \det[f(x_j) w(x_j) x_j^{i-1}]_{i,j=1,\ldots,N}$$

$$= \det[f(x_j) w(x_j) R_{i-1}(x_j)]_{i,j=1,\ldots,N}, \tag{5.5.2}$$

where $R_i(x)$, $i = 0, 1, \ldots, N-1$, are any set of N linearly independent polynomials each of degree less than N, suitably normalized to have equality in the above equation.

As x_1 occurs only in the first column, we can integrate over it, replacing the column $f(x_1) w(x_1) R_i(x_1)$ by the column

$$g_i(x_2) := \int_a^{x_2} f(x) w(x) R_i(x) \, dx. \tag{5.5.3}$$

Now x_2 occurs in two columns and we can't do much about it. But we can integrate over x_3 replacing the third column by the column

$$\int_{x_2}^{x_4} f(x)w(x)R_i(x)\,dx = g_i(x_4) - g_i(x_2). \tag{5.5.4}$$

Adding the first column to the third does not change the determinant and we can eliminate the variable x_2 from the third column, so that the variables x_2 and x_4 appear each in two columns.

Then integrate over x_5, add the third column to the resulting fifth, so as to make it $g_i(x_6)$, and so on. This integration over the alternate variables $x_1, x_3, x_5, \ldots$ replaces the columns $f(x_{2j-1})w(x_{2j-1})R_i(x_{2j-1})$ successively by the columns $g_i(x_{2j})$, for $j = 1, 2, \ldots$. If $N = 2s+1$ is odd, then the last column is replaced by the column of constants $g_i(b)$.

We still have to integrate over $x_2, x_4, \ldots$ with the restrictions $a \leqslant x_2 \leqslant x_4 \leqslant \cdots \leqslant b$. As each variable occurs now in two columns, the integrand is completely symmetric in them and we can ignore their ordering. Thus if $N = 2s$,

$$\begin{aligned}\mathcal{I}_1(f) &= N! \int_{a \leqslant x_2 \leqslant x_4 \leqslant \cdots \leqslant x_{2s} \leqslant b} \det\big[f(x_{2j})w(x_{2j})R_i(x_{2j}) \quad g_i(x_{2j})\big]\,dx_2\,dx_4 \cdots dx_{2s} \\ &= \frac{N!}{s!} \int_a^b \det\big[f(x_{2j})w(x_{2j})R_i(x_{2j}) \quad g_i(x_{2j})\big]\,dx_2\,dx_4 \cdots dx_{2s},\end{aligned} \tag{5.5.5}$$

and if $N = 2s+1$, one has an extra column $g_i(b)$ in the integrand above at the extreme right.

Now each variable occurs in two columns and the situation is similar to the case $\beta = 4$. Expand the integrand by pairs of columns in the Laplace manner, and make similar observations:

The integral is a sum of products of s (or $s+1$, in case $N = 2s+1$) factors

$$\begin{aligned}\alpha_{i,j} &= \int_a^b f(x)w(x)(R_i(x)g_j(x) - R_j(x)g_i(x))\,dx \\ &= \int_{a \leqslant y \leqslant x \leqslant b} f(x)w(x)f(y)w(y)(R_i(x)R_j(y) - R_j(x)R_i(y))\,dx\,dy \\ &= \int_a^b \int_a^b f(x)w(x)f(y)w(y)R_i(x)R_j(y)\varepsilon(x-y)\,dx\,dy,\end{aligned} \tag{5.5.6}$$

$$\varepsilon(x) = \begin{cases} 1/2, & x > 0, \\ -1/2, & x < 0, \\ 0, & x = 0, \end{cases} \tag{5.5.7}$$

(and eventually $g_i(b)$, in case $N = 2s + 1$) with all indices distinct, they are the indices $0, 1, \ldots, N-1$ taken in some order, each term in the sum has a plus or minus sign, sign of the permutation of the indices. Thus the result is a Pfaffian

$$\mathcal{I}_1(f) = (2s)!\,\mathrm{pf}[\alpha_{i,j}]_{i,j=0,\ldots,2s-1}, \tag{5.5.8}$$

if $N = 2s$ is even, and

$$\mathcal{I}_1(f) = (2s+1)!\,\mathrm{pf}\begin{bmatrix} \alpha_{i,j} & g_i(b) \\ -g_j(b) & 0 \end{bmatrix}_{i,j=0,\ldots,2s}, \tag{5.5.9}$$

if $N = 2s + 1$ is odd, with $\alpha_{i,j}$ given by Eq. (5.5.6).

If $N = 2s$ is even, we choose $R_i(x)$ monic of precise degree i and skew-orthogonal of "real" type with the weight $w(x)$,

$$\int w(x)w(y)\varepsilon(x-y)R_{2i}(x)R_{2j}(y)\,dx\,dy = 0, \tag{5.5.10}$$

$$\int w(x)w(y)\varepsilon(x-y)R_{2i+1}(x)R_{2j+1}(y)\,dx\,dy = 0, \tag{5.5.11}$$

$$\int w(x)w(y)\varepsilon(x-y)R_{2i}(x)R_{2j+1}(y)\,dx\,dy = r_i\delta_{i,j}. \tag{5.5.12}$$

Then taking $f(x) = 1$ in Eq. (5.2.1) we get the "normalization constant" or the "partition function"

$$A_1 = \mathcal{I}_1(1) = (2s)!\prod_{i=0}^{s-1} r_i. \tag{5.5.13}$$

Choosing $R_i(x)$ and r_i as above and taking $f(x) = 1 + a(x)$, we get for $N = 2s$,

$$\left\langle \prod_{i=1}^{N}(1+a(x_i)) \right\rangle_1 := \mathcal{I}_1(f)/\mathcal{I}_1(1) = \mathrm{pf}[Z_{i,j} + \Psi_{i,j}(a)]_{i,j=0,1,\ldots,2s-1}$$

$$= \mathrm{pf}\begin{bmatrix} \Psi_{2i,2j}(a) & \delta_{i,j} + \Psi_{2i,2j+1}(a) \\ -\delta_{i,j} + \Psi_{2i+1,2j}(a) & \Psi_{2i+1,2j+1}(a) \end{bmatrix}_{i,j=0,1,\ldots,s-1}, \tag{5.5.14}$$

$$\Psi_{i,j}(a) := \frac{1}{\sqrt{r_i r_j}} \int a(x)a(y)w(x)w(y)\varepsilon(x-y)R_i(x)R_j(y)\,dx\,dy. \tag{5.5.15}$$

If $N = 2s + 1$ is odd, we choose $R_0(x), R_1(x), \ldots, R_{2s}(x)$ as linearly independent polynomials each of degree less than or equal to $2s$ satisfying skew-orthogonality relations (5.5.10)–(5.5.12) and in addition

$$\int w(x) R_i(x)\, dx = \delta_{i,2s}. \tag{5.5.16}$$

To see that such a choice is possible, though not with $R_i(x)$ a monic polynomial of precise degree i, see e.g. Mehta (1989MT), Section 8.12. With such a choice taking $f(x) = 1$, one has the "normalization constant" or the "partition function"

$$A_1 = \mathcal{I}_1(1) = (2s+1)! \int_a^b w(x) R_{2s}(x)\, dx \prod_{i=0}^{s-1} r_i. \tag{5.5.17}$$

And taking $f(x) = 1 + a(x)$ one has

$$\left\langle \prod_{i=1}^{N} (1 + a(x_i)) \right\rangle_1 := \mathcal{I}_1(f)/\mathcal{I}_1(1)$$

$$= \mathrm{pf} \begin{bmatrix} \Psi_{2i,2j} & \delta_{i,j} + \Psi_{2i,2j+1} & \Psi_{2i,2s} & g_{2i} \\ -\delta_{i,j} + \Psi_{2i+1,2j} & \Psi_{2i+1,2j+1} & \Psi_{2i+1,2s} & g_{2i+1} \\ \Psi_{2s,2j} & \Psi_{2s,2j+1} & 0 & g_{2s} \\ -g_{2j} & -g_{2j+1} & -g_{2s} & 0 \end{bmatrix}_{i,j=0,1,\ldots,s-1} \tag{5.5.18}$$

with $\Psi_{i,j} = \Psi_{i,j}(a)$ given by Eq. (5.5.15) and

$$g_i := g_i(a) = \int_a^b a(x) w(x) R_i(x)\, dx. \tag{5.5.19}$$

We get the m-point correlation function by functionally differentiating m times the above Pfaffians with respect to $a(x)$; which is the same thing as expanding in powers of a and taking $m!$ times the coefficient of $\prod_{i=1}^{s} a(x_i)$. However, as in the case $\beta = 4$, it is much simpler to introduce the quaternions and treat the Pfaffians as quaternion determinants. Hence we will not pursue this line any further.

Equations (5.5.6)–(5.5.9) are valid for any N linearly independent polynomials $R_i(x)$ each of degree less than N. We can choose them as we like. Choices other than skew-orthogonal ones may be more convenient depending on the function $f(x)$ and/or the weight $w(x)$. For example, if $w(-x) = w(x)$, the domain of integration is symmetric about the origin, $b = -a$, $f(-x) = \pm f(x)$, and $R_i(-x) = (-1)^i R_i(x)$, say $R_i(x) = x^i$, then alternate $\alpha_{i,j}$ and $g_i(b)$ in Eqs. (5.5.8) and (5.5.9) are zero and the computation may be simpler.

5.6 Average Value of $\prod_{j=1}^{N} \psi(x_j, y_j)$; Bi-Orthogonal Polynomials

Consider in particular the multiple integral

$$\left\langle \prod_{j=1}^{N} \psi(x_j, y_j) \right\rangle := \text{const} \cdot \int \left[\prod_{j=1}^{N} \psi(x_j, y_j) w(x_j, y_j) \right] \times \Delta_N(x)\Delta_N(y)\, dx_1 \cdots dx_N\, dy_1 \cdots dy_N, \tag{5.6.1}$$

with the normalization constant such that $\langle 1 \rangle = 1$. As in Eq. (5.3.3) we write $\Delta_N(x)$ as a polynomial alternant

$$\begin{aligned} \Delta_N(x)\Delta_N(y) &= \det[x_j^{i-1}]_{i,j=1,\dots,N} \det[y_j^{i-1}]_{i,j=1,\dots,N} \\ &= \det[P_{i-1}(x_j)]_{i,j=1,\dots,N} \det[\Pi_{i-1}(y_j)]_{i,j=1,\dots,N}, \end{aligned} \tag{5.6.2}$$

where $P_i(x)$ and $\Pi_i(x)$ are any monic polynomials of degree i. Expanding the determinants one has

$$\Delta_N(x)\Delta_N(y) = \sum_{(i)} \sum_{(j)} \sigma(i)\sigma(j) P_{i_1}(x_1) \cdots P_{i_N}(x_N) \Pi_{j_1}(y_1) \cdots \Pi_{j_N}(y_N), \tag{5.6.3}$$

where $\sigma(i)$ is the sign of the permutation $(i) = \binom{0 \ \dots \ N-1}{i_1 \ \dots \ i_N}$, the sum (i) is over all the $N!$ permutations (i), and similarly for the permutations (j).

If $\Psi(\mathbf{x}, \mathbf{y}) = \prod_{j=1}^{N} \psi(x_j, y_j)$, then the average value of $\Psi(\mathbf{x}, \mathbf{y})$ is

$$\begin{aligned} \langle \Psi(\mathbf{x}, \mathbf{y}) \rangle &= \text{const} \cdot \sum_{(i)} \sum_{(j)} \sigma(i)\sigma(j) \prod_{k=1}^{N} \int w(x_k, y_k)\, dx_k\, P_{i_k}(x_k) \Pi_{j_k}(y_k) \psi(x_k, y_k) \\ &= \text{const} \cdot \sum_{(i)} \sum_{(j)} \sigma(i)\sigma(j) \Psi_{i_1, j_1} \cdots \Psi_{i_N, j_N} \\ &= \text{const} \cdot N! \det[\Psi_{i,j}]_{i,j=0,\dots,N-1}, \end{aligned} \tag{5.6.4}$$

where

$$\Psi_{i,j} := \int P_i(x) \Pi_j(y) w(x, y) \psi(x, y)\, dx\, dy. \tag{5.6.5}$$

Here $P_i(x)$ and $\Pi_i(x)$ are arbitrary monic polynomials of order i. We can choose them as we like. If we choose them such that $\Psi_{i,j}$ is zero for $i \neq j$, then the determinant in Eq. (5.6.4) is diagonal and can easily be evaluated. For example, if we choose $P_i(x)$ and

$\Pi_i(x)$ bi-orthogonal with the weight function $w(x, y)$, i.e.

$$\int w(x, y) P_i(x) \Pi_j(y)\, dx\, dy = h_j \delta_{ij}, \tag{5.6.6}$$

then the normalization constant is

$$\int \prod_{j=1}^{N} w(x_j, y_j) \Delta_N(x) \Delta_N(y)\, dx_1 \cdots dx_N\, dy_1 \cdots dy_N = N! \prod_{j=0}^{N-1} h_j. \tag{5.6.7}$$

For the case when only some, not all, of the variables are integrated ("correlation functions") one may choose the polynomials $P_i(x)$ and $\Pi_i(x)$ bi-orthogonal as above, set $\psi(x, y) = (1 + a(x))(1 + b(y))$ and differentiate functionally several times with respect to $a(x)$ and $b(y)$. However, it is simpler to use a theorem similar to Theorem 5.1.4 to get them. The result is a single determinant involving four functions

$$K_{11}(x, y) = \sum_{k=0}^{N-1} \frac{1}{h_k} P_k(x) P_k(y), \qquad K_{12}(x, y) = \sum_{k=0}^{N-1} \frac{1}{h_k} P_k(x) \Pi_k(y),$$

$$K_{21}(x, y) = \sum_{k=0}^{N-1} \frac{1}{h_k} \Pi_k(x) P_k(y), \qquad K_{22}(x, y) = \sum_{k=0}^{N-1} \frac{1}{h_k} \Pi_k(x) \Pi_k(y), \tag{5.6.8}$$

of the remaining variables. In Chapter 23 we will study a more general problem which includes this as a particular case.

5.7 Correlation Functions

In this section the quantities of interest, to be calculated, are the "m-point correlation functions" defined by

$$X_m^{(\beta)}(x_1, \ldots, x_m) := A_\beta^{-1} \frac{N!}{(N-m)!} \int |\Delta_N(x)|^\beta \left[\prod_{k=1}^{N} w(x_k) \right] dx_{m+1} \cdots dx_N, \tag{5.7.1}$$

where the normalization constant A_β is the "partition function"

$$A_\beta = \int |\Delta_N(x)|^\beta \left[\prod_{k=1}^{N} w(x_k) \right] dx_1 \cdots dx_N. \tag{5.7.2}$$

For this purpose we will express the integrand in Eq. (5.7.1) or (5.7.2) as an $N \times N$ determinant with quaternion elements $K_\beta(x, y)$ satisfying the conditions (5.1.21) and (5.1.22) of Theorem 5.1.4

$$A_\beta^{-1}|\Delta_N(x)|^\beta \prod_{i=1}^{N} w(x_i) = \frac{1}{N!} \det[K_\beta(x_i, x_j)]_{i,j=1,2,\ldots,N}. \tag{5.7.3}$$

Define three series of monic polynomials, $C_k(x)$ and $R_k(x)$ for $0 \leqslant k < N$ each of degree less than N and $Q_k(x)$ for $0 \leqslant k < 2N$ each of degree less than $2N$. Recall that a polynomial is called monic when the coefficient of the highest power is one. The polynomial with an index k may be of precise degree k, but that is not necessary. What is essential is that the polynomials $C_0(x), C_1(x), \ldots, C_{N-1}(x)$ be linearly independent. Similarly, the polynomials $Q_0(x), Q_1(x), \ldots, Q_{2N-1}(x)$ or the polynomials $R_0(x), R_1(x), \ldots, R_{N-1}(x)$ should be linearly independent. Let these polynomials satisfy the orthogonality relations

$$\langle C_k, C_l\rangle_2 = c_k \delta_{kl}, \tag{5.7.4}$$

$$\langle Q_k, Q_l\rangle_4 = q_{[k/2]} Z_{kl}, \tag{5.7.5}$$

$$\langle R_k, R_l\rangle_1 = r_{[k/2]} Z_{kl}, \tag{5.7.6}$$

where c_k, $q_{[k/2]}$ and $r_{[k/2]}$ are normalization constants, $[x]$ is the largest integer not greater than x and Z_{kl} are given by Eq. (5.1.12); $Z_{2k,2k+1} = -Z_{2k+1,2k} = 1, k \geqslant 0$, and $Z_{j,k} = 0$ otherwise. The scalar product $\langle f, g\rangle_2$ and the two skew scalar products $\langle f, g\rangle_4$ and $\langle f, g\rangle_1$ of the functions $f(x)$ and $g(x)$ are defined by

$$\langle f, g\rangle_2 := \int f(x)g(x)w(x)\,dx, \tag{5.7.7}$$

$$\langle f, g\rangle_4 := \int [f(x)g'(x) - f'(x)g(x)]w(x)\,dx, \tag{5.7.8}$$

$$\langle f, g\rangle_1 := \int\int f(x)g(y)w(x)w(y)\varepsilon(y-x)\,dx\,dy, \tag{5.7.9}$$

with $\varepsilon(x)$ equal to $1/2$, $-1/2$ or 0 according as $x > 0$, $x < 0$ or $x = 0$.

When $N = 2s + 1$ is odd, for the monic polynomials $R_i(x)$ one has to take a set of N linearly independent polynomials $R_i(x)$ each of degree less than N satisfying the skew-orthogonality relations (5.7.6) and some more relations

$$\int w(x)R_i(x)\,dx = \delta_{i,N-1}. \tag{5.7.10}$$

Define the quaternions $\chi_k(x)$ and $\varpi_k(x)$, the 2×2 matrix representations of which are

$$\Theta[\chi_k(x)] := \begin{bmatrix} Q_{2k}(x) & -Q'_{2k}(x) \\ -Q_{2k+1}(x) & Q'_{2k+1}(x) \end{bmatrix}, \tag{5.7.11}$$

$$\Theta[\varpi_k(x)] := \begin{bmatrix} \psi_{2k}(x) & -\psi'_{2k}(x) \\ -\psi_{2k+1}(x) & \psi'_{2k+1}(x) \end{bmatrix}, \tag{5.7.12}$$

with

$$\psi_k(x) := \int \varepsilon(x-y) R_k(y) w(y)\, dy. \tag{5.7.13}$$

Define the real $K_2(x, y)$ and the two quaternions $K_4(x, y)$ and $K_1(x, y)$

$$K_2(x, y) := \sqrt{w(x)w(y)} \sum_{k=0}^{N-1} \frac{1}{c_k} C_k(x) C_k(y), \tag{5.7.14}$$

$$K_4(x, y) := \sqrt{w(x)w(y)} \sum_{k=0}^{N-1} \frac{1}{q_k} \overline{\chi}_k(x) \chi_k(y) \tag{5.7.15}$$

$$= \begin{bmatrix} S_4(x, y) & D_4(x, y) \\ I_4(x, y) & S_4(y, x) \end{bmatrix}, \tag{5.7.16}$$

$$K_1(x, y) := \sum_{k=0}^{N/2-1} \frac{1}{r_k} \overline{\varpi}_k(x) \varpi_k(y) - \begin{bmatrix} 0 & 0 \\ \varepsilon(x-y) & 0 \end{bmatrix} + K_{\text{odd}}(x, y), \tag{5.7.17}$$

$$K_{\text{odd}}(x, y) := \begin{bmatrix} \alpha(x) & 0 \\ u(x) - u(y) & \alpha(y) \end{bmatrix}, \tag{5.7.18}$$

or

$$K_1(x, y) = \begin{bmatrix} S_1(x, y) + \alpha(x) & D_1(x, y) \\ J_1(x, y) & S_1(y, x) + \alpha(y) \end{bmatrix}, \tag{5.7.19}$$

with

$$S_4(x, y) = \sqrt{w(x)w(y)} \sum_{k=0}^{N-1} \frac{1}{q_k} (Q'_{2k+1}(x) Q_{2k}(y) - Q'_{2k}(x) Q_{2k+1}(y)), \tag{5.7.20}$$

$$D_4(x, y) = \sqrt{w(x)w(y)} \sum_{k=0}^{N-1} \frac{1}{q_k} (-Q'_{2k+1}(x) Q'_{2k}(y) + Q'_{2k}(x) Q'_{2k+1}(y)), \tag{5.7.21}$$

$$I_4(x,y) = \sqrt{w(x)w(y)} \sum_{k=0}^{N-1} \frac{1}{q_k} (Q_{2k+1}(x)Q_{2k}(y) - Q_{2k}(x)Q_{2k+1}(y)), \tag{5.7.22}$$

$$S_1(x,y) = \sum_{k=0}^{N-1} \frac{1}{r_k} (\psi'_{2k+1}(x)\psi_{2k}(y) - \psi'_{2k}(x)\psi_{2k+1}(y)), \tag{5.7.23}$$

$$D_1(x,y) = \sum_{k=0}^{N-1} \frac{1}{r_k} (-\psi'_{2k+1}(x)\psi'_{2k}(y) + \psi'_{2k}(x)\psi'_{2k+1}(y)) \tag{5.7.24}$$

$$= -\frac{\partial}{\partial y} S_1(x,y), \tag{5.7.25}$$

$$J_1(x,y) = \sum_{k=0}^{N-1} \frac{1}{r_k} (\psi_{2k+1}(x)\psi_{2k}(y) - \psi_{2k}(x)\psi_{2k+1}(y)) - \varepsilon(x-y) + u(x) - u(y) \tag{5.7.26}$$

$$= \int \varepsilon(x-\xi) S_1(\xi,y)\, d\xi - \varepsilon(x-y) + u(x) - u(y), \tag{5.7.27}$$

where $\alpha(x) = u(x) = 0$, if N is even and

$$\alpha(x) = w(x)R_{2s}(x) \div \int w(y)R_{2s}(y)\, dy, \tag{5.7.28}$$

$$u(x) = \int \varepsilon(x-y)\alpha(y)\, dy, \tag{5.7.29}$$

if $N = 2s+1$ is odd.

Then Eq. (5.7.3) is valid and $K_\beta(x,y)$ satisfies both conditions (5.1.21) and (5.1.22) of Theorem 5.1.4. Our final result is the general formula

Theorem 5.7.1. *With $K_\beta(x,y)$ defined by Eqs.* (5.7.14)–(5.7.18) *one has*

$$X_m^{(\beta)}(x_1,\ldots,x_m) = \det[K_\beta(x_i,x_j)]_{i,j=1,\ldots,m}, \tag{5.7.30}$$

valid for the three values 1, 2 *and* 4 *of* β.

We also have the three partition functions A_β

$$A_2 = N! \prod_{k=0}^{N-1} c_k, \qquad A_4 = N! \prod_{k=0}^{N-1} q_k. \tag{5.7.31}$$

The result for $\beta = 1$ is slightly different according as N is even or odd

$$A_1 = \begin{cases} 2^s (2s)! \prod_{k=0}^{s-1} r_k, & N = 2s, \text{ even}, \quad (5.7.32) \\ 2^s (2s+1)! \prod_{k=0}^{s-1} r_k, & N = 2s+1, \text{ odd}. \quad (5.7.33) \end{cases}$$

From the positivity of these functions for every N, we deduce that c_k, q_k and r_k are positive numbers for every k.

5.8 Proof of Theorem 5.7.1

5.8.1 Case $\beta = 2$. Writing $\Delta_N(x)$ as an $N \times N$ determinant

$$\Delta_N(x) = \det[x_i^{j-1}]_{i,j=1,\dots,N} = \det[C_{j-1}(x_i)]_{i,j=1,\dots,N}, \tag{5.8.1}$$

one has

$$\begin{aligned} \Delta_N^2(x) &= \left(\prod_{j=0}^{N-1} c_j\right) \det\left[\frac{1}{c_{j-1}} C_{j-1}(x_k)\right]_{j,k=1,\dots,N} \det[C_{j-1}(x_k)]_{j,k=1,\dots,N} \\ &= \left(\prod_{k=0}^{N-1} c_k\right) \det\left[\sum_{k=0}^{N-1} \frac{1}{c_k} C_k(x_i) C_k(x_j)\right]_{i,j=1,\dots,N}, \end{aligned} \tag{5.8.2}$$

where $C_k(x)$ are the orthogonal polynomials

$$\int C_k(x) C_l(x) w(x)\, dx = c_k \delta_{kl}. \tag{5.8.3}$$

Now set

$$K_2(x,y) := \sqrt{w(x)w(y)} \sum_{k=0}^{N-1} \frac{1}{c_k} C_k(x) C_k(y). \tag{5.8.4}$$

Then

$$\det[K_2(x_i,x_j)]_{i,j=1,\dots,N} = \left(\prod_{k=0}^{N-1} c_k\right)^{-1} \Delta_N^2(x) \prod_{i=1}^{N} w(x_i), \tag{5.8.5}$$

$K_2(x, y) = K_2(y, x) = K_2^*(x, y)$ and from the orthogonality relation (5.8.3) one has

$$\int K_2(x, y) K_2(y, z)\, dy = K_2(x, z). \tag{5.8.6}$$

Thus $K_2(x, y)$ satisfies both conditions (5.1.21) and (5.1.22) of Theorem 5.1.4. The constant c defined by Eq. (5.1.24) is equal to N, so that Eq. (5.1.23) writes

$$\int \det[K_2(x_i, y_j)]_{i,j=1,\ldots,p+1}\, dx_{p+1} = (N - p) \det[K_2(x_i, y_j)]_{i,j=1,\ldots,p}. \tag{5.8.7}$$

The m-point correlation function takes the form

$$X_m^{(2)}(x_1, \ldots, x_m)$$
$$= A_2^{-1} \frac{N!}{(N - m)!} \left(\prod_{k=0}^{N-1} c_k \right) \int dx_{m+1} \cdots dx_N \det[K_2(x_i, y_j)]_{i,j=1,\ldots,N}. \tag{5.8.8}$$

Successive applications of Eq. (5.8.7) allow us to perform the integrations over the variables x_{m+1} to x_N. The calculation of the normalization constant A_2 along the same lines is straightforward, and the final result is Eq. (5.7.30) for $\beta = 2$.

5.8.2 Case $\beta = 4$. We write $\Delta_N^4(x)$ as a confluent alternant (see e.g. Mehta (1989MT), Section 7.1)

$$\Delta_N^4(x) = \det[\, x_j^{i-1} \quad (i-1)x_j^{i-2}\,] = \det[\, Q_{i-1}(x_j) \quad Q'_{i-1}(x_j)\,]$$
$$(i = 1, 2, \ldots, 2N;\ j = 1, 2, \ldots, N), \tag{5.8.9}$$

where the $Q_j(x)$ are monic polynomials of degree j, and prime denotes the derivation. Define the quaternion $\chi_k(x)$, which for short we identify with its 2×2 matrix representation

$$\chi_k(x) = \begin{bmatrix} Q_{2k}(x) & -Q'_{2k}(x) \\ -Q_{2k+1}(x) & Q'_{2k+1}(x) \end{bmatrix}. \tag{5.8.10}$$

Its dual is

$$\overline{\chi}_k(x) = \begin{bmatrix} Q'_{2k+1}(x) & Q'_{2k}(x) \\ Q_{2k+1}(x) & Q_{2k}(x) \end{bmatrix}. \tag{5.8.11}$$

Then $\Delta_N^4(x)$ can be written as follows

$$\Delta_N^4(x) = \det \Theta[Q] \tag{5.8.12}$$
$$= \det(Q\overline{Q}) = \det(\overline{Q}Q), \tag{5.8.13}$$

where Q is the $N \times N$ quaternion matrix with elements

$$Q_{ij} = \chi_{i-1}(x_j). \tag{5.8.14}$$

To write Eq. (5.8.13), use has been made of Corollary 5.1.3. Let us set

$$K_4(x, y) := \sqrt{w(x)w(y)} \sum_{k=0}^{N-1} \frac{1}{q_k} \overline{\chi}_k(x) \chi_k(y), \tag{5.8.15}$$

where the q_k are non-zero constants. Then

$$\begin{aligned} \Delta_N^4(x) \prod_{k=1}^{N} w(x_k) &= \prod_{k=1}^{N} w(x_k) \det(\overline{Q}Q) \\ &= \left(\prod_{k=0}^{N-1} q_k \right) \det[K_4(x_i, x_j)]_{i,j=1,\ldots,N}. \end{aligned} \tag{5.8.16}$$

Obviously $K_4(x, y)$ satisfies the first condition, Eq. (5.1.21), of Theorem 5.1.4. Let us now see for the second condition, Eq. (5.1.22). From Eqs. (5.8.10) and (5.8.11) one readily obtains

$$\int \chi_l(x) \overline{\chi}_k(x) w(x)\, dx = \begin{bmatrix} \langle Q_{2l}, Q_{2k+1} \rangle_4 & \langle Q_{2l}, Q_{2k} \rangle_4 \\ \langle Q_{2k+1}, Q_{2l+1} \rangle_4 & \langle Q_{2k}, Q_{2l+1} \rangle_4 \end{bmatrix}, \tag{5.8.17}$$

where we have used the skew scalar product $\langle\, , \rangle_4$ defined in Eq. (5.7.9). Now choose the $Q_k(x)$ as the skew orthogonal polynomials satisfying

$$\langle Q_{2l}, Q_{2k+1} \rangle_4 = \langle Q_{2k}, Q_{2l+1} \rangle_4 = q_k \delta_{kl}, \tag{5.8.18}$$

$$\langle Q_{2l}, Q_{2k} \rangle_4 = \langle Q_{2k+1}, Q_{2l+1} \rangle_4 = 0. \tag{5.8.19}$$

Equation (5.8.17) then becomes

$$\int \chi_l(x) \overline{\chi}_k(x) w(x)\, dx = q_k \delta_{kl} \mathbf{1}, \tag{5.8.20}$$

(recall that $\mathbf{1}$ designates the quaternion unity, or the unit 2×2 matrix). As a consequence

$$\int K_4(x, y) K_4(y, z)\, dy = K_4(x, z). \tag{5.8.21}$$

The end of the calculation goes exactly as in the case $\beta = 2$. Here again the constant c defined by Eq. (5.1.24) is equal to N, and the final result is Eq. (5.7.30) for $\beta = 4$.

5.8.3 Case $\beta = 1$, Even Number of Variables. This time the integrand we have to deal with is

$$X = \left[\prod_{k=1}^{N} w(x_k)\right] |\Delta_N(x)|. \tag{5.8.22}$$

We note that the sign of the Vandermonde determinant $\Delta_N(x)$ is given (de Bruijn, 1955) by the Pfaffian of the $N \times N$ matrix $[\mathrm{sign}(x_j - x_i)]_{i,j=1,\dots,N}$, so that we can write

$$X = 2^{[N/2]} \left[\prod_{k=1}^{N} w(x_k)\right] \Delta_N(x)\, \mathrm{pf}[\varepsilon(x_j - x_i)]_{i,j=1,\dots,N}, \tag{5.8.23}$$

where $\varepsilon(x)$, defined in the previous section, is half the sign function.

To start with, the case of an even number of variables, $N = 2s$, is easier. We replace the Vandermonde determinant $\Delta_{2s}(x)$ by the determinant of the $2s \times 2s$ matrix $[R_{j-1}(x_i)]_{i,j=1,\dots,2s}$, where the $R_k(x)$ are monic polynomials of degree k. Define the quaternion $\varpi_k(x)$, which for short we identify with its 2×2 matrix representation

$$\varpi_k(x) = \begin{bmatrix} \psi_{2k}(x) & -\psi'_{2k}(x) \\ -\psi_{2k+1}(x) & \psi'_{2k+1}(x) \end{bmatrix}, \tag{5.8.24}$$

where

$$\psi_k(x) = \int \varepsilon(x - y) w(y) R_k(y)\, dy, \quad \psi'_k(x) = w(x) R_k(x). \tag{5.8.25}$$

Define also the two self-dual quaternions

$$g(x, y) = \sum_{k=0}^{s-1} \frac{1}{r_k} \overline{\varpi}_k(x) \varpi_k(y), \tag{5.8.26}$$

$$K_1(x, y) = g(x, y) - \begin{bmatrix} 0 & 0 \\ \varepsilon(x - y) & 0 \end{bmatrix}, \tag{5.8.27}$$

where the r_k are non-zero constants. Obviously, the first condition (5.1.21) of Theorem 5.1.4 is satisfied by both $g(x, y)$ and $K_1(x, y)$.

From the above definitions, one readily obtains

$$\int \varpi_l(x) \overline{\varpi}_k(x)\, dx = 2 \begin{bmatrix} \langle R_{2l}, R_{2k+1} \rangle_1 & \langle R_{2l}, R_{2k} \rangle_1 \\ \langle R_{2k+1}, R_{2l+1} \rangle_1 & \langle R_{2k}, R_{2l+1} \rangle_1 \end{bmatrix}, \tag{5.8.28}$$

where we have used the skew scalar product $\langle\, ,\, \rangle_1$ defined by Eq. (5.7.9). Now choose the $R_k(x)$ as the skew orthogonal polynomials satisfying

$$\langle R_{2l}, R_{2k+1}\rangle_1 = \langle R_{2k}, R_{2l+1}\rangle_1 = r_k\delta_{kl}, \tag{5.8.29}$$

$$\langle R_{2k}, R_{2l}\rangle_1 = \langle R_{2l+1}, R_{2k+1}\rangle_1 = 0. \tag{5.8.30}$$

Equation (5.8.28) becomes

$$\int \varpi_k(x)\overline{\varpi}_l(x)\, dx = 2r_k\delta_{kl}\mathbf{1}, \tag{5.8.31}$$

which entails that $g(x, y)$ satisfies Eq. (5.1.22), up to a factor 2

$$\int g(x, y)g(y, z)\, dy = 2g(x, z). \tag{5.8.32}$$

Furthermore, a straightforward calculation shows that

$$\int \begin{bmatrix} 0 & 0 \\ \varepsilon(x-y) & 0 \end{bmatrix} \overline{\varpi}_k(y)\, dy = \begin{bmatrix} 0 & 0 \\ 0 & 1 \end{bmatrix} \overline{\varpi}_k(x), \tag{5.8.33}$$

$$\int \varpi_k(y) \begin{bmatrix} 0 & 0 \\ \varepsilon(y-z) & 0 \end{bmatrix} dy = \varpi_k(z) \begin{bmatrix} 1 & 0 \\ 0 & 0 \end{bmatrix}. \tag{5.8.34}$$

Putting together all these results, one obtains

$$\int K_1(x, y)K_1(y, z)\, dy = K_1(x, z) + \lambda K_1(x, z) - K_1(x, z)\lambda, \tag{5.8.35}$$

where λ is the constant quaternion $\begin{bmatrix} 1/2 & 0 \\ 0 & -1/2 \end{bmatrix}$. This is the second condition (5.1.22) of Theorem 5.1.4. The constant c of the same theorem is easily calculated with the help of Eq. (5.8.31), and found to be equal to $2s$.

Consider now the $2s \times 2s$ self-dual quaternion matrix $\mathcal{M}$ with elements

$$\mathcal{M}_{ij} = K_1(x_i, x_j). \tag{5.8.36}$$

Note that $[g(x_i, x_j)]_{2s}$ is the product of two rectangular $2s \times s$ and $s \times 2s$ quaternion matrices, with (i, j) elements equal to $r_{j-1}^{1/2}\overline{\varpi}_{j-1}(x_i)$ and $r_{i-1}^{1/2}\varpi_{i-1}(x_j)$. Its $4s \times 4s$ matrix representation is a rank $2s$ matrix with a vanishing determinant.

Let us be more specific. $\mathcal{M}$ being self-dual, $Z\Theta[\mathcal{M}]$ is anti-symmetric. It is convenient to reorder the rows and columns of $Z\Theta[\mathcal{M}]$ by writing first the $2s$ odd rows (and columns) before the $2s$ even rows (and columns). Let us call $\Xi[\mathcal{M}]$ the new anti-symmetric matrix thus obtained. It takes the form

$$\Xi[\mathcal{M}] = \begin{bmatrix} M & M\alpha^T \\ \alpha M & \alpha M\alpha^T - \mathcal{E} \end{bmatrix}, \tag{5.8.37}$$

where $M, \mathcal{E}$ and α are $2s \times 2s$ matrices. M and $\mathcal{E}$ are anti-symmetric

$$M_{ij} = \sum_{k,l=0}^{2s-1} \frac{1}{r_{[k/2]}} \psi_k'(x_i) Z_{kl} \psi_l'(x_j), \tag{5.8.38}$$

$$\mathcal{E}_{ij} = \varepsilon(x_j - x_i), \tag{5.8.39}$$

and α is defined, for almost all x_i by

$$\psi_k(x_i) = -\sum_{j=1}^{2s} \alpha_{ij} \psi_k'(x_j) \quad (i = 1, 2, \ldots, 2s;\ k = 0, 1, \ldots, 2s-1). \tag{5.8.40}$$

One easily finds that

$$\det \Theta[\mathcal{M}] = \det \Xi[\mathcal{M}], \tag{5.8.41}$$

$$\det \mathcal{M} = \operatorname{pf} Z\Theta[\mathcal{M}] = (-1)^s \operatorname{pf} \Xi[\mathcal{M}]. \tag{5.8.42}$$

Obviously, if we forget the matrix $\mathcal{E}$, the $2s$ last rows (columns) of the matrix (5.8.37) are linear combinations of the $2s$ first rows (columns). As a result, the determinant of $\Xi[\mathcal{M}]$ is independent of α:

$$\det \Xi[\mathcal{M}] = (\det M)(\det \mathcal{E}). \tag{5.8.43}$$

Its Pfaffian, a linear function of the elements of α, indeed is also independent of α, since squared it coincides with the determinant. Thus

$$\operatorname{pf} \Xi[\mathcal{M}] = (-1)^s (\operatorname{pf} M)(\operatorname{pf} \mathcal{E}). \tag{5.8.44}$$

Next, we remark that the matrix M appears in Eq. (5.8.38) as the product NZN^T, where N is the $2s \times 2s$ matrix with (i, k) element equal to $r_{[k/2]}^{-1/2} \psi_k'(x_i)$. Its Pfaffian is thus given by

$$\operatorname{pf} M = (\det N)(\operatorname{pf} Z) \tag{5.8.45}$$

$$= \left(\prod_{k=0}^{s-1} \frac{1}{r_k} \right) \left[\prod_{k=1}^{2s} w(x_k) \right] \det[R_{i-1}(x_j)]_{2s}. \tag{5.8.46}$$

Finally, from Eqs. (5.8.23), (5.8.36), (5.8.42), (5.8.44), (5.8.46), we obtain

$$X = 2^s \left(\prod_{k=0}^{s-1} r_k \right) \det[K_1(x_i, x_j)]_{i,j=1,\ldots,2s}. \tag{5.8.47}$$

Now, the constant A_1 defined by Eq. (5.1.24) can be calculated by applying $2s$ times Theorem 5.1.4

$$\begin{aligned} A_1 &= 2^s \left(\prod_{k=0}^{s-1} r_k \right) \int \det[K_1(x_i, x_j)]_{i,j=1,\ldots,2s}\, dx_1 \cdots dx_{2s} \\ &= 2^s (2s)! \left(\prod_{k=0}^{s-1} r_k \right). \end{aligned} \tag{5.8.48}$$

The correlation functions defined by Eq. (5.7.1) are then given by

$$X_{2s}^{(1)}(x_1, \ldots, x_{2s}) = \det[K_1(x_i, x_j)]_{i,j=1,\ldots,2s}, \tag{5.8.49}$$

$$\begin{aligned} X_m^{(1)}(x_1, \ldots, x_m) &= \frac{(2s)!}{(2s-m)!} \int \det[K_1(x_i, x_j)]_{i,j=1,\ldots,2s}\, dx_{m+1} \cdots dx_{2s} \\ &= \det[K_1(x_i, x_j)]_{i,j=1,\ldots,m}. \end{aligned} \tag{5.8.50}$$

This is Eq. (5.7.30) for $\beta = 1$ and $N = 2s$.

5.8.4 Case $\beta = 1$, Odd Number of Variables. When the number of variables $N = 2s + 1$ is odd, the kernel $K_1(x, y)$ contains one more term $K_{\text{odd}}(x, y)$. It is evidently self-dual, and so satisfies condition (5.1.21) of Theorem 5.1.4. As the polynomials $R_i(x)$, though no longer of precise degree i, still satisfy the skew-orthogonality relations (5.7.6), Eqs. (5.8.29), (5.8.30) are still valid. Also

$$\begin{aligned} \int \varpi_k(y) K_{\text{odd}}(y, z)\, dy &= \int \begin{bmatrix} \psi_{2k}(y) & -\psi'_{2k}(y) \\ -\psi_{2k+1}(y) & \psi'_{2k+1}(y) \end{bmatrix} \begin{bmatrix} \alpha(y) & 0 \\ u(y) - u(z) & \alpha(z) \end{bmatrix} dy \\ &= 0, \end{aligned} \tag{5.8.51}$$

$$\begin{aligned} \int K_{\text{odd}}(x, y) \overline{\varpi}_k(y)\, dy &= \int \begin{bmatrix} \alpha(x) & 0 \\ u(x) - u(y) & \alpha(y) \end{bmatrix} \begin{bmatrix} \psi'_{2k+1}(y) & \psi'_{2k}(y) \\ \psi_{2k+1}(y) & \psi_{2k}(y) \end{bmatrix} dy \\ &= 0, \end{aligned} \tag{5.8.52}$$

because of the additional restrictions (5.7.10) the polynomials $R_i(x)$ satisfy. Finally a straightforward calculation gives

$$\int \begin{bmatrix} 0 & 0 \\ \varepsilon(x-y) & 0 \end{bmatrix} K_{\text{odd}}(y,z)\,dy + \int K_{\text{odd}}(x,y) \begin{bmatrix} 0 & 0 \\ \varepsilon(y-z) & 0 \end{bmatrix} dy$$

$$= \begin{bmatrix} 0 & 0 \\ u(x)-u(z) & 0 \end{bmatrix} = K_{\text{odd}}(x,z)\lambda - \lambda K_{\text{odd}}(y,z), \tag{5.8.53}$$

and

$$\int K_{\text{odd}}(x,y) K_{\text{odd}}(y,z)\,dz = K_{\text{odd}}(x,z). \tag{5.8.54}$$

Putting together these results one verifies Eq. (5.8.35), i.e. $K_1(x,y)$ satisfies the second condition (5.1.22) of Theorem 5.1.4 in this case as well.

It remains to verify that indeed

$$\det[K_1(x_i,x_j)]_{i,j=1,\ldots,2s+1} = |\Delta_{2s+1}(x)| \prod_{i=1}^{2s+1} w(x_i). \tag{5.8.55}$$

For this, one considers a matrix slightly different from $[K_1(x_i,x_j)]$, that is

$$\det[\Theta(K_\eta(x_i,x_j))]_{i,j=1,\ldots,N}, \tag{5.8.56}$$

with

$$\Theta(K_\eta(x,y)) = \begin{bmatrix} S_1(x,y)+\alpha(x) & D_1(x,y)+\eta u(x)u(y) \\ J_1(x,y)+1/\eta & S_1(y,x)+\alpha(y) \end{bmatrix}, \tag{5.8.57}$$

computes its determinant which is simpler, and take the limit as $\eta \to 0$. For details see e.g. Mehta (1989MT), Appendix A.13.

Note. One should note that the expressions for the kernels $K_\beta(x,y)$ given above are not unique. Any expression for them will be acceptable as long as

$$\det[K_\beta(x_i,x_j)]_{i,j=1,\ldots,N} = |\Delta_N(x)|^\beta \prod_{i=1}^{N} w(x_i) \tag{5.8.58}$$

and the conditions of Theorem 5.1.4 are satisfied,

$$K * K = K + \lambda K - K\lambda. \tag{5.1.22$'$}$$

5.9 Spacing Functions

The spacing function $E_\beta(m, I)$ is defined as

$$E_\beta(m, I) := \frac{N!}{m!(N-m)!} \int_I dx_1 \cdots dx_m \int_J dx_{m+1} \cdots dx_N \prod_{i=1}^{N} w(x_i) |\Delta_N(x)|^\beta, \tag{5.9.1}$$

where in the integral the first m variables vary over the interval I and the other $N-m$ variables over the remaining part J of the range of integration (a, b) outside I. If $\prod_{i=1}^{N} w(x_i)|\Delta_N(x)|^\beta$ is the joint probability density of the N variables x_i, $i = 1, \ldots, N$, in (a, b), then $E_\beta(m, I)$ is the probability that exactly m of those lie in the interval I while the $N-m$ others lie outside of I.

The simplest of the spacing functions is $E_\beta(0, I)$, the probability that the interval I contains none of the x_i. To calculate it one sets in Eq. (5.2.1) $f(x) = 1 - \xi(I)$, where $\xi(I)$ is the characteristic function of I, i.e. $\xi(x) = 1$ if $x \in I$ and $\xi(x) = 0$ if $x \notin I$. If one sets instead $f(x) = 1 - z\xi(I)$, computes $\langle f(\mathbf{x})\rangle_\beta$ of Eq. (5.2.2) as a function of z, then differentiates the result partially m times with respect to z and sets $z = 1$, one will get $E_\beta(m, I)$ apart from a sign and a symmetry factor. However, to get the final result in a simple tractable form, one has to know the theory of linear algebraic equations (for the cases $\beta = 4$ and $\beta = 1$ with quaternion coefficients). Hence we will postpone its discussion for later chapters.

5.10 Determinantal Representations

Let the positive weight function $w(x)$ be such that all its integer moments are finite, $m_j = \int w(x)x^j\, dx < \infty$. Then the uniquely determined monic orthogonal polynomials $C_N(x)$ are obtained by Schmidt's process of orthogonalization of the successive powers of x and can be written as a determinant

$$C_N(x) = G_N^{-1} \det \begin{bmatrix} m_0 & m_1 & m_2 & \ldots & m_N \\ m_1 & m_2 & m_3 & \ldots & m_{N+1} \\ \ldots & \ldots & \ldots & \ldots & \ldots \\ m_{N-1} & m_N & m_{N+1} & \ldots & m_{2N-1} \\ 1 & x & x^2 & \ldots & x^N \end{bmatrix}, \tag{5.10.1}$$

with

$$G_N = \det[m_{i+j}]_{i,j=0,1,\ldots,N-1}. \tag{5.10.2}$$

Only the skew orthogonal polynomials $Q_{2N}(x)$ of even order are uniquely determined. To the polynomial $Q_{2N+1}(x)$ one can add an arbitrary constant multiple of $Q_{2N}(x)$ without changing the skew orthogonality relations. For example, one can choose this constant multiple so that the coefficient of x^{2N} in $Q_{2N+1}(x)$ is zero. The polynomials $Q_N(x)$ are then unique. They can be written as a determinant

$$Q_{2N}(x) = G_N^{-1} \det \begin{bmatrix} m_{0,0} & m_{0,1} & \cdots & m_{0,2N-1} & 1 \\ m_{1,0} & m_{1,1} & \cdots & m_{1,2N-1} & x \\ \cdots & \cdots & \cdots & \cdots & \cdots \\ m_{2N,0} & m_{2N,1} & \cdots & m_{2N,2N-1} & x^{2N} \end{bmatrix}, \tag{5.10.3}$$

$$G_N = \det[m_{i,j}]_{i,j=0,1,\ldots,2N-1}, \tag{5.10.4}$$

and

$$Q_{2N+1}(x) = H_N^{-1} \det \begin{bmatrix} m_{0,0} & m_{0,1} & \cdots & m_{0,2N+1} \\ m_{1,0} & m_{1,1} & \cdots & m_{1,2N+1} \\ \cdots & \cdots & \cdots & \cdots \\ m_{2N-1,0} & m_{2N-1,1} & \cdots & m_{2N-1,2N+1} \\ 1 & x & \cdots & x^{2N+1} \\ m_{2N+1,0} & m_{2N+1,1} & \cdots & m_{2N+1,2N+1} \end{bmatrix} + aQ_{2N}(x), \tag{5.10.5}$$

where the constant a is arbitrary, $m_{i,j} = (j-i) \int x^{j+i-1} w(x)\, dx$ and

$$H_N = \det[m_{i,j}]_{i=0,1,\ldots,2N-1,2N+1;\, j=0,1,\ldots,2N}. \tag{5.10.6}$$

Similarly, only the polynomials $R_{2N}(x)$ of even order are uniquely determined. To the polynomial $R_{2N+1}(x)$ one can add an arbitrary constant multiple of $R_{2N}(x)$ without changing the skew orthogonality relations. For example, one can choose this constant multiple so that the coefficient of x^{2N} in $R_{2N+1}(x)$ is zero. The polynomials $R_N(x)$ are then unique. They have similar determinantal expressions (5.10.3) and (5.10.5), where now $m_{i,j} = \int\int x^i y^j \varepsilon(y-x) w(x) w(y)\, dx\, dy$.

Bi-orthogonal polynomials do not exist for arbitrary positive weights $w(x,y)$. They exist when the moment matrix $M_{i,j}$, $i, j = 0, 1, \ldots, N-1$, is non-singular for every positive integer N, where $M_{i,j} = \int w(x,y) x^i y^j\, dx\, dy$. They have the determinantal expressions

$$\Pi_N(x) = G_N^{-1} \det \begin{bmatrix} M_{0,0} & M_{0,1} & \cdots & M_{0,N-1} & 1 \\ M_{1,0} & M_{1,1} & \cdots & M_{1,N-1} & x \\ \cdots & \cdots & \cdots & \cdots & \cdots \\ M_{N,0} & M_{N,1} & \cdots & M_{N,N-1} & x^N \end{bmatrix}, \tag{5.10.7}$$

and

$$P_N(x) = G_N^{-1} \det \begin{bmatrix} M_{0,0} & M_{0,1} & \cdots & M_{0,N} \\ M_{1,0} & M_{1,1} & \cdots & M_{1,N} \\ \cdots & \cdots & \cdots & \cdots \\ M_{N-1,0} & M_{N-1,1} & \cdots & M_{N-1,N} \\ 1 & x & \cdots & x^N \end{bmatrix}, \tag{5.10.8}$$

$$G_N = \det[M_{i,j}]_{i,j=0,1,\ldots,N-1}. \tag{5.10.9}$$

If $w(x, y) = w(y, x)$ then $M_{i,j} = M_{j,i}$ and $P_N(x) = \Pi_N(x)$.

All these determinantal forms are almost evident. In fact, if we take the appropriate scalar or skew scalar product of one of these polynomials with a lower power of the variable we get a determinant with either two identical rows or two identical columns, thus verifying the needed orthogonality, skew-orthogonality or bi-orthogonality.

5.11 Integral Representations

It has long been known that the orthogonal polynomial $C_N(x)$ can be expressed as a multiple integral (see e.g. Szegö, 1959)

$$C_N(x) \propto \int \cdots \int \Delta_N^2(y) \prod_{j=1}^{N} [(x - y_j) w(y_j)\, dy_j]. \tag{5.11.1}$$

Similar multiple integral expressions for the skew-orthogonal polynomials $Q_N(x)$ and $R_N(x)$ and the bi-orthogonal polynomials $P_i(x)$ and $\Pi_i(x)$ can be written down (Eynard, 2001; Ghosh and Pandey, 2002)

$$Q_{2N}(x) \propto \int \cdots \int \Delta_N^4(y) \prod_{j=1}^{N} [(x - y_j)^2 w(y_j)\, dy_j], \tag{5.11.2}$$

$$Q_{2N+1}(x) \propto \int \cdots \int \Delta_N^4(y) \left(x + 2 \sum_{j=1}^{N} y_j \right) \prod_{j=1}^{N} [(x - y_j)^2 w(y_j)\, dy_j], \tag{5.11.3}$$

$$R_{2N}(x) \propto \int \cdots \int |\Delta_{2N}(y)| \prod_{j=1}^{2N} [(x - y_j) w(y_j)\, dy_j], \tag{5.11.4}$$

$$R_{2N+1}(x) \propto \int \cdots \int |\Delta_{2N}(y)| \left(x + \sum_{j=1}^{N} y_j \right) \prod_{j=1}^{2N} [(x - y_j) w(y_j)\, dy_j], \tag{5.11.5}$$

$$P_N(x) \propto \int \cdots \int \Delta_N(x) \Delta_N(y) \prod_{j=1}^{N} [(x - x_j) w(x_j, y_j)\, dx_j\, dy_j], \tag{5.11.6}$$

$$\Pi_N(y) \propto \int \cdots \int \Delta_N(x) \Delta_N(y) \prod_{j=1}^{N} [(y - y_j) w(x_j, y_j)\, dx_j\, dy_j]. \tag{5.11.7}$$

Proof. Writing $\Delta_N(y)\prod_{i=1}^N(x-y_i)$ as an alternant

$$\Delta_N^2(y)\prod_{i=1}^N(x-y_i) = \det[y_j^{i-1}]_{i,j=1,\ldots,N}\det[y_j^i, x^i]_{i=0,1,\ldots,N;\,j=1,2,\ldots,N}$$
$$= \det[C_{i-1}(y_j)]_{i,j=1,\ldots,N}\det[C_i(y_j), C_i(x)]_{i=0,1,\ldots,N;\,j=1,2,\ldots,N}$$
$$= \sum_{(i)}\sigma(i)C_{i_1}(y_1)\cdots C_{i_N}(y_N)$$
$$\times\sum_{(j)}\sigma(j)C_{j_1}(y_1)\cdots C_{j_N}(y_N)C_{j_{N+1}}(x), \tag{5.11.8}$$

where $\sigma(i)$ is the sign of the permutation $(i)=\binom{0\ \ldots\ N-1}{i_1\ \ldots\ i_N}$, the sum (i) is over all the $N!$ permutations (i); $\sigma(j)$ is the sign of the permutation $(j)=\binom{0\ \ldots\ N}{j_1\ \ldots\ j_{N+1}}$ and the sum (j) is over all the $(N+1)!$ permutations (j). Since the $C_j(y)$ are orthogonal with the weight $w(y)$, if we multiply both sides of Eq. (5.11.8) with $\prod_{i=1}^N w(y_i)$ and integrate over $y_1,\ldots,y_N$, the only terms which give a non-zero contribution are those with $\sigma(i_k)=\sigma(j_k)$, $k=1,\ldots,N$. And each term gives the same contribution $c_0c_1\cdots c_{N-1}$. This proves Eq. (5.11.1).

The proofs of Eqs. (5.11.6) and (5.11.7) can be derived by similar arguments; expand the determinants on the right-hand side of the equalities

$$\Delta_N(x)\prod_{j=1}^N(x-x_j) = \det[x_j^i, x^i]_{i=0,1,\ldots,N;\,j=1,2,\ldots,N}$$
$$= \det[P_i(x_j), P_i(x)]_{i=0,1,\ldots,N;\,j=1,2,\ldots,N}, \tag{5.11.9}$$

$$\Delta_N(y)\prod_{j=1}^N(y-y_j) = \det[y_j^i, y^i]_{i=0,1,\ldots,N;\,j=1,2,\ldots,N}$$
$$= \det[\Pi_i(y_j), \Pi_i(y)]_{i=0,1,\ldots,N;\,j=1,2,\ldots,N}, \tag{5.11.10}$$

multiply both sides by $\prod_{j=1}^N w(x_j,y_j)$, integrate over x_j, y_j, $1\leqslant j\leqslant N$, and use the bi-orthogonality of the polynomials $P_i(x)$ and $\Pi_i(y)$.

To prove Eq. (5.11.2) write $\Delta_N^4(y)\prod_{i=1}^N(x-y_i)^2$ as a confluent alternant (see e.g. Mehta (1989MT), Section 7.1)

$$\Delta_N^4(y)\prod_{i=1}^N(x-y_i)^2$$
$$= \det[y_j^i,\ iy_j^{i-1},\ x^i]_{i=0,1,\ldots,2N;\,j=1,2,\ldots,N}$$

$$= \det[Q_i(y_j), Q_i'(y_j), Q_i(x)]_{i=0,1,\dots,2N;\, j=1,2,\dots,N} \tag{5.11.11}$$

$$= \sum_{(i)} \sigma(i) \det\begin{bmatrix} Q_{i_1}(y_1) & Q'_{i_1}(y_1) \\ Q_{i_2}(y_1) & Q'_{i_2}(y_1) \end{bmatrix} \cdots$$

$$\det\begin{bmatrix} Q_{i_{2N-1}}(y_N) & Q'_{i_{2N-1}}(y_N) \\ Q_{i_{2N}}(y_N) & Q'_{i_{2N}}(y_N) \end{bmatrix} Q_{i_{2N+1}}(x), \tag{5.11.12}$$

where, as above, $\sigma(i)$ is the sign of the permutation $(i) = \binom{0 \ \cdots \ 2N}{i_1 \ \cdots \ i_{2N+1}}$, and the sum (i) is over all those of the $(2N+1)!$ permutations of the indices $0, 1, \dots, 2N$ which satisfy the inequalities $i_1 < i_2,\ i_3 < i_4, \dots, i_{2N-1} < i_{2N}$. The $Q_i(y)$ are the skew-orthogonal polynomials

$$\int [Q_{i_1}(y)Q'_{i_2}(y) - Q'_{i_1}(y)Q_{i_2}(y)]w(y)\,dy = \begin{cases} q_k, & \text{if } i_1 = i_2 - 1 = 2k, \\ -q_k, & \text{if } i_2 = i_1 - 1 = 2k, \\ 0, & \text{otherwise.} \end{cases} \tag{5.11.13}$$

Therefore multiplying both sides of Eq. (5.11.12) with $\prod_{k=1}^{N} w(y_k)$ and integrating over $y_1, \dots, y_N$, the only terms which survive are $i_{2j+1} = 2k$, $i_{2j+2} = 2k+1$, $i_{2N+1} = 2N$, and each of them has the same coefficient, namely $q_0 q_1 \cdots q_{N-1}$. This proves Eq. (5.11.2).

The proof of Eq. (5.11.3) depends on the equality

$$\Delta_N^4(y) \prod_{i=1}^{N} (x - y_i)^2 \left(x + 2\sum_{i=1}^{N} y_i \right) = \det[y_j^i, i y_j^{i-1}, x^i]_{i=0,1,\dots,2N-1,2N+1;\, j=1,2,\dots,N}. \tag{5.11.14}$$

Multiply both sides by $\prod_{j=1}^{N} w(y_j)$ and integrate over $y_1, \dots, y_N$. Expanding the determinant by blocks of two columns as in Eq. (5.11.12), multiplying with the weight function and integrating one sees from the skew-orthogonality of the polynomials $Q_i(y)$ that the only terms which survive are $i_{2j+1} = 2k$, $i_{2j+2} = 2k+1$, $i_{2N+1} = 2N+1$, and each of them has the same coefficient $q_0 q_1 \cdots q_{N-1}$.

The proofs of Eqs. (5.11.4) and (5.11.5) are similar and depend on the equalities

$$\Delta_{2N}(y) \prod_{i=1}^{2N} (x - y_i) = \det[y_j^i, x^i]_{i=0,1,\dots,2N;\, j=1,2,\dots,2N}$$

$$= \det[R_i(y_j), R_i(x)]_{i=0,1,\dots,2N;\, j=1,2,\dots,2N}, \tag{5.11.15}$$

and

$$\Delta_{2N}(y) \prod_{i=1}^{2N} (x - y_i) \left(x + \sum_{i=1}^{2N} y_i \right)$$

$$= \det[y_j^i, x^i]_{i=0,1,\ldots,2N-1,2N+1;\, j=1,2,\ldots,2N}$$

$$= \det[R_i(y_j), R_i(x)]_{i=0,1,\ldots,2N-1,2N+1;\, j=1,2,\ldots,2N}. \tag{5.11.16}$$

Multiply both sides by $\prod_{i=1}^{2N} w(y_j)$, and integrate over $y_1, \ldots, y_{2N}$. To take care of the absolute value sign of $\Delta_{2N}(y)$, order the variables as $y_1 \leqslant y_2 \leqslant \cdots \leqslant y_{2N}$, integrate over alternate ones $y_1, y_3, \ldots, y_{2N-1}$, so as to replace the columns $R_i(y_{2j+1})$ by $\int \varepsilon(y_{2j+2} - y) R_i(y) w(y)\, dy$. Then in the integration over the remaining variables $y_2, y_4, \ldots$ the ordering can be removed. The proofs can now be completed by similar arguments; expand the determinants by blocks of pairs of columns and use the skew-orthogonality of the polynomials $R_j(y)$. □

5.12 Properties of the Zeros

Several properties of the zeros of the orthogonal polynomials $C_N(x)$ have been known for a long time. For example,

(1) The zeros of $C_N(x)$ are real, simple and lie in the support of the weight function $w(x)$.
(2) The zeros interlace for successive polynomials, i.e. between every pair of consecutive zeros of $C_{N+1}(x)$ there is one and only one zero of $C_N(x)$ for every $N > 0$.

What can one say about the zeros of $Q_N(x)$, $R_N(x)$, $P_N(x)$ or of $\Pi_N(x)$? Not much.

As we said earlier in Section 5.10, if the positive weight function $w(x, y)$ is such that all the moments $M_{i,j} = \int w(x, y) x^i y^j\, dx\, dy$ exist for all $i, j \geqslant 0$ and the moment matrix $[M_{i,j}]$, $i, j = 0, \ldots, N-1$ is non-singular for every positive integer N, then unique monic polynomials $P_i(x)$ and $\Pi_i(x)$, bi-orthogonal with the weight $w(x, y)$, exist.

From limited numerical evidence for the weights

(i) $w(x, y) = \sin(\pi x y)$, $0 \leqslant x, y \leqslant 1$;
(ii) $w(x, y) = |x - y|$, $-1 \leqslant x, y \leqslant 1$;
(iii) $w(x, y) = [1/(x + y)] \exp[-x - y]$, $0 \leqslant x, y < \infty$;
(iv) $w(x, y) = \exp(-x^2 - y^2 - cxy)$, $-\infty < x, y < \infty$, $0 < c < 2$;

one might think that the zeros of the bi-orthogonal polynomials are real, simple, lie respectively in the x or y-support of $w(x, y)$, interlace for successive $N, \ldots$.

Alas, this is not true in general as seen by the following examples due to P. Deligne. If one takes

$$w(x, y) = u(x, y) + v(x, y), \tag{5.12.1}$$

$$u(x, y) = \begin{cases} \delta(x - y), & -1 \leqslant x, y \leqslant 1, \\ 0, & \text{otherwise}, \end{cases} \tag{5.12.2}$$

$$v(x, y) = \frac{1}{8}[\delta(x-1)\delta(y+2) + \delta(x+1)\delta(y-2)]. \tag{5.12.3}$$

Then $P_3(x) = 5x^3 + 3x$ and $\Pi_3(x) = 5x^3 + 48x$ have complex zeros.

As a second example, if one takes

$$w(x, y) = \frac{1}{\sqrt{\pi}} e^{-x^2} \delta(x-y) + \frac{5}{16}[\delta(x-1)\delta(y+1) + \delta(x+1)\delta(y-1)], \tag{5.12.4}$$

then $P_3(x) = \Pi_3(x) = x^3 + x$ have complex zeros.

However, an argument of Ercolani and McLaughlin (2001) shows that if the moments $M_{i,j}$ of the weight function $w(x, y)$ exist for all $i, j \geqslant 0$ and if $\det[w(x_i, y_j)]_{i,j=1}^{N} > 0$ for $x_1 < x_2 < \cdots < x_N$, $y_1 < y_2 < \cdots < y_N$, then bi-orthogonal polynomials exist and their zeros are real, simple and lie in the respective supports of the weight $w(x, y)$.

This is a sufficient condition, not a necessary one, as indicated by the weights (i) and (ii) above. We will come back to this question in Chapter 23 when we study a linear chain of coupled Hermitian matrices.

5.13 Orthogonal Polynomials and the Riemann–Hilbert Problem

Orthogonal polynomials also arise as a solution of the following Riemann–Hilbert problem. Let a 2×2 matrix function $Y(z)$ satisfy the three conditions:

(1) $Y(z)$ is analytic in the entire complex plane z except for the real line.

(2) $Y_\pm(x) = \lim_{\varepsilon \to 0+} Y(x \pm i\varepsilon)$, x and ε real, are related as

$$Y_+(x) = Y_-(x) \begin{bmatrix} 1 & w(x) \\ 0 & 1 \end{bmatrix},$$

where $w(x) \geqslant 0$ and all its moments are finite, $\int w(x) x^j \, dx < \infty$.

(3) As $|z| \to \infty$,

$$Y(z) = (I + \mathrm{O}(z^{-1})) \begin{bmatrix} z^n & 0 \\ 0 & z^{-n} \end{bmatrix},$$

with I the 2×2 unit matrix.

Then $Y(z)$ is uniquely given by

$$Y(z) \equiv Y^{(n)}(z) = \begin{bmatrix} P_n(z) & \dfrac{1}{2\pi i} \displaystyle\int_{-\infty}^{\infty} \frac{w(x) P_n(x)\, dx}{x - z} \\ \gamma_{n-1} P_{n-1}(z) & \dfrac{\gamma_{n-1}}{2\pi i} \displaystyle\int_{-\infty}^{\infty} \frac{w(x) P_{n-1}(x)\, dx}{x - z} \end{bmatrix},$$

where $P_n(x)$ is the nth monic orthogonal polynomial with the weight function $w(x)$ and γ_n is a normalization constant.

One may expand $Y^{(n)}(z)$ as

$$\{I + Y_1^{(n)} z^{-1} + Y_2^{(n)} z^{-2} + \cdots\} \begin{bmatrix} z^n & 0 \\ 0 & z^{-n} \end{bmatrix},$$

and determine successively the coefficient matrices $Y_1^{(n)}, Y_2^{(n)}, \ldots$. Powerful methods of complex variable theory can be applied to get, for example, the asymptotic behaviour of the orthogonal polynomials $P_n(z)$ for large n. For details see e.g. Deift (2000) or Fokas et al. (1990).

5.14 A Remark (Balian)

Define the scalar product of $f(x)$ and $g(x)$ depending on an auxiliary parameter z as

$$\langle f, g \rangle_z = \int \int f(x) g(y) \sqrt{w(x) w(y)} \varepsilon(y - x + z) \, dx \, dy, \tag{5.14.1}$$

where as usual $\varepsilon(x)$ is $1/2$, $-1/2$ or 0 according as x is positive, negative or zero. The power series expansion of this scalar product starts as

$$\langle f, g \rangle_z = \langle f, g \rangle_1 + z \langle f, g \rangle_2 + \frac{z^2}{2} \langle f, g \rangle_4 + \cdots. \tag{5.14.2}$$

The first three terms are just the scalar products (with the weight function replaced by its square root in case $\beta = 1$), Eqs. (5.7.7)–(5.7.9), we need to deal with the three values 1, 2 and 4 of β.

Can one find an application of the later terms in the series? For example, the next term will be (subscript 8 for "octonions" !!)

$$\langle f, g \rangle_8 = \int w(x) \left(f(x) g''(x) + f''(x) g(x) - \frac{1}{2} \left[\frac{w'(x)}{w(x)} \right]^2 f(x) g(x) \right) dx. \tag{5.14.3}$$

Summary of Chapter 5

A general method of evaluating integrals of $|\Delta_N(x)|^\beta \prod_{i=1}^N w(x_i)$ is described. Here $\Delta_N(x)$ is the product of differences of the N variables $x_1, \ldots, x_N$, β takes one of the values 1, 2 or 4 and integration is done over any number of variables. For the simplest case $\beta = 2$, a knowledge of determinants and orthogonal polynomials based on a symmetric scalar product is needed (Section 5.3). For the other two values 1 and 4 of β properties of Pfaffians are needed. In analogy to the case $\beta = 2$, determinants of matrices with quaternion elements and two kinds of skew-orthogonal polynomials based on anti-symmetric scalar products are introduced (Sections 5.4, 5.5 and 5.6). The method is to express the integrand as an $N \times N$ determinant in such a way that each variable occurs in one row and one column and integration over that variable amounts to erasing that row and column thus decreasing the size of the determinant by one and multiplying by a known constant. This way one can integrate over as many variables as one likes by simply erasing the corresponding rows and columns. The final result, the m-point correlation function, is thus expressed as a single $m \times m$ determinant.

Similar integrals over two sets of variables lead us naturally to introduce bi-orthogonal polynomials.

All these polynomials, orthogonal, skew-orthogonal and bi-orthogonal, can be represented either as determinants (Section 5.10) or as multiple integrals (Section 5.11). Finally some interrogatory observations are made as to the nature and location of the zeros and the relation among the three kinds of scalar products.

6

GAUSSIAN UNITARY ENSEMBLE

In Chapter 2 it was suggested that if the zero and the unit of the energy scale are properly chosen, then the statistical properties of the fluctuations of energy levels of a sufficiently complicated system will behave as if they were the eigenvalues of a matrix taken from one of the Gaussian ensembles. In Chapter 3 we derived the joint probability density function for the eigenvalues of matrices from orthogonal, symplectic or unitary ensembles. To examine the behaviour of a few eigenvalues, one has to integrate out the others, and this we will undertake now.

In this chapter we take up the Gaussian unitary ensemble, which is the simplest from the mathematical point of view, and derive expressions for the probability of a few eigenvalues to lie in certain specified intervals. For other ensembles the same thing will be undertaken in later chapters.

In Section 6.1 we define correlation, cluster and spacing probability functions. These definitions are general and apply to any of the ensembles. In Section 6.2 we derive the correlation and cluster functions by the method described in Chapter 5. Section 6.3 explains Gaudin's way of using Fredholm theory of integral equations to write the probability $E_2(0; s)$, that an interval of length s contains none of the eigenvalues, as an infinite product. In Section 6.4 we derive an expression for $E_2(n; s)$, the probability that an interval of length s contains exactly n eigenvalues. Finally Section 6.5 contains a few remarks about the convergence and numerical computation of $E_2(n; s)$ as well as some other equalities.

The two-level cluster function is given by Eq. (6.2.14), its Fourier transform by (6.2.17); the probability $E_2(n; s)$ for an interval of length s to contain exactly n levels is given by Eq. (6.4.30) while the nearest neighbor spacing density $p_2(0; s)$ is given by Eq. (6.4.32) or (6.4.33). These various functions are tabulated in Appendices A.14 and A.15.

6.1 Generalities

The joint probability density function of the eigenvalues (or levels) of a matrix from one of the Gaussian invariant ensembles is (cf. Theorem 3.3.1, Eq. (3.3.8))

$$P_{N\beta}(x_1, \ldots, x_N) = \text{const} \cdot \exp\left(-\frac{\beta}{2}\sum_{j=1}^{N} x_j^2\right) \prod_{1\leqslant j<k\leqslant N} |x_j - x_k|^\beta, \quad -\infty < x_i < \infty, \tag{6.1.1}$$

where $\beta = 1, 2$ or 4 according as the ensemble is orthogonal, unitary or symplectic. The considerations of this section are general, they apply to all the three ensembles. From Section 6.2 onwards we will concentrate our attention to the simplest case $\beta = 2$. For simplicity of notation the index β will sometimes be suppressed; for example, we will write P_N instead of $P_{N\beta}$.

6.1.1 About Correlation and Cluster Functions. The n-point correlation function is defined by (Dyson, 1962a, III)

$$R_n(x_1, \ldots, x_n) = \frac{N!}{(N-n)!}\int_{-\infty}^{\infty} \cdots \int_{-\infty}^{\infty} P_N(x_1, \ldots, x_N)\, dx_{n+1} \cdots dx_N, \tag{6.1.2}$$

which is the probability density of finding a level (regardless of labeling) around each of the points $x_1, x_2, \ldots, x_n$, the positions of the remaining levels being unobserved. In particular, $R_1(x)$ will give the overall level density. Each function R_n for $n > 1$ contains terms of various kinds describing the grouping of n levels into various subgroups or clusters. For practical purposes it is convenient to work with the n-level cluster function (or the cumulant) defined by

$$T_n(x_1, \ldots, x_n) = \sum_G (-1)^{n-m}(m-1)! \prod_{j=1}^{m} R_{G_j}(x_k, \text{ with } k \text{ in } G_j). \tag{6.1.3}$$

Here G stands for any division of the indices $(1, 2, \ldots, n)$ into m subgroups $(G_1, G_2, \ldots, G_m)$. For example,

$$\begin{aligned} T_1(x) &= R_1(x), \\ T_2(x_1, x_2) &= -R_2(x_1, x_2) + R_1(x_1)R_1(x_2), \\ T_3(x_1, x_2, x_3) &= R_3(x_1, x_2, x_3) - R_1(x_1)R_2(x_2, x_3) - \cdots - \cdots \\ &\quad + 2R_1(x_1)R_1(x_2)R_1(x_3), \end{aligned}$$

$$\begin{aligned}T_4(x_1,x_2,x_3,x_4) = &-R_4(x_1,x_2,x_3,x_4)\\&+[R_1(x_1)R_3(x_2,x_3,x_4)+\cdots+\cdots+\cdots]\\&+[R_2(x_1,x_2)R_2(x_3,x_4)+\cdots+\cdots]\\&-2\big[R_2(x_1,x_2)R_1(x_3)R_1(x_4)+\cdots+\cdots+\cdots+\cdots+\cdots\big]\\&+6R_1(x_1)R_1(x_2)R_1(x_3)R_1(x_4),\end{aligned}$$

where in the last equation the first bracket contains four terms, the second contains three terms, and the third contains six terms. Equation (6.1.3) is a finite sum of products of the R functions, the first term in the sum being $(-1)^{n-1}R_n(x_1,x_2,\ldots,x_n)$ and the last term being $(n-1)!R_1(x_1)\cdots R_1(x_n)$.

We would be particularly pleased if these functions R_n and T_n turn out to be functions only of the differences $|x_i - x_j|$. Unfortunately, this is not true in general. Even the level density R_1 will turn out to be a semi-circle rather than a constant (cf. Section 6.2). However, as long as we remain in the region of maximum (constant) density, we will see that the functions R_n and T_n satisfy this requirement. Outside the central region of maximum density one can always approximate the overall level density to be a constant in a small region ("small" meaning that the number of levels n in this interval is such that $1 \ll n \ll N$). And one believes that the statistical properties of these levels are independent of where this interval is chosen. However, if the interval is chosen not symmetrically about the origin, the mathematics is more difficult.

It was this unsatisfactory feature of the Gaussian ensembles that led Dyson to define the circular ensembles discussed in Chapters 10, 11 and 12.

The inverse of (6.1.3) (cf. Appendix A.7) is

$$R_n(x_1,\ldots,x_n) = \sum_G (-1)^{n-m} \prod_{j=1}^{m} T_{G_j}(x_k, \text{ with } k \text{ in } G_j). \tag{6.1.4}$$

Thus each set of functions R_n and T_n is easily determined in terms of the other. The advantage of the cluster functions is that they have the property of vanishing when any one (or several) of the separations $|x_i - x_j|$ becomes large in comparison to the local mean level spacing. The function T_n describes the correlation properties of a single cluster of n levels, isolated from the more trivial effects of lower order correlations.

Of special interest for comparison with experiments are those features of the statistical model that tend to definite limits as $N \to \infty$. The cluster functions are convenient also from this point of view. While taking the limit $N \to \infty$, we must measure the energies in units of the mean level spacing α and introduce the variables

$$y_j = x_j/\alpha. \tag{6.1.5}$$

The y_j then form a statistical model for an infinite series of energy levels with mean spacing $\alpha = 1$. The cluster functions

$$Y_n(y_1, y_2, \ldots, y_n) = \lim_{N\to\infty} \alpha^n T_n(x_1, x_2, \ldots, x_n) \tag{6.1.6}$$

are well defined and finite everywhere. In particular,

$$Y_1(y) = 1,$$

whereas $Y_2(y_1, y_2)$ defines the shape of the neutralizing charge cloud induced by each particle around itself when the model is interpreted as a classical Coulomb gas (see Section 6.2).

We will calculate all correlation and cluster functions. The R_n will be exhibited as a certain determinant and T_n as a sum over $(n-1)!$ permutations. The *two level form-factor* or the Fourier transform of Y_2

$$b(k) = \int_{-\infty}^{\infty} Y_2(r)\exp(2\pi i k r)\,dr, \quad r = |y_1 - y_2|, \tag{6.1.7}$$

is of special interest, as many important properties of the level distribution, such as mean square values, depend only on it. The Fourier transform of Y_n, or the *n-level form-factor*

$$\int Y_n(y_1, \ldots, y_n)\exp\left(2\pi i \sum_{j=1}^{n} k_j y_j\right) dy_1 \cdots dy_n = \delta(k_1 + \cdots + k_n) b(k_1, \ldots, k_n), \tag{6.1.8}$$

will be given as a single integral. The presence of δ reflects the fact that $Y_n(y_1, \ldots, y_n)$ depends only on the differences of the y_j.

6.1.2 About Level-Spacings. The probability density for several consecutive levels inside an interval is defined by

$$A_\beta(\theta; x_1, \ldots, x_n) = \frac{N!}{n!(N-n)!}\int \cdots \int_{\text{out}} P_{N\beta}(x_1, \ldots, x_N)\,dx_{n+1} \cdots dx_N, \tag{6.1.9}$$

where the subscript "out" means that integrations over $x_{n+1}, \ldots, x_N$ are taken outside the interval $(-\theta, \theta)$, $|x_j| \geqslant \theta$, $j = n+1, \ldots, N$, while the remaining variables are inside that interval, $|x_j| \leqslant \theta$, $j = 1, \ldots, n$.

The quantities of interest are again those which remain finite in the limit $N \to \infty$. We put

$$t = \theta/\alpha, \tag{6.1.10}$$

and define

$$B_\beta(t; y_1, \ldots, y_n) = \lim_{N\to\infty} \alpha^n A_\beta(\theta; x_1, \ldots, x_n). \tag{6.1.11}$$

This is the probability density that in a series of eigenvalues with mean spacing unity an interval of length $2t$ contains exactly n levels at positions around the points $y_1, \ldots, y_n$. To get the probability that a randomly chosen interval of length $s = 2t$ contains exactly n levels one has to integrate $B_\beta(t; y_1, \ldots, y_n)$ over $y_1, \ldots, y_n$ from $-t$ to t,

$$E_\beta(n; s) = \int_{-t}^{t} \cdots \int_{-t}^{t} B_\beta(t; y_1, \ldots, y_n)\, dy_1 \cdots dy_n. \tag{6.1.12}$$

If we put $y_1 = -t$, and integrate over $y_2, y_3, \ldots, y_n$ from $-t$ to t, we get

$$\tilde{F}_\beta(0; s) = B_\beta(t; -t), \tag{6.1.13a}$$

$$\tilde{F}_\beta(n-1; s) = n \int_{-t}^{t} \cdots \int_{-t}^{t} B_\beta(t; -t, y_2, \ldots, y_n)\, dy_2 \cdots dy_n, \quad n > 1. \tag{6.1.13b}$$

If we put $y_1 = -t$, $y_2 = t$ and integrate over $y_3, \ldots, y_n$ from $-t$ to t, we get,

$$p_\beta(0; s) = 2B_\beta(t; -t, t), \tag{6.1.14a}$$

$$p_\beta(n-2; s) = n(n-1) \int_{-t}^{t} \cdots \int_{-t}^{t} B_\beta(t; -t, t, y_3, \ldots, y_n)\, dy_3 \cdots dy_n, \quad n > 2. \tag{6.1.14b}$$

Omitting the sub-script β for the moment, $E(n; s)$ is the probability that an interval of length $s(= 2t)$ chosen at random contains exactly n levels. If we choose a level at random and measure a distance s from it, say to the right, then the probability that this distance contains exactly n levels, not counting the level at the beginning, is $\tilde{F}(n; s)$. Lastly, let us choose a level at random, denote it by A_0, move to the right, say, and denote the successive levels by $A_1, A_2, A_3, \ldots$. Then $p(n; s)\, ds$ is the probability that the distance between levels A_0 and A_{n+1} lies between s and $s + ds$. Let us also write

$$I(n) = \int_0^\infty E(n; s)\, ds. \tag{6.1.15}$$

The functions $E(n; s)$, $\tilde{F}(n; s)$ and $p(n; s)$ are of course related (cf. Appendix A.8),

$$-\frac{d}{ds} E(0; s) = \tilde{F}(0; s), \qquad -\frac{d}{ds} \tilde{F}(0; s) = p(0; s), \tag{6.1.16a}$$

$$-\frac{d}{ds} E(n; s) = \tilde{F}(n; s) - \tilde{F}(n-1; s), \quad n \geqslant 1, \tag{6.1.16b}$$

and

$$-\frac{d}{ds}\tilde{F}(n;s) = p(n;s) - p(n-1;s), \quad n \geqslant 1, \tag{6.1.16c}$$

or

$$\tilde{F}(n;s) = -\frac{d}{ds}\sum_{j=0}^{n} E(j;s), \quad n \geqslant 0, \tag{6.1.17}$$

and

$$\begin{aligned} p(n;s) &= -\frac{d}{ds}\sum_{j=0}^{n} \tilde{F}(j;s), \\ &= \frac{d^2}{ds^2}\sum_{j=0}^{n}(n-j+1)E(j;s), \quad n \geqslant 0. \end{aligned} \tag{6.1.18}$$

The functions $E(n;s)$ and $\tilde{F}(n;s)$ satisfy a certain number of constraints. Evidently, they are non-negative, and

$$\sum_{n=0}^{\infty} E(n;s) = \sum_{n=0}^{\infty} \tilde{F}(n;s) = 1, \tag{6.1.19}$$

for any s, since the number of levels in the interval s is either 0, or 1, or 2, or .... As we have chosen the average level density to be unity, we have

$$\sum_{n=0}^{\infty} nE(n;s) = s. \tag{6.1.20}$$

If the levels never overlap, then

$$E(n;0) = \tilde{F}(n;0) = \delta_{n0} \equiv \begin{cases} 1, & \text{if } n = 0, \\ 0, & \text{if } n > 0. \end{cases} \tag{6.1.21}$$

Moreover, the $\tilde{F}(n;s)$ and $p(n;s)$ are probabilities, so that the right-hand sides of Eqs. (6.1.17), (6.1.18) are non-negative for n and s. Also $p(n;s)$ are normalized and $E(n;s)$ and $\tilde{F}(n;s)$ decrease fast to zero as $s \to \infty$, so that

$$\int_0^s p(n;x)\,dx = 1 - \sum_{j=0}^{n} \tilde{F}(j;s) = 1 + \frac{d}{ds}\sum_{j=0}^{n}(n-j+1)E(j;s) \tag{6.1.22}$$

increases from 0 to 1 as s varies from 0 to ∞. Thus

$$\sum_{j=0}^{n} \tilde{F}(j;s) \quad \text{and} \quad \sum_{j=0}^{n} (n-j+1)E(j;s) \tag{6.1.23}$$

are non-increasing functions of s.

The numbers $I(n)$, Eq. (6.1.15), are also not completely arbitrary. For example, consider the mean square scatter of s_n, the distance between the levels A_0 and A_n, separated by $n-1$ other levels,

$$\left\langle (s_n - \langle s_n \rangle)^2 \right\rangle = \langle s_n^2 \rangle - \langle s_n \rangle^2 \geqslant 0. \tag{6.1.24}$$

Integrating by parts and using Eqs. (6.1.16)–(6.1.18), we get

$$\langle s_n^2 \rangle \equiv \int_0^\infty s^2 p(n-1;s)\,ds = 2\sum_{j=0}^{n-1}(n-j)I(j), \tag{6.1.25}$$

$$\langle s_n \rangle \equiv \int_0^\infty s p(n-1;s)\,ds = n, \tag{6.1.26}$$

where we have dropped the integrated end terms. From Eqs. (6.1.24)–(6.1.26) one gets the inequalities

$$2\sum_{j=0}^{n-1}(n-j)I(j) \geqslant n^2, \quad n \geqslant 1. \tag{6.1.27}$$

In particular, $I(0) \geqslant 1/2$. Also $I(0) \geqslant I(n)/2$ (see Section 16.4).

For illustration purposes, let us consider 2 or 3 examples.

Example 1. The positions of the various levels are independent random variables with no correlation whatsoever. This case is known as the Poisson distribution. For all $n \geqslant 0$, one has

$$E(n;s) = \tilde{F}(n;s) = p(n;s) = s^n e^{-n}/n!, \tag{6.1.28}$$

$$I(n) = 1. \tag{6.1.29}$$

Example 2. Levels are regularly spaced, i.e. they are located at integer coordinates. In this case

$$E(n;s) = \begin{cases} 1 - |s-n|, & \text{if } |s-n| \leqslant 1, \\ 0, & \text{if } |s-n| \geqslant 1, \end{cases} \tag{6.1.30}$$

$$\tilde{F}(n; s) = \begin{cases} 1, & \text{if } n < s < n+1, \\ 0, & \text{otherwise,} \end{cases} \tag{6.1.31}$$

$$p(n; s) = \delta(n + 1 - s), \tag{6.1.32}$$

$$I(0) = 1/2, \qquad I(n) = 1, \quad n \geqslant 1. \tag{6.1.33}$$

The δ in Eq. (6.1.32) above is the Dirac delta function.

Example 3. Levels occur regularly at intervals a, b and c apart with $a + b + c = 3$, so that the average density is one. In this case

$$\begin{aligned} I(0) &= \frac{1}{2} I(3) = \frac{1}{6}(a^2 + b^2 + c^2), \\ I(1) &= I(2) = \frac{1}{3}(bc + ca + ab), \\ I(n) &= I(n-3), \quad n \geqslant 4. \end{aligned} \tag{6.1.34}$$

If we take $a \leqslant b < a + b \leqslant c$, for example, $a = b = 1/4$, $c = 5/2$, then $I(0) > 1$, $I(1) < 1$, $I(3) > 2$, etc. This example shows that $I(n)$ are not necessarily monotone.

To get the probability that $n - 1$ consecutive spacings have values $s_1, \ldots, s_{n-1}$ it is sufficient first to order the y_j as

$$-t \leqslant y_1 \leqslant \cdots \leqslant y_n \leqslant t, \tag{6.1.35}$$

in Eq. (6.1.11) and then to substitute

$$t = -y_1 = \frac{1}{2} \sum_{j=1}^{n-1} s_j, \tag{6.1.36}$$

$$y_i = \sum_{j=1}^{i-1} s_j - \frac{1}{2} \sum_{j=1}^{n-1} s_j, \quad i = 2, \ldots, n; \tag{6.1.37}$$

(thus $y_n = t$); in the expression for $B_\beta(t; y_1, \ldots, y_n)$.

In particular, let $B_\beta(t; y) = \mathcal{B}_\beta(x_1, x_2)$ and $B_\beta(t; -t, t, y) = \mathcal{P}_\beta(x_1, x_2)$ with $x_1 = t + y$, $x_2 = t - y$. Then $\mathcal{B}_\beta(x_1, x_2)$ is the probability that no eigenvalues lie for a distance x_1 on one side and x_2 on the other side of a given eigenvalue. Similarly, $\mathcal{P}_\beta(x_1, x_2)$ is the joint probability density function for two adjacent spacings x_1 and x_2; the distances are measured in units of mean spacing. Also $\mathcal{P}_\beta(x_1, x_2) = \partial^2 \mathcal{B}_\beta(x_1, x_2)/(\partial x_1 \partial x_2)$.

Explicit expressions will be derived for $B_\beta(t; y_1, \ldots, y_n)$, and $E_\beta(n; s)$, for $\beta = 2$ in this chapter, and for $\beta = 1$ and 4 in later chapters.

6.1.3 Spacing Distribution. The functions E_β, $\tilde{F}_\beta$ and p_β are of particular interest for small values of n. Thus the probability that the distance between a randomly chosen pair of consecutive eigenvalues lies between s and $s + ds$ is $p_\beta(0; s)\,ds$; $p_\beta(0; s)$ is known as the probability density for the spacings between the nearest neighbors. The probability that this distance (between the nearest neighbors) be less than or equal to s is

$$\Psi_\beta(s) = \int_0^s p_\beta(0; x)\,dx = 1 - \tilde{F}_\beta(0; s). \tag{6.1.38}$$

This $\Psi_\beta(s)$ is known as the distribution function of the spacings (between the nearest neighbors). It increases from 0 to 1 as s varies from 0 to ∞ (cf. Eq. (6.1.22) and what follows).

6.1.4 Correlations and Spacings. From the interpretation of $Y_2(s)$, $s = |y_1 - y_2|$, and $p(n; s)$, Eqs. (6.1.6) and (6.1.14), as probability densities one has for all s,

$$Y_2(s) = \sum_{n=0}^{\infty} p(n; s). \tag{6.1.39}$$

The E_β and $R_{n\beta}$ or $T_{n\beta}$ are also related. From this relation and the expression for $R_{n\beta}$ or $T_{n\beta}$, one can derive an expression for $E_\beta(n; s)$. See Appendix A.7.

6.2 The n-Point Correlation Function

From now on in the rest of this chapter we will consider the Gaussian unitary ensemble corresponding to $\beta = 2$ in Eq. (6.1.1).

For the calculation of $R_{n2}(x_1, \ldots, x_n)$ it is convenient to use the results of Section 5.3. The orthogonal polynomials corresponding to the weight function $\exp(-x^2)$ over $(-\infty, \infty)$ are the Hermite polynomials

$$H_j(x) = \exp(x^2)\left(-\frac{d}{dx}\right)^j \exp(-x^2) = j!\sum_{i=0}^{[j/2]} (-1)^i \frac{(2x)^{j-2i}}{i!(j-2i)!}. \tag{6.2.1}$$

Introducing the "oscillator wave functions",

$$\begin{aligned}\varphi_j(x) &= (2^j j!\sqrt{\pi})^{-1/2} \exp(-x^2/2) H_j(x) \\ &= (2^j j!\sqrt{\pi})^{-1/2} \exp(x^2/2)(-d/dx)^j \exp(-x^2),\end{aligned} \tag{6.2.2}$$

orthogonal over $(-\infty, \infty)$,

$$\int_{-\infty}^{\infty} \varphi_j(x)\varphi_k(x)\,dx = (2^{j+k}j!k!\pi)^{-1/2}\int_{-\infty}^{\infty} H_j(x)H_k(x)\exp(-x^2)\,dx$$

$$= \delta_{jk} = \begin{cases} 1, & \text{if } j = k, \\ 0, & \text{otherwise,} \end{cases} \tag{6.2.3}$$

we write according to the prescriptions of Section 5.3

$$P_{N2}(x_1, \ldots, x_N) = \frac{1}{N!}(\det[\varphi_{j-1}(x_i)])^2 \tag{6.2.4}$$

$$= \frac{1}{N!}\det[K_N(x_i, x_j)]_{i,j=1,\ldots,N}, \tag{6.2.5}$$

$$K_N(x, y) = \sum_{k=0}^{N-1} \varphi_k(x)\varphi_k(y). \tag{6.2.6}$$

This kernel $K_N(x, y)$ satisfies the conditions of Theorem 5.1.4 and hence integration over $N - n$ variables gives the n-point correlation function

$$R_n = \det[K_N(x_i, x_j)]_{i,j=1,\ldots,n}, \tag{6.2.7}$$

with $K_N(x, y)$ given by (6.2.6).

We may expand the determinant in Eq. (6.2.7),

$$R_n = \sum_P (-1)^{n-m} \prod_1^m K_N(x_a, x_b)K_N(x_b, x_c)\cdots K_N(x_d, x_a), \tag{6.2.8}$$

where the permutation P is a product of m exclusive cycles of lengths $h_1, \ldots, h_m$ of the form $(a \to b \to c \to \cdots \to d \to a)$, $\sum_1^m h_j = n$. Taking h_j as the number of indices in G_j, and comparing with Eq. (6.1.4), we get

$$T_n(x_1, \ldots, x_n) = \sum_P K_N(x_1, x_2)K_N(x_2, x_3)\cdots K_N(x_n, x_1), \tag{6.2.9}$$

where the sum is over all the $(n - 1)!$ distinct cyclic permutations of the indices in $(1, \ldots, n)$.

Putting $n = 1$ in (6.2.7), we get the level density

$$\sigma_N(x) = K_N(x, x) = \sum_{j=0}^{N-1} \varphi_j^2(x). \tag{6.2.10}$$

As $N \to \infty$, this goes to the semi-circle law (cf. Appendix A.9)

$$\sigma_N(x) \to \sigma(x) = \begin{cases} \frac{1}{\pi}(2N - x^2)^{1/2}, & |x| < (2N)^{1/2}, \\ 0, & |x| > (2N)^{1/2}. \end{cases} \tag{6.2.11}$$

The mean spacing at the origin is thus

$$\alpha = 1/\sigma(0) = \pi/(2N)^{1/2}. \tag{6.2.12}$$

Figure 6.1 shows $\sigma_N(x)$ for a few values of N.

Putting $n = 2$ in (6.2.9) we get the two-level cluster function

$$T_2(x_1, x_2) = (K_N(x_1, x_2))^2 = \left(\sum_{j=0}^{N-1} \varphi_j(x_1)\varphi_j(x_2) \right)^2. \tag{6.2.13}$$

Taking the limit as $N \to \infty$, this equation, the definition (6.1.6) and (6.2.12) (cf. Appendix A.10) give

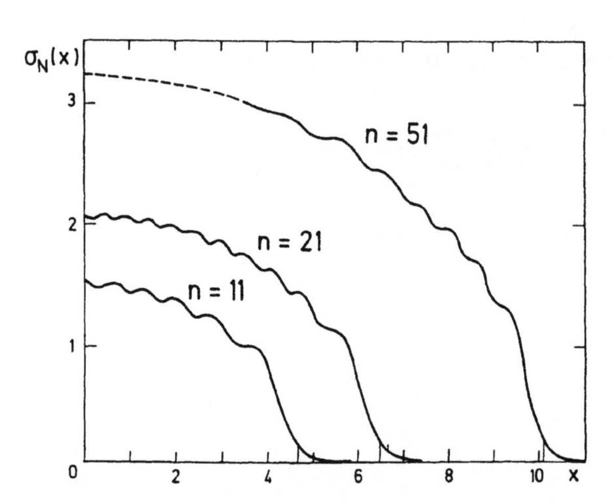

Figure 6.1. The level density $\sigma_N(x)$, Eq. (6.2.10), for $N = 11$, 21 and 51. The oscillations are noticeable even for $N = 51$. The "semi-circle", Eq. (6.2.11), ends at points marked at $\sqrt{22} \approx 4.7$, $\sqrt{42} \approx 6.5$ and $\sqrt{102} \approx 10.1$. Reprinted with permission from E.P. Wigner, "Distribution laws for roots of a random Hermitian matrix" (1962), in *Statistical theories of spectra: fluctuations*, ed. C.E. Porter, Academic Press, New York (1965).

$$Y_2(x_1, x_2) = \lim_{N\to\infty} \left(\frac{\pi}{(2N)^{1/2}} \sum_{j=0}^{N-1} \varphi_j(x_1)\varphi_j(x_2) \right)^2$$

$$= \left(s(r)\right)^2 \equiv \left(\frac{\sin(\pi r)}{\pi r} \right)^2, \tag{6.2.14}$$

with

$$r = |y_1 - y_2|, \quad y_1 = (2N)^{1/2} x_1/\pi, \; y_2 = (2N)^{1/2} x_2/\pi. \tag{6.2.15}$$

Figure 6.2 shows the limiting two-level correlation function $1 - (\sin(\pi r)/(\pi r))^2$. Note the oscillations around integer values of r.

In the limit $N \to \infty$, the n-level cluster function is

$$Y_n(y_1, \ldots, y_n) = \sum_P s(r_{12}) s(r_{23}) \cdots s(r_{n1}), \tag{6.2.16}$$

the sum being taken over the $(n-1)!$ distinct cyclic permutations of the indices $(1, 2, \ldots, n)$, and $r_{ij} = |y_i - y_j| = (2N)^{1/2} |x_i - x_j|/\pi$.

The two-level form factor is (cf. Appendix A.11)

$$b(k) = \int_{-\infty}^{\infty} Y_2(r) \exp(2\pi i k r)\, dr = \begin{cases} 1 - |k|, & |k| \leqslant 1, \\ 0, & |k| \geqslant 1. \end{cases} \tag{6.2.17}$$

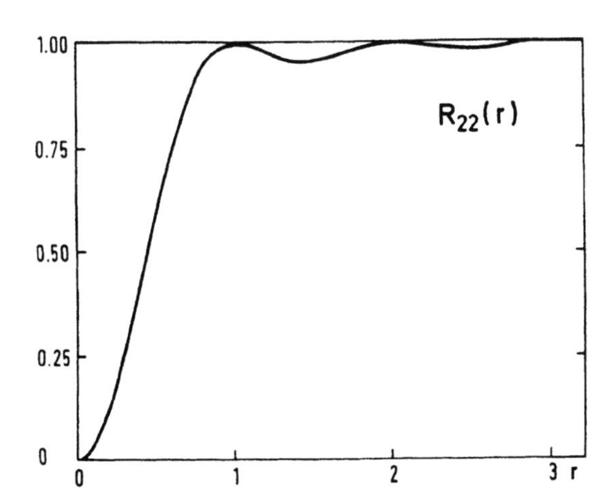

Figure 6.2. Two point correlation function, $1 - (\sin(\pi r)/(\pi r))^2$, for the unitary ensemble.

The n-level form-factor or the Fourier transform of Y_n is

$$\int Y_n(y_1, \ldots, y_n) \exp\left(2\pi i \sum_{j=1}^{n} k_j y_j\right) dy_1 \cdots dy_n$$

$$= \delta(k_1 + \cdots + k_n) \cdot \int_{-\infty}^{\infty} dk \sum_P f_2(k) f_2(k + k_1) \cdots$$

$$f_2(k + k_1 + \cdots + k_{n-1}), \tag{6.2.18}$$

with

$$f_2(k) = \begin{cases} 1, & \text{if } |k| < 1/2, \\ 0, & \text{if } |k| > 1/2. \end{cases} \tag{6.2.19}$$

It must be noted that the three and higher order correlation functions can all be expressed in terms of the level density and the two-level function $K_N(x, y)$, for every finite N, as is evident from (6.2.7). This is a particular feature of the unitary ensemble.

6.3 Level Spacings

In this section we will express $A_2(\theta)$, the probability that the interval $(-\theta, \theta)$ does not contain any of the points $x_1, \ldots, x_N$, as a Fredholm determinant of a certain integral equation. In the limit of large N the solutions of this integral equation are the so called spheroidal functions satisfying a second order differential equation. The limit of $A_2(\theta)$ will thus be expressed as a fast converging infinite product.

To start with, let us calculate the simplest level spacing function

$$A_2(\theta) = \int \cdots \int_{\text{out}} P_{N2}(x_1, \ldots, x_N)\, dx_1 \cdots dx_N, \tag{6.3.1}$$

case $n = 0$, of Eq. (6.1.9). Substituting from (6.2.4) for P_{N2}, we get

$$A_2(\theta) = \frac{1}{N!} \int \cdots \int_{\text{out}} \left(\det[\varphi_{j-1}(x_k)]\right)^2 dx_1 \cdots dx_N. \tag{6.3.2}$$

At this point we apply Gram's result (cf. Appendix A.12) to get

$$A_2(\theta) = \det G, \tag{6.3.3}$$

where the elements of the matrix G are given by

$$g_{jk} = \int_{\text{out}} \varphi_{j-1}(x)\varphi_{k-1}(x)\, dx = \delta_{jk} - \int_{-\theta}^{\theta} \varphi_{j-1}(x)\varphi_{k-1}(x)\, dx. \tag{6.3.4}$$

To diagonalize G, consider the integral equation

$$\lambda\psi(x) = \int_{-\theta}^{\theta} K(x, y)\psi(y)\, dy, \tag{6.3.5}$$

$$K(x, y) \equiv K_N(x, y) = \sum_{i=0}^{N-1} \varphi_i(x)\varphi_i(y). \tag{6.2.6}$$

As the kernel $K(x, y)$ is a sum of separable ones, the eigenfunction $\psi(x)$ is necessarily of the form

$$\psi(x) = \sum_{i=0}^{N-1} c_i\varphi_i(x). \tag{6.3.6}$$

Substituting (6.3.6) in (6.3.5) and remembering that $\varphi_i(x)$ for $i = 0, 1, \ldots, N-1$, are linearly independent functions, we obtain the system of linear equations

$$\lambda c_i = \sum_{j=0}^{N-1} c_j \int_{-\theta}^{\theta} \varphi_i(y)\varphi_j(y)\, dy, \quad i = 0, 1, \ldots, N-1. \tag{6.3.7}$$

This system will have a non-zero solution if and only if λ satisfies the following equation

$$\det\left[\lambda\delta_{ij} - \int_{-\theta}^{\theta} \varphi_i(y)\varphi_j(y)\, dy\right]_{i,j=0,\ldots,N-1} = 0. \tag{6.3.8}$$

This algebraic equation in λ has N real roots; let them be $\lambda_0, \ldots, \lambda_{N-1}$, so that

$$\det\left[\lambda\delta_{ij} - \int_{-\theta}^{\theta} \varphi_i(y)\varphi_j(y)\, dy\right] \equiv \prod_{i=0}^{N-1} (\lambda - \lambda_i). \tag{6.3.9}$$

Comparing with (6.3.3) and (6.3.4), we see that

$$A_2(\theta) = \prod_{i=0}^{N-1} (1 - \lambda_i), \tag{6.3.10}$$

where λ_i are the eigenvalues of the integral Eq. (6.3.5).

As we are interested in large N, we take the limit $N \to \infty$. In this limit the quantity which is finite is not $K_N(x, y)$, but

$$\mathcal{K}_N(\xi, \eta) = \left(\pi t/(2N)^{1/2}\right) K_N(x, y), \tag{6.3.11}$$

with

$$\pi t = (2N)^{1/2}\theta, \quad \pi t\xi = (2N)^{1/2}x, \quad \pi t\eta = (2N)^{1/2}y, \tag{6.3.12}$$

and (cf. Appendix A.10)

$$\lim \mathcal{K}_N(\xi, \eta) = \mathcal{K}(\xi, \eta) = \frac{\sin(\xi - \eta)\pi t}{(\xi - \eta)\pi}. \tag{6.3.13}$$

The limiting integral equation is then

$$\lambda f(\xi) = \int_{-1}^{1} \mathcal{K}(\xi, \eta) f(\eta)\, d\eta, \tag{6.3.14}$$

where

$$f(\xi) = \psi\big(\pi t\xi/(2N)^{1/2}\big),$$

and

$$2t = 2\theta(2N)^{1/2}/\pi = \text{spacing}/(\text{mean spacing at the origin}). \tag{6.3.15}$$

As $\mathcal{K}(\xi, \eta) = \mathcal{K}(-\xi, -\eta)$, and the interval of integration in Eq. (6.3.14) is symmetric about the origin, if $f(\xi)$ is a solution of the integral Eq. (6.3.14) with the eigenvalue λ, then so is $f(-\xi)$ with the same eigenvalue. Hence every solution of (6.3.14) is (or can be chosen to be) either even or odd. Consequently, instead of $\sin(\xi - \eta)\pi t/((\xi - \eta)\pi)$, one can as well study the kernel $\sin(\xi + \eta)\pi t/((\xi + \eta)\pi)$.

Actually, from the orthogonality of the $\varphi_j(x)$, $g_{ij} = 0$ whenever $i + j$ is odd; the matrix G has a checker board structure, every alternate element being zero; $\det G$ is a product of two determinants, one containing only even functions $\varphi_{2i}(x)$, and the other only odd functions $\varphi_{2i+1}(x)$; and $A_2(\theta)$ in Eq. (6.3.10) is a product of two factors, one containing the λ_i corresponding to the even solutions of Eq. (6.3.5) and the other containing those corresponding to the odd solutions.

The kernel $\mathcal{K}(x, y)$ is the square of another symmetric kernel $(t/2)^{1/2}\exp(\pi ixyt)$,

$$\int_{-1}^{1} \big((t/2)^{1/2} e^{\pi ixzt}\big)\big((t/2)^{1/2} e^{\pi izyt}\big)^* \, dz = \frac{\sin(x - y)\pi t}{(x - y)\pi}. \tag{6.3.16}$$

Therefore Eq. (6.3.14) may be replaced by the integral equation

$$\mu f(x) = \int_{-1}^{1} \exp(\pi ixyt) f(y)\, dy, \tag{6.3.17}$$

with

$$\lambda = \frac{1}{2} t|\mu|^2. \tag{6.3.18}$$

Taking the limit $N \to \infty$, Eqs. (6.3.10) and (6.3.18) then give

$$B_2(t) = \prod_{i=0}^{\infty} \left(1 - \frac{1}{2} t |\mu_i|^2 \right), \tag{6.3.19}$$

where μ_i are determined from Eq. (6.3.17).

The integral equation (6.3.17) can be written as a pair of equations corresponding to even and odd solutions

$$\mu_{2j} f_{2j}(x) = 2 \int_0^1 \cos(\pi x y t) f_{2j}(y)\, dy, \tag{6.3.20}$$

$$\mu_{2j+1} f_{2j+1}(x) = 2i \int_0^1 \sin(\pi x y t) f_{2j+1}(y)\, dy. \tag{6.3.21}$$

The eigenvalues corresponding to the even solutions are real and those corresponding to odd solutions are pure imaginary.

A careful examination of (6.3.17) shows that its solutions are the spheroidal functions (Robin, 1959) that depend on the parameter t. These functions are defined as the solutions of the differential equation

$$(L - \ell) f(x) \equiv \left((x^2 - 1) \frac{d^2}{dx^2} + 2x \frac{d}{dx} + \pi^2 t^2 x^2 - \ell \right) f(x) = 0, \tag{6.3.22}$$

which are regular at the points $x = \pm 1$. In fact, one can verify that the self-adjoint operator L commutes with the kernel $\exp(\pi i x y t)$ defined over the interval $(-1, 1)$; that is,

$$\int_{-1}^1 \exp(\pi i x y t) L(y) f(y)\, dy = L(x) \int_{-1}^1 \exp(\pi i x y t) f(y)\, dy, \tag{6.3.23}$$

provided

$$(1 - x^2) f(x) = 0 = (1 - x^2) f'(x), \quad x \to \pm 1. \tag{6.3.24}$$

Equation (6.3.24) implies that $f(x)$ is regular at $x = \pm 1$. Hence (6.3.17) and (6.3.22) both have the same set of eigenfunctions. Once the eigenfunctions are known, the corresponding eigenvalues can be computed from (6.3.17). For example, for even solutions put $x = 0$, to get

$$\mu_{2i} = (f_{2j}(0))^{-1} \cdot \int_{-1}^1 f_{2j}(y)\, dy, \tag{6.3.25}$$

while for the odd solutions, differentiate first with respect to x and then put $x = 0$,

$$\mu_{2j+1} = i\pi t\big(f'_{2j+1}(0)\big)^{-1} \int_{-1}^{1} y f_{2j+1}(y)\, dy. \tag{6.3.26}$$

The spheroidal functions $f_j(x)$, $j = 0, 1, 2, \ldots$, form a complete set of functions integrable over $(-1, 1)$. They are therefore the complete set of solutions of the integral equation (6.3.17), or that of (6.3.14). We therefore have

$$\begin{aligned} E_2(0; s) = B_2(t) &= \lim_{N\to\infty} A_2\big(\theta(2N)^{1/2}/\pi\big) \\ &= \prod_{i=0}^{\infty}(1 - \lambda_i) = \prod_{i=0}^{\infty}\left(1 - \frac{1}{2}t|\mu_i|^2\right), \end{aligned} \tag{6.3.27}$$

$s = 2t$, where μ_i are given by Eqs. (6.3.25) and (6.3.26), the functions $f_i(x)$ being the spheroidal functions, solutions of the differential equation (6.3.22).

Figure 6.3 shows the eigenvalues λ_i for some values of i and s.

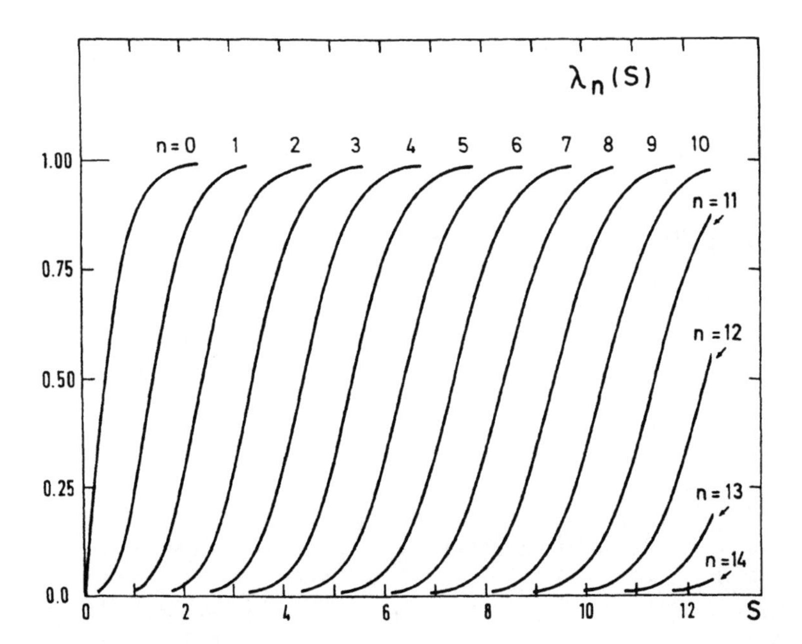

Figure 6.3. The eigenvalues λ_i; Eq. (6.3.14). Reprinted with permission from M.L. Mehta and J. des Cloizeaux, The probabilities for several consecutive eigenvalues of a random matrix, *Indian J. Pure Appl. Math.* 3, 329–351 (1972).

6.4 Several Consecutive Spacings

Next we take the case of several consecutive spacings, integral (6.1.9), with $n > 0$. From Eqs. (6.1.1) and (6.2.4) one has

$$P_{N2}(x_1, \ldots, x_N) = \frac{1}{N!} \det \left[\sum_{k=1}^{N} \varphi_i(x_k)\varphi_j(x_k) \right]_{i,j=0,1,\ldots,N-1}. \tag{6.4.1}$$

We want to calculate

$$A_2(\theta; x_1, \ldots, x_n) = \frac{N!}{n!(N-n)!} \int \cdots \int_{\text{out}} P_{N2}(x_1, \ldots, x_N)\, dx_{n+1} \cdots dx_N. \tag{6.4.2}$$

The subscript "out" means that the variables $x_1, \ldots, x_N$ belong to the intervals

$$\begin{aligned} |x_j| &\le \theta, \quad \text{if } 1 \leqslant j \leqslant n, \\ |x_j| &\ge \theta, \quad \text{if } n+1 \le j \le N. \end{aligned} \tag{6.4.3}$$

In Eq. (6.4.1) the row with index i can be considered as a sum of N rows with elements $\varphi_i(x_k)\varphi_j(x_k)$, $k = 1, \ldots, N$. Expressing each row as such sums, we write the determinant in Eq. (6.4.1) as a sum of determinants. If a variable occurs in two or more rows, these rows are proportional and therefore the corresponding determinant is zero. The non-zero determinants are those in which all the indices k are different.

$$P_{N2}(x_1, \ldots, x_N) = \frac{1}{N!} \sum_{(k_i)} \det \left[\varphi_i(x_{k_i})\varphi_j(x_{k_i}) \right]_{i,j=0,\ldots,N-1}. \tag{6.4.4}$$

The indices $k_0, \ldots, k_{N-1}$ are obtained by a permutation of $1, \ldots, N$ and in Eq. (6.4.4) the summation is over all permutations. As each variable $x_1, \ldots, x_N$ occurs in one and only one row of any of the determinants of Eq. (6.4.4), one may easily integrate over as many variables as one likes. After integrating over $x_{n+1}, \ldots, x_N$ we expand each determinant in the Laplace manner (see for example, Mehta (1989MT)) according to the n rows containing the variables $x_1, \ldots, x_n$. Introducing the matrix G with elements

$$g_{ij} = \delta_{ij} - \int_{-\theta}^{\theta} \varphi_i(x)\varphi_j(x)\, dx, \tag{6.3.4}$$

we obtain the result:

$$A_2(\theta; x_1, \ldots, x_n) = \frac{1}{n!} \sum_{(i,j)} (\det[\varphi_{i_k}(x_k)\varphi_{j_\ell}(x_k)]_{k,\ell=1,\ldots,n}) G'(i_1, \ldots, i_n; j_1, \ldots, j_n). \tag{6.4.5}$$

The indices $i_1, \ldots, i_n$ are chosen from $0, \ldots, N-1$ as also the indices $j_1, \ldots, j_n$. The indices $i_1, \ldots, i_n$ are not ordered, while the indices $j_1, \ldots, j_n$ are ordered $j_1 < \cdots < j_n$. The summation in (6.4.5) is extended over all possible choices of indices satisfying the above conditions. The cofactor $G'(i_1, \ldots, i_n; j_1, \ldots, j_n)$ is apart from a sign the determinant of the $(N-n) \times (N-n)$ matrix obtained from G by omitting the rows $i_1, \ldots, i_n$ and the columns $j_1, \ldots, j_n$. The sign is plus or minus according as $\sum_{k=1}^{n}(i_k + j_k)$ is even or odd. Therefore (cf. Mehta (1989MT), Section 3.9)

$$G'(i_1, \ldots, i_n; j_1, \ldots, j_n) = \det G \cdot \det\left(\left[\left(G^{-1}\right)_{ji}\right]_{i=i_1,\ldots,i_n;\, j=j_1,\ldots,j_n}\right). \tag{6.4.6}$$

If we integrate over $x_1, \ldots, x_n$ in the interval $(-\theta, \theta)$, Eq. (6.4.5), we get

$$C_2(n;\theta) \equiv \int_{-\theta}^{\theta} \cdots \int_{-\theta}^{\theta} A_2(\theta; x_1, \ldots, x_n)\, dx_1 \cdots dx_n = \sum_{(i,j)} \det \gamma(i; j) \cdot G'(i; j), \tag{6.4.7}$$

where $\gamma(i; j) \equiv \gamma(i_1, \ldots, i_n; j_1, \ldots, j_n)$ is the $n \times n$ matrix formed from the rows $i_1, \ldots, i_n$ and columns $j_1, \ldots, j_n$ of the matrix γ with elements

$$\gamma_{pq} = \int_{-\theta}^{\theta} \varphi_p(x)\varphi_q(x)\, dx, \tag{6.4.8}$$

and as above

$$G'(i; j) = \det G \cdot \det[G^{-1}(j; i)]. \tag{6.4.6}$$

The summation in Eq. (6.4.7) is over all possible choices of indices with $0 \leqslant i_1 < i_2 < \cdots < i_n \leqslant N-1$; $0 \leqslant j_1 < j_2 < \cdots < j_n \leqslant N-1$. (The ordering of the $i_1, \ldots, i_n$ removes the factor $n!$ between (6.4.5) and (6.4.7).) Hence

$$C_2(n;\theta) = \det G \cdot \sum_{(i,j)} \det \gamma(i; j) \cdot \det G^{-1}(j; i), \tag{6.4.9}$$

which in the limit $N \to \infty$, $\theta \to 0$, $s = 2\theta(2N)^{1/2}/\pi$ finite, goes to $E_2(n; s)$,

$$E_2(n; s) = \lim \alpha^n C_2(n; \theta). \tag{6.4.10}$$

Recalling the diagonalization of G from Section 6.3 above, Eq. (6.4.5) reads for $n = 0$,

$$A_2(\theta) = \det G = \prod_{i=0}^{N-1} (1 - \lambda_i), \tag{6.4.11}$$

where the λ_i, $i = 0, \ldots, N-1$, are the eigenvalues of the matrix $[\gamma]$ defined by equation (6.4.8), i.e.

$$\sum_{j=0}^{N-1} \gamma_{ij} h_{jk} = h_{ik} \lambda_k. \tag{6.4.12}$$

Equivalently, the λ_i are the eigenvalues of the integral equation (6.3.5),

$$\int_{-\theta}^{\theta} K(x, y) \psi_i(y)\, dy = \lambda_i \psi_i(x), \tag{6.4.13}$$

with the kernel $K(x, y)$, Eq. (6.2.6),

$$K(x, y) = \sum_{i=0}^{N-1} \varphi_i(x) \varphi_i(y), \tag{6.4.14}$$

and the eigenfunctions, Eq. (6.3.6),

$$\psi_i(x) = \sum_{j=0}^{N-1} h_{ji} \varphi_j(x). \tag{6.4.15}$$

The normalization of the $\psi_i(x)$ depends on that of the eigenvectors of γ. As γ is real and symmetric we may choose its eigenvectors to be real, orthogonal and normalized to unity (cf. Mehta, 1989MT), so that $h = [h_{ij}]$ is a real orthogonal matrix:

$$\sum_{j=0}^{N-1} h_{ij} h_{kj} = \sum_{j=0}^{N-1} h_{ji} h_{jk} = \delta_{ik}. \tag{6.4.16}$$

From Eq. (6.4.14)–(6.4.16) we deduce

$$K(x, y) = \sum_{i=0}^{N-1} \psi_i(x) \psi_i(y). \tag{6.4.17}$$

Since K is real and symmetric, its eigenfunctions $\psi_i(x)$ are orthogonal

$$\int_{-\theta}^{\theta} \psi_i(x) \psi_j(x)\, dx = 0, \quad i \neq j. \tag{6.4.18}$$

However, their normalization is not unity. From Eqs. (6.4.13), (6.4.17) and (6.4.18) one gets

$$\int_{-\theta}^{\theta} \psi_i^2(x)\,dx = \lambda_i. \tag{6.4.19}$$

To take the limits $N \to \infty$, let us put as before

$$\alpha = \pi/(2N)^{1/2}, \quad \theta = \alpha t, \quad x = \alpha t\xi, \quad y = \alpha t\eta, \tag{6.4.20}$$

and keep ξ, η, t finite. Then

$$K(x, y) \to (\alpha t)^{-1}\mathcal{K}(\xi, \eta), \tag{6.4.21}$$

$$\psi_i(x) \to a_i f_i(\xi), \tag{6.4.22}$$

where $\mathcal{K}(\xi, \eta)$ is given by Eq. (6.3.13) and $f_i(\xi)$ are the spheroidal functions, depending on the parameter t, solutions of the integral equation (6.3.14) or of (6.3.17). If we normalize the spheroidal functions as

$$\int_{-1}^{1} f_i(\xi) f_j(\xi)\,d\xi = \delta_{ij}, \tag{6.4.23}$$

then the constants a_i are given by

$$\lambda_i = \int_{-\theta}^{\theta} \psi_i^2(x)\,dx = a_i^2 \alpha t \int_{-1}^{1} f_i^2(\xi)\,d\xi,$$

or

$$a_i = (\lambda_i/\alpha t)^{1/2}. \tag{6.4.24}$$

After this revision of Section 6.3, let us come back to Eqs. (6.4.5) and (6.4.6). The matrix h diagonalizes G, and hence also G^{-1}, while it transforms the functions $\varphi_i(x)$ into $\psi_i(x)$. Using Eqs. (6.4.12), (6.4.15) and (6.4.16) we can write Eqs. (6.4.5) and (6.4.6) as

$$A_2(\theta; x_1, \ldots, x_n) = \frac{1}{n!} \sum_{(i;j)} \det[\psi_{i_k}(x_k)\psi_{j_\ell}(x_k)]_{k,\ell=1,\ldots,n}$$

$$\times \det[1 - \Lambda] \cdot \det[(1 - \Lambda)_{ji}^{-1}]_{j=j_1,\ldots,j_n; i=i_1,\ldots,i_n}, \tag{6.4.25}$$

where Λ is the diagonal matrix with diagonal elements λ_i. As $[1 - \Lambda]$ is diagonal, non-vanishing terms result if and only if the indices i can be obtained by a permutation P of the indices j:

$$j_1 < \cdots < j_n, \quad i_k = j_{Pk}, \; k = 1, \ldots, n. \tag{6.4.26}$$

Thus Eq. (6.4.25) can be written as

$$A_2(\theta; x_1, \ldots, x_n) = \frac{1}{n!} \prod_{\rho=0}^{N-1} (1-\lambda_\rho) \cdot \sum_{(j)} (1-\lambda_{j_1})^{-1} \cdots (1-\lambda_{j_n})^{-1}$$
$$\times \sum_P \varepsilon_P \det[\psi_{j_{Pk}}(x_k)\psi_{j_\ell}(x_k)]_{k,\ell=1,\ldots,n}, \tag{6.4.27}$$

where P is a permutation of the indices $(1, 2, \ldots, n)$ and ε_P its sign. Equation (6.4.27) coincides with

$$A_2(\theta; x_1, \ldots, x_n) = \frac{1}{n!} \prod_{\rho=0}^{N-1} (1-\lambda_\rho) \cdot \sum_{(j)} (1-\lambda_{j_1})^{-1} \cdots (1-\lambda_{j_n})^{-1} \cdot (\det[\psi_{j_\ell}(x_k)])^2. \tag{6.4.28}$$

When $N \to \infty$, $\theta \to 0$ while t and y_j given by Eqs. (6.1.5), (6.1.10) are finite, we get from Eqs. (6.1.9), (6.1.11), (6.4.28), (6.4.22) and (6.4.24):

$$B_2(t; y_1, \ldots, y_n)$$
$$= \frac{t^{-n}}{n!} \prod_\rho (1-\lambda_\rho) \sum_{(j)} \frac{\lambda_{j_1}}{1-\lambda_{j_1}} \cdots \frac{\lambda_{j_n}}{1-\lambda_{j_n}} \cdot \left(\det([f_{j_\ell}(y_k/t)]_{\ell,k=1,\ldots,n})\right)^2. \tag{6.4.29}$$

Let us recall that the eigenvalues λ_i and the functions $f_i(\xi)$ depend on t as a parameter and that the indices $0 \leqslant j_1 < \cdots < j_n$ are integers.

Due to the orthonormality of the $f_i(\xi)$ we may easily integrate over $y_1, \ldots, y_n$ in (6.4.29). Actually, for $j = j_1, \ldots, j_n$; $k = 1, \ldots, n$;

$$\left(\det[f_j(x_k)]\right)^2 = \sum_{P,Q} \varepsilon_P \varepsilon_Q f_{P1}(x_1) \cdots f_{Pn}(x_n) f_{Q1}(x_1) \cdots f_{Qn}(x_n),$$

where P and Q are permutations of the indices $j_1, \ldots, j_n$ and ε_P, ε_Q their signs. On integration one gets 1 when $P = Q$ and 0 when $P \neq Q$. Hence

$$E_2(n; s) = \int_{-t}^{t} \cdots \int_{-t}^{t} B_2(t; y_1, \ldots, y_n)\, dy_1 \cdots dy_n$$
$$= \prod_\rho (1-\lambda_\rho) \sum_{(j)} \frac{\lambda_{j_1}}{1-\lambda_{j_1}} \cdots \frac{\lambda_{j_n}}{1-\lambda_{j_n}}. \tag{6.4.30}$$

In the same way, from (6.1.14), (6.4.29) and (6.4.23), we get

$$p_2(n-2;s)=\frac{4}{s^2}\prod_\rho(1-\lambda_\rho)\sum_{(j)}\frac{\lambda_{j_1}}{1-\lambda_{j_1}}\cdots\frac{\lambda_{j_n}}{1-\lambda_{j_n}}$$
$$\times\sum_{\ell,m=1}^{n} f_{j_\ell}(1)f_{j_m}(-1)\big(f_{j_\ell}(1)f_{j_m}(-1)-f_{j_\ell}(-1)f_{j_m}(1)\big). \quad (6.4.31)$$

On the other hand, to obtain the probability density of $(n-1)$ consecutive spacings $s_1,\ldots,s_{n-1}$ one has only to make the substitutions (6.1.36), (6.1.37) in the expression (6.4.29) as explained at the end of Section 6.1.

The case $n=2$ is of special interest. Putting $s=2t$, we get the probability for a single spacing s as:

$$p_2(0;s)=2B_2(s/2;-s/2,s/2)$$
$$=\frac{16}{s^2}\prod_\rho(1-\lambda_\rho)\cdot\sum_j\frac{\lambda_{2j}}{1-\lambda_{2j}}f_{2j}^2(1)\sum_k\frac{\lambda_{2k+1}}{1-\lambda_{2k+1}}f_{2k+1}^2(1). \quad (6.4.32)$$

Using Eqs. (6.1.18) and (6.4.30) we may also write

$$p_2(0;s)=\frac{d^2}{ds^2}E_2(0;s)=\frac{d^2}{ds^2}\prod_\rho(1-\lambda_\rho). \quad (6.4.33)$$

Introducing

$$F_2(z,t):=\prod_{\rho=0}^{\infty}(1-z\lambda_\rho), \quad (6.4.34)$$

one can write Eq. (6.4.30) as

$$E_2(n;2t)=\frac{1}{n!}\left(-\frac{\partial}{\partial z}\right)^n F_2(z,t)|_{z=1}. \quad (6.4.35)$$

This equation can also be derived from the relation between $E(n;s)$ and the n-level correlation or cluster functions. See Appendix A.7.

Let us write

$$F_+\equiv F_+(z,t)=\prod_{j=0}^{\infty}(1-z\lambda_{2j}), \quad (6.4.36)$$

$$F_- \equiv F_-(z,t) = \prod_{j=0}^{\infty}(1 - z\lambda_{2j+1}), \tag{6.4.37}$$

so that they are the products over the eigenvalues corresponding respectively to the even and odd solutions of the integral equation (6.3.14). In other words $F_\pm(z,t)$ are the Fredholm determinants of the kernels

$$\mathcal{K}_\pm(\xi,\eta;z) = \frac{z}{2}\left(\frac{\sin(\xi-\eta)\pi t}{(\xi-\eta)\pi} \pm \frac{\sin(\xi+\eta)\pi t}{(\xi+\eta)\pi}\right) \tag{6.4.38}$$

over the interval $(-1,1)$. These functions $F_\pm(z,t)$ have some remarkable relations studied in Chapters 20, 21 and Appendix A.16. There among other things one computes the logarithmic derivatives of $F_\pm(z,t)$ to be

$$\frac{d}{dt}\ln F_+(z,t) = -\frac{2}{t}\sum_{i=0}^{\infty}\frac{z\lambda_{2i}}{1-z\lambda_{2i}} f_{2i}^2(1), \tag{6.4.39}$$

and

$$\frac{d}{dt}\ln F_-(z,t) = -\frac{2}{t}\sum_{i=0}^{\infty}\frac{z\lambda_{2i+1}}{1-z\lambda_{2i+1}} f_{2i+1}^2(1). \tag{6.4.40}$$

The equality of the expressions (6.4.32) and (6.4.33) is then expressed as

$$\frac{d^2F_2}{dt^2} = \frac{d^2}{dt^2}(F_+F_-) = 4\left(\frac{dF_+}{dt}\right)\left(\frac{dF_-}{dt}\right), \tag{6.4.41}$$

or, indicating derivatives by primes, as

$$\frac{d}{dt}\left(\frac{F_+'}{F_+} + \frac{F_-'}{F_-}\right) = -\left(\frac{F_+'}{F_+} - \frac{F_-'}{F_-}\right)^2, \tag{6.4.42}$$

for $z=1$. A direct proof of relation (6.4.42) is given in Chapter 21.

We will encounter $F_+(z,t)$ and $F_-(z,t)$ later in connection with (Gaussian or circular) orthogonal and symplectic ensembles.

Figure 6.4 gives a graphical representation of $E_2(n;s)$ for small values of n. For comparison the corresponding probabilities

$$E_0(n;s) = s^n \exp(-s)/n! \tag{6.4.43}$$

for a set of independent random levels, Poisson process, are drawn in Figure 6.5, and the probabilities $E_\infty(n;s)$ corresponding to equally spaced levels in Figure 6.6. From

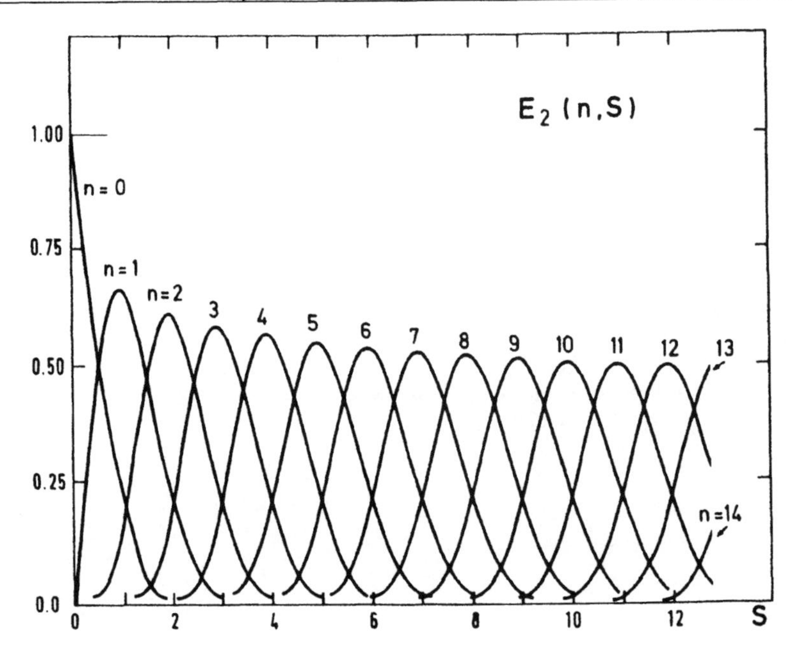

Figure 6.4. The n-level spacings $E_2(n, s)$ for the Gaussian unitary ensemble. Reprinted with permission from M.L. Mehta and J. des Cloizeaux, The probabilities for several consecutive eigenvalues of a random matrix, *Indian J. Pure Appl. Math.* 3, 329–351 (1972).

these figures one sees that the set of eigenvalues of a matrix from the Gaussian unitary ensemble is more or less equally spaced, each individual peak is quite isolated, it looses its height and gains in width only slowly as n increases. Figures 6.7 and 6.8 are the contour maps for two consecutive spacings.

For the empirical probability density of the nearest neighbor spacings of the zeros of the Riemann zeta function on the critical line see Figures 1.12 to 1.15 in Chapter 1. Figures 6.9 and 6.10 represent those for the next nearest neighbor spacings in comparison to $p_2(1; s)$ for the Gaussian unitary ensemble.

6.5 Some Remarks

A few remarks about the analysis of the previous two sections are in order.

6.5.1 We have put aside the question of convergence. In fact, for fixed ξ and η, $\mathcal{K}_N(\xi, \eta)$ tends to $\mathcal{K}(\xi, \eta)$ uniformly (Goursat, 1956) with respect to ξ and η in any finite interval $|\xi|, |\eta| \leqslant 1$. Hence the Fredholm determinant of the kernel $\mathcal{K}_N(\xi, \eta)$ converges (Goursat, 1956) to the Fredholm determinant of the limiting kernel $\mathcal{K}(\xi, \eta)$; that

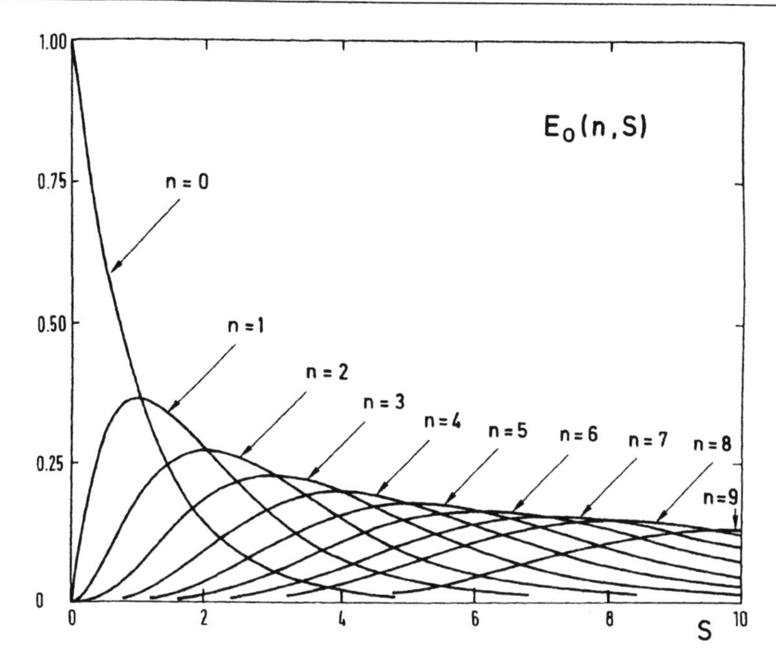

Figure 6.5. The n-level spacings $E_0(n; s)$ for the Poisson process. Reprinted with permission from M.L. Mehta and J. des Cloizeaux, The probabilities for several consecutive eigenvalues of a random matrix, *Indian J. Pure Appl. Math.* 3, 329–351 (1972).

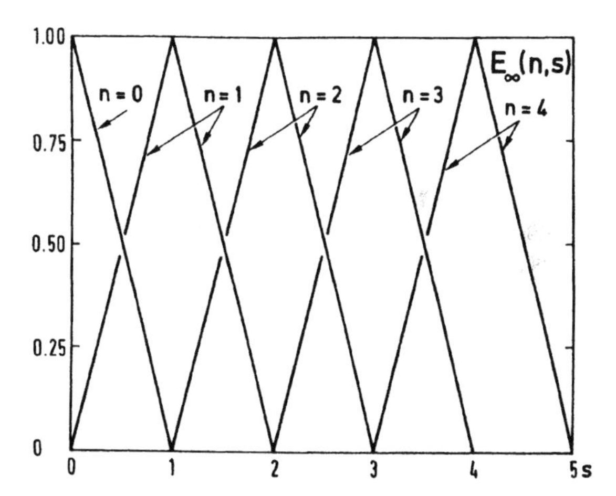

Figure 6.6. The same as Figure 6.5, but for equally spaced levels. Reprinted with permission from M.L. Mehta and J. des Cloizeaux, The probabilities for several consecutive eigenvalues of a random matrix, *Indian J. Pure Appl. Math.* 3, 329–351 (1972).

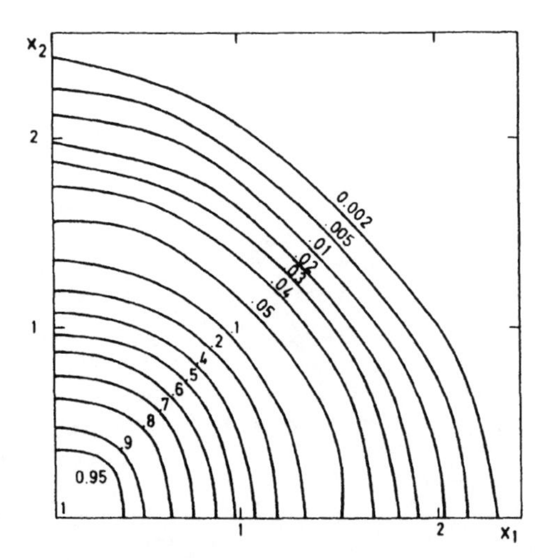

Figure 6.7. Contour map of the probability $\mathcal{B}_2(x_1, x_2)$ that no eigenvalues (of a random matrix chosen from the Gaussian unitary ensemble) lie for a distance x_1 on one side and x_2 on the other side of a given eigenvalue, the distances being measured in units of the mean spacing.

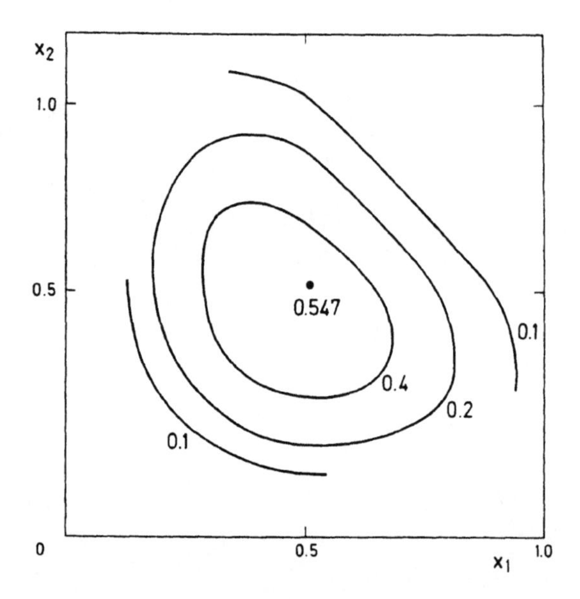

Figure 6.8. Contour map of $B_2(t; -t, t, y)$ as a function of $x_1 = t + y$, $x_2 = t - y$, or of the function $\mathcal{P}_2(x_1, x_2) = \partial^2 \mathcal{B}_2(x_1, x_2)/\partial x_1 \partial x_2$, the joint probability density function for the two adjacent spacings x_1 and x_2 measured in units of the mean spacing.

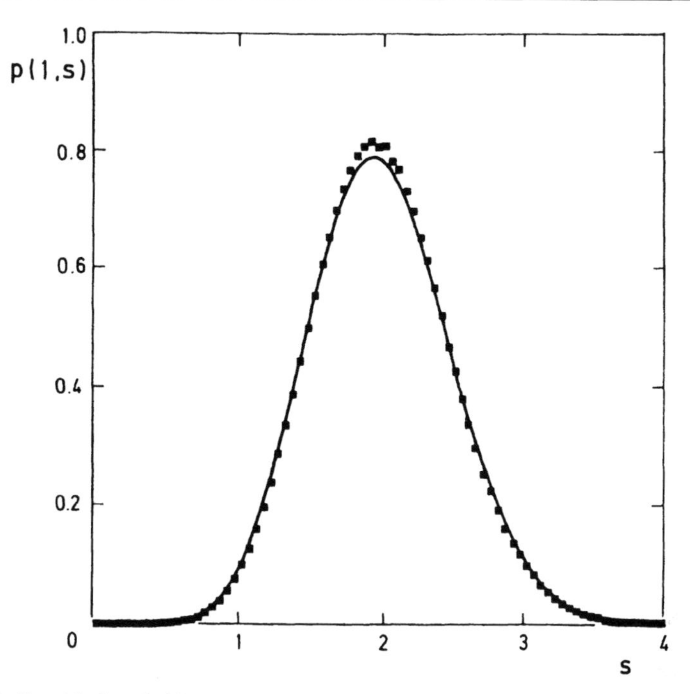

Figure 6.9. Empirical probability density $p(1;s)$ of the next nearest neighbor spacings for the zeros $\frac{1}{2}+i\gamma_n$ of the Riemann zeta function for $10^{12} < n < 10^{12}+10^5$ compared to that for the Gaussian unitary ensemble. From A. Odlyzko (1989). Copyright 1989, American Telephone and Telegraph Company, reprinted with permission.

is,

$$\lim A_2(\theta) = B_2(t) = E_2(0; s), \tag{6.5.1}$$

and in general

$$\lim A_2(\theta; x_1, \ldots, x_n) = B_2(t; y_1, \ldots, y_n). \tag{6.5.2}$$

6.5.2 For $t=0$ the spheroidal functions $f_i(\xi)$ are proportional to the Legendre polynomials

$$f_i(\xi) = \left(\frac{2i+1}{2}\right)^{1/2} P_i(\xi) \quad (t=0), \tag{6.5.3}$$

and for small t, they can be expanded in terms of them (Stratton et al., 1956; Robin, 1959):

$$f_i(x) = \sum_j d_j(t,i) P_j(x). \tag{6.5.4}$$

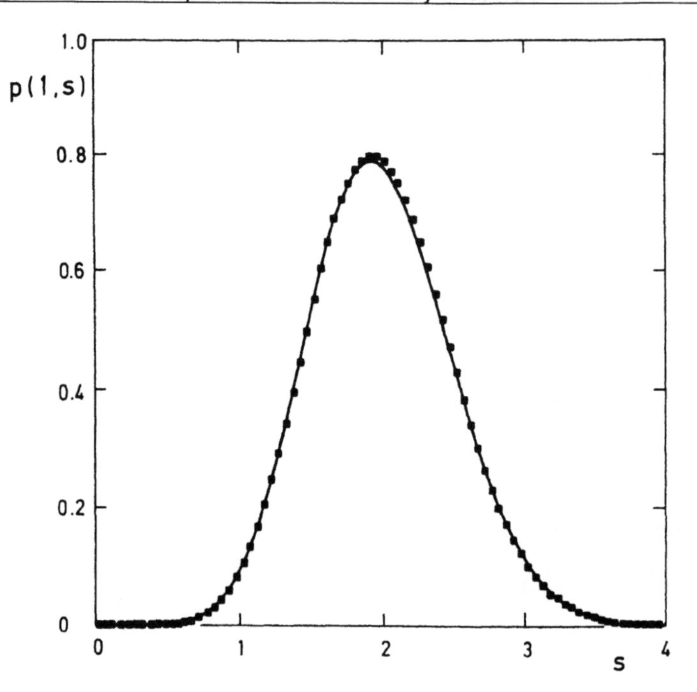

Figure 6.10. Same as Figure 6.9, but for the 79 million zeros around $n \approx 10^{20}$. From A. Odlyzko (1989). Copyright 1989, American Telephone and Telegraph Company, reprinted with permission.

Because of parity only those j occur in the summation for which $j-i$ is an even integer. For example, to the smallest order we get from (6.3.25) and (6.3.26)

$$\mu_{2j} = \int_{-1}^{1} P_{2j}(y)\,dy / P_{2j}(0) = 2\delta_{j0}, \tag{6.5.5}$$

and

$$\mu_{2j+1} = i\pi t \int_{-1}^{1} y P_{2j+1}(y)\,dy / P'_{2j+1}(0) = \frac{2}{3} i\pi t \delta_{j0}. \tag{6.5.6}$$

A few terms in such power series expansions are given in Appendix A.13. For the power series expansions of $E_2(n,s)$ see Chapters 20 and 21.

6.5.3 Extensive numerical tables of expansion coefficients $d_j(t,i)$, Eq. (6.5.4), are available (Stratton et al., 1956) or can be computed (Van Buren, 1976). Using them one can calculate $f_i(x)$, μ_i, λ_i, and so on. From (6.3.25), (6.3.26), we get e.g.

$$\mu_{2k} = 2d_0(t, 2k)\left(\sum_j (-1)^j \frac{1\cdot 3\cdots(2j-1)}{2\cdot 4\cdots(2j)} d_{2j}(t, 2k)\right)^{-1}, \tag{6.5.7}$$

and

$$\mu_{2k+1} = \frac{2}{3} i\pi t d_1(t, 2k+1)\left(\sum_j (-1)^j \frac{3\cdot 5\cdots(2j+1)}{2\cdot 4\cdots(2j)} d_{2j+1}(t, 2k+1)\right)^{-1}. \tag{6.5.8}$$

Let us note that for $p_2(0; s)$, expression (6.4.32) is better than (6.4.33) for the following reason. If a function is known only numerically, its derivative is known with a lesser precision. Equation (6.4.33) involves two numerical differentiations, hence its precision is considerably less than that of (6.4.32) which involves no numerical differentiation. Good numerical precision can also be obtained by using the power series expansion for $s \leqslant 2$ and the asymptotic expansion for $s \geq 1$. (See Chapters 20 and 21.)

6.5.4 In Section 6.2 we saw that the level density is a semi-circle. And still all our considerations about correlations and spacings were restricted to the central part of the spectrum, i.e. all the eigenvalues were supposed to lie near the origin where the level density is flat and equals $(2N)^{1/2}/\pi$. This was done only for simplicity. Actually the p-point correlation function for any finite p is stationary under translations over the spectrum provided we measure distances in terms of the local mean spacing. For example, Eqs. (6.3.11) and (6.3.12) can be replaced by

$$\mathcal{K}_N(\xi, \eta) = \pi t K_N(x, y)(2N - a^2)^{-1/2}, \tag{6.5.9}$$

and

$$\pi t = \theta(2N - a^2)^{1/2}, \qquad \pi t\xi = x(2N - a^2)^{1/2}, \qquad \pi t\eta = y(2N - a^2)^{1/2}, \tag{6.5.10}$$

without changing Eq. (6.3.13); the points x and y are now near the point a and not near the origin. Similarly, the 2-point cluster function for levels not near the origin is again given by (cf. Eq. (6.2.14))

$$\begin{aligned} Y_2(x_1, x_2) &= \lim_{N\to\infty} \frac{x^2}{(2N - x_1^2)^{1/2}(2N - x_2^2)^{1/2}} \left(\sum_{j=0}^{N-1} \varphi_j(x_1)\varphi_j(x_2)\right)^2 \\ &= \left(\frac{\sin(\pi r)}{\pi r}\right)^2, \end{aligned} \tag{6.5.11}$$

with $r = |y_1 - y_2|$, $\pi y_1 = (2N - x_1^2)^{1/2} x_1$, $\pi y_2 = (2N - x_2^2)^{1/2} x_2$, with $y_1 - y_2$ finite. The stationarity property of the p-point correlation function then follows from the fact that it can be expressed in terms of the 2-point function.

Expressing all distances in terms of the local mean spacing for comparison with the theory has usually been called "unfolding".

Another property of some conceptual importance is the so called "ergodicity". The basic ergodic theorem, familiar in the statistical mechanics and in the theory of random processes, states that the time average of a physical quantity equals its ensemble average for "almost all" members of the ensemble in some reasonable limit. For random matrices we have an ordering of the eigenvalues instead of time, the spectrum is discrete rather than continuous, and some physical quantities, such as level density, may not be stationary. By the ergodic theorem for random matrices one means that in the limit of large matrices the spectral average of a fluctuation measure equals in "almost all" cases its ensemble average. For a detailed discussion and a proof of this theorem see e.g. Pandey (1979).

6.5.5 From the explicit expressions for, say, the correlation functions or the level spacings, one can compute them and represent them as graphs. However, the question of the probable (or mean square) thickness of such curves was not asked. The reason lies in the difficulty of giving a precise definition of this "thickness" and then in estimating it. Similar other questions arise when one tries to compare the empirical or experimental data with the predictions of a theory. We will come back to it in Chapter 16.

6.5.6 For every integer m, let the sides of a fixed $2m$-gon be labeled $s_1, s_2, \ldots, s_{2m}$ consecutively around its boundary. Let $\varepsilon_g(m)$ be the number of ways of putting these sides into m pairs, each side belonging to one and only one pair, and such that if the sides in each pair are identified we obtain an orientable surface of genus g. (An orientable surface of genus g is topologically equivalent to a sphere with g handles.) The generating function

$$C(m, N) = \sum_{g=0}^{\infty} \varepsilon_g(m) N^{m+1-2g} \tag{6.5.12}$$

was studied by Harer and Zagier (1986). They showed from combinatorial arguments that

$$C(m, N) = 2^m \left\langle \operatorname{tr} A^{2m} \right\rangle \equiv 2^m \int \operatorname{tr} A^{2m} e^{-\operatorname{tr} A^2} dA \div \int e^{-\operatorname{tr} A^2} dA, \tag{6.5.13}$$

where A is an $N \times N$ Hermitian matrix, $A_{jk} = A_{jk}^{(0)} + i A_{jk}^{(1)} = A_{kj}^{(0)} - i A_{kj}^{(1)}$, $1 \leqslant j \leqslant k \leqslant N$, $A_{kj}^{(0)}$, $A_{kj}^{(1)}$ real and

$$dA = \prod_{j\leqslant k} dA_{jk}^{(0)} \prod_{j<k} dA_{jk}^{(1)}. \tag{2.5.1}$$

Here and in what follows, all the integrals are taken from $-\infty$ to $+\infty$.

Taking the eigenvalues of A as new variables, one has

$$C(m,N) = 2^m \left\langle \sum_{i=1}^{N} x_i^{2m} \right\rangle = 2^m N \langle x_1^{2m} \rangle, \tag{6.5.14}$$

with the notation

$$\langle f(x) \rangle = \int f(x)\Phi(x)\,dx \div \int \Phi(x)\,dx, \tag{6.5.15}$$

$$\Phi(x) \equiv \exp\left(-\sum_{i=1}^{N} x_i^2\right) \prod_{1\leqslant i<k\leqslant N} (x_i - x_j)^2, \quad dx \equiv dx_1 \cdots dx_N. \tag{6.5.16}$$

From Eqs. (6.2.1), (6.2.7), and (6.2.10) we can therefore write

$$\begin{aligned} C(m,N) &= 2^m N \int x^{2m} \sum_{j=0}^{N-1} \varphi_j^2(x)\,dx \div \int \sum_{j=0}^{N-1} \varphi_j^2(x)\,dx \\ &= 2^m \int x^{2m} \left\{ N\varphi_N^2(x) - \sqrt{N(N+1)}\varphi_{N-1}(x)\varphi_{N+1}(x) \right\} dx, \end{aligned} \tag{6.5.17}$$

where we have used the orthonormality of the functions $\varphi_j(x)$, Eq. (6.2.3), and the Christoffel–Darboux formula (cf. Bateman, 1953b)

$$\sum_{j=0}^{N-1} \varphi_j^2(x) = N\varphi_N^2(x) - \sqrt{N(N+1)}\varphi_{N-1}(x)\varphi_{N+1}(x). \tag{6.5.18}$$

Now using the relations

$$(2x)^k = \sum_{j=0}^{[k/2]} \frac{k!}{j!(k-2j)!} H_{k-2j}(x), \tag{6.5.19}$$

and (Bateman, 1954)

$$\int e^{-x^2} H_j(x) H_k(x) H_\ell(x)\, dx$$
$$= \begin{cases} \dfrac{j!k!\ell!}{(s-j)!(s-k)!(s-\ell)!} 2^s \sqrt{\pi}, & \text{if } j+k+\ell = 2s \text{ is even,} \\ 0, & \text{otherwise,} \end{cases} \tag{6.5.20}$$

we can express $C(m, N)$ as a finite sum

$$\begin{aligned} C(m, N) &= \sum_{j=0}^{m} \frac{(2m)!N!2^{-j}}{j!(m-j)!(m-j+1)!(N-m+j)!} \\ &\quad \times (N(m-j+1) - (N+1)(m-j)) \\ &= \frac{(2m)!}{2^m m!} \sum_{j=0}^{m} \binom{m}{j} \binom{N}{m-j+1} 2^{m-j} \\ &= \frac{(2m)!}{2^m m!} c(m, N), \end{aligned} \tag{6.5.21}$$

with

$$c(m, N) = \sum_{j=0}^{m} \binom{m}{j} \binom{N}{m-j+1} 2^{m-j} = \sum_{j=0}^{m} \binom{m}{j} \binom{N}{j+1} 2^j. \tag{6.5.22}$$

Actually, the summation over j can be formally extended to all integers since the binomial coefficient is zero outside the allowed range. There is a nice recurrence relation for $c(m, N)$, which can be derived as follows. From

$$\begin{aligned} c(m, N) - c(m, N-1) &= \sum_j \binom{m}{j} \left(\binom{N}{j+1} - \binom{N-1}{j+1} \right) 2^j \\ &= \sum_j \binom{m}{j} \binom{N-1}{j} 2^j, \end{aligned} \tag{6.5.23}$$

and

$$c(m-1, N) - c(m-1, N-1) = \sum_j \binom{m-1}{j} \binom{N-1}{j} 2^j, \tag{6.5.24}$$

one has by subtraction

$$
\begin{aligned}
&c(m,N)-c(m,N-1)-c(m-1,N)+c(m-1,N-1)\\
&=\sum_j\left(\binom{m}{j}-\binom{m-1}{j}\right)\binom{N-1}{j}2^j\\
&=\sum_j\binom{m-1}{j-1}\binom{N-1}{j}2^j=\sum_j\binom{m-1}{j}\binom{N-1}{j+1}2^{j+1}\\
&=2c(m-1,N-1),
\end{aligned}
\tag{6.5.25}
$$

or

$$c(m,N)=c(m,N-1)+c(m-1,N)+c(m-1,N-1). \tag{6.5.26}$$

This relation is symmetric in m and N. The initial values

$$c(m,1)=1, \qquad c(0,N)=N, \tag{6.5.27}$$

computed directly from Eq. (6.5.22), are not symmetric. Equation (6.5.22) or Eqs. (6.5.26) and (6.5.27) determine completely the $c(m,N)$. From Eqs. (6.5.26) and (6.5.27) one can also derive the generating function

$$1+2\sum_{m=0}^{\infty}c(m,N)x^{m+1}=\left(\frac{1+x}{1-x}\right)^N, \tag{6.5.28}$$

and by expanding the binomials $(1+x)^N(1-x)^{-N}$ one can get still another form for the $c(m,N)$:

$$c(m,N)=\frac{1}{2}\sum_{j_1+j_2=m+1}\binom{N}{j_1}\binom{N+j_2-1}{j_2}. \tag{6.5.29}$$

6.5.7 Instead of the $2m$-gon of Section 6.5.6 above, consider now a compact surface of genus 0 with k boundary components, and divide the ith boundary component in n_i edges. Let $f_g(n_1,\ldots,n_k)$ be the number of ways of identifying these edges in pairs to obtain a closed orientable connected surface of genus g. Clearly $f_g(n_1,\ldots,n_k)$ is symmetric in the variables, $f_g(n_1,\ldots,n_k)=0$ unless $n_1+\cdots+n_k$ is even, and $f_g(2m)=\varepsilon_g(m)$.

An equation similar to Eq. (6.5.13) seems to hold here,

$$\sum_{g=0}^{\infty}N^{m+1-2g}f_g(n_1,\ldots,n_k)=2^m\langle\operatorname{tr}A^{n_1}\cdots\operatorname{tr}A^{n_k}\rangle, \quad n_1+\cdots+n_k=2m. \tag{6.5.30}$$

One can compute the average

$$\left\langle (\operatorname{tr} A)^{j_1} (\operatorname{tr} A^2)^{j_2} \right\rangle \tag{6.5.31}$$

for any non-negative integers j_1, j_2 by partial differentiation with respect to a and b at $a=1, b=0$ of

$$\begin{aligned}\int \exp(-a \operatorname{tr} A^2 - 2b \operatorname{tr} A) &= e^{Nb^2/a} \int \exp(-a \operatorname{tr}(A + b/a)^2)\, dA \\ &= \text{const} \times \exp(Nb^2/a) a^{-N^2/2}.\end{aligned} \tag{6.5.32}$$

Also, the average $\langle x_1^{j_1} \cdots x_k^{j_k} \rangle$ with $j_1 + \cdots + j_k = 2m$ can be expressed as a linear combination of similar averages with $j_1 + \cdots + j_k = 2m - 2$, by an argument of Aomoto (see Chapter 17, Section 17.8). A general formula for such averages is not known.

Summary of Chapter 6

For the Gaussian unitary ensemble with the joint probability density of the eigenvalues

$$P_{N2}(x_1, \ldots, x_N) \propto \exp\left(-\sum_{j=1}^{N} x_j^2 \right) \prod_{1 \leqslant j < k \leqslant N} (x_j - x_k)^2, \tag{6.1.1}$$

the asymptotic two level cluster function is

$$Y_2(r) = \left(\frac{\sin(\pi r)}{\pi r} \right)^2, \tag{6.2.14}$$

and its Fourier transform is

$$b(k) = \begin{cases} 1 - |k|, & |k| \leqslant 1, \\ 0, & |k| \geqslant 1. \end{cases} \tag{6.2.17}$$

The n-level cluster function is

$$Y_n(y_1, \ldots, y_n) = \sum_P \left(\frac{\sin(\pi r_{12})}{\pi r_{12}} \right) \left(\frac{\sin(\pi r_{23})}{\pi r_{23}} \right) \cdots \left(\frac{\sin(\pi r_{n1})}{\pi r_{n1}} \right), \tag{6.2.16}$$

where $r_{ij} = |y_i - y_j|$, and the sum is taken over all $(n-1)!$ cyclic permutation of the indices $(1, 2, \ldots, n)$.

The "sine kernel" $\sin(x-y)\pi t/(x-y)\pi$ is the square of another kernel $\exp(i\pi xyt)$ over the interval $(-1, 1)$; it commutes with a second order linear differential operator

$$L := \frac{d}{dx}(x^2-1)\frac{d}{dx} + (\pi t x)^2. \tag{6.3.22}$$

The probability $E_2(n; s)$ that a randomly chosen interval of length s contains exactly n levels is given by the formulas

$$E_2(0; s) = \prod_{i=0}^{\infty}(1-\lambda_i), \tag{6.3.27}$$

and for $n > 0$,

$$E_2(n; s) = E_2(0; s) \cdot \sum_{0 \leqslant j_1 < j_2 < \cdots < j_n} \frac{\lambda_{j_1}}{1-\lambda_{j_1}} \cdots \frac{\lambda_{j_n}}{1-\lambda_{j_n}}, \tag{6.4.30}$$

where $\lambda_i = s|\mu_i|^2/4$, and μ_i and $f_i(x)$ are the eigenvalues and eigenfunctions of the integral equation

$$\mu f(x) = \int_{-1}^{1} \exp(i\pi xys/2) f(y)\, dy. \tag{6.3.17}$$

The eigenfunctions $f_i(x)$ also satisfy the differential equation $Lf(x) = \ell f(x)$ with L of Eq. (6.3.22) and are known as the prolate spheroidal functions.

One can also write

$$E_2(n, s) = \frac{1}{n!}\left(-\frac{\partial}{\partial z}\right)^n F_2(z, s/2)|_{z=1}, \tag{6.4.35}$$

where $F_2(z, t)$ is the Fredholm determinant of the kernel

$$z\mathcal{K}(x, y) = z\frac{\sin(x-y)\pi t}{(x-y)\pi}$$

over the interval $(-1, 1)$.

7

GAUSSIAN ORTHOGONAL ENSEMBLE

The Gaussian unitary ensemble studied in Chapter 6 is the simplest from the mathematical point of view. However, for physical applications the most important one is the Gaussian orthogonal ensemble (GOE). The eigenvalues of a real symmetric matrix chosen at random from the GOE are real and have the joint probability density, Theorem 3.3.1, Eq. (3.3.8)

$$P_{N1}(x_1, \ldots, x_N) = \text{const} \cdot \left(-\frac{1}{2}\sum_{i=1}^{N} x_i^2\right) \prod_{1 \leqslant i < j \leqslant N} |x_i - x_j|. \tag{7.0.1}$$

As we will be dealing with the GOE throughout this chapter, we will some times omit the subscript 1.

It was suggested in Chapter 2 that if the energy scale is properly chosen, then the fluctuation properties of the series of points $x_1, \ldots, x_N$ should provide a good model for the eigenvalues of a complicated system having the time reversal invariance and rotational symmetry.

The main objective of this chapter is to derive expressions for the n-level correlation function R_n, cluster function T_n, the form-factor Y_n, and the spacing distributions $B(n; y_1, \ldots, y_n)$, $E(n; s)$ and $p(n; s)$ corresponding to Eqs. (6.2.7), (6.2.9), (6.2.16), (6.4.29), (6.2.30) and (6.4.31) of the preceding chapter.

7.1 Generalities

The first serious difficulty with the function (7.0.1) is its unfavorable symmetry caused by the presence of the absolute value sign. This will be taken care of by the method of integration over alternate variables and the introduction of matrices whose elements are no longer ordinary complex numbers, but quaternions. As explained in Chapter 5, once this is understood, one can quietly follow step by step all the manipulations of Chapter 6 and arrive at the corresponding expressions for all the quantities pertaining to the GOE. The method of integration over alternate variables to modify the symmetry of the integrand, though explained in detail in Chapter 5, will be somewhat repeated.

We will follow the plan of Chapter 6. Thus Section 7.2 deals with correlation functions, Sections 7.3, 7.4 and 7.5 with the spacing probabilities and Section 7.6 with bounds of the spacing distribution. The two-level cluster function is given by Eq. (7.2.41), its Fourier transform by Eq. (7.2.46); the probability $E_1(n; s)$ that an interval of length s contains exactly n levels is given by Eqs. (7.3.19), (7.4.12) and (7.5.21), while the nearest neighbor spacing probability density $p_1(0; s)$ is given by Eq. (7.4.18) or (7.4.19). These various functions are shown graphically on Figures 7.1 and 7.3 and are tabulated in Appendices A.14 and A.15.

As in Chapter 6, we can write the integrand in (7.0.1) as a determinant containing "oscillator wave functions" (cf. Eq. (6.2.4))

$$\exp\left(-\frac{1}{2}\sum_{i=1}^{N} x_i^2\right)\prod_{j<i}(x_i - x_j) = \text{const}\cdot\det[\varphi_{i-1}(x_j)]_{i,j=1,\ldots,N}, \tag{7.1.1}$$

where

$$\begin{aligned}\varphi_j(x) &= (2^j j!\sqrt{\pi})^{-1/2}\exp(-x^2/2)H_j(x)\\ &= (2^j j!\sqrt{\pi})^{-1/2}\exp(x^2/2)\left(-\frac{d}{dx}\right)^j\exp(-x^2),\end{aligned} \tag{7.1.2}$$

are the "oscillator wave functions" orthogonal over $(-\infty, \infty)$

$$\int_{-\infty}^{\infty}\varphi_j(x)\varphi_k(x)\,dx = \delta_{jk}. \tag{6.2.3}$$

From the recurrence relations for the Hermite polynomials $H_j(x)$,

$$H_{j+1}(x) = 2xH_j(x) - 2jH_{j-1}(x), \qquad H_j'(x) = 2jH_{j-1}(x), \tag{7.1.3}$$

where prime denotes the derivative, one deduces

$$\sqrt{2}\,\varphi_j'(x) = \sqrt{j}\varphi_{j-1}(x) - \sqrt{j+1}\varphi_{j+1}(x), \tag{7.1.4}$$

i.e. $\varphi_{j+1}(x)$ is a linear combination of $\varphi'_j(x)$ and $\varphi_{j-1}(x)$. Therefore one can, for example, replace everywhere $\varphi_{2j+1}(x)$ by $\varphi'_{2j}(x)$ in the determinant (7.1.2). Thus for N even, we have

$$\exp\left(-\frac{1}{2}\sum_{i=1}^{N} x_i^2\right)\prod_{j<i}(x_i - x_j) = c\det[\varphi_{2j-2}(x_i), \varphi'_{2j-2}(x_i)],$$
$$i = 1, \ldots, N;\ j = 1, \ldots, N/2, \tag{7.1.5}$$

where c is a constant. If N is odd, there is an extra column $\varphi_{N-1}(x_i)$. For $N = 2m + 1$, we may replace $\varphi_{2j}(x)$ by $\varphi'_{2j+1}(x)$ for $j = 0, 1, \ldots, m - 1$, using the last column $\varphi_{2m}(x)$ to eliminate the extra terms,

$$\exp\left(-\frac{1}{2}\sum_{i=1}^{N} x_i^2\right)\prod_{j<i}(x_i - x_j) = c\det[\varphi'_{2j-1}(x_i), \varphi_{2j-1}(x_i), \varphi_{2m}(x_i)],$$
$$i = 1, \ldots, N;\ j = 1, \ldots, (N-1)/2, \tag{7.1.6}$$

where c is again a constant, not necessarily the same.

In Eq. (7.1.5) the $(2j - 1)$th column is $\varphi_{2j-2}(x_i)$ and the $2j$th column is $\varphi'_{2j-2}(x_i)$, i standing for the ith component of the column. In Eq. (7.1.6) the $(2j - 1)$th column is $\varphi'_{2j-1}(x_i)$, the $2j$th column is $\varphi_{2j-1}(x_i)$, while the last, i.e. the Nth or the $(2m + 1)$th column is $\varphi_{2m}(x_i)$. Such a convenient self-explanatory notation will often be used in what follows.

7.2 Correlation and Cluster Functions

In view of the relations (5.1.14) and (5.1.15) it will be sufficient to form an ordinary $2N \times 2N$ matrix $\Theta[K]$ such that $Z\Theta[K]$ is anti-symmetric (or the $N \times N$ matrix K with quaternion elements is self-dual) and $\det\Theta[K]$ is the square of the function (7.1.1). We will follow the procedure outlined in Section 5.7.

One can directly verify that the monic skew-orthogonal polynomials of the real type with the weight $w(x) = \exp(-x^2/2)$ over $(-\infty, \infty)$ are

$$R_{2j}(x) = 2^{-2j} H_{2j}(x), \tag{7.2.1}$$

$$R_{2j+1}(x) = 2^{-2j}\left(x H_{2j}(x) - H'_{2j}(x)\right), \tag{7.2.2}$$

$$r_j = 2^{1-2j}(2j)!\sqrt{\pi}, \tag{7.2.3}$$

where $H_j(x)$ are the Hermite polynomials, Eq. (6.2.1),

$$H_j(x) = \exp(x^2)\left(-\frac{d}{dx}\right)^j \exp(-x^2) = j! \sum_{i=0}^{[j/2]} (-1)^i \frac{(2x)^{j-2i}}{i!(j-2i)!} \tag{6.2.1}$$

and the prime denotes differentiation. The $\psi_k(x)$ of Eq. (5.7.13) are therefore

$$\psi'_{2j}(x) = \sqrt{r_j}\varphi_{2j}(x), \tag{7.2.4}$$

$$\psi_{2j+1}(x) = \sqrt{r_j}\varphi_{2j}(x). \tag{7.2.5}$$

Hence for $N = 2m$ we will write Eq. (5.8.26) in its 2×2 matrix form as

$$g(x, y) = \begin{bmatrix} S_N(x, y) & D_N(x, y) \\ I_N(x, y) & S_N(y, x) \end{bmatrix}, \tag{7.2.6}$$

where

$$S_N(x, y) = \sum_{i=0}^{m-1} \left(\varphi_{2i}(x)\varphi_{2i}(y) - \varphi'_{2i}(x) \int_{-\infty}^{\infty} \varepsilon(y-t)\varphi_{2i}(t)\, dt\right), \tag{7.2.7}$$

$$D_N(x, y) = \sum_{i=0}^{m-1} \left(-\varphi_{2i}(x)\varphi'_{2i}(y) + \varphi'_{2i}(x)\varphi_{2i}(y)\right), \tag{7.2.8}$$

$$I_N(x, y) = \sum_{i=0}^{m-1} \left(\varphi_{2i}(y) \int_{-\infty}^{\infty} \varepsilon(x-t)\varphi_{2i}(t)\, dt - \varphi_{2i}(x) \int_{-\infty}^{\infty} \varepsilon(y-t)\varphi_{2i}(t)\, dt\right). \tag{7.2.9}$$

For $N = 2m + 1$, we will choose the skew-orthogonal polynomials $R_j(x)$ for $j = 0, 1, \ldots, m-1$, as

$$R_{2j}(x) = 2^{-2j-1} H_{2j+1}(x), \tag{7.2.10}$$

$$R_{2j+1}(x) = 2^{-2j-1}(x H_{2j+1}(x) - H'_{2j+1}(x)), \tag{7.2.11}$$

and

$$R_{2m}(x) = H_{2m}(x) \div \int_{-\infty}^{\infty} H_{2m}(y) e^{-y^2}\, dy. \tag{7.2.12}$$

These $2m+1$ polynomials, each of degree less than $2m+1$, are linearly independent, are skew-orthogonal and satisfy the additional constraints (5.7.10). Hence for $N = 2m+1$, we have Eq. (7.2.6) with

$$S_N(x,y) = \sum_{i=0}^{m-1} \left(\varphi_{2i+1}(x)\varphi_{2i+1}(y) - \varphi'_{2i+1}(x) \int_{-\infty}^{\infty} \varepsilon(y-t)\varphi_{2i+1}(t)\, dt \right), \tag{7.2.13}$$

$$D_N(x,y) = \sum_{i=0}^{m-1} (-\varphi_{2i+1}(x)\varphi'_{2i+1}(y) + \varphi'_{2i+1}(x)\varphi_{2i+1}(y)), \tag{7.2.14}$$

$$\begin{aligned} I_N(x,y) = \sum_{i=0}^{m-1} \bigg(& \varphi_{2i+1}(y) \int_{-\infty}^{\infty} \varepsilon(x-t)\varphi_{2i+1}(t)\, dt \\ & - \varphi_{2i+1}(x) \int_{-\infty}^{\infty} \varepsilon(y-t)\varphi_{2i+1}(t)\, dt \bigg). \end{aligned} \tag{7.2.15}$$

Note that from Eq. (7.1.4) denoting the derivative by a prime

$$\begin{aligned} -\varphi'_{2i}(x)\varphi_{2i}(y) &= \left(\sqrt{i+1/2}\varphi_{2i+1}(x) - \sqrt{i}\varphi_{2i-1}(x)\right)\varphi_{2i}(y) \\ &= \varphi_{2i+1}(x)(\varphi'_{2i+1}(y) + \sqrt{i+1}\varphi_{2i+2}(y)) - \sqrt{i}\varphi_{2i-1}(x)\varphi_{2i}(y) \\ &= \varphi_{2i+1}(x)\varphi'_{2i+1}(y) + \sqrt{i+1}\varphi_{2i+1}(x)\varphi_{2i+2}(y) \\ &\quad - \sqrt{i}\varphi_{2i-1}(x)\varphi_{2i}(y). \end{aligned} \tag{7.2.16}$$

Summing this for $i = 0$ to $i = m-1$, we get

$$-\sum_{i=0}^{m-1} \varphi'_{2i}(x)\varphi_{2i}(y) = \sum_{i=0}^{m-1} \varphi_{2i+1}(x)\varphi'_{2i+1}(y) + \sqrt{m}\varphi_{2m-1}(x)\varphi_{2m}(y). \tag{7.2.17}$$

Or interchanging x and y and using once more Eq. (7.1.4),

$$-\sum_{i=0}^{m-1} \varphi'_{2i+1}(x)\varphi_{2i+1}(y) = \sum_{i=0}^{m} \varphi_{2i}(x)\varphi'_{2i}(y) + \sqrt{m+1/2}\varphi_{2m}(x)\varphi_{2m+1}(y). \tag{7.2.18}$$

Thus whether $N = 2m$ or $N = 2m+1$, Eqs. (7.2.7) and (7.2.13) agree with

$$S_N(x,y) = \sum_{j=0}^{N-1} \varphi_j(x)\varphi_j(y) + \left(\frac{N}{2}\right)^{1/2} \varphi_{N-1}(x) \int_{-\infty}^{\infty} \varepsilon(y-t)\varphi_N(t)\, dt. \tag{7.2.19}$$

So we finally write

$$K_{N1}(x, y) = g(x, y) - \begin{bmatrix} 0 & 0 \\ \varepsilon(x-y) & 0 \end{bmatrix} + K_{\text{odd}}(x, y) \tag{7.2.20}$$

$$= \begin{bmatrix} S_N(x, y) + \alpha(x) & D_N(x, y) \\ J_N(x, y) & S_N(y, x) + \alpha(y) \end{bmatrix}, \tag{7.2.21}$$

with

$$S_N(x, y) = \sum_{j=0}^{N-1} \varphi_j(x)\varphi_j(y) + \left(\frac{N}{2}\right)^{1/2} \varphi_{N-1}(x) \int_{-\infty}^{\infty} \varepsilon(y-t)\varphi_N(t)\, dt, \tag{7.2.22}$$

$$D_N(x, y) = -\frac{\partial}{\partial y} S_N(x, y), \tag{7.2.23}$$

$$I_N(x, y) = \int_{-\infty}^{\infty} \varepsilon(x-t) S_N(t, y)\, dt, \tag{7.2.24}$$

$$J_N(x, y) = I S_N(x, y) - \varepsilon(x-y) + u(x) - u(y), \tag{7.2.25}$$

$$u(x) = \int_{-\infty}^{\infty} \varepsilon(x-y)\alpha(y)\, dy, \tag{7.2.26}$$

where $\varepsilon(x) = (1/2)\text{sign}(x)$ is $1/2$, $-1/2$ or 0 according as $x > 0$, $x < 0$ or $x = 0$,

$$\alpha(x) = \varphi_{2m}(x) \div \int_{-\infty}^{\infty} \varphi_{2m}(t)\, dt, \tag{7.2.27}$$

if $N = 2m + 1$ and

$$\alpha(x) = 0, \tag{7.2.28}$$

if $N = 2m$. Note the interchange of x and y in the lower right corner of Eq. (7.2.21); also that $D_N(x, y)$ and $J_N(x, y)$ change sign under the exchange of x and y.

According to Chapter 5, Sections 5.7 and 5.8

$$P_{N1}(x_1, \ldots, x_N) = \frac{1}{N!} \det[K_{N1}(x_i, x_j)]_{i,j=1,2,\ldots,N}, \tag{7.2.29}$$

and $K_{N1}(x, y)$ satisfies the two conditions of Theorem 5.1.4. Hence one immediately has the n-level correlation function

$$R_n(x_1, \ldots, x_n) = \frac{N!}{(N-n)!} \int \cdots \int P_N(x_1, \ldots, x_N)\, dx_{n+1} \cdots dx_N$$

$$= \det[K_{N1}(x_j, x_k)]_{j,k=1,\ldots,n}. \tag{7.2.30}$$

And as in Section 6.2, the n-level cluster function is

$$T_n(x_1,\ldots,x_n)=\sum_P K_{N1}(x_1,x_2)K_{N1}(x_2,x_3)\cdots K_{N1}(x_n,x_1), \tag{7.2.31}$$

the sum being taken over all $(n-1)!$ distinct cyclic permutations of the indices $(1,2,\ldots,n)$.

Setting $n=1$, we get the level density

$$R_1(x)=S_N(x,x)+\alpha(x). \tag{7.2.32}$$

In the limit $N\to\infty$, this goes to the "semi-circle law" (cf. Appendix A.9),

$$R_1(x)=\begin{cases}\frac{1}{\pi}(2N-x^2)^{1/2}, & |x|<(2N)^{1/2},\\ 0, & |x|>(2N)^{1/2}.\end{cases} \tag{7.2.33}$$

The mean spacing at the origin is thus

$$\alpha=1/R_1(0)=\pi/(2N)^{1/2}. \tag{7.2.34}$$

Setting $n=2$, we get the two-level cluster function

$$T_2(x,y)=K_{N1}(x,y)K_{N1}(y,x). \tag{7.2.35}$$

In the limit $N\to\infty$ (cf. Appendix A.10),

$$\lim(\pi/(2N))^{1/2}S_N(x,y)=s(r), \tag{7.2.36}$$

$$\lim(\pi/(2N))\,\mathrm{sign}(x-y)D_N(x,y)=\frac{d}{dr}s(r)\equiv D(r), \tag{7.2.37}$$

$$\lim\mathrm{sign}(x-y)\cdot I_N(x,y)=-\int_0^r s(t)\,dt\equiv I(r), \tag{7.2.38}$$

with

$$x=\pi\xi/(2N)^{1/2},\qquad y=\pi\eta/(2N)^{1/2},\qquad r=|\xi-\eta|, \tag{7.2.39}$$

and

$$s(r)=\sin(\pi r)/(\pi r). \tag{7.2.40}$$

Hence

$$Y_2(r)=\lim(\pi/2N)T_2(x,y)=K_1(r)K_1(-r), \tag{7.2.41}$$

with

$$K_1(r) = \begin{bmatrix} s(r) & D(r) \\ I(r) - 1/2 & s(r) \end{bmatrix}, \tag{7.2.42}$$

or

$$\begin{aligned} Y_2(r) &= \left(\frac{1}{2} - \int_0^r s(t)\,dt\right)\left(\frac{d}{dr}s(r)\right) + (s(r))^2 \\ &= \left(\int_r^\infty s(t)\,dt\right)\left(\frac{d}{dr}s(r)\right) + (s(r))^2. \end{aligned} \tag{7.2.43}$$

Figure 7.1 shows the limiting two level correlation function. Note that the oscillations are almost imperceptible in contrast to Figure 6.1.

The behaviour of $Y_2(r)$ for small and large r is given by

$$Y_2(r) = 1 - \frac{\pi^2 r}{6} + \frac{\pi^4 r^3}{60} - \frac{\pi^4 r^4}{135} + \cdots \tag{7.2.44}$$

and

$$Y_2(r) = \frac{1}{\pi^2 r^2} - \frac{1 + \cos^2 \pi r}{\pi^4 r^4} + \cdots \tag{7.2.45}$$

respectively. The two-level form factor is (cf. Appendix A.11)

$$b(k) = \int_{-\infty}^{\infty} Y_2(r) \exp(2\pi i k r)\,dr$$

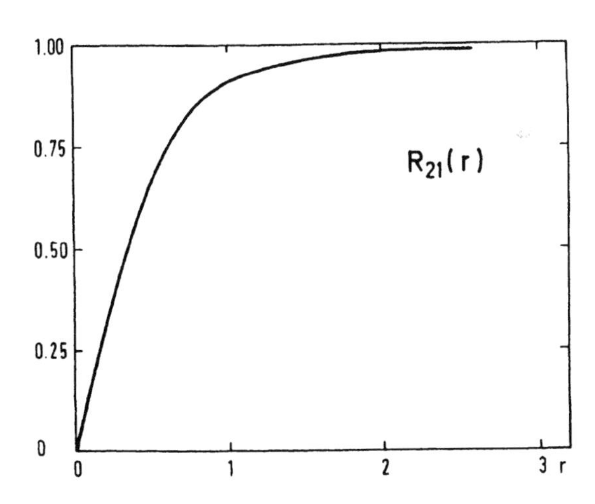

Figure 7.1. Two level correlation function for the orthogonal ensemble.

$$
= \begin{cases} 1 - 2|k| + |k| \ln(1 + 2|k|), & |k| \leqslant 1, \\ -1 + |k| \ln\left(\dfrac{2|k|+1}{2|k|-1}\right), & |k| \geqslant 1. \end{cases} \tag{7.2.46}
$$

This has the behaviour

$$
b(k) = 1 - 2|k| + 2k^2 + \cdots, \tag{7.2.47}
$$

$$
b(k) = \frac{1}{12k^2} + \frac{1}{80k^4} + \cdots, \tag{7.2.48}
$$

for small and large k respectively.

In the limit $N \to \infty$, the n-level cluster function is

$$
Y_n(\xi_1, \ldots, \xi_n) = \sum_P K_1(r_{12}) K_1(r_{23}) \cdots K_1(r_{n1}), \tag{7.2.49}
$$

the sum being taken over the $(n-1)!$ distinct cyclic permutations of the indices $(1, 2, \ldots, n)$, and $r_{ij} = |\xi_i - \xi_j| = (2N)^{1/2}|x_i - x_j|/\pi$. The n-level form-factor or the Fourier transform of Y_n is

$$
\begin{aligned}
&\int Y_n(\xi_1, \ldots, \xi_n) \exp\left(2\pi i \sum_{j=1}^{n} k_j \xi_j\right) d\xi_1 \cdots d\xi_n \\
&\quad = \delta(k_1 + \cdots + k_n) \\
&\quad\quad \times \int_{-\infty}^{\infty} dk \sum_P f_1(k) f_1(k + k_1) \cdots f_1(k + k_1 + \cdots + k_{n-1}),
\end{aligned} \tag{7.2.50}
$$

with

$$
f_1(k) = \begin{bmatrix} f_2(k) & k f_2(k) \\ (f_2(k) - 1)/k & f_2(k) \end{bmatrix}, \qquad f_2(k) = \begin{cases} 1, & |k| < 1/2, \\ 0, & |k| > 1/2. \end{cases} \tag{7.2.51}
$$

The sum in Eq. (7.2.49) is over all $(n-1)!$ distinct cyclic permutations of the indices $(1, \ldots, n)$ and that in Eq. (7.2.50) is over all the $(n-1)!$ permutations of the indices $(1, \ldots, n-1)$.

7.3 Level Spacings. Integration Over Alternate Variables

To begin with, we calculate the level-spacing function

$$
A_1(\theta) = \int_{\text{out}} \cdots \int P_{N1}(x_1, \ldots, x_N)\, dx_1 \cdots dx_N, \tag{7.3.1}
$$

case $n=0$ of Eq. (6.1.9). To take care of the absolute value sign in (7.1.1), we order the variables as $-\infty < x_1 \leqslant x_2 \leqslant \cdots \leqslant x_N < \infty$, and multiply the result by $N!$. Substituting from Eq. (7.1.5) we can write

$$A_1(\theta) = c \cdot N! \int_{R(\text{out})} \cdots \int \det \begin{bmatrix} \varphi_{2j-2}(x_i) \\ \varphi'_{2j-2}(x_i) \end{bmatrix} dx_1 \cdots dx_N, \tag{7.3.2}$$

where the region of integration $R(\text{out})$ is $-\infty < x_1 \leqslant \cdots \leqslant x_N < \infty$, and $|x_j| > \theta$, $j = 1, \ldots, N$; If we integrate over x_1 we replace the first column by the column $[F_j(x_2)]$, $j = 0, \ldots, [N/2]-1$, where the functions $F_j(x)$ are defined by

$$F_{2j}(x) = \int_{-\infty}^{x} u(y)\varphi_{2j}(y)\,dy, \qquad F_{2j+1}(x) = \int_{-\infty}^{x} u(y)\varphi'_{2j}(y)\,dy, \tag{7.3.3}$$

with

$$u(x) = \begin{cases} 0, & |x| \leqslant \theta, \\ 1, & |x| > \theta. \end{cases} \tag{7.3.4}$$

Now x_2 occurs in two columns and we cannot integrate over it. However, integrating over x_3 we replace the third column by the column $[F_j(x_4) - F_j(x_2)]$. But we have already a column $[F_j(x_2)]$, so we may drop the terms with a negative sign. Thus at each integration over $x_3, x_5, \ldots$, the corresponding column is replaced by a column of an F_j. In case N is odd, the last column is replaced by a column of pure numbers $F_j(\infty)$.

Thus an integration over the alternate variables $x_1, x_3, x_5, \ldots$, gives

$$A_1(\theta) = c \cdot N! \int_{R(\text{out})} \cdots \int \det \begin{bmatrix} F_{2j}(x_{2i}) & \varphi_{2j}(x_{2i}) \\ F_{2j+1}(x_{2i}) & \varphi'_{2j}(x_{2i}) \end{bmatrix} dx_2\, dx_4 \cdots dx_{2m}, \tag{7.3.5}$$

where either $N = 2m$ or $N = 2m+1$. In the latter case, there is an extra column $[F_j(\infty)]$ at the right end. To avoid minor complications we take in the rest of the chapter N even, $N = 2m$.

The integrand in (7.3.5) is now symmetric in the remaining variables, therefore one can integrate over them independently and divide by $(N/2)!$; the result is a determinant (cf. Appendix A.17),

$$\begin{aligned} A_1(\theta) &= c \cdot N![(N/2)!]^{-1} \cdot \int_{\text{out}} \cdots \int \det \begin{bmatrix} F_{2j}(x_{2i}) & \varphi_{2j}(x_{2i}) \\ F_{2j+1}(x_{2i}) & \varphi'_{2j}(x_{2i}) \end{bmatrix} dx_2\, dx_4 \cdots dx_{2m} \\ &= c \cdot N!(-2)^{N/2} \det[\overline{g}_{ij}]_{i,j=0,1,\ldots,m-1}, \end{aligned} \tag{7.3.6}$$

where the region of integration is now the entire real line outside the interval $(-\theta, \theta)$ for each of the variables $x_2, x_4, \ldots, x_{2m}$, and

$$\begin{aligned}\overline{g}_{ij} &= -\frac{1}{2}\int_{-\infty}^{\infty}\big(F_{2i}(x)\varphi'_{2j}(x) - F_{2j+1}(x)\varphi_{2i}(x)\big)u(x)\,dx \\ &= \delta_{ij} - \int_{-\theta}^{\theta}\varphi_{2i}(x)\varphi_{2j}(x)\,dx.\end{aligned} \tag{7.3.7}$$

We can fix the constant by observing that $A_1(0) = 1$, so that

$$c \cdot (-2)^{N/2} N! = 1, \tag{7.3.8}$$

$$A_1(\theta) = \det \overline{G} = \det[\overline{g}_{ij}], \tag{7.3.9}$$

with the elements $\overline{g}_{ij}$ of $\overline{G}$ given by Eq. (7.3.7).

From this point on, the analysis follows Section 6.3. The diagonalization of $\overline{G}$ and the passage to the limit $m \to \infty$, $\theta \to 0$, proceed exactly as in Sections 6.3, 6.4 except that we now need only the even part in x of $K(x, y)$, and only the even functions $\psi_{2i}(x)$ and $f_{2i}(\xi)$. Relations corresponding to Eqs. (6.4.8) and (6.4.11) to (6.4.24) are

$$\overline{\gamma}_{ij} = \int_{-\theta}^{\theta}\varphi_{2i}(x)\varphi_{2j}(x)\,dx, \tag{7.3.10}$$

$$A_1(\theta) = \prod_{i=0}^{m-1}(1 - \lambda_{2i}), \tag{7.3.11}$$

$$\sum_{j=0}^{m-1}\overline{\gamma}_{ij}\overline{h}_{jk} = \overline{h}_{ik}\lambda_{2k}, \tag{7.3.12}$$

$$\int_{-\theta}^{\theta}\overline{K}(x, y)\psi_{2i}(y)\,dy = \lambda_{2i}\psi_{2i}(x), \tag{7.3.13}$$

$$\overline{K}(x, y) = \sum_{i=0}^{m-1}\varphi_{2i}(x)\varphi_{2i}(y) = \sum_{i=0}^{m-1}\psi_{2i}(x)\psi_{2i}(y), \tag{7.3.14}$$

$$\psi_{2i}(x) = \sum_{j=0}^{m-1}\overline{h}_{ji}\varphi_{2j}(x), \tag{7.3.15}$$

$$\sum_{j=0}^{m-1}\overline{h}_{ij}\overline{h}_{kj} = \sum_{j=0}^{m-1}\overline{h}_{ji}\overline{h}_{jk} = \delta_{ik}, \tag{7.3.16}$$

$$\alpha = \pi/(2N)^{1/2}, \quad \theta = \alpha t, \quad x = \alpha t \xi, \quad y = \alpha t \eta, \quad s = 2t, \tag{7.3.17}$$

$$\overline{K}(x, y) = (\alpha t)^{-1} \overline{\mathcal{K}}(\xi, \eta), \tag{7.3.18}$$

$$E_1(0; s) = B_1(t) = \lim A_1(\theta) = \prod_i (1 - \lambda_{2i}) = \prod_i \left(1 - \frac{1}{2} t \mu_{2i}^2\right), \tag{7.3.19}$$

$$\overline{\mathcal{K}}(\xi, \eta) = \frac{1}{2}[\mathcal{K}(\xi, \eta) + \mathcal{K}(\xi, -\eta)]. \tag{7.3.20}$$

The symbols $\mathcal{K}(\xi, \eta)$, $\psi_i(x)$, $f_i(\xi)$, and λ_i have the same meaning as in Section 6.4. The μ_{2i} are the eigenvalues of the integral equation (6.3.20)

$$\mu_{2i} f(\xi) = 2 \int_0^1 \cos(\pi \xi \eta t) f(\eta) \, d\eta. \tag{7.3.21}$$

Equations (6.4.22), (6.4.24) may be completed by

$$\psi_{2i}^{(j)}(\xi) \to (\alpha t)^{-j-1/2} \lambda_{2i}^{1/2} f_{2i}^{(j)}(\xi), \tag{7.3.22}$$

where the super-script j denotes the jth derivative.

Note that the constant in Eq. (7.1.1) is fixed by Eqs. (7.1.5), (7.3.2) and (7.3.8),

$$P_{N1}(x_1, \ldots, x_{2m}) = \frac{(-2)^{-m}}{(2m)!} \det \begin{bmatrix} \varphi_{2i}(x_j) \\ \varphi_{2i}'(x_j) \end{bmatrix}_{i=0,\ldots,m-1;\ j=1,\ldots,2m}; \tag{7.3.23}$$

or written in full,

$$P_{N1}(x_1, \ldots, x_{2m}) = \frac{(-2)^{-m}}{(2m)!} \det \begin{bmatrix} \varphi_0(x_1) & \varphi_0(x_2) & \cdots & \varphi_0(x_{2m}) \\ \varphi_0'(x_1) & \varphi_0'(x_2) & \cdots & \varphi_0'(x_{2m}) \\ \cdots & \cdots & \cdots & \cdots \\ \varphi_{2m-2}(x_1) & \varphi_{2m-2}(x_2) & \cdots & \varphi_{2m-2}(x_{2m}) \\ \varphi_{2m-2}'(x_1) & \varphi_{2m-2}'(x_2) & \cdots & \varphi_{2m-2}'(x_{2m}) \end{bmatrix}. \tag{7.3.24}$$

7.4 Several Consecutive Spacings: $n = 2r$

Next we take the case of several consecutive spacings, the integral

$$A_1(\theta; x_1, \ldots, x_n) = \frac{N!}{n!(N-n)!} \int_{\text{out}} \cdots \int P_{N1}(x_1, \ldots, x_N) \, dx_{n+1} \cdots dx_N, \tag{7.4.1}$$

with $n > 0$. In writing the expression (7.1.5) we took the ordering $x_1 \leqslant x_2 \leqslant \cdots \leqslant x_{2m}$. However, the same expression is valid also when $-\theta \leqslant x_1 \leqslant \cdots \leqslant x_{2r} \leqslant \theta$, $x_{2r+1} \leqslant$

$x_{2r+2} \leqslant \cdots \leqslant x_{2m}$, and $|x_j| \geqslant \theta$ for $j = 2r+1, 2r+2, \ldots, 2m$; in other words, none or some of the $x_{2r+1}, \ldots, x_{2m}$ are less than $-\theta$ and others are greater than θ. This pertains to the fact that the determinant in Eq. (7.1.5) does not change sign when the columns containing the variables $x_{2r+1}, \ldots, x_{2m}$ are passed one by one over the $2r$ columns containing the variables $x_1, \ldots, x_{2r}$. Let us therefore take $n = 2r$, $r \geqslant 1$. We expand the determinant in Eq. (7.1.5) by the first $2r$ columns in the Laplace manner and integrate every term so obtained over $x_{2r+1}, \ldots, x_{2m}$ outside the interval $(-\theta, \theta)$, while $-\theta \leqslant x_1 \leqslant \cdots \leqslant x_{2r} \leqslant \theta$, using the method of integration over alternate variables (Section 7.3). Let us remark that the $2r \times 2r$ determinants formed from the first $2r$ columns of (7.1.5) not all will have a non-zero coefficient. Only those containing r rows of functions φ and another r rows of functions φ' will survive. This is so because

$$\begin{aligned} \int_{(\text{out})\, y \leqslant x} \cdots \int (\varphi_{2i}(y)\varphi_{2j}(x) - \varphi_{2i}(x)\varphi_{2j}(y))\, dx\, dy &= 0, \\ \int_{(\text{out})\, y \leqslant x} \cdots \int (\varphi'_{2i}(y)\varphi'_{2j}(x) - \varphi'_{2i}(x)\varphi'_{2j}(y))\, dx\, dy &= 0. \end{aligned} \tag{7.4.2}$$

We also have

$$\int_{(\text{out})\, y \leqslant x} \cdots \int (\varphi_{2i}(y)\varphi'_{2j}(x) - \varphi_{2i}(x)\varphi'_{2j}(y))\, dx\, dy = 2\overline{g}_{ij}, \tag{7.4.3}$$

the $\overline{g}_{ij}$ given by Eq. (7.3.7). The result is

$$A_1(\theta; x_1, \ldots, x_{2r}) = (-2)^{-r} \sum_{(i,j)} \det\left(\begin{bmatrix} \varphi_{2i_k}(x_\ell) \\ \varphi'_{2j_k}(x_\ell) \end{bmatrix}_{1 \leqslant \ell \leqslant 2r}^{1 \leqslant k \leqslant r} \right) \cdot \bar{G}'(i_1, \ldots, i_r; j_1, \ldots, j_r), \tag{7.4.4}$$

where the summation is extended over all possible choices of the indices $i_1 < \cdots < i_r$, $j_1 < \cdots < j_r$ from $0, \ldots, m-1$. The $\overline{G}'(i_1, \ldots, i_r; j_1, \ldots, j_r)$ is apart from a sign the $(m-r) \times (m-r)$ determinant obtained from $\overline{G}$ by omitting the rows $i_1, \ldots, i_r$ and the columns $j_1, \ldots, j_r$. The sign is plus or minus according as $\sum_{k=1}^{r}(i_k + j_k)$ is even or odd. Thus $\bar{G}'(i_1, \ldots, i_r; j_1, \ldots, j_r)$ is equal to (cf. Mehta, 1989MT, Section 3.9), the determinant of $\overline{G}$ multiplied by an $r \times r$ determinant from the elements of the inverse of $\overline{G}$

$$\overline{G}'(i_1, \ldots, i_r; j_1, \ldots, j_r) = \det[\overline{G}] \cdot \det\left([(\overline{G}^{-1})_{ji}]_{i=i_1,\ldots,i_r}^{j=j_1,\ldots,j_r} \right). \tag{7.4.5}$$

Diagonalizing $\overline{G}$ as explained above in Section 7.3, and using Eqs. (7.3.9), (7.3.11) and (7.3.15)

$$A_1(\theta; x_1, \ldots, x_{2r}) = (-2)^{-r} \prod_{\rho=0}^{m-1} (1 - \lambda_{2\rho}) \cdot \sum_{(i)} (1 - \lambda_{2i_1})^{-1} \cdots (1 - \lambda_{2i_r})^{-1}$$

$$\times \det \left(\begin{bmatrix} \psi_{2i_k}(x_j) \\ \psi'_{2i_k}(x_j) \end{bmatrix}_{1 \leqslant j \leqslant 2r}^{1 \leqslant k \leqslant r} \right). \tag{7.4.6}$$

With

$$\alpha = \pi/(2\sqrt{m}), \qquad \theta = \alpha t, \qquad x_j = \alpha y_j, \tag{7.4.7}$$

we get in the limit $m \to \infty$,

$$B_1(t; y_1, \ldots, y_{2r}) = \lim \alpha^{2r} A(\theta; x_1, \ldots, x_{2r}) = (-2)^{-r} t^{-2r} \prod_{\rho} (1 - \lambda_{2\rho})$$

$$\times \sum_{(i)} \frac{\lambda_{2i_1}}{1 - \lambda_{2i_1}} \cdots \frac{\lambda_{2i_r}}{1 - \lambda_{2i_r}} \cdot \det \left(\begin{bmatrix} f_{2i_k}(y_j/t) \\ f'_{2i_k}(y_j/t) \end{bmatrix}_{1 \leqslant j \leqslant 2r}^{1 \leqslant k \leqslant r} \right). \tag{7.4.8}$$

The variables y_j satisfy of course the inequalities

$$-t \leqslant y_1 \leqslant \cdots \leqslant y_{2r} \leqslant t, \tag{7.4.9}$$

the λ_{2i} and the $f_{2i}(\xi)$ depend on t as a parameter and the indices $0 \leqslant i_1 < \cdots < i_r$ are non-negative integers.

To get $E_1(2r; s)$, Eq. (6.1.12), from (7.4.8) one may again use the method of integration over alternate variables (Chapter 5, Section 5.5 and Appendix A.17). From parity one sees that

$$\int_{-1 \leqslant y \leqslant x \leqslant 1} \cdots \int dy\, dx\, (f_{2i}(y) f_{2j}(x) - f_{2i}(x) f_{2j}(y)) = 0,$$

$$\int_{-1 \leqslant y \leqslant x \leqslant 1} \cdots \int dy\, dx\, (f'_{2i}(y) f'_{2j}(x) - f'_{2i}(x) f'_{2j}(y)) = 0, \tag{7.4.10}$$

and that

$$\int_{-1\leqslant y\leqslant x\leqslant 1}\cdots\int dy\,dx\,(f_{2i}(y)f'_{2j}(x)-f_{2i}(x)f'_{2j}(y))$$
$$=-2\left(\delta_{ij}-f_{2j}(1)\int_{-1}^{1}f_{2i}(y)\,dy\right). \tag{7.4.11}$$

Therefore, the method of integration over alternate variables gives us

$$E_1(2r;s)=\prod_{\rho}(1-\lambda_{2\rho})\cdot\sum_{i_1<\cdots<i_r}\frac{\lambda_{2i_1}}{1-\lambda_{2i_1}}\cdots\frac{\lambda_{2i_r}}{1-\lambda_{2i_r}}$$
$$\times\det\left[\delta_{ij}-f_{2i}(1)\int_{-1}^{1}f_{2j}(y)\,dy\right]_{i,j=i_1,i_2,\ldots,i_r}, \tag{7.4.12}$$

or (cf. Appendix A.18)

$$E_1(2r;s)=\prod_{\rho}(1-\lambda_{2\rho})\cdot\sum_{i_1<\cdots<i_r}\left\{\frac{\lambda_{2i_1}}{1-\lambda_{2i_1}}\cdots\frac{\lambda_{2i_r}}{1-\lambda_{2i_r}}\right.$$
$$\left.\times\left(1-\sum_{j=1}^{r}f_{2i_j}(1)\int_{-1}^{1}f_{2i_j}(y)\,dy\right)\right\}. \tag{7.4.13}$$

As we obtained (7.4.4) by integrating the expression (7.3.22) so we obtain from (7.4.8) the following expression for $p_1(2r-2;s)$

$$p_1(2r-2;s)=-\frac{2}{s^2}\prod_{\rho}(1-\lambda_{2\rho})\cdot\sum_{i_1<\cdots<i_r}\left(\frac{\lambda_{2i_1}}{1-\lambda_{2i_1}}\cdots\frac{\lambda_{2i_r}}{1-\lambda_{2i_r}}\sum_{j,k}a_{jk}b_{jk}\right), \tag{7.4.14}$$

where the summation over j,k in the above equation is over the indices $i_1,\ldots,i_r$, while

$$a_{jk}=\det\begin{bmatrix}f_{2j}(-1) & f_{2j}(1)\\ f'_{2k}(-1) & f'_{2k}(1)\end{bmatrix}=2f_{2j}(1)f'_{2k}(1), \tag{7.4.15}$$

and b_{jk} is the cofactor of the element (j,k) in

$$\det\left[\delta_{ij}-f_{2j}(1)\int_{-1}^{1}f_{2i}(y)\,dy\right]_{i,j=i_1,\ldots,i_r}. \tag{7.4.16}$$

That is to say (cf. Appendix A.18)

$$b_{jk} = f_{2j}(1) \int_{-1}^{1} f_{2k}(y)\, dy + \delta_{jk}\left(1 - \sum_{\ell} f_{2\ell}(1) \int_{-1}^{1} f_{2\ell}(y)\, dy\right), \tag{7.4.17}$$

where the index ℓ takes all the values $i_1, \ldots, i_r$.

A case of special interest is $r = 1$,

$$p_1(0; s) = -\frac{4}{s^2} \prod_{\rho} (1 - \lambda_{2\rho}) \cdot \sum_{i} \frac{\lambda_{2i}}{1 - \lambda_{2i}} f_{2i}(1) f_{2i}'(1), \tag{7.4.18}$$

which from (6.1.18) and (7.3.19) can also be written as

$$p_1(0; s) = \frac{d^2}{ds^2} \prod_{\rho} (1 - \lambda_{2\rho}). \tag{7.4.19}$$

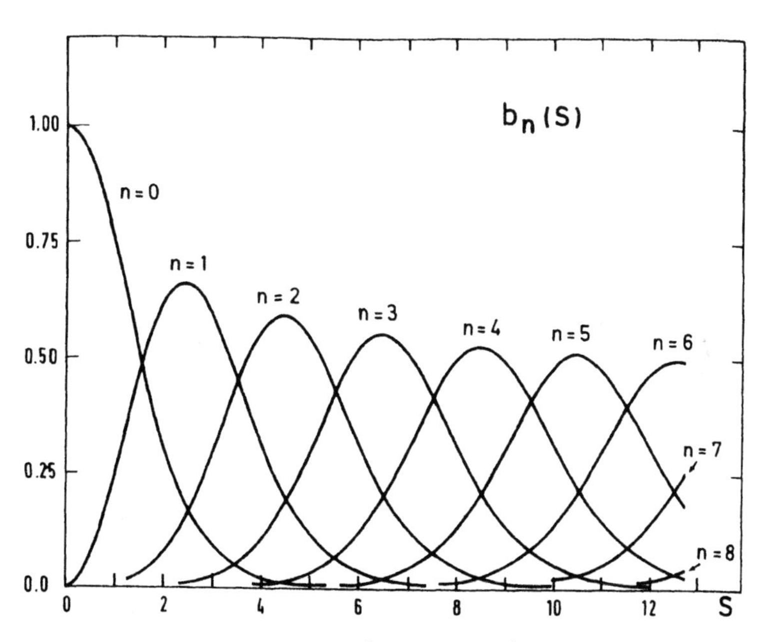

Figure 7.2. The functions $b_n(s) = f_{2n}(1) \int_{-1}^{1} f_{2n}(x)\, dx / \int_{-1}^{1} f_{2n}^2(x)\, dx$, where $f_n(x)$ is the prolate spheroidal function.

As we remarked in Section 6.5.3, for numerical computations Eq. (7.4.18) is better suited than Eq. (7.4.19). Still better will be to use the power series expansion for $s \leqslant 2$ and the asymptotic expansion for $s \geqslant 1$ (cf. Chapters 20 and 21).

Figure 7.2 shows the functions $b_j(s) = f_{2j}(1)\int_{-1}^{1} f_{2j}(x)\,dx$, for small values of j and s.

7.5 Several Consecutive Spacings: $n = 2r - 1$

The case $n = 2r - 1$ is more complicated for the following reason. If we order the variables as $-\infty < x_{2r} \leqslant x_{2r+1} \leqslant \cdots \leqslant x_{2m} < \infty$; $|x_j| \geqslant \theta$, $j = 2r, 2r+1, \ldots, 2m$; then the determinant in (7.3.22) changes sign each time a variable x_j, $2r \leqslant j \leqslant 2m$, passes over the excluded interval $(-\theta, \theta)$. To overcome this difficulty one divides the integral into two parts according to whether an odd or an even number of variables are greater than θ;

$$A_1(\theta; x_1, \ldots, x_{2r-1}) = \mathcal{O}(\theta; x_1, \ldots, x_{2r-1}) + \mathcal{E}(\theta; x_1, \ldots, x_{2r-1}), \tag{7.5.1}$$

with

$$\mathcal{O}(\theta; x_1, \ldots, x_{2r-1}) = \sum_{k=0}^{m-r} \int \cdots \int P_{N1}(x_1, \ldots, x_{2m})\, dx_{2r}\, dx_{2r+1} \cdots dx_{2m}, \tag{7.5.2}$$

where the integration limits are

$$\begin{aligned} &x_j < -\theta, \quad && 2r \leqslant j \leqslant 2r + 2k - 1, \\ &x_j > \theta, \quad && 2r + 2k \leqslant j \leqslant 2m; \end{aligned} \tag{7.5.3}$$

and $-\mathcal{E}(\theta; x_1, \ldots, x_{2r-1})$ is the same sum of integrals (7.5.2) but with the integration limits

$$\begin{aligned} &x_j < -\theta, \quad && 2r \leqslant j \leqslant 2r + 2k, \\ &x_j > \theta, \quad && 2r + 2k + 1 \leqslant j \leqslant 2m. \end{aligned} \tag{7.5.4}$$

If for some k, the upper limit of variation of j is less than its lower limit, the corresponding line in (7.5.3) or (7.5.4) is ignored.

In $\mathcal{O}$ we integrate over alternate variables $x_{2r+1}, x_{2r+3}, \ldots, x_{2m-1}$; introducing functions $u(x)$, $F_j(x)$, Eqs. (7.3.3), (7.3.4) and

$$\varepsilon(x) = \begin{cases} 1, & \text{if } x > \theta, \\ 0, & \text{if } x < \theta, \end{cases} \tag{7.5.5}$$

so that

$$\mathcal{O}(\theta; x_1, \ldots, x_{2r-1}) = (-2)^{-m} \int \cdots \int \prod_{j=r}^{m} u(x_{2j})\, dx_{2j}$$

$$\times \det \begin{bmatrix} 0 & \varepsilon(x_{2j}) & 0 \\ \varphi_{2i}(x_k) & F_{2i}(x_{2j}) & \varphi_{2i}(x_{2j}) \\ \varphi'_{2i}(x_k) & F_{2i+1}(x_{2j}) & \varphi'_{2i}(x_{2j}) \end{bmatrix}, \tag{7.5.6}$$

$$i = 0, 1, \ldots, m-1; \quad k = 1, 2, \ldots, 2r-1; \quad j = r, r+1, \ldots, m. \tag{7.5.7}$$

In (7.5.6) one has $-\theta \leqslant x_1 \leqslant x_2 \leqslant \cdots \leqslant x_{2r-1} \leqslant \theta$, and integrations are carried over the domain $-\infty < x_{2r} \leqslant x_{2r+1} \leqslant \cdots \leqslant x_{2m} < \infty$.

Now one can drop the ordering and integrate independently over the remaining variables. One expands the determinant according to its first $(2r-1)$ columns and uses the orthonormality of the functions $\varphi_i(x)$.

In $\mathcal{E}$ we integrate over the other alternate variables, $x_{2m}, x_{2m-2}, \ldots, x_{2r}$, introducing similar functions

$$F'_{2i}(x) = \int_x^\infty u(t)\varphi_{2i}(t)\, dt, \qquad F'_{2i+1}(x) = \int_x^\infty u(t)\varphi'_{2i}(t)\, dt, \tag{7.5.8}$$

and

$$\varepsilon'(x) = \begin{cases} 1, & \text{if } x < -\theta, \\ 0, & \text{if } x > -\theta; \end{cases} \tag{7.5.9}$$

change variables $x_1, \ldots, x_{2r-1}$ to their negatives and compare with $\mathcal{O}$;

$$\mathcal{E}(\theta; x_1, \ldots, x_{2r-1}) = \mathcal{O}(\theta; -x_{2r-1}, \ldots, -x_1). \tag{7.5.10}$$

For clarity of the exposition we take in greater detail the case $n = 1$. The general case is similar, though more cumbersome.

7.5.1 Case $n = 1$. Equation (7.5.6) gives now (see Appendix A.17),

$$\mathcal{O}(\theta; x) = (-2)^{-m} \int \cdots \int \prod_{j=1}^{m} u(x_{2j})\, dx_{2j} \cdot \det \begin{bmatrix} 0 & \varepsilon(x_{2j}) & 0 \\ \varphi_{2i}(x) & F_{2i}(x_{2j}) & \varphi_{2i}(x_{2j}) \\ \varphi'_{2i}(x) & F_{2i+1}(x_{2j}) & \varphi'_{2i}(x_{2j}) \end{bmatrix} \tag{7.5.11}$$

$$= (-2)^{-m} \left(\det \begin{bmatrix} 0 & \rho_{2j+1}(\theta) \\ \varphi_{2i}(x) & -2\overline{g}_{ij} \end{bmatrix} + \det \begin{bmatrix} 0 & -\rho_{2j}(\theta) \\ \varphi'_{2i}(x) & -2\overline{g}_{ij} \end{bmatrix} \right), \tag{7.5.12}$$

the range of integration in (7.5.11) being $x_2 \leqslant x_4 \leqslant \cdots \leqslant x_{2m}$,

$$\rho_{2j}(\theta) = \int_{-\infty}^{\infty} \varepsilon(x)u(x)\varphi_{2j}(x)\,dx = \int_{\theta}^{\infty} \varphi_{2j}(x)\,dx, \tag{7.5.13}$$

$$\rho_{2j+1}(\theta) = \int_{-\infty}^{\infty} \varepsilon(x)u(x)\varphi'_{2j}(x)\,dx = -\varphi_{2j}(\theta), \tag{7.5.14}$$

and $\overline{g}_{ij}$ given by (7.3.7). From (7.5.1), (7.5.10) and (7.5.12), one has

$$A_1(\theta; x) = -\det\begin{bmatrix} 0 & \varphi_{2j}(\theta) \\ \varphi_{2i}(x) & \overline{g}_{ij} \end{bmatrix}_{i,j=0,1,\ldots,m-1} \tag{7.5.15}$$

$$= \sum_{i,j=0}^{m-1} \varphi_{2i}(x)\varphi_{2j}(\theta)\overline{G}(i; j). \tag{7.5.16}$$

Following the diagonalization of $\overline{G}$ as in Section 7.3 above, we make the replacements

$$\varphi_{2i}(\theta) \to \psi_{2i}(t) \to (\alpha t)^{-1/2}\lambda_{2i}^{1/2} f_{2i}(1),$$

$$\varphi_{2i}(x) \to \psi_{2i}(y) \to (\alpha t)^{-1/2}\lambda_{2i}^{1/2} f_{2i}(y/t), \tag{7.5.17}$$

to get

$$\begin{aligned} B_1(t; y) &= \lim \alpha \cdot A_1(\theta; x) \\ &= -\frac{1}{t}\det\begin{bmatrix} 0 & \sqrt{\lambda_{2j}} f_{2j}(1) \\ \sqrt{\lambda_{2i}} f_{2i}(y/t) & (1-\lambda_{2i})\delta_{ij} \end{bmatrix} \\ &= \frac{1}{t}\prod_{\rho}(1-\lambda_{2\rho}) \cdot \sum_j \frac{\lambda_{2j}}{1-\lambda_{2j}} f_{2j}(1) f_{2j}(y/t). \end{aligned} \tag{7.5.18}$$

7.5.2 Case $n = 2r - 1$. Equation (7.5.18) is now replaced by

$$\begin{aligned} B_1(t; y_1, \ldots, y_{2r-1}) &= \lim \alpha^{2r-1} A_1(\theta; x_1, \ldots, x_{2r-1}) \\ &= (-2)^{1-r} t^{-2r+1} \prod_{\rho}(1-\lambda_{2\rho}) \\ &\quad \times \sum_{(i)} \frac{\lambda_{2i_1}}{1-\lambda_{2i_1}} \cdots \frac{\lambda_{2i_r}}{1-\lambda_{2i_r}} \sum_{(j)} f_{2j}(1) \cdot \det M_j(i_1, \ldots, i_r) \end{aligned} \tag{7.5.19}$$

where the $(2r-1)\times(2r-1)$ matrix $M_j(i_1,\dots,i_r)$ depends on the variables $y_1,\dots,y_{2r-1}$ as

$$M_j(i_1,\dots,i_r)$$
$$=\begin{bmatrix} f_{2j}\left(\frac{y_1}{t}\right) & f_{2i_1}\left(\frac{y_1}{t}\right) & f'_{2i_1}\left(\frac{y_1}{t}\right) & \cdots & f_{2i_{r-1}}\left(\frac{y_1}{t}\right) & f'_{2i_{r-1}}\left(\frac{y_1}{t}\right) \\ \cdots & \cdots & \cdots & \cdots & \cdots & \cdots \\ f_{2j}\left(\frac{y_{2r-1}}{t}\right) & f_{2i_1}\left(\frac{y_{2r-1}}{t}\right) & f'_{2i_1}\left(\frac{y_{2r-1}}{t}\right) & \cdots & f_{2i_{r-1}}\left(\frac{y_{2r-1}}{t}\right) & f'_{2i_{r-1}}\left(\frac{y_{2r-1}}{t}\right) \end{bmatrix}, \tag{7.5.20}$$

j is one of the indices $i_1,\dots,i_r$ and the summation in (7.5.19) is over all possible choices of non-negative integers $0\leqslant i_1<\cdots<i_r$ and of j among these integers. The variables y_j are supposed to be ordered $-t\leqslant y_1\leqslant\cdots\leqslant y_{2r-1}\leqslant t$.

Integration over alternate variables $y_1, y_3,\dots,y_{2r-1}$ and then independently over the rest from $-t$ to t, using the orthonormality of the functions f_j gives

$$E_1(2r-1;s)=\prod_\rho(1-\lambda_{2\rho})\cdot\sum_{(i)}\frac{\lambda_{2i_1}}{1-\lambda_{2i_1}}\cdots\frac{\lambda_{2i_r}}{1-\lambda_{2i_r}}\cdot\left(\sum_{j=1}^{r} f_{2i_j}(1)\int_{-1}^{1} f_{2i_j}(y)\,dy\right). \tag{7.5.21}$$

The expression for $p_1(2r-3;s)$ is obtained by putting $y_1=-t$, $y_{2r-1}=t$ and integrating over the other variables. We get

$$p_1(2r-3;s)=-\frac{2}{s^2}\prod_\rho(1-\lambda_{2\rho})\cdot\sum_{(i)}\frac{\lambda_{2i_1}}{1-\lambda_{2i_1}}\cdots\frac{\lambda_{2i_r}}{1-\lambda_{2i_r}}\cdot\sum_{(j)} f_{2j}(1)a_{j_1j_2j_3}b_{j_2j_3;j_1j}. \tag{7.5.22}$$

By definition

$$a_{j_1j_2j_3}=2f'_{2j_1}(1)\left(f_{2j_2}(1)\int_{-1}^{1} f_{2j_3}(t)\,dt-f_{2j_3}(1)\int_{-1}^{1} f_{2j_2}(t)\,dt\right), \tag{7.5.23}$$

and $b_{j_2j_3;j_1j}$ is, apart from a sign, the determinant obtained from (7.4.16) by omitting the rows corresponding to the values j_2, j_3 and the columns corresponding to the values j_1, j. The sign can be fixed, as always, by bringing this minor to the leading position in the upper left hand corner. The indices j_1, j_2, j_3 and j are chosen from $i_1,\dots,i_r$, and the summation in (7.5.22) above is over all such choices and then over all choices of the non-negative integers $0\leqslant i_1<i_2<\cdots<i_r$.

Figure 7.3 shows $E_1(n;s)$ for a few values of n. Figures 7.4 and 7.5 are the contour maps for two consecutive spacings. As in Figure 6.4, we notice that the set of eigenvalues of a matrix from the Gaussian orthogonal ensemble is more or less equally spaced,

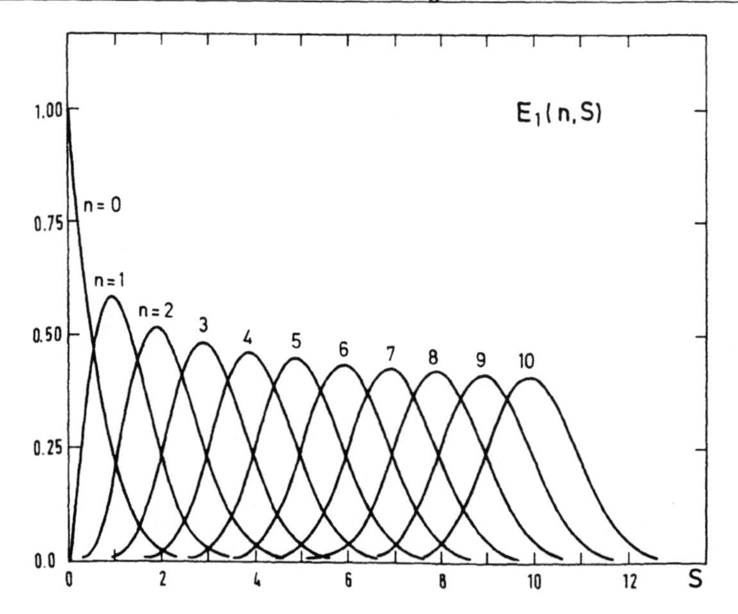

Figure 7.3. The n-level spacings $E_1(n, s)$ for the Gaussian orthogonal ensemble. Reprinted with permission from M.L. Mehta and J. des Cloizeaux, The probabilities for several consecutive eigenvalues of a random matrix, *Indian J. Pure Appl. Math.* 3 (1972) 329–351.

though not as good, since here the individual peaks are less high and a little wider than for the Gaussian unitary case.

According to Eqs. (6.4.36)–(6.4.37) we will introduce

$$F_+(z, t) := \prod_{j=0}^{\infty}(1 - z\lambda_{2j}), \tag{7.5.24}$$

$$F_-(z, t) := \prod_{j=0}^{\infty}(1 - z\lambda_{2j+1}), \tag{7.5.25}$$

and

$$E_\pm(r, 2t) = \frac{1}{r!}\left(-\frac{\partial}{\partial z}\right)^r F_\pm(z, t)|_{z=1}. \tag{7.5.26}$$

Then from Eqs. (7.4.13) and (7.5.21) one sees that

$$E_+(0, s) = E_1(0, s), \tag{7.5.27a}$$

$$E_+(r, s) = E_1(2r - 1, s) + E_1(2r, s), \quad r \geqslant 1. \tag{7.5.27b}$$

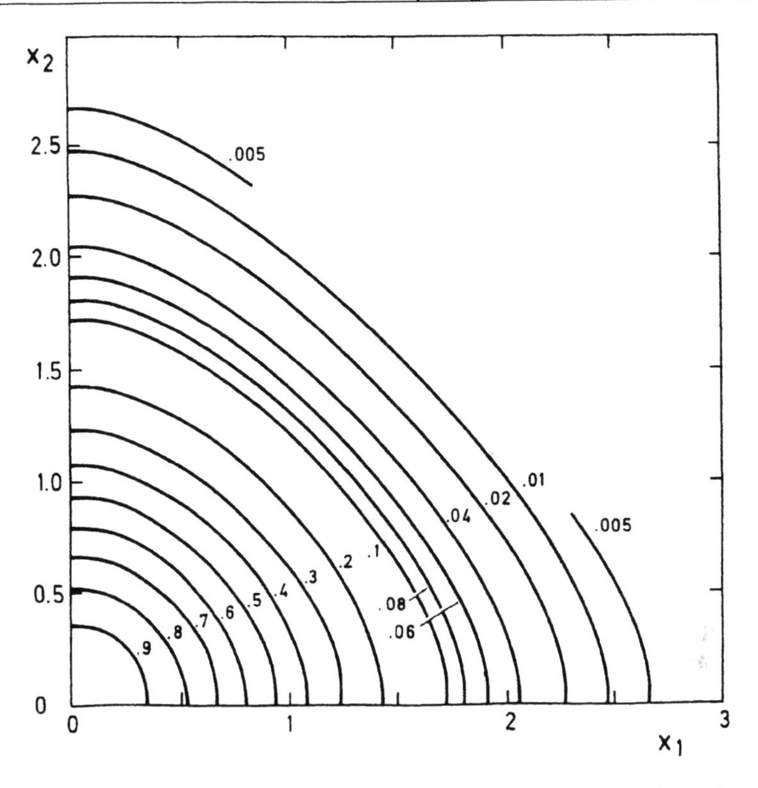

Figure 7.4. The contour map of $\mathcal{B}_1(x, y)$, the same as Figure 6.7 but this time the random matrix is chosen from the Gaussian orthogonal ensemble.

Moreover from Eq. (20.2.14) or (21.1.17) we get

$$\prod_{j=0}^{\infty}(1 - z\lambda_{2j+1}) = \prod_{j=0}^{\infty}(1 - z\lambda_{2j})\left\{1 + \sum_{i=0}^{\infty}\frac{z\lambda_{2i}}{1 - z\lambda_{2i}} f_{2i}(1)\int_{-1}^{1} f_{2i}(y)\,dy\right\}. \tag{7.5.28}$$

If we differentiate Eq. (7.5.28) r times with respect to z and then set $z = 1$, we get the remarkable set of identities

$$E_{-}(r, s) = E_1(2r, s) + E_1(2r + 1, s), \quad r \geqslant 0. \tag{7.5.29}$$

More about this in Chapter 20.

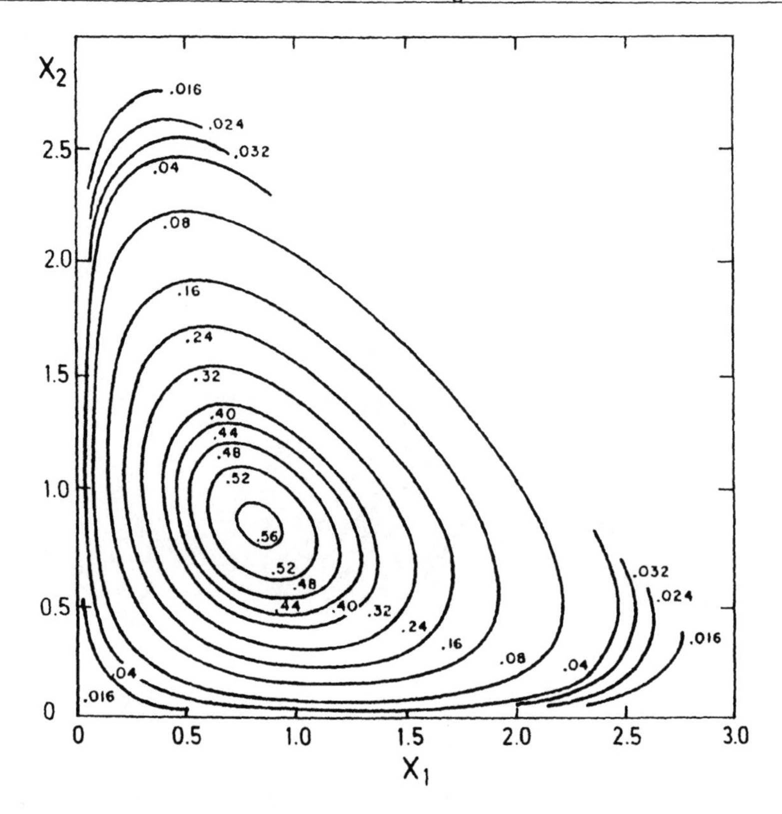

Figure 7.5. The contour map of $\mathcal{P}_1(x, y)$, the same as Figure 6.8 but this time the random matrix is chosen from the Gaussian orthogonal ensemble.

7.6 Bounds for the Distribution Function of the Spacings

Equations (7.3.9) and (7.3.7) can also be written as

$$A_1(\theta) = \det\left[2\int_\theta^\infty \varphi_{2i}(x)\varphi_{2j}(x)\,dx\right]_{i,j=0,1,\ldots,m-1}. \tag{7.6.1}$$

We then apply Gram's result (cf. Appendix A.12) to write

$$\det\left[\int_\theta^\infty \varphi_{2i}(x)\varphi_{2j}(x)\,dx\right] = \frac{1}{m!}\int_\theta^\infty \cdots \int_\theta^\infty (\det[\varphi_{2i-2}(x_j)]_{i,j=1,\ldots,m})^2\,dx_1 \cdots dx_m. \tag{7.6.2}$$

In Section 6.2 we wrote $\exp(-\frac{1}{2}\sum_i x_i^2)\prod_{i<j}(x_i - x_j)$ as a determinant containing oscillator functions $\varphi_j(x)$. Following step by step the same procedure but in the reverse

direction we can convert the integrand of (7.6.2) back to the form

$$\exp\left(-\sum_{i=1}^{m} x_i^2\right) \prod_{1\leqslant i<j\leqslant m} (x_i^2 - x_j^2)^2, \tag{7.6.3}$$

except for some multiplicative factors. Since Eq. (7.6.2) contains only even functions $\varphi_j(x)$, we have in Eq. (7.6.3) the product of differences of the x_j^2 instead of those of the x_j. Thus

$$A_1(\theta) = \text{const} \cdot \int_\theta^\infty \cdots \int_\theta^\infty dx_1 \cdots dx_m \exp\left(-\sum_{i=1}^{m} x_i^2\right) \prod_{1\leqslant i<j\leqslant m} (x_i^2 - x_j^2)^2. \tag{7.6.4}$$

Differentiation with respect to θ gives

$$\frac{dA_1(\theta)}{d\theta} = \text{const} \cdot e^{-\theta^2} \int_\theta^\infty \cdots \int_\theta^\infty \exp\left(-\sum_{i=1}^{m-1} x_i^2\right)$$

$$\times \prod_{1\leqslant i<j\leqslant m-1} (x_i^2 - x_j^2)^2 \prod_{i=1}^{m-1} (x_i^2 - \theta^2)^2 \, dx_i. \tag{7.6.5}$$

Introducing new variables y_i defined by

$$x_i^2 = y_i^2 + \theta^2, \tag{7.6.6}$$

one may write

$$\frac{dA_1(\theta)}{d\theta} = -I_m(\theta) \exp(-m\theta^2), \tag{7.6.7}$$

with

$$I_m(\theta) = \text{const} \cdot \int_0^\infty \cdots \int_0^\infty \exp\left(-\sum_{i=1}^{m-1} y_i^2\right)$$

$$\times \prod_{1\leqslant i<j\leqslant m-1} (y_i^2 - y_j^2)^2 \cdot \prod_{i=1}^{m-1} y_i^5 (y_i^2 + \theta^2)^{-1/2} \, dy_i. \tag{7.6.8}$$

Applying Gram's result once again (cf. Appendix A.12) we write $I_m(\theta)$ as a determinant

$$I_m(\theta) = \text{const} \cdot \det[\eta_{i+j}]_{i,j=1,\ldots,m-1}, \tag{7.6.9}$$

where

$$\eta_j(\theta) = 2\int_0^\infty e^{-y^2} y^{2j+1}(y^2+\theta^2)^{-1/2}\,dy. \tag{7.6.10}$$

We may expand $\eta_j(\theta)$ and hence $I_m(\theta)$ in a power series in θ (cf. Appendix A.19),

$$\eta_j(\theta) = \Gamma\left(j+\frac{1}{2}\right) - \frac{1}{2}\theta^2\Gamma\left(j-\frac{1}{2}\right) + \frac{3}{8}\theta^2\Gamma\left(j-\frac{3}{2}\right) - \cdots, \tag{7.6.11}$$

$$I_m(\theta) = I_m(0)\left(1 - \frac{1}{3}(m-1)\theta^2 + \frac{1}{30}(m-1)(7m+1)\theta^4 + \cdots\right). \tag{7.6.12}$$

Finally we may take the limit $m \to \infty$, $2\theta\sqrt{m} = \pi t$ finite,

$$\frac{dE_1(0;2t)}{2dt} = \lim \frac{\sqrt{m}}{\pi}\frac{dA_1(\theta)}{d\theta} = -I(t)\exp\left(-(\pi t/2)^2\right). \tag{7.6.13}$$

As $dE_1(0;s)/ds = -1$, at $s = 0$ (cf. Section 6.1.2), we get from (7.6.12) the power series expansion of $I(t)$,

$$I(t) = \lim \frac{\sqrt{m}}{\pi} I_m(\theta) = 1 - \frac{1}{3}\left(\frac{\pi t}{2}\right)^2 + \frac{7}{30}\left(\frac{\pi t}{2}\right)^4 + \cdots. \tag{7.6.14}$$

The form (7.6.9) expressing $I_m(\theta)$ as a determinant is convenient for calculations, as in arriving at (7.6.12) and (7.6.14), whereas the integral form (7.6.8) is useful to find bounds for $I(t)$.

One can prove (cf. Appendix A.20) that for all positive values of y_i

$$1 - \frac{1}{2}\sum_{i=1}^m \frac{\theta^2}{y_i^2} \leqslant \prod_{i=1}^m \left(y_i(y_i^2+\theta^2)^{-1/2}\right) \leqslant 1. \tag{7.6.15}$$

The expansion (7.6.12) in the limit $m \to \infty$ gives then the inequalities

$$1 - \frac{1}{3}\left(\frac{\pi t}{2}\right)^2 \leqslant I(t) \leqslant 1. \tag{7.6.16}$$

Hence we get rigorous lower and upper bounds for the distribution function $\Psi(s)$ ($s = 2t$) of the spacings (cf. Section 6.1.3)

$$\Psi_L(s) \leqslant \Psi(s) \leqslant \Psi_U(s), \tag{7.6.17}$$

where

$$\Psi_L(s) = 1 - \exp\left(-\frac{\pi^2 s^2}{16}\right), \tag{7.6.18}$$

and

$$\Psi_U(s) = 1 - \left(1 - \frac{1}{3}\frac{\pi^2 s^2}{16}\right)\exp\left(-\frac{\pi^2 s^2}{16}\right). \tag{7.6.19}$$

Because the differences $\Psi - \Psi_L$ and $\Psi_U - \Psi$ are every where non negative, the difference between unity and the approximate mean values $\langle s\rangle_L$ and $\langle s\rangle_U$ obtained by substituting Ψ_L and Ψ_U for Ψ in

$$\langle s\rangle = \int_0^\infty s p_1(0; s)\, ds = \int_0^\infty (1 - \Psi(s))\, ds \tag{7.6.20}$$

(cf. Eq. (6.1.26)) provides a good estimation of the accuracy of the corresponding approximations to Ψ. One has

$$\langle s\rangle_L - 1 = \frac{2}{\sqrt{\pi}} - 1 \cong 0.1284, \qquad 1 - \langle s\rangle_U = 1 - \frac{5}{3\sqrt{\pi}} \cong 0.0597. \tag{7.6.21}$$

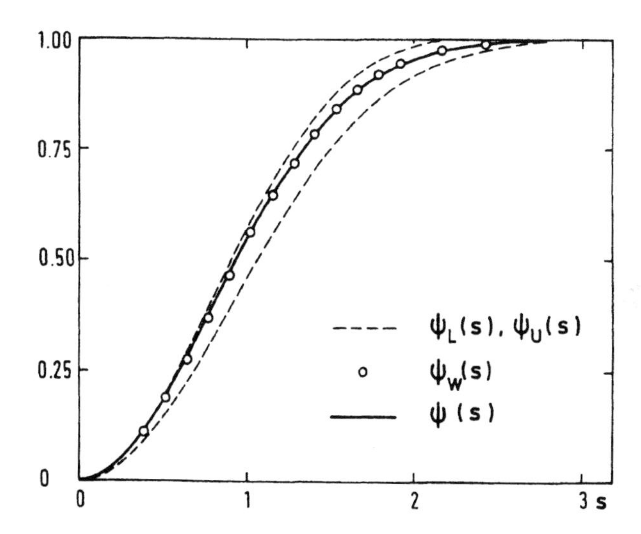

Figure 7.6. The distribution function of the spacings $\Psi(s)$, the lower and upper bounds $\Psi_L(s)$, $\Psi_U(s)$ and the Wigner surmise $\Psi_W(s)$. Reprinted with permission from Elsevier Science Publishers, Gaudin M., Sur la loi limite de l'espacement des valeurs propres d'une matrice aléatoire, *Nucl. Phys.* 25, 447–458 (1961).

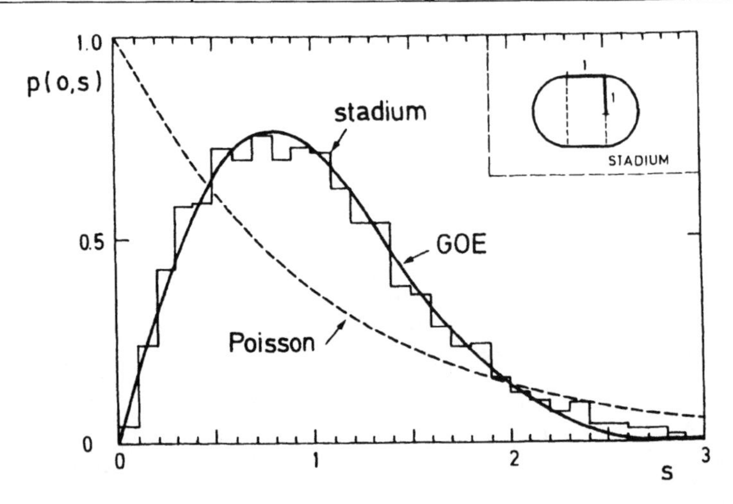

Figure 7.7. Empirical probability density of the nearest neighbor spacings of the possible energies of a particle free to move on the stadium consisting of a rectangle of size 1×2 with semi-circular caps of radius 1, depicted in the right upper corner. The stadium can be defined by the inequalities $|y| \leqslant 1$, and either $|x| \leqslant 1/2$ or $(x \pm 1/2)^2 + y^2 \leqslant 1$. The solid curve represents Eq. (7.3.19) corresponding to the Gaussian orthogonal ensemble (GOE), while the dashed curve is for the Poisson process corresponding to no correlations. Supplied by O. Bohigas, from Bohigas et al. (1984a).

For visual comparison, Figure 7.6 is a plot of the functions Ψ_L, Ψ, Ψ_U and the Wigner surmise

$$\Psi_W(s) = 1 - \exp\left(-\frac{\pi s^2}{4}\right). \tag{7.6.22}$$

It is a surprise that Wigner surmise is so close to the real distribution.

Figures 1.3, 1.4, 1.6–1.8 in Chapter 1, represent histograms of the nearest neighbor spacings of the nuclear and atomic levels and Figures 7.7 and 7.8 represent those for chaotic systems. Finally, Figure 7.9 corresponds to the ultrasonic resonance frequencies of an aluminium block.

Summary of Chapter 7

For the Gaussian orthogonal ensemble with the joint probability density of the eigenvalues

$$P_{N1}(x_1, \ldots, x_N) \propto \exp\left(-\frac{1}{2}\sum_{j=1}^{N} x_j^2\right) \prod_{1 \leqslant j < k \leqslant N} |x_j - x_k|, \tag{7.0.1}$$

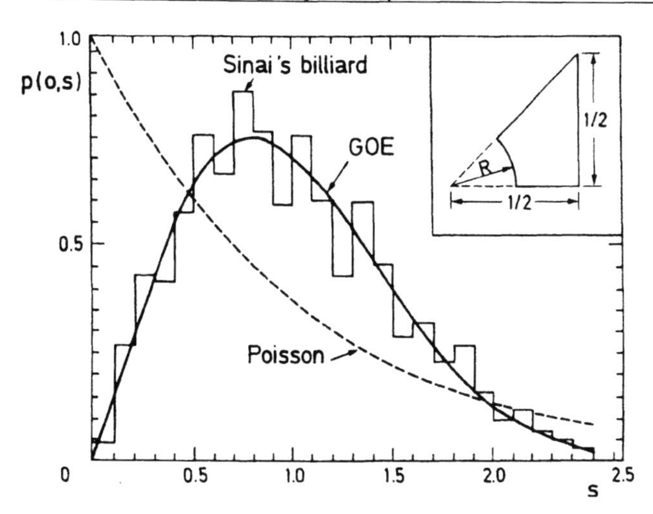

Figure 7.8. Same as Figure 7.7 but when the particle moves on Sinai's billiard table consisting of $1/8$ of a square cut by a circular arc, depicted in the right upper corner. One may define it by the inequalities $y \geqslant 0$, $x \geqslant y$, $x \leqslant 1$ and $x^2 + y^2 \geqslant r$. Only 1/8th of the square is taken so that all obvious symmetries of the square are disposed of. Supplied by O. Bohigas, from Bohigas et al. (1984a).

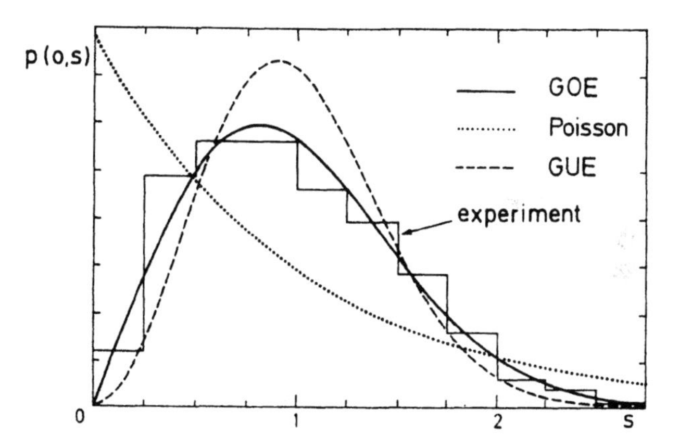

Figure 7.9. Empirical probability density of the nearest neighbor spacings for the ultrasonic frequencies of an aluminium block. The three curves correspond respectively to the Poisson process with no correlations, to the Gaussian orthogonal ensemble (GOE) and to the Gaussian unitary ensemble (GUE). Reprinted with permission from American Institute of Physics, Weaver R.L., Spectral statistics in elastodynamics, *J. Acoust. Soc. Amer.* 85 (1989) 1005–1013.

the asymptotic two level cluster function is

$$Y_2(r) = \left(\frac{\sin(\pi r)}{\pi r}\right)^2 + \left(\int_r^\infty \left(\frac{\sin(\pi t)}{\pi t}\right) dt\right)\left(\frac{d}{dr}\left(\frac{\sin(\pi r)}{\pi r}\right)\right). \tag{7.2.43}$$

The n-level cluster function is a little complicated, see Eq. (7.2.49).

The Fourier transform of $Y_2(r)$ is

$$b(k) = \begin{cases} 1 - 2|k| + |k| \ln(1 + 2|k|), & |k| \leqslant 1, \\ -1 + |k| \ln\left(\dfrac{2|k|+1}{2|k|-1}\right), & |k| \geqslant 1. \end{cases} \tag{7.2.46}$$

The probability that a randomly chosen interval of length s contains exactly n levels is given by the formulas

$$E_1(0; s) = \prod_{i=0}^{\infty} (1 - \lambda_{2i}), \tag{7.3.19}$$

and for $r > 0$,

$$E_1(2r; s) = E_1(0; s) \cdot \sum_{0 \leqslant j_1 < j_2 < \cdots < j_r} \prod_{i=1}^{r} \left(\frac{\lambda_{j_i}}{1 - \lambda_{j_i}}\right)[1 - (b_{j_1} + \cdots + b_{j_r})], \tag{7.4.13}$$

$$E_1(2r - 1; s) = E_1(0; s) \cdot \sum_{0 \leqslant j_1 < j_2 < \cdots < j_r} \prod_{i=1}^{r} \left(\frac{\lambda_{j_i}}{1 - \lambda_{j_i}}\right)(b_{j_1} + \cdots + b_{j_r}), \tag{7.5.21}$$

where

$$b_j = f_{2j}(1) \int_{-1}^{1} f_{2j}(x)\, dx \div \int_{-1}^{1} f_{2j}^2(x)\, dx,$$

$\lambda_j = s|\mu_j|^2/4$, and μ_j and $f_j(x)$ are the eigenvalues and the eigenfunctions of the integral equation

$$\mu f(x) = \int_{-1}^{1} \exp(i\pi xys/2) f(y)\, dy \tag{6.3.17}$$

or μ_{2j} and $f_{2j}(x)$ are the eigenvalues and the eigenfunctions of the integral equation

$$\mu f(x) = 2\int_0^1 \cos(\pi xys/2) f(y)\, dy. \tag{6.3.20}$$

8

GAUSSIAN SYMPLECTIC ENSEMBLE

In this chapter we will consider briefly the Gaussian symplectic ensemble. The joint probability density function of the eigenvalues, Eq. (3.2.14), or (5.0.1) with $\beta = 4$, will be expressed as a determinant of an $N \times N$ quaternion self-dual matrix satisfying the requirements of Theorem 5.1.4. The correlation and cluster functions will then follow from that theorem. In particular, the two-level cluster function is given by Eq. (8.2.6) and its Fourier transform by Eq. (8.2.9). The question of level spacings is more complicated and will not receive any detailed attention here. In fact, in Chapter 11 we will show the equivalence of the Gaussian and circular ensembles and a profound relation between orthogonal and symplectic ensembles. This will allow us to deduce the level spacings for the symplectic ensemble from that of the orthogonal ensemble.

8.1 A Quaternion Determinant

For simplicity of writing we will take the weight function $w(x)$ to be $\exp(-x^2)$ instead of $\exp(-2x^2)$. This is just a scale change and has no effect on our reasoning or the final result if one measures distances in terms of the local mean spacing.

For the weight function $w(x) = \exp(-x^2)$ over $(-\infty, \infty)$ the monic skew-orthogonal polynomials of quaternion type are

$$Q_{2k}(x) = \sum_{j=0}^{k} \frac{k!}{j!} 2^{-2j} H_{2j}(x), \tag{8.1.1}$$

$$Q_{2k+1}(x) = 2^{-2k-1} H_{2k+1}(x), \tag{8.1.2}$$

$$q_k = 2^{-2k}(2k+1)!\sqrt{\pi}, \tag{8.1.3}$$

where $H_k(x)$ are the Hermite polynomials

$$\begin{aligned} H_k(x) &= e^{x^2}\left(-\frac{d}{dx}\right)^k e^{-x^2} \\ &= k! \sum_{j=0}^{[k/2]} (-1)^j \frac{(2x)^{k-2j}}{j!(k-2j)!}. \end{aligned} \tag{8.1.4}$$

Introducing the oscillator wave functions

$$\varphi_k(x) = (2^k k! \sqrt{\pi})^{-1/2} H_k(x) e^{-x^2/2}, \tag{8.1.5}$$

we can write

$$\frac{1}{\sqrt{q_k}} Q_{2k+1}(x) e^{-x^2/2} = \frac{1}{\sqrt{2}} \varphi_{2k+1}(x). \tag{8.1.6}$$

Also using the recurrence relation for Hermite polynomials

$$\begin{aligned} H_{k+1}(x) &= 2x H_k(x) - H_k'(x) \\ &= 2x H_k(x) - 2k H_{k-1}(x), \end{aligned} \tag{8.1.7}$$

an integration by parts gives

$$\begin{aligned} \phi_{2k+1}(x) &:= \int_{-\infty}^{x} \varphi_{2k+1}(t)\, dt \\ &= -\sqrt{\frac{2}{2k+1}} \varphi_{2k}(x) + \sqrt{\frac{2k}{2k+1}} \phi_{2k-1}(x). \end{aligned} \tag{8.1.8}$$

Iterating the last equation and using Eqs. (8.1.1), (8.1.3) and (8.1.5) one can write

$$\frac{1}{\sqrt{q_k}} Q_{2k}(x) e^{-x^2/2} = -\frac{1}{\sqrt{2}} \int_{-\infty}^{x} \varphi_{2k+1}(t)\, dt. \tag{8.1.9}$$

Hence from Chapter 5, Eqs. (5.7.15), (5.7.16), (5.7.20)–(5.7.22) we have

$$K_4(x, y) = \begin{bmatrix} S(x, y) & D(x, y) \\ I(x, y) & S(y, x) \end{bmatrix}, \tag{8.1.10}$$

with

$$\sqrt{2}S(x,y) = e^{-(x^2+y^2)/2}\sum_{k=0}^{N-1}\frac{1}{q_k}[Q'_{2k+1}(x)Q_{2k}(y) - Q'_{2k}(x)Q_{2k+1}(y)] \quad (8.1.11)$$

$$= S_{2N+1}(x,y) + xI_{2N+1}(x,y), \quad (8.1.12)$$

$$\sqrt{2}D(x,y) = e^{-(x^2+y^2)/2}\sum_{k=0}^{N-1}\frac{1}{q_k}[-Q'_{2k+1}(x)Q'_{2k}(y) + Q'_{2k}(x)Q'_{2k+1}(y)] \quad (8.1.13)$$

$$= -\sqrt{2}D(y,x)$$

$$= D_{2N+1}(x,y) + xS_{2N+1}(y,x) - yS_{2N+1}(x,y) + xyI_{2N+1}(x,y), \quad (8.1.14)$$

$$\sqrt{2}I(x,y) = e^{-(x^2+y^2)/2}\sum_{k=0}^{N-1}\frac{1}{q_k}[Q_{2k+1}(x)Q_{2k}(y) - Q_{2k}(x)Q_{2k+1}(y)]$$

$$= I_{2N+1}(x,y)$$

$$= -\sqrt{2}I(y,x). \quad (8.1.15)$$

Here we have used the quantities $S_{2N+1}(x,y)$, $D_{2N+1}(x,y)$ and $I_{2N+1}(x,y)$ encountered in Chapter 7, Eqs. (7.2.13)–(7.2.15). The kernel $K_4(x,y)$ of Eq. (8.1.10) is by construction self-dual and satisfies all the conditions of Theorem 5.1.4. However, we will modify it a little to have a more familiar form

$$K_{N4}(x,y) = \frac{1}{\sqrt{2}}\begin{bmatrix} S_{2N+1}(x,y) & D_{2N+1}(x,y) \\ I_{2N+1}(x,y) & S_{2N+1}(y,x) \end{bmatrix}. \quad (8.1.16)$$

The change from (8.1.10) to (8.1.16) can be achieved by linear operations on rows and columns, not affecting the self-duality, the determinant or other conditions of Theorem 5.1.4, in particular the relation $K * K = K$ as can be verified directly.

That the two kernels (8.1.10) and (8.1.16) are different can be ascertained, for example, by writing them explicitly for the case $N = 2$.

8.2 Correlation and Cluster Functions

From Eq. (5.8.16) and Theorem 5.1.4 integrating over all the variables one can fix the constants,

$$P_{N4}(x_1,\ldots,x_N) = \frac{1}{N!}\det[K_{N4}(x_i,x_j)]_{i,j=1,\ldots,N}. \quad (8.2.1)$$

By the same Theorem 5.1.4 the n-level correlation function is

$$\begin{aligned} R_n(x_1,\ldots,x_n) &= \frac{N!}{(N-n)!}\int\cdots\int P_{N4}(x_1,\ldots,x_N)\,dx_{n+1}\cdots dx_N \\ &= \det[K_{N4}(x_i,x_j)]_{i,j=1,\ldots,n}. \end{aligned} \tag{8.2.2}$$

And as in Section 7.2, the n-level cluster function is

$$T_n(x_1,\ldots,x_n) = \sum_P K_{N4}(x_1,x_2)K_{N4}(x_2,x_3)\cdots K_{N4}(x_n,x_1), \tag{8.2.3}$$

the sum being taken over all $(n-1)!$ distinct cyclic permutations of the indices $(1,2,\ldots,n)$.

Setting $n=1$, we get the level density which in the limit $N\to\infty$ is given again by Eq. (6.2.11) or (7.2.33). Setting $n=2$, we get the two-level cluster function

$$\begin{aligned} T_2(x,y) &= K_{N4}(x,y)K_{N4}(y,x) \\ &= \frac{1}{2}[S_{2N+1}(x,y)S_{2N+1}(y,x) - D_{2N+1}(x,y)\cdot I_{2N+1}(x,y)]. \end{aligned} \tag{8.2.4}$$

In the limit $N\to\infty$,

$$\begin{aligned} Y_2(r) &= \lim \frac{\pi}{2N}T_2(x,y) = K_4(r)K_4(-r), \\ K_4(r) &= \begin{bmatrix} s(2r) & D(2r) \\ I(2r) & s(2r) \end{bmatrix}, \end{aligned} \tag{8.2.5}$$

with the functions s, D and I given by Eqs. (7.2.40), (7.2.37) and (7.2.38). Or

$$Y_2(r) = \left(\frac{\sin(2\pi r)}{2\pi r}\right)^2 - \frac{d}{dr}\left(\frac{\sin(2\pi r)}{2\pi r}\right)\cdot\int_0^r\left(\frac{\sin(2\pi t)}{2\pi t}\right)dt. \tag{8.2.6}$$

The behaviour of $Y_2(r)$ for small and large r is given by

$$Y_2(r) = 1 - \frac{(2\pi r)^4}{135} + \cdots, \tag{8.2.7}$$

and

$$Y_2(r) = -\frac{\pi}{2}\frac{\cos(2\pi r)}{2\pi r} + \frac{1+(\pi/2)\sin(2\pi r)}{(2\pi r)^2} - \frac{1+\cos^2(2\pi r)}{(2\pi r)^4} + \cdots, \tag{8.2.8}$$

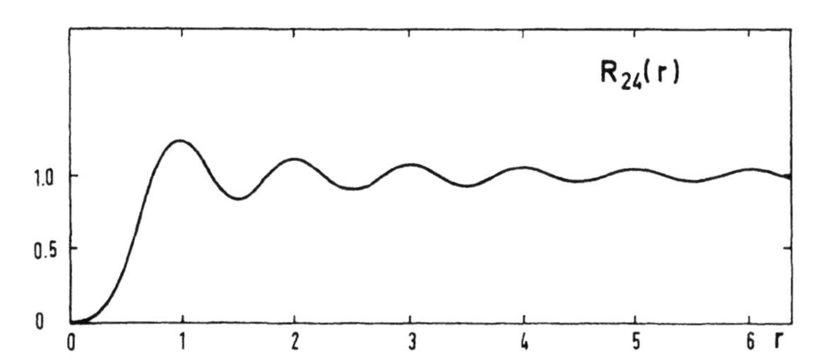

Figure 8.1. Two-level correlation function for the symplectic ensemble.

respectively. The two-level form factor is (cf. Appendix A.11)

$$b(k) = \int_{-\infty}^{\infty} Y_2(r) \exp(2\pi ikr)\, dr$$

$$= \begin{cases} 1 - \frac{1}{2}|k| + \frac{1}{4}|k| \ln(1 - |k|), & |k| < 2, \\ 0, & |k| > 2. \end{cases} \tag{8.2.9}$$

In the limit $N \to \infty$, the n-level cluster function is given by Eq. (6.2.16) or (7.2.49), where $s(r)$ or $K_1(r)$ is now replaced by $K_4(r)$ of Eq. (8.2.5). The n-level form factor is given by Eq. (6.2.18) or (7.2.50), where $f_2(k)$ or $f_1(k)$ is now replaced by

$$f_4(k) = \frac{1}{2} f_2(k) \begin{bmatrix} 1 & k \\ k^{-1} & 1 \end{bmatrix}, \tag{8.2.10}$$

with $f_2(k)$ given by Eq. (6.2.19)

$$f_2(k) = \begin{cases} 1, & |k| < 1/2, \\ 0, & |k| > 1/2. \end{cases} \tag{8.2.11}$$

Figure 8.1 shows the limiting two-level correlation function. Note that the oscillations persist much longer than for the Gaussian unitary ensemble, Figure 6.2.

8.3 Level Spacings

The joint probability density function for the eigenvalues of a matrix taken from the Gaussian symplectic ensemble can be written by Eqs. (8.2.1), (8.1.5) and (8.1.9) as

$$P_{N4}(x_1, \ldots, x_N) = \text{const} \cdots \det[\varphi_i(x_j) \quad \varphi_i'(x_j)],$$

$$i = 0, 1, \ldots, 2N-1; \quad j = 1, 2, \ldots, N. \tag{8.3.1}$$

Note that each variable occurs in two columns, we do not need to order the variables and one can integrate over any number of them by using Pfaffians, cf. Appendix A.17. Moreover, since our integration limits are $(-\infty, -\theta)$ and (θ, ∞), the excluded interval $(-\theta, \theta)$ situated symmetrically about the origin, we have

$$\int_{\text{out}} \varphi_{2i}(x)\varphi'_{2j}(x)\,dx = \int_{\text{out}} \varphi_{2i+1}(x)\varphi'_{2j+1}(x)\,dx = 0. \tag{8.3.2}$$

Therefore the Pfaffians reduce to determinants. For example, the probability that the interval $(-\theta, \theta)$ contains none of the eigenvalues is

$$\begin{aligned} A_4(\theta) &= \int \cdots \int_{\text{out}} P_{N4}(x_1, \dots, x_N)\,dx_1 \cdots dx_N \\ &= \text{const} \cdot \det[B_{ij}]_{i,j=0,1,\dots,N-1}, \end{aligned} \tag{8.3.3}$$

with

$$B_{ij} = \int_{\text{out}} (\varphi_{2i}(x)\varphi'_{2j+1}(x) - \varphi'_{2i}(x)\varphi_{2j+1}(x))\,dx, \tag{8.3.4}$$

where the subscript "out" means, as usual, that we integrate outside the interval $(-\theta, \theta)$.

Similarly, the probability that the interval $(-\theta, \theta)$ contains exactly n eigenvalues can be written as

$$\text{const} \cdot \sum_{(i;j)} B(i;j) \cdot \det \begin{bmatrix} \varphi_{2i_p}(x_k) & \varphi'_{2i_p}(x_k) \\ \varphi_{2j_p+1}(x_k) & \varphi'_{2j_p+1}(x_k) \end{bmatrix}_{p,k=1,\dots,n}, \tag{8.3.5}$$

where the sum is taken over all choices of the indices satisfying $0 \leqslant i_1 < \cdots < i_n \leqslant N-1$; $0 \leqslant j_1 < \cdots < j_n \leqslant N-1$; and $B(i;j)$ is the $(N-n) \times (N-n)$ minor of B obtained by removing the rows $i_1, \dots, i_n$ and the columns $j_1, \dots, j_n$. The x_k lie in the interval $(-\theta, \theta)$, $|x_k| \leqslant \theta$.

However, taking the limit $N \to \infty$ is difficult, since the element B_{ij} is a sum of two terms of different nature, one involving the integration of the product of two even functions φ, and the other of two odd functions φ.

Another expression for $E_4(n;s)$ is given, for example, by Eq. (A.7.21), Appendix A.7, where λ_i are the eigenvalues of the integral equation (A.7.23), with $\phi(x,y)$ replaced now by $K_4(x-y)$, K_4 given by Eq. (8.2.5).

Instead of either of these we will deduce these level spacing probabilities from the equivalence of Gaussian and circular ensembles and a profound relation between orthogonal and symplectic ensembles, proved in Chapter 11 (cf. Sections 11.6 and 11.7).

Summary of Chapter 8

For the Gaussian symplectic ensemble with the joint probability density of the eigenvalues

$$P_{N4}(x_1, \ldots, x_N) \propto \exp\left(-\sum_{j=1}^{N} x_j^2\right) \prod_{1 \leqslant j < k \leqslant N} (x_j - x_k)^4,$$

the asymptotic two-level cluster function is

$$Y_2(r) = \left(\frac{\sin(2\pi r)}{2\pi r}\right)^2 - \frac{d}{dr}\left(\frac{\sin(2\pi r)}{2\pi r}\right) \cdot \int_0^r \left(\frac{\sin(2\pi t)}{2\pi t}\right) dt, \tag{8.2.6}$$

with the Fourier transform

$$b(k) = \begin{cases} 1 - \frac{1}{2}|k| + \frac{1}{4}|k| \ln|1 - |k||, & |k| \leqslant 2, \\ 0, & |k| \geqslant 2. \end{cases} \tag{8.2.9}$$

The probability $E_4(n; s)$ that a randomly chosen interval of length s contains exactly n levels will be given in Chapter 11.

9

GAUSSIAN ENSEMBLES: BROWNIAN MOTION MODEL

9.1 Stationary Ensembles

In Chapter 4 we exploited the idea that the probability $P(x_1, \ldots, x_N)$, Eq. (3.3.8), for the eigenvalues of a random matrix to lie in unit intervals around the points $x_1, \ldots, x_N$,

$$P(x_1, \ldots, x_N) = C_{N\beta} e^{-\beta W}, \tag{9.1.1}$$

$$W = \frac{1}{2} \sum_{1}^{N} x_j^2 - \sum_{i<j} \ln |x_i - x_j|, \tag{9.1.2}$$

is identical with the probability density of the positions of N unit charges free to move on the infinite straight line $-\infty < x < \infty$ under the influence of forces derived from the potential energy (9.1.2), according to the laws of classical mechanics, in a state of thermodynamical equilibrium at a temperature given by

$$kT = \beta^{-1}. \tag{9.1.3}$$

This system of point charges in thermodynamical equilibrium is called the stationary Coulomb gas model or simply the Coulomb gas model, which corresponds to the Gaussian ensembles.

9.2 Nonstationary Ensembles

In this chapter we present an idea of Dyson, generalizing the notion of a matrix ensemble in such a way that the Coulomb gas model acquires meaning not only as a static model in timeless thermodynamical equilibrium but as a dynamical system that may be in an arbitrary nonequilibrium state changing with time. The word "time" in this chapter always refers to a fictitious time which is a property of the mathematical model and has nothing to do with real physical time.

When we try to interpret Coulomb gas as a dynamical system, we naturally consider it first as an ordinary conservative system in which the charges move as Newtonian particles and exchange energy with one another only through the electric forces arising from the potential (9.1.2). We then have to give meaning to the velocity of each particle and to regulate the behavior of the random matrix H in such a way that the eigenvalues have the normal Newtonian property of inertia. No reasonable way of doing this has yet been found. Perhaps there is no such way.

After considerable and fruitless efforts to develop a Newtonian theory of ensembles, Dyson (1962b) discovered that the correct procedure is quite different and much simpler. The x_j should be interpreted as positions of particles in Brownian motion (Chandrasekhar, 1943; Uhlenbeck and Ornstein, 1930; Wang and Uhlenbeck, 1945). This means that the particles have no well-defined velocities nor do they possess inertia. Instead, they feel frictional forces resisting their motion. The gas is not a conservative system, for it is constantly exchanging energy with its surroundings through these frictional forces. The potential (9.1.2) still operates on the particles in the following way. The particle at x_j experiences an external electric force

$$E(x_j) = -\frac{\partial W}{\partial x_j} = -x_j + \sum_{i(\neq j)} \frac{1}{x_j - x_i} \tag{9.2.1}$$

in addition to the local frictional force and the constantly fluctuating force giving rise to the Brownian motion. (The symbol $\sum_{i(\neq j)}$ means that the sum is over all i except j, in particular there is no summation over the index j; while the symbol $\sum_{i\neq j}$ will mean that the sum is over both the indices i and j except when $i = j$.)

The equation of motion of the Brownian particle at x_j may be written as

$$\frac{d^2 x_j}{dt^2} = -f\frac{dx_j}{dt} + E(x_j) + A(t), \tag{9.2.2}$$

where f is the friction coefficient and $A(t)$ is a rapidly fluctuating force. For $A(t)$ we postulate the usual properties (Uhlenbeck and Ornstein, 1930)

$$\langle A(t_1)A(t_2)\cdots A(t_{2n+1})\rangle = 0, \tag{9.2.3}$$

$$\langle A(t_1)A(t_2)\cdots A(t_{2n})\rangle = \sum_{\text{pairs}} \langle A(t_i)A(t_j)\rangle\langle A(t_k)A(t_\ell)\rangle\cdots, \tag{9.2.4}$$

and

$$\langle A(t_1)A(t_2)\rangle = \frac{2kT}{f}\delta(t_1 - t_2), \tag{9.2.5}$$

where the summation in (9.2.4) extends over all distinct ways in which the $2n$ indices can be divided into n pairs.

There is nothing new in the integration of the Langevin equation (9.2.2). After long enough time for the effect of the initial velocities to become negligible, let $x_1, x_2, \ldots, x_N$ be the positions of the particles at time t. At a later time $t + \delta t$, let these positions be changed to $x_1 + \delta x_1, x_2 + \delta x_2, \ldots, x_N + \delta x_N$. The δx_j, $j = 1, 2, \ldots, N$, will in general be different for every member of the ensemble. They are random variables. Using Eqs. (9.2.3) to (9.2.5) we find that to the first order in the small quantities

$$f\langle\delta x_j\rangle = E(x_j)\delta t, \tag{9.2.6}$$

$$f\langle(\delta x_j)^2\rangle = 2kT\,\delta t, \tag{9.2.7}$$

and all other ensemble averages, for example, $\langle\delta x_j \delta x_\ell\rangle$, $\langle(\delta x_j)^2\delta x_\ell\rangle$, $\langle(\delta x_j)^3\rangle$, are of a higher order in δt.

An alternative description of the Brownian motion is obtained by deriving the Fokker–Planck or Smoluchowski equation. Let $P(x_1, x_2, \ldots, x_N; t)$ be the time-dependent joint probability density that the particles will be at the positions x_j at time t. Assuming that the future evolution of the system is completely determined by its present state, with no reference to its past (that is, the process is a Markov process), we obtain

$$\begin{aligned} P(x_1, \ldots, x_N; t + \delta t) &\\ = \int \cdots \int & P(x_1 - \delta x_1, \ldots, x_N - \delta x_N; t) \\ & \times \psi(x_1 - \delta x_1, \ldots, x_N - \delta x_N; \delta x_1, \ldots, \delta x_N; \delta t)\, d(\delta x_1) \cdots d(\delta x_N), \end{aligned} \tag{9.2.8}$$

where ψ under the integral sign is the probability that the positions of the particles will change from $x_1 - \delta x_1, \ldots, x_N - \delta x_N$ to $x_1, \ldots, x_N$ in a time interval δt. Expanding both sides of (9.2.8) in a power series of δx_j, δt, using (9.2.6) and (9.2.7), and going to the limit $\delta t \to 0$, we get (Uhlenbeck and Ornstein, 1930)

$$f\frac{\partial P}{\partial t} = \sum_{j=1}^{N}\left\{kT\frac{\partial^2 P}{\partial x_j^2} - \frac{\partial}{\partial x_j}(E(x_j)P)\right\}. \tag{9.2.9}$$

Equation (9.2.9) describes the evolution of the Coulomb gas with time. If we start from an arbitrary initial probability density P at time $t = t_0$, a unique solution of (9.2.9) will exist for all $t \geqslant t_0$. Any solution of this sort we call a *time-dependent Coulomb gas model*.

Equation (9.2.9) implies in turn (9.2.6) and (9.2.7). To see this we multiply both sides of (9.2.9) by x_j and integrate over all x_i. Making the usual assumptions that $P(x_1, \ldots, x_N; t)$, as well as its derivatives, vanish quite fast on the boundary, we get on partial integration

$$f \frac{d}{dt} \langle x_j \rangle = \langle E(x_j) \rangle, \tag{9.2.10}$$

where

$$\langle F \rangle = \int F(x_1, \ldots, x_N) P(x_1, \ldots, x_N; t) \, dx_1 \cdots dx_N$$

is the ensemble average of F. Starting at the positions $x_1, \ldots, x_N$ and executing the motion for a small time interval δt, we find that (9.2.10) is the same as (9.2.6). Similarly, by multiplying by x_j^2 and integrating (9.2.9) we have

$$f \frac{d}{dt} \langle x_j^2 \rangle = 2kT + 2\langle x_j E(x_j) \rangle,$$

which together with $\langle (\delta x_j)^2 \rangle = \langle x_j^2 \rangle - \langle x_j \rangle^2$ yields (9.2.7).

Thus the descriptions of the motion by (9.2.6) and (9.2.7), and by (9.2.9) are equivalent. Also there exists a unique solution to (9.2.9) which is independent of time, and this time independent solution is given by (9.1.1) and (9.1.2).

A Brownian motion model can also be constructed for the matrix H, of which x_j are the eigenvalues. The independent real parameters H_{ii}, $1 \leqslant i \leqslant N$; $H_{ij}^{(\lambda)}$; $1 \leqslant i < j \leqslant N$, $0 \leqslant \lambda \leqslant \beta - 1$, which determine all the matrix elements of H, are $p = N + N(N-1)\beta/2$ in number. Let us denote them by H_μ, where μ is a single index that runs from 1 to p and replaces the three indices i, j and λ. Suppose that the parameters H_μ have the values $H_1, H_2, \ldots, H_p$ at time t and $H_1 + \delta H_1, \ldots, H_p + \delta H_p$ at a later time $t + \delta t$. Brownian motion of H is defined by requiring that each δH_μ be a random variable with the ensemble averages

$$f \langle \delta H_\mu \rangle = -H_\mu \delta t, \tag{9.2.11}$$

$$f \langle (\delta H_\mu)^2 \rangle = g_\mu k T \delta t, \tag{9.2.12}$$

where

$$g_\mu = g_{ij}^{(\lambda)} = 1 + \delta_{ij} = \begin{cases} 2, & \text{if } i = j, \\ 1, & \text{if } i \neq j. \end{cases} \tag{9.2.13}$$

All other averages are of a higher order in δt. This is a Brownian motion of the simplest type, the various components H_μ being completely uncoupled and each being subject to a fixed simple harmonic force. The Smoluchowski equation which corresponds to (9.2.11) and (9.2.12) is

$$f\frac{\partial P}{\partial t} = \sum_{\mu}\left(\frac{1}{2}g_{\mu}kT\frac{\partial^2 P}{\partial H_{\mu}^2} + \frac{\partial}{\partial H_{\mu}}(H_{\mu}P)\right), \tag{9.2.14}$$

where $P(H_1, \ldots, H_p; t)$ is the time dependent joint probability density of the H_{μ}. The solution of (9.2.14) which corresponds to a given initial condition $H = H'$ at $t = 0$, is known explicitly (Uhlenbeck and Ornstein, 1930)

$$P(H,t) = C(1-q^2)^{-p/2}\exp\left(-\frac{\operatorname{tr}(H-qH')^2}{2kT(1-q^2)}\right), \tag{9.2.15}$$

where

$$q = \exp(-t/f). \tag{9.2.16}$$

The solution shows that the Brownian process is invariant under symmetry preserving unitary transformations of the matrix H; in fact, the awkward looking factor g_{μ} in (9.2.12) is put in to ensure this invariance. When $t \to \infty$, $q \to 0$, and the probability density (9.2.15) tends to the stationary form,

$$P(H_1, \ldots, H_p) = \text{const} \cdot \exp(-\operatorname{tr} H^2/2kT), \tag{9.2.17}$$

which is the unique time independent solution of (9.2.14). Note that with the relation (9.1.3) between β and the temperature kT, (9.2.17) is essentially the same as (2.6.18).

We are now in a position to state the main result of this chapter.

Theorem 9.2.1. *When the matrix H executes a Brownian motion according to the simple harmonic law* (9.2.11), (9.2.12), *starting from any initial conditions, its eigenvalues $x_1, x_2, \ldots, x_N$ execute a Brownian motion that obeys the equations of motion* (9.2.6) *and* (9.2.7) *of the time dependent Coulomb gas.*

To prove the theorem we need only show that (9.2.6) and (9.2.7) follow from (9.2.11) and (9.2.12). Suppose, then, that (9.2.11) and (9.2.12) hold. We have seen that the process described by (9.2.11) and (9.2.12) is independent of the representation of H. Therefore we may choose the representation so that H is diagonal at time t. The instantaneous values of H_{μ} at time t are then

$$H_{jj}^{(0)} = x_j, \quad j = 1, 2, \ldots, N, \tag{9.2.18}$$

and all other components are zero. At a later time $t + \delta t$ the matrix $H + \delta H$ is no longer diagonal and its eigenvalues $x_j + \delta x_j$ must be calculated by perturbation theory. We have to the second order in δH

$$\delta x_j = \delta H_{jj}^{(0)} + \sum_{i(\neq j)} \sum_{\lambda=0}^{\beta-1} \frac{(\delta H_{ij}^{(\lambda)})^2}{x_j - x_i}. \tag{9.2.19}$$

Higher terms in the perturbation series will not contribute to the first order in δt. When we take the ensemble average on each side of (9.2.19) and use (9.2.11), (9.2.18), (9.2.12), (9.1.3) and (9.2.1), the result is (9.2.6). When we take the ensemble average of $(\delta x_j)^2$, only the first term on the right side of (9.2.19) contributes to the order δt, and this term gives (9.2.7) by virtue of (9.2.12) and (9.2.13). The theorem is thus proved.

When the limit $t \to \infty$ is taken, Theorem 9.2.1 reduces to Theorem 3.3.1. This new proof of Theorem 3.3.1 is in some respects more illuminating. It shows how the repulsive Coulomb potential (9.1.2), pushing apart each pair of eigenvalues, arises directly from the perturbation formula (9.2.19). It has long been known that perturbations generally split levels that are degenerate in an unperturbed system. We now see that this splitting effect of perturbations is quantitatively identical with the repulsive force of the Coulomb gas model.

Theorem 9.2.1 is a much stronger statement than Theorem 3.3.1. It shows that the electric force (9.2.1), acting on the eigenvalues x_j, has a concrete meaning for any matrix H whatever, not only for an ensemble of matrices in stationary thermal equilibrium. The force $E(x_j)$ is precisely proportional to the mean rate of drift of x_j which occurs when the matrix H is subjected to a random perturbation.

9.3 Some Ensemble Averages

We now describe a general property of the time-dependent Coulomb gas model which may be used to calculate a few ensemble averages. Dyson observed that if $G = G(x_1, \ldots, x_N)$ is any function of the positions of the charges, not depending explicitly on time, then the time variation of $\langle G \rangle$, the ensemble average of G, is governed by the equation

$$f \frac{d}{dt} \langle G \rangle = -\sum_j \left\langle \frac{\partial W}{\partial x_j} \frac{\partial G}{\partial x_j} \right\rangle + kT \sum_j \left\langle \frac{\partial^2 G}{\partial x_j^2} \right\rangle. \tag{9.3.1}$$

This equation is obtained by multiplying (9.2.9) throughout by G and partial integrations; W is given by (9.1.2).

As a first example, choose

$$R = \sum_j x_j^2 \tag{9.3.2}$$

for G so that

$$\frac{\partial W}{\partial x_j}\frac{\partial R}{\partial x_j} = -2\sum_{i(\neq j)} \frac{x_j}{x_j - x_i} + 2x_j^2,$$

$$\frac{\partial^2 R}{\partial x_j^2} = 2,$$

and (9.3.1) becomes

$$f\frac{\partial\langle R\rangle}{\partial t} = -2\langle R\rangle + N(N-1) + 2kTN = 2(R_\infty - \langle R\rangle), \tag{9.3.3}$$

with

$$R_\infty = \frac{1}{2}N(N-1) + kTN. \tag{9.3.4}$$

The solution of (9.3.3) is

$$\langle R\rangle = R_0 q^2 + R_\infty(1-q^2), \tag{9.3.5}$$

where q is given by (9.2.16) and R_0 is the value of $\langle R\rangle$ at $t=0$. Equation (9.3.5) shows that the ensemble average $\langle R\rangle$ approaches its equilibrium value R_∞ with exponential speed as $t \to \infty$.

Next, take $G = W$ in (9.3.1), so that

$$\left(\frac{\partial W}{\partial x_j}\right)^2 = x_j^2 + \sum_{i(\neq j)}\left[\left(\frac{1}{x_j - x_i}\right)^2 - \frac{2x_j}{x_j - x_i}\right] + \sum_{\substack{i,\ell \\ (i,j,\ell \text{ all different})}} \left[(x_j - x_i)(x_j - x_\ell)\right]^{-1}, \tag{9.3.6}$$

and

$$\frac{\partial^2 W}{\partial x_j^2} = 1 + \sum_{i(\neq j)} (x_j - x_i)^{-2}. \tag{9.3.7}$$

On performing a summation over j the last term in (9.3.6) drops out (cf. Appendix A.5, paragraph preceding (A.5.14), whereas the second term in the first bracket gives $-N(N-1)$. Substituting in (9.3.1) and simplifying, we get

$$f\frac{\partial\langle W\rangle}{\partial t} = (kT-1)\sum_{i\neq j}\left\langle (x_j - x_i)^{-2}\right\rangle + (N^2 - N + NkT) - \sum_j \langle x_j^2\rangle. \tag{9.3.8}$$

For the stationary Coulomb gas at temperature kT the left side of (9.3.8) vanishes and (9.3.4) may be used on the right. Thus we find a "virial theorem" for the stationary gas:

$$\sum_{i \neq j} \left\langle (x_j - x_i)^{-2} \right\rangle = \frac{N(N-1)}{2(1-kT)}. \tag{9.3.9}$$

The probability density of eigenvalues becomes proportional to $|x_j - x_i|^{\beta}$, when two eigenvalues x_i, x_j come close together. The ensemble average of $(x_j - x_i)^{-2}$ is therefore defined only for $\beta > 1$ and (9.3.8) and (9.3.9) hold only for $kT < 1$.

An especially interesting case $\beta = 1$, requires a passage to the limit in (9.3.8). As $kT \to 1$, we have for any fixed value of Δ

$$\lim(kT - 1) \int_{-\Delta}^{\Delta} |y|^{\beta - 2} \, dy = \lim(kT - 1)(\beta - 1)^{-1} 2\Delta^{\beta - 1} = -2. \tag{9.3.10}$$

We obtain the correct limit in (9.3.8) if we replace

$$(kT - 1)(x_j - x_i)^{-2}$$

with

$$-2(x_j - x_i)^{-1} \delta(x_j - x_i), \tag{9.3.11}$$

which has a well-defined meaning as an ensemble average when $kT = 1$, for the probability density then contains a factor $|x_j - x_i|$; (9.3.8) thus becomes in the limit

$$f \frac{d\langle W \rangle}{dt} = -2 \sum_{i \neq j} \left\langle |x_j - x_i|^{-1} \delta(x_j - x_i) \right\rangle + N^2 - \sum_j \langle x_j^2 \rangle, \quad kT = 1. \tag{9.3.12}$$

The corresponding "virial theorem" is

$$\sum_{i \neq j} \left\langle (x_j - x_i)^{-1} \delta(x_j - x_i) \right\rangle = \frac{1}{4} N(N-1), \quad kT = 1 \tag{9.3.13}$$

for the stationary gas.

Summary of Chapter 9

Suppose that all the real parameters entering the definition of a Hermitian matrix H execute Brownian motion with the averages

$$\langle \delta H_{ij} \rangle = -H_{ij} \delta t, \tag{9.2.11}$$

$$\langle(\delta H_{ij})^2\rangle = (1+\delta_{ij})\delta t/\beta. \tag{9.2.12}$$

Then starting at whatever initial values, after a long enough time the matrix elements H_{ij} reach an equilibrium distribution given by

$$P(H) \propto \exp(-\beta \operatorname{tr} H^2/2). \tag{9.2.17}$$

Some ensemble averages can also be computed, such as

$$\sum_{i\neq j}\langle(x_j - x_i)^{-2}\rangle = \frac{N(N-1)}{2(1-kT)}, \quad \beta > 1, \tag{9.3.9}$$

or

$$\sum_{i\neq j}\langle(x_j - x_i)^{-1}\delta(x_j - x_i)\rangle = \frac{1}{4}N(N-1), \quad \beta = 1. \tag{9.3.13}$$

10

CIRCULAR ENSEMBLES

In the preceding chapters we presented a detailed study of the Gaussian ensembles. We pointed out at the end of Section 2.6 that the requirements of invariance and the statistical independence of various independent components seem to overrestrict the possible choices. In particular, the various values of the matrix elements are not equally weighted. Rosenzweig has tried to answer this question in part with a consideration of the "fixed-strength ensemble" in which the joint probability density function for the matrix elements is taken to be proportional to the Dirac delta function $\delta(\operatorname{tr} H^2 - r^2)$, where r is a fixed number; see Chapter 27. However, a uniform probability density cannot be defined on the infinite real line.

Because of this unsatisfactory feature Dyson (1962a, I) introduced his circular ensembles, as follows.

Suppose that the system is characterized not by its Hamiltonian H but by a unitary matrix S, whose elements give the transition probabilities between the various states. The matrix S is unitary; its eigenvalues are therefore of the form $e^{i\theta_j}$, where the angles θ_j are real and can be taken to lie between 0 and 2π. The matrix S is a function of the Hamiltonian H of the system. This functional dependence need not be specified. All that is needed is that for small ranges of variation, the θ_j be linear functions of the eigenvalues x_j of H. To help the reader's imagination he may think of a relation such as

$$S = e^{i\tau H} \quad \text{or} \quad S = \frac{1 - i\tau H}{1 + i\tau H}. \tag{10.0.1}$$

However, such a definite relation between S and H cannot be correct except in a limited range of energy. We will deliberately leave this relation vague because we are going to restrict the order of our matrices to $N \times N$, where N is very large but finite. And this cannot represent, say, a nucleus, for the real nucleus has an infinite number of energy levels. Like the Gaussian ensembles, the circular ensembles are gross mutilations of the over-all actual situation. The most we can expect of such models is that in any energy region that is small compared to the total excitation energy the statistical distribution of the levels will be correctly reproduced. With no further apologies we make the following fundamental assumption:

The statistical behavior of n consecutive levels of an actual system, whenever n is small compared with the total number of levels, is the same as that of n consecutive angles $\theta_1, \theta_2, \ldots, \theta_n$, where n is small compared with N.

10.1 Orthogonal Ensemble

According to the analysis of Chapter 2, a system having time-reversal invariance and rotational symmetry or having time-reversal invariance and integral spin will be characterized by a symmetric S. Following Dyson, we define the orthogonal circular ensemble E_{1c} of symmetric unitary matrices S by assigning the probabilities in the following way.

Every symmetric unitary S (cf. Appendix A.3) can be written as

$$S = U^T U, \tag{10.1.1}$$

where U is unitary. Define a small neighborhood of S by

$$S + dS = U^T (1 + i\, dM) U, \tag{10.1.2}$$

where dM is a real symmetric matrix with elements dM_{ij} and the elements dM_{ij} for $i \leqslant j$ vary independently in some small intervals of length $d\mu_{ij}$. The "volume" of this neighborhood is defined by

$$\mu_1(dS) = \prod_{i \leqslant j} d\mu_{ij}. \tag{10.1.3}$$

The ensemble E_{1c} is defined by the statement: *The probability that a matrix from the ensemble E_{1c} lies between S and $S + dS$ is proportional to*

$$P(S)\, dS = \frac{1}{V_1} \mu_1(dS), \tag{10.1.4}$$

where V_1 is the total volume of the space T_{1c} of unitary symmetric matrices of order $N \times N$ and therefore, a normalization constant.

For this definition to have a meaning one must be sure that $\mu_1(dS)$ does not depend on the choice of U in (10.1.1). The fact that it is so can be verified as follows. Let

$$S = U^T U = V^T V, \tag{10.1.5}$$

where both U and V are unitary. The matrix

$$R = VU^{-1} \tag{10.1.6}$$

is unitary and also satisfies

$$R^T R = (U^T)^{-1} V^T V U^{-1} = (U^T)^{-1} U^T U U^{-1} = 1. \tag{10.1.7}$$

Therefore R is real and orthogonal. Let

$$\mu_1'(dS) = \prod_{i \leqslant j} d\mu_{ij}' \tag{10.1.8}$$

be the volume derived from V as $\mu_1(dS)$ was derived from U. We now have

$$S + dS = V^T(1 + i\,dM')V, \tag{10.1.9}$$

with

$$dM' = R\,dM\,R^{-1}. \tag{10.1.10}$$

To prove that $\mu_1(dS) = \mu_1'(dS)$ we need to show that the Jacobian

$$J = \det\left[\frac{\partial(dM_{ij}')}{\partial(dM_{ij})}\right] \tag{10.1.11}$$

has absolute value unity when dM, dM' are real symmetric matrices related by (10.1.10). A proof of this is given in Appendix A.21. Thus the volume $\mu_1(dS)$ is unique. Incidentally, we have established that for a fixed S the unitary matrix U in (10.1.1) is undetermined precisely to the extent of a transformation

$$U \to RU, \tag{10.1.12}$$

where R is an arbitrary real orthogonal matrix.

The motivation for the choice of the ensemble E_{1c} will be made clearer by the following theorem.

Theorem 10.1.1. *The orthogonal ensemble E_{1c} is uniquely defined in the space of unitary symmetric matrices of order $N \times N$ by the property of being invariant under every automorphism*

$$S \to W^T S W \tag{10.1.13}$$

of T_{1c} into itself where W is any $N \times N$ unitary matrix.

Theorem 10.1.1 comprises two statements: (a) *that E_{1c} is invariant under the automorphisms* (10.1.13) and (b) *that it is unique.* To prove (a) we suppose that a neighborhood $S + dS$ of S is transformed into a neighborhood $S' + dS'$ of S' by the automorphism (10.1.13). Equations (10.1.1) and (10.1.2) then hold and

$$S' = W^T S W = V^T V, \quad V = UW, \tag{10.1.14}$$

$$S' + dS' = V^T (1 + i\, dM) V. \tag{10.1.15}$$

The volumes $\mu_1(dS)$ and $\mu_1(dS')$ are then identical by definition as the same dM is occurring in (10.1.2) and (10.1.15). This proof of (a) is trivial, for we could choose a convenient unitary matrix V in (10.1.14) and it was already shown that the value of $\mu_1(dS)$ does not depend on the choice of V. To prove (b) let E_1' be any ensemble invariant under (10.1.13). The probability density $P'(S)\, dS$ associated with E_1' will define a certain volume $\mu_1'(dS)$ of the neighborhood of S in the space T_{1c}. The ratio

$$\varphi(S) = \frac{\mu_1'(dS)}{\mu_1(dS)} \tag{10.1.16}$$

is a function of S defined on T_{1c} and invariant under the transformations (10.1.13). If $S = U^T U$, we choose $W = U^{-1}$ in (10.1.13) so that S is transformed to unity and therefore $\varphi(S)$ is a constant. Thus the probability densities in E_{1c} and E_{1c}' are proportional. Also they are both normalized to unity and are therefore identical.

10.2 Symplectic Ensemble

Next we consider systems with half-integral spin and time-reversal invariance but no rotational symmetry. In this section we use the quaternion notation developed in Chapter 2. The systems are described by self-dual unitary quaternion matrices (cf. Chapter 2)

$$S^R \equiv -Z S^T Z = S, \quad S^\dagger = S^{-1}. \tag{10.2.1}$$

Once again we have to assign a probability that a matrix chosen randomly from the space T_{4c} of self-dual unitary quaternion matrices of order $N \times N$ will lie between S and $S + dS$. This is done as follows:

Every matrix S in T_{4c} can be written as

$$S = U^R U, \tag{10.2.2}$$

where U is unitary. To see that this is possible, observe that in the ordinary language without quaternions SZ is an antisymmetric unitary matrix and can be reduced to the canonical form

$$SZ = VZV^T, \tag{10.2.3}$$

where V is unitary. Choosing $U = (ZV)^T$ then gives (10.2.2) (see also Appendix A.3). For a given S, the unitary matrix U in (10.2.2) is precisely undetermined to the extent of a transformation

$$U \to BU, \tag{10.2.4}$$

where B is an arbitrary symplectic matrix. For a proof it is sufficient to observe that the dual of a product of matrices is the product of their duals taken in the reverse order; and a symplectic matrix is, by definition, one that satisfies

$$B^R B = BB^R = 1. \tag{10.2.5}$$

A small neighborhood of S in T_{4c} is defined by

$$S + dS = U^R(1 + i\, dM)U, \tag{10.2.6}$$

where dM is a self-dual quaternion real matrix with elements

$$dM_{ij} = dM_{ij}^{(0)} + \sum_{\alpha=1}^{3} dM_{ij}^{(\alpha)} e_\alpha. \tag{10.2.7}$$

The real coefficients $dM_{ij}^{(\alpha)}$ satisfy

$$dM_{ij}^{(0)} = dM_{ji}^{(0)}, \qquad dM_{ij}^{(\alpha)} = -dM_{ji}^{(\alpha)}, \quad \alpha = 1, 2, 3. \tag{10.2.8}$$

There are $N(2N-1)$ independent real variables $dM_{ij}^{(\alpha)}$ and they are allowed to vary over some small intervals of lengths $d\mu_{ij}^{(\alpha)}$. The neighborhood of S, thus defined, is assigned a volume

$$\mu_4(dS) = \prod_{i<j} \prod_{\alpha=1}^{3} d\mu_{ij}^{(\alpha)} \prod_{i \leqslant j} d\mu_{ij}^{(0)}. \tag{10.2.9}$$

In terms of this volume the symplectic ensemble E_{4c} is defined in exactly the same way as E_{1c} was defined in terms of the volume (10.1.3). The statistical weight of the neighborhood dS in T_{4c} is

$$P(S)\,dS = \frac{1}{V_4}\mu_4(dS), \tag{10.2.10}$$

where V_4 is the total volume of the space T_{4c} of self-dual unitary quaternion matrices of order $N \times N$.

We can now repeat almost without change the arguments in Section 10.1. We must first prove that the volume $\mu_4(dS)$ is independent of the choice of U in (10.2.2). This involves showing that the Jacobian J has absolute value unity, where

$$J = \det\left[\frac{\partial(dM_{ij}'^{(\alpha)})}{\partial(dM_{ij}^{(\alpha)})}\right], \tag{10.2.11}$$

$$dM' = B\,dM\,B^{-1}, \tag{10.2.12}$$

and B is symplectic. As before Appendix A.21 contains a proof of this.

The analog of Theorem 10.1.1 is the following.

Theorem 10.2.1. *The symplectic ensemble E_{4c} is uniquely defined in the space T_{4c} of self-dual unitary quaternion matrices of order $N \times N$ by the property of being invariant under every automorphism*

$$S \to W^R S W \tag{10.2.13}$$

of T_{4c} into itself, where W is any $N \times N$ unitary quaternion matrix.

Theorem 10.2.1 can be proved by following word for word the proof of Theorem 10.1.1, the operation of transposition being replaced by that of taking the dual.

10.3 Unitary Ensemble

A system without time-reversal symmetry is associated with an arbitrary unitary matrix S not restricted to be symmetric or self-dual. A neighborhood of S in the space T_{2c} of all unitary $N \times N$ matrices is defined by

$$S + dS = U(1 + i\,dM)V, \tag{10.3.1}$$

where U and V are any unitary matrices that satisfy the equation $S = UV$ and dM is an infinitesimal Hermitian matrix with elements $dM_{ij} = dM_{ij}^{(0)} + i\,dM_{ij}^{(1)}$. The real components $dM_{ij}^{(0)}$, $dM_{ij}^{(1)}$ are N^2 in number and are allowed to vary independently over small intervals of lengths $d\mu_{ij}^{(0)}$, $d\mu_{ij}^{(1)}$. The volume $\mu_2(dS)$ is defined by the equation

$$\mu_2(dS) = \prod_{i \leqslant j} d\mu_{ij}^{(0)} \prod_{i<j} d\mu_{ij}^{(1)}, \tag{10.3.2}$$

and is independent of the choice of U and V. The ensemble E_{2c} gives to each neighborhood dS the statistical weight

$$P(S)\, dS = \frac{1}{V_2} \mu_2(dS), \tag{10.3.3}$$

where V_2 is the total volume of the space T_{2c}.

The invariance property of E_{2c}, analogous to Theorems 10.1.1 and 10.2.1, is stated in Theorem 10.3.1.

Theorem 10.3.1. *The unitary ensemble E_{2c} is uniquely defined in the space T_{2c} of all $N \times N$ unitary matrices by the property of being invariant under every automorphism*

$$S \to USV \tag{10.3.4}$$

of T_{2c} into itself, where U and V are any two $N \times N$ unitary matrices.

This theorem merely expresses the well-known result that $\mu_2(dS)$ is the invariant group-measure of the N-dimensional unitary group $U(N)$.

10.4 The Joint Probability Density of the Eigenvalues

We give below a few lemmas which will be used subsequently. For their proofs see Appendix A.3.

Lemma 10.4.1. *Let S be any unitary symmetric $N \times N$ matrix. Then there exists a real orthogonal matrix R which diagonalizes S; that is*

$$S = R^{-1} E R, \tag{10.4.1}$$

where E is diagonal. The diagonal elements of E are complex numbers $e^{i\theta_j}$ lying on the unit circle.

Lemma 10.4.2. *Let S be a unitary self-dual quaternion matrix of order $N \times N$. Then there exists a symplectic matrix B such that*

$$S = B^{-1} E B, \tag{10.4.2}$$

where E is diagonal and scalar (cf. Section 2.4). The diagonal elements of E are N complex numbers $e^{i\theta_j}$ on the unit circle; each is repeated twice.

Lemma 10.4.3. *Let S be a unitary matrix. Then there exists a unitary matrix U such that*

$$S = U^{-1} E U, \tag{10.4.3}$$

where E is diagonal. The diagonal elements of E are N complex numbers $e^{i\theta_j}$ on the unit circle.

Though this result is well known, we stated it here for completeness.

We are now in a position to prove the main result of this chapter.

Lemma 10.4.4. *In the ensemble $E_{\beta c}$ the probability of finding the eigenvalues $e^{i\phi_j}$ of S with an angle in each of the intervals $(\theta_j, \theta_j + d\theta_j)$, $j = 1, \ldots, N$, is given by*

$$P_{N\beta}(\theta_1, \ldots, \theta_N)\, d\theta_1 \cdots d\theta_N, \tag{10.4.4}$$

where

$$P_{N\beta}(\theta_1, \ldots, \theta_N) = C'_{N\beta} \prod_{1 \leqslant \ell < j \leqslant N} |e^{i\theta_\ell} - e^{i\theta_j}|^\beta. \tag{10.4.5}$$

Here $\beta = 1$ for orthogonal, $\beta = 4$ for symplectic, and $\beta = 2$ for unitary circular ensembles. The constant $C'_{N\beta}$ is fixed by normalization.

Proof. (1) Let $\beta = 1$. By Lemma 10.4.1 every S in T_{1c} can be diagonalized in the form

$$S = R^{-1} E R, \tag{10.4.6}$$

with R orthogonal. We now wish to express the volume $\mu_1(dS)$ in terms of the volumes $\mu(dE)$ and $\mu(dR)$, defined for the neighborhoods of the matrices E and R, respectively. A small neighborhood of E is given by

$$dE = i E\, d\theta, \tag{10.4.7}$$

where $d\theta$ means a diagonal matrix with elements $d\theta_1, \ldots, d\theta_N$. To find the neighborhood of R we differentiate

$$R R^T = 1, \tag{10.4.8}$$

thus getting

$$R(dR)^T + (dR)R^T = 0, \tag{10.4.9}$$

showing that the infinitesimal matrix

$$dA \equiv (dR)R^T = -R(dR)^T \tag{10.4.10}$$

is a real antisymmetric matrix with elements dA_{ij}. The volumes $\mu(dE)$ and $\mu(dR)$ are given by

$$\mu(dE) = \prod_{j=1}^{N} d\theta_j, \tag{10.4.11}$$

$$\mu(dR) = \prod_{1 \leqslant \ell < j \leqslant N} dA_{\ell j}. \tag{10.4.12}$$

The volume $\mu(dS)$ is defined by (10.1.3), where dM is given by (10.1.2) and U is any unitary matrix satisfying (10.1.1). Differentiating (10.4.6) and using (10.1.1), (10.1.2), (10.4.7), (10.4.8), and (10.4.10), we obtain

$$iRU^T\,dM\,U R^{-1} = -dA\,E + iE\,d\theta + E\,dA, \tag{10.4.13}$$

which is the relation between $dM, d\theta$, and dA. Since E is a diagonal unitary matrix, it has a square root F which is also diagonal with elements $\pm e^{i\theta_j/2}$. There is an ambiguity in the sign of each element, but it does not matter how these signs are chosen. A convenient choice for U satisfying (10.1.1) is then

$$U = FR \tag{10.4.14}$$

by virtue of (10.4.6). With this choice of U, (10.4.13) reduces to

$$iF\,dM F = -dAF^2 + iF^2\,d\theta + F^2\,dA, \tag{10.4.15}$$

or

$$dM = d\theta + i(F^{-1}\,dAF - F\,dAF^{-1}). \tag{10.4.16}$$

Equation (10.4.16) gives $dM_{\ell j}$ in terms of the quantities $d\theta_j$, $dA_{\ell j}$, and θ_j for each pair of indices ℓ, j; namely

$$dM_{jj} = d\theta_j, \tag{10.4.17}$$

$$dM_{\ell j} = 2\sin\left[\frac{1}{2}(\theta_\ell - \theta_j)\right]dA_{\ell j}, \quad \ell \neq j. \tag{10.4.18}$$

Assembling the definitions (10.1.3), (10.4.11), and (10.4.12), we deduce from (10.4.17) and (10.4.18)

$$\mu(dS) = \prod_{\ell<j} \left| 2\sin\left(\frac{\theta_\ell - \theta_j}{2}\right)\right| \mu(dE)\mu(dR)$$

$$= \prod_{\ell<j} |e^{i\theta_\ell} - e^{i\theta_j}| \mu(dE)\mu(dR). \tag{10.4.19}$$

Now keep the angles $\theta_1, \ldots, \theta_N$ fixed and integrate (10.4.19) with respect to the parameters $dA_{\ell j}$ over the entire allowed range. This will give (10.4.4) with P_{N1} given by (10.4.5). Thus the theorem is proved for the orthogonal case.

(2) Next let $\beta = 4$. The matrix S is now diagonalized with the help of a symplectic matrix

$$S = B^{-1}EB \tag{10.4.20}$$

(Lemma 10.4.2). The infinitesimal matrix

$$dA = dB\, B^R \tag{10.4.21}$$

is quaternion real and anti-Hermitian. The components of dA are $dA_{\ell j}^{(\alpha)}$, which are real. They are antisymmetric in ℓ, j for $\alpha = 0$ and symmetric in ℓ, j for $\alpha = 1, 2$, and 3. The volume $\mu(dB)$ is now given by

$$\mu(dB) = \prod_{\ell, j, \alpha} dA_{\ell j}^{(\alpha)}. \tag{10.4.22}$$

The volume $\mu(dS)$ is given by (10.2.9) with dM given by (10.2.6). The matrix dM is Hermitian and quaternion real. The algebra leading up to (10.4.16) goes exactly as before. Equation (10.4.17) still holds, the diagonal elements dM_{jj} being real scalar quaternions with only one independent component. Equation (10.4.18) now holds separately for each of the four quaternion components $\alpha = 0, 1, 2, 3$. The equation analogous to (10.4.19) is

$$\mu(dS) = \left(\prod_{\ell<j} |e^{i\theta_\ell} - e^{i\theta_j}|^4\right)\mu(dE)C, \tag{10.4.23}$$

where C does not depend on the θ_j. The power 4 in (10.4.23) arises from the fact that every nondiagonal element $dM_{\ell j}$, $\ell < j$, gives according to (10.4.18) four factors corresponding to the four components $\alpha = 0, 1, 2$ and 3. Note also that the C in (10.4.23) is not equal to $\mu(dB)$, for the diagonal components $dA_{jj}^{(\alpha)}$ with $\alpha = 1, 2, 3$ do not occur in (10.4.17), whereas $\mu(dB)$ contains their product. For our purposes it is sufficient to know that C does not depend on the θ_j.

The rest of the proof proceeds as in the case $\beta = 1$.

(3) Lastly let $\beta = 2$. In this case (10.4.6) holds with a unitary R. The infinitesimal matrix dA is now anti-Hermitian, and the diagonal elements dA_{jj} are pure imaginary. The real part $dA_{\ell j}^{(0)}$ of the non-diagonal elements $dA_{\ell j}$ is antisymmetric in ℓ, j, whereas the imaginary part $dA_{\ell j}^{(1)}$ is symmetric in ℓ, j. Equation (10.4.17) holds in this case as well; (10.4.18) holds separately for the real and imaginary components $dM_{\ell j}^{(0)}$ and $dM_{\ell j}^{(1)}$ of the nondiagonal elements $dM_{\ell j}$. The equation analogous to (10.4.19) is therefore

$$\mu(dS) = \left\{ \prod_{\ell<j} |e^{i\theta_\ell} - e^{i\theta_j}|^2 \right\} \mu(dE)C, \tag{10.4.24}$$

where C does not depend on the θ_j. As in the case $\beta = 4$, this is sufficient for our purposes.

The theorem is thus established for all three cases. □

As in the Gaussian ensembles, it is possible to account for the different powers of the product of differences appearing in Eqs. (10.4.19), (10.4.23) and (10.4.24) by a dimensional argument. The dimension of the space T_{1c} of all symmetric unitary matrices is $N(N+1)/2$, whereas the dimension of the subspace T_{1c}' composed of all symmetric unitary matrices with two equal eigenvalues is $N(N+1)/2 - 2$ (see Appendix A.4). Because of the single restriction, the equality of two eigenvalues, the dimension should normally have decreased by one; as it is decreased by 2, it indicates a factor in Eq. (10.4.19) linear in $|\exp(i\theta_j) - \exp(i\theta_k)|$. Similarly, for unitary matrices the dimension decreases from N^2 to $N^2 - 3$ on equating two eigenvalues, and for quaternion self-dual unitary matrices it decreases from $N(2N-1)$ to $N(2N-1) - 5$. See the end of Section 3.3 and Appendix A.4.

Summary of Chapter 10

Three ensembles of unitary matrices S are considered; (i) that of unitary symmetric matrices S invariant under the transformations $S \to W^T SW$, known as the orthogonal (circular) ensemble; (ii) that of unitary matrices S invariant under the transformations $S \to USV$, known as the unitary (circular) ensemble; and (iii) that of self-dual unitary matrices S invariant under the transformations $S \to W^R SW$, known as the symplectic (circular) ensemble. Here W, U and V are any unitary matrices, W^T is the transpose of W, and W^R is the dual of W.

The joint probability density of the eigenvalues $\exp(i\theta_j)$, $j = 1, \ldots, N$, is found to be

$$P(\theta_1, \ldots, \theta_N) \propto \prod_{1 \leqslant j < k \leqslant N} |\exp(i\theta_j) - \exp(i\theta_k)|^\beta, \tag{10.4.5}$$

where $\beta = 1$ for the orthogonal, $\beta = 2$ for the unitary and $\beta = 4$ for the symplectic (circular) ensemble.

11

CIRCULAR ENSEMBLES (CONTINUED)

The joint probability density of the eigenvalues of a matrix taken from one of the three circular ensembles (orthogonal, unitary or symplectic) was derived in Chapter 10. We will now evaluate their correlation and cluster functions. For this purpose we will again use the theory of quaternion matrices. In the limit of large N, these will be seen to be identical to those for the corresponding Gaussian ensembles. It will be shown that alternate eigenvalues of a matrix from the orthogonal ensemble behave as if they were the eigenvalues of a matrix from the symplectic ensemble. This will allow us in Section 11.7 to deduce the spacing probabilities for the symplectic ensemble from those of the orthogonal ensemble. As for the Gaussian ensembles, a Brownian motion model can be constructed for the circular ensembles as well.

11.1 Unitary Ensemble. Correlation and Cluster Functions

Systems with no time reversal invariance may be characterized by unitary random matrices. The joint probability density of the eigenvalue angles of such matrices taken from the unitary ensemble (cf. Chapter 10, Theorem 10.4.4) is

$$P_{N2}(\theta_1, \ldots, \theta_N) = \text{const} \cdot \prod_{j<k} |e^{i\theta_j} - e^{i\theta_k}|^2. \quad (11.1.1)$$

Calculating averages with the probability density P_{N2} is mathematically the simplest. One may write

$$\prod_{j<k} |e^{i\theta_j} - e^{i\theta_k}|^2 = \prod_{j<k} (e^{i\theta_j} - e^{i\theta_k}) \cdot \prod_{j<k} (e^{-i\theta_j} - e^{-i\theta_k})$$

$$= \det[e^{ip\theta_j}] \cdot \det[e^{-ip\theta_j}], \tag{11.1.2}$$

$$p = \frac{1}{2}(1-N), \frac{1}{2}(3-N), \ldots, \frac{1}{2}(N-3), \frac{1}{2}(N-1), \tag{11.1.3}$$

$$j, k = 1, 2, \ldots, N. \tag{11.1.4}$$

Or

$$\prod_{j<k} |e^{i\theta_j} - e^{i\theta_k}|^2 = (2\pi)^N \det[S_N(\theta_j - \theta_k)]_{j,k=1,\ldots,N}, \tag{11.1.5}$$

with

$$S_N(\theta) = \frac{1}{2\pi} \sum_p e^{ip\theta} = \frac{1}{2\pi} \frac{\sin(N\theta/2)}{\sin(\theta/2)}, \tag{11.1.6}$$

the sum on p being over the values as in Eq. (11.1.3).

Now

$$\int_0^{2\pi} S_N(0)\, d\theta = N, \tag{11.1.7}$$

and

$$\int_0^{2\pi} S_N(\theta_j - \theta_k) S_N(\theta_k - \theta_\ell)\, d\theta_k = S_N(\theta_j - \theta_\ell), \tag{11.1.8}$$

so that all the conditions of Theorem 5.1.4 are satisfied.

Integrating over all the variables, and using that theorem, we fix the constants,

$$P_{N2}(\theta_1, \ldots, \theta_N) = \frac{1}{N!} \det[S_N(\theta_j - \theta_k)]_{j,k=1,\ldots,N}. \tag{11.1.9}$$

According to the same Theorem 5.1.4, the n-level correlation function is

$$R_n(\theta_1, \ldots, \theta_n) = \det[S_N(\theta_j - \theta_k)]_{j,k=1,\ldots,n}. \tag{11.1.10}$$

The level density is

$$R_1(\theta) = S_N(0) = \frac{N}{2\pi}, \tag{11.1.11}$$

as it should be. The two-level correlation function is

$$R_2(\theta, \varphi) = (S_N(0))^2 - (S_N(\theta - \varphi))^2, \tag{11.1.12}$$

and the two-level cluster function is

$$T_2(\theta - \varphi) = (S_N(\theta - \varphi))^2 = \left(\frac{\sin(N(\theta - \varphi)/2)}{2\pi \sin((\theta - \varphi)/2)} \right)^2. \tag{11.1.13}$$

On taking the limit $N \to \infty$, while keeping $N\theta = 2\pi\xi$ and $N\varphi = 2\pi\eta$ finite, we have

$$Y_2(\xi, \eta) = \lim \left(\frac{2\pi}{N} \right)^2 T_2(\theta, \varphi) = \left(\frac{\sin(\pi r)}{\pi r} \right)^2, \tag{11.1.14}$$

$$r = |\xi - \eta|. \tag{11.1.15}$$

This result is identical to the two-point cluster function for the case of the unitary Gaussian ensemble, Eq. (6.2.14).

More generally, in the limit $N \to \infty$,

$$\lim \frac{2\pi}{N} S_N(\theta - \varphi) = \frac{\sin(\pi r)}{\pi r} \equiv s(r), \tag{11.1.16}$$

so that in this limit the n-level correlation and cluster functions for any finite n are identical to those for the Gaussian unitary ensemble, Section 6.2.

11.2 Unitary Ensemble. Level Spacings

For the spacing distribution one first finds the probability $A_2(\alpha)$ that a randomly chosen interval of length 2α will contain none of the angles θ_j:

$$A_2(\alpha) = \int_\alpha^{2\pi-\alpha} \cdots \int_\alpha^{2\pi-\alpha} P_{N2}(\theta_1, \ldots, \theta_N)\, d\theta_1 \cdots d\theta_N \tag{11.2.1}$$

$$= \text{const} \cdot \int_\alpha^{2\pi-\alpha} \cdots \int_\alpha^{2\pi-\alpha} |\det[e^{ip\theta_j}]|^2\, d\theta_1 \cdots d\theta_N.$$

$$= \text{const} \cdot \int_\alpha^{2\pi-\alpha} \cdots \int_\alpha^{2\pi-\alpha} \det\left[\frac{1}{2\pi} \sum_{j=1}^{N} e^{i(p-q)\theta_j} \right] d\theta_1 \cdots d\theta_N \tag{11.2.2}$$

from (11.1.2), where p and q take values as in (11.1.3); or (cf. Appendix A.12)

$$A_2(\alpha) = \det\left[\frac{1}{2\pi}\int_\alpha^{2\pi-\alpha} d\theta\, e^{i(p-q)\theta}\right] \tag{11.2.3}$$

$$= \det\left[\delta_{pq} - \frac{1}{2\pi}\int_{-\alpha}^{\alpha} d\theta \cos(p-q)\theta\right]. \tag{11.2.4}$$

The constant was fixed by the requirement that $A_2(0) = 1$. Since

$$\sum_q (\delta_{pq} - \cos(p\theta)\cos(q\theta))(\delta_{qr} - \sin(q\theta)\sin(r\theta)) = \delta_{pr} - \cos(p-r)\theta, \tag{11.2.5}$$

we may factorize the expression (11.2.4),

$$A_2(\alpha) = A_1(\alpha)\cdot A_1'(\alpha), \tag{11.2.6}$$

with

$$A_1(\alpha) = \det\left[\delta_{pq} - \frac{1}{2\pi}\int_{-\alpha}^{\alpha} d\theta \cos(p\theta)\cos(q\theta)\right]$$

$$= \det\left[\delta_{pq} - \frac{1}{2\pi}\left(\frac{\sin(p-q)\alpha}{p-q} + \frac{\sin(p+q)\alpha}{p+q}\right)\right], \tag{11.2.7}$$

$$A_1'(\alpha) = \det\left[\delta_{pq} - \frac{1}{2\pi}\int_{-\alpha}^{\alpha} d\theta \sin(p\theta)\sin(q\theta)\right]$$

$$= \det\left[\delta_{pq} - \frac{1}{2\pi}\left(\frac{\sin(p-q)\alpha}{p-q} - \frac{\sin(p+q)\alpha}{p+q}\right)\right], \tag{11.2.8}$$

$$p, q = -\frac{1}{2}(N-1), -\frac{1}{2}(N-3), \ldots, \frac{1}{2}(N-1). \tag{11.2.9}$$

In the limit $N \to \infty$, $N\alpha = 2\pi t$, $Np\alpha = 2\pi\xi t$, $Nq\alpha = 2\pi\eta t$, all finite, $A_1(\alpha)$ becomes $B_1(t)$, the Fredholm determinant of the integral equation

$$\lambda f(\xi) = \int_{-1}^{1} \bar{\mathcal{K}}(\xi, \eta) f(\eta)\, d\eta, \tag{11.2.10}$$

where

$$\bar{\mathcal{K}}(\xi, \eta) = \frac{1}{2}(\mathcal{K}(\xi, \eta) + \mathcal{K}(\xi, -\eta)), \tag{11.2.11}$$

is the even part of

$$\mathcal{K}(\xi, \eta) = \frac{\sin(\xi - \eta)\pi t}{(\xi - \eta)\pi}. \tag{11.2.12}$$

In other words,

$$A_1(\alpha) \to B_1(t) = \prod_{i=0}^{\infty}(1 - \lambda_{2i}), \tag{11.2.13}$$

where λ_{2i} are the eigenvalues of the integral equation (6.3.14) corresponding to the even solutions.

Similarly,

$$A_1'(\alpha) \to B_1'(t) = \prod_{i=0}^{\infty}(1 - \lambda_{2i+1}), \tag{11.2.14}$$

where λ_{2i+1} are the eigenvalues of the same integral equation (6.3.14) corresponding to the odd solutions.

Thus in the limit of large N,

$$A_2(\alpha) \to E_2(0; s) = E_1(0; s) \cdot E_1'(s) = \prod_{i=0}^{\infty}(1 - \lambda_i), \tag{11.2.15}$$

which is identical to the spacing distribution for the Gaussian unitary ensemble.

This result was expected. Actually, the limiting n-point correlation function for any finite n being identical for the circular and Gaussian unitary ensembles, all their statistical properties involving a finite number of levels will coincide in the limit $N \to \infty$.

11.3 Orthogonal Ensemble. Correlation and Cluster Functions

The orthogonal ensemble is the most important from a practical point of view. It is also one of the most complicated from the mathematical point of view. We can use the identity

$$|e^{i\theta_j} - e^{i\theta_k}| = i^{-1} \exp\left(-\frac{i}{2}(\theta_j + \theta_k)\right)(e^{i\theta_k} - e^{i\theta_j}), \quad \text{if } \theta_j \leqslant \theta_k, \tag{11.3.1}$$

to write

$$\prod_{1 \leqslant j < k \leqslant N} |e^{i\theta_j} - e^{i\theta_k}| = i^{-N(N-1)/2} \exp\left(-\frac{i}{2}(N-1)\sum_1^N \theta_j\right) \prod_{1 \leqslant j < k \leqslant N} (e^{i\theta_k} - e^{i\theta_j}), \tag{11.3.2}$$

where the θ_j are supposed to be ordered

$$-\pi \leqslant \theta_1 \leqslant \theta_2 \leqslant \cdots \leqslant \theta_N \leqslant \pi. \tag{11.3.3}$$

Writing the product of differences in (11.3.2) as a Vandermonde determinant and multiplying the column containing the powers of $e^{i\theta_j}$ by $\exp(-i(N-1)\theta_j/2)$, we have

$$\prod_{1\leqslant j<k\leqslant N} |e^{i\theta_j} - e^{i\theta_k}| = i^{-N(N-1)/2} \det[e^{ip\theta_j}], \tag{11.3.4}$$

$$p = \frac{1}{2}(1-N), \frac{1}{2}(3-N), \ldots, \frac{1}{2}(N-3), \frac{1}{2}(N-1),$$
$$j = 1, 2, \ldots, N \tag{11.3.5}$$

if the θ_j are ordered as in (11.3.3).

We want to write (11.3.4) as an $N \times N$ quaternion determinant. For this purpose let $S_N(\theta)$ be as in (11.1.6), and let

$$D_N(\theta) = \frac{d}{d\theta} S_N(\theta) = \frac{1}{2\pi} \sum_p ipe^{ip\theta}, \tag{11.3.6}$$

$$I_N(\theta) = \int_0^\theta S_N(\varphi)\, d\varphi, \tag{11.3.7}$$

so that

$$I_N(\theta) = \frac{1}{2\pi i} \sum_p p^{-1} e^{ip\theta}, \quad N \text{ even}, \tag{11.3.8}$$

$$= \frac{1}{2\pi i} \sum_p{}' p^{-1} e^{ip\theta} + \frac{\theta}{2\pi}, \quad N \text{ odd}. \tag{11.3.9}$$

the term $p = 0$ being excluded in the last sum. In addition, let

$$J_N(\theta) = -\frac{1}{2\pi i} \sum_q q^{-1} e^{iq\theta}, \tag{11.3.10}$$

where q takes the values

$$q = \pm\frac{1}{2}(N+1), \pm\frac{1}{2}(N+3), \ldots. \tag{11.3.11}$$

Then

$$I_N(\theta) - J_N(\theta) = \varepsilon_N(\theta) \tag{11.3.12}$$

is a step function whose character depends only on the parity of N. In fact, for any integer m with

$$2\pi m < \theta < 2\pi(m+1), \tag{11.3.13}$$

we have

$$\varepsilon_N(\theta) = \frac{1}{2}(-1)^m, \quad N \text{ even}; \qquad \varepsilon_N(\theta) = m + \frac{1}{2}, \quad N \text{ odd}. \tag{11.3.14}$$

At the points of discontinuity $\theta = 2\pi m$,

$$\varepsilon_N(\theta) = 0, \quad N \text{ even}; \qquad \varepsilon_N(\theta) = m, \quad N \text{ odd}. \tag{11.3.15}$$

The lack of uniform convergence of the series defining J_N will not cause any difficulty. The function $S_N(\theta)$ is even in θ, while D_N, I_N, J_N and ε_N are all odd in θ.

Define the quaternion K_{N1} by its 2×2 matrix representation

$$\Theta(K_{N1}(\theta)) = \begin{bmatrix} S_N(\theta) & D_N(\theta) \\ J_N(\theta) & S_N(\theta) \end{bmatrix}, \tag{11.3.16}$$

and the $N \times N$ quaternion matrix

$$[K_{N1}(\theta_j - \theta_k)]_{j,k=1,\dots,N}. \tag{11.3.17}$$

We will now verify that $P_{N1}(\theta_1, \dots, \theta_N)$, Eq. (11.3.4), is proportional to $\det[K_{N1}(\theta_j - \theta_k)]$, or that $\det[\Theta(K_{N1}(\theta_j - \theta_k))]$ is proportional to the square of $P_{N1}(\theta_1, \dots, \theta_N)$ (cf. Section 5.1). Moreover, we will also verify that this K_{N1} satisfies all the conditions of Theorem 5.1.4, allowing us to write down immediately the n-level correlation and cluster functions. We treat separately the cases N even and N odd. These considerations parallel those of Chapter 7.

11.3.1 The Case $N = 2m$, Even. The $2N \times 2N$ matrix

$$\begin{bmatrix} S_N(\theta_j - \theta_k) & D_N(\theta_j - \theta_k) \\ I_N(\theta_j - \theta_k) & S_N(\theta_j - \theta_k) \end{bmatrix} \tag{11.3.18}$$

is the product of two matrices

$$\begin{bmatrix} e^{ip\theta_j} \\ (ip)^{-1} e^{ip\theta_j} \end{bmatrix} \quad \text{and} \quad \frac{1}{2\pi}[e^{-ip\theta_k} \quad ipe^{-ip\theta_k}], \tag{11.3.19}$$

of orders $2N \times N$ and $N \times 2N$ respectively. The rank of (11.3.18) is thus N. The N rows

$$[I_N(\theta_j - \theta_k) \quad S_N(\theta_j - \theta_k)] \tag{11.3.20}$$

are linear combinations of the N rows

$$[S_N(\theta_j - \theta_k) \quad D_N(\theta_j - \theta_k)]. \tag{11.3.21}$$

Therefore the determinant of $\Theta([K_{N1}])$ is not changed when we subtract the rows (11.3.20) from the corresponding rows of $\Theta([K_{N1}])$. This gives

$$\det[\Theta(K_{N1})] = \det[D_N(\theta_j - \theta_k)] \cdot \det[\varepsilon_N(\theta_j - \theta_k)]. \tag{11.3.22}$$

Now by virtue of (11.3.4) and (11.3.6),

$$\begin{aligned}\det[D_N(\theta_j - \theta_k)] &= (2\pi)^{-N} \det[e^{ip\theta_j}] \cdot \det[ipe^{-ip\theta_j}] \\ &= \text{const} \cdot (P_{N1}(\theta_1, \ldots, \theta_N))^2. \end{aligned} \tag{11.3.23}$$

The second determinant in (11.3.22), $\det[\varepsilon_N(\theta_j - \theta_k)]$ is (i) piecewise constant with possible discontinuities only at places where $\theta_j - \theta_k = 2\pi m$ with integer m, (ii) periodic with period 2π in each variable θ_j, and (iii) a symmetric function of $(\theta_1, \ldots, \theta_N)$. It follows from these three properties that this determinant is a constant independent of $\theta_1, \ldots, \theta_N$, except at the points of discontinuity where $P_{N1}(\theta_1, \ldots, \theta_N) = 0$.

11.3.2 The Case $N = 2m + 1$, Odd. In this case zero appears as a value of p in (11.3.19). Replacing

$$(ip)^{-1} e^{ip\theta_j} \to \delta^{-1}, \quad ipe^{ip\theta_k} \to \delta, \tag{11.3.24}$$

when $p = 0$, consider now instead of (11.3.18), the matrix

$$\begin{bmatrix} S_N(\theta_j - \theta_k) & D_N(\theta_j - \theta_k) + \delta/2\pi \\ I_N(\theta_j - \theta_k) + (2\pi\delta)^{-1} - \dfrac{\theta_j - \theta_k}{2\pi} & S_N(\theta_j - \theta_k) \end{bmatrix}, \tag{11.3.25}$$

of rank N. The determinant of

$$\Theta_\delta(K_{N1}) = \begin{bmatrix} S_N(\theta_j - \theta_k) & D_N(\theta_j - \theta_k) + \delta/2\pi \\ J_N(\theta_j - \theta_k) & S_N(\theta_j - \theta_k) \end{bmatrix} \tag{11.3.26}$$

is not changed by subtracting the N rows of (11.3.25) from the corresponding N rows of (11.3.26). Therefore,

$$\begin{aligned}\det[\Theta_\delta(K_{N1})] = {} & \det[D_N(\theta_j - \theta_k) + \delta/2\pi] \\ & \times \det[\varepsilon_N(\theta_j - \theta_k) + (2\pi\delta)^{-1} + (\theta_k - \theta_j)/(2\pi)]. \end{aligned} \tag{11.3.27}$$

The first factor is

$$\text{const} \cdot \delta \cdot (\det[e^{ip\theta_j}])^2. \tag{11.3.28}$$

In the second factor we subtract the first column from each of the remaining columns, obtaining

$$(2\pi\delta)^{-1}\det[1_N + \mathrm{O}(\delta), \varepsilon_N(\theta_j - \theta_k) - \varepsilon_N(\theta_j - \theta_1) + ((\theta_k - \theta_1)/2\pi)], \quad (11.3.29)$$

where 1_N means a single column of unit elements, and k labels the remaining columns from 2 to N. Passing to the limit $\delta \to 0$ in (11.3.27), (11.3.28) and (11.3.29),

$$\det[\Theta(K_{N1})] = \text{const} \cdot (\det[e^{ip\theta_j}])^2 \cdot d_N, \quad (11.3.30)$$

where now

$$d_N = \det[1_N, \varepsilon_N(\theta_j - \theta_k) - \varepsilon_N(\theta_j - \theta_1)]. \quad (11.3.31)$$

The terms $((\theta_k - \theta_1)/2\pi)$ in Eq. (11.3.29) contributing nothing to the determinant. By the same argument as was used for N even; d_N is independent of $\theta_1, \ldots, \theta_N$ except at places where $P_{N1}(\theta_1, \ldots, \theta_N) = 0$.

11.3.3 Conditions of Theorem 5.1.4. From the parity of the functions S_N, D_N and J_N and Eq. (2.4.17), it is evident that $[K_{N1}]$ is self-dual. Adopting the notation

$$(f * g)(\theta) = \int_0^{2\pi} f(\theta - \varphi) g(\varphi)\, d\varphi, \quad (11.3.32)$$

it is straightforward to verify that

$$S_N * S_N = S_N, \quad (11.3.33)$$

$$D_N * S_N = S_N * D_N = D_N, \quad (11.3.34)$$

$$J_N * S_N = S_N * J_N = 0, \quad (11.3.35)$$

$$J_N * D_N = D_N * J_N = 0. \quad (11.3.36)$$

Hence

$$K_{N1} * K_{N1} = K_{N1} + \lambda K_{N1} - K_{N1}\lambda, \quad (11.3.37)$$

with

$$\lambda = \frac{1}{2}\begin{bmatrix} 1 & 0 \\ 0 & -1 \end{bmatrix} = -\frac{i}{2} e_1. \quad (11.3.38)$$

Also

$$\int_0^{2\pi} K_{N1}(0)\, d\theta = \int_0^{2\pi} S_N(0)\, d\theta = N. \quad (11.3.39)$$

Thus all the conditions of Theorem 5.1.4 are satisfied. Integrating over all the variables, one can fix the constants:

$$P_{N1}(\theta_1, \ldots, \theta_N) = \frac{1}{N!} \det[K_{N1}(\theta_j - \theta_k)]_{j,k=1,\ldots,N}. \tag{11.3.40}$$

11.3.4 Correlation and Cluster Functions. As in Chapter 5, Theorem 5.1.4 now gives us with (11.3.40) the n-level correlation function

$$R_n(\theta_1, \ldots, \theta_n) = \det[K_{N1}(\theta_j - \theta_k)]_{j,k=1,\ldots,n}, \tag{11.3.41}$$

and the n-level cluster function

$$T_n(\theta_1, \ldots, \theta_n) = \sum_P K_{N1}(\theta_1 - \theta_2) K_{N1}(\theta_2 - \theta_3) \cdots K_{N1}(\theta_n - \theta_1), \tag{11.3.42}$$

the sum is taken over all $(n-1)!$ distinct cyclic permutations of the indices $(1, \ldots, n)$.

The level density is

$$R_1(\theta) = K_{N1}(0) = S_N(0) = \frac{N}{2\pi}, \tag{11.3.43}$$

as was expected. In the limit $N \to \infty$, $N\theta = 2\pi\xi$, $N\varphi = 2\pi\eta$ finite, one gets

$$\lim \frac{2\pi}{N} S_N(\theta - \varphi) = s(r) = \frac{\sin(\pi r)}{\pi r}, \tag{11.3.44}$$

$$\lim \left(\frac{2\pi}{N}\right)^2 D_N(\theta - \varphi) = D(r) = \frac{d}{dr} s(r), \tag{11.3.45}$$

$$\lim I_N(\theta - \varphi) = I(r) = \int_0^r s(t)\, dt, \tag{11.3.46}$$

$$\lim J_N(\theta - \varphi) = J(r) = I(r) - \frac{1}{2}, \tag{11.3.47}$$

$$\lim \frac{2\pi}{N} K_{N1}(\theta - \varphi) \approx K_1(r) = \begin{bmatrix} s(r) & D(r) \\ J(r) & s(r) \end{bmatrix}. \tag{11.3.48}$$

Thus in the limit of large N, the n-level correlation and cluster functions for the circular orthogonal ensemble are identical to those for the Gaussian orthogonal ensemble.

11.4 Orthogonal Ensemble. Level Spacings

To avoid unnecessary complications let us take N even, $N = 2m$. Let $u(\theta)$, $v(\theta)$ be functions defined over $(-\pi, \pi)$ and consider the average

$$H = \left\langle \prod_{\text{alt}} u(\theta_j) \prod_{\text{alt}}{}' v(\theta_k) \right\rangle$$
$$= \int_0^{2\pi} \cdots \int_0^{2\pi} \prod_{\text{alt}} u(\theta_j) \prod_{\text{alt}}{}' v(\theta_k) P_{N1}(\theta_1, \ldots, \theta_N)\, d\theta_1 \cdots d\theta_N, \quad (11.4.1)$$

taken with respect to the orthogonal ensemble $\beta = 1$, defined by Theorem 10.4.4. Here $\prod_{\text{alt}}$ denotes a product taken over a set of m alternate points θ_j as they lie on the unit circle and $\prod'_{\text{alt}}$ a product over the remaining m points. This average can again be calculated by integration over alternate variables, using Eq. (11.3.4). We define

$$f'_{pq} = \int_{-\pi}^{\pi} \int_{-\pi}^{\pi} u(\theta) v(\varphi) \varepsilon(\theta - \varphi) (e^{ip\varphi + iq\theta} - e^{ip\theta + iq\varphi})\, d\theta\, d\varphi, \quad (11.4.2)$$

with

$$\varepsilon(\theta) = \frac{1}{2}\text{sign}(\theta) = \begin{cases} |\theta|/(2\theta), & \theta \neq 0, \\ 0, & \text{otherwise}, \end{cases} \quad (11.4.3)$$

and do integrations step by step as in Section 5.5 or 7.3. The result is

$$H^2 = \text{const} \cdot \det[f'_{pq}], \quad p, q = -m + \frac{1}{2}, -m + \frac{3}{2}, \ldots, m - \frac{1}{2}. \quad (11.4.4)$$

By reversing the order of the columns we can write

$$H^2 = \text{const} \cdot \det[f_{pq}], \quad (11.4.5)$$

with

$$f_{pq} = \frac{ip}{4\pi} f'_{p,-q} = \frac{ip}{4\pi} \int_{-\pi}^{\pi} \int_{-\pi}^{\pi} u(\theta) v(\varphi) \varepsilon(\theta - \varphi) (e^{ip\varphi - iq\theta} - e^{ip\theta - iq\varphi})\, d\theta\, d\varphi. \quad (11.4.6)$$

If the functions $u(\theta)$ and $v(\varphi)$ satisfy the relation

$$u(-\theta) v(-\varphi) = u(\theta) v(\varphi), \quad (11.4.7)$$

then

$$f_{-p,-q} = f_{p,q}, \quad (11.4.8)$$

and there are further simplifications. As can be verified, we now have

$$\det[f_{p,q}+f_{-p,q}]=\det[f_{p,q}-f_{-p,q}], \quad p,q=\frac{1}{2},\frac{3}{2},\ldots,m-\frac{1}{2}, \tag{11.4.9}$$

and H may be written as a determinant

$$H=\text{const}\cdot\det[F_{pq}], \quad p,q=\frac{1}{2},\frac{3}{2},\ldots,m-\frac{1}{2}; \tag{11.4.10}$$

with

$$\begin{aligned} F_{pq} &= f_{p,q}+f_{-p,q} \\ &= \frac{p}{\pi}\int_{\pi}^{\pi}\int_{-\pi}^{\pi} u(\theta)v(\varphi)\varepsilon(\theta-\varphi)(\cos(p\varphi)\sin(q\theta)-\cos(p\theta)\sin(q\varphi))\,d\theta\,d\varphi. \end{aligned} \tag{11.4.11}$$

The constant in (11.4.10) above can be fixed to be 1 by the condition that when $u(\theta)=v(\varphi)=1$, then $H=1$.

To get the probability that a randomly chosen interval of length 2α is empty of eigenvalues, put

$$u(\theta)=v(\theta)=\begin{cases}1, & \text{if } -\pi+\alpha<\theta<\pi-\alpha,\\ 0, & \text{otherwise,}\end{cases} \tag{11.4.12}$$

in (11.4.1). We have chosen the center of the excluded interval to be at $\theta=\pi$ so that (11.4.7) is satisfied. From (11.4.10) and (11.4.11) we then have

$$A_1(\alpha)=\det[F_{pq}], \quad p,q=\frac{1}{2},\frac{3}{2},\ldots,m-\frac{1}{2}, \tag{11.4.13}$$

where

$$\begin{aligned} F_{pq} &= \delta_{pq}-\frac{1}{\pi}\int_{-\alpha}^{\alpha}\cos(p\theta)\cos(q\theta)\,d\theta \\ &= \delta_{pq}-\frac{\sin(p-q)\alpha}{(p-q)\pi}-\frac{\sin(p+q)\alpha}{(p+q)\pi}. \end{aligned} \tag{11.4.14}$$

If we choose

$$\begin{aligned} &u(\theta)=v(\theta)=1, \quad \text{if } -\pi+\alpha<\theta<\pi-\alpha, \\ &u(\theta)=0,\ v(\theta)=2, \quad \text{if } \pi-\alpha<\theta<\pi+\alpha, \end{aligned} \tag{11.4.15}$$

in (11.4.1), then we get the probability that a randomly chosen interval of length 2α will contain at most one eigenvalue. The choice $v(\theta) = 2$, rather than 1, in the interval $(\pi - \alpha, \pi + \alpha)$ arises from the fact that while ordering $-\pi \leqslant \theta_1 \leqslant \cdots \leqslant \theta_{2m} \leqslant \pi$, the interval $(\pi - \alpha, \pi + \alpha)$ becomes unattainable for half the levels belonging to the alternate series. Equations (11.4.10) and (11.4.11) then give

$$A_1'(\alpha) = \det[F_{pq}'], \quad p, q = \frac{1}{2}, \frac{3}{2}, \ldots, m - \frac{1}{2}; \tag{11.4.16}$$

$$\begin{aligned} F_{pq}' &= \delta_{pq} - \frac{1}{\pi} \int_{-\alpha}^{\alpha} \sin(p\theta) \sin(q\theta)\, d\theta \\ &= \delta_{pq} - \frac{\sin(p-q)\alpha}{(p-q)\pi} + \frac{\sin(p+q)\alpha}{(p+q)\pi}. \end{aligned} \tag{11.4.17}$$

When $m \to \infty$, while $m\alpha = \pi t$ remains finite, the limits of $A_1(\alpha)$ and $A_1'(\alpha)$ are $B_1(t)$ and $B_1'(t)$ respectively. These limits are obtained exactly as in Section 11.2.

Note that although $B_1'(t)$ is the probability that a randomly chosen interval of length $2t$ will not contain any of the eigenvalues belonging to the same alternate series, its second derivative is not a probability. The probability density for spacings between pairs of next nearest neighbors (i.e. pairs of levels having one level in between) is given instead (cf. Appendix A.8) by

$$p_1(1; s) = \frac{d^2}{ds^2}(B_1(t) + B_1'(t)), \quad s = 2t, \tag{11.4.18}$$

where $B_1(t) = E_1(0; s)$, the probability that the interval of length $2t$ will contain none of the eigenvalues, is given by (7.3.11) or (11.2.13) and $B_1'(t)$ is given by (11.2.14). Comparing with (7.3.19) and (7.5.21) we get the identity

$$B_1'(t) - B_1(t) = \prod_{\rho=0}^{\infty} (1 - \lambda_{2\rho}) \cdot \sum_{i=0}^{\infty} \frac{\lambda_{2i}}{1 - \lambda_{2i}} f_{2i}(1) \int_{-1}^{1} f_{2i}(y)\, dy, \tag{11.4.19}$$

or with (11.2.13) and (11.2.14),

$$\prod_{i=0}^{\infty} \left(\frac{1 - \lambda_{2i+1}}{1 - \lambda_{2i}} \right) = 1 + \sum_{i=0}^{\infty} \frac{\lambda_{2i}}{1 - \lambda_{2i}} f_{2i}(1) \int_{-1}^{1} f_{2i}(y)\, dy, \tag{11.4.20}$$

where $f_i(x)$ are the normalized spheroidal functions defined by Eqs. (6.3.17) or (6.3.22) and (6.4.23). A proof of (11.4.20) is given in Chapter 20 and another one in Chapter 21.

Proceeding as in Section 7.8, we can find lower and upper bounds to $B_1'(t)$,

$$\exp\left(-\left(\frac{\pi t}{2}\right)^2\right) \leqslant B_1'(t) \leqslant \left(1+\left(\frac{\pi t}{2}\right)^2\right)\exp\left(-\left(\frac{\pi t}{2}\right)^2\right). \tag{11.4.21}$$

11.5 Symplectic Ensemble. Correlation and Cluster Functions

The joint probability density of the eigenvalue angles of a unitary self-dual random matrix taken from the symplectic ensemble was derived in Chapter 10 as

$$P_{N4}(\theta_1, \dots, \theta_N) = \text{const} \cdot \prod_{j<k} |e^{i\theta_j} - e^{i\theta_k}|^4. \tag{11.5.1}$$

To deal with integrals containing such an expression we write (11.5.1) as a confluent alternant. As in (11.3.2)

$$\prod_{j<k} |e^{i\theta_j} - e^{i\theta_k}|^4 = \exp\left(-2i(N-1)\sum_1^N \theta_j\right) \cdot \prod_{j<k} (e^{i\theta_j} - e^{i\theta_k})^4. \tag{11.5.2}$$

Note that because the power 4 is an even integer, ordering of the variables is no longer necessary. The fourth power of the product of differences expressed as a determinant (cf. Mehta, 1989MT, Section 7.1) is

$$\prod_{1\leqslant j<k\leqslant N} (e^{i\theta_j} - e^{i\theta_k})^4 = \det[e^{ik\theta_j}, ke^{i(k-1)\theta_j}]_{1\leqslant j\leqslant N,\ 0\leqslant k\leqslant 2N-1}. \tag{11.5.3}$$

If we multiply the $(2j-1)$th and the $(2j)$th columns by $\exp(-(2N-1)i\theta_j/2)$ and $\exp(-(2N-3)i\theta_j/2)$, respectively, we obtain

$$\begin{aligned}\exp\left(-2i(N-1)\sum_1^N \theta_j\right)\prod_{j<k}(e^{i\theta_j} - e^{i\theta_k})^4 &= \det\left[e^{ip\theta_j} \quad \left(p+N-\frac{1}{2}\right)e^{ip\theta_j}\right] \\ &= \det[e^{ip\theta_j} \quad pe^{ip\theta_j}],\end{aligned} \tag{11.5.4}$$

where p varies over the half odd integers

$$-\left(N-\frac{1}{2}\right), -\left(N-\frac{3}{2}\right), \dots, \left(N-\frac{1}{2}\right). \tag{11.5.5}$$

We construct next an $N \times N$ quaternion self-dual matrix $K_{N4}(\theta_1, \dots, \theta_N)$ satisfying the conditions of Theorem 5.1.4 and such that its determinant be proportional to

$P_{N4}(\theta_1, \ldots, \theta_N)$. With S_N, D_N and I_N defined as in Eqs. (11.1.6), (11.3.6) and (11.3.7), let the 2×2 matrix representation of $K_{N4}(\theta)$ be

$$\Theta[K_{N4}(\theta)] = \frac{1}{2}\begin{bmatrix} S_{2N}(\theta) & D_{2N}(\theta) \\ IS_{2N}(\theta) & S_{2N}(\theta) \end{bmatrix}. \tag{11.5.6}$$

With Eqs. (11.3.33), (11.3.34) and

$$D_{2N} * I_{2N} = I_{2N} * D_{2N} = S_{2N}, \tag{11.5.7}$$

it is straightforward to verify that

$$K_{N4} * K_{N4} = K_{N4}. \tag{11.5.8}$$

Also

$$\int_0^{2\pi} K_{N4}(0)\, d\theta = \frac{1}{2}\int S_{2N}(0)\, d\theta = \frac{N}{2\pi}. \tag{11.5.9}$$

Thus all the conditions of Theorem 5.1.4 are satisfied. It remains to verify that

$$\det[\Theta(K_{N4})] = \text{const} \cdot \prod_{1 \leqslant j < k \leqslant N} |e^{i\theta_j} - e^{i\theta_k}|^8. \tag{11.5.10}$$

For this, note that $\Theta(K_{N4})$ is the product of two matrices,

$$\Theta(K_{N4}) = \frac{1}{4\pi}\begin{bmatrix} e^{ip\theta_j} \\ (ip)^{-1}e^{ip\theta_j} \end{bmatrix}\begin{bmatrix} e^{-ip\theta_k} & ipe^{-ip\theta_k} \end{bmatrix}, \tag{11.5.11}$$

and the two matrices on the right have, apart from a constant, the same determinant, the confluent alternant of Eq. (11.5.4).

We are now prepared to write down the correlation and cluster functions. Integrating over all the variables, we fix the constants,

$$P_{N4}(\theta_1, \ldots, \theta_N) = \frac{1}{N!}\det[K_{N4}(\theta_j - \theta_k]_{j,k=1,\ldots,N}. \tag{11.5.12}$$

The n-level correlation function is (cf. Theorem 5.4.1)

$$R_n(\theta_1, \ldots, \theta_n) = \det[K_{N4}(\theta_j - \theta_k)]_{j,k=1,\ldots,n}; \tag{11.5.13}$$

and the n-level cluster function is

$$T_n(\theta_1, \ldots, \theta_n) = \sum_P K_{N4}(\theta_1 - \theta_2)K_{N4}(\theta_2 - \theta_3)\cdots K_{N4}(\theta_n - \theta_1), \tag{11.5.14}$$

the sum being taken, as usual, over all $(n-1)!$ distinct cyclic permutations of the indices $(1, 2, \ldots, n)$.

Putting $n = 1$, we get the level density

$$R_1(\theta) = K_{N4}(0) = \frac{1}{2} S_{2N}(0) = \frac{N}{2\pi}, \tag{11.5.15}$$

as expected. In the limit $N \to \infty$, $N\theta = 2\pi\xi$ finite,

$$\lim \frac{2\pi}{N} K_{N4}(\theta) \approx K_4(\xi) = \frac{1}{2} \begin{bmatrix} s(2\xi) & D(2\xi) \\ I(2\xi) & s(2\xi) \end{bmatrix}, \tag{11.5.16}$$

with s, Ds and Is given by Eqs. (11.3.44)–(11.3.46) or (7.2.36)–(7.2.39).

A comparison with Section 7.2 will show that in the limit of large N the n-level correlation and cluster functions for the circular symplectic ensemble coincide with those of the Gaussian symplectic ensemble.

11.6 Relation Between Orthogonal and Symplectic Ensembles

With Eqs. (11.3.4) and (11.5.4) we will now prove the important result:

11.6.1. Theorem. *The statistical properties of N alternate angles θ_j, where $e^{i\theta_j}$ are the eigenvalues of a symmetric unitary random matrix of order $2N \times 2N$ taken from the orthogonal ensemble, are identical to those of the N angles φ_j, where $e^{i\varphi_j}$ are the eigenvalues of an $N \times N$ quaternion self-dual unitary random matrix taken from the symplectic ensemble.*

Proof. Suppose that $\theta_1 \leqslant \theta_2 \leqslant \cdots \leqslant \theta_{2N} \leqslant \theta_1 + 2\pi$. We write the joint probability density

$$P_{2N,1}(\theta_1, \ldots, \theta_{2N}) = \text{const} \cdot \prod_{1 \leqslant j < k \leqslant 2N} |e^{i\theta_j} - e^{i\theta_k}| = \text{const} \cdot \det[e^{ip\theta_j}], \tag{11.6.1}$$

$$p = -\left(N - \frac{1}{2}\right), -\left(N - \frac{3}{2}\right), \ldots, \left(N - \frac{1}{2}\right), \quad j = 1, 2, \ldots, 2N, \tag{11.6.2}$$

as in (11.3.4) and integrate over alternate variables $\theta_1, \theta_3, \ldots, \theta_{2N-1}$, as in Section 7.3 or 11.4. The limits of integration for θ_{2j-1} are $(\theta_{2j-2}, \theta_{2j})$, except when $j = 1$. For θ_1 these limits are $(\theta_{2N} - 2\pi, \theta_2)$. Thus the integration over the odd-indexed variables replaces the θ_1 column with

$$\int_{\theta_{2N}-2\pi}^{\theta_2} d\theta \, e^{ip\theta} = (ip)^{-1}(e^{ip\theta_2} + e^{ip\theta_{2N}}), \tag{11.6.3}$$

and the θ_{2j-1} column, for $j > 1$, with

$$\int_{\theta_{2j-2}}^{\theta_{2j}} d\theta\, e^{ip\theta} = (ip)^{-1}(e^{ip\theta_{2j}} - e^{ip\theta_{2j-2}}). \tag{11.6.4}$$

This later column can be changed, by adding other columns, to

$$(ip)^{-1}(e^{ip\theta_{2j}} + e^{ip\theta_{2N}}). \tag{11.6.5}$$

The $(2N-1)$th column is now simply $(ip)^{-1}2e^{ip\theta_{2N}}$, which allows us to drop the $e^{ip\theta_{2N}}$ term from every other column. The final result (the limits of integrations on the left-hand side being $\theta_1 \leqslant \theta_2 \leqslant \cdots \leqslant \theta_{2N} \leqslant \theta_1 + 2\pi$) is

$$\begin{aligned}\int \cdots \int d\theta_1\, d\theta_3 \cdots d\theta_{2N-1} P_{2N,1}(\theta_1, \ldots, \theta_{2N}) \\ = \text{const} \cdot \det[\, e^{ip\theta_{2j}} \quad ipe^{ip\theta_{2j}}\,] \\ = \text{const} \cdot P_{N4}(\theta_2, \theta_4, \ldots, \theta_{2N}),\end{aligned} \tag{11.6.6}$$

which establishes the theorem. □

11.7 Symplectic Ensemble. Level Spacings

We did not study in detail the level spacings for the Gaussian symplectic ensemble in Section 8.3. We have seen in Sections 11.1, 11.3 and 11.5 that in the limit of large N Gaussian ensembles are statistically equivalent to the corresponding circular ensembles. We have also seen in Section 11.6 that for any N, the eigenvalues of a random $N \times N$ matrix from the circular symplectic ensemble are statistically equivalent to the alternate eigenvalues of a random $2N \times 2N$ matrix from the circular orthogonal ensemble. We now use this equivalence to deduce the spacing probabilities for the symplectic ensemble, from those of the orthogonal one.

Let us denote by $B_4(t; y_1, \ldots, y_r)$ the probability that a randomly chosen interval of length $2t$ (measured in units of the local mean spacing) contains one level each at positions $y_1 < \cdots < y_r$ inside this interval and none others, the levels being the eigenvalues of a random matrix taken from the (Gaussian or circular) symplectic ensemble; let $B_1(t; y_1, \ldots, y_r)$ denote the same probability when the random matrix is taken from the (Gaussian or circular) orthogonal ensemble. The relation between symplectic and orthogonal ensembles, proved in Section 11.6 above, then tells us that $B_4(t; y_1, \ldots, y_r)$ is a sum of four terms, arising from the various possibilities of alternating $y_1, \ldots, y_r$ with the other eigenvalues $z_1, \ldots, z_s$, $s = r-1, r$ or $r+1$,

$$
\begin{aligned}
2B_4(t; y_1, \ldots, y_r) = & \int B_1(2t; y_1, z_1, y_2, z_2, \ldots, z_{r-1}, y_r)\, dz_1 \cdots dz_{r-1} \\
& + \int B_1(2t; y_1, z_1, y_2, z_2, \ldots, y_r, z_r)\, dz_1 \cdots dz_r \\
& + \int B_1(2t; z_0, y_1, z_1, y_2, \ldots, z_{r-1}, y_r)\, dz_0\, dz_1 \cdots dz_{r-1} \\
& + \int B_1(2t; z_0, y_1, z_1, y_2, \ldots, y_r, z_r)\, dz_0\, dz_1 \cdots dz_r,
\end{aligned}
\tag{11.7.1}
$$

where the integrations on the right-hand side are carried with the restrictions

$$
-2t < z_0 < y_1 < z_1 < y_2 < \cdots y_r < z_r < 2t. \tag{11.7.2}
$$

Note that on the right-hand side we have $2t$ instead of t and an extra factor 2 on the left-hand side, because the local mean spacings for orthogonal and symplectic ensembles differ by a factor 2.

It is straightforward, though somewhat cumbersome, to substitute the expressions of B_1 from Sections 7.4 and 7.5 and integrate over the alternate variables z_j. The result is much simpler if we integrate over all the variables in the region (11.7.2). Thus the probability $E_4(r; s)$, that a randomly chosen interval of length s contains exactly r eigenvalues of a self-dual quaternion random matrix taken from the symplectic ensemble is given by

$$
E_4(0; s) = E_1(0; 2s) + \frac{1}{2} E_1(1; 2s), \tag{11.7.3}
$$

$$
E_4(r; s) = E_1(2r; 2s) + \frac{1}{2} E_1(2r-1; 2s) + \frac{1}{2} E_1(2r+1; 2s), \quad r > 0, \tag{11.7.4}
$$

where $E_1(r; s)$ is the same probability when the random matrix is chosen from the orthogonal ensemble. The $E_1(r; s)$ are given by Eqs. (7.4.13) and (7.5.21). Equations (11.7.3) and (11.7.4) can also be written as

$$
\begin{aligned}
E_4(0; s/4) &= \frac{1}{2}\left(E_+(0, t) + E_-(0, t)\right) \\
&= \frac{1}{2}\left(F_+(1, t) + F_-(1, t)\right) \\
&= \frac{1}{2}\left(\prod_i (1 - \lambda_{2i}) + \prod_i (1 - \lambda_{2i+1})\right),
\end{aligned}
\tag{11.7.5}
$$

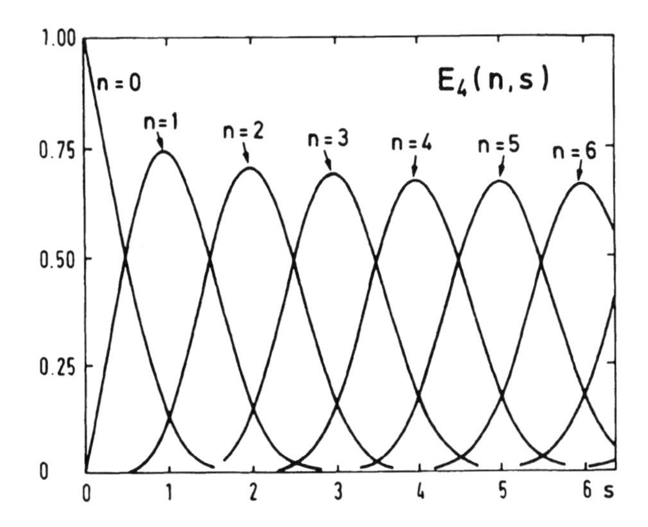

Figure 11.1. The n-level spacings $E_4(n, s)$ for the Gaussian symplectic ensemble.

$$E_4(r; s/4) = \frac{1}{2}(E_+(r; t) + E_-(r, t)), \tag{11.7.6}$$

$s = 2t$; where λ_i are the eigenvalues of the integral equation (6.3.14) and $E_\pm(r, t)$ are defined by Eqs. (7.5.24)–(7.5.26).

Figure 11.1 shows the functions $E_4(n; s)$ for some values of n. Note that the various peaks are higher and narrower than the corresponding ones of $E_2(n; s)$ for the unitary ensemble.

11.8 Brownian Motion Model

Just as in Chapter 9, we can construct a Brownian motion model for the elements of our unitary matrices. Every matrix U taken from the ensemble $E_{\beta c}$ can be written as

$$U = VV^D, \tag{11.8.1}$$

where V is unitary and V^D is the transpose or the dual (cf. Chapter 2 and Appendix A.3) of V, according as β is 1 or 4. For the unitary ensemble, $\beta = 2$, V^D is unrelated to V, except that V^D is unitary and (11.8.1) holds. A permissible small change in U is then given by

$$\delta U = V(i\delta M)V^D, \tag{11.8.2}$$

where δM is an infinitesimal Hermitian matrix which is symmetric and hence real if $\beta = 1$ and self-dual if $\beta = 4$. Let us denote the independent real components of δM by

δM_μ ($\equiv \delta M_{jk}^{(\lambda)}$); $\mu = 1, 2, \ldots, p$; $p = N + \frac{1}{2}N(N-1)\beta$ (or $\lambda = 0$ for $1 \leqslant j = k \leqslant N$, and $0 \leqslant \lambda \leqslant \beta - 1$ for $1 \leqslant j < k \leqslant N$). The isotropic and representation independent Brownian motion of U is defined by the statement that U at time t moves to $U + \delta U$ at time $t + \delta t$, where δU is given by (11.8.2) and the real parameters δM_μ are independent random variables with the moments

$$\langle \delta M_\mu \rangle = 0, \tag{11.8.3}$$

$$f\langle (\delta M_\mu)^2 \rangle = g_\mu kT\delta t, \tag{11.8.4}$$

where g_μ is given by (9.2.13).

The effect of the Brownian motion of U on its eigenvalues $\exp(i\theta_j)$ may again be found by choosing U to be diagonal at time t and calculating $\exp(i\theta_j + i\delta\theta_j)$ at time $t + \delta t$ by perturbation theory:

$$\delta\theta_j = \delta M_{jj}^{(0)} + \sum_{\substack{\ell \\ (\ell \neq j)}} \sum_{\lambda=0}^{\beta-1} (\delta M_{\ell j}^{(\lambda)})^2 \left(\frac{1}{2} \cot\left(\frac{\theta_j - \theta_\ell}{2} \right) \right) + \cdots. \tag{11.8.5}$$

Equations (11.8.3) and (11.8.4) imply that the angles θ_j execute a Brownian motion with

$$f\langle \delta\theta_j \rangle = E(\theta_j)\delta t, \tag{11.8.6}$$

$$f\langle (\delta\theta_j)^2 \rangle = 2kT\delta t, \tag{11.8.7}$$

where

$$E(\theta_j) = \sum_{\substack{\ell \\ (\ell \neq j)}} \frac{1}{2} \cot\left(\frac{\theta_j - \theta_\ell}{2} \right). \tag{11.8.8}$$

This force $E(\theta_j)$ is exactly the component, tangential to the circle, of the electric field produced at $\exp(i\theta_j)$ by unit charges placed at all the other points $\exp(i\theta_\ell)$ at which U has eigenvalues. Thus

$$E(\theta_j) = -\frac{\partial W}{\partial \theta_j}, \tag{11.8.9}$$

$$W = -\sum_{\ell < j} \ln |e^{i\theta_\ell} - e^{i\theta_j}|. \tag{11.8.10}$$

One may write the corresponding Focker–Planck equation and find that the unique stationary probability density for the eigenvalue angles which corresponds to the completely diffused probability density of U is

$$P(\theta_1, \ldots, \theta_N) = c' \prod_{\ell<j} |e^{i\theta_\ell} - e^{i\theta_j}|^\beta. \tag{11.8.11}$$

This is again a new proof of Theorem 10.4.4.

Summary of Chapter 11

Starting with the joint probability density $P(\theta_1, \ldots, \theta_N)$ of the eigenvalues $\exp(i\theta_j)$, $j = 1, \ldots, N$; for each of the three ensembles, orthogonal, unitary and symplectic, expressions are derived for the n-level correlation and cluster functions and for the spacings. The asymptotic results are identical to those for the corresponding Gaussian ensembles. (See the summaries of Chapters 6, 7 and 8.) The alternate eigenvalues of a matrix taken from the orthogonal (circular) ensemble behave as if they were the eigenvalues of a matrix from the symplectic (circular) ensemble, Section 11.6. From this one deduces the following expressions for the probability $E_4(n; s)$ that a randomly chosen interval of length s contains exactly n levels of a matrix from the symplectic ensemble:

$$E_4(0; s) = E_1(0; 2s) + \frac{1}{2} E_1(1; 2s), \tag{11.7.3}$$

and for $n > 0$,

$$\begin{aligned} E_4(n; s) &= E_1(2n; 2s) + \frac{1}{2} E_1(2n-1; 2s) + \frac{1}{2} E_1(2n+1; 2s) \\ &= \frac{1}{2}[E_+(n, 2s) + E_-(n, 2s)], \end{aligned} \tag{11.7.4}$$

where $E_1(n; s)$ is the same probability for a matrix from the orthogonal ensemble.

A Brownian motion model starting with any initial values of the matrix elements and ultimately settling to their equilibrium distributions is considered in Section 11.8. The joint probability density for the eigenvalues also settles to

$$P(\theta_1, \ldots, \theta_N) \propto \prod_{1\leqslant j<k\leqslant N} |\exp(i\theta_j) - \exp(i\theta_k)|^\beta, \tag{11.8.11}$$

where $\beta = 1$ for unitary symmetric matrices, $\beta = 2$ for unitary matrices and $\beta = 4$ for unitary symplectic matrices.

12

CIRCULAR ENSEMBLES. THERMODYNAMICS

In Chapter 11 we studied the correlation functions, cluster functions, and spacing distribution of the eigenvalues of a unitary matrix taken from Dyson's circular ensembles. In this chapter we calculate the partition function, the energy, the free energy, the entropy, and the specific heat of the energy levels defined in complete analogy with the classical mechanics. We then apply thermodynamic ideas to "derive" the leading terms in the asymptotic behavior of the spacing distribution for large spacings.

12.1 The Partition Function

As in Chapter 4, consider a thin circular conducting wire of radius unity; let N point charges be free to move on this wire. The universe is supposed to be two-dimensional. The charges repel one another according to the two-dimensional Coulomb law, so that the potential energy due to this electrostatic repulsion is

$$W = -\sum_{1 \leqslant \ell < j \leqslant N} \log |e^{i\theta_\ell} - e^{i\theta_j}|. \tag{12.1.1}$$

As in Chapter 4, we discard the trivial velocity-dependent contributions. The positional partition function at a temperature kT is given by

$$\Psi_N(\beta) = (2\pi)^{-N} \int_0^{2\pi} \cdots \int_0^{2\pi} e^{-\beta W} \, d\theta_1 \cdots d\theta_N, \tag{12.1.2}$$

$$\beta = (kT)^{-1} \tag{12.1.3}$$

where W is given by (12.1.1). The aim of this section is to prove the following theorem.

Theorem 12.1.1. *For any positive integer N and a real or complex β the partition function $\Psi_N(\beta)$ is given by*

$$\Psi_N(\beta) = \Gamma(1 + \beta N/2)[\Gamma(1 + \beta/2)]^{-N}. \tag{12.1.4}$$

Proof. The proof may be divided into three parts:

(1) The argument depends on the fact that W is bounded from below. It is intuitively clear (for a proof see Appendix A.22) that this minimum is attained when the N charges are situated at the corners of a regular polygon of N sides inscribed in the unit circle. The value of the minimum can be calculated and gives

$$W \geqslant W_0 = -(N/2) \log N. \tag{12.1.5}$$

Therefore we can write

$$\Psi_N(\beta) = \int_0^Y P(y) y^\beta \, dy, \tag{12.1.6}$$

where

$$Y = \exp(-W_0) = N^{N/2}, \tag{12.1.7}$$

and $P(y)$ is a positive weight function. In other words, $\Psi_N(\beta)$ is a moment function defined over a finite interval $(0, Y)$, and thus has special analytical properties (Shohat and Tamarkin, 1943). It is analytic in the upper half plane, $\operatorname{Re}\beta > 0$, and in this region it satisfies the inequality

$$|\Psi_N(\beta)| < C|Y^\beta|. \tag{12.1.8}$$

Now the function

$$\psi_N(\beta) = \Gamma(1 + \beta N/2)[\Gamma(1 + \beta/2)]^{-N} \tag{12.1.9}$$

satisfies all these conditions. It is singular only on the negative real axis and for large $|\beta|$,

$$\psi_N(\beta) \approx N^{1/2} (\pi\beta)^{-(N-1)/2} Y^\beta. \tag{12.1.10}$$

The function

$$\Delta(\beta) = e^{2\beta W_0}[\Psi_N(2\beta) - \psi_N(2\beta)] \tag{12.1.11}$$

is therefore regular and bounded in the upper half plane, $\operatorname{Re}\beta > 0$. At this point we can use a theorem of Carlson (Titchmarsh, 1939).

Carlson's theorem. *If a function of β is regular and bounded in the upper half plane,* $\operatorname{Re}\beta > 0$, *and is zero for* $\beta = 1, 2, 3, \ldots$, *then it is identically zero.*

Applying this theorem to the function $\Delta(\beta)$ we conclude that if the expressions (12.1.2) and (12.1.4) are equal for all even integers β, then they are identically equal.

(2) The integral

$$(2\pi)^{-N} \int_0^{2\pi} \cdots \int_0^{2\pi} \prod_{\ell<j} |e^{i\theta_\ell} - e^{i\theta_j}|^{2k} \, d\theta_1 \cdots d\theta_N \tag{12.1.12}$$

is equal to the constant term in the series expansion, involving positive as well as negative powers of z_j, $j = 1, 2, \ldots, N$, of the product

$$\prod_{\substack{\ell,j \\ (\ell \neq j)}} \left(1 - \frac{z_\ell}{z_j}\right)^k. \tag{12.1.13}$$

To see this let $z_j = \exp(i\theta_j)$, so that

$$|z_j - z_\ell|^{2k} = (z_j - z_\ell)^k (z_j^{-1} - z_\ell^{-1})^k = \left(1 - \frac{z_\ell}{z_j}\right)^k \left(1 - \frac{z_j}{z_\ell}\right)^k, \tag{12.1.14}$$

and note that any power other than zero of z_j, $j = 1, 2, \ldots, N$, vanishes on integration.

(3) The constant term in the expansion of

$$\prod_{\substack{\ell,j \\ (\ell \neq j)}} \left(1 - \frac{z_j}{z_\ell}\right)^{a_j} \tag{12.1.15}$$

is given by

$$\frac{(a_1 + \cdots + a_N)!}{a_1! \cdots a_N!}. \tag{12.1.16}$$

A proof of this is reproduced in Appendix A.23 (Good, 1970).

If we put $a_1 = a_2 = \cdots = a_N = k$ in (12.1.15) and (12.1.16), we get the constant term in the expansion of (12.1.13) as

$$(Nk)!(k!)^{-N} = \Gamma(1+kN)[\Gamma(1+k)]^{-N}. \tag{12.1.17}$$

The proof of Theorem 12.1.1 is now complete. □

(4) The fact that the constant term in (12.1.14) is given by (12.1.17) can also be deduced from Selberg's integral, Section 17.7. But the statement in 3 above is stronger.

12.2 Thermodynamic Quantities

Theorem 12.1.1 specifies completely the thermodynamic properties of a finite Coulomb gas of N charges on the unit circle. For applications to the energy level series we are interested only in the special case of a very large N, $N \to \infty$. In this section we study the statistical mechanics of an infinite Coulomb gas or, equivalently, that of an infinitely long series of eigenvalues.

The partition function (12.1.2) is normalized in a way that the energy of the gas is zero at infinite temperature ($\beta = 0$). The potential energy at zero temperature is then the ground-state energy

$$W_0 = -(N/2)\log N. \tag{12.2.1}$$

To obtain finite limits for the thermodynamic variables as $N \to \infty$ we must first change the zero of the energy to the position W_0. By definition the gas then has zero energy at zero temperature and positive energy at any positive temperature. The partition function defined on the new energy scale is

$$\Phi_N(\beta) = (2\pi)^{-N} \int_0^{2\pi} \cdots \int_0^{2\pi} \exp(-\beta(W - W_0))\, d\theta_1 \cdots d\theta_N. \tag{12.2.2}$$

The free energy per particle $F_N(\beta)$ is

$$F_N(\beta) = -(\beta N)^{-1} \log \Phi_N(\beta) \tag{12.2.3}$$

$$= (1/2)\log N - (\beta N)^{-1} \log \Gamma(1+\beta N/2) + \beta^{-1} \log \Gamma(1+\beta/2), \tag{12.2.4}$$

where we have used Theorem 12.1.1. Taking the limit $N \to \infty$, we obtain the following theorem:

Theorem 12.2.1. *As $N \to \infty$ the free energy per particle of the Coulomb gas at temperature $kT = \beta^{-1}$ tends to the limiting value*

$$F(\beta) = \beta^{-1} L(\beta/2) + [1 - \log(\beta/2)]/2, \tag{12.2.5}$$

$$L(z) = \log \Gamma(1+z). \tag{12.2.6}$$

The values of other thermodynamic quantities follow from (12.2.5):
Energy per particle:

$$U(\beta) = F + \beta \frac{\partial F}{\partial \beta} = (1/2)[L'(\beta/2) - \log(\beta/2)]. \tag{12.2.7}$$

Entropy per particle:

$$S(\beta) = \beta^2 \frac{\partial F}{\partial \beta} = (\beta/2)[L'(\beta/2) - 1] - L(\beta/2). \tag{12.2.8}$$

Specific heat per particle:

$$C(\beta) = -\beta^2 \frac{\partial U}{\partial \beta} = -(\beta^2/4)L'(\beta/2) + \beta/2. \tag{12.2.9}$$

To calculate the values of these thermodynamic quantities for physically interesting values of β the following formulas (cf. Bateman (1953a) or Abramowitz and Stegun (1965)) may be used:

$$L(z) = -\gamma z + \sum_2^\infty (-1)^n \frac{\zeta(n)}{n} z^n, \quad |z| \leqslant 1, \tag{12.2.10}$$

$$L(z) = z \log z - z + \frac{1}{2} \log z + \frac{1}{2} \log(2\pi) + \frac{1}{12z} + \mathrm{O}\left(\frac{1}{z^2}\right), \quad |z| \to \infty, \tag{12.2.11}$$

$$L(z+n) - L(z) = \sum_{r=1}^n \log(z+r), \tag{12.2.12}$$

$$L'(z) = -\gamma + \sum_2^\infty (-1)^n \zeta(n) z^{n-1}, \quad |z| < 1, \tag{12.2.13}$$

$$L'(z+n) - L'(z) = \sum_{r=1}^n (z+r)^{-1}, \tag{12.2.14}$$

$$L'(1/2) = 2 - \gamma - 2\log 2, \tag{12.2.15}$$

$$L''(z) = \sum_{n=1}^\infty (z+n)^{-2}, \quad z \neq -1, -2, \ldots, \tag{12.2.16}$$

where γ is Euler's constant $\gamma = 0.5772\ldots$ and $\zeta(k)$ are the sums of the inverse powers of the integers

$$\zeta(k) = \sum_{n=1}^{\infty} n^{-k}. \tag{12.2.17}$$

In particular,

$$\zeta(2) = \sum_{1}^{\infty} n^{-2} = \frac{\pi^2}{6}, \tag{12.2.18}$$

and

$$\sum_{1}^{\infty} (2n-1)^{-2} = \frac{\pi^2}{8}. \tag{12.2.19}$$

Table 12.1 summarizes this calculation.

12.3 Statistical Interpretation of U and C

If we denote the ensemble average by $\langle\,\rangle$,

$$\langle f \rangle = \frac{\int_0^{2\pi} \cdots \int_0^{2\pi} f(\theta_1, \ldots, \theta_N) e^{-\beta(W-W_0)}\, d\theta_1 \cdots d\theta_N}{\int_0^{2\pi} \cdots \int_0^{2\pi} e^{-\beta(W-W_0)}\, d\theta_1 \cdots d\theta_N}, \tag{12.3.1}$$

then from (12.1.4), (12.1.5), (12.2.2), (12.2.7) and (12.2.9) we have

$$\langle W - W_0 \rangle = -[\Phi_N(\beta)]^{-1} \frac{\partial}{\partial \beta} \Phi_N(\beta) = NU, \tag{12.3.2}$$

and

$$\begin{aligned} \left\langle (W - \langle W \rangle)^2 \right\rangle &= \left\langle (W - W_0)^2 \right\rangle - (\langle W - W_0 \rangle)^2 \\ &= [\Phi_N(\beta)]^{-1} \frac{\partial^2}{\partial \beta^2} \Phi_N(\beta) - (NU)^2, \\ &= \beta^{-2} NC, \end{aligned} \tag{12.3.3}$$

where W is the electrostatic energy given by (12.1.1) and $W_0 = -(1/2)N \log N$ is the minimum value of W when the charges are uniformly spaced. Thus U is, apart from normalization, the ensemble average of the logarithm of the geometric mean of all distances between pairs of eigenvalues, and $\beta^{-2}C$ is the statistical mean square fluctuation of the same quantity.

Table 12.1.

	$\beta\to 0$	$\beta=1$	$\beta=2$
$L\left(\frac{\beta}{2}\right)$	$-\frac{\gamma\beta}{2}+\frac{\pi^2\beta^2}{48}+\cdots$	$\frac{1}{2}\log\left(\frac{\pi}{4}\right)=-0.121$	0
$L'\left(\frac{\beta}{2}\right)$	$-\gamma+\frac{\pi^2\beta}{12}+\cdots$	$2-\gamma-\log 4=0.036$	$1-\gamma=0.423$
$L''\left(\frac{\beta}{2}\right)$	$\frac{\pi^2}{6}+\cdots$	$\frac{\pi^2}{2}-4=0.935$	$\frac{\pi^2}{6}-1=0.645$
F	$12\left(1-\gamma-\log\left(\frac{\beta}{2}\right)\right)+\frac{\pi^2\beta}{48}+\cdots$	$\frac{1}{2}\left(1+\log\left(\frac{\pi}{2}\right)\right)=0.726$	$\frac{1}{2}=0.500$
U	$\frac{1}{2}\left(-\gamma-\log\left(\frac{\beta}{2}\right)\right)+\frac{\pi^2\beta}{24}+\cdots$	$1-\frac{1}{2}(\gamma+\log 2)=0.365$	$\frac{1}{2}(1-\gamma)=0.211$
S	$-\frac{\beta}{2}+\frac{\pi^2\beta^2}{48}+\cdots$	$\frac{1}{2}(1-\gamma-\log\pi)=-0.361$	$-\gamma=-0.577$
C	$\frac{\beta}{2}-\frac{\pi^2\beta^2}{24}+\cdots$	$\frac{3}{2}-\frac{\pi^2}{8}=0.266$	$2-\frac{\pi^2}{6}=0.355$

	$\beta=4$	$\beta\to\infty$
$L\left(\frac{\beta}{2}\right)$	$\log 2=0.693$	$\frac{\beta}{2}\log\left(\frac{\beta}{2}\right)-\frac{\beta}{2}+\frac{1}{2}\log(\pi\beta)+\frac{1}{6\beta}+\cdots$
$L'\left(\frac{\beta}{2}\right)$	$\frac{3}{2}-\gamma=0.923$	$\log\left(\frac{\beta}{2}\right)+\frac{1}{\beta}-\frac{1}{3\beta^2}+\cdots$
$L''\left(\frac{\beta}{2}\right)$	$\frac{\pi^2}{6}-\frac{5}{4}=0.395$	$\frac{2}{\beta}-\frac{2}{\beta^2}+\frac{4}{3\beta^3}+\cdots$
F	$\frac{1}{2}-\frac{\log 2}{4}=0.327$	$\frac{1}{2\beta}\log(\pi\beta)+\frac{1}{6\beta^2}+\cdots$
U	$\frac{3}{4}-\frac{1}{2}(\gamma+\log 2)=0.115$	$\frac{1}{2\beta}-\frac{1}{6\beta^2}+\cdots$
S	$1-2\gamma-\log 2=-0.848$	$\frac{1}{2}(1-\log(\pi\beta))-\frac{1}{3\beta}+\cdots$
C	$7-\frac{2}{3}\pi^2=0.420$	$\frac{1}{2}-\frac{1}{3\beta}+\cdots$

For analyzing the properties of observed eigenvalue series, W seems to be a good quantity. It has two great advantages over the other quantities such as F and S:

(1) W can be computed from the eigenvalue pair-correlation function alone without analyzing higher order correlations.

(2) The statistical uncertainty of W is known from the value of C.

Theorem 12.3.1. *Let $z_1, z_2, \ldots, z_N$ be the eigenvalues of a random unitary matrix taken from one of the ensembles E_1, E_2 or E_4. The quantity*

$$W - W_0 = \frac{1}{2} N \log N - \sum_{i<j} \log|z_i - z_j| \tag{12.3.4}$$

has the average value NU and the root mean square deviation $\beta^{-1}(NC)^{1/2}$ with the values of U and C listed in Table 12.1.

12.4 Continuum Model for the Spacing Distribution

In this section we exploit an argument of classical statistical mechanics to arrive at the asymptotic form of spacing distribution for large spacings.

As in Section 12.1, we write the joint probability density of the eigenvalues $e^{i\theta_j}$, $j = 1, 2, \ldots, N$, of a random unitary $N \times N$ matrix as

$$P_{N\beta}(\theta_1, \ldots, \theta_N) = C'_{N\beta} e^{-\beta W}, \tag{12.4.1}$$

with

$$W = -\sum_{j<k} \log |e^{i\theta_j} - e^{i\theta_k}|. \tag{12.4.2}$$

The probability that an arc of length 2α will contain none of the angles θ_j, that is, the $E_m(\alpha)$ of (11.2.1) or (11.4.13), is given by

$$E_\beta(0, \alpha) = \Psi_\beta(\alpha) / \Psi_\beta(0), \tag{12.4.3}$$

where

$$\Psi_\beta(\alpha) = \int_\alpha^{2\pi-\alpha} \cdots \int_\alpha^{2\pi-\alpha} e^{-\beta W} \, d\theta_1 \cdots d\theta_N \tag{12.4.4}$$

is the partition function of the analogous Coulomb gas of N charges compressed in a circular arc of length $2\pi - 2\alpha$, whereas $\Psi_\beta(0)$ is the partition function of the same gas on the whole unit circle. We write

$$E_\beta(0, \alpha) = \exp(-\beta[F_N(\beta, \alpha) - F_N(\beta, 0)]), \tag{12.4.5}$$

where $F_N(\beta, \alpha)$ is the free energy of the Coulomb gas on the arc $2\pi - 2\alpha$.

The hypothesis that for large N the Coulomb gas forms a continuous electric fluid obeying the laws of thermodynamics may be put in the form of the following three assumptions:

(1) There is a macroscopic charge density; that is there exists a smooth function $\sigma_\alpha(\theta)$ such that the average number of charges on the arc $(\theta, \theta + d\theta)$ is $\sigma_\alpha(\theta)\, d\theta$.
(2) For a given density $\sigma_\alpha(\theta)$ the free energy is the sum of the two terms

$$F = V_1 + V_2, \tag{12.4.6}$$

where V_1 is the macroscopic potential energy

$$V_1 = -\frac{1}{2} \iint_{\alpha}^{2\pi-\alpha} \sigma_\alpha(\theta)\sigma_\alpha(\phi) \log |e^{i\theta} - e^{i\phi}| \, d\theta \, d\phi, \tag{12.4.7}$$

and V_2 is the contribution from the individual arcs, depending only on the local density

$$V_2 = \int_{\alpha}^{2\pi-\alpha} \sigma_\alpha(\theta) f_\beta [\sigma_\alpha(\theta)] \, d\theta, \tag{12.4.8}$$

$f_\beta[\sigma]$ being the free energy per particle of a Coulomb gas having uniform density σ on the whole unit circle. The factor $1/2$ in (12.4.7) is there because the interaction between two arc elements is counted twice.

(3) In almost all cases the density σ_α adjusts itself in such a way that the free energy $F_N(\beta, \alpha)$ given by (12.4.6), (12.4.7) and (12.4.8) is a minimum, subject to the condition

$$\int_{\alpha}^{2\pi-\alpha} \sigma_\alpha(\theta) \, d\theta = N. \tag{12.4.9}$$

This last equation expresses the fact that the total number of charges is fixed to N.

There is no rigorous mathematical justification for the above assumptions. But they are so much accepted by tradition that we make no apologies for adopting them.

The functional $f_\beta(\sigma)$ remains to be specified, and we write

$$f_\beta(\sigma) = U_\beta(\sigma) - \beta^{-1} S_\beta(\sigma), \tag{12.4.10}$$

where $U_\beta(\sigma)$ is the energy and $S_\beta(\sigma)$ is the entropy per particle for a uniform gas of

$$N' = 2\pi\sigma \tag{12.4.11}$$

charges on the whole unit circle.

As in (12.2.7), the energy per particle is

$$U_\beta(\sigma) = -\frac{1}{2} \log N' + U(\beta). \tag{12.4.12}$$

The term $-\frac{1}{2} \log N'$ is included, for we now have to take the total energy, including the ground-state energy $-\frac{1}{2} N' \log N'$.

The entropy $S_\beta(\sigma)$, if calculated as in Section 12.2, is independent of N' for large N'. However, one thing should be noted. The calculation of the entropy in Section 12.2 was made for a gas of N distinguishable particles. The entropy so defined is not an extensive quantity. To make it extensive we must subtract $N!$ from the classical entropy, which amounts to treating the particles as indistinguishable. As we need the $S_\beta(\sigma)$ in (12.4.10) to be an extensive quantity, we write

$$S_\beta(\sigma) = \log\left(\frac{N}{N'}\right) + S(\beta), \tag{12.4.13}$$

where $S(\beta)$ is given by (12.2.8). Putting

$$\sigma_\alpha(\theta) = \frac{N}{2\pi}\rho_\alpha(\theta), \tag{12.4.14}$$

and collecting (12.4.6) to (12.4.13), we have

$$\beta F_N(\beta, \alpha) = G_2 + G_1 + G_0, \tag{12.4.15}$$

where

$$G_2 = -\frac{\beta}{2}\left(\frac{N}{2\pi}\right)^2 \iint_\alpha^{2\pi-\alpha} \rho_\alpha(\theta)\rho_\alpha(\phi)\log|e^{i\theta} - e^{i\phi}|\,d\theta\,d\phi, \tag{12.4.16}$$

$$G_1 = (1-\beta/2)\left(\frac{N}{2\pi}\right)\int_\alpha^{2\pi-\alpha} \rho_\alpha(\theta)\log\rho_\alpha(\theta)\,d\theta, \tag{12.4.17}$$

$$G_0 = \beta N\left[F(\beta) - \frac{1}{2}\log N\right]; \tag{12.4.18}$$

and $F(\beta)$ given by (12.2.5). One has to minimize the quantity (12.4.15) under the restriction

$$\int_\alpha^{2\pi-\alpha} \rho_\alpha(\theta)\,d\theta = 2\pi. \tag{12.4.19}$$

When $\alpha = 0$, the equilibrium density $\rho_\alpha(\theta) = 1$, and $G_2 = G_1 = 0$, so that

$$\beta F_N(\beta, 0) = G_0 = \beta N\left[F(\beta) - \frac{1}{2}\log N\right], \tag{12.4.20}$$

and from (12.4.5)

$$E_\beta(0, \alpha) = \exp[-\min_\rho(G_2 + G_1)]. \tag{12.4.21}$$

Using Lagrange's method to minimize $(G_2 + G_1)$ under the restriction (12.4.19), we get for $\beta \neq 2$

$$-\beta\left(\frac{N}{2\pi}\right)^2 \int_\alpha^{2\pi-\alpha} \rho_\alpha(\phi)\log|e^{i\theta} - e^{i\phi}|\,d\phi + \left(\frac{N}{2\pi}\right)(1-\beta/2)\log\rho_\alpha(\theta)$$

$$+\left(\frac{N}{2\pi}\right)(1-\beta/2)-\lambda=0, \quad \alpha<\theta<2\pi-\alpha, \tag{12.4.22}$$

where λ is the undetermined constant. Letting

$$V_\alpha(\theta)=-\int_\alpha^{2\pi-\alpha}\rho_\alpha(\phi)\log|e^{i\theta}-e^{i\phi}|\,d\phi, \tag{12.4.23}$$

we have

$$\beta\left(\frac{N}{2\pi}\right)(1-\beta/2)^{-1}V_\alpha(\theta)+\log\rho_\alpha(\theta)=\text{constant},$$

or

$$\rho_\alpha(\theta)=Ae^{-\nu V_\alpha(\theta)}, \tag{12.4.24}$$

$$\nu=\beta\left(\frac{N}{2\pi}\right)(1-\beta/2)^{-1}. \tag{12.4.25}$$

When $\beta=2$, G_1 is zero, and the minimization leads to

$$V_\alpha(\theta)=-\int_\alpha^{2\pi-\alpha}\rho_\alpha(\phi)\log|e^{i\theta}-e^{i\phi}|\,d\phi=\text{constant}. \tag{12.4.26}$$

The $V_\alpha(\theta)$ is the electrostatic potential at the angle θ produced by all the other charges. If $\beta\neq 2$, these charges are in thermal equilibrium at an effective temperature

$$kT_\beta=\frac{1}{\beta}(1-\beta/2)=\frac{1}{\beta}-\frac{1}{2}, \tag{12.4.27}$$

under the potential $V_\alpha(\theta)$ generated by themselves. If $\beta=2$, the potential $V_\alpha(\theta)$ is constant and the charges are in electrostatic equilibrium on a conducting circular arc of length $2\pi-2\alpha$.

The problem in the case of unitary ensemble $\beta=2$ is the easiest to handle. The classical problem of charge distribution on a slotted conducting cylinder is well known. Here we give only the result (Smythe, 1950). The solution of (12.4.19) and (12.4.26) is

$$\rho_\alpha(\theta)=\sin\frac{\theta}{2}\left(\sin^2\frac{\theta}{2}-\sin^2\frac{\alpha}{2}\right)^{-1/2}, \tag{12.4.28}$$

$$V_\alpha=2\pi\log\left(\cos\frac{\alpha}{2}\right). \tag{12.4.29}$$

Equation (12.4.21) therefore gives

$$\log E_2(0, \alpha) = -\min_{\rho} G_2 = N^2 \log\left(\cos\frac{\alpha}{2}\right). \tag{12.4.30}$$

In the limit $N \to \infty$, $s = 2t = 2\alpha N/(2\pi)$ finite, we obtain

$$\begin{aligned}\log E_2(0; s) &= \lim(N^2 \log\cos(\alpha/2)) \\ &= \lim\left(-\frac{1}{8}N^2\alpha^2\right) = -\frac{\pi^2}{2}t^2.\end{aligned} \tag{12.4.31}$$

When $\beta \neq 2$, one can apply perturbation theory to expand F in powers of αN. Since G_2 is of order $(\alpha N)^2$ and G_1 is of order αN, one can treat G_1 as a "small perturbation" over G_2. The first order contribution is

$$\beta F_1 = (1 - \beta/2)\left(\frac{N}{2\pi}\right)\int_{\alpha}^{2\pi-\alpha} \bar{\rho}_\alpha(\theta)\log\bar{\rho}_\alpha(\theta)\,d\theta, \tag{12.4.32}$$

where $\bar{\rho}_\alpha(\theta)$ is the unperturbed charge density (12.4.28).

To evaluate (12.4.32) make the following change of variables,

$$\sin^2\frac{\theta}{2} = \sin^2\frac{\alpha}{2} + \cos^2\frac{\alpha}{2}\sin^2\varphi. \tag{12.4.33}$$

Hence

$$\begin{aligned}\int_{\alpha}^{2\pi-\alpha} \bar{\rho}_\alpha(\theta)\log\bar{\rho}_\alpha(\theta)\,d\theta &= 2\int_{\alpha}^{\pi} \bar{\rho}_\alpha(\theta)\log\bar{\rho}_\alpha(\theta)\,d\theta, \\ &= 4\int_0^{\pi/2} d\varphi\left[\frac{1}{2}\log\left(\sin^2\varphi + \tan^2\frac{\alpha}{2}\right) - \log\sin\varphi\right] \\ &= 2\pi\log\left(\tan\frac{\alpha}{2} + \sec\frac{\alpha}{2}\right).\end{aligned} \tag{12.4.34}$$

(For the last step one can look in a table of integrals, e.g. Gradshteyn and Rizhik, 1965.) In the limit $N \to \infty$, $s = 2t = 2\alpha N/(2\pi)$ finite, this gives

$$\beta F_1 = (1 - \beta/2)\pi t, \tag{12.4.35}$$

and

$$\log E_\beta(0; s) = -\frac{\pi^2}{4}\beta t^2 - (1 - \beta/2)\pi t. \tag{12.4.36}$$

In Chapters 18 and 21 we will derive an asymptotic series for $E_\beta(0; s)$ valid for large s. It will then be seen that the terms derived above from thermodynamic arguments are correct.

Summary of Chapter 12

Considering the eigenvalues of matrices from the circular ensembles as electric charges confined to the circumference of the unit circle, various thermodynamic quantities, such as the partition function, the free energy, the specific heat, ..., are calculated and their statistical interpretation is given (Sections 12.1–12.3). From such thermodynamic considerations the dominant terms in the asymptotic behaviour of $E(0; s)$ is derived

$$\log E_\beta(0; s) = -\beta(\pi s)^2/16 - (1 - \beta/2)\pi s/2, \tag{12.4.36}$$

where $E_\beta(0; s)$ is the probability that a randomly chosen interval of length s contains no eigenvalues.

13

GAUSSIAN ENSEMBLE OF ANTI-SYMMETRIC HERMITIAN MATRICES

This short chapter deals with the statistical properties of the eigenvalues of an anti-symmetric Hermitian matrix whose elements are Gaussian random variables. While of no immediate physical interest, the precise analytical results obtained merit some mention for their mathematical elegance.

13.1 Level Density. Correlation Functions

In Section 3.4 we saw that the eigenvalues of an anti-symmetric Hermitian matrix always come in pairs $\pm x_i$, x_i real. Let $x_1, \ldots, x_N$ be the positive eigenvalues of H. The probability density $\exp(-\operatorname{tr} H^2/2)$ for H implies for $x_1, \ldots, x_N$ the joint probability density (cf. Section 3.4),

$$P_{2N}(x_1, \ldots, x_N) = C \prod_{j=1}^{N} \exp(-x_j^2) \prod_{1 \leqslant j < k \leqslant N} (x_j^2 - x_k^2)^2, \tag{13.1.1}$$

if H is of even order $2N$, and

$$P_{2N+1}(x_1,\ldots,x_N)=C\prod_{j=1}^{N}x_j^2\exp(-x_j^2)\prod_{1\leqslant j<k\leqslant N}(x_j^2-x_k^2)^2, \tag{13.1.2}$$

if H is of odd order, $2N+1$. The constant C in the above two equations is not the same. As in Chapters 6 and 7, it is convenient to introduce the oscillator functions

$$\varphi_j(x)=(2^j j!\sqrt{\pi})^{-1/2}\exp\left(\frac{x^2}{2}\right)\left(-\frac{d}{dx}\right)^j e^{-x^2}, \tag{13.1.3}$$

so that (13.1.1) and (13.1.2) can be written as

$$P_{2N}(x_1,\ldots,x_N)=\frac{2^N}{N!}(\det[\varphi_{2j-2}(x_k)])^2=\frac{1}{N!}\det[K_N^+(x_j,x_k)]_{j,k=1,\ldots,N}, \tag{13.1.4}$$

$$P_{2N+1}(x_1,\ldots,x_N)=\frac{2^N}{N!}(\det[\varphi_{2j-1}(x_k)])^2=\frac{1}{N!}\det[K_N^-(x_j,x_k)]_{j,k=1,\ldots,N}, \tag{13.1.5}$$

where

$$K_N^+(x,y)=2\sum_{j=0}^{N-1}\varphi_{2j}(x)\varphi_{2j}(y), \tag{13.1.6}$$

$$K_N^-(x,y)=2\sum_{j=0}^{N-1}\varphi_{2j+1}(x)\varphi_{2j+1}(y), \tag{13.1.7}$$

and the constants have been chosen such that

$$\int_0^\infty\cdots\int_0^\infty P_{2N}(x_1,\ldots,x_N)\,dx_1\cdots dx_N$$
$$=\int_0^\infty\cdots\int_0^\infty P_{2N+1}(x_1,\ldots,x_N)\,dx_1\cdots dx_N=1. \tag{13.1.8}$$

The $K_N^\pm(x,y)$ satisfy the conditions of Theorem 5.1.4, namely

$$\int_0^\infty K_N(x,x)\,dx=N, \tag{13.1.9}$$

$$\int_0^{\infty} K_N(x,y)K_N(y,z)\,dy = K_N(x,z). \tag{13.1.10}$$

Hence the n-level correlation function is

$$R_n(x_1,\ldots,x_N) = \det[K_N(x_j,x_k)]_{j,k=1,\ldots,n}. \tag{13.1.11}$$

In particular the level-density is

$$\begin{aligned}\sigma_{2N}(x) &= 2\sum_{j=0}^{N-1}\varphi_{2j}^2(x)\\ &= 4\sqrt{N}\left(A_N(x) - \frac{1}{4x}\varphi_{2N}(x)\varphi_{2N-1}(x)\right),\end{aligned} \tag{13.1.12}$$

$$\begin{aligned}\sigma_{2N+1}(x) - \delta(x) &= 2\sum_{j=0}^{N-1}\varphi_{2j+1}^2(x)\\ &= 4\sqrt{N}\left(A_N(x) + \frac{1}{4x}\varphi_{2N}(x)\varphi_{2N-1}(x)\right),\end{aligned} \tag{13.1.13}$$

where

$$A_N(x) = \frac{1}{4\sqrt{N}}\sum_{j=0}^{2N-1}\varphi_j^2(x). \tag{13.1.14}$$

In the limit $N \to \infty$ (cf. Appendices A.9 and A.10),

$$A_N(x) \approx \frac{1}{2\pi}\left(1 - \frac{x^2}{4N}\right)^{1/2}, \tag{13.1.15}$$

$$\frac{1}{4x}\varphi_{2N}(x)\varphi_{2N-1}(x) \approx -\frac{1}{2\pi}\frac{\sin(2\xi)}{2\xi}, \tag{13.1.16}$$

$$\begin{aligned}K_N^{\pm}(x,y) &= \sqrt{N}\left\{\frac{\varphi_{2N}(x)\varphi_{2N-1}(y) - \varphi_{2N}(y)\varphi_{2N-1}(x)}{x-y}\right.\\ &\qquad \left.\mp \frac{\varphi_{2N}(x)\varphi_{2N-1}(y) + \varphi_{2N}(y)\varphi_{2N-1}(x)}{x+y}\right\}\\ &\approx \frac{2}{\pi}\sqrt{N}\left(\frac{\sin(\xi-\eta)}{\xi-\eta} \pm \frac{\sin(\xi+\eta)}{\xi+\eta}\right),\end{aligned} \tag{13.1.17}$$

where $\xi = 2x\sqrt{N}$, $\eta = 2y\sqrt{N}$.

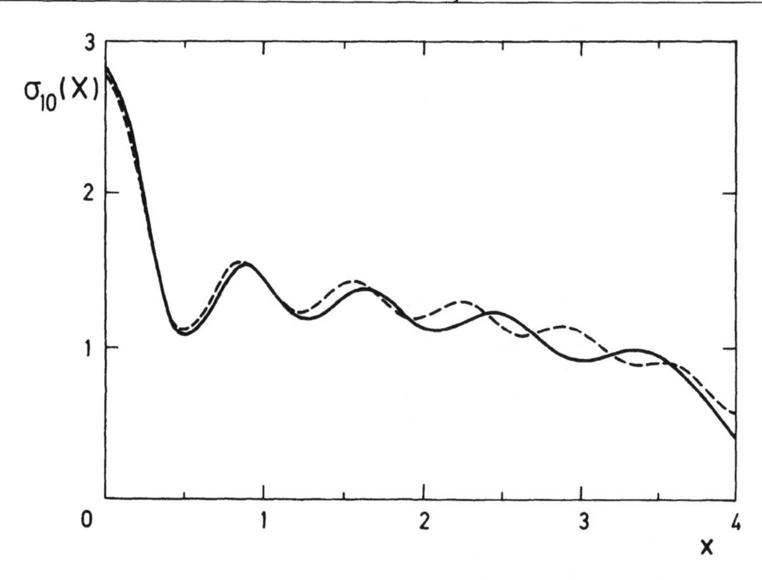

Figure 13.1. Level density for 10×10 anti-symmetric Hermitian matrices. The solid curve is the exact result, Eq. (13.1.12), while the dashed curve is its asymptotic expression, Eqs. (13.1.15) and (13.1.16). Reprinted with permission from Elsevier Science Publishers, Mehta M.L. and Rosenzweig N., Distribution laws for the roots of a random anti-symmetric matrix, *Nucl. Phys. A* 109, 449–456 (1968).

It is curious to note that near the origin the level density contains a term $\sin(2\xi)/(2\xi)$, and therefore has large oscillations. Thus the average spacing has to be defined in an artificial way. Figure 13.1 is a plot of $\sigma_{2N}(x)$ for $N = 5$ and its asymptotic expression.

13.2 Level Spacings

In considering level spacings, one has to distinguish between two situations: (i) central spacings, i.e. the probability of having the smallest positive eigenvalue at $s/2$, and the largest negative one therefore at $-s/2$, and (ii) non-central spacings, i.e. the probability of having two adjacent eigenvalues at α and β, $\beta \geqslant \alpha > 0$, and simultaneously therefore two other adjacent ones at $-\alpha$ and $-\beta$. This non-central spacing distribution will depend not only on $(\beta - \alpha)$ but very strongly on $(\beta + \alpha)/2$ as well. This dependence is expected from the oscillatory behavior of the level density. As we move away from the origin, the dependence on $(\beta + \alpha)/2$ decreases and at about 20 to 30 mean spacings from the origin becomes negligible.

13.2.1 Central Spacings. For definiteness we shall take the matrices to be of even order $2N$. The probability that all positive eigenvalues are $\geqslant \alpha$, and therefore all negative ones $\leqslant -\alpha$, is

$$\int_{\alpha}^{\infty} \cdots \int_{\alpha}^{\infty} \frac{2^N}{N!} (\det[\varphi_{2j-2}(x_k)])^2 \, dx_1 \cdots dx_N$$

$$= \det\left[2\int_{\alpha}^{\infty} \varphi_{2j-2}(x)\varphi_{2k-2}(x)\, dx\right]_{j,k=1,2,\ldots,N} \tag{13.2.1}$$

$$= \det\left[\delta_{jk} - \int_{-\alpha}^{\alpha} \varphi_{2j-2}(x)\varphi_{2k-2}(x)\, dx\right]_{j,k=1,2,\ldots,N} \tag{13.2.2}$$

$$= F_N(\alpha) \underset{N\to\infty}{\approx} F_+\left(\frac{2}{\pi}\alpha N^{1/2}\right), \tag{13.2.3}$$

where the function $F_+(t)$ is given by Eq. (6.4.36) or (7.3.19) with $z = 1$; i.e.

$$F_+(t) = \prod_j (1 - \lambda_{2j}), \tag{13.2.4}$$

and the λ_{2j} are the eigenvalues of the integral equation (6.3.14) corresponding to the even solutions, i.e.

$$2\pi\lambda f(x) = \int_{-t}^{t} \left[\frac{\sin(x+y)\pi}{x+y} + \frac{\sin(x-y)\pi}{x-y}\right] f(y)\, dy. \tag{13.2.5}$$

The probability that there is one eigenvalue at α ($\alpha \geqslant 0$) and none in the interval $(0, \alpha)$, and therefore one at $-\alpha$ and none in $(-\alpha, 0)$, is obtained by differentiation. The result is

$$P(-\alpha, \alpha) = -\frac{1}{2}\frac{d}{d\alpha} F_N(\alpha) \underset{N\to\infty}{\approx} -\frac{1}{\pi} N^{1/2} F'_+\left(\frac{1}{\pi} S N^{1/2}\right), \tag{13.2.6}$$

where $S = 2\alpha$ is the spacing. The average value of the central spacings is

$$\langle S \rangle = \int_0^{\infty} 2\alpha \cdot P(-\alpha, \alpha)\, d\alpha = -\int_0^{\infty} \alpha \frac{d}{d\alpha} F_N(\alpha)\, d\alpha$$

$$\approx -\frac{\pi}{2\sqrt{N}} \int_0^{\infty} t \frac{d}{dt} F_+(t)\, dt = C N^{-1/2}, \tag{13.2.7}$$

where the constant C is

$$C = \frac{\pi}{2} \int_0^{\infty} F_+(t)\, dt \approx 1.0097. \tag{13.2.8}$$

The cumulative distribution function of the central spacings is

$$F(S) = \int_0^S P(S_1)\,dS_1 = 1 - F_+\left(\frac{C}{\pi}\frac{S}{\langle S\rangle}\right). \tag{13.2.9}$$

Note that the occurrence of levels at α and $-\alpha$ are not independent events. Thus we should not divide the result by the probability of having a level at $-\alpha$. Also a factor $1/2$ is introduced in Eq. (13.2.6) to take this into account.

13.2.2 Non-Central Spacings. The probability that there are no eigenvalues in the interval (α, β), $\beta \geqslant \alpha > 0$, and hence none in the interval $(-\beta, -\alpha)$ is given by

$$\int_{\text{out}(\alpha,\beta)} \cdots \int \frac{2^N}{N!}\{\det[\varphi_{2j-2}(x_k)]\}^2 dx_1 \cdots dx_N \tag{13.2.10}$$

$$= \det\left[\delta_{jk} - 2\int_\alpha^\beta \varphi_{2j-2}(x)\varphi_{2k-2}(x)\,dx\right]_{j,k=1,2,\ldots,N}, \tag{13.2.11}$$

where out (α, β) means that the integrations are carried from 0 to α and from β to ∞, the interval (α, β) being left out. The probability of having one level at α, another at β, and none in between can then be obtained (cf. Section 6.1) by two differentiations of the expression (13.2.11). In the limit of large N, expression (13.2.11) simplifies to

$$A\left(\frac{4}{\pi}\alpha N^{1/2}, \frac{4}{\pi}\beta N^{1/2}\right), \tag{13.2.12}$$

where

$$A(a,b) = \prod_j (1 - \lambda_j), \tag{13.2.13}$$

and the λ_j are the eigenvalues of the integral equation

$$\pi\lambda f(x) = \int_a^b \left[\frac{\sin(x+y)\pi}{x+y} + \frac{\sin(x-y)\pi}{x-y}\right] f(y)\,dy. \tag{13.2.14}$$

However, if a and b are far removed from the origin, we may neglect $\sin[(x+y)\pi]/(x+y)$ in comparison with $\sin[(x-y)\pi]/(x-y)$. Thus for large values of $(a+b)/2$, Eq. (13.2.14) reduces to

$$\pi\lambda f(x) = \int_{-(b-a)/2}^{(b-a)/2} \frac{\sin(x-y)\pi}{x-y} f(y)\,dy, \tag{13.2.15}$$

whose solutions are known to be spheroidal functions (cf. Section 6.3). The final result is

$$A(a,b) \approx F_+(b-a)F_-(b-a), \tag{13.2.16}$$

where $F_\pm(b-a)$ are the $F_\pm(1, b-a)$ of Eqs. (6.4.36) and (6.4.37).

Summary of Chapter 13

For a Gaussian ensemble of anti-symmetric Hermitian matrices we give the level density, the n-point correlation function and the probability density for the nearest neighbor spacings. The results depend sensitively on whether the order of the matrices is even or odd and in what region of the level density lie the spacings considered.

14

A GAUSSIAN ENSEMBLE OF HERMITIAN MATRICES WITH UNEQUAL REAL AND IMAGINARY PARTS

In Chapters 6, 7 and 8 we studied three particular Gaussian ensembles.

(i) The ensemble of Hermitian matrices with equally probable real and imaginary parts (for quaternion matrices, equally probable self-dual and anti-self-dual parts), known as the Gaussian unitary ensemble (GUE).
(ii) The ensemble of real symmetric matrices, known as the Gaussian orthogonal ensemble (GOE).
(iii) The ensemble of self-dual quaternion matrices, known as the Gaussian symplectic ensemble (GSE).

These three ensembles are basic models for energy level fluctuations of complex systems. For time-reversal invariant systems, GOE or GSE is appropriate, depending on the properties of the Hamiltonian. On the other hand in the absence of time-reversal symmetry, the GUE is appropriate. For nuclear spectra GOE appears to be a good model, while for the zeros of the Riemann zeta function on the critical line GUE seems to apply.

Gaussian ensembles with an arbitrary ratio of the mean square values of their real and imaginary parts (of time reversal invariant and non-invariant parts) are of interest due to a suggestion of Wigner, that the analysis of data in comparison with such ensembles may give an upper bound on the time-reversal breaking part of the nuclear forces. As we will see, the transition in the fluctuation properties for the ensemble going from GOE to GUE, adequate for the above purpose, is very rapid and is in fact, discontinuous for infinite order matrices. The situation is similar for ensembles going from GSE to GUE.

In this chapter we will study two ensembles of Hermitian matrices, the elements of which are Gaussian random variables. In one case the real and imaginary parts of the matrix elements have mean square values $2v^2$ and $2v^2\alpha^2$ respectively. In the other case the self-dual and anti-self-dual parts of the (quaternion) matrix elements have mean square values $2v^2$ and $2v^2\alpha^2$ respectively. In particular, we will derive all correlation and cluster functions of the eigenvalues for finite or infinite order matrices. Though only the case $\alpha^2 \ll 1$ is relevant for immediate physical applications, these results are obtained with little extra effort and it is worthwhile to record them for their mathematical elegance.

In Sections 14.1 and 14.2 we describe the ensembles and give a summary of the results. Later sections deal with the proofs; in Section 14.3 we derive the joint probability density for the eigenvalues and in Section 14.4 the correlation and cluster functions for finite and infinite order matrices. Recurrence relations and integrals which are frequently used in this chapter will be found in Appendix A.24; the normalization integrals for the joint probability density of the eigenvalues is given in Appendices A.25 and A.26, Appendices A.27 and A.28 contain the verification of some equations of Section 14.4 while Appendix A.29 deals with some asymptotic forms when the order N of the matrices is large.

14.1 Summary of Results. Matrix Ensembles From GOE to GUE and Beyond

Consider an ensemble of $N \times N$ Hermitian matrices $[H_{jk}] = [R_{jk} + iS_{jk}]$, with R real symmetric and S real anti-symmetric, i.e.

$$R_{jk} = R^*_{jk} = R_{kj}, \qquad S_{jk} = S^*_{jk} = -S_{kj}. \tag{14.1.1}$$

The joint probability density for the matrix elements is taken to be

$$P(H) = \mathrm{c} \cdot \exp\left[-\sum_{j,k}\left(\frac{R^2_{jk}}{4v^2} + \frac{S^2_{jk}}{4v^2\alpha^2}\right)\right], \tag{14.1.2}$$

$$dH = \prod_{j\leqslant k} dR_{jk} \prod_{j<k} dS_{jk}, \tag{14.1.3}$$

where the normalization constant c is

$$\mathrm{c} = 2^{-N/2}\alpha^{-N(N-1)/2}(2\pi v^2)^{-N^2/2}. \tag{14.1.4}$$

On the average

$$\frac{\langle \|\mathrm{Im} H\|^2 \rangle}{\langle \|\mathrm{Re} H\|^2 \rangle} = \frac{\langle \|\mathrm{S}\|^2 \rangle}{\langle \|\mathrm{R}\|^2 \rangle} = \frac{N-1}{N+1}\alpha^2 \approx \alpha^2 \quad \text{for large } N. \tag{14.1.5}$$

We shall choose the scale v^2 such that

$$2v^2(1+\alpha^2) = 1. \tag{14.1.6}$$

As special cases we have (i) $\alpha^2 = 0$, so that $S = 0$ with probability one and the matrices H form the GOE; (ii) $\alpha^2 = 1$, on the average R and S have the same magnitude for large N and the ensemble is GUE; (iii) $\alpha^2 \to \infty$, S dominates R, and the ensemble of H may be referred to as the anti-symmetric Gaussian orthogonal ensemble (cf. Chapter 13).

The joint probability density for the eigenvalues $x_1, \ldots, x_N$ of H is

$$P(x_1, \ldots, x_N) = C_N \exp\left[-\frac{1}{2}(1+\alpha^2)\sum x_j^2\right] \Delta(x_1, \ldots, x_N)\, \mathrm{Pf}[F_{ij}], \tag{14.1.7}$$

where

$$\Delta(x) \equiv \Delta(x_1, \ldots, x_N) = \prod_{1 \leqslant i < j \leqslant N} (x_i - x_j), \tag{14.1.8}$$

and $\mathrm{Pf}[F_{ij}]$ is the Pfaffian of a $2m \times 2m$ matrix defined below. If N is even, $N = 2m$, we define

$$F_{ij} = f(x_i - x_j), \quad i, j = 1, 2, \ldots, N. \tag{14.1.9}$$

If N is odd, $N = 2m + 1$, we use the preceding definition and in addition

$$F_{i,N+1} = -F_{N+1,i} = 1, \quad i = 1, 2, \ldots, N, \qquad F_{N+1,N+1} = 0, \tag{14.1.9$'$}$$

with

$$f(x) = \mathrm{erf}\left[\left(\frac{1-\alpha^4}{4\alpha^2}\right)^{1/2} x\right] \equiv \left(\frac{1-\alpha^4}{\pi\alpha^2}\right)^{1/2} \int_0^x \exp\left(-\frac{(1-\alpha^4)y^2}{4\alpha^2}\right) dy. \tag{14.1.10}$$

The $[F_{ij}]$ is an anti-symmetric matrix of even order and its Pfaffian is, apart from a sign, the square root of its determinant. (Note: Let $m = [N/2]$ be the integral part of $N/2$. The Pfaffian (see e.g. Mehta, 1989MT) of any $N \times N$ anti-symmetric matrix $A = [a_{ij}]$ is the alternating multi-linear form in the elements a_{ij} with $i < j$,

$$\mathrm{pf}\, A \equiv \mathrm{pf}\,[a_{ij}] = \sum \pm a_{i_1 i_2} a_{i_3 i_4} \cdots a_{i_{2m-1} i_{2m}},$$

where the sum is taken over all permutations $(i_1, i_2, \ldots, i_N)$ of $(1, 2, \ldots, N)$ with the restrictions $i_1 < i_2, i_3 < i_4, \ldots, i_{2m-1} < i_{2m}, i_1 < i_3 < \cdots < i_{2m-1}$, and the sign is $+$ or $-$ according to whether the permutation is even or odd. With this general definition $\text{Pf}\,[F_{ij}]$ is just the Pfaffian of the $N \times N$ anti-symmetric matrix $[f(x_i - x_j)]$.) The normalization constant C_N is

$$C_N^{-1} = 2^{3N/2}(1-\alpha^2)^{N(N-1)/4}(1+\alpha^2)^{-N(N+1)/4}\prod_{j=1}^{N}\Gamma(1+j/2). \tag{14.1.11}$$

Note that when $\alpha^2 > 1$, $f(x)$ and C_N are both pure imaginary but $p(x_1, \ldots, x_N)$ remains real positive. Moreover, for $\alpha^2 = 0$, 1 or ∞, $p(x_1, \ldots, x_N)$ and all the quantities derived below have well defined limits.

All the eigenvalue correlations can be expressed in terms of functions of two variables. The n-level correlation function (cf. Section 6.1) is

$$R_n(x_1, \ldots, x_n) = \left\{\det[\phi(x_i, x_j)]_{i,j=1,\ldots,n}\right\}^{1/2}, \tag{14.1.12}$$

and the n-level cluster function (cumulant of the preceding) is

$$T_n(x_1, \ldots, x_n) = \frac{1}{2}\text{tr}\sum \phi(x_1, x_2)\phi(x_2, x_3)\cdots\phi(x_n, x_1), \tag{14.1.13}$$

where

$$\phi(x, y) = \begin{bmatrix} \xi_N(x, y) + S_N(x, y) & D_N(x, y) \\ J_N(x, y) & \xi_N(y, x) + S_N(y, x) \end{bmatrix}, \tag{14.1.14}$$

and the sum in (14.1.13) is taken over all $(n-1)!$ distinct cyclic permutations of the indices $(1, 2, \ldots, n)$. Note the interchange of x and y in the lower right-hand corner of (14.1.14). The two-point functions ξ_N, S_N, D_N and J_N are given in terms of other two-point functions $I_N(x, y)$, $g(x, y)$, $\mu_N(x, y)$ and the one-point functions $\varphi_j(x)$, $\psi_j(x)$, $A_j(x)$ and $\varepsilon(x)$ defined below.

$$\xi_N(x, y) = \begin{cases} \varphi_{2m}(x)\exp\left(-\dfrac{\alpha^2 y^2}{2}\right) \div \displaystyle\int_{-\infty}^{\infty}\varphi_{2m}(t)\exp\left(-\dfrac{\alpha^2 t^2}{2}\right)dt, & N = 2m+1 \text{ odd}, \\ 0, & N = 2m \text{ even}, \end{cases} \tag{14.1.15}$$

$$S_N(x, y) = \sum_{j=0}^{N-1}\varphi_j(x)\varphi_j(y) + (1-\alpha^2)$$

$$\times \left(\frac{1+\alpha^2}{1-\alpha^2}\right)^N \left(\frac{N}{2}\right)^{1/2} \varphi_{N-1}(x) A_N(y) \tag{14.1.16}$$

$$= \begin{cases} \sum_{j=0}^{m-1} [\varphi_{2j}(x)\varphi_{2j}(y) - \psi_{2j}(x)A_{2j}(y)], & N = 2m \text{ even}, \\ \sum_{j=0}^{m-1} [\varphi_{2j+1}(x)\varphi_{2j+1}(y) - \psi_{2j+1}(x)A_{2j+1}(y)], & N = 2m+1 \text{ odd}, \end{cases} \tag{14.1.17}$$

$$D_N(x,y) = \sum_{j=0}^{N-1} \varphi_j(x)\psi_j(y) + (1-\alpha^2)\left(\frac{1+\alpha^2}{1-\alpha^2}\right)^N \left(\frac{N}{2}\right)^{1/2} \varphi_{N-1}(x)\varphi_N(y) \tag{14.1.18}$$

$$= \begin{cases} \sum_{j=0}^{m-1} [\varphi_{2j}(x)\psi_{2j}(y) - \psi_{2j}(x)\varphi_{2j}(y)], & N = 2m \text{ even}, \\ \sum_{j=0}^{m-1} [\varphi_{2j+1}(x)\psi_{2j+1}(y) - \psi_{2j+1}(x)\varphi_{2j+1}(y)], & N = 2m+1 \text{ odd}, \end{cases} \tag{14.1.19}$$

$$J_N(x,y) = I_N(x,y) + g(x,y) + \mu_N(x,y) - \mu_N(y,x), \tag{14.1.20}$$

where

$$I_N(x,y) = \sum_{j=0}^{N-1} \varphi_j(x)A_j(y) + (1-\alpha^2)\left(\frac{1+\alpha^2}{1-\alpha^2}\right)^N \left(\frac{N}{2}\right)^{1/2} A_N(x)A_{N-1}(y) \tag{14.1.21}$$

$$= \begin{cases} \sum_{j=0}^{m-1} [\varphi_{2j}(x)A_{2j}(y) - A_{2j}(x)\varphi_{2j}(y)], & N = 2m \text{ even}, \\ \sum_{j=0}^{m-1} [\varphi_{2j+1}(x)A_{2j+1}(y) - A_{2j+1}(x)\varphi_{2j+1}(y)], & N = 2m+1 \text{ odd}, \end{cases} \tag{14.1.22}$$

$$g(x,y) = \frac{1}{2}\left(\frac{1+\alpha^2}{1-\alpha^2}\right)^{1/2} \exp\left[-\frac{1}{2}\alpha^2(x^2+y^2)\right] f(x-y) \tag{14.1.23}$$

$$= \sum_{j=0}^{\infty} A_j(x)\varphi_j(y) = \frac{1}{2}\sum_{j=0}^{\infty} [A_j(x)\varphi_j(y) - \varphi_j(x)A_j(y)]$$

$$= \sum_{j=0}^{\infty} [A_{2j}(x)\varphi_{2j}(y) - \varphi_{2j}(x)A_{2j}(y)]$$

$$= \sum_{j=0}^{\infty} [A_{2j+1}(x)\varphi_{2j+1}(y) - \varphi_{2j+1}(x)A_{2j+1}(y)], \tag{14.1.24}$$

$$\mu_N(x, y) = \begin{cases} \exp\left(-\frac{\alpha^2 x^2}{2}\right) A_{2m}(y) \div \int_{-\infty}^{\infty} \varphi_{2m}(t) \exp\left(-\frac{\alpha^2 t^2}{2}\right) dt, & N = 2m+1 \text{ odd}, \\ 0, & N = 2m \text{ even}, \end{cases} \tag{14.1.25}$$

$$\varphi_j(x) = (2^j j! \sqrt{\pi})^{-1/2} \exp\left(\frac{x^2}{2}\right)\left(-\frac{d}{dx}\right)^j \exp(-x^2), \tag{14.1.26}$$

$$\psi_j(x) = \left(\frac{1+\alpha^2}{1-\alpha^2}\right)^j \exp\left(\frac{\alpha^2 x^2}{2}\right)\left(\frac{d}{dx}\right)\left[\exp\left(-\frac{\alpha^2 x^2}{2}\right)\varphi_j(x)\right], \tag{14.1.27}$$

$$A_j(x) = \left(\frac{1-\alpha^2}{1+\alpha^2}\right)^j \exp\left(-\frac{\alpha^2 x^2}{2}\right) \int_{-\infty}^{\infty} \exp\left(\frac{\alpha^2 t^2}{2}\right) \varepsilon(x-t)\varphi_j(t)\, dt, \tag{14.1.28}$$

$$\varepsilon(x) = \frac{1}{2} \operatorname{sign} x = \begin{cases} 1/2, & x > 0, \\ 0, & x = 0, \\ -1/2, & x < 0. \end{cases} \tag{14.1.29}$$

To see the equivalence of the pairs of Eqs. (14.1.16)–(14.1.19); (14.1.21), (14.1.22) and (14.1.23), (14.1.24) see Appendix A.24.

Note that $\varphi_{2j}(x)$, $\psi_{2j+1}(x)$ and $A_{2j+1}(x)$ are even functions, while $\varphi_{2j+1}(x)$, $\psi_{2j}(x)$ and $A_{2j}(x)$ are odd functions of x.

For large N, the level density $R_1(x)$ becomes a semi-circle for all α

$$R_1(x) \approx \pi^{-1}(2N - x^2)^{1/2}. \tag{14.1.30}$$

The n-level correlation and cluster functions for $n > 1$ are discontinuous in α^2 at $\alpha^2 = 0$. We have the GOE results for $\alpha^2 = 0$ and the GUE results for $\alpha^2 > 0$. On the other hand if (RMS being the abbreviation of Root Mean Square)

$$\rho \equiv \rho(x) = \frac{\text{RMS value of } (\operatorname{Im} H_{ij})}{\text{Local average spacing at } x} \approx \frac{\alpha}{\sqrt{2}} R_1(x), \tag{14.1.31}$$

remains finite when $\alpha^2 \to 0$ and $N \to \infty$, the n-level correlation and cluster functions have well defined limits for all ρ. With $x_i - x_j \to 0$ and $(x_i - x_j)R(x_i) \to r_{ij} = r_i - r_j$, the n-level correlation function becomes

$$R_n(r_1, \ldots, r_n; \rho) \equiv \lim_{N\to\infty} \{R_1(x_1), \ldots, R_1(x_n)\}^{-1} R_n(x_1, \ldots, x_n)$$
$$= \left\{\det[K(r_{ij}; \rho)]_{i,j=1,\ldots,n}\right\}^{1/2}, \tag{14.1.32}$$

and the corresponding n-level cluster function is

$$Y_n(r_1, \ldots, r_n; \rho) \equiv \lim_{N\to\infty} \{R_1(x_1), \ldots, R_1(x_n)\}^{-1} T_n(x_1, \ldots, x_n)$$
$$= \frac{1}{2} \operatorname{tr} \sum K(r_{12}; \rho) K(r_{23}; \rho), \ldots, K(r_{n1}; \rho), \tag{14.1.33}$$

where the sum is taken, as in (14.1.13), over all $(n-1)!$ distinct cyclic permutations of the indices $(1, 2, \ldots, n)$ and

$$K(r; \rho) = \begin{bmatrix} \dfrac{\sin(\pi r)}{\pi r} & D(r; \rho) \\ J(r; \rho) & \dfrac{\sin(\pi r)}{\pi r} \end{bmatrix}, \tag{14.1.34}$$

is considered as an ordinary 2×2 matrix (not a quaternion!), with

$$D(r; \rho) = -\frac{1}{\pi} \int_0^{\pi} k \sin kr \exp(2\rho^2 k^2)\, dk, \tag{14.1.35}$$

$$J(r; \rho) = -\frac{1}{\pi} \int_{\pi}^{\infty} \frac{\sin kr}{k} \exp(-2\rho^2 k^2)\, dk. \tag{14.1.36}$$

Note that as $\rho \to \infty$, $D \to \infty$, $J \to 0$, while the product $JD \to 0$, so that in the expressions for R_n and Y_n one may replace D and J by zeros. For the asymptotic forms of J and D see Appendix A.29.

If we put $\alpha = 0$, we get back the results of Chapter 7. Putting $\alpha = 1$ is more delicate, since it involves a limiting process. Similarly, $\alpha \to \infty$ is also delicate.

14.2 Matrix Ensembles From GSE to GUE and Beyond

As in Section 14.1, consider an ensemble of $N \times N$ Hermitian quaternion matrices

$$[\mathbf{H}_{jk}] = [\mathbf{R}_{jk} + i\mathbf{S}_{jk}], \tag{14.2.1}$$

where $\mathbf{R}$ and $\mathbf{S}$ are real self-dual and real anti-self-dual quaternion matrices respectively, i.e. with

$$\mathbf{R} = \mathbf{R}^0 + \sum_{\mu=1}^{3} \mathbf{R}^\mu e_\mu, \qquad \mathbf{S} = \mathbf{S}^0 + \sum_{\mu=1}^{3} \mathbf{S}^\mu e_\mu; \tag{14.2.2}$$

$\mathbf{R}^0$ and $\mathbf{S}^\mu$ are real symmetric while $\mathbf{R}^\mu$ and $\mathbf{S}^0$ are real anti-symmetric. The probability density for the matrix elements will be taken as

$$\mathbf{P}(\mathbf{H}) = c \exp\left[-\sum_{j,k}\sum_{\mu=0}^{3}\left(\frac{(\mathbf{R}^\mu_{jk})^2}{4v^2} + \frac{(\mathbf{S}^\mu_{jk})^2}{4v^2\alpha^2}\right)\right], \tag{14.2.3}$$

$$d\mathbf{H} = \prod_{j\leqslant k}\left(d\mathbf{R}^0_{jk}\prod_{\mu=1}^{3} d\mathbf{S}^\mu_{jk}\right)\prod_{j<k}\left(d\mathbf{S}^0_{jk}\prod_{\mu=1}^{3} d\mathbf{R}^\mu_{jk}\right), \tag{14.2.4}$$

with the normalization constant

$$c = 2^{-2N}\alpha^{-N(2N+1)}(2\pi v^2)^{-2N^2}. \tag{14.2.5}$$

On the average the ratio of the anti-self-dual and self-dual parts of $\mathbf{H}$ is

$$\frac{\langle\|\mathbf{S}\|^2\rangle}{\langle\|\mathbf{R}\|^2\rangle} = \left(\frac{2N+1}{2N-1}\right)\alpha^2 \approx \alpha^2 \quad \text{for large } N. \tag{14.2.6}$$

We shall again choose the scale $2v^2(1+\alpha^2) = 1$ as in (14.1.6).

As previously, we have the particular cases: (i) $\alpha^2 = 0$, so that $\mathbf{S} = 0$ with probability one and the matrices $\mathbf{H}$ form the GSE; (ii) $\alpha^2 = 1$, on the average $\mathbf{R}$ and $\mathbf{S}$ have the same magnitude and the ensemble is GUE (of $2N \times 2N$ matrices); (iii) $\alpha^2 \to \infty$, $\mathbf{S}$ dominates $\mathbf{R}$ and the ensemble of $\mathbf{H}$ may be referred to as the anti-self-dual Gaussian symplectic ensemble (AGSE).

The joint probability density for the eigenvalues $x_1, \ldots, x_{2N}$ of $\mathbf{H}$ is

$$\mathbf{P}(x_1, \ldots, x_{2N}) = \mathbf{C}_N \exp\left(-\frac{1+\alpha^2}{2}\sum_{j=1}^{2N} x_j^2\right)\Delta(x_1, \ldots, x_{2N})\,\mathrm{Pf}[\mathbf{F}(x_i - x_j)]_{i,j=1,\ldots,2N}, \tag{14.2.7}$$

where Δ is the product of differences of the $2N$ variables $x_1, \ldots, x_{2N}$, Eq. (14.1.8),

$$\mathbf{F}(x) = x\exp[-(1-\alpha^4)x^2/4\alpha^2], \tag{14.2.8}$$

and

$$\mathbf{C}_N^{-1} = 2^{-N(N-1)}\pi^N \alpha^{3N} \frac{(1-\alpha^2)^{N(N-1)}}{(1+\alpha^2)^{N(N+1)}} \frac{(2N)!}{N!} \prod_1^N (2j)!. \tag{14.2.9}$$

The limits in the three cases $\alpha^2 = 0$, 1 and ∞ are again well defined.

The n-level correlation and cluster functions are again given by (14.1.12) and (14.1.13) with ϕ replaced by

$$\mathbf{\Phi}(x, y) = \begin{bmatrix} \mathbf{S}_N(x, y) & \mathbf{D}_N(x, y) \\ \mathbf{I}_N(x, y) & \mathbf{S}_N(y, x) \end{bmatrix}, \tag{14.2.10}$$

with the new functions

$$\mathbf{S}_N(x, y) = \sum_{j=0}^{N-1} (\varphi_{2j+1}(x)\varphi_{2j+1}(y) - \mathbf{\Psi}_{2j+1}(x)\mathbf{A}_{2j+1}(y)) \tag{14.2.11}$$

$$= \sum_{j=0}^{2N} \varphi_j(x)\varphi_j(y) + (1+\alpha^2) \times \left(\frac{1-\alpha^2}{1+\alpha^2}\right)^{2N+1} (N+1/2)^{1/2} \varphi_{2N}(x)\mathbf{A}_{2N+1}(y), \tag{14.2.12}$$

$$\mathbf{I}_N(x, y) = \sum_{j=0}^{N-1} [\varphi_{2j+1}(x)\mathbf{A}_{2j+1}(y) - \mathbf{A}_{2j+1}(x)\varphi_{2j+1}(y)] \tag{14.2.13}$$

$$= \sum_{j=0}^{2N} \varphi_j(x)\mathbf{A}_j(y) + (1+\alpha^2) \times \left(\frac{1-\alpha^2}{1+\alpha^2}\right)^{2N+1} (N+1/2)^{1/2} \mathbf{A}_{2N+1}(x)\mathbf{A}_{2N}(y), \tag{14.2.14}$$

$$\mathbf{D}_N(x, y) = \overline{\mathbf{D}}_N(x, y) + \mathbf{g}(x, y), \tag{14.2.15}$$

$$\overline{\mathbf{D}}_N(x, y) = \sum_{j=0}^{N-1} [\varphi_{2j+1}(x)\mathbf{\Psi}_{2j+1}(y) - \mathbf{\Psi}_{2j+1}(x)\varphi_{2j+1}(y)] \tag{14.2.16}$$

$$= \sum_{j=0}^{2N} \varphi_j(x)\mathbf{\Psi}_j(y) + (1+\alpha^2)$$

$$\times \left(\frac{1-\alpha^2}{1+\alpha^2}\right)^{2N+1} \left(N+\frac{1}{2}\right)^{1/2} \varphi_{2N}(x)\varphi_{2N+1}(y), \tag{14.2.17}$$

$$\mathbf{g}(x,y) = -\frac{(1+\alpha^2)(1-\alpha^4)}{4\alpha^3\sqrt{\pi}} \exp\left[-\frac{1}{2}\alpha^2(x^2+y^2)\right]\mathbf{F}(x-y) \tag{14.2.18}$$

$$= \sum_{j=0}^{\infty} \mathbf{\Psi}_j(x)\varphi_j(y) = \frac{1}{2}\sum_{j=0}^{\infty}[\mathbf{\Psi}_j(x)\varphi_j(y) - \varphi_j(x)\mathbf{\Psi}_j(y)]$$

$$= \sum_{j=0}^{\infty}[\mathbf{\Psi}_{2j}(x)\varphi_{2j}(y) - \varphi_{2j}(x)\mathbf{\Psi}_{2j}(y)]$$

$$= \sum_{j=0}^{\infty}[\mathbf{\Psi}_{2j+1}(x)\varphi_{2j+1}(y) - \varphi_{2j+1}(x)\mathbf{\Psi}_{2j+1}(y)], \tag{14.2.19}$$

where $\varphi_j(x)$ and $\varepsilon(x)$ are given by (14.1.26), (14.1.29),

$$\mathbf{\Psi}_j(x) = \left(\frac{1-\alpha^2}{1+\alpha^2}\right)^j \exp\left(-\frac{1}{2}\alpha^2x^2\right)\frac{d}{dx}\left(\exp\left(\frac{1}{2}\alpha^2x^2\right)\varphi_j(x)\right), \tag{14.2.20}$$

$$\mathbf{A}_j(x) = \left(\frac{1+\alpha^2}{1-\alpha^2}\right)^j \exp\left(\frac{1}{2}\alpha^2x^2\right)\int_{-\infty}^{\infty} \exp\left(-\frac{1}{2}\alpha^2y^2\right)\varepsilon(x-y)\varphi_j(y)\,dy. \tag{14.2.21}$$

For large N, we again have a semi-circular level density

$$\mathbf{R}_1(x) \approx \pi^{-1}(4N-x^2)^{1/2}, \tag{14.2.22}$$

whereas the n-level functions for $n > 1$ are discontinuous in α^2 at $\alpha^2 = 0$. On the other hand when $\alpha \to 0$, $N \to 0$ and

$$\rho \equiv \rho(x) \approx (\alpha/\sqrt{2})\mathbf{R}_1(x) \tag{14.2.23}$$

is finite, the functions have well defined limits. Equations (14.1.32) and (14.1.33) are unchanged but with K replaced by

$$\mathbf{K}(r;\rho) = \begin{bmatrix} (\sin \pi r)/\pi r & \mathbf{D}(r;\rho) \\ \mathbf{I}(r;\rho) & (\sin \pi r)/\pi r \end{bmatrix}, \tag{14.2.24}$$

where

$$\mathbf{I}(r;\rho) = -\pi^{-1}\int_0^{\pi} \frac{\sin kr}{k} \exp(2\rho^2k^2)\,dk, \tag{14.2.25}$$

$$\mathbf{D}(r;\rho) = -\pi^{-1}\int_{\pi}^{\infty} k\sin kr\exp(-2\rho^2k^2)\,dk. \tag{14.2.26}$$

Note that the limit of $\mathbf{D}(r;\rho)$ at $\rho=0$ contains a term in $\delta(r)/r$. This is due to the fact that in the GSE the eigenvalues are doubly degenerate. For $\rho\to\infty$, the product ID being zero, I and D can be replaced by zeros (see the remark following (14.1.36)).

14.3 Joint Probability Density for the Eigenvalues

In this section we derive the joint probability density for the eigenvalues, Eqs. (14.1.7) and (14.2.7), from that of the matrix elements, Eqs. (14.1.2) and (14.2.3). The matrix element probability densities depend on the eigenvalues and the angular variables characterizing the eigenvectors, and one has to integrate over these angular variables. For $\alpha^2=0,1,\infty$ in both ensembles the matrix element probability densities depend only on the eigenvalues. Also the Jacobian separates into a product of two functions, one involving only the eigenvalues and the other involving only the eigenvectors (cf. Chapter 3); therefore the integral over the angular variables, giving only a constant, need not be calculated. For arbitrary α^2 this simplification is not there, but we know (cf. Appendix A.5) an integral over the group of unitary matrices U,

$$\int \exp\left[c\,\mathrm{tr}(A-U^{\dagger}BU)^2\right]dU = \mathrm{const}\cdot\left[\Delta(a)\Delta(b)\right]^{-1}\det\left[\exp\{c(a_i-b_j)^2\}\right], \tag{14.3.1}$$

valid for arbitrary Hermitian matrices A and B having eigenvalues (a_i) and (b_i) respectively, where Δ is the product of differences, Eq. (14.1.8). We also have two other results at our disposal.

(i) The convolution of two independent Gaussian distributions with variances σ_1^2 and σ_2^2 is again a Gaussian with the variance $\sigma_1^2+\sigma_2^2$:

$$\int_{-\infty}^{\infty}(2\pi\sigma_1^2)^{-1/2}\exp\left(-\frac{x^2}{2\sigma_1^2}\right)(2\pi\sigma_2^2)^{-1/2}\exp\left(-\frac{(y-x)^2}{2\sigma_2^2}\right)dx$$
$$= \left[2\pi(\sigma_1^2+\sigma_2^2)\right]^{-1/2}\exp\left[-\frac{y^2}{2\sigma_1^2+2\sigma_2^2}\right]. \tag{14.3.2}$$

(ii) For any sets of functions $\theta_i(x)$, $\tau_i(x)$ and $\chi_i(x)$ and for a suitable measure $d\mu(x)$, integrals of the form

$$I_1 = \int\cdots\int\prod_1^N d\mu(x_i)\det[\theta_i(x_j)]\,\mathrm{sign}\Delta(x), \tag{14.3.3}$$

$$I_2 = \int\cdots\int\prod_1^m d\mu(x_i)\det[\theta_i(x_j),\tau_i(x_j)]_{i=1,\ldots,2m;\ j=1,\ldots,m}, \tag{14.3.4}$$

$$I_3 = \int \cdots \int \prod_1^m d\mu(x_i) \det[\theta_i(x_j), \tau_i(x_j), \chi_i(x_m)]_{i=1,\dots,2m-1;\ j=1,\dots,m-1}, \tag{14.3.5}$$

can be evaluated by the method of integration over alternate variables and the theory of Pfaffians (cf. Appendix A.17). The result is

$$I_1 = N!\,\mathrm{Pf}\,[a_{ij}]_{i,j=1,\dots,2m}, \tag{14.3.6}$$

$$I_2 = m!\,\mathrm{Pf}\,[b_{ij}]_{i,j=1,\dots,2m}, \tag{14.3.7}$$

$$I_3 = (m-1)!\,\mathrm{Pf}\,[c_{ij}]_{i,j=1,\dots,2m}, \tag{14.3.8}$$

where $2m = N$ if N is even and $2m = N+1$ if N is odd,

$$a_{ij} = \iint_{x \leqslant y} d\mu(x)\,d\mu(y)\,[\theta_i(x)\theta_j(y) - \theta_j(x)\theta_i(y)], \tag{14.3.9}$$

for $i, j = 1, 2, \dots, N$. When N is odd, we have in addition

$$a_{i,N+1} = -a_{N+1,i} = \int d\mu(x)\,\theta_i(x), \tag{14.3.10}$$

for $i = 1, 2, \dots, N$, and $a_{N+1,N+1} = 0$. Also

$$b_{ij} = \int d\mu(x)\,[\theta_i(x)\tau_j(x) - \tau_i(x)\theta_j(x)], \tag{14.3.11}$$

for $i, j = 1, 2, \dots, 2m$,

$$c_{ij} = b_{ij}, \quad i, j = 1, 2, \dots, 2m-1, \tag{14.3.12}$$

$$c_{i,2m} = -c_{2m,i} = \int d\mu(x)\,\chi_i(x), \tag{14.3.13}$$

$$c_{2m,2m} = 0. \tag{14.3.14}$$

We therefore proceed in three steps.

Firstly we write H as a sum, $H = A + B$. This convenient breaking will be different according to whether $\alpha^2 < 1$ or $\alpha^2 > 1$ and has nothing to do with the previous writing of H as a sum of its real and imaginary parts.

Secondly we use Eq. (14.3.1) to integrate over the unitary matrix, diagonalizing B. The result is some determinant containing the eigenvalues of H and of A, and the product of differences of these eigenvalues.

Finally we integrate over the eigenvalues of A using (14.3.3)–(14.3.8).

The treatment of the ensemble of Hermitian quaternion matrices will be parallel. Constants will be ignored in the intermediate steps, the final constants will be fixed by the normalization condition (see Appendices A.25 and A.26). The detailed working is given in the following subsections.

14.3.1 Matrices From GOE to GUE and Beyond. We write H as a sum of two Hermitian matrices

$$H = A + B, \tag{14.3.15}$$

and choose B so that its real and imaginary parts have the same variance. If $\alpha^2 < 1$, the real part of H has a larger variance than its imaginary part and we choose A to be real symmetric. If $\alpha^2 > 1$ it is the imaginary part of H which has a larger variance and we choose A to be pure imaginary anti-symmetric. In either case we adjust the variance of A and the common variance of the real and imaginary parts of B in a proper way. Thus our choices are:

$$P_1(A) \propto \exp\left\{-\frac{\operatorname{tr} A^2}{4v^2(1-\alpha^2)}\right\},$$

$$P_2(B) \propto \exp\left(-\frac{\operatorname{tr} B^2}{4v^2\alpha^2}\right),$$

if $\alpha^2 < 1$, and

$$P_1(A) \propto \exp\left\{-\frac{\operatorname{tr} A^2}{4v^2(\alpha^2-1)}\right\},$$

$$P_2(B) \propto \exp\left(-\frac{\operatorname{tr} B^2}{4v^2}\right),$$

if $\alpha^2 > 1$. We combine these equations as

$$A^T = A^* = A \operatorname{sign}(1-\alpha^2),$$

$$P_1(A) \propto \exp\left[-\frac{\operatorname{tr} A^2}{4v^2|1-\alpha^2|}\right], \tag{14.3.16}$$

$$P_2(B) \propto \exp\left(-\frac{\operatorname{tr} B^2}{4v^2\gamma^2}\right), \tag{14.3.17}$$

where $\gamma^2 = \min(1, \alpha^2)$ and $\operatorname{sign}(x)$, the sign of x, is $+1$ or -1 according as x is positive or negative; for $x = 0$, $\operatorname{sign}(x) = 0$. Though Eqs. (14.3.16) and (14.3.17) look alike, they have an important difference, $\operatorname{tr} B^2$ contains two sums of squares, those of the real and imaginary parts of B, whereas $\operatorname{tr} A^2$ contains only one of them. Equation (14.1.2) is now written in the equivalent form

$$P(H) = \int P_1(A) P_2(H - A)\, dA. \tag{14.3.18}$$

Let $x_1, \ldots, x_N$ be the eigenvalues of H

$$H = U X U^\dagger, \quad U^\dagger U = 1, \tag{14.3.19}$$

so that (see Chapter 3)

$$dH \propto \Delta^2(x)\, dU\, dx, \tag{14.3.20}$$

and the joint probability density for the $x_1, \ldots, x_N$ is

$$P(x) \equiv p(x_1, \ldots, x_N) \propto \int P_1(A) P_2(H - A) \Delta^2(x)\, dU\, dA. \tag{14.3.21}$$

We consider separately the cases of A symmetric and anti-symmetric.

When A is real symmetric, its real eigenvalues $a_1, \ldots, a_N$ are in general distinct and from (14.3.16), (14.3.17), (14.3.21) and (14.3.1) we have

$$P(x) \propto \int \exp\left(-\sum_1^N \frac{a_i^2}{4v^2(1-\alpha^2)}\right) \exp\left(-\frac{\operatorname{tr}(UXU^\dagger - A)^2}{4v^2}\right) \Delta^2(x)\, dU\, dA \tag{14.3.22}$$

$$\propto \Delta(x) \int \exp\left(-\sum_1^N \frac{a_i^2}{4v^2(1-\alpha^2)}\right) \det\left[\exp\left(-\frac{(x_i - a_j)^2}{4v^2}\right)\right] \frac{1}{\Delta(a)}\, dA. \tag{14.3.23}$$

As far as the dependence on the eigenvalues is concerned, one has (cf. Chapter 3)

$$dA \propto |\Delta(a)|\, da_1 \cdots da_N, \tag{14.3.24}$$

so that

$$P(x) \propto \Delta(x) \int \exp\left(-\sum_1^N \frac{a_i^2}{4v^2(1-\alpha^2)}\right) \times \det\left[\exp\left(-\frac{(x_i - a_j)^2}{4v^2}\right)\right] \operatorname{sign} \Delta(a)\, da_1 \cdots da_N. \tag{14.3.25}$$

Using Eqs. (14.3.3), (14.3.6) and (14.3.9), we get

$$P(x) \propto \Delta(x) \exp\left(-\sum_{1}^{N} \frac{x_i^2}{4v^2}\right) \mathrm{Pf}\,[b_{ij}], \tag{14.3.26}$$

where

$$b_{ij} = \iint_{-\infty \leqslant z_1 \leqslant z_2 \leqslant \infty} dz_1\, dz_2 (A(z_1, x_i) A(z_2, x_j) - A(z_1, x_j) A(z_2, x_i)), \tag{14.3.27}$$

$$A(z, x) = \exp\left(\frac{(z - (1-\alpha^2)x)^2}{4v^2\alpha^2(1-\alpha^2)}\right), \tag{14.3.27'}$$

for $i, j = 1, 2, \ldots, N$, when the order N of the matrices is even. For N odd, we have one more row and column;

$$b_{N+1,N+1} = 0,$$

and for $i = 1, \ldots, N$,

$$b_{i,N+1} = -b_{N+1,i} = \int_{-\infty}^{\infty} dz\, A(z, x_i). \tag{14.3.28}$$

On introducing the new variables $s = (z_2 + z_1)/\sqrt{2}$, $t = (z_2 - z_1)/\sqrt{2}$, the integration on s in (14.3.27) can be performed. A little algebra then gives

$$b_{ij} = \mathrm{const} \cdot \mathrm{erf}\left[\left(\frac{1-\alpha^2}{8v^2\alpha^2}\right)^{1/2} (x_i - x_j)\right], \quad i, j = 1, \ldots, N. \tag{14.3.29}$$

Equations (14.3.26), (14.3.28) and (14.3.29) give (14.1.7) when A is real symmetric.

When A is anti-symmetric pure imaginary, its eigenvalues are real and come in pairs $\pm a_i$; zero is necessarily an eigenvalue if the order N of A is odd. As far as the dependence on the eigenvalues is concerned (cf. Chapter 3)

$$dA \propto \prod_{1 \leqslant i < j \leqslant m} (a_i^2 - a_j^2)^2\, da_1 \cdots da_m, \tag{14.3.30}$$

$$\Delta(a) = 2^m \prod_{1}^{m} a_i \prod_{1 \leqslant i < j \leqslant m} (a_i^2 - a_j^2)^2, \tag{14.3.31}$$

for $N = 2m$ even; while

$$dA \propto \prod_1^m a_i^2 \prod_{1 \leqslant i < j \leqslant m} (a_i^2 - a_j^2)^2, \tag{14.3.32}$$

$$\Delta(a) = 2^m \prod_1^m a_i^3 \prod_{1 \leqslant i < j \leqslant m} (a_i^2 - a_j^2)^2, \tag{14.3.33}$$

for $N = 2m + 1$ odd. (We take $a_1, \ldots, a_m$ as distinct positive numbers.) Thus from (14.3.16)–(14.3.18), (14.3.21), (14.3.1), (14.3.30) and (14.3.31) we get for $N = 2m$ even

$$P(x) \propto \Delta(x) \exp\left(-\sum_1^N \frac{x_i^2}{4v^2\alpha^2}\right) \int_0^\infty \cdots \int_0^\infty \frac{da_1 \cdots da_m}{a_1 \cdots a_m}$$
$$\times \det\left[\exp\left(-\frac{\alpha^2 - 1}{4v^2\alpha^2}\left(\frac{\alpha^2 a_j}{\alpha^2 - 1} - x_i\right)^2\right) \cdot \exp\left(-\frac{\alpha^2 - 1}{4v^2\alpha^2}\left(\frac{\alpha^2 a_j}{\alpha^2 - 1} + x_i\right)^2\right)\right],$$
$$i = 1, \ldots, N; \; j = 1, \ldots, m. \tag{14.3.34}$$

For $N = 2m + 1$ odd, instead of (14.3.30) and (14.3.31) we use (14.3.32) and (14.3.33). The result is again Eq. (14.3.34) in which the determinant contains one more column

$$\exp\left[-\frac{\alpha^2}{4v^2(\alpha^2 - 1)}\left(\frac{\alpha^2 - 1}{\alpha^2} x_i\right)^2\right] = \exp\left(-\frac{\alpha^2 - 1}{4v^2\alpha^2} x_i^2\right). \tag{14.3.35}$$

For N even, we use Eqs. (14.3.4), (14.3.7) and (14.3.11), to get from (14.3.34)

$$P(x) \propto \Delta(x) \exp\left(-\sum_1^N \frac{x_i^2}{4v^2\alpha^2}\right) \mathrm{Pf}[b_{ij}], \tag{14.3.36}$$

with

$$b_{ij} = \int_0^\infty \frac{dz}{z}\left(\exp\left\{-\frac{\alpha^2}{4v^2(\alpha^2 - 1)}\left[\left(z - \frac{\alpha^2 - 1}{\alpha^2} x_i\right)^2 + \left(z + \frac{\alpha^2 - 1}{\alpha^2} x_j\right)^2\right]\right\}\right.$$
$$\left. - \exp\left\{-\frac{\alpha^2}{4v^2(\alpha^2 - 1)}\left[\left(z + \frac{\alpha^2 - 1}{\alpha^2} x_i\right)^2 + \left(z - \frac{\alpha^2 - 1}{\alpha^2} x_j\right)^2\right]\right\}\right). \tag{14.3.37}$$

For N odd, we use (14.3.5), (14.3.8), (14.3.12) and (14.3.13); the result is again (14.3.36) in which the Pfaffian has an additional row and an additional column

$$b_{i,N+1} = -b_{N+1,i} = \exp\left(-\frac{\alpha^2 - 1}{4v^2\alpha^2}x_i^2\right). \tag{14.3.38}$$

A little algebra gives for (14.3.37)

$$\begin{aligned} b_{ij} &\propto \exp\left(-\frac{\alpha^2-1}{4v^2\alpha^2}(x_i^2+x_j^2)\right)\int_0^\infty \frac{dt}{t}e^{-t^2}\sinh\left[\left(\frac{\alpha^2-1}{2v^2\alpha^2}\right)^{1/2}(x_i-x_j)t\right] \\ &\propto \exp\left(-\frac{\alpha^2-1}{4v^2\alpha^2}(x_i^2+x_j^2)\right)\operatorname{erf}\left[\left(\frac{\alpha^2-1}{8v^2\alpha^2}\right)^{1/2}(x_i-x_j)\right]. \end{aligned} \tag{14.3.39}$$

Equations (14.3.36), (14.3.39) and (14.3.38) give Eq. (14.1.7) for anti-symmetric A.

Equation (14.1.7) is therefore established in all cases except for the overall constant. This constant is evaluated in Appendix A.25.

14.3.2 Matrices From GSE to GUE and Beyond. We write the $N \times N$ quaternion matrix $\mathbf{H}$ as a sum

$$\mathbf{H} = \mathbf{A} + \mathbf{B}, \qquad \bar{\mathbf{A}} = \mathbf{A}\,\mathrm{sign}(1-\alpha^2). \tag{14.3.40}$$

The quaternion matrices $\mathbf{A}$ and $\mathbf{B}$ are Hermitian; the self-dual and anti-self-dual parts of $\mathbf{B}$ have the same variance; $\mathbf{A}$ is self-dual quaternion real if $\alpha^2 < 1$, and anti-self-dual quaternion pure imaginary if $\alpha^2 > 1$. Due to Eq. (14.3.2), Eq. (14.2.3) is equivalent to Eq. (14.3.18) with

$$P_1(\mathbf{A}) \propto \exp\left[-\operatorname{tr}\mathbf{A}^2/(4v^2|1-\alpha^2|)\right], \tag{14.3.41}$$

$$P_2(\mathbf{B}) \propto \exp(-\operatorname{tr}\mathbf{B}^2/4v^2\gamma^2), \tag{14.3.42}$$

where, as before, $\gamma^2 = \min(1, \alpha^2)$. Let

$$\mathbf{H} = UXU^\dagger, \quad UU^\dagger = 1, \tag{14.3.43}$$

where $\mathbf{H}$, X and U are $2N \times 2N$ ordinary matrices, U is unitary, X is diagonal and the diagonal elements $x_1, \ldots, x_{2N}$ of X are the eigenvalues of $\mathbf{H}$. As before (Chapter 3)

$$d\mathbf{H} \propto \Delta^2(x_1, \ldots, x_{2N})\,dx\,dU, \tag{14.3.44}$$

$$dx \equiv dx_1, \ldots, dx_{2N}. \tag{14.3.45}$$

We consider separately, as before, the cases of $\mathbf{A}$ self-dual and $\mathbf{A}$ anti-self-dual.

When **A** is self-dual, its eigenvalues are real numbers $a_1, \ldots, a_N$ each repeated twice. Since the right-hand side of Eq. (14.3.1) is now of the form 0/0, we cannot use it directly, but have to take limits when the eigenvalues of **A** become equal in pairs. The calculation gives

$$\int \exp\left[-\frac{\operatorname{tr}(\mathbf{A} - UXU^\dagger)^2}{4v^2\alpha^2}\right] dU \propto (\Delta(x)\Delta^4(a))^{-1}$$
$$\times \det\left[\exp\left(-\frac{(a_i - x_j)^2}{4v^2\alpha^2}\right) \quad (a_i - x_j)\exp\left(-\frac{(a_i - x_j)^2}{4v^2\alpha^2}\right)\right], \tag{14.3.46}$$

where $\Delta(x) \equiv \Delta(x_1, \ldots, x_{2N})$ and $\Delta(a) \equiv \Delta(a_1, \ldots, a_N)$, Eq. (14.1.8).

Now as far as the dependence on the eigenvalues is concerned (cf. Chapter 3)

$$d\mathbf{A} \propto \Delta^4(a)\, da_1, \ldots, da_N \equiv \Delta^4(a)\, da, \tag{14.3.47}$$

so that

$$\mathbf{P}(x) \equiv \mathbf{P}(x_1, \ldots, x_{2N}) \propto \Delta(x) \int da \exp\left(-\sum_1^N \frac{a_i^2}{2v^2(1-\alpha^2)}\right)$$
$$\times \det\left[\exp\left(-\frac{(a_i - x_j)^2}{4v^2\alpha^2}\right) \quad (a_i - x_j)\exp\left(-\frac{(a_i - x_j)^2}{4v^2\alpha^2}\right)\right]. \tag{14.3.48}$$

Using Eqs. (14.3.4), (14.3.7) and (14.3.11)

$$\mathbf{P}(x) \propto \Delta(x) \exp\left(-\sum_1^{2N} \frac{x_i^2}{4v^2}\right) \operatorname{Pf}[h_{ij}], \tag{14.3.49}$$

where

$$h_{ij} = (x_i - x_j)\exp\left(\frac{x_i^2 + x_j^2}{4v^2}\right) \int_{-\infty}^{\infty} dz \exp\left(-\frac{z^2}{2v^2(1-\alpha^2)}\right)$$
$$\times \exp\left(-\frac{(z - x_i)^2}{4v^2\alpha^2} - \frac{(z - x_j)^2}{4v^2\alpha^2}\right). \tag{14.3.50}$$

A little algebra now gives

$$h_{ij} \propto (x_i - x_j)\exp\left(-\frac{1-\alpha^2}{8v^2\alpha^2}(x_i - x_j)^2\right). \tag{14.3.51}$$

From (14.3.49) and (14.3.51) we get Eq. (14.2.7) for the case $\alpha^2 < 1$.

When $\mathbf{A}$ is anti-self-dual, its eigenvalues are the real numbers $\pm a_1, \pm a_2, \ldots, \pm a_N$, and from Eq. (14.3.1)

$$\int \exp\left(-\frac{\operatorname{tr}(\mathbf{A} - UXU^\dagger)^2}{4v^2}\right) dU \propto [\Delta(x)\Delta(\pm a_1, \ldots, \pm a_N)]^{-1} \times \det\left[\exp\left(-\frac{(a_i - x_j)^2}{4v^2}\right) \quad \exp\left(-\frac{(a_i + x_j)^2}{4v^2}\right)\right]. \tag{14.3.52}$$

However,

$$\Delta(\pm a_1, \ldots, \pm a_N) = 2^N \prod_1^N a_i [\Delta(a_1^2, \ldots, a_N^2)]^2, \tag{14.3.53}$$

and as far as dependence on the eigenvalues is concerned (cf. Section 3.4)

$$d\mathbf{A} \propto \prod_1^N a_i^2 [\Delta(a_1^2, \ldots, a_N^2)]^2, \tag{14.3.54}$$

so that

$$\begin{aligned} \mathbf{P}(x) &\propto \Delta(x) \int_0^\infty \cdots \int_0^\infty da_1 \cdots da_N \prod_1^N a_i \exp\left(-\sum_1^N \frac{a_i^2}{2v^2(\alpha^2 - 1)}\right) \\ &\quad \times \det\left[\exp\left(\frac{(a_i - x_j)^2}{4v^2}\right) \quad \exp\left(-\frac{(a_i + x_j)^2}{4v^2}\right)\right] \\ &\propto \Delta(x) \exp\left(-\sum_1^{2N} \frac{x_i^2}{4v^2}\right) \operatorname{Pf}[h_{ij}], \end{aligned} \tag{14.3.55}$$

where in the last step we have used Eqs. (14.3.4), (14.3.7), (14.3.11), and

$$\begin{aligned} h_{ij} &= \exp\left(\frac{x_i^2 + x_j^2}{4v^2}\right) \int_0^\infty dz\, z \exp\left(-\frac{z^2}{2v^2(\alpha^2 - 1)}\right) \\ &\quad \times \left[\exp\left(-\frac{(z - x_i)^2}{4v^2} - \frac{(z + x_j)^2}{4v^2}\right) - \exp\left(-\frac{(z + x_i)^2}{4v^2} - \frac{(z - x_j)^2}{4v^2}\right)\right]. \end{aligned} \tag{14.3.56}$$

A little algebra now gives

$$h_{ij} \propto (x_i - x_j) \exp\left(-\frac{\alpha^2 - 1}{8v^2\alpha^2}(x_i - x_j)^2\right). \tag{14.3.57}$$

Equations (14.3.55) and (14.3.57) give (14.2.7) for $\alpha^2 > 1$.

Equation (14.2.7) is therefore established for all values of α^2, except for the overall constant, which is evaluated in Appendix A.26.

14.4 Correlation and Cluster Functions

To derive the correlation functions we shall use Theorem 5.1.4 with determinants of quaternion matrices.

Let us consider the first ensemble. Using Theorem 5.1.4 several times one can get the correlation function

$$R_n(x_1, \ldots, x_n) = \frac{N!}{(N-n)!} \int_{-\infty}^{\infty} \cdots \int_{-\infty}^{\infty} P(x_1, \ldots, x_N)\, dx_{n+1} \cdots dx_N, \tag{14.4.1}$$

as a determinant of an $n \times n$ quaternion matrix; the derivation relies on the fact that the $p(x)$ of Eq. (14.1.7) can be written as an $N \times N$ quaternion determinant

$$P(x_1, \ldots, x_N) = (1/N!) \det[\phi(x_i, x_j)]_{i,j=1,\ldots,N}, \tag{14.4.2}$$

where the quaternion $\phi(x, y)$ (Eq. (14.1.14)) satisfies the equality

$$\int_{-\infty}^{\infty} \phi(x, y)\phi(y, z)\, dy = \phi(x, z) + \lambda\phi(x, z) - \phi(x, z)\lambda, \tag{14.4.3}$$

in which λ is the constant quaternion

$$\lambda = \frac{1}{2}\begin{bmatrix} 1 & 0 \\ 0 & -1 \end{bmatrix}. \tag{14.4.4}$$

Equation (14.1.12) then follows from Theorem 5.1.4. The proof of Eq. (14.4.2) is given in Appendix A.27 and that of (14.4.3) in Appendix A.28.

Similarly one can get the correlation functions for the other ensemble, since

$$\mathbf{P}(x_1, \ldots, x_{2N}) = [(2N)!]^{-1} \det[\mathbf{\Phi}(x_i, x_j)]_{i,j=1,\ldots,2N}, \tag{14.4.5}$$

where $\mathbf{P}$ is now given by (14.2.7) and $\mathbf{\Phi}$ by (14.2.10), and since this $\mathbf{\Phi}$ satisfies Eq. (14.4.3).

The proof of Eq. (14.4.5) is given in Appendix A.27. The $\mathbf{\Phi}$ satisfies Eq. (14.4.3) with a quaternion τ which is the negative of (14.4.4), see Appendix A.28.

Equation (14.1.13) is obtained from the observation that (cf. Eq. (6.1.3)) the cluster function T_n is the cumulant of the correlation function R_n and the expression on the right-hand side of (14.1.13) is the cumulant of that on the right-hand side of (14.1.12) (consider it as a determinant of an $n \times n$ self-dual matrix with quaternion elements, Section 5.1).

To get the limits for large N of the correlation and cluster functions it suffices then to take the limits of $\phi(x, y)$ and $\mathbf{\Phi}(x, y)$. The limits of the functions $S_N(x, y)$, $D_N(x, y)$, $I_N(x, y)$ and $J_N(x, y)$ are derived as follows (Pandey and Mehta, 1983). For large N

$$S_N(x, y) \approx \sum_{j=0}^{N-1} \varphi_j(x)\varphi_j(y). \tag{14.4.6}$$

Thus from Appendix A.9 the asymptotic level density is

$$R_1(x) = \lim_{N\to\infty} S_N(x, x) \approx \frac{1}{\pi}(2N - x^2)^{1/2}. \tag{14.1.30}$$

Also, as $N \to \infty$, $x - y \to 0$, while

$$r = (x - y)R_1(x) \tag{14.4.7}$$

remains finite, we have from Eq. (14.4.6) and Appendix A.10

$$S_N(x, y) \to R_1(x)\sin(\pi r)/(\pi r). \tag{14.4.8}$$

The limits of $I_N(x, y)$ and $D_N(x, y)$ are derived in Appendix A.29. Those of $\mathbf{S}_N$, $\overline{\mathbf{D}}_N$ and $\mathbf{I}_N$ are obtained from those of S_N, $\overline{D}_N$ and I_N by changing ρ^2 to $-\rho^2$ and replacing N by $2N$. For $\mathbf{g}$ we have

$$\begin{aligned}[\mathbf{R}_1(x)]^{-2} g(x, y) &\approx -(r/8\rho^3\sqrt{\pi})\exp(-r^2/8\rho^2) \\ &\approx -\pi^{-1}\int_0^\infty k\sin kr\exp(-2\rho^2k^2)\,dk,\end{aligned} \tag{14.4.9}$$

from which we get the limit of $\mathbf{D}_N$.

This completes the proof of the statements listed in Section 14.1. Some technical details are collected in Appendices A.24 to A.30.

Summary of Chapter 14

In this chapter we derive all correlation and cluster functions for the eigenvalues of the Hermitian matrices $H = H_1 + H_2$ when the probability density of the matrix elements is proportional to the product of $\exp(-\operatorname{tr} H_1^2/(2v^2))$ and $\exp(-\operatorname{tr} H_2^2/(2v^2\alpha^2))$, where H_1 and H_2 are either the symmetric and anti-symmetric parts of H or they are the self-dual and anti-self-dual parts of H.

15

MATRICES WITH GAUSSIAN ELEMENT DENSITIES BUT WITH NO UNITARY OR HERMITIAN CONDITIONS IMPOSED

An ensemble of matrices whose elements are complex, quaternion, or real numbers, but with no other restrictions as to their Hermitian or unitary character, is of no immediate physical interest, for their eigenvalues may lie anywhere on the complex plane. However, efforts have been made (Ginibre, 1965; Edelman et al., 1994) to investigate them and the results are interesting in their own right.

To define a matrix ensemble one has to specify two things: the space T on which the matrices vary and a probability density function over T. For the case to be tractable, these choices have to be reasonable. Following Ginibre (1965) we will take T to be the set of all $N \times N$ matrices and a Gaussian probability density for the matrix elements.

15.1 Complex Matrices

Let T be the set of all $N \times N$ complex matrices. The probability that a matrix from the set T will lie in $(A, A + dA)$ is $P(A)\mu(dA)$, where $\mu(dA)$ is the linear measure

$$\mu(dA) = \prod_{j,k} dA_{jk}^{(0)} \, dA_{jk}^{(1)}, \tag{15.1.1}$$

and $A_{jk}^{(0)}$, $A_{jk}^{(1)}$ are the real and imaginary parts of the matrix element

$$A_{jk} = A_{jk}^{(0)} + i A_{jk}^{(1)}. \tag{15.1.2}$$

For the function $P(A)$ we may choose, for example (Ginibre, 1965)

$$P(A) = \exp\left[-\operatorname{tr}(A^{\dagger} A)\right]. \tag{15.1.3}$$

We denote the ensemble so defined as T_c. It is visibly invariant under all unitary transformations.

To get any information about the eigenvalues, we must first find their joint probability density. This can be done, as in Chapter 3, by changing the variables from A_{jk} to the (complex) eigenvalues z_j of A and the auxiliary variables p_j. Since $\operatorname{tr}(A^{\dagger} A)$ is not only a function of z_j but contains other variables p_j as well, these variables have to be chosen carefully to facilitate later integrations. Let the eigenvalues of A be distinct; the case when A has multiple eigenvalues need not be considered for the same reasons as in Chapter 3. Also, let X be the $N \times N$ matrix whose columns are the eigenvectors of A so that X is nonsingular and $X^{-1} A X = E$ is diagonal. From $A = X E X^{-1}$ we obtain by differentiation

$$dA = X(dE + dB\,E - E\,dB)X^{-1}, \tag{15.1.4}$$

with

$$dB = X^{-1}\,dX. \tag{15.1.5}$$

Equation (15.1.4) reads in terms of its components:

$$(X^{-1}\,dAX)_{jj}^{(0)} = dz_j^{(0)} = dx_j, \qquad (X^{-1} dAX)_{jj}^{(1)} = dz_j^{(1)} = dy_j, \tag{15.1.6}$$

$$\begin{aligned}
(X^{-1}\,dAX)_{jk}^{(0)} &= (x_k - x_j)\,dB_{jk}^{(0)} - (y_k - y_j)\,dB_{jk}^{(1)}, \\
(X^{-1}\,dAX)_{jk}^{(1)} &= (y_k - y_j)\,dB_{jk}^{(0)} + (x_k - x_j)\,dB_{jk}^{(1)}, \quad j \neq k,
\end{aligned} \tag{15.1.7}$$

where x_j, y_j are the real and imaginary parts of z_j, the diagonal elements of E, whereas $dB_{jk}^{(0)}$ and $dB_{jk}^{(1)}$ are the real and imaginary parts of dB_{jk}. Whenever any set of differentials is expressed linearly in terms of those of the others, the ratio of the volume elements is equal to the Jacobian. The volume element in (15.1.1) is therefore given by

$$\begin{aligned}
\mu(dA) &= \mu(X^{-1}\,dAX) \\
&= \prod_{j \neq k} |z_k - z_j|^2\,dB_{jk}^{(0)}\,dB_{jk}^{(1)} \prod_i dx_i\,dy_i.
\end{aligned} \tag{15.1.8}$$

One still has to evaluate the integral

$$J = \int \exp[-\operatorname{tr}(A^\dagger A)] \prod_{j\neq k} dB_{jk}^{(0)}\, dB_{jk}^{(1)}, \tag{15.1.9}$$

which can be done by a careful choice of the new variables of integration and by using properties of determinant expansions (see Appendix A.30 or A.33). The result is as follows.

The joint probability density for the eigenvalues of A belonging to the ensemble T_c of all complex matrices is given by

$$P_c(z_1, z_2, \ldots, z_N) = K_c \exp\left(-\sum_1^N |z_i|^2\right) \prod_{1\leqslant i<j\leqslant N} |z_i - z_j|^2, \tag{15.1.10}$$

where K_c is the normalization constant given later by (15.1.17).

With this joint probability density function one can determine various quantities of interest as easily as in Chapters 6 or 11. For example, the probability that all the eigenvalues z_i will lie outside a circle of radius α centered at $z = 0$ is

$$E_{N_c}(\alpha) = \int \cdots \int_{|z_i|\geqslant\alpha} P_c(z_1, \ldots, z_N) \prod_i dx_i\, dy_i. \tag{15.1.11}$$

By writing

$$\prod_{i<j} |z_i - z_j|^2 = \prod_{i<j} (z_i - z_j)(z_i^* - z_j^*)$$

$$= \det\begin{bmatrix} 1 & \ldots & 1 \\ z_1 & \ldots & z_N \\ \ldots & \ldots & \ldots \\ z_1^{N-1} & \ldots & z_N^{N-1} \end{bmatrix} \det\begin{bmatrix} 1 & \ldots & 1 \\ z_1^* & \ldots & z_N^* \\ \ldots & \ldots & \ldots \\ z_1^{*N-1} & \ldots & z_N^{*N-1} \end{bmatrix},$$

and multiplying the two determinants row by row we get

$$E_{N_c}(\alpha) = K_c \int \cdots \int_{|z_i|\geqslant\alpha} \left(\prod_i dx_i\, dy_i\right) \exp\left(-\sum_1^N |z_i|^2\right)$$

$$\times \det \begin{bmatrix} N & \sum_i z_i & \cdots & \sum_i z_i^{N-1} \\ \sum_i z_i^* & \sum_i z_i^* z_i & \cdots & \sum_i z_i^* z_i^{N-1} \\ \cdots & \cdots & \cdots & \cdots \\ \sum_i z_i^{*N-1} & \sum_i z_i^{*N-1} z_i & \cdots & \sum_i z_i^{*N-1} z_i^{N-1} \end{bmatrix}. \tag{15.1.12}$$

Since the integrand is symmetric in all the z_i, we can replace the first row with $1, z_1, z_1^2, \ldots, z_1^{N-1}$ and multiply the result by N; z_1 can now be eliminated from the other rows by subtracting a suitable multiple of the first row. The resulting determinant is symmetric in the $N-1$ variables $z_2, z_3, \ldots, z_N$; therefore we replace the second row with $z_2^*, z_2^* z_2, \ldots, z_2^* z_2^{N-1}$ and multiply the result by $N-1$. The process can be repeated and we get

$$E_{N_c}(\alpha) = K_c N! \int \cdots \int_{|z_i| \geqslant \alpha} \left\{ \prod_i dx_i\, dy_i \right\} \exp\left(-\sum_1^N |z_i|^2 \right)$$
$$\times \det \begin{bmatrix} 1 & z_1 & \cdots & z_1^{N-1} \\ z_2^* & z_2^* z_2 & \cdots & z_2^* z_2^{N-1} \\ \cdots & \cdots & \cdots & \cdots \\ z_N^{*N-1} & z_N^{*N-1} z_N & \cdots & z_N^{*N-1} z_N^{N-1} \end{bmatrix}. \tag{15.1.13}$$

Since the various rows now depend on distinct variables, we can integrate them separately with the exponential factor. By changing to polar coordinates and performing the angular integrations first we see that

$$\int_{|z| \geqslant \alpha} e^{-|z|^2} z^{*j} z^k\, dx\, dy = \pi \delta_{jk} \Gamma(j+1, \alpha^2), \tag{15.1.14}$$

so that

$$E_{N_c}(\alpha) = K_c N! \pi^N \prod_{j=1}^N \Gamma(j, \alpha^2), \tag{15.1.15}$$

where $\Gamma(j, \alpha^2)$ is the incomplete gamma function

$$\Gamma(j, \alpha^2) = \int_{\alpha^2}^{\infty} e^{-x} x^{j-1}\, dx = \Gamma(j) e^{-\alpha^2} \sum_{\ell=0}^{j-1} \frac{\alpha^{2\ell}}{\ell!}. \tag{15.1.16}$$

Since $E_{N_c}(0) = 1$, the constant K_c can be determined from (15.1.15) as

$$K_c^{-1} = \pi^N \prod_{j=1}^{N} j!, \tag{15.1.17}$$

and therefore

$$E_{N_c}(\alpha) = \prod_{j=1}^{N} \left(e^{-\alpha^2} \sum_{\ell=0}^{j-1} \frac{\alpha^{2\ell}}{\ell!} \right). \tag{15.1.18}$$

One can convince oneself that $E_{N_c}(\alpha)$ tends to a well-defined limit as $N \to \infty$. For small values of α one may expand $E_{N_c}(\alpha)$ in a power series:

$$E_{N_c}(\alpha) = 1 - \alpha^2 + \frac{1}{2}\alpha^6 - \frac{5}{12}\alpha^8 + \frac{7}{24}\alpha^{10} - \cdots. \tag{15.1.19}$$

To get the coefficient of α^{2i} in the above power series one may replace $e^{-\alpha^2} \sum_{\ell=0}^{j-1} \alpha^{2\ell}/\ell!$ by unity for all $j > i$. In fact, one can even get for $E_c(\alpha) = \lim_{N\to\infty} E_{N_c}(\alpha)$ a series of upper bounds and a series of lower bounds converging toward each other. We have the obvious inequality

$$0 < \prod_{j=r}^{N-1} [e^{-\alpha^2} a_j(\alpha^2)] \leqslant 1, \quad r \geqslant 0, \tag{15.1.20}$$

where

$$a_j(x) = \sum_{\ell=0}^{j} \frac{x^\ell}{\ell!}, \tag{15.1.21}$$

is the truncated exponential series. On the other hand, the identity

$$e^{-\alpha^2} a_j(\alpha^2) = \exp\left(- \int_0^{\alpha^2} \frac{a_j(x) - a_{j-1}(x)}{a_j(x)}\, dx \right), \tag{15.1.22}$$

and the inequality $a_j(x) \geqslant a_\ell(x)$ for $j \geqslant \ell$ give us

$$\begin{aligned} \prod_{j=r}^{N-1} [e^{-\alpha^2} a_j(\alpha^2)] &= \exp\left[- \sum_{j=r}^{N-1} \int_0^{\alpha^2} \frac{a_j(x) - a_{j-1}(x)}{a_j(x)}\, dx \right] \\ &\geqslant \exp\left[- \sum_{j=r}^{N-1} \int_0^{\alpha^2} \frac{a_j(x) - a_{j-1}(x)}{a_r(x)}\, dx \right] \end{aligned}$$

$$= \exp\left[-\int_0^{\alpha^2} \frac{a_{N-1}(x) - a_{r-1}(x)}{a_r(x)}\, dx\right]. \tag{15.1.23}$$

Taking the limit $N \to \infty$, the inequalities (15.1.20) and (15.1.23) give us finally

$$0 < F_s(\alpha^2) f_s(\alpha^2) \leqslant F_r(\alpha^2) f_r(\alpha^2) \leqslant E_c(\alpha) \leqslant F_{r'}(\alpha^2) \leqslant F_{s'}(\alpha^2) \leqslant 1,$$
$$\text{if } r > s > 0, \quad r' > s' > 0, \tag{15.1.24}$$

where

$$F_r(\alpha^2) = \prod_{j=0}^{r-1} \left[e^{-\alpha^2} a_j(\alpha^2)\right] \equiv E_{rc}(\alpha), \tag{15.1.25}$$

and

$$f_r(\alpha^2) = \exp\left[-\int_0^{\alpha^2} \frac{e^x - a_{r-1}(x)}{a_r(x)}\, dx\right]. \tag{15.1.26}$$

Equation (15.1.24) gives us in particular

$$\exp\left(-\alpha^2 - \int_0^{\alpha^2} \frac{e^x - 1}{1+x}\, dx\right) \leqslant E_c(\alpha) \leqslant (1+\alpha^2) e^{-2\alpha^2}. \tag{15.1.27}$$

To get the n-point correlation function

$$R_n(z_1, \ldots, z_n) = \frac{N!}{(N-n)!} \int \cdots \int P_c(z_1, \ldots, z_N) \prod_{i=n+1}^{N} dx_i\, dy_i, \tag{15.1.28}$$

we proceed exactly as in Section 6.2 or 11.1. Equation (15.1.14) corresponds to the orthogonality property of φ_k in Section 6.2, Eq. (6.2.3). The result is

$$R_n(z_1, \ldots, z_n) = \pi^{-n} \exp\left(-\sum_{j=1}^{n} |z_j|^2\right) \det\left[K_N(z_i, z_j)\right]_{i,j=1,\ldots,n}, \tag{15.1.29}$$

where

$$K_N(z_i, z_j) = \sum_{\ell=0}^{N-1} \frac{(z_i z_j^*)^\ell}{\ell!}. \tag{15.1.30}$$

As $N \to \infty$ the correlation functions tend to well-defined limits:

$$R_n(z_1, \ldots, z_N) \simeq \pi^{-n} \exp\left(-\sum_1^n |z_i|^2\right) \det\left[e^{z_i z_j^*}\right]_{i,j=1,2,\ldots,n}. \tag{15.1.31}$$

From Eqs. (15.1.29) and (15.1.30) the density of the eigenvalues is

$$R_1(z) = \pi^{-1} e^{-|z|^2} \sum_{\ell=0}^{N-1} \frac{|z|^{2\ell}}{\ell!}. \tag{15.1.32}$$

This density is isotropic and depends only on $|z| = r$, which was to be expected. It is constant $R_1(z) \approx 1/\pi$ for $r^2 \ll N$ and $R_1(z) \approx 0$ for $r^2 \gg N$. The sum in (15.1.32) can be estimated in an elementary way. From the inequalities

$$\begin{aligned} e^{r^2} - \sum_0^{N-1} \frac{r^{2\ell}}{\ell!} &= \sum_N^{\infty} \frac{r^{2\ell}}{\ell!} \leqslant \frac{r^{2N}}{N!} \sum_0^{\infty} \frac{r^{2\ell}}{(N+1)^{\ell}} \\ &= \frac{r^{2N}}{N!} \frac{N+1}{N+1-r^2} \quad \text{for } r^2 < N, \end{aligned} \tag{15.1.33}$$

and

$$\begin{aligned} \sum_0^{N-1} \frac{r^{2\ell}}{\ell!} &\leqslant \frac{r^{2(N-1)}}{(N-1)!} \sum_0^{N-1} \left(\frac{N-1}{r^2} \right)^{\ell} \\ &\leqslant \frac{r^{2(N-1)}}{(N-1)!} \frac{r^2}{r^2 - N + 1} \quad \text{for } r^2 > N, \end{aligned} \tag{15.1.34}$$

we get

$$1 - \pi R_1(z) \leqslant e^{-r^2} \frac{r^{2N}}{N!} \frac{N+1}{N+1-r^2} \quad \text{for } r^2 < N, \tag{15.1.35}$$

and

$$\pi R_1(z) \leqslant e^{-r^2} \frac{r^{2N}}{N!} \frac{N}{r^2+1-N} \quad \text{for } r^2 > N. \tag{15.1.36}$$

One can also estimate how fast the eigenvalue density falls from $1/\pi$ to 0 around $r^2 = N$. Putting $r = N^{1/2} \pm u$, $0 \leqslant u \leqslant 1 \ll N$, the leading term in (15.1.35) and (15.1.36) is $e^{-u^2}/2u\sqrt{\pi}$.

The two-point correlation function in the limit $N \to \infty$ is

$$R_2(z_1, z_2) = \pi^{-2}\left[1 - \exp(-|z_1 - z_2|^2)\right], \tag{15.1.37}$$

and depends only on the distance between the eigenvalues.

15.2 Quaternion Matrices

In this section we consider matrices whose elements are real quaternions (cf. Chapter 2). All four quaternion components of each matrix element are random variables. To proceed any further one has to know about the diagonalization of these matrices. The eigenvalue equation may be written as

$$AY = Y\lambda, \tag{15.2.1}$$

where Y is a vector with N quaternion components (the eigenvector) and λ is a quaternion number (the eigenvalue). There is no reason a priori for (real quaternion) solutions to (15.2.1) to exist. Fortunately, they do and in sufficient number (Appendix A.31). Writing (15.2.1) as

$$AY\mu = Y\mu(\mu^{-1}\lambda\mu), \tag{15.2.2}$$

we see that if λ is an eigenvalue then so is $\mu^{-1}\lambda\mu$ for arbitrary μ. Thus the eigenvalues of a given matrix are not just discrete points but describe closed curves, and one has to talk about the distribution of these eigencurves in the four-dimensional space. Even if one chooses to describe these curves by some fixed point or points on them, only one-sided linear independence of the corresponding eigenvector rays can be established by the usual methods. Although, for a given quaternion real matrix A another such X can be found (in the favorable circumstance of distinct eigencurves) which diagonalizes it,

$$A = XEX^{-1}. \tag{15.2.3}$$

(E diagonal and real (Appendix A.31)), it seems difficult to establish it by purely quaternion means.

In view of these difficulties, from now on we shall employ the matrix representation of quaternions (cf. Chapter 2), thus doubling the size of the matrix A, and use well-known results on matrices with complex elements. Thus, in reality, this section does not deal with the quaternion matrices as such but with even-order complex matrices having a special structure; the elements of A satisfy the relations

$$A_{2i,2j} = A^*_{2i-1,2j-1}, \qquad A_{2i-1,2j} = -A^*_{2i,2j-1}, \tag{15.2.4}$$

or, in the matrix notation,

$$AZ = ZA^*, \tag{15.2.5}$$

where Z is the antisymmetric, real, unitary matrix (2.4.1).

If X is the $2N \times 2N$ matrix whose columns are the eigenvectors of A, the eigenvalues being all distinct, then

$$A = XEX^{-1}, \tag{15.2.6}$$

where E is diagonal. The diagonal elements of E occur in complex conjugate pairs z_j, z_j^*; $j = 1, 2, \ldots, N$. The linear measure is

$$\mu(dA) = \prod_{\substack{1 \leqslant i \leqslant j \leqslant N \\ \lambda=0,1}} dA_{2i,2j}^{(\lambda)} \, dA_{2i-1,2j}^{(\lambda)}, \tag{15.2.7}$$

where $dA_{ij}^{(0)}$ and $dA_{ij}^{(1)}$ are the real and imaginary parts of dA_{ij}. For $P(A)$ we take

$$P(A) = \exp\left[-\frac{1}{2}\operatorname{tr}(A^\dagger A)\right]; \tag{15.2.8}$$

the factor $1/2$ is there to compensate for the artificial doubling of the size of A. We denote the ensemble so defined by T_Q. Equations (15.1.4) and (15.1.5) are valid. If we write (15.1.4) in terms of the various components, the volume element of dA is

$$\mu(dA) = \prod_i |z_i - z_i^*|^2 \prod_{i \neq j} \left(|z_i - z_j|^2 |z_i - z_j^*|^2\right) \prod_{\substack{i<j \\ \lambda=0,1}} dB_{2i,2j}^{(\lambda)} \, dB_{2i-1,2j}^{(\lambda)}. \tag{15.2.9}$$

The integration corresponding to (15.1.9) for this case is carried out in Appendix A.32 or A.33. The result is as follows.

The joint probability density function for the eigenvalues of A belonging to the ensemble T_Q of all complex matrices satisfying (15.2.4) is given by

$$P_Q(z_1, \ldots, z_N) = K_Q \exp\left(-\sum_1^N |z_i|^2\right) \prod_i |z_i - z_i^*|^2 \prod_{i<j} (|z_i - z_j|^2 |z_i - z_j^*|^2), \tag{15.2.10}$$

where K_Q is the normalization constant given by (15.2.15).

With this joint probability density function one can determine the various quantities of interest with almost the same ease as in the unitary case. The method to be followed, in all such calculations is to express $P_Q(z_1, \ldots, z_N)$ as a quaternion determinant and use the integration method developed in Chapter 5 and Appendix A.17.

Let us write a $2N \times 2N$ Vandermonde determinant of the variables z_j, z_j^*, $j = 1, 2, \ldots, N$; that is, the determinant whose $(2j-1)$th column consists of the successive powers $(1, z_j, z_j^2, \ldots, z_j^{2N-1})$ of z_j and whose $2j$th column consists of the successive powers $(1, z_j^*, z_j^{*2}, \ldots, z_j^{*2N-1})$ of z_j^* for $j = 1, 2, \ldots, N$. We can clearly see that this determinant is nothing but

$$\prod_i (z_i^* - z_i) \prod_{i<j} \left(|z_i - z_j|^2 |z_i - z_j^*|^2\right). \tag{15.2.11}$$

Thus we are led to define with $z = x + iy$,

$$f_{ij}(u) = \int\int e^{-|z|^2}(z - z^*)u(z)(z^i z^{*j} - z^j z^{*i})\,dx\,dy, \tag{15.2.12}$$

and the average value of $\prod_i u(z_i)$ (see Appendix A.17) is

$$\left\langle \prod_i u(z_i) \right\rangle = \int \cdots \int P_Q(z_1, \ldots, z_N) \prod_i u(z_i)\,dx_i\,dy_i$$

$$= K_Q N!(\det[f_{ij}]_{i,j=0,1,\ldots,2N-1})^{1/2}. \tag{15.2.13}$$

Putting $u(z) = 1$ and equating the average (15.2.13) to unity, we get the value of K_Q:

$$f_{ij}(1) = 2\pi[j!\delta_{i+1,j} - i!\delta_{j+1,i}], \tag{15.2.14}$$

$$K_Q^{-1} = N!(2\pi)^N \prod_1^N \Gamma(2j). \tag{15.2.15}$$

Next we put

$$u(z) = \begin{cases} 0, & \text{if } |z| < \alpha, \\ 1, & \text{if } |z| \geqslant \alpha, \end{cases} \tag{15.2.16}$$

and obtain an expression for $E_{NQ}(\alpha)$, the probability that no eigenvalue will lie inside a circle of radius α centered at the origin

$$E_{NQ}(\alpha) = \prod_{j=1}^N \frac{\Gamma(2j, \alpha^2)}{\Gamma(2j, 0)}, \tag{15.2.17}$$

where the incomplete gamma functions $\Gamma(j, \alpha^2)$ are defined by (15.1.16). Corresponding to (15.1.19), (15.1.20), and (15.1.26), we now have

$$E_Q(\alpha) = \lim_{N\to\infty} E_{NQ}(\alpha) = 1 - \frac{1}{2}\alpha^4 + \frac{1}{3}\alpha^6 - \frac{1}{6}\alpha^8 + \frac{1}{15}\alpha^{10} - \cdots, \tag{15.2.18}$$

$$f_s(\alpha) \leqslant f_r(\alpha^2) \leqslant E_Q(\alpha) \leqslant F_{r'}(\alpha^2) \leqslant F_{s'}(\alpha^2) \quad \text{for } r > s, r' > s' > 0, \tag{15.2.19}$$

where now

$$F_r(\alpha^2) = \prod_{j=0}^{r-1} \left(e^{-\alpha^2} a_{2j+1}(\alpha^2)\right), \tag{15.2.20}$$

and

$$f_r(\alpha^2) = F_r(\alpha^2) \exp\left[- \int_0^{\alpha^2} \frac{(e^x - e^{-x})/2 - a_{2r-1}(x)}{a_{2r+1}} \, dx \right], \tag{15.2.21}$$

with $a_j(x)$ given by (15.1.21).

To get the correlation functions $R_n(z_1, \ldots, z_n)$, Eq. (6.1.2), we proceed as in Section 5.1. Consider the $N \times N$ quaternion matrix

$$Q(z_1, \ldots, z_N) = [K_N(z_i, z_j)]_{i,j=1,\ldots,N}, \tag{15.2.22}$$

with

$$\Theta(K_N(z, \zeta)) = \begin{bmatrix} \phi_N(\zeta, z^*) & \phi_N(\zeta^*, z^*) \\ \phi_N(z, \zeta) & \phi_N(z, \zeta^*) \end{bmatrix}, \tag{15.2.23}$$

and

$$\begin{aligned} \phi_N(u, v) &= -\phi_N(v, u) \\ &= \frac{1}{2\pi} \sum_{0 \leqslant i \leqslant j \leqslant N-1} \frac{2^j j!}{2^i i!(2j+1)!} (u^{2i} v^{2j+1} - v^{2i} u^{2j+1}). \end{aligned} \tag{15.2.24}$$

15.2.1. Lemma. *The matrix Q has the following properties*:

(i) *Q is self-dual, its (i, j) element $K_N(z_i, z_j)$ depends only on z_i and z_j*;

(ii)

$$\begin{aligned} &\int K_N(z, z)(z - z^*) \exp(-|z|^2) \, d^2 z \\ &= \int K_N(z, z)(z - z^*) \exp(-|z|^2) \, dx \, dy = N; \end{aligned} \tag{15.2.25}$$

(iii)

$$\int K_N(z, \zeta) K_N(\zeta, \xi)(\zeta - \zeta^*) \exp(-|\zeta|^2) \, d^2\zeta = K_N(z, \xi), \tag{15.2.26}$$

(iv)

$$\det Q = \text{const} \cdot \prod_{i=1}^{N} (z_i^* - z_i) \prod_{1 \leqslant i < j \leqslant N} \left(|z_i - z_j|^2 |z_i - z_j^*|^2\right), \tag{15.2.27}$$

or

$$\det \Theta(Q) = \text{const} \cdot \prod_{i=1}^{N} (z_i^* - z_i)^2 \prod_{1 \leqslant i < j \leqslant N} (|z_i - z_j|^4 |z_i - z_j^*|^4), \quad (15.2.28)$$

where $\Theta(Q)$ *is the* $2N \times 2N$ *ordinary matrix corresponding to the* $N \times N$ *quaternion matrix* Q *(cf. Section* 5.1).

Proof. Property (i) is almost obvious. Verification of (ii) is straightforward using the orthogonality property (15.1.14)

$$\int e^{-|z|^2} z^i z^{*j} d^2 z = \pi j! \delta_{ij}. \quad (15.2.29)$$

For (iii) it is convenient first to note that

$$\begin{aligned} &\int \phi_N(z, \zeta^*) \phi_N(\zeta, \xi)(\zeta - \zeta^*) e^{-|\zeta|^2} d^2\zeta \\ &= - \int \phi_N(z, \zeta) \phi_N(\zeta^*, \xi)(\zeta - \zeta^*) e^{-|\zeta|^2} d^2\zeta \\ &= \int \phi_N(z, \zeta) \phi_N(\xi, \zeta^*) e^{-|\zeta|^2} d^2\zeta = \frac{1}{2} \phi_N(z, \xi). \end{aligned} \quad (15.2.30)$$

To verify (iv) requires a little more work. Consider the $2N \times 2N$ Vandermonde determinant $\det[u_i^{j-1}]$. In this determinant if we replace the rows of even powers as

$$u^{2j} \to \xi_{2j}(u) = \sum_{i=0}^{j} \frac{2^j j!}{2^i i!} u^{2i}, \quad (15.2.31)$$

and the rows of odd powers as

$$u^{2j+1} \to \xi_{2j+1}(u) = \frac{u^{2j+1}}{(2j+1)!}, \quad (15.2.32)$$

the determinant is multiplied by a known constant. Therefore,

$$\prod_{1 \leqslant i < j \leqslant 2N} (u_i - u_j)^2 = \text{const} \cdot \det [\, \xi_{2j}(u_i) \quad \xi_{2j+1}(u_i) \,] \begin{bmatrix} \xi_{2j+1}(u_k) \\ -\xi_{2j}(u_k) \end{bmatrix},$$

$$i, k = 1, \ldots, 2N, \quad j = 0, 1, \ldots, N-1,$$

or

$$\prod_{i<j}(u_i - u_j)^2 = \text{const}\cdot\det\left[\sum_{j=0}^{N-1}(\xi_{2j}(u_i)\xi_{2j+1}(u_k) - \xi_{2j+1}(u_i)\xi_{2j}(u_k))\right]$$
$$= \text{const}\cdot\det[\phi_N(u_i, u_k)]. \tag{15.2.33}$$

This is essentially Eq. (15.2.28). □

Hence the n-point correlation function, according to Theorem 5.1.4, is

$$R_n(z_1,\ldots,z_n) = \prod_{i=1}^{n}(e^{-|z_i|^2}(z_i - z_i^*))\det[K_N(z_i,z_j)]_{i,j=1,\ldots,n}, \tag{15.2.34}$$

with the quaternion K_N given by Eq. (15.2.23). The constant is fixed by the requirement that

$$\int P_Q(z_1,\ldots,z_N)\,dz_1,\ldots,dz_N = 1.$$

In the limit $N \to \infty$,

$$\phi_N(u,v) \to \phi(u,v) = \frac{1}{(2\pi)^{1/2}}\sum_{0\leqslant k\leqslant i<\infty}[k!\Gamma(i+3/2)]^{-1}$$
$$\times\left\{\left(\frac{u^2}{2}\right)^k\left(\frac{v^2}{2}\right)^{i+1/2} - \left(\frac{u^2}{2}\right)^{i+1/2}\left(\frac{v^2}{2}\right)^k\right\}. \tag{15.2.35}$$

Taking k and $i - k = k_1$ as independent summation indices,

$$\phi(u,v) = (2\pi)^{-1/2}\sum_{k_1=0}^{\infty} I_{k_1+1/2}(uv)\left\{\left(\frac{v}{u}\right)^{k_1+1/2} - \left(\frac{u}{v}\right)^{k_1+1/2}\right\}, \tag{15.2.36}$$

where $I_{k_1+1/2}$ is the Bessel function

$$I_{k_1+1/2}(x) = \sum_{k=0}^{\infty}(k!\Gamma(k+k_1+3/2))^{-1}(x/2)^{2k+k_1+1/2}. \tag{15.2.37}$$

By using the recurrence relation for Bessel functions, we obtain the following simpler equation (cf. Appendix A.34)

$$\phi(u,v) = \frac{1}{2\pi}(v-u)e^{uv}\int_0^1 \exp\left(\frac{1}{2}(u-v)^2 x\right)\frac{dx}{\sqrt{1-x}}. \tag{15.2.38}$$

Equation (15.2.34) now becomes

$$R_n(z_1,\ldots,z_n) = \prod_{i=1}^{n}(e^{-|z_i|^2}(z_i - z_i^*))\left\{\det\begin{bmatrix}\phi(z_i,z_j) & \phi(z_i,z_j^*)\\ \phi(z_i^*,z_j) & \phi(z_i^*,z_j^*)\end{bmatrix}\right\}^{1/2}. \tag{15.2.39}$$

15.3 Real Matrices

A matrix with real elements does not necessarily possess a sufficient number of real solutions to the eigenvalue equation (15.2.1). This is perhaps the reason for the great difficulties experienced in the investigation of random matrices with real elements. We will take

$$P(A)\mu(dA) = \pi^{-N^2/2}e^{-\operatorname{tr}(A^TA)}\prod_{i,j} dA_{i,j}, \tag{15.3.1}$$

so that the elements of the real $N\times N$ matrix A are independent random Gaussian (normal) variables. One may ask the questions: (i) What is the probability $P_{N,k}$ that exactly k eigenvalues out of N are real. (ii) What is the average number of real eigenvalues, i.e. $\sum_{k=1}^{N} kP_{N,k} =$? (iii) How the eigenvalues are distributed in the complex plane. Partial or complete answers to these questions are now known. In the rest of this section we will follow Edelman et al. (1994) to arrive at the following results.

The average number E_N of real eigenvalues of an $N\times N$ real matrix whose elements are independent Gaussian random variables is

$$E_N = \frac{1}{2}(1-(-1)^N) + \sqrt{\frac{2}{\pi}}\sum_{k=1}^{[N/2]}\frac{\Gamma(N-2k+1/2)}{\Gamma(N-2k+1)}, \tag{15.3.2}$$

where $[N/2]$ is the integer part of $N/2$. For large N we quote from Edelman et al. (1994) the asymptotic series

$$E_N \approx \sqrt{\frac{2N}{\pi}}\left(1-\frac{3}{8N}-\frac{3}{128N^2}+\frac{27}{1024N^3}+\frac{499}{32768N^4}+\mathrm{O}(N^{-5})\right)+\frac{1}{2}. \tag{15.3.3}$$

The generating function of the E_N is astonishingly simple

$$\sum_{N=1}^{\infty} E_N z^N = \frac{z}{1-z^2} + \sqrt{\frac{2}{\pi}} \frac{z^2}{1-z^2} \sum_{k=0}^{\infty} \Gamma(k+1/2) \frac{z^k}{k!}$$

$$= \frac{z}{1-z^2} + \frac{\sqrt{2} z^2}{1-z^2} (1-z)^{-1/2}. \tag{15.3.4}$$

For the details see Appendix A.36.

Let the $N \times N$ real matrix A with the probability density (15.3.1) has k real eigenvalues λ_j, $1 \leqslant j \leqslant k$, and $m = (N-k)/2$ pairs of complex conjugate eigenvalues $x_j \pm i y_j$, $1 \leqslant j \leqslant m$, the probability of which we will denote by $P_{N,k}$, then the joint probability density of its eigenvalues is

$$P(\lambda_1, \ldots, \lambda_k, x_1 \pm i y_1, \ldots, x_m \pm i y_m)$$

$$= P_{N,k}^{-1} C_{N,k} \Delta_0 \exp\left[- \sum_{i=1}^{k} \lambda_i^2 + 2 \sum_{i=1}^{m} (y_i^2 - x_i^2) \right] \prod_{i=1}^{m} \left[y_i \operatorname{erfc}(2 y_i) \right], \tag{15.3.5}$$

where

$$C_{N,k}^{-1} = 2^{-2m} k! m! \prod_{j=1}^{N} \Gamma\left(\frac{j}{2}\right), \tag{15.3.6}$$

$$\Delta_0 = |\Delta_k(\lambda)| \Delta_{k,m}(\lambda, x, y) \Delta_m^+(x, y) \Delta_m^-(x, y), \tag{15.3.7}$$

$$\Delta_k(\lambda) = \prod_{1 \leqslant i < j \leqslant k} (\lambda_i - \lambda_j), \tag{15.3.8}$$

$$\Delta_{k,m}(\lambda, x, y) = \prod_{i=1}^{k} \prod_{j=1}^{m} \left[(\lambda_i - x_j)^2 + y_j^2 \right], \tag{15.3.9}$$

$$\Delta_m^+(x, y) = \prod_{1 \leqslant i < j \leqslant m} \left[(x_i - x_j)^2 + (y_i - y_j)^2 \right] \tag{15.3.10}$$

$$\Delta_m^-(x, y) = \prod_{1 \leqslant i < j \leqslant m} \left[(x_i - x_j)^2 + (y_i + y_j)^2 \right], \tag{15.3.11}$$

$$\operatorname{erfc}(z) = \frac{2}{\sqrt{\pi}} \int_z^{\infty} e^{-\xi^2} d\xi. \tag{15.3.12}$$

By convention, a null sum is interpreted as zero and a null product as one. For details see Appendix A.37.

When all the eigenvalues are real, $m = 0$, then

$$\Delta_{k,m}(\lambda, x, y) = \Delta_m^+(x, y) = \Delta_m^-(x, y) = 1, \tag{15.3.13}$$

and the probability density function (15.3.5) for the eigenvalues is identical to that of Gaussian orthogonal ensembles (3.1.17). This fact was known to Ginibre (1965).

The probability $P_{N,k}$ that an $N \times N$ real matrix (with independent Gaussian random elements) has exactly k real eigenvalues and $m = (N-k)/2$ pairs of complex conjugate eigenvalues is given by the multiple integral

$$P_{N,k} = C_{N,k} \int_{-\infty}^{\infty} d\lambda_1 \cdots d\lambda_k \int_{-\infty}^{\infty} dx_1 \cdots dx_m \int_0^{\infty} dy_1 \cdots dy_m \, \Delta_0$$
$$\times \exp\left[-\sum_{i=1}^{k} \lambda_i^2 + 2\sum_{i=1}^{m} (y_i^2 - x_i^2) \right] \prod_{i=1}^{m} [y_i \operatorname{erfc}(2y_i)], \tag{15.3.14}$$
$$C_{N,k} = 2^{N-k} \left[k!m! \prod_{j=1}^{N} \Gamma(j/2) \right]^{-1}. \tag{15.3.15}$$

Edelman (1997) evaluated $P_{N,k}$ for a few small values of N and k and we quote their results here as table 15.3.1. For $k = N$, the integral (15.3.14) can be deduced from that of Selberg (see e.g. Chapter 17, Eq. (17.6.7)) and one has

$$P_{N,N} = 2^{-N(N-1)/4}. \tag{15.3.16}$$

For arbitrary values of N and k the multiple integral (15.3.14) has not yet been simplified.

From numerical evidence and heuristic reasoning it seems reasonable to assume that the eigenvalues in general are spread evenly in the complex plane over a circular region around the origin. If this were so then one might think that the real line being a set of measure zero, the average number of real eigenvalues will be finite for large N. But this is not the case; they increase as $\sqrt{N}$. The eigenvalues of a real matrix seem to prefer the real line.

15.4 Determinants: Probability Densities

The determinant of random real matrices is an old subject (see e.g. Kullback, 1934; Nyquist et al., 1954), while that of random complex or quaternion matrices is relatively recent.

For quaternions, multiplication being not commutative, it is not possible to define a determinant having the usual three properties; namely, (i) $\det A = 0$ if and only if

Table 15.3.1. Probability $P_{N,k}$ that a real $N \times N$ matrix with Gaussian random elements has exactly k real eigenvalues for a few small values of N and k

N	k	$P_{N,k}$	
1	1	1	1
2	2	$\frac{1}{\sqrt{2}}$	0.70711
2	0	$1-\frac{1}{\sqrt{2}}$	0.29289
3	3	$\frac{1}{2\sqrt{2}}$	0.35355
3	1	$1-\frac{1}{2\sqrt{2}}$	0.64645
4	4	$\frac{1}{r8}$	0.125
4	2	$-\frac{1}{4}+\frac{11}{8\sqrt{2}}$	0.72227
4	0	$\frac{9}{8}-\frac{11}{8\sqrt{2}}$	0.15273
5	5	$\frac{1}{32}$	0.03125
5	3	$-\frac{1}{16}+\frac{13}{16\sqrt{2}}$	0.51202
5	1	$\frac{33}{32}-\frac{13}{16\sqrt{2}}$	0.45673
6	6	$\frac{1}{128\sqrt{2}}$	0.00552
6	4	$\frac{271}{1024}-\frac{3}{128\sqrt{2}}$	0.24808
6	2	$-\frac{271}{512}+\frac{107}{64\sqrt{2}}$	0.65290
6	0	$\frac{1295}{1024}-\frac{53}{32\sqrt{2}}$	0.09350
7	7	$\frac{1}{1024\sqrt{2}}$	0.00069
7	5	$\frac{355}{4096}-\frac{3}{1024\sqrt{2}}$	0.08460
7	3	$-\frac{355}{2048}+\frac{1087}{1024\sqrt{2}}$	0.57727
7	1	$\frac{4451}{4096}-\frac{1085}{1024\sqrt{2}}$	0.33744
8	8	$\frac{1}{16384}$	0.00006
8	6	$-\frac{1}{4096}+\frac{3851}{131072\sqrt{2}}$	0.02053
8	4	$\frac{53519}{131072}-\frac{11553}{131072\sqrt{2}}$	0.34599
8	2	$-\frac{53487}{65536}+\frac{257185}{131072\sqrt{2}}$	0.57131
8	0	$\frac{184551}{131072}-\frac{249483}{131072\sqrt{2}}$	0.06210
9	9	$\frac{1}{262144}$	0.00000
9	7	$-\frac{1}{65536}+\frac{5297}{1048576\sqrt{2}}$	0.00356
9	5	$\frac{82347}{524288}-\frac{15891}{1048576\sqrt{2}}$	0.14635
9	3	$-\frac{82339}{262144}+\frac{1345555}{1048576\sqrt{2}}$	0.59328
9	1	$\frac{606625}{524288}-\frac{1334961}{1048576\sqrt{2}}$	0.25681

(continued)

Notice that $P_{N,k}$ is of the form $\frac{a}{b}+\frac{c}{d}\sqrt{2}$ with a, b, c, d integers, and the denominators b and d are powers of 2.

$Ax = 0$ has a non-zero solution $x \neq 0$, (ii) $\det(AB) = \det A \cdot \det B$, (iii) $\det A$ is multilinear in the rows of A. So the definition of a determinant varies according to which of the property or properties one wants to keep. We will adopt the following definition due to Dieudonné or Artin (cf. Mehta, 1989MT, Chapter 8).

Any $N \times N$ matrix A is either singular (i.e. $Ax = yA = 0$ have non-zero solutions) or has an inverse (i.e. $AB = BA = I$). If A is singular, define $\det A = 0$. If A has an inverse, define $\det A$ by recurrence on N as follows. If $N = 1$, define $\det A = |a_{11}|$, where $|x|$ means the norm of (the quaternion) x. If $N > 1$, then let A_{ij} be the $(N-1) \times (N-1)$

matrix obtained by removing the ith row and the jth column of A. The matrix elements of B, the inverse of A, are written as b_{ij}. Not all b_{ij} are zero. One shows (see e.g. Artin (1966) that whenever $b_{ij} = 0$, $\det A_{ji} = 0$, and whenever $b_{ij} \neq 0$, $\det A_{ji} \neq 0$ and $|b_{ij}^{-1} \det A_{ji}|$ is independent of i or j. One then defines $\det A = |b_{ij}^{-1} \det A_{ji}|$. Thus for a quaternion matrix A, $\det A$ is a non-negative real number. (For real or complex A this definition also gives a non-negative real number, the absolute value of the usual ordinary determinant. For quaternion self-dual A this definition gives the absolute value of the definition adopted in Section 5.1.) Note that this determinant is not linear in the rows of A, but has the other two properties. Also that a quaternion matrix A may be singular while its transpose has an inverse.

The eigenvalues and eigenvectors of a matrix A are defined as the solutions of $A\varphi = \varphi x$, where φ is an $N \times 1$ matrix and x is a number. For a real or complex A one can eliminate φ to get $\det(A - xI) = 0$, where I is the unit matrix. For quaternion A, if x is an eigenvalue with the eigenvector φ and μ any constant quaternion, then $\mu^{-1} x \mu$ is an eigenvalue with the eigenvector $\varphi\mu$. Thus x and $\mu^{-1} x \mu$ are not essentially distinct as eigenvalues. It is not evident that an $N \times N$ quaternion matrix should have N (quaternion) eigenvalues, but it has. One can actually put them in one to one correspondence with complex numbers (see Appendix A.31). Here we will only note that the norm of the product of eigenvalues gives the determinant defined above.

As we said in Section 15.2 above, if all the eigenvalues of A are essentially distinct, then one can diagonalize A by a non-singular matrix. To make things clearer, we give an example:

$$\begin{bmatrix} 1 & e_2 \\ e_1 & e_3 \end{bmatrix} \begin{bmatrix} 1 & 1 \\ e_2 & e_1 \end{bmatrix} = \begin{bmatrix} 1 & 1 \\ e_2 & e_1 \end{bmatrix} \begin{bmatrix} 0 & 0 \\ 0 & 1 - e_3 \end{bmatrix}, \tag{15.4.1}$$

$$\begin{bmatrix} 1 & e_1 \\ e_2 & e_3 \end{bmatrix} \begin{bmatrix} 1 & 1 \\ a & b \end{bmatrix} = \begin{bmatrix} 1 & 1 \\ a & b \end{bmatrix} \begin{bmatrix} x_1 & 0 \\ 0 & x_2 \end{bmatrix} \tag{15.4.2}$$

with

$$a = \frac{1}{2}(1 - \sqrt{3})(e_1 - e_2), \qquad b = \frac{1}{2}(1 + \sqrt{3})(e_1 - e_2), \tag{15.4.3}$$

$$x_1 = \frac{1}{2}(1 + \sqrt{3}) - \frac{1}{2}(1 - \sqrt{3})e_3, \qquad x_2 = \frac{1}{2}(1 - \sqrt{3}) - \frac{1}{2}(1 + \sqrt{3})e_3, \tag{15.4.4}$$

showing that the eigenvalues of $\left[\begin{smallmatrix} 1 & e_2 \\ e_1 & e_3 \end{smallmatrix}\right]$ are 0 and $1 - e_3$, while those of its transpose $\left[\begin{smallmatrix} 1 & e_1 \\ e_2 & e_3 \end{smallmatrix}\right]$ are x_1 and x_2. Their determinants are, respectively, 0 and 2.

If all the eigenvalues x_i are real and positive (respectively, real and non-negative), one says that A is positive definite (respectively, positive semi-definite). Denote by $A^\dagger$ the transpose, hermitian conjugate or the dual of A according as A is real, complex or quaternion. For any matrix A, the product $AA^\dagger$ (or $A^\dagger A$) is positive semi-definite, its eigenvalues are real and non-negative. The positive square roots of the eigenvalues of $AA^\dagger$ (or of $A^\dagger A$, they are the same) are known as the singular values of A. The eigenvalues and singular values of A have, in general, nothing in common, except that

$$\prod_{i=1}^{N} \lambda_i^2 = \det(AA^\dagger) = \det A \cdot \det A^\dagger = |\det A|^2 = \prod_{i=1}^{N} |x_i|^2, \tag{15.4.5}$$

where λ_i are the singular values and x_i are the eigenvalues of A.

We start with the joint probability density of the singular values and calculate the Mellin transform of the probability density $p(|y|)$ of the (absolute value) of the determinant y of a random matrix A. From symmetry, when A is real, y is real and $p(y)$ is even in y; when A is complex, y is complex and $p(y)$ depends only on $|y|$. When A is quaternion, y is by definition real and non-negative. One can therefore recover $p(y)$ from $p(|y|)$ when A is real or complex. Our results derived below and confirming those in the known cases are as follows.

$$p_1(y) = \prod_{j=1}^{N} [\Gamma(j/2)]^{-1} G_{0,N}^{N,0}\left(y^2 \mid 0, \frac{1}{2}, \frac{2}{2}, \frac{3}{2}, \ldots, \frac{N-1}{2}\right), \quad y \text{ real}, \tag{15.4.6}$$

$$p_2(y) = \frac{1}{\pi} \prod_{j=1}^{N} [\Gamma(j)]^{-1} G_{0,N}^{N,0}(|y|^2 \mid 0, 1, 2, \ldots, N-1), \quad y \text{ complex}, \tag{15.4.7}$$

$$p_4(y) = 2 \prod_{j=1}^{N} [\Gamma(2j)]^{-1} G_{0,N}^{N,0}\left(y^2 \mid \frac{3}{2}, \frac{7}{2}, \frac{11}{2}, \ldots, 2N - \frac{1}{2}\right), \quad y \text{ real non negative}. \tag{15.4.8}$$

Here $G_{0,N}^{N,0}$ is a Meijer G function. In the above results the Gaussian probability distribution $P(A)$ for the matrix A was taken $P(A) \propto \exp(-\operatorname{tr} A^\dagger A)$.

The joint probability density of the singular values can conveniently be derived in two steps from the two observations

(i) Any matrix A can almost uniquely be written as $U\Lambda V$ where Λ is a diagonal matrix with real non-negative diagonal elements, while U and V are real orthogonal, complex unitary or quaternion symplectic matrices according as A is real, complex or quaternion; "almost uniquely" referring to the fact that either U or V is undetermined up to multiplication by a diagonal matrix.

(ii) Any positive semi-definite matrix $H = AA^\dagger$ can be written uniquely as $H = TT^\dagger$, where T is a triangular matrix with real non-negative diagonal elements.

As a result the Gaussian joint probability density $\exp(-\operatorname{tr} AA^{\dagger})$ for the matrix elements of A gets transformed to (see e.g. Hua, 1963)

$$F(\Lambda) \equiv F(\lambda_1, \ldots, \lambda_N) = \text{const} \cdot \exp\left(-\sum_{j=1}^{N} \lambda_j^2\right) |\Delta(\lambda^2)|^{\beta} \prod_{j=1}^{N} \lambda_j^{\beta-1}, \tag{15.4.9}$$

where $\lambda_1, \ldots, \lambda_N$ are the singular values of A, Δ is the product of differences

$$\Delta(\lambda^2) = \prod_{1 \leqslant j < k \leqslant N} (\lambda_k^2 - \lambda_j^2), \tag{15.4.10}$$

and $\beta = 1, 2$ or 4 according as A is real, complex or quaternion.

The Mellin transform of the product of the λ's is

$$\begin{aligned}
\mathcal{M}_N(s) &= \text{const} \cdot \int_0^{\infty} \eta^{s-1} \delta(\eta - \lambda_1 \ldots \lambda_N) F(\Lambda)\, d\lambda_1 \cdots d\lambda_N\, d\eta \\
&= \text{const} \cdot \int_0^{\infty} \exp\left(-\sum_{j=1}^{N} \lambda_j^2\right) |\Delta(\lambda^2)|^{\beta} \prod_{j=1}^{N} \lambda_j^{\beta+s-2}\, d\lambda_j \\
&= \text{const} \cdot \int_0^{\infty} \exp\left(-\sum_{j=1}^{N} t_j\right) |\Delta(t)|^{\beta} \prod_{j=1}^{N} t_j^{(\beta+s-3)/2}\, dt_j \\
&= \prod_{j=1}^{N} \left[\frac{\Gamma((s-1)/2 + j\beta/2)}{\Gamma(j\beta/2)}\right].
\end{aligned} \tag{15.4.11}$$

In the last line we have used a result derived from Selberg's integral (see e.g. Chapter 17, Eq. (17.6.5). The constant, independent of s, is fixed from the requirement that $\mathcal{M}_N(1) = 1$.

In particular, setting $s = q + 1$, we get integer moments of $|\det A|$ as

$$\left\langle |\det A|^q \right\rangle := \frac{\int \exp(-\operatorname{tr} A^{\dagger} A) |\det A|^q\, dA}{\int \exp(-\operatorname{tr} A^{\dagger} A)\, dA} = \prod_{j=1}^{N} \frac{\Gamma[(q + j\beta)/2)]}{\Gamma[(j\beta)/2)]}. \tag{15.4.12}$$

The inverse Mellin transform of the expression (15.4.11) is (see e.g. Bateman, 1954)

$$p_{\beta}(|y|) = 2 \prod_{j=1}^{N} [\Gamma(j\beta/2)]^{-1} G_{0,N}^{N,0}\left(|y|^2 \,\middle|\, \frac{\beta - 1}{2}, \frac{2\beta - 1}{2}, \ldots, \frac{N\beta - 1}{2}\right). \tag{15.4.13}$$

When $\beta = 1$, the matrix A is real, its determinant y is real, from symmetry the probability density $p_1(y)$ is an even function of y and we have

$$p_1(y) = \frac{1}{2} p_1(|y|), \tag{15.4.14}$$

giving Eq. (15.4.6). When $\beta = 2$, A is complex, y is complex, from symmetry $p_2(y)$ depends only on the absolute value $|y|$ of y, and one has

$$\begin{aligned} p_2(y) &= \frac{1}{2\pi|y|} p_2(|y|) \\ &= \frac{1}{\pi} \prod_{j=1}^{N} [\Gamma(j)]^{-1} \frac{1}{|y|} G_{0,N}^{N,0}\left(|y|^2 \,\middle|\, \frac{1}{2}, \frac{3}{2}, \ldots, N - \frac{1}{2}\right) \\ &= \frac{1}{\pi} \prod_{j=1}^{N} [\Gamma(j)]^{-1} G_{0,N}^{N,0}(|y|^2 \,|\, 0, 1, \ldots, N-1), \end{aligned} \tag{15.4.15}$$

which is Eq. (15.4.7). When $\beta = 4$, A is quaternion, y is, by definition, real positive, and $p_4(y) = p_4(|y|)$, giving Eq. (15.4.8).

Summary of Chapter 15

Three ensembles of random matrices A are considered; (i) the elements of A are complex numbers, (ii) they are real quaternions, and (iii) they are real numbers. The joint probability density of the matrix elements is taken to be proportional to $\exp(-\operatorname{tr}(A^\dagger A))$ in each case. An explicit expression for the r-point correlation function is derived when the elements of A are either complex numbers or real quaternions, Eqs. (15.1.31) and (15.2.39), respectively. When the elements of A are real numbers, the average number of real eigenvalues is known, Eqs. (15.3.2)–(15.3.4). When there are k real and m complex conjugate pairs of eigenvalues the joint probability density of the eigenvalues is given, Eqs. (15.3.6)–(15.3.12). The probability that exactly k eigenvalues out of N are real is expressed as a complicated multiple integral which can be evaluated only for small values of N and k. Using the Mellin transform the probability density of the determinant in each case is calculated and found to be a Meijer G function.

16

STATISTICAL ANALYSIS OF A LEVEL-SEQUENCE

Experimentally one observes a finite stretch of energy levels of a system, giving rise to a list of numbers. This may be the set of resonance levels of a nucleus, the possible energies of a free particle on an odd shaped billiard table or the zeros of the Riemann zeta function on the critical line. Suppose these numbers are $(E_1 \leqslant E_2 \leqslant \cdots \leqslant E_n)$ occupying an interval of length $2L$. The question is how well these numbers agree with the predictions of the statistical theory.

As we said in Section 1.4, one may draw for example, a histogram of the nearest neighbor spacings. For this one computes $S_i = E_{i+1} - E_i$, the average $D = \langle S_i \rangle$ of the S_i, the normalized spacings $s_i = S_i/\langle S_i \rangle = (E_{i+1} - E_i)/D$, then sorts these spacings according to their size, and draws a graph whose height between i/m and $(i+1)/m$ is proportional to the number of s_j such that $i/m \leqslant s_j < (i+1)/m$ for some convenient integer m, say, 10 or 20 depending on n, and for $i = 0, 1, 2, \ldots$. Or one may draw a histogram of the two level correlation function. And compare such histograms with the curves given by the theory. If n is fairly large, one may choose a reasonably big integer m and the histogram gets finer. But as French said once (1971, private conversation), "the statistical theory gives us the limiting curve of, say, the nearest neighbor spacings, but not its thickness. And one cannot give a quantitative estimate of how good a histogram fits a given curve".

The statistical theory makes the hypothesis that there exists a mean level spacing D and a very large number N, such that the statistical behaviour of $x_j = \pi E_j/(D\sqrt{2N})$, $j = 1, \ldots, n$, is the same as that of n consecutive real numbers $(x_1, \ldots, x_n)$ all much smaller than $(2N)^{1/2}$ of a much longer series $(x_1, \ldots, x_N)$. The $(x_1, \ldots, x_N)$ are distributed on the real line with the joint probability density

$$P(x_1, \ldots, x_N) = \text{const} \cdot \exp\Big(-\beta \sum_j x_j^2\Big) \prod_{i<j} |x_i - x_j|^\beta, \tag{16.0.1}$$

$\beta = 1$, 2 or 4. Or another theory makes the hypothesis that there exists a mean level spacing D and a very large N, such that the statistical properties of the n angles $\theta_j = 2\pi E_j/(ND)$, $j = 1, \ldots, n$, are the same as that of n consecutive angles of a much longer series $(\theta_1, \ldots, \theta_N)$. The $(\theta_1, \ldots, \theta_N)$ are distributed on the whole circle $0 \leqslant \theta_j \leqslant 2\pi$ with the probability density

$$P(\theta_1, \ldots, \theta_N) = \text{const} \cdot \prod_{j<k} |e^{i\theta_j} - e^{i\theta_k}|^\beta, \tag{16.0.2}$$

$\beta = 1, 2$ or 4. We did prove in Chapter 11 that the statistical behaviour of the n numbers $(x_1, \ldots, x_n)$ is identical to that of the n angles $(\theta_1, \ldots, \theta_n)$ in the limit $N \to \infty$. For the analysis of the experiments, one needs to know the statistical properties of such a partial series.

For small n we already have at our disposal the cluster functions and level-spacings studied earlier, though the "thickness" of the curves is lacking. In this chapter we will assume that $1 \ll n \ll N$. That is to say, the number n of observed levels is large enough to be described in statistical terms, but is still very small compared with the number of unobserved levels. The procedure will be to search for convenient statistics of the observed level series. The word "statistic" is here used in the technical sense customary among statisticians. A "statistic" is a number W which can be computed from an observed sequence of levels alone without other information, and for which the average value $\langle W \rangle$ and the variance

$$V_W = \big\langle (W - \langle W \rangle)^2 \big\rangle, \tag{16.0.3}$$

are known from the theoretical model. A convenient statistic should satisfy two conditions: (i) the computation of W from the observed data should be simple, and (ii) the theoretical figure of merit

$$\Phi_W = \big[V_W / \langle W \rangle^2\big], \tag{16.0.4}$$

should be as small as possible. This Φ_W is the mean square fractional deviation which is to be expected between the observed W and the theoretical $\langle W \rangle$, if the theoretical model is a valid one.

A statistic may serve either of two purposes. If $\langle W \rangle$ involves some unknown parameter contained in the theoretical model, then W provides a measurement of this parameter with fractional error proportional to $\Phi_W^{1/2}$. If $\langle W \rangle$ is independent of parameters, then W provides a test of the theory. In the second case, the theory has clearly failed if $(W - \langle W \rangle)^2$ is found to be much larger than V_W.

In the theoretical model described above, there is only one parameter, namely the "ideal level-spacing" D. The integer N is not an effective parameter, since all properties of the model become independent of N as $N \to \infty$. In the analysis of observations, one needs only one statistic to measure D, and any additional statistic will provide a test of the model.

In practice one often has level series which are mixtures of two or more superimposed series with different values of angular momentum and parity. For the analysis of such series one must consider a generalized theoretical model. The generalized model consists of m uncorrelated level series superimposed on the same interval of length $2L$, their theoretical level spacings being $[D_1, \ldots, D_m]$. This generalized model contains m free parameters D_μ. It is more convenient to take for the parameters the overall mean level-spacing

$$D = \left[\sum D_\mu^{-1}\right]^{-1}, \tag{16.0.5}$$

and the fractions

$$a_\mu = (D/D_\mu), \qquad \sum a_\mu = 1, \tag{16.0.6}$$

a_μ being the proportion of levels belonging to the μth series. Equation (16.0.5) expresses the fact that the overall density of the levels is the sum of the partial level densities.

When one is dealing with a double series ($m = 2$), it is possible with two statistics to measure D and $a_1 = 1 - a_2$, and with a third statistic to obtain a meaningful test of the model. When the series is multiple ($m > 2$), the information that can be obtained in this way is necessarily more fragmentary.

We describe now the various types of statistics found useful in applications. They have been extensively applied in the analysis of nuclear resonance levels, energy levels of quantum chaotic systems and the zeros of the Riemann zeta function on the critical line. More recently, they are finding applications in analyzing the elastodynamic properties of structural materials, architectural acoustics, and so on. (Cf. Weaver, 1989.) Any article on the statistical properties of such sequences will contain a few graphs of the number variance, of the least square statistic Δ and values of the covariance of the nearby spacings in addition to the histograms of the two level correlation function and of the probability densities $p(k; s)$ of the kth neighbor spacings for small k, (cf. Section 6.1). The F-statistic, Section 16.5, has sometimes been used to detect slight imperfections in the experimental nuclear data (see Liou et al., 1972b). The energy statistic Q, Section 16.3, the Λ- and other statistics, Sections 16.6, 16.7 did not receive much popularity.

16.1 Linear Statistic or the Number Variance

The simplest statistic for the measurement of the average spacing D is

$$W = n, \qquad \langle W \rangle = \frac{2L}{D} = s. \tag{16.1.1}$$

The variance of this W, $V_n \equiv \mu_2 = \langle (n - \langle n \rangle)^2 \rangle = \langle n^2 \rangle - \langle n \rangle^2 = (\delta n)^2$, from Appendix A.38 is

$$\begin{aligned} V_n &= \frac{2}{\pi^2}\left(\ln(2\pi s) + \gamma - \frac{\pi^2}{8} - \mathrm{Ci}(2\pi s)\right) \\ &\quad + \frac{4s}{\pi}\int_{\pi s}^{\infty}\left(\frac{\sin\xi}{\xi}\right)^2 d\xi + \frac{1}{\pi^2}\left(\int_{\pi s}^{\infty}\frac{\sin\xi}{\xi}\,d\xi\right)^2 \\ &= \frac{2}{\pi^2}\left(\ln(2\pi s) + 1 + \gamma - \frac{\pi^2}{8}\right) + \mathrm{O}(s^{-1}), \quad \beta = 1, \end{aligned} \tag{16.1.2}$$

$$\begin{aligned} V_n &= \frac{1}{\pi^2}(\ln(2\pi s) + \gamma - \mathrm{Ci}(2\pi s)) + \frac{2s}{\pi}\int_{\pi s}^{\infty}\left(\frac{\sin\xi}{\xi}\right)^2 d\xi \\ &= \frac{1}{\pi^2}(\ln(2\pi s) + 1 + \gamma) + \mathrm{O}(s^{-1}), \quad \beta = 2, \end{aligned} \tag{16.1.3}$$

and

$$\begin{aligned} V_n &= \frac{1}{2\pi^2}(\ln(4\pi s) + \gamma - \mathrm{Ci}(4\pi s)) + \frac{2s}{\pi}\int_{2\pi s}^{\infty}\left(\frac{\sin\xi}{\xi}\right)^2 d\xi + \frac{1}{4\pi^2}\left(\int_0^{2\pi s}\frac{\sin\xi}{\xi}\,d\xi\right)^2 \\ &= \frac{1}{2\pi^2}\left(\ln(4\pi s) + 1 + \gamma + \frac{\pi^2}{8}\right) + \mathrm{O}(s^{-1}), \quad \beta = 4, \end{aligned} \tag{16.1.4}$$

where $\mathrm{Ci}(x)$ is the cosine integral

$$\mathrm{Ci}(x) = \ln x + \gamma - \int_0^x \frac{1 - \cos\xi}{\xi}\,d\xi = -\int_x^{\infty}\frac{\cos\xi}{\xi}\,d\xi. \tag{16.1.5}$$

The figure of merit for this statistic n is $\Phi_n = V_n/s^2$, and is quite small in practice.
This n is one of a general class of linear statistics

$$W = \sum_{i=1}^{n} f(E_i), \qquad \langle W \rangle = \frac{1}{D}\int_{-L}^{L} f(E)\,dE, \tag{16.1.6}$$

where $f(x)$ is any function defined on the interval $(-L, L)$, the zero of energy is chosen for convenience at the center of the observed interval. The choice $f(x) = 1$ gives the statistic (16.1.1). As $f(x) = 1$ has discontinuities at the end points, this is not the linear statistic giving a minimum Φ_W. In fact any smooth function $f(x)$ will give a Φ_W lower than that for an $f(x)$ having discontinuities. And

$$f(x) = (1 - (x/L)^2)^{1/2}, \tag{16.1.7}$$

gives the minimum having a constant variance independent of s (cf. Appendix A.39). But in the practical situations with about a hundred observed levels, the precision gained is not worth the extra trouble.

Equations (16.1.1) and (16.1.6) hold both for simple and multiple series. In the case of multiple series, the variance of a linear statistic is the sum of the variances for each individual series. Thus, for example, Eq. (16.1.2) becomes

$$V_n = \frac{2m}{\pi^2}\left(\ln(2\pi s) + 1 + \gamma - \frac{\pi^2}{8}\right) + \frac{2}{\pi^2}\sum_{\mu} a_\mu \ln a_\mu + \mathrm{O}(s^{-1}), \quad \beta = 1, \tag{16.1.8}$$

where m is the number of independent series and a_μ is the fraction of the levels belonging to the μth series. Thus a rare series contributes as much as a dense series to the error in the measurement of D. This effect of a rare series may be of some practical importance.

Figures 16.1–16.6 show the number variance for the nuclear levels, for energies of a chaotic system, for ultrasonic resonance frequencies of an aluminium block and for the zeros of the Riemann zeta function $\zeta(z)$ on the critical line $\mathrm{Re}\, z = 1/2$. Note that for energy intervals less than or equal to about 10 times the mean spacing, the number variance agrees quite well with Eq. (16.1.2) or (16.1.3). For larger energy intervals, the number variance does not rise as required by those equations, but rather oscillates indefinitely around a roughly constant value, thus announcing the failure of the random matrix theory. This seems to happen for energy intervals when the classical periodic orbits gain importance on the classical chaotic motions. We do not yet have a clear understanding of the transition from an integrable system to a chaotic one and the consequent changes in the energy level statistics. But surprisingly enough, in many cases one can apparently devise an ad hoc formula which reproduces the oscillations in the number variance. For example, for the Riemann zeta zeros, Berry (1988) proposed from heuristic arguments the formula

$$V(L) = V_2(L) + \frac{1}{\pi^2}(\mathrm{Ci}(4\pi L\tau) - \ln(4\pi L\tau) - \gamma)$$
$$+ \frac{2}{\pi^2}\sum_{p,r} \frac{\sin^2(2\pi Lr \ln p / \ln(E/2\pi))}{r^2 p^r}, \tag{16.1.9}$$

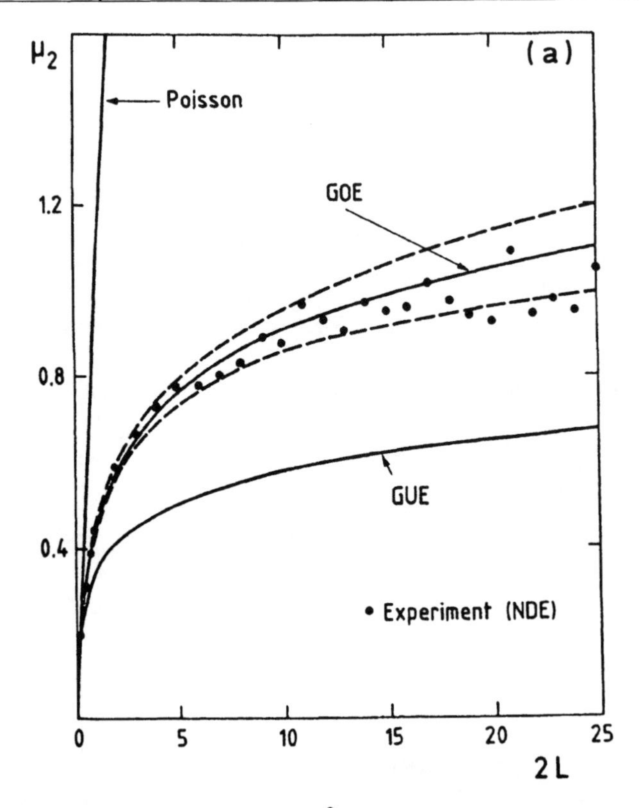

Figure 16.1. Number variance $\mu_2 = \langle (n - \langle n \rangle)^2 \rangle$ for the nuclear data ensemble. For a detailed explanation of the data analyzed here see the caption of Figure 1.4, Chapter 1. The solid curves correspond to a Poisson process, the Gaussian orthogonal ensemble (GOE) and the Gaussian unitary ensemble (GUE). The dashed curves correspond to one standard deviation from the GOE curve. Points represent the experimental data. From Bohigas et al. (1983). Reprinted with permission of the American Physical Society.

where $V_2(L)$ is given by Eq. (16.1.3), τ is a constant of the order of $1/4$, Ci is the cosine integral, Eq. (16.1.5), and the sum is taken over all primes p and all positive integers r such that $p^r < (E/2\pi)^\tau$. Here the zeros of $\zeta(z)$ taken into consideration are $z = 1/2 + i\gamma$ with $E - L \leqslant \gamma \leqslant E + L$.

Formula (16.1.9) reproduces the empirical variance of the zeros of $\zeta(z)$ very well for all values of L (see Figures 16.4–16.6) and is quite insensitive to small changes in the value of τ. This only shows that the zeros of $\zeta(z)$ have a deep relation with the primes and powers of primes, which we do not yet fully understand.

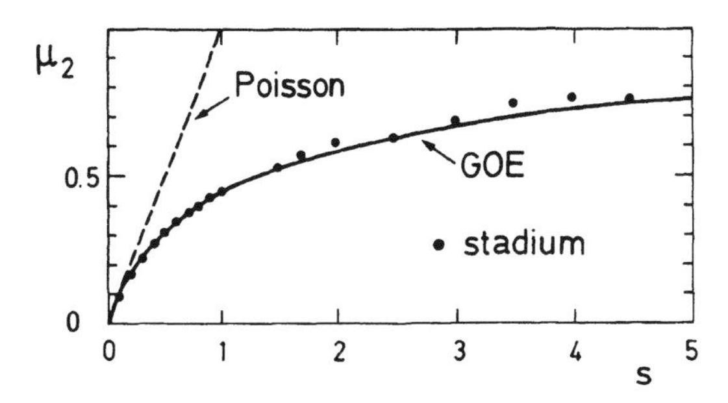

Figure 16.2. Number variance for the possible energies of a particle free to move on the stadium consisting of a rectangle of size 1×2 with semi-circular caps of radius 1. The stadium can be defined by the inequalities $|y| \leqslant 1$, and either $|x| \leqslant 1/2$ or $(x \pm 1/2)^2 + y^2 \leqslant 1$; cf. Figure 7.7. The solid curve corresponds to the Gaussian orthogonal ensemble (GOE). Supplied by O. Bohigas, from Bohigas et al. (1984a).

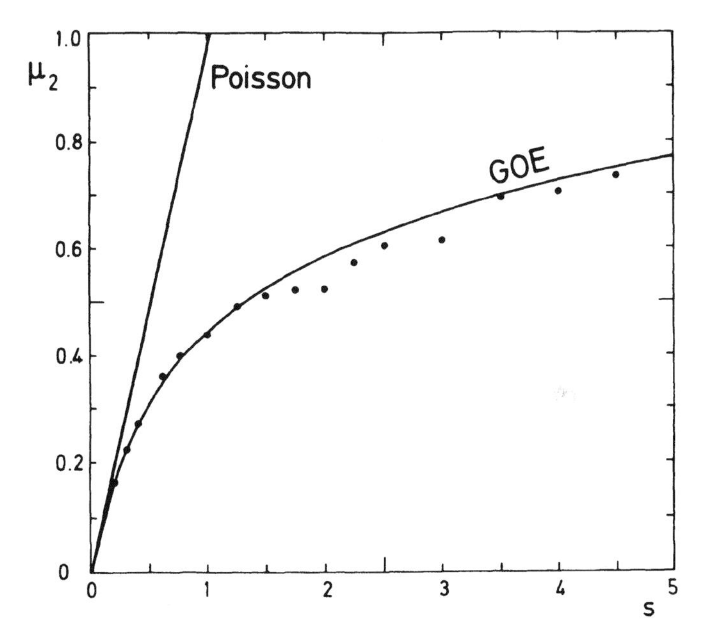

Figure 16.3. Number variance for the ultrasonic resonance frequencies of an aluminium block. The linearly rising curve corresponds to the Poisson process, while the other one corresponds to the Gaussian orthogonal ensemble (GOE). From Weaver (1989). Reprinted with permission from American Institute of Physics, Weaver R.L., Spectral statistics in elastodynamics, *J. Acoust. Soc. Amer.* 85, 1005–1013 (1989).

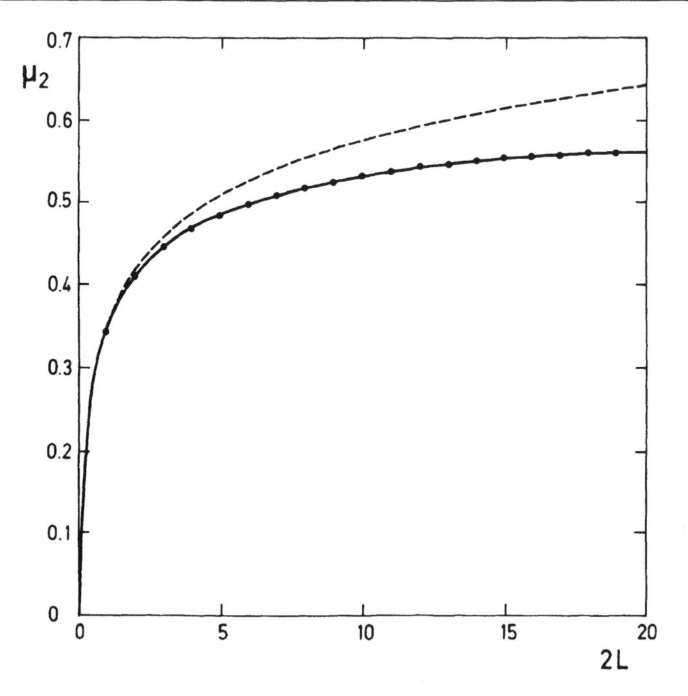

Figure 16.4. Number variance for the zeta zeros. The dashed curve corresponds to the Gaussian unitary ensemble (GUE). The solid curve passes smoothly through the numerical data. From Odlyzko (1989). Copyright 1989, American Telephone and Telegraph Company, reprinted with permission.

16.2 Least Square Statistic

It is customary to represent the experimental level distributions graphically by a plot of $N(E)$ against E, where $N(E)$ is the number of levels having energy between zero and E. The resulting graphs are in appearance like staircases, and the staircase gives a good visual impression of the overall regularity of the level series (cf. Figure 16.7). It is customary to measure the average level spacing by drawing a line having the same average slope as the staircase. Fluctuations in the level series then appear as deviations of the staircase from the straight line.

The aim of this section is to make a quantitative study of this "staircase" method of analysis of data, and to find out precisely what can be learned from it. It should be possible to construct, from the deviations between the staircase and the straight line, a statistic which will serve to test whether the overall irregularity of the observed level series is in agreement with the theoretical model or not.

It is convenient to take the observed energy interval as $[-L, +L]$, with zero in the center. For negative E, $N(E)$ is defined as *minus* the number of levels between E and

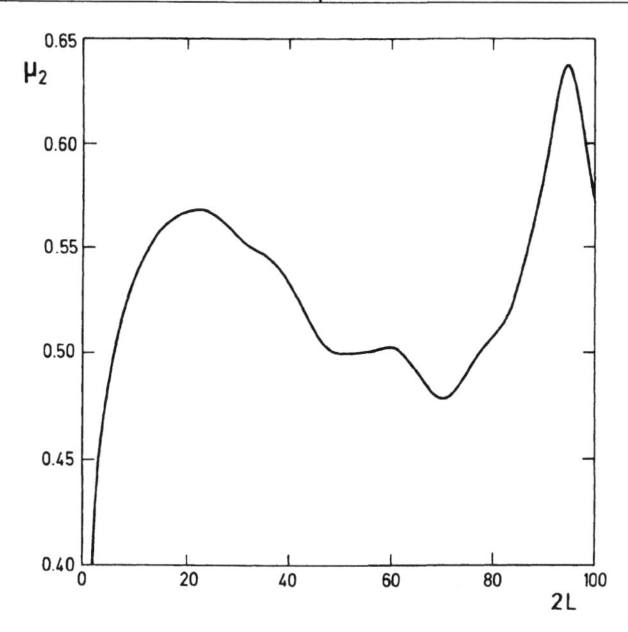

Figure 16.5. Number variance for the zeta zeros (continued). The difference between the curve (16.1.9) and the numerical data is imperceptible on this figure and Figure 16.6. From Odlyzko (1989). Copyright 1989, American Telephone and Telegraph Company, reprinted with permission.

zero. The problem is to analyze the deviation of the staircase graph

$$y = N(E), \tag{16.2.1}$$

from a suitably chosen straight line

$$y = AE + B. \tag{16.2.2}$$

A suitable method of analysis is to fit Eq. (16.2.2) to (16.2.1) by a least-square criterion. The mean-square deviation of Eq. (16.2.1) from the best fit (16.2.2) is

$$\Delta = \operatorname*{Min}_{A,B}\left(\frac{1}{2L}\int_{-L}^{L}[N(E) - AE - B]^2\,dE\right). \tag{16.2.3}$$

We choose this Δ as the statistic which measures the irregularity of the level series.

Originally three possible ways of choosing the straight line were considered (Dyson and Mehta, 1963), but the experience shows that it is better to let A and B vary inde-

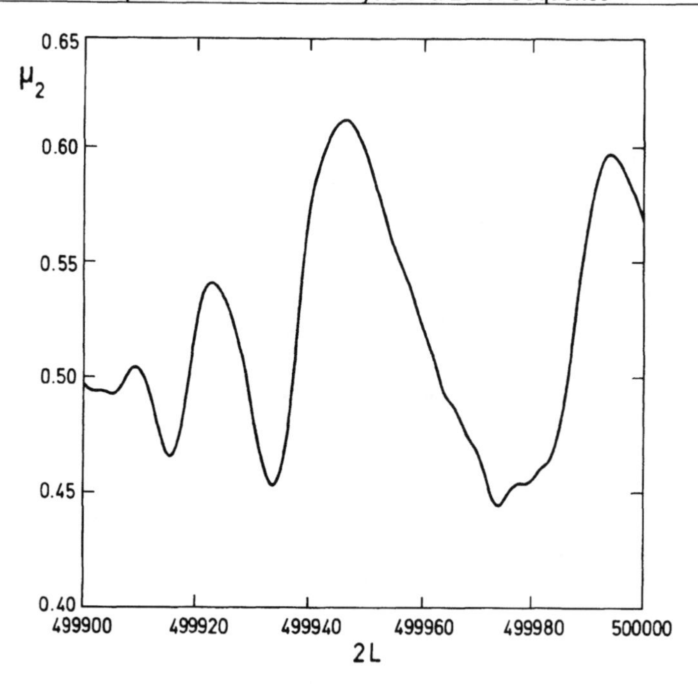

Figure 16.6. Number variance for the zeta zeros (continued). From Odlyzko (1989). Copyright 1989, American Telephone and Telegraph Company, reprinted with permission.

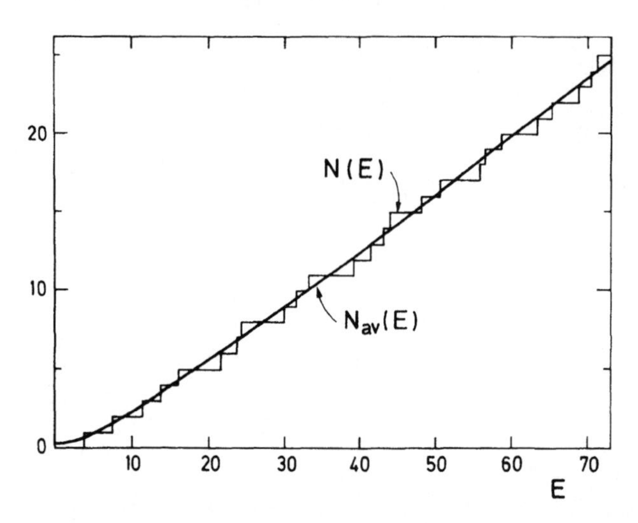

Figure 16.7. $N(E)$ for the stadium, cf. Figure 16.2. Supplied by O. Bohigas, from Bohigas et al. (1984a).

pendently and make a two parameter fit. For applications, one should know the mean and the variance of Δ.

The mean value of Δ is (cf. Appendix A.40)

$$\langle\Delta\rangle = \frac{1}{\pi^2}\left(\ln(2\pi s) + \gamma - \frac{5}{4} - \frac{\pi^2}{8}\right) + O(s^{-1}), \quad \beta = 1, \tag{16.2.4}$$

$$= \frac{1}{2\pi^2}\left(\ln(2\pi s) + \gamma - \frac{5}{4}\right) + O(s^{-1}), \quad \beta = 2, \tag{16.2.5}$$

$$= \frac{1}{4\pi^2}\left(\ln(4\pi s) + \gamma - \frac{5}{4} + \frac{\pi^2}{8}\right) + O(s^{-1}), \quad \beta = 4, \tag{16.2.6}$$

It is instructive to compare these values with the average Δ for a random sequence without correlations. For such a sequence (Poisson process),

$$\langle\Delta\rangle = s/15. \tag{16.2.7}$$

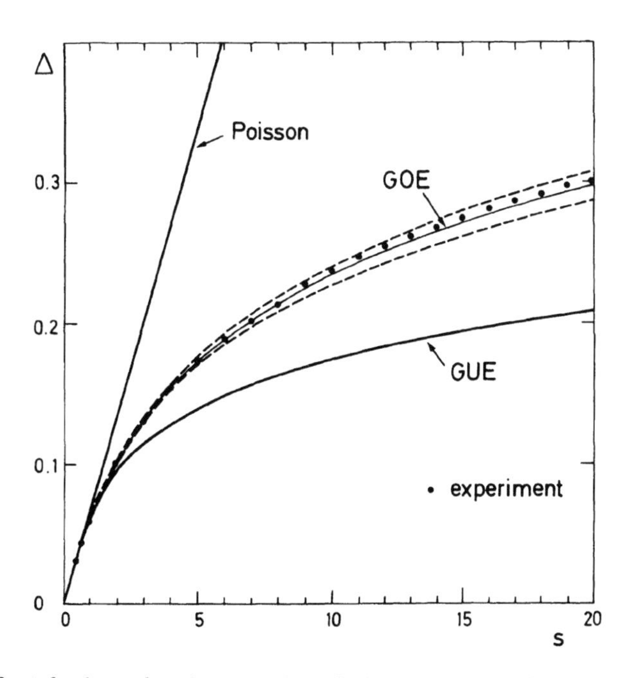

Figure 16.8. Δ for the nuclear data ensemble, cf. Figure 16.1 or 1.4. From Haq et al. (1982).

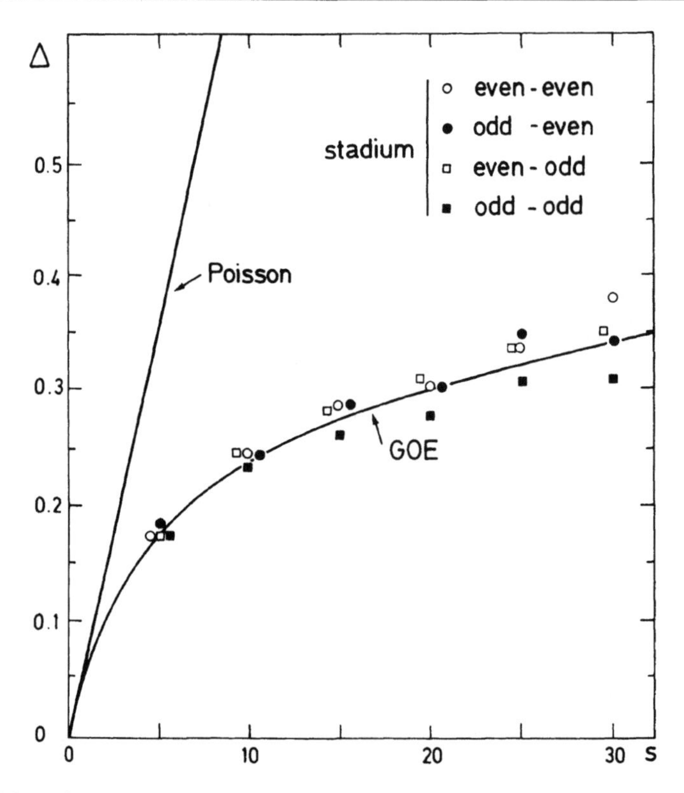

Figure 16.9. Δ for the stadium, cf. Figure 16.2. The computed possible energies of the particle were divided into four categories according to the symmetry of its wave function and each category was analyzed separately.

To calculate the variance of Δ is more lengthy, as it involves 3 and 4 level correlation functions as well. We will not do it, but content ourselves to say that the variance is a small constant independent of s.

Figures 16.8 to 16.11 give graphs of Δ for the nuclear energy levels, for energies of chaotic systems, and for the ultrasonic resonance frequencies of aluminium blocks.

16.3 Energy Statistic

We saw in Section 12.3 that the total energy

$$W = \sum_{1 \leqslant j < k \leqslant N} \ln |e^{i\theta_j} - e^{i\theta_k}|$$

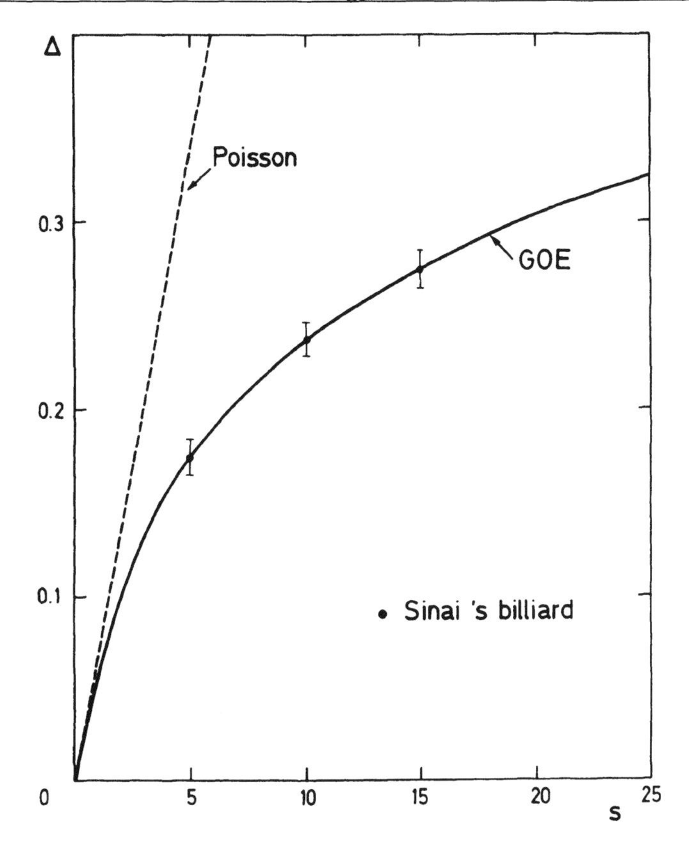

Figure 16.10. Δ for the Sinai billiard, cf. Figure 7.8.

is a good statistic, since its mean and variance are known from the theory. By means of W one could decide how accurately a given sequence of angles $0 \leqslant \theta_1, \theta_2, \ldots, \theta_N < 2\pi$ is or is not in agreement with the theoretical probability density (10.4.5). However, in practice we are given only a partial sequence of levels $(E_1, \ldots, E_n)$ lying in the interval $(-L, L)$. One could construct a statistic Q for this partial sequence similar to W and calculate its mean and variance. Take for example,

$$Q = -\sum_{i<j} f(E_i, E_j) \ln |(E_i - E_j)/R| + \frac{2L}{D} \sum_i U(E_i),$$

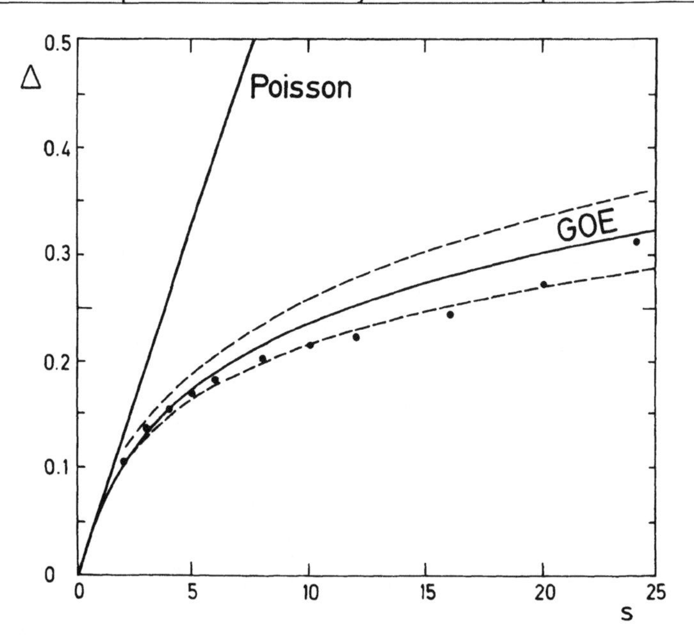

Figure 16.11. Δ for the ultrasonic resonance frequencies of an aluminium block, cf. Figure 16.3. Dashed curves are one standard deviation for the GOE. Reprinted with permission from American Institute of Physics, Weaver R.L., Spectral statistics in elastodynamics, *J. Acoust. Soc. Amer.* 85, 1005–1013 (1989).

where

$$f(E, E') = \begin{cases} 1, & \text{if } |E - E'| < R,\ |E| < L,\ |E'| < L, \\ 0, & \text{otherwise,} \end{cases}$$

$$U(E) = \begin{cases} -R/L, & \text{if } |E| < L - R, \\ -\dfrac{R}{2L} + \dfrac{L - |E|}{R}\left(1 - \ln\left(\dfrac{L - |E|}{R}\right)\right), & \text{if } L - R < |E| < L, \\ 0, & \text{otherwise,} \end{cases}$$

and R is chosen to be 2 or 3 times the mean spacing D. In this example, apart from the first term similar to the free energy, an extra term has been added so that its variation with respect to each E_i is quite small. The mean value of Q can be evaluated using the two-level correlation function. To find its variance is more difficult, since it involves 3- and 4-level correlations as well. We will not do these calculations here, but refer the interested reader to the original paper (Dyson and Mehta, 1963).

16.4 Covariance of Two Consecutive Spacings

To have some idea of how consecutive spacings are correlated, one sometimes computes their covariance. Let $s_1, s_2, s_3, \ldots$ be the distances between successive eigenvalues (measured in units of the local mean spacing), $E(n; s)$ the probability that a randomly chosen interval of length s contains exactly n eigenvalues and $p(n; s)\,ds$ the probability that the sum $s_1 + \cdots + s_{n+1}$ has a value between s and $s + ds$, i.e. $p(n; s)\,ds$ is the probability that two eigenvalues separated by n other eigenvalues be at a distance between s and $s + ds$. From

$$(s_1 + \cdots + s_{n+1})^2 = \sum_{j=1}^{n+1} s_j^2 + 2 \sum_{1 \leqslant j < k \leqslant n+1} s_j s_k, \tag{16.4.1}$$

taking averages and using Eqs. (6.1.16)–(6.1.18) we get on two partial integrations

$$\langle s_1^2 \rangle = 2 \int_0^\infty E(0; s)\,ds, \tag{16.4.2}$$

and

$$\langle s_1 s_{n+1} \rangle = \int_0^\infty E(n; s)\,ds, \quad n > 0, \tag{16.4.3}$$

provided that the first two derivatives of the $E(n; s)$ decrease fast for large s, i.e. provided that $s\,dE(n; s)/ds$ and $s^2 d^2 E(n; s)/ds^2$ tend to 0 as $s \to \infty$, for all n. Also from $\langle (s_1 - s_{n+1})^2 \rangle \geqslant 0$, one has the inequality

$$\int_0^\infty E(n; s)\,ds \leqslant 2 \int_0^\infty E(0; s)\,ds. \tag{16.4.4}$$

The covariance of s_1, s_{n+1} is defined by

$$\begin{aligned} \operatorname{cov}(s_1, s_{n+1}) &= \frac{\langle (s_1 - \langle s_1 \rangle)(s_{n+1} - \langle s_{n+1} \rangle) \rangle}{(\langle (s_1 - \langle s_1 \rangle)^2 \rangle \langle (s_{n+1} - \langle s_{n+1} \rangle)^2 \rangle)^{1/2}} \\ &= \frac{\langle s_1 s_{n+1} \rangle - 1}{\langle s_1^2 \rangle - 1} = \frac{\int_0^\infty E(n; s)\,ds - 1}{2 \int_0^\infty E(0; s)\,ds - 1}. \end{aligned} \tag{16.4.5}$$

From the tables of the functions $E(n; s)$, Appendix A.14, one can estimate the integrals of $E(n; s)$ numerically, getting

$$2 \int_0^\infty E_\beta(0; s)\,ds = \begin{cases} 1.286, & \beta = 1, \\ 1.180, & \beta = 2, \\ 1.104, & \beta = 4; \end{cases} \tag{16.4.6}$$

$$\int_0^\infty E_\beta(1;s)\,ds = \begin{cases} 0.922, & \beta=1, \\ 0.944, & \beta=2, \\ 0.964, & \beta=4. \end{cases} \tag{16.4.7}$$

Hence,

$$\operatorname{cov}(s_1,s_2) = \begin{cases} -0.27, & \beta=1 \text{ (orthogonal ensemble)}, \\ -0.31, & \beta=2 \text{ (unitary ensemble)}, \\ -0.34, & \beta=4 \text{ (symplectic ensemble)}. \end{cases} \tag{16.4.8}$$

16.5 The F-Statistic

In order to detect a few missing or spurious levels in an otherwise pure sequence, Dyson introduced the so called F-statistic. It is defined by

$$F_i = \sum_{j(\neq i)} f((x_i - x_j)/mD), \tag{16.5.1}$$

where $x_1, x_2, \ldots$ are the positions of the successive levels, D their average spacing and m is a free parameter. The "optimum" function $f(x)$ having the smallest figure of merit is found to be

$$f(x) = \begin{cases} \frac{1}{2}\ln\left(\frac{1+\sqrt{1-x^2}}{1-\sqrt{1-x^2}}\right), & \text{if } |x| \leqslant 1, \\ 0, & \text{if } |x| \geqslant 1. \end{cases} \tag{16.5.2}$$

Thus the sum in Eq. (16.5.1) extends over all x_j within $x_i \pm mD$. The average value of F depends only on the two point cluster function,

$$\begin{aligned} \langle F\rangle &= \int_0^\infty (1-Y_2(r)) f(r/m)\,dr \\ &= \int_0^m (1-Y_2(r)) \ln\left(\frac{m+\sqrt{m^2-r^2}}{m-\sqrt{m^2-r^2}}\right) dr \\ &= m\pi - 2\int_0^m dx \int_0^x dr\, \frac{Y_2(r)}{\sqrt{x^2-r^2}}. \end{aligned} \tag{16.5.3}$$

Thus for a level from the Gaussian orthogonal ensemble

$$\langle F_i\rangle = m\pi - \ln(8m\pi) + 2 - \gamma, \quad \gamma = 0.577\ldots, \text{ Euler's constant},$$

and

$$\langle F_i\rangle = m\pi,$$

for a spurious level. The variance of F involves three and four point correlations as well, and has been estimated to be $\ln(m\pi)$ for the Gaussian orthogonal ensemble. Choosing a reasonable value of m, say, about 10, the sequence of the numbers F_j will roughly be constant if the level series is pure. For a spurious level x_i, F_i will rise to a peak; if a level is missing, then F_i will dip at the two adjacent levels. This statistic seems to be good for an almost pure level series; as the number of missing or spurious levels increases, it gets certainly unreliable.

16.6 The Λ-Statistic

Another statistic sometimes used to detect missing or spurious levels was introduced by Monahan and Rosenzweig (1972). Comparing the theoretical level spacing distribution $\Psi(s)$, Section 6.1.3, with the experimental staircase graph $\Psi^*(s)$ deduced from a set of $(n+1)$ levels or n consecutive spacings, they define

$$\Lambda(n) = n \int_0^\infty (\Psi(x) - \Psi^*(x))^2 \, dx. \tag{16.6.1}$$

Actually, they use the Wigner surmise

$$\Psi(x) = 1 - \exp(-\pi x^2/4),$$

which is simpler and close enough to the $\Psi(x)$ of the Gaussian orthogonal ensemble (GOE). Monahan and Rosenzweig computed the average and the variance of $\Lambda(n)$ by Monte Carlo calculations of randomly generated matrices of a certain order from the GOE. If the successive spacings were statistically independent, each of them following the Wigner curve, then the average of $\Lambda(n)$ will be a constant, 0.293, independent of n. The long range order of the levels makes the average of $\Lambda(n)$ a decreasing function of n.

16.7 Statistics Involving Three and Four Level Correlations

Most of the statistics described yet depend exclusively on the two point correlations. To test whether a given sequence of levels have higher correlations as required by the theory, one has to consider statistics involving these correlations. Two such statistics are the skewness and the excess of the number of levels in a given energy interval.

Let $\mu_j = \langle (n - \langle n \rangle)^j \rangle$, which involves ν-level correlations for all $\nu \leqslant j$. We have already considered the number variance μ_2 above, Section 16.1. The skewness γ_1 and the excess γ_2 are defined as

$$\gamma_1 = \mu_3 \mu_2^{-3/2}, \qquad \gamma_2 = \mu_4 \mu_2^{-2} - 3. \tag{16.7.1}$$

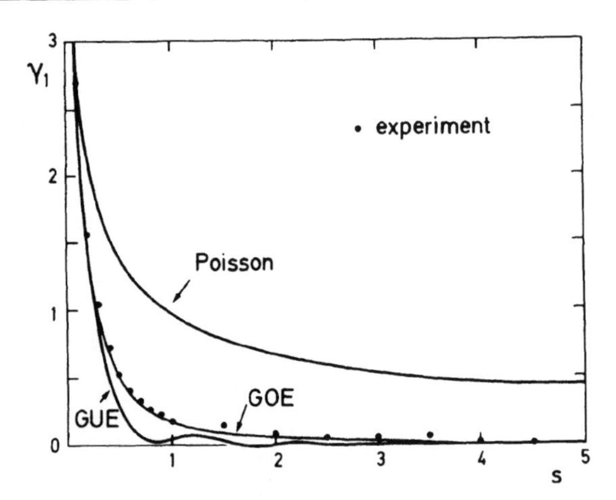

Figure 16.12a. The skewness γ_1 for nuclear data ensemble, cf; Figure 16.1 or 1.4. From Bohigas et al. (1985).

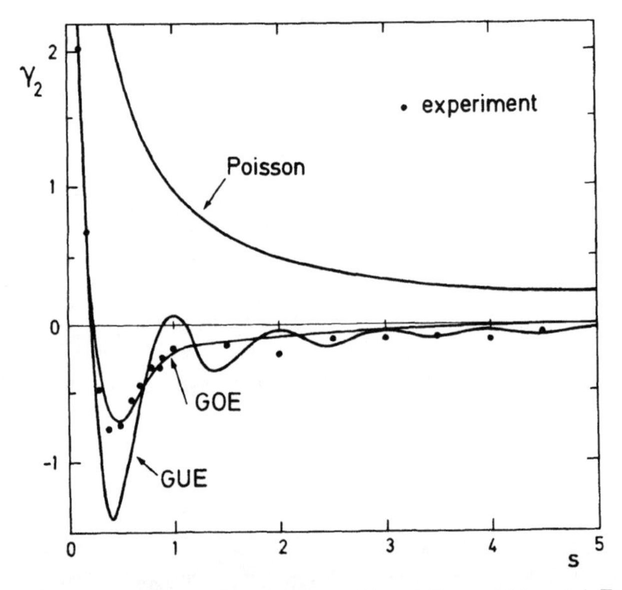

Figure 16.12b. The excess γ_2 for nuclear data ensemble, cf; Figure 16.1 or 1.4. From Bohigas et al. (1985).

One can express μ_2, μ_3 and μ_4 in terms of the averages

$$\langle n^2 \rangle = \iint R_2(x, y)\, dx\, dy + \int R_1(x)\, dx,$$

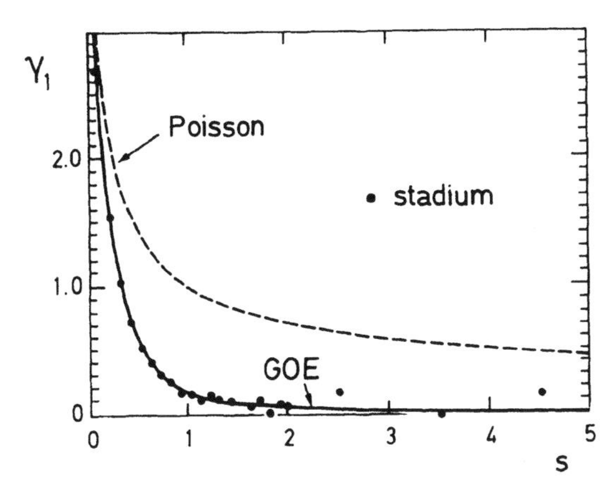

Figure 16.13a. The skewness γ_1 for chaotic systems, stadium or Sinai's billiard.

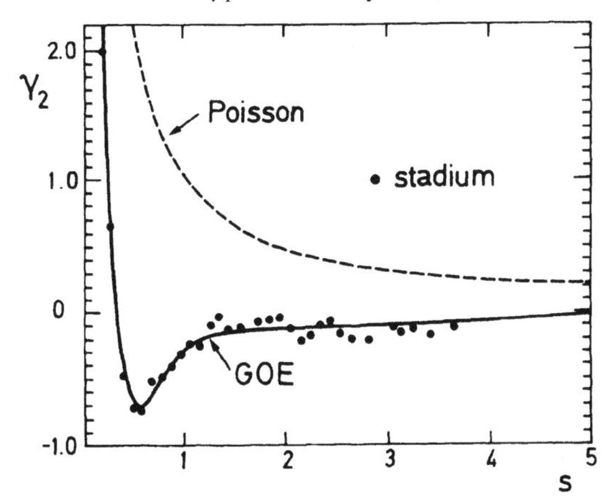

Figure 16.13b. The excess γ_2 for chaotic systems, stadium or Sinai's billiard.

$$\langle n^3 \rangle = \iiint R_3(x,y,z)\,dx\,dy\,dz + 3\iint R_2(x,y)\,dx\,dy + \int R_1(x)\,dx,$$

$$\langle n^4 \rangle = \iiiint R_4(x,y,z,t)\,dx\,dy\,dz\,dt + 6\iiint R_3(x,y,z)\,dx\,dy\,dz$$
$$+7\iint R_2(x,y)\,dx\,dy + \int R_1(x)\,dx, \tag{16.7.2}$$

all integrals being taken from $-L$ to L. (The various coefficients $1, 3, 6, 7$, etc. are the Stirling numbers of the first kind; they appear in the expansions $n^j = n_j + c_1 n_{j-1} + \cdots + c_{j-1} n_1 + c_j$, where $n_k = n(n-1)\cdots(n-k+1)$.)

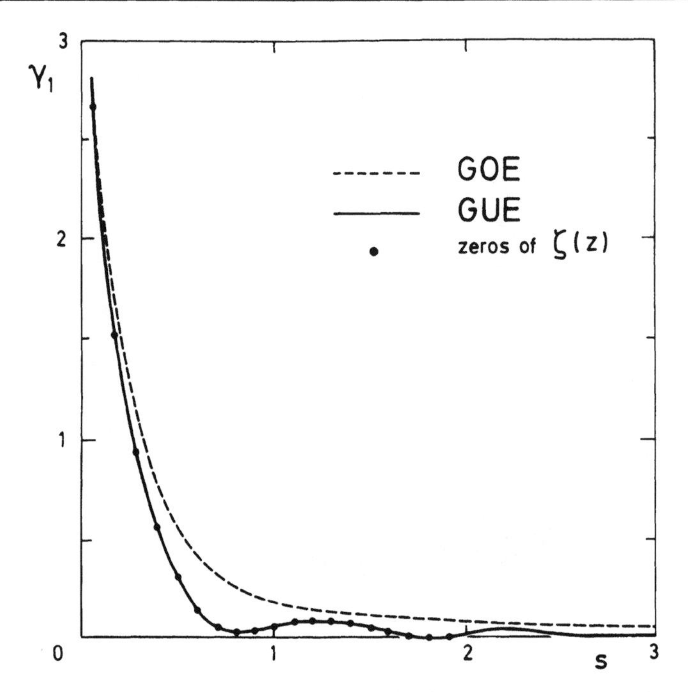

Figure 16.14a. The skewness γ_1 for the zeros of the Riemann zeta function on the critical line $\text{Re}\, z = 1/2$.

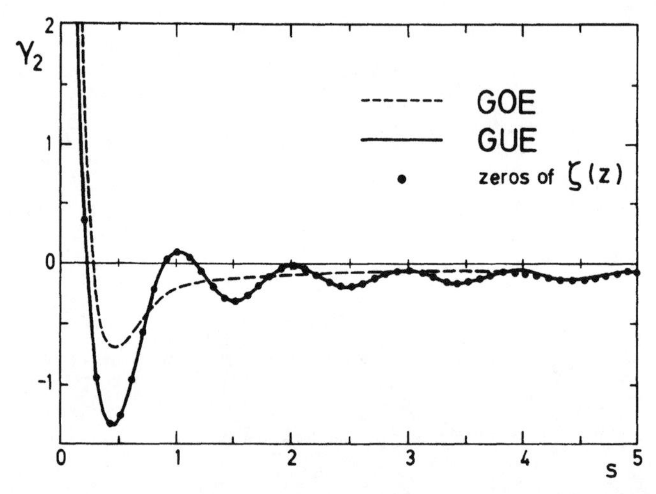

Figure 16.14b. The excess γ_2 for the zeros of the Riemann zeta function on the critical line $\text{Re}\, z = 1/2$. Supplied by A. Odlyzko and O. Bohigas. Figure 16.14 is based on the work of Odlyzko.

For a sequence of levels without correlations, Poisson process, $\mu_2 = \mu_3 = s$ and $\mu_4 = 3s^2 + s$, so that $\gamma_1 = s^{-1/2}$ and $\gamma_2 = s^{-1}$. For matrix ensembles it is easier to compute them using numerical tables of $E(n; s)$,

$$\mu_j = \sum_{n=0}^{\infty} (n - s)^j E(n; s). \tag{16.7.3}$$

The γ_1 and γ_2 for the nuclear data ensemble are shown on Figure 16.12, for chaotic systems on Figure 16.13 and for the zeros of the zeta function on Figure 16.14. For comparison we have also given the curves of γ_1 and γ_2 for the Poisson process, for the Gaussian orthogonal ensemble (GOE) and for the Gaussian unitary ensemble (GUE). One sees that the nuclear data ensemble and energies of chaotic systems are consistent with GOE while the zeta zeros are consistent with GUE. Though not shown on the figures, the results for $5 \leqslant s \leqslant 25$ were also calculated and the above consistencies persist. In other words, the three and four level correlations of the considered level sequences are in good agreement with those required by the theory.

16.8 Other Statistics

From various other statistics considered in the literature, we will mention two more. One is the variance of the rth neighbor spacing probability density, $p(r-1; s)$, Section 6.1.2. Actually as r increases, $p(r-1; s)$ tends to a Gaussian centered at r and only its width $\sigma(r)$ has any significance. This $\sigma(r)$ is closely related to the number variance $(\delta n)^2 = \langle n^2 \rangle - \langle n \rangle^2$, Section 16.1 (see Brody et al., 1981). The other one is the repulsion parameter. Generalizing the Poisson distribution and the Wigner surmise, Brody (1973) writes the spacing probability density $p(0; s)$ as

$$p(0; s) = A s^{\omega} \exp(-B s^{\omega+1}), \tag{16.8.1}$$

the two constants A and B are fixed in terms of ω by the requirement that $p(0; s)$ is normalized and the mean value of s is unity;

$$A = B(1 + \omega), \qquad B = \left(\Gamma\left(\frac{\omega + 2}{\omega + 1} \right) \right)^{1+\omega}. \tag{16.8.2}$$

The kth moment of s is

$$\mu_k = \int_0^{\infty} s^k \cdot A s^{\omega} \exp(-B s^{\omega+1}) \, ds = \Gamma\left(\frac{\omega + k + 1}{\omega + 1} \right) \left(\Gamma\left(\frac{\omega + 2}{\omega + 1} \right) \right)^{-k}. \tag{16.8.3}$$

The parameter ω measures the degree of repulsion between adjacent levels. It is determined either by comparing some moment of s, say μ_2, to that given by the data, or better by fitting the curve (16.8.1) to the experimental histogram by least squares.

Summary of Chapter 16

To estimate how well a given sequence of levels agrees with the eigenvalues of a random matrix, several statistics are considered. A statistic is a quantity that can be computed for the given sequence of levels and whose average and mean square scatter are known from the statistical theory of random matrices. Most popular of these statistics are the number variance and the best straight-line-fit Δ statistic along with the two-point correlation function $R_2(r)$ and the nearest neighbor spacing probability density $p(0; s)$. The F statistic has sometimes been used to detect small impurities in an otherwise pure sequence. The other statistics like the Q, Λ, or the repulsion parameter ω have not been so widely used.

17

SELBERG'S INTEGRAL AND ITS CONSEQUENCES

Selberg's integral is a generalization of Euler's beta-integral. Aomoto found a slight generalization of Selberg's integral, thus providing an alternative proof of it. We give here both derivations essentially following the original papers. When Selberg's integral was not so widely known, some of its consequences were thought of as conjectures by various authors. Several new proofs and generalizations have since been found as described in Andrews et al. (1993), Chapter 8. Macdonald established a connection between this integral, simple Lie algebras and reflection groups. Most of the considerations of Macdonald were later proved. Here we describe these developments in some detail.

17.1 Selberg's Integral

Theorem. *For any positive integer* n, *let* $dx \equiv dx_1 \cdots dx_n$,

$$\Delta(x) \equiv \Delta(x_1, \ldots, x_n) = \prod_{1 \leqslant j < \ell \leqslant n} (x_j - x_\ell), \quad \textit{if } n > 1,$$

$$\Delta(x) = 1, \quad \textit{if } n = 1, \tag{17.1.1}$$

and

$$\Phi(x) \equiv \Phi(x_1, \ldots, x_n) = |\Delta(x)|^{2\gamma} \prod_{j=1}^{n} x_j^{\alpha-1}(1 - x_j)^{\beta-1}. \tag{17.1.2}$$

Then

$$I(\alpha,\beta,\gamma,n) \equiv \int_0^1 \cdots \int_0^1 \Phi(x)\,dx = \prod_{j=0}^{n-1} \frac{\Gamma(1+\gamma+j\gamma)\Gamma(\alpha+j\gamma)\Gamma(\beta+j\gamma)}{\Gamma(1+\gamma)\Gamma(\alpha+\beta+(n+j-1)\gamma)}, \tag{17.1.3}$$

and for $1 \leqslant m \leqslant n$,

$$\int_0^1 \cdots \int_0^1 x_1 x_2 \cdots x_m \Phi(x)\,dx = \prod_{j=1}^{m} \frac{\alpha+(n-j)\gamma}{\alpha+\beta+(2n-j-1)\gamma} \int_0^1 \cdots \int_0^1 \Phi(x)\,dx, \tag{17.1.4}$$

valid for integer n and complex α, β, γ *with*

$$\operatorname{Re}\alpha > 0, \quad \operatorname{Re}\beta > 0, \quad \operatorname{Re}\gamma > -\min\left(\frac{1}{n}, \frac{\operatorname{Re}\alpha}{(n-1)}, \frac{\operatorname{Re}\beta}{(n-1)}\right). \tag{17.1.5}$$

Equation (17.1.3) *is Selberg's integral and* (17.1.4) *is Aomoto's extension of it.*

Equation (17.1.4) has the extra factors $x_1 \cdots x_m$ over those in Eq. (17.1.3). Can one introduce extra factors $(1-x_j)$ as well? When there is no overlap in the two sets of factors, the result is

$$\begin{aligned} B(m_1,m_2) &:= \int_0^1 \cdots \int_0^1 \prod_{i_1=1}^{m_1} x_{i_1} \prod_{i_2=m_1+1}^{m_1+m_2} (1-x_{i_2})\Phi(x)\,dx \\ &= \frac{\prod_{i_1=1}^{m_1}(\alpha+(n-i_1)\gamma)\prod_{i_2=1}^{m_2}(\beta+(n-i_2)\gamma)}{\prod_{i=1}^{m_1+m_2}(\alpha+\beta+(2n-i-1)\gamma)} I(\alpha,\beta,\gamma,n), \\ &\quad m_1, m_2 \geqslant 0, \quad m_1+m_2 \leqslant n, \end{aligned} \tag{17.1.6}$$

and when there is overlap

$$\begin{aligned} C(m_1,m_2,m_3) &:= \int_0^1 \cdots \int_0^1 \prod_{i_1=1}^{m_1} x_{i_1} \prod_{l_2=m_1+1-m_3}^{m_1+m_2-m_3} (1-x_{i_2})\Phi(x)\,dx \\ &= \prod_{i=1}^{m_3} \frac{\alpha+\beta+(n-i-1)\gamma)}{\alpha+\beta+1+(2n-i-1)\gamma} B(m_1,m_2), \\ &\quad m_1, m_2, m_3 \geqslant 0, \quad m_1+m_2-m_3 \leqslant n. \end{aligned} \tag{17.1.7}$$

Equation (17.1.6) is easy to deduce from Eq. (17.1.4); one verifies that both sides satisfy the recurrence relation

$$B(m_1.m_2) = B(m_1, m_2 - 1) - B(m_1 + 1, m_2 - 1), \tag{17.1.8}$$

and the initial condition that $B(m, 0)$ gives back Eq. (17.1.4). To deduce Eq. (17.1.7) is a little longer; for details see Andrews et al. (1993), Chapter 8.

17.2 Selberg's Proof of Eq. (17.1.3)

(i) Let γ be a positive integer.

We will use the facts that $|\Delta(x)|^{2\gamma}$ is a homogeneous polynomial in $x_1, \dots, x_n$ of degree $\gamma n(n-1)$ and that $|\Delta(x_1, \dots, x_n)|^{2\gamma}$ is divisible by $|\Delta(x_1, \dots, x_m)|^{2\gamma}$ for $1 \leqslant m \leqslant n$.

For $n = 1$, $\Delta(x) = 1$, and (17.1.3) is just Euler's integral

$$\int_0^1 x^{\alpha-1}(1-x)^{\beta-1}\, dx = \frac{\Gamma(\alpha)\Gamma(\beta)}{\Gamma(\alpha+\beta)}. \tag{17.2.1}$$

For $n > 1$, let

$$|\Delta(x)|^{2\gamma} = \sum_{(j)} c_{j_1 \cdots j_n} x_1^{j_1} x_2^{j_2} \cdots x_n^{j_n}, \tag{17.2.2}$$

where $c_{(j)} \equiv c_{j_1 \cdots j_n}$ are integers and the integers $j_1, \dots, j_n$ satisfy $j_1 + \cdots + j_n = \gamma n(n-1)$. Without loss of generality we may suppose that $j_1 \leqslant j_2 \leqslant \cdots \leqslant j_n$, so that $j_n \geqslant \gamma(n-1)$. Also for $1 \leqslant m \leqslant n$,

$$|\Delta(x_1, \dots, x_n)|^{2\gamma} = |\Delta(x_1, \dots, x_m)|^{2\gamma} P(x_1, \dots, x_n),$$

with P a polynomial. Therefore,

$$\sum_{(j)} c_{j_1 \cdots j_n} x_1^{j_1} x_2^{j_2} \cdots x_n^{j_n} = P(x_1, \dots, x_n) \cdot \sum_{(q)} c'_{q_1 \cdots q_m} x_1^{q_1} x_2^{q_2} \cdots x_m^{q_m}.$$

As $P(x)$ is a polynomial, one has $j_1 \geqslant q_1$, $j_2 \geqslant q_2, \dots, j_m \geqslant q_m$. Also counting powers, $q_1 + \cdots + q_m = \gamma m(m-1)$; and at least one of the q, say q_i, is $\geqslant \gamma(m-1)$. Hence for $1 \leqslant m \leqslant n$, one has

$$j_m \geqslant j_i \geqslant q_i \geqslant \gamma(m-1). \tag{17.2.3}$$

On the other hand

$$|\Delta(x)|^{2\gamma} = (x_1 \cdots x_n)^{2\gamma(n-1)} \left| \Delta\left(\frac{1}{x}\right) \right|^{2\gamma},$$

so that

$$\sum_{(j)} c_{j_1\cdots j_n} x_1^{j_1} x_2^{j_2} \cdots x_n^{j_n} = (x_1 \cdots x_n)^{2\gamma(n-1)} \sum_{(p)} c_{p_1\ldots p_n} x_1^{-p_1} x_2^{-p_2} \cdots x_n^{-p_n},$$

Examine the set $p_1 = 2\gamma(n-1) - j_n$, $p_2 = 2\gamma(n-1) - j_{n-1}, \ldots, p_n = 2\gamma(n-1) - j_1$. For this set $p_1 \leqslant p_2 \leqslant \cdots \leqslant p_n$, if $j_1 \leqslant j_2 \leqslant \cdots \leqslant j_n$. Therefore, as above, for $1 \leqslant m \leqslant n$, $p_m \geqslant \gamma(m-1)$, or $2\gamma(n-1) - j_m \geqslant \gamma(n-m)$. Combining with (17.2.3),

$$\gamma(m-1) \leqslant j_m \leqslant 2\gamma(n-1) - \gamma(n-m) = \gamma(n+m-2). \tag{17.2.4}$$

Now substitute (17.2.2) on the left-hand side of (17.1.3), integrate term by term and use (17.2.1), to obtain

$$\begin{aligned} I(\alpha, \beta, \gamma, n) &= \sum_{(j)} c_{j_1\cdots j_n} \prod_{m=1}^{n} \frac{\Gamma(\alpha + j_m)\Gamma(\beta)}{\Gamma(\alpha + \beta + j_m)} \\ &= \prod_{m=1}^{n} \frac{\Gamma(\alpha + \gamma(m-1))\Gamma(\beta)}{\Gamma(\alpha + \beta + \gamma(n+m-2))} \sum_{(j)} c_{j_1\cdots j_n} Q_{(j)}(\alpha, \beta), \end{aligned}$$

where

$$Q_{(j)}(\alpha, \beta) = \prod_{m=1}^{n} \frac{\Gamma(\alpha + j_m)}{\Gamma(\alpha + \gamma(m-1))} \frac{\Gamma(\alpha + \beta + \gamma(n+m-2))}{\Gamma(\alpha + \beta + j_m)}$$

is, from (17.2.4), a polynomial in α and β. The degree of $Q_{(j)}$ in β is

$$\sum_{m=1}^{n} (\gamma(n+m-2) - j_m) = \gamma(n-2)n + \gamma n(n+1)/2 - \gamma n(n-1) = \gamma n(n-1)/2.$$

Thus

$$I(\alpha, \beta, \gamma, n) = \prod_{m=1}^{n} \frac{\Gamma(\alpha + \gamma(m-1))\Gamma(\beta + \gamma(m-1))}{\Gamma(\alpha + \beta + \gamma(n+m-2))} \frac{R(\alpha, \beta)}{S(\beta)},$$

with

$$R(\alpha, \beta) = \sum_{(j)} c_{j_1\cdots j_n} Q_{(j)}(\alpha, \beta) \quad \text{and} \quad S(\beta) = \prod_{m=1}^{n} \frac{\Gamma(\beta + \gamma(m-1))}{\Gamma(\beta)}.$$

The $R(\alpha, \beta)$ is a polynomial in α and β; its degree in β being at most $\gamma n(n-1)/2$, while $S(\beta)$ is a polynomial in β of degree $\sum_{m=1}^{n} \gamma(m-1) = \gamma n(n-1)/2$.

As $\Delta(1-x)=\pm\Delta(x)$, the integral I is symmetric in α and β, so that $R(\alpha,\beta)/S(\beta)=R(\beta,\alpha)/S(\alpha)$ is a polynomial in β. Therefore $R(\alpha,\beta)$ is divisible by $S(\beta)$. Since the degree of $R(\alpha,\beta)$ in β is at most equal to that of $S(\beta)$, $R(\alpha,\beta)/S(\beta)$ is independent of β. By symmetry it is also independent of α. It may still depend on γ and n. Therefore

$$I(\alpha,\beta,\gamma,n)=c(\gamma,n)\cdot\prod_{m=1}^{n}\frac{\Gamma(\alpha+\gamma(m-1))\Gamma(\beta+\gamma(m-1))}{\Gamma(\alpha+\beta+\gamma(n+m-2))}. \tag{17.2.5}$$

To determine $c(\gamma,n)$, put $\alpha=\beta=1$;

$$I(1,1,\gamma,n)=\int_0^1\cdots\int_0^1|\Delta(x)|^{2\gamma}\,dx_1\cdots dx_n=c(\gamma,n)\cdot\prod_{m=1}^{n}\frac{(\Gamma(1+\gamma(m-1))^2}{\Gamma(2+\gamma(n+m-2))}. \tag{17.2.6}$$

Let y be the largest of the $x_1,\ldots,x_n$ and replace the other x_j by $x_j=yt_j$, $0\leqslant t_j\leqslant 1$. Then

$$|\Delta(x_1,\ldots,x_n)|^{2\gamma}=y^{\gamma n(n-1)}\cdot|\Delta(t_1,\ldots,t_{n-1})|^{2\gamma}\cdot\prod_{j=1}^{n-1}(1-t_j)^{2\gamma},$$

and

$$\begin{aligned}I(1,1,\gamma,n)&=n\int_0^1 dy\cdot y^{n-1}\cdot y^{\gamma n(n-1)}\cdot I(1,2\gamma+1,\gamma,n-1)\\&=\frac{c(\gamma,n-1)}{\gamma(n-1)+1}\cdot\prod_{m=1}^{n-1}\frac{\Gamma(1+\gamma(m-1))\Gamma(2\gamma+1+\gamma(m-1))}{\Gamma(1+2\gamma+1+\gamma(n-1+m-2))}.\end{aligned} \tag{17.2.7}$$

Comparing (17.2.6) and (17.2.7) one gets on simplification,

$$\frac{c(\gamma,n)}{c(\gamma,n-1)}=\frac{\Gamma(1+\gamma n)}{\Gamma(1+\gamma)},\quad\text{or}\quad c(\gamma,n)=c(\gamma,1)\cdot\prod_{m=2}^{n}\frac{\Gamma(1+\gamma m)}{\Gamma(1+\gamma)},$$

and from (17.2.1) and (17.2.5), $c(\gamma,1)=1$.

This completes the proof when γ is a non-negative integer.

(ii) When γ is a complex number, consider

$$f(\gamma)=\int_0^1\cdots\int_0^1|\Delta(x)|^{2\gamma}\prod_{j=1}^{n}x_j^{\alpha-1}(1-x_j)^{\beta-1}\,dx_j,$$

and

$$g(\gamma) = \prod_{j=0}^{n-1} \frac{\Gamma(1+\gamma+j\gamma)\Gamma(\alpha+j\gamma)\Gamma(\beta+j\gamma)}{\Gamma(1+\gamma)\Gamma(\alpha+\beta+(n+j-1)\gamma)}$$

$$= \prod_{j=1}^{n} \frac{\Gamma(1+j\gamma)\Gamma(\alpha+(j-1)\gamma)\Gamma(\beta+(n-j)\gamma)}{\Gamma(1+\gamma)\Gamma(\alpha+\beta+(n+j-2)\gamma)},$$

as functions of the complex variable γ, for fixed α and β, $\operatorname{Re}\alpha > 0$, $\operatorname{Re}\beta > 0$. The functions $f(\gamma)$ and $g(\gamma)$ are regular and analytic in the upper half plane. For $0 \leqslant x_j \leqslant 1$, $|\Delta(x)| \leqslant 1$, so that $f(\gamma)$ is bounded for $\operatorname{Re}\gamma \geqslant 0$. As for $g(\gamma)$ we may use Stirling's formula

$$\Gamma(z) \to \sqrt{2\pi} \cdot z^{z-1/2} e^{-z}, \quad |z| \to \infty, \quad \operatorname{Re} z \geqslant 0,$$

to estimate it when $|\gamma| \to \infty$. After simplification we have

$$\frac{\Gamma(1+j\gamma)\Gamma(\alpha+(j-1)\gamma)\Gamma(\beta+(n-j)\gamma)}{\Gamma(1+\gamma)\Gamma(\alpha+\beta+(n+j-2)\gamma)}$$
$$\approx \left(\frac{2\pi j}{\gamma}\right)^{1/2} \cdot \frac{(j-1)^{\alpha-1/2}(n-j)^{\beta-1/2}}{(n+j-2)^{\alpha+\beta-1/2}}$$
$$\times \left(\frac{j^j (j-1)^{j-1} (n-j)^{n-j}}{(n+j-2)^{n+j-2}}\right)^{\gamma},$$

when $|\gamma| \to \infty$, $\operatorname{Re}\gamma \geqslant 0$. As $j^j(j-1)^{j-1}(n-j)^{n-j}(n+j-2)^{-(n+j-2)} \leqslant 1$ for $2 \leqslant j \leqslant n-1$, $|g(\gamma)|$ is bounded for $\operatorname{Re}\gamma \geqslant 0$ and goes to 0 as $|\gamma| \to \infty$. Thus in the upper half γ-plane, $\operatorname{Re}\gamma \geqslant 0$, the function $h(\gamma) = f(\gamma) - g(\gamma)$ is regular, bounded, and takes the value 0 at non-negative integer values of γ. At this point we can apply a theorem of Carlson (Titchmarsh, 1939).

Carlson's theorem. *If a function of γ is regular and bounded in the half plane $\operatorname{Re}\gamma \geqslant 0$, and is zero for $\gamma = 0, 1, 2, 3, \ldots$, then it is identically zero.*

Applying this theorem to the function $h(\gamma)$, we deduce that (17.1.3) holds for all complex values of γ, $\operatorname{Re}\gamma \geqslant 0$.

(iii) To extend the validity of (17.1.3) to slightly negative values of $\operatorname{Re}\gamma$, as indicated in (17.1.5), one needs to examine in detail the convergence and analytic nature of the integral f(γ) and the function $g(\gamma)$ in that region. We will not do it here and refer to the original paper (Selberg, 1944). End of proof.

17.3 Aomoto's Proof of Eqs. (17.1.4) and (17.1.3)

For brevity let us write

$$\langle f(x_1,\dots,x_n)\rangle = \int_0^1 \cdots \int_0^1 f(x_1,\dots,x_n)\Phi(x)\,dx \Big/ \int_0^1 \cdots \int_0^1 \Phi(x)\,dx. \quad (17.3.1)$$

Integrating between 0 and 1 the identity

$$\frac{d}{dx_1}(x_1^a x_2\cdots x_m\Phi) = (a+\alpha-1)x_1^{a-1}x_2\cdots x_m\Phi - (\beta-1)\frac{x_1^a x_2\cdots x_m}{1-x_1}\Phi$$
$$+2\gamma\sum_{j=2}^{n}\frac{x_1^a x_2\cdots x_m}{x_1-x_j}\Phi, \quad (17.3.2)$$

we get, for $a=1$ and $a=2$,

$$0=\alpha\langle x_2\cdots x_m\rangle-(\beta-1)\left\langle\frac{x_1x_2\cdots x_m}{1-x_1}\right\rangle+2\gamma\sum_{j=2}^{n}\left\langle\frac{x_1x_2\cdots x_m}{x_1-x_j}\right\rangle, \quad (17.3.3)$$

and

$$0=(\alpha+1)\langle x_1x_2\cdots x_m\rangle-(\beta-1)\left\langle\frac{x_1^2x_2\cdots x_m}{1-x_1}\right\rangle+2\gamma\sum_{j=2}^{n}\left\langle\frac{x_1^2x_2\cdots x_m}{x_1-x_j}\right\rangle. \quad (17.3.4)$$

Now

$$\left\langle\frac{x_1x_2\cdots x_m}{1-x_1}\right\rangle-\left\langle\frac{x_1^2x_2\cdots x_m}{1-x_1}\right\rangle=\langle x_1x_2\cdots x_m\rangle; \quad (17.3.5)$$

interchanging x_1 and x_j and observing the symmetry,

$$\left\langle\frac{x_1x_2\cdots x_m}{x_1-x_j}\right\rangle=-\left\langle\frac{x_jx_2\cdots x_m}{x_1-x_j}\right\rangle=\begin{cases}0, & \text{if } 2\leqslant j\leqslant m,\\ \frac{1}{2}\langle x_2\cdots x_m\rangle, & \text{if } m<j\leqslant n,\end{cases} \quad (17.3.6)$$

and

$$\left\langle\frac{x_1^2x_2\cdots x_m}{x_1-x_j}\right\rangle=-\left\langle\frac{x_j^2x_2\cdots x_m}{x_1-x_j}\right\rangle=\begin{cases}\frac{1}{2}\langle x_1x_2\cdots x_m\rangle, & \text{if } 2\leqslant j\leqslant m,\\ \langle x_1x_2\cdots x_m\rangle, & \text{if } m<j\leqslant n.\end{cases} \quad (17.3.7)$$

Subtracting (17.3.4) from (17.3.3) and using the above, we get

$$(\alpha+1+\gamma(m-1)+2\gamma(n-m)+\beta-1)\langle x_1x_2\cdots x_m\rangle=(\alpha+\gamma(n-m))\langle x_2\cdots x_m\rangle, \tag{17.3.8}$$

or repeating the process

$$\begin{aligned}\langle x_1x_2\cdots x_m\rangle &= \frac{\alpha+\gamma(n-m)}{\alpha+\beta+\gamma(2n-m-1)}\langle x_1\cdots x_{m-1}\rangle\\ &= \prod_{j=1}^{m}\frac{\alpha+\gamma(n-j)}{\alpha+\beta+\gamma(2n-j-1)}.\end{aligned} \tag{17.3.9}$$

which is Eq. (17.1.4).

For $m=n$, Eq. (17.1.4) or (17.3.9) can be written as

$$\frac{I(\alpha+1,\beta,\gamma,n)}{I(\alpha,\beta,\gamma,n)}=\langle x_1\cdots x_n\rangle=\prod_{j=1}^{n}\frac{\alpha+\gamma(n-j)}{\alpha+\beta+\gamma(2n-j-1)}, \tag{17.3.10}$$

or for a positive integer α,

$$I(\alpha,\beta,\gamma,n)=\prod_{j=1}^{n}\frac{\Gamma(\alpha+\gamma(n-j))\Gamma(1+\beta+\gamma(2n-j-1))}{\Gamma(\alpha+\beta+\gamma(2n-j-1))\Gamma(1+\gamma(n-j))}I(1,\beta,\gamma,n). \tag{17.3.11}$$

As

$$I(\alpha,\beta,\gamma,n)=I(\beta,\alpha,\gamma,n), \tag{17.3.12}$$

we get

$$I(\alpha,\beta,\gamma,n)=\prod_{j=1}^{n}\frac{\Gamma(\alpha+\gamma(n-j))\Gamma(\beta+\gamma(n-j))}{\Gamma(\alpha+\beta+\gamma(2n-j-1))}c(\gamma,n), \tag{17.3.13}$$

where $c(\gamma,n)$ is independent of α and β. This is Eq. (17.2.5). The rest of the proof follows Selberg's reasoning.

Note that Eq. (17.3.13) is derived here for integer α, β and complex γ, while Selberg gets (17.2.5) for complex α, β and integer γ. One then completes the argument by analytical continuation using Carlson's theorem.

A slight change of reasoning due to Askey gives Eq. (17.3.13) directly for complex α, β and γ as follows. Equations (17.3.10) and (17.3.12) give the ratio of $I(\alpha,\beta,\gamma,n)$

and $I(\alpha, \beta + m, \gamma, n)$ for any integer m,

$$I(\alpha, \beta, \gamma, n) = \prod_{j=1}^{n} \frac{(\alpha + \beta + \gamma(2n - j - 1))_m}{(\beta + \gamma(n - j))_m} I(\alpha, \beta + m, \gamma, n), \tag{17.3.14}$$

where we have used the notation

$$(a)_m = \Gamma(a + m)/\Gamma(a), \quad m \geqslant 0; \tag{17.3.15}$$

i.e.

$$(a)_0 = 1,$$

and

$$(a)_m = a(a + 1) \cdots (a + m - 1), \quad m \geqslant 1.$$

Now

$$\begin{aligned} I(\alpha, \beta + m, \gamma, n) &= \int_0^1 \cdots \int_0^1 |\Delta(x)|^{2\gamma} \prod_{j=1}^{n} x_j^{\alpha-1} (1 - x_j)^{\beta+m-1} \, dx_j \\ &= m^{-\alpha n - \gamma n(n-1)} \int_0^m \cdots \int_0^m |\Delta(x)|^{2\gamma} \prod_{j=1}^{n} x_j^{\alpha-1} \left(1 - \frac{x_j}{m}\right)^{\beta+m-1} dx_j, \end{aligned} \tag{17.3.16}$$

and

$$\frac{(a)_m}{(b)_m} m^{b-a} = \frac{\Gamma(b)}{\Gamma(a)} \cdot \frac{\Gamma(a + m)}{\Gamma(b + m)} m^{b-a}, \tag{17.3.17}$$

so in the limit $m \to \infty$,

$$I(\alpha, \beta, \gamma, n) = \mathcal{I}(\alpha, \gamma, n) \cdot \prod_{j=1}^{n} \frac{\Gamma(\beta + \gamma(n - j))}{\Gamma(\alpha + \beta + \gamma(2n - j - 1))}, \tag{17.3.18}$$

where

$$\mathcal{I}(\alpha, \gamma, n) = \int_0^\infty \cdots \int_0^\infty |\Delta(x)|^{2\gamma} \prod_{j=1}^{n} x_j^{\alpha-1} \exp(-x_j) \, dx_j. \tag{17.3.19}$$

Equations (17.3.12) and (17.3.18) give (17.3.13).

17.4 Other Averages

Taking polynomials other than $x_1^a x_2 \cdots x_m$ in Eq. (17.3.2), and/or differentiating with respect to x_2 or x_3, one can evaluate other integrals. For example, for $n \geqslant m > 1$,

$$(\alpha+\beta+\gamma(2n-m-2))\langle x_1^2 x_2 \cdots x_m\rangle = (\alpha+\gamma(n-m))\langle x_1^2 x_2 \cdots x_{m-1}\rangle + \gamma\langle x_1 x_2 \cdots x_m\rangle,$$

and

$$(\alpha+\beta+1+2\gamma(n-1))\langle x_1^2\rangle = (\alpha+1+2\gamma(n-1))\langle x_1\rangle - \gamma(n-1)\langle x_1 x_2\rangle, \quad n \geqslant 1,$$

these together with Eq. (17.3.9) give the value of $\langle x_1^2 x_2 \cdots x_m\rangle$.

17.5 Other Forms of Selberg's Integral

Selberg's integral can also be written in various other forms. We note two more of them here:

$$I(\alpha,\beta,\gamma,n) = \int_0^\infty \cdots \int_0^\infty |\Delta(x)|^{2\gamma} \prod_{j=1}^n x_j^{\alpha-1}(1+x_j)^{-\alpha-\beta-2\gamma(n-1)}\, dx_j, \quad (17.5.1)$$

obtained from (17.1.3) by the change of variables $x_j = y_j/(1+y_j)$; and

$$J(a,b,\alpha,\beta,\gamma,n) = \int_{-\infty}^\infty \cdots \int_{-\infty}^\infty |\Delta(x)|^{2\gamma} \prod_{j=1}^n (a+ix_j)^{-\alpha}(b-ix_j)^{-\beta}\, dx_j$$
$$= \frac{(2\pi)^n}{(a+b)^{(\alpha+\beta)n-\gamma n(n-1)-n}} \times \prod_{j=0}^{n-1} \frac{\Gamma(1+\gamma+j\gamma)\Gamma(\alpha+\beta-(n+j-1)\gamma-1)}{\Gamma(1+\gamma)\Gamma(\alpha-j\gamma)\Gamma(\beta-j\gamma)}. \quad (17.5.2)$$

Equations corresponding to (17.1.4) are

$$\int_0^\infty \cdots \int_0^\infty |\Delta(x)|^{2\gamma} \prod_{j=1}^m (x_j(1+x_j)^{-1}) \cdot \prod_{j=1}^n x_j^{\alpha-1}(1+x_j)^{-\alpha-\beta-2\gamma(n-1)}\, dx_j$$
$$= \prod_{j=n-m}^{n-1} \frac{\alpha+j\gamma}{\alpha+\beta+(n+j-1)\gamma} \cdot I(\alpha,\beta,\gamma,n), \quad (17.5.3)$$

and

$$\int_{-\infty}^{\infty}\cdots\int_{-\infty}^{\infty}\prod_{j=1}^{m}(a+ix_j)\cdot|\Delta(x)|^{2\gamma}\prod_{j=1}^{n}(a+ix_j)^{-\alpha}(b-ix_j)^{-\beta}\,dx_j$$

$$=(a+b)^m\prod_{j=n-m}^{n-1}\left(\frac{\alpha-j\gamma-1}{\alpha+\beta-(n+j-1)\gamma-2}\right)\cdot J(a,b,\alpha,\beta,\gamma,n),\tag{17.5.4}$$

where m and n are integers with $1\leqslant m\leqslant n$ while a,b,α,β and γ are complex numbers. Equations (17.5.1) and (17.5.3) are valid for α,β,γ satisfying (17.1.5), while Eqs. (17.5.2) and (17.5.4) are valid for $\operatorname{Re}a$, $\operatorname{Re}b$, $\operatorname{Re}\alpha$, $\operatorname{Re}\beta>0$, $\operatorname{Re}(\alpha+\beta)>1$, and

$$-\frac{1}{n}<\operatorname{Re}\gamma<\min\left(\frac{\operatorname{Re}\alpha}{n-1},\frac{\operatorname{Re}\beta}{n-1},\frac{\operatorname{Re}(\alpha+\beta+1)}{2(n-1)}\right).\tag{17.5.5}$$

(Here $i=\sqrt{-1}$, and not an index!)

In (17.5.2) and (17.5.4) one can make sure of the dependence on a and b by the change of variables, $2x_j=i(a-b)+(a+b)y_j$. A proof of Eq. (17.5.2) can be supplied following Selberg's reasoning, Section 17.2, by starting with the Cauchy integral

$$\int_{-\infty}^{\infty}(a+ix)^{-\alpha}(b-ix)^{-\beta}\,dx=\frac{2\pi}{(a+b)^{\alpha+\beta-1}}\frac{\Gamma(\alpha+\beta-1)}{\Gamma(\alpha)\Gamma(\beta)},\tag{17.5.6}$$

instead of the Euler integral (17.2.1). Or one can first establish (17.5.4) following Aomoto's reasoning, Section 17.3. Then taking $m=n$ in Eq. (17.5.4) and iterating, one gets from the symmetry in $(a,\alpha)\Leftrightarrow(b,\beta)$ the equation corresponding to (17.2.5) or (17.3.13) as

$$(2\pi)^{-n}\cdot J(a,b,\alpha,\beta,\gamma,n)=C(\gamma,n)\cdot(a+b)^{-(\alpha+\beta)n+\gamma n(n-1)+n}$$
$$\times\prod_{j=0}^{n-1}\frac{\Gamma(\alpha+\beta-(n+j-1)\gamma-1)}{\Gamma(\alpha-j\gamma)\Gamma(\beta-j\gamma)},\tag{17.5.7}$$

where $C(\gamma,n)$ does not depend on a,b,α or β. To determine $C(\gamma,n)$, put $a=b=1$, $\alpha=\beta=\gamma(n-1)+1$, and make the change of variables $x_j=\operatorname{tg}(\theta_j/2)$. This gives

$$C(\gamma,n)=(2\pi)^{-n}2^{\gamma n(n-1)+n}\cdot\prod_{j=0}^{n-1}\Gamma(1+j\gamma)J(1,1,\gamma(n-1)+1,\gamma(n-1)+1,\gamma,n)$$

$$=(2\pi)^{-n}2^{\gamma n(n-1)}$$

$$\times \prod_{j=0}^{n-1} \Gamma(1+j\gamma) \cdot \int_{-\pi}^{\pi} \cdots \int_{-\pi}^{\pi} \prod_{1 \leqslant j < k \leqslant n} |\sin(\theta_j - \theta_k)/2|^{2\gamma} \prod_{j=1}^{n} d\theta_j.$$

The integral in the last line above is (see for example, Sections 11.1 or 17.7)

$$\int \cdots \int \prod |\sin(\theta_j - \theta_k)/2|^{2\gamma} \prod d\theta_j = \int \cdots \int \prod |e^{i\theta_j} - e^{i\theta_k}|^{2\gamma} \prod d\theta_j$$
$$= (2\pi)^n \Gamma(1+n\gamma)/(\Gamma(1+\gamma))^n. \quad (17.5.8)$$

This gives

$$C(\gamma, n) = 2^{\gamma n(n-1)} \prod_{j=1}^{n} \frac{\Gamma(1+j\gamma)}{\Gamma(1+\gamma)}. \quad (17.5.9)$$

For more details see Morris, 1982.

17.6 Some Consequences of Selberg's Integral

By changing variables of integration, choosing various appropriate values for α and β or taking limits we may derive a few Selberg type integrals related to all classical orthogonal polynomials. Thus

Jacobi,

$$x_j = (1-y_j)/2, \quad 1-x_j = (1+y_j)/2,$$
$$|\Delta(x)| = 2^{-n(n-1)/2}|\Delta(y)|, \quad dx_j = dy_j/2,$$
$$\int_{-1}^{1} \cdots \int_{-1}^{1} |\Delta(x)|^{2\gamma} \cdot \prod_{j=1}^{n} (1-x_j)^{\alpha-1}(1+x_j)^{\beta-1}\, dx_j$$
$$= 2^{\gamma n(n-1)+n(\alpha+\beta-1)} \prod_{j=0}^{n-1} \frac{\Gamma(1+\gamma+j\gamma)\Gamma(\alpha+\gamma j)\Gamma(\beta+\gamma j)}{\Gamma(1+\gamma)\Gamma(\alpha+\beta+\gamma(n+j-1))}. \quad (17.6.1)$$

Gegenbauer, $\alpha = \beta = \lambda + 1/2$ in (17.6.1),

$$\int_{-1}^{1} \cdots \int_{-1}^{1} |\Delta(x)|^{2\gamma} \cdot \prod_{j=1}^{n} (1-x_j^2)^{\lambda-1/2}\, dx_j$$
$$= 2^{\gamma n(n-1)+2\lambda n} \cdot \prod_{j=0}^{n-1} \frac{\Gamma(1+\gamma+j\gamma)(\Gamma(\lambda+\gamma j+1/2)^2}{\Gamma(1+\gamma)\Gamma(2\lambda+\gamma(n+j-1)+1)}. \quad (17.6.2)$$

Legendre, $\alpha = \beta = 1$ in (17.6.1),

$$\int_{-1}^{1} \cdots \int_{-1}^{1} |\Delta(x)|^{2\gamma} \cdot \prod_{j=1}^{n} dx_j = 2^{\gamma n(n-1)+n} \cdot \prod_{j=0}^{n-1} \frac{\Gamma(1+\gamma+j\gamma)(\Gamma(\gamma j+1)^2}{\Gamma(1+\gamma)\Gamma(\gamma(n+j-1)+2)}. \tag{17.6.3}$$

Tchebichev, $\alpha = \beta = 3/2$ or $\alpha = \beta = 1/2$ in (17.6.1),

$$\int_{-1}^{1} \cdots \int_{-1}^{1} |\Delta(x)|^{2\gamma} \cdot \prod_{j=1}^{n} (1-x_j^2)^{\pm 1/2} \, dx_j$$

$$= 2^{\gamma n(n-1)+n\pm n} \cdot \prod_{j=0}^{n-1} \frac{\Gamma(1+\gamma+j\gamma)(\Gamma(\gamma j+1\pm 1/2)^2}{\Gamma(1+\gamma)\Gamma(\gamma(n+j-1)+2\pm 1)}. \tag{17.6.4}$$

Laguerre, put $x_j = y_j/L$, $\beta = L+1$, and take the limit $L \to \infty$ in Eq. (17.1.3);

$$\int_{0}^{\infty} \cdots \int_{0}^{\infty} |\Delta(x)|^{2\gamma} \cdot \prod_{j=1}^{n} x_j^{\alpha-1} \exp(-x_j) \, dx_j = \prod_{j=0}^{n-1} \frac{\Gamma(1+\gamma+j\gamma)\Gamma(\alpha+\gamma j)}{\Gamma(1+\gamma)}. \tag{17.6.5}$$

Set $x_j = y_j^2/2$ in (17.6.5) to get

$$\int_{-\infty}^{\infty} \cdots \int_{-\infty}^{\infty} \prod_{1 \leqslant j < \ell \leqslant n} |x_j^2 - x_\ell^2|^{2\gamma} \prod_{j=1}^{n} |x_j|^{2\alpha-1} \exp(-x_j^2/2) \, dx_j$$

$$= 2^{\alpha n+\gamma n(n-1)} \prod_{j=1}^{n} \frac{\Gamma(1+j\gamma)\Gamma(\alpha+\gamma(j-1))}{\Gamma(1+\gamma)}. \tag{17.6.6}$$

Hermite, put $x_j = y_j/L, \alpha = \beta = aL^2+1$, and take the limit $L \to \infty$ in (17.6.1);

$$\int_{-\infty}^{\infty} \cdots \int_{-\infty}^{\infty} |\Delta(x)|^{2\gamma} \cdot \prod_{j=1}^{n} \exp(-ax_j^2) \, dx_j$$

$$= (2\pi)^{n/2}(2a)^{-n(\gamma(n-1)+1)/2} \cdot \prod_{j=1}^{n} \frac{\Gamma(1+j\gamma)}{\Gamma(1+\gamma)}. \tag{17.6.7}$$

For integer γ this last equation can also be written as a finite algebraic identity. From

$$\int_{-\infty}^{\infty} \exp(-a^2x^2 - 2iax\lambda)x^n\,dx = \left(\frac{i}{2a}\frac{d}{d\lambda}\right)^n \int_{-\infty}^{\infty} \exp(-a^2x^2 - 2iax\lambda)\,dx$$
$$= \left(\frac{i}{2a}\frac{d}{d\lambda}\right)^n \exp(-\lambda^2) \int_{-\infty}^{\infty} \exp(-(ax+i\lambda)^2)\,dx$$
$$= \frac{\sqrt{\pi}}{a}\left(\frac{i}{2a}\frac{d}{d\lambda}\right)^n \exp(-\lambda^2),$$

we can replace the integration by a differentiation at the point $x_j = 0$,

$$\int_{-\infty}^{\infty} \exp(-ax_j^2)x_j^n\,dx_j = \sqrt{\frac{\pi}{a}}\left(\frac{i}{2\sqrt{a}}\frac{\partial}{\partial x_j}\right)^n \exp(-x_j^2)|_{x_j=0}.$$

Equation (17.6.7) therefore takes the form

$$\left(\frac{i}{2\sqrt{a}}\right)^{\gamma n(n-1)} \prod_{1\leqslant p<q\leqslant n} \left(\frac{\partial}{\partial x_p} - \frac{\partial}{\partial x_q}\right)^{2\gamma} \exp\left(-\sum_{j=1}^{n} x_j^2\right)\Bigg|_{x_j=0}$$
$$= (2a)^{-\gamma n(n-1)/2} \prod_{j=1}^{n} \frac{(j\gamma)!}{\gamma!}. \tag{17.6.8}$$

Replacing the exponential by its power series expansion, one notes that the term $(-\sum x_j^2)^\ell$ gives zero on differentiation if $\ell < \gamma n(n-1)/2$, and leaves a homogeneous polynomial of order $\ell - \gamma n(n-1)/2$ in the variables $x_1, \ldots, x_n$, if $\ell > \gamma n(n-1)/2$. On setting $x_j = 0$, $j = 1, \ldots, n$, one sees therefore that there is only one term, corresponding to $\ell = \gamma n(n-1)/2$, which gives a non zero contribution. Equation (17.6.8) is now

$$\prod_{1\leqslant p<q\leqslant n} \left(\frac{\partial}{\partial x_p} - \frac{\partial}{\partial x_q}\right)^{2\gamma} \left(\sum_{j=1}^{n} x_j^2\right)^{\ell} = 2^\ell \ell! \prod_{j=1}^{n} \frac{(j\gamma)!}{\gamma!}, \quad \ell = \gamma n(n-1)/2.$$

If $P(x) \equiv P(x_1, \ldots, x_n)$ and $Q(x) \equiv Q(x_1, \ldots, x_n)$ are homogeneous polynomials in $x \equiv (x_1, \ldots, x_n)$ of the same degree, then a little reflection shows that $P(\partial/\partial x)Q(x)$ is a constant which is also equal to $Q(\partial/\partial x)P(x)$. Thus one can interchange the roles of x_j and $\partial/\partial x_j$ to get

$$\left(\sum_{j=1}^{n} \frac{\partial^2}{\partial x_j^2}\right)^{\ell} \prod_{1\leqslant i<j\leqslant n} (x_i - x_j)^{2\gamma} = 2^\ell \ell! \prod_{j=1}^{n} \frac{(j\gamma)!}{\gamma!}, \quad \ell = \gamma n(n-1)/2. \tag{17.6.9}$$

17.7 Normalization Constant for the Circular Ensembles

Put $\alpha = -\gamma(n-1)+\varepsilon$ in (17.1.3) and take the limit $\varepsilon \to 0$. When one substitutes the expansion (17.2.2) in the left-hand side integrand of (17.1.3), the dominant (i.e. the most divergent) contribution comes from the term $(x_1 x_2 \cdots x_n)^{\gamma(n-1)}$. Using $\Gamma(-j+\varepsilon) = \Gamma(\varepsilon)\prod_{p=1}^{j}(-p+\varepsilon)^{-1}$, one gets the following result.

If one expands $(\Delta(x))^{2\gamma}$ in a power series of the variables $x_1, \ldots, x_n$, the coefficient of $(x_1 \cdots x_n)^{\gamma(n-1)}$ is $(-1)^{\gamma n(n-1)/2}(n\gamma)!/(\gamma!)^n$. Or, if one expands $(\Delta(x)\Delta(x^{-1}))^{\gamma}$ in positive and negative powers of $x_1, \ldots, x_n$, the constant term is $(n\gamma)!/(\gamma!)^n$.

One can express this result also as

$$\begin{aligned}
&\int_0^{2\pi} \cdots \int_0^{2\pi} \prod_{1\leqslant j<k\leqslant n} |e^{i\theta_j} - e^{i\theta_k}|^{2\gamma} \prod_{j=1}^{n} d\theta_j/(2\pi) \\
&= \int_0^{2\pi} \cdots \int_0^{2\pi} \prod_{1\leqslant j<k\leqslant n} \{e^{i\theta_j} - e^{i\theta_k}\}^{\gamma}\{e^{-i\theta_j} - e^{-i\theta_k}\}^{\gamma} \prod_{j=1}^{n} d\theta_j/(2\pi) \\
&= (n\gamma)!/(\gamma!)^n;
\end{aligned} \tag{17.7.1}$$

since if one expands the integrand in the above equation in positive and negative powers of $\exp(i\theta_j)$, only the constant term survives on integration.

By computing the sub-dominant terms one can get a few more coefficients in such expansions.

17.8 Averages With Laguerre or Hermite Weights

Let us write

$$\langle f(x)\rangle = \int_0^{\infty} \cdots \int_0^{\infty} f(x)\Phi(x)\,dx \Big/ \int_0^{\infty} \cdots \int_0^{\infty} \Phi(x)\,dx, \tag{17.8.1}$$

where

$$\Phi(x) \equiv \Phi(x_1, \ldots, x_n) = |\Delta(x)|^{2\gamma} \cdot \prod_{j=1}^{n} x_j^{\alpha-1} \exp(-x_j)\,dx_j.$$

Integrating

$$\frac{\partial}{\partial x_1} f(x)\Phi(x) = \frac{\partial f}{\partial x_1}\Phi(x) + f(x)\left(\frac{\alpha-1}{x_1} - 1 + 2\gamma \sum_{j=2}^{n} \frac{1}{x_1 - x_j}\right)\Phi(x),$$

choosing various polynomials for $f(x)$, and applying Aomoto's argument one gets, for example,

$$\langle x_1 \cdots x_m \rangle = (\alpha + \gamma(n-m))\langle x_1 \cdots x_{m-1} \rangle = \prod_{i=1}^{m} (\alpha + \gamma(n-i)), \tag{17.8.2}$$

$$\begin{aligned} \langle x_1^2 \cdots x_j^2 x_{j+1} \cdots x_m \rangle &= (\alpha + 1 + \gamma(2n - m - j))\langle x_1^2 \cdots x_{j-1}^2 x_j \cdots x_m \rangle \\ &= \prod_{i=1}^{j} (\alpha + 1 + \gamma(2n - m - i)) \cdot \prod_{k=1}^{m} (\alpha + \gamma(n-k)), \end{aligned} \tag{17.8.3}$$

$$\begin{aligned} &\langle x_1^3 \cdots x_i^3 x_{i+1}^2 \cdots x_j^2 x_{j+1} \cdots x_m \rangle \\ &\quad = (\alpha + 2 + \gamma(2n - j - i))\langle x_1^3 \cdots x_{i-1}^3 x_i^2 \cdots x_j^2 x_{j+1} \cdots x_m \rangle \\ &\qquad + \gamma(n-m)\langle x_1^3 \cdots x_{i-1}^3 x_i^2 \cdots x_{j-1}^2 x_j \cdots x_m x_{m+1} \rangle, \end{aligned} \tag{17.8.4}$$

etc.

Taking $m = n$ in (17.8.2) one can express the (Laguerre) integral (17.6.5) with parameter α as a known multiple of that with parameter $\alpha \pm j$, j a positive integer. Instead of thus decreasing α, Askey and Richards (1989) increased it and took the limit $\alpha \to \infty$. In this way they related the integral (17.6.5) to the (Hermite) integral (17.6.7) and with a fine analysis determined them both, without recourse to (17.1.3). We will not reproduce the details here.

If in the above we replace $\Phi(x)$ by

$$\Phi(x) = |\Delta(x)|^{2\gamma} \prod_{j=1}^{n} \exp(-a x_j^2),$$

then similar arguments give

$$\langle x_1 \cdots x_{2m} \rangle = \left(-\frac{\gamma}{2a}\right)(2m-1)\langle x_1 \cdots x_{2m-2} \rangle = \left(-\frac{\gamma}{2a}\right)^m \frac{(2m)!}{2^m m!}, \tag{17.8.5}$$

(a result known to Ullah (1986) for $\gamma = 1/2$),

$$\begin{aligned} 2a\langle x_1^2 \cdots x_j^2 x_{j+1} \cdots x_m \rangle &= (1 + \gamma(n-m))\langle x_1^2 \cdots x_{j-1}^2 x_j \cdots x_{m-1} \rangle \\ &\quad - \gamma(j-1)\langle x_1^2 \cdots x_{j-2}^2 x_{j-1} \cdots x_m \rangle, \end{aligned} \tag{17.8.6}$$

$$\begin{aligned} 2a\langle x_1^3 \cdots x_j^3 x_{j+1} \cdots x_m \rangle &= (2 + \gamma(2n - m - j))\langle x_1^3 \cdots x_{j-1}^3 x_j \cdots x_m \rangle \\ &\quad - \gamma(j-1)\langle x_1^3 \cdots x_{j-2}^3 x_{j-1}^2 x_j^2 x_{j+1} \cdots x_m \rangle, \end{aligned} \tag{17.8.7}$$

etc., so that, e.g.,

$$\langle x_1^2 x_2 \cdots x_{2m+1}\rangle = \frac{1}{2a}(1+\gamma(n-2m-1))\langle x_1 \cdots x_{2m}\rangle$$
$$= \frac{1}{2a}(1+\gamma(n-2m-1))\left(-\frac{\gamma}{2a}\right)^m \frac{(2m)!}{2^m m!}, \tag{17.8.8}$$

$$\langle x_1^2 x_2^2 x_3 \cdots x_{2m}\rangle = \left(\frac{1}{2a}\right)^2 (1+\gamma(n-2m))(1+\gamma(n-2m+1))$$
$$\times \left(-\frac{\gamma}{2a}\right)^{m-1} \frac{(2m-2)!}{2^{m-1}(m-1)!}$$
$$+\left(-\frac{\gamma}{2a}\right)^{m+1} \frac{(2m)!}{2^m m!}, \tag{17.8.9}$$

$$\langle x_1^3 x_2 \cdots x_{2m}\rangle = \frac{1}{2a}(2+\gamma(2n-2m-1))\left(-\frac{\gamma}{2a}\right)^m \frac{(2m)!}{2^m m!}, \tag{17.8.10}$$

etc.

17.9 Connection With Finite Reflection Groups

Macdonald (1982) observed that $x_i - x_j = 0,\ 1 \leqslant i < j \leqslant n$, are the equations of the reflecting (hyper-)planes of the finite group A_n. There are other finite groups generated by reflections. In the n-dimensional Euclidean space $\mathcal{R}^n$, consider a certain number of planes all passing through the origin. If the angles between these planes are properly chosen, then the group generated by reflections in these planes is finite. These groups were enumerated and classified by Coxeter as $A_n (n \geqslant 1)$, $B_n (n \geqslant 2)$, $D_n (n \geqslant 4)$, E_6, E_7, E_8, F_4, G_2, H_3, H_4, and $I_2(p)$, $(p \geqslant 5)$ $(G_2 \equiv I_2(6))$. The first three sequences are known as the classical groups, while the rest are the exceptional groups. Let G be one of these groups and let $P(x)$ be the product of distances of the point $x \equiv (x_1, \ldots, x_n)$ from all the reflecting planes belonging to G. In other words, if the equations of the planes are $\sum_{i=1}^n a_i^{(\alpha)} x_i = 0$ with $\sum_{i=1}^n (a_i^{(\alpha)})^2 = 1$, then $P(x) = \prod_\alpha (\sum_i a_i^{(\alpha)} x_i)$. Let N be the degree of $P(x)$, i.e. the number of reflecting planes in G.

Not knowing Selberg's integral, Macdonald conjectured (1982) that

$$\int_{-\infty}^{\infty} \cdots \int_{-\infty}^{\infty} \exp\Big(-\sum x_i^2/2\Big) |P(x)|^{2\gamma} dx_1 \cdots dx_n = 2^{-N\gamma} (2\pi)^{n/2} \prod_{j=1}^{n} \frac{\Gamma(1+\gamma d_i)}{\Gamma(1+\gamma)}, \tag{17.9.1}$$

where d_i are the degrees of a basic set of polynomials invariant under G. The n integers $m_i = d_i - 1, i = 1, \ldots, n$, are known also as the (Coxeter) indices of G and have a few other remarkable properties (cf. Mehta, 1989MT, Section 7.11).

For the group A_n Eq. (17.9.1) is the same as (17.6.7). Note the difference in normalization; the reflecting planes are $(x_i - x_j)/\sqrt{2} = 0$ in (17.9.1) and not $x_i - x_j = 0$ as in (17.6.7). For the groups B_n and D_n, Eq. (17.9.1) is the same as (17.6.6) with $\alpha = \gamma + 1/2$ and $\alpha = 1/2$ respectively. For the two-dimensional discrete rotation group $I_2(p)$,

$$P(x) = \prod_{j=0}^{p-1} (x_1 \sin(j\pi/p) - x_2 \cos(j\pi/p)), \tag{17.9.2}$$

and Eq. (17.9.1) can be directly verified by introducing polar coordinates. For other exceptional groups Eq. (17.9.1) was verified for $\gamma = 1$ by Macdonald himself. Opdam (1989) gave a uniform proof of Eq. (17.9.1) for the groups A_n, B_n, D_n, E_6, E_7, E_8, F_4 and G_2 (or $I_2(6)$); the proof depends on the existence of a crystallographic lattice corresponding to each of these groups. Garvan (1989) gave a computer proof of Eq. (17.9.1) for the groups H_3 and F_4. Opdam (1993) gave a proof for the remaining groups H_3 and H_4. Thus Eq. (17.9.1) is no longer a conjecture but a theorem.

It is instructive to recall the computer proof of Garvan. Denoting the two sides of Eq. (17.9.1) by $f(\gamma)$ and $F(\gamma)$ respectively, it will be sufficient to show that

$$\frac{f(\gamma+1)}{f(\gamma)} = \frac{F(\gamma+1)}{F(\gamma)},$$

since starting with the evident equality $f(0) = F(0)$, the above equation will imply $f(\gamma) = F(\gamma)$ for any positive integer γ and then Carlson's theorem will extend its validity for complex γ. Now $F(\gamma+1)/F(\gamma)$ is a polynomial in γ of degree $\sum_{i=1}^{n} d_i = N + n$. From Aomoto's argument, Section 17.8, one can show that $f(\gamma+1)/f(\gamma)$ is also a polynomial of degree $N + n$. If two polynomials each of degree $N + n$ are equal at $N + n + 1$ distinct points, then they are identically equal. Garvan could verify the equality of the polynomials with the help of a computer for small values of N and n, i.e. for H_3 and F_4.

We give here a convenient list of reflecting (hyper-)planes for the exceptional (non-factorizable finite reflection) groups mentioned above, together with the degree of $P(x)$ and the list of the integers d_i. The entries are arranged as follows.

(i) Coxeter symbol of the group, the subscript indicating the dimension of the space in which it operates, i.e. the number n of the variables in the integral (17.9.1);
(ii) Number N of the reflecting (hyper-)planes or the degree of the polynomial $P(x)$ in (17.9.1);
(iii) The integers d_i, the degrees of the polynomials in a basic invariant set (invariant under reflections of the group);
(iv) A list of reflecting (hyper-)planes or the linear factors of the polynomial $P(x)$, not normalized in general.

Group E_6: $N=36$; $d_i=(2,5,6,8,9,12)$; factors of $P(x)$: x_1, x_2, x_3, x_4, $x_1\pm x_2\pm x_3\pm x_4$, $x_1\pm x_2\pm\sqrt{2}x_5$, $x_3\pm x_4\pm\sqrt{2}x_5$, $x_1\pm x_3\pm\frac{1}{\sqrt{2}}(x_5-\sqrt{3}x_6)$, $x_2\pm x_4\pm\frac{1}{\sqrt{2}}(x_5-\sqrt{3}x_6)$, $x_1\pm x_4\pm\frac{1}{\sqrt{2}}(x_5+\sqrt{3}x_6)$, $x_2\pm x_3\pm\frac{1}{\sqrt{2}}(x_5-\sqrt{3}x_6)$.

Group E_7: $N=63$; $d_i=(2,6,8,10,12,14,18)$; factors of $P(x)$: x_i, $i=1,2,\ldots,7$, and $x_i\pm x_j\pm x_k\pm x_\ell$ with $(i,j,k,\ell)=(1,2,3,4)$, $(1,2,5,6)$, $(1,3,5,7)$, $(1,4,6,7)$, $(2,3,6,7)$, $(2,4,5,7)$ and $(3,4,5,6)$.

Group E_8: $N=120$; $d_i=(2,8,12,14,18,20,24,30)$; factors of $P(x)$: x_i, $i=1,2,\ldots,8$, and $x_i\pm x_j\pm x_k\pm x_\ell$ with $(i,j,k,\ell)=(1,2,3,4)$, $(1,2,5,6)$, $(1,3,5,7)$, $(1,4,6,7)$, $(2,3,6,7)$, $(2,4,5,7)$, $(3,4,5,6)$, $(5,6,7,8)$, $(3,4,7,8)$, $(2,4,6,8)$, $(2,3,5,8)$, $(1,4,5,8)$, $(1,3,6,8)$ and $(1,2,7,8)$.

Group F_4: $N=24$; $d_i=(2,6,8,12)$; factors of $P(x)$: x_1, x_2, x_3, x_4, $x_1\pm x_2\pm x_3\pm x_4$, $x_i\pm x_j$, $1\leqslant i<j\leqslant 4$.

Group H_3: $N=10$; $d_i=(2,6,10)$; factors of $P(x)$: x_1, x_2, x_3, $ax_1\pm bx_2\pm cx_3$ and those obtained by the cyclic permutations of (a,b,c), i.e. $ax_2\pm bx_3\pm cx_1$ and $ax_3\pm bx_1\pm cx_2$, where $a=\cos(\pi/5)=(\sqrt{5}+1)/4$, $b=\cos(2\pi/5)=(\sqrt{5}-1)/4$ and $c=\cos(\pi/3)=1/2$.

Group H_4: $N=60$; $d_i=(2,12,20,30)$; factors of $P(x)$: x_1, x_2, x_3, x_4, $x_1\pm x_2\pm x_3\pm x_4$, $ax_1\pm bx_2\pm cx_3$, $ax_1\pm bx_3\pm cx_4$, $ax_1\pm bx_4\pm cx_2$, $ax_2\pm bx_4\pm cx_3$, and those obtained by the cyclic permutations of (a,b,c) in the last four sets. The numbers a,b,c are respectively the cosines of the angles $\pi/5$, $2\pi/5$ and $\pi/3$ as for the group H_3.

17.10 A Second Generalization of the Beta Integral

A second generalization of the beta-integral by Selberg is as follows.

$$\int\cdots\int y_1\cdots y_m\cdot|\Delta(y)|^{2\gamma}\left(1-\sum_{i=1}^n y_i\right)^{\beta-1}\prod_{j=1}^n y_j^{\alpha-1}dy_j$$
$$=\frac{\Gamma(\beta)}{\Gamma(\beta+m+\alpha n+\gamma n(n-1))}\cdot\prod_{i=1}^m(\alpha+\gamma(n-i))$$
$$\times\prod_{j=1}^n\frac{\Gamma(\alpha+\gamma(n-j))\Gamma(1+\gamma j)}{\Gamma(1+\gamma)},\tag{17.10.1}$$

where the integral is taken over $0\leqslant y_i$, $y_1+\cdots+y_n\leqslant 1$. This is to be compared with the long known generalization

$$\int\cdots\int\left(1-\sum_{i=1}^n y_i\right)^{\beta-1}\prod_{j=1}^n y_j^{\alpha_j-1}dy_j=\Gamma(\beta)\prod_{j=1}^n\Gamma(\alpha_j)/\Gamma(\alpha_1+\cdots+\alpha_n+\beta).\tag{17.10.2}$$

the integral being again taken over $0\leqslant y_i$, $y_1+\cdots+y_n\leqslant 1$.

Proof. From (17.6.5) and (17.8.2) one has by a change of variables,

$$\int_0^\infty \cdots \int_0^\infty x_1 \cdots x_m \cdot |\Delta(x)|^{2\gamma} \prod_{j=1}^n x_j^{\alpha-1} \exp(-\lambda x_j)\, dx_j$$

$$= \lambda^{-m-\alpha n-\gamma n(n-1)} \prod_{i=1}^m (\alpha+\gamma(n-i)) \cdot \prod_{j=1}^n \frac{\Gamma(\alpha+\gamma(n-j))\Gamma(1+\gamma j)}{\Gamma(1+\gamma)}.$$

Multiplying both sides by $\lambda^{\beta-1}e^{-\lambda}$ and integrating over λ from 0 to ∞, one gets

$$\int_0^\infty \cdots \int_0^\infty x_1 \cdots x_m \cdot |\Delta(x)|^{2\gamma} \left(1+\sum_{i=1}^n x_i\right) \cdot \prod_{j=1}^n x_j^{\alpha-1}\, dx_j$$

$$= \frac{\Gamma(\beta-m-\alpha n-\gamma n(n-1))}{\Gamma(\beta)} \prod_{i=1}^m (\alpha+\gamma(n-i))$$

$$\times \prod_{j=1}^n \frac{\Gamma(\alpha+\gamma(n-j))\Gamma(1+\gamma j)}{\Gamma(1+\gamma)},$$

provided that the variables appearing in the gamma functions all have positive real parts. Now make a change of variables,

$$x_i = y_i \left(1-\sum_{i=1}^n y_i\right)^{-1}, \qquad y_i = x_i \left(1+\sum_{i=1}^n x_i\right)^{-1}.$$

Setting $u = (1-\sum_{i=1}^n y_i)^{-1}$, the Jacobian is

$$\det\left[\frac{\partial x_i}{\partial y_j}\right] = \det[u\delta_{ij} + y_i u^2].$$

This determinant can be calculated by bordering it with an extra row and a column,

$$\det[u\delta_{ij} + u^2 y_i] = \det\begin{bmatrix} 1 & 0 \\ y_i & u\delta_{ij}+u^2 y_i \end{bmatrix} = \det\begin{bmatrix} 1 & -u^2 \\ y_i & u\delta_{ij} \end{bmatrix}$$

$$= u^n + u^2 \cdot u^{n-1} \cdot \sum_{i=1}^n y_i = u^{n+1}.$$

The limits of integration for y_i are $0 \leqslant y_i, \sum_{i=1}^{n} y_i \leqslant 1$. Hence

$$\int_0^\infty \cdots \int_0^\infty x_1 \cdots x_m \cdot |\Delta(x)|^{2\gamma} \left(1 + \sum_{i=1}^{n} x_i\right)^{-\beta} \prod_{j=1}^{n} x_j^{\alpha-1}\, dx_j$$

$$= \int \cdots \int y_1 \cdots y_m \cdot |\Delta(y)|^{2\gamma} \left(1 - \sum_{i=1}^{n} y_i\right)^{\beta-m-\alpha n-\gamma n(n-1)-1} \prod_{j=1}^{n} y_j^{\alpha-1}\, dy_j,$$

where the integral is taken over $0 \leqslant y_i$, $y_1 + \cdots + y_n \leqslant 1$. Or redefining β, we get

$$\int \cdots \int y_1 \cdots y_m |\Delta(y)|^{2\gamma} \left(1 - \sum_{i=1}^{n} y_i\right)^{\beta-1} \cdot \prod_{j=1}^{n} y_j^{\alpha-1}\, dy_j$$

$$= \frac{\Gamma(\beta)}{\Gamma(\beta + m + \alpha n + \gamma n(n-1))} \prod_{i=1}^{m} (\alpha + \gamma(n-i))$$

$$\times \prod_{j=1}^{n} \frac{\Gamma(\alpha + \gamma(n-j))\Gamma(1+\gamma j)}{\Gamma(1+\gamma)}. \quad \square \tag{17.10.3}$$

17.11 Some Related Difficult Integrals

Following Macdonald one may replace $\Delta(x)$ in (17.1.2) by the polynomial $P(x)$ of Section 17.9; no evaluation or guess is known for

$$\int_0^1 \cdots \int_0^1 |P(x)|^{2\gamma} \prod_{i=1}^{n} x_i^{\alpha-1} (1-x_i)^{\beta-1}\, dx_i. \tag{17.11.1}$$

As other possible extensions consider the integrals

$$A(n, s, \gamma) = \pi^{-ns/2} \int \cdots \int |\Delta(\vec{r})|^{2\gamma} \exp\left(\sum_{i=1}^{n} [-\vec{r}_i^{\,2}]\right) \prod_{i=1}^{n} d\vec{r}_i, \tag{17.11.2}$$

$$B(n, s, \gamma) = \pi^{-ns/2} \prod_{j=1}^{n} \Gamma(j + s/2) \int \cdots \int_{|\vec{r}_i| \leqslant 1} |\Delta(\vec{r})|^{2\gamma}\, d\vec{r}_1 \cdots d\vec{r}_n, \tag{17.11.3}$$

with

$$\Delta(\vec{r}) = \prod_{1 \leqslant i < j \leqslant n} |\vec{r}_i - \vec{r}_j|, \tag{17.11.4}$$

where $\vec{r}_1, \ldots, \vec{r}_n$ are vectors in the s-dimensional Euclidean space $\mathcal{R}^s$; they vary over all the space $\mathcal{R}^s$ in (17.11.2) and inside the unit sphere $|\vec{r}_i| \leqslant 1$, $i = 1, \ldots, n$, in (17.11.3). Or the integral

$$C(n,s,\gamma) = 2^{-n(n+1)/2}\pi^{-n(s+1)/2}\left(\Gamma\left(\frac{s+1}{2}\right)\right)^n \int \cdots \int_{|\vec{r}_i|=1} |\Delta(\vec{r})|^{2\gamma}\, d\Omega_1 \cdots d\Omega_n, \tag{17.11.5}$$

where $\vec{r}_1, \ldots, \vec{r}_n$ are vectors in $\mathcal{R}^{s+1}$ and they vary over the s-dimensional surface of the sphere $|\vec{r}_i| = 1$, $i = 1, \ldots, n$, in $\mathcal{R}^{s+1}$. Note that here s is the dimension of the surface of the unit sphere in $s+1$ dimensions.

The case $s = 1$ of these integrals was considered up to now (cf. Eqs. (17.6.7), (17.6.1) and (11.2.1)). For $s = 2$ and $\gamma = 1$, the integrals (17.11.2) and (17.11.3) can be dealt with as in Chapter 15 with the result

$$A(n,2,1) = B(n,2,1) = \prod_{j=1}^{n} j!. \tag{17.11.6}$$

The integral (17.11.5) for $s = 2, \gamma = 1$ was evaluated by Caillol (1981) using a similar method, as explained below.

In terms of the Cayley–Klein parameters

$$\alpha_j = \cos(\theta_j/2)\exp(i\varphi_j/2), \qquad \beta_j = \sin(\theta_j/2)\exp(-i\varphi_j/2),$$

where θ_j, φ_j are the polar angles of $\vec{r}_j$, one has

$$|\vec{r}_j - \vec{r}_k|^2 = 4|\alpha_j\beta_k - \alpha_k\beta_j|^2 = 4|\beta_j\beta_k|^2\left|\frac{\alpha_j}{\beta_j} - \frac{\alpha_k}{\beta_k}\right|^2,$$

so that

$$|\Delta(\vec{r})|^2 = 2^{n(n-1)}\prod_{j=1}^{n}|\beta_j|^{2(n-1)}\prod_{1\leqslant j<k\leqslant n}\left|\frac{\alpha_j}{\beta_j} - \frac{\alpha_k}{\beta_k}\right|^2.$$

Also

$$\int d\Omega\left(\frac{\alpha}{\beta}\right)^j\left(\frac{\alpha}{\beta}\right)^{*k}|\beta|^{2(n-1)}$$
$$= \int_0^{2\pi} d\varphi \int_0^{\pi} d\theta\,\sin\theta\cdot(\cos(\theta/2))^{j+k}(\sin(\theta/2))^{2(n-1)-j-k}\exp(i(j-k)\varphi)$$
$$= 2\pi\cdot\delta_{jk}\int_0^{\pi} d\theta\,\sin\theta\cdot(\cos(\theta/2))^{2j}(\sin(\theta/2))^{2(n-1-j)}$$

$$= 4\pi \frac{j!(n-j-1)!}{n!} \delta_{jk}. \tag{17.11.7}$$

Therefore, if one writes

$$K_n(\vec{r}_j, \vec{r}_k) = \sum_{\ell=0}^{n-1} \frac{n!}{4\pi \cdot \ell!(n-\ell-1)!} \left(\frac{\alpha_j}{\beta_j}\right)^{\ell} \left(\frac{\alpha_k}{\beta_k}\right)^{*\ell} (\beta_j \beta_k^*)^{n-1}, \tag{17.11.8}$$

then

$$|\Delta(\vec{r})|^2 = 2^{n(n-1)} (4\pi)^n \prod_{\ell=0}^{n-1} \frac{\ell!(n-\ell-1)!}{n!} \det[K_n(\vec{r}_j, \vec{r}_k)]_{j,k=1,\ldots,n},$$

$$= 2^{n(n-1)} (4\pi)^n \prod_{\ell=0}^{n-1} \frac{(\ell!)^2}{n!} \det[K_n(\vec{r}_j, \vec{r}_k)]_{j,k=1,\ldots,n},$$

$$\int K_n(\vec{r}_j, \vec{r}_j)\, d\Omega_j = n,$$

$$\int K_n(\vec{r}_j, \vec{r}_k) K_n(\vec{r}_k, \vec{r}_p)\, d\Omega_k = K_n(\vec{r}_j, \vec{r}_p).$$

So that from Theorem 5.1.4,

$$\int \cdots \int_{|\vec{r}_j|=1} |\Delta(\vec{r})|^2 d\Omega_1 \cdots d\Omega_n = 2^{n(n-1)} n! (4\pi)^n \prod_{\ell=0}^{n-1} \frac{(\ell!)^2}{n!}. \tag{17.11.9}$$

If one writes

$$P(\vec{r}_1, \ldots, \vec{r}_n) = \frac{1}{n!} \det[K_n(\vec{r}_j, \vec{r}_k)]_{j,k=1,\ldots,n},$$

then this P is normalized and can be interpreted as a probability density. The m-point correlation function, from Theorem 5.1.4, is

$$\frac{n!}{(n-m)!} \int \cdots \int P(\vec{r}_1, \ldots, \vec{r}_n)\, d\Omega_{m+1} \cdots d\Omega_n = \det[K_n(\vec{r}_j, \vec{r}_k)]_{j,k=1,\ldots,m}. \tag{17.11.10}$$

Still another integral of interest is the average value of the product of traces of the matrix in the circular ensembles. For example,

$$I(k, n, \gamma) = \frac{1}{(2\pi)^n} \frac{(\gamma!)^n}{(n\gamma)!} \int_0^{2\pi} \cdots \int_0^{2\pi} \left| \sum_{j=1}^{n} e^{i\theta_j} \right|^{2k} \prod_{1 \leqslant j < \ell \leqslant n} |e^{i\theta_j} - e^{i\theta_\ell}|^{2\gamma} \prod_{j=1}^{n} d\theta_j. \tag{17.11.11}$$

It is known (cf. Chapter 25) that $I(k,n,1)$ gives the number of permutations of $(1,\ldots,k)$ in which the length of the longest increasing subsequence is less than or equal to n. One has in particular,

$$I(k,n,1)=k!,\quad 0\leqslant k\leqslant n. \tag{17.11.12}$$

Integrals (17.11.2), (17.11.3) and (17.11.5) for $s>2$ or $s=2,\gamma>1$ are not known. Of course, one can compute them with some effort for $\gamma=1$, any s and small values of n, say, $n=2,3$, or 4. For example,

$$A(1,s,1)=B(1,s,1)=C(1,s,1)=C(2,s,1)=1; \tag{17.11.13}$$

$$A(2,s,1)=B(2,s,1)=s;$$

$$C(3,s,1)=s(s+2)(s+1)^{-2}; \tag{17.11.14}$$

$$A(3,s,1)=s(s+2)(2s-1)/2;$$

$$A(4,s,1)=s(s+2)^2(2s^3+4s^2-9s+4)/2; \tag{17.11.15}$$

$$B(3,s,1)=s(s^2+2s-2);$$

$$C(4,s,1)=s(s^2+3s-1)(s+1)^{-3}; \tag{17.11.16}$$

etc.

For applications in polymer theory one would like to know the value of

$$\int\cdots\int|\Delta(\vec{r})|^{2\gamma}\exp\left[-\sum_{j=1}^{n}(\vec{r}_{j-1}-\vec{r}_j)^2\right]\prod_{j=1}^{n}|\vec{r}_j|^{2\gamma}\,d\vec{r}_j, \tag{17.11.17}$$

with $\vec{r}_0=0$, specially in $s=3$ dimensions. But this has a different structure.

It is curious to note that with a constant vector $\vec{a}$ say of unit length, the integral

$$I(\alpha_0,\alpha_1,\ldots,\alpha_n)=\int\cdots\int|\vec{a}-\vec{r}_1-\cdots-\vec{r}_n|^{-2\alpha_0}\prod_{j=1}^{n}|\vec{r}_j|^{-2\alpha_j}d\vec{r}_j, \tag{17.11.18}$$

can be evaluated without much difficulty, viz.,

$$I(\alpha_0,\ldots,\alpha_n)=\pi^{ns/2}\frac{\Gamma(\alpha_0+\cdots+\alpha_n-ns/2)}{\Gamma((n+1)s/2-\alpha_0-\cdots-\alpha_n)}\prod_{j=0}^{n}\frac{\Gamma(s/2-\alpha_j)}{\Gamma(\alpha_j)}, \tag{17.11.19}$$

valid for values of α_j such that the arguments of all the gamma functions have a positive real part.

Proof (*Selberg*). Taking $\vec{r}_1 + \vec{r}_2$ instead of $\vec{r}_2$ as the new variable,

$$I(\alpha_0, \ldots, \alpha_n) = \int \cdots \int |\vec{a} - \vec{r}_2 - \cdots - \vec{r}_n|^{-2\alpha_0} |\vec{r}_1|^{-2\alpha_1} |\vec{r}_2 - \vec{r}_1|^{-2\alpha_2} \ldots$$

$$|\vec{r}_n|^{-2\alpha_n} \, d\vec{r}_1 \cdots d\vec{r}_n. \tag{17.11.20}$$

Setting here $\vec{r}_1 = |\vec{r}_2| \cdot \vec{r}_1'$, we get

$$I(\alpha_0, \ldots, \alpha_n) = I(\alpha_2, \alpha_1) \cdot I(\alpha_0, \alpha_1 + \alpha_2 - s/2, \alpha_3, \ldots, \alpha_n). \tag{17.11.21}$$

Repeating the process we can express $I(\alpha_0, \ldots, \alpha_n)$ in terms of $I(\beta_1, \beta_2)$.

For $I(\alpha_0, \alpha_1)$, writing

$$|\vec{r}|^{-2\alpha} = (\Gamma(\alpha))^{-1} \int_0^\infty t^{\alpha-1} \exp(-t|\vec{r}|^2) \, dt, \tag{17.11.22}$$

one has

$$I(\alpha_0, \alpha_1) = \int |\vec{r}|^{-2\alpha_1} |\vec{a} - \vec{r}|^{-2\alpha_0} \, d\vec{r}$$

$$= \frac{1}{\Gamma(\alpha_0)\Gamma(\alpha_1)} \int_0^\infty dt_0 \int_0^\infty dt_1 \cdot t_0^{\alpha_0-1} t_1^{\alpha_1-1} \int d\vec{r} \, \exp\left[-t_0|\vec{a} - \vec{r}|^2 - t_1|\vec{r}|^2\right]. \tag{17.11.23}$$

The Gaussian integral over $\vec{r}$ gives

$$\pi^{s/2}(t_0 + t_1)^{-s/2} \exp(-t_0 t_1/(t_0 + t_1)). \tag{17.11.24}$$

Now make a change of variables

$$t_0 = uv, \quad t_1 = (1-u)v, \quad \frac{\partial(t_0, t_1)}{\partial(u, v)} = v, \tag{17.11.25}$$

so that

$$\Gamma(\alpha_0)\Gamma(\alpha_1) I(\alpha_0, \alpha_1)$$

$$= \pi^{s/2} \int_0^1 du \int_0^\infty dv \cdot u^{\alpha_0-1}(1-u)^{\alpha_1-1} v^{\alpha_0+\alpha_1-s/2-1} \exp(-vu(1-u)). \tag{17.11.26}$$

Integration over v leaves us with a beta integral.

Collecting the results we get (17.11.19). □

Summary to Chapter 17

The main object of this chapter is to present the evaluation of certain definite multiple integrals. These integrals were encountered in different contexts by various investigators and the discovery of their similarity was a pleasant surprise. The most significant of them is Eq. (17.1.3) or its slight generalizations Eqs. (17.1.4), (17.1.6) and (17.1.7). The other important related integrals or consequences being Eqs. (17.6.1) to (17.6.7), (17.7.1), (17.8.5) to (17.8.10), (17.9.1) and (17.10.1).

It is desirable to know the integrals (17.11.1)–(17.11.3) and (17.11.11) in particular.

18

ASYMPTOTIC BEHAVIOUR OF $E_\beta(0, s)$ BY INVERSE SCATTERING

In Chapters 6, 7 and 11 we derived expressions for the probability $E_\beta(0, s)$ that an arbitrary interval of length s contains none of the eigenvalues of a random matrix chosen from the unitary ($\beta = 2$), orthogonal ($\beta = 1$), or symplectic ($\beta = 4$) ensemble. They were expressed in terms of the infinite products $F_+(t) = \prod_{j=0}^{\infty}(1 - \lambda_{2j}(t))$, and $F_-(t) = \prod_{j=0}^{\infty}(1 - \lambda_{2j+1}(t))$, $s = 2t$, where $\lambda_j(t)$ are the eigenvalues of a certain integral equation. In other words, $F_\pm(t)$ are the Fredholm determinants of certain kernels over the interval $(-1, 1)$, or with a change of variables, of other kernels over the interval $(0, t)$. The eigenfunctions of the integral equation in question also satisfy a certain second order linear differential equation depending on the parameter t. They are known as the prolate spheroidal functions in the mathematical literature. Their asymptotic behaviour for large t can be ascertained from the differential equation and from this knowledge one can derive the asymptotic behaviour of $\lambda_j(t)$ and hence of $F_\pm(t)$ for large t.

Another way to get the asymptotic behaviour of $F_\pm(t)$ is to find the same for the Toeplitz determinant (cf. Eq. (11.2.3))

$$F_N(\alpha) = \det[C_{j-k}]_{j,k=1,\ldots,N}, \tag{18.0.1}$$

with

$$C_j \equiv C_j(\alpha) = \frac{1}{2\pi}\int_\alpha^{2\pi-\alpha} \exp(ij\theta)\, d\theta, \tag{18.0.2}$$

$$F_+(t)F_-(t) = \lim_{N\to\infty} F_N\left(\frac{2\pi t}{N}\right). \tag{18.0.3}$$

The behaviour of $F_+(t)$ or $F_-(t)$ can be deduced from that of their product as we will see.

A third way to find the asymptotic behaviour of $F_\pm(t)$ is to express them in terms of other Fredholm determinants $\Delta_\pm(t)$ defined over (t,∞). To construct these new convenient kernels over (t,∞), Dyson used the mathematical methods of the inverse scattering theory. The quantum theory of scattering deals with finding the phase shifts starting from a given potential, while the problem of inverse scattering consists in constructing the potential from the phase shifts. These problems can be solved by studying the wave functions either near the origin or near infinity and as Dyson noticed, involves certain Fredholm determinants. From the knowledge of $F_\pm(t)$ near $t \approx 0$, its dominant behaviour for $t \to \infty$ and from the considerations of inverse scattering theory near the origin one finds a potential corresponding to the Fredholm determinants $F_\pm(t)$. For this potential one then constructs from the considerations near infinity other Fredholm determinants $\Delta_\pm(t)$ defined this time over (t,∞). Actually near the origin or at infinity one considers always a pair of potentials, one of them being the object of study while the other "comparison" potential is simple enough so that the corresponding solutions are either known or easily calculable. One can expand these new Fredholm determinants $\Delta_\pm(t)$ in inverse powers of t.

As a result of the above three methods one finally ends up with an asymptotic series for $\log F_\pm(t)$, whose successive terms can in principle be computed with more and more effort. This is how the first few terms of the asymptotic series of $F_\pm(t)$ were initially obtained and we will describe it in this chapter for historical and pedagogical reasons.

Still another, and by far the fastest, method to get the asymptotic behavior of $F_\pm(t)$ is to use the second order non-linear differential equations satisfied by them. This method also gives the power series expansions of $F_\pm(t)$ near $t=0$. We will take it up later in Chapter 21.

18.1 Asymptotics of $\lambda_n(t)$

From Eqs. (6.3.10), (7.3.11) and (11.7.5) we have

$$E_2(0,s) = F(s) \equiv F_+(t)F_-(t), \qquad E_1(0,s) = F_+(t), \tag{18.1.1}$$

and

$$E_4(0,t) = \frac{1}{2}\big(F_+(t) + F_-(t)\big), \tag{18.1.2}$$

with

$$F_+(t)=\prod_{n=0}^{\infty}\big(1-\lambda_{2n}(t)\big),\qquad F_-(t)=\prod_{n=0}^{\infty}\big(1-\lambda_{2n+1}(t)\big),\tag{18.1.3}$$

$s=2t$, and where λ_n are the eigenvalues of the integral equation

$$\lambda f(x)=\int_{-1}^{1}\frac{\sin(x-y)\pi t}{(x-y)\pi}f(y)\,dy,\tag{18.1.4}$$

arranged in decreasing order of magnitude

$$1\geqslant\lambda_0\geqslant\lambda_1\geqslant\cdots\geqslant 0.\tag{18.1.5}$$

Though Eqs. (18.1.1)–(18.1.5) as they stand are complete, it is helpful to know that for certain discrete values of χ (depending on t, but independent of x), the solutions of the differential equation (6.3.22),

$$\frac{d}{dx}\big(1-x^2\big)\frac{df}{dx}+\big(\chi-\pi^2t^2x^2\big)f(x)=0\tag{18.1.6}$$

are regular at $x=\pm 1$ and then they are also the solutions of the integral equation (18.1.4). They are called prolate spheroidal functions, they have been studied extensively in the literature, they have been tabulated for t not too large, and their asymptotic behavior has been derived. The asymptotic formulas for $\lambda_n(t)$ with t large and any n are a little involved, since the order of n relative to that of t is arbitrary. They take a simpler form when n is finite and t is large. These formulas read as follows (cf. des Cloizeaux and Mehta, 1972).

First determine b_n by the implicit relation

$$\left(n+\frac{1}{2}\right)\frac{\pi}{2}=\pi t\int_0^{\min(1,\sqrt{\varepsilon})}\left(\frac{\varepsilon-y^2}{1-y^2}\right)^{1/2}dy+\eta(b),\tag{18.1.7}$$

where

$$\varepsilon=1-\frac{2b}{\pi t},\tag{18.1.8}$$

$$\eta(b)=\varphi(b)-\frac{b}{2}\left(\log\left|\frac{b}{2}\right|-1\right),\tag{18.1.9}$$

and $\varphi(b)$ is the phase of $\Gamma[\frac{1+ib}{2}]$,

$$\Gamma\left[\frac{1+ib}{2}\right]=\left(\frac{\pi}{\cosh(\pi b/2)}\right)^{1/2}e^{i\varphi(b)}.\tag{18.1.10}$$

Then

$$1-\lambda_n = \left(\frac{2}{\pi}\right)^{1/2}\left(\frac{u}{2e}\right)^{u/2}\left[\Gamma\left(\frac{u+1}{2}\right)\right]^{-1} e^{-2\pi t\delta(\varepsilon)}\left[1+e^{\pi b}\right]^{-1}, \tag{18.1.11}$$

where

$$u = \pi t\varepsilon = \pi t - 2b, \tag{18.1.12}$$

and the function $\delta(\varepsilon)$ is defined by

$$\begin{aligned} \delta(\varepsilon) &= (1-\varepsilon)\left(\int_0^{\pi/2} \frac{\sin^2\alpha}{(\sin^2\alpha+\varepsilon\cos^2\alpha)^{1/2}}\,d\alpha - \frac{\pi}{4}\right), \quad \text{if } \varepsilon \leqslant 1, \\ \delta(\varepsilon) &= 0, \quad \text{if } \varepsilon \geqslant 1. \end{aligned} \tag{18.1.13}$$

When n is finite and $t \gg 1$, these formulas take the simpler form (cf. Slepian, 1965)

$$1-\lambda_n(t) = \frac{\pi^{n+1}}{n!} 2^{3n+2} t^{n+1/2} \exp(-2\pi t)\left\{1+O(t^{-1})\right\}, \tag{18.1.14}$$

so that

$$\frac{(1-\lambda_{2n}(t))^{1/4}(1-\lambda_{2n+2}(t))^{3/4}}{(1-\lambda_{2n+1}(t))^{3/4}(1-\lambda_{2n+3}(t))^{1/4}} \approx \left(1-\frac{1}{(2n+2)^2}\right)^{1/4}. \tag{18.1.15}$$

Also when $n \approx t \gg 1$, $b_n \approx \pi(n-2t)/(2\log(2\sqrt{\pi t}))$, $1-\lambda_n \approx (1+e^{-\pi b_n})^{-1} \approx 1$, and relation (18.1.15) is again valid. Thus Eq. (18.1.15) can be taken to be correct for $t \gg 1$ and any n. Hence

$$\begin{aligned} \frac{F_+(t)}{F_-(t)} &= \frac{(1-\lambda_0)^{3/4}}{(1-\lambda_1)^{1/4}} \prod_{n=0}^{\infty} \frac{(1-\lambda_{2n})^{1/4}(1-\lambda_{2n+2})^{3/4}}{(1-\lambda_{2n+1})^{3/4}(1-\lambda_{2n+3})^{1/4}} \\ &\approx \frac{(1-\lambda_0)^{3/4}}{(1-\lambda_1)^{1/4}} \prod_{n=0}^{\infty}\left(1-\frac{1}{(2n+2)^2}\right)^{1/4}. \end{aligned} \tag{18.1.16}$$

From Eq. (18.1.14) and $\sin\theta = \theta\prod_{n=1}^{\infty}(1-\theta^2/(n^2\pi^2))$ for $\theta = \pi/2$, this gives

$$\frac{F_+(t)}{F_-(t)} \approx \left(4\pi\sqrt{t}\,e^{-2\pi t}\right)^{3/4}\left(32\pi^2 t^{3/2} e^{-2\pi t}\right)^{-1/4}(2/\pi)^{1/4} \approx \sqrt{2}\,e^{-\pi t}. \tag{18.1.17}$$

Thus we get a relation between the asymptotic behaviours of $F_+(t)$, $F_-(t)$ and their product $F(t) = F_+(t)F_-(t)$ for large t, namely

$$\log F_\pm(t) \approx \frac{1}{2}\log F(t) \mp \frac{\pi}{2}t. \tag{18.1.18}$$

Using Eqs. (18.1.7) to (18.1.13) and doing some analysis one can actually find the coefficients of t^2, t and $\log t$ in the asymptotic expansion of $\log F(t)$. The details of this tedious analysis will not be reproduced here (cf. des Cloizeaux and Mehta, 1973), as we will be finding these terms in the next section from a theorem of Widom.

18.2 Asymptotics of Toeplitz Determinants

A theorem of Szegö states that (Szegö, 1939)

Theorem 18.2.1. *If $f(\theta)$ is a positive function over $0 \leqslant \theta \leqslant 2\pi$, its derivative satisfies a Lipschitz condition and $F_N(f)$ is the $N \times N$ Toeplitz determinant*

$$F_N(f) = \det\left[\frac{1}{2\pi}\int_0^{2\pi} f(\theta)\exp\bigl(i(j-k)\theta\bigr)\,d\theta\right]_{j,k=1,\ldots,N}. \tag{18.2.1}$$

Then as $N \to \infty$

$$\log F_N(f) \approx Nf_0 + \frac{1}{4}\sum_{k=1}^{\infty} kf_k f_{-k} + \mathrm{O}(1), \tag{18.2.2}$$

where f_k are the Fourier coefficients of $\log f(\theta)$,

$$\log f(\theta) = \sum_{k=-\infty}^{\infty} f_k e^{ik\theta}. \tag{18.2.3}$$

Widom (1971) extended this theorem for functions $f(\theta)$ which are positive only on an arc of the unit circle.

Theorem 18.2.2. *Let $f(\theta) = 1$, if $\alpha \leqslant \theta \leqslant 2\pi - \alpha$, and $f(\theta) = 0$ if either $\theta < \alpha$ or $\theta > 2\pi - \alpha$. Then as $N \to \infty$*

$$\log F_N(f) \approx N^2 \log\cos\frac{\alpha}{2} - \frac{1}{4}\log\left(N\sin\frac{\alpha}{2}\right) + \frac{1}{12}\log 2 + 3\zeta'(-1) \tag{18.2.4}$$

where $\zeta'(z)$ is the derivative of the Riemann zeta function.

Taking the limit $\alpha N = 2\pi t \gg 1$, one has $(s = 2t)$,

$$\log F(t) \approx -\frac{1}{2}(\pi t)^2 - \frac{1}{4}\log(\pi t) + \frac{1}{12}\log 2 + 3\zeta'(-1). \tag{18.2.5}$$

This together with Eq. (18.1.17) gives

$$\log F_\pm(t) \approx -\frac{(\pi t)^2}{4} \mp \frac{\pi t}{2} - \frac{1}{8}\log(\pi t) + \left(\frac{1}{24} \pm \frac{1}{4}\right)\log 2 + \frac{3}{2}\zeta'(-1). \tag{18.2.6}$$

18.3 Fredholm Determinants and the Inverse Scattering Theory

With the new variables $\xi = x\pi t$, $\eta = y\pi t$, Eq. (18.1.4) takes the form

$$\lambda f\left(\frac{\xi}{\pi t}\right) = \frac{1}{\pi}\int_{-\pi t}^{\pi t} \frac{\sin(\xi - \eta)}{\xi - \eta} f\left(\frac{\eta}{\pi t}\right) d\eta.$$

And as $f(-x) = \pm f(x)$, the λ_j are the eigenvalues of the integral equations

$$\lambda_{2j} g(x) = \int_0^{\pi t} K_+(x, y) g(y)\, dy, \tag{18.3.1}$$

$$\lambda_{2j+1} g(x) = \int_0^{\pi t} K_-(x, y) g(y)\, dy, \tag{18.3.2}$$

with

$$K_\pm(x, y) = \frac{1}{\pi}\left(\frac{\sin(x - y)}{x - y} \pm \frac{\sin(x + y)}{x + y}\right). \tag{18.3.3}$$

Hence $F_\pm(t)$ are the Fredholm determinants of the kernels (18.3.3) or of

$$K_+(x, y) = \frac{2}{\pi}\int_0^1 \cos(kx)\cos(ky)\, dk, \tag{18.3.4}$$

$$K_-(x, y) = \frac{2}{\pi}\int_0^1 \sin(kx)\sin(ky)\, dk, \tag{18.3.5}$$

over the interval $(0, \tau)$, $F_\pm(t) = \det[1 - K_\pm]_0^\tau$, $\tau = \pi t$.

Fredholm determinants occur also in the theory of inverse scattering. In the quantum theory of scattering the potential is given and phase shifts are calculated. The problem of inverse scattering deals with reconstructing the potential given the phase shifts. Two methods are available for this purpose. The Gel'fand–Levitan method constructs the potential by working from $x = 0$ upwards, while Marchenko method does it by working downwards from $x = \infty$. The first method is related to a Fredholm determinant over the interval $(0, \tau)$, while the second involves that over the interval (τ, ∞). Both methods are equivalent and any one of them is sufficient to solve the inverse scattering problem.

Dyson took advantage of this equivalence and related the Fredholm determinants $F_\pm(t)$ defined over $(0, \pi t)$ to an inverse scattering problem applying the Gel'fand–Levitan method. He then applied Marchenko method to the same problem and expressed $F_\pm(t)$ in terms of other Fredholm determinants $\Delta_\pm(t)$ defined this time over $(\pi t, \infty)$. These later can be expanded in inverse powers of t.

As we are unable to do better than Dyson, we will reproduce his entire analysis. Sections 18.3–18.6 are taken almost verbatim from his paper "Fredholm determinants and inverse scattering problems" in *Comm. Math. Phys.* **47** (1976) 171–183.

In the problem of inverse scattering we are dealing with two potential $V_1(x)$, $V_2(x)$ on the half-line $0 < x < \infty$, V_1 being supposed unknown and V_2 known. We have two corresponding families of wave-functions $u_1(k, x)$, $u_2(k, x)$ satisfying the wave-equations

$$\left[\frac{d^2}{dx^2} + k^2 - V_j(x)\right] u_j(k, x) = 0, \quad j = 1, 2, \tag{18.3.6}$$

and generating spectral kernels K_1, K_2 defined by

$$\delta(x - y) - K_j(x, y) = \int u_j(k, x) u_j(k, y)\, dk. \tag{18.3.7}$$

In the Gel'fand–Levitan version (Gel'fand and Levitan, 1951) of inverse scattering theory, the wave-functions $u_j(k, x)$ are subject to boundary conditions at $x = 0$, namely

$$u_1(k, 0) = u_2(k, 0) = g q(k), \tag{18.3.8}$$

$$u_j'(k, 0) = h_j q(k), \tag{18.3.9}$$

where $q(k)$ is a given function and g, h_1, h_2 are given coefficients. The Fredholm determinants

$$F_j(\tau) = \det[1 - K_j]_0^\tau, \quad \tau = \pi t, \tag{18.3.10}$$

are defined on the finite interval $[0, \tau]$, and satisfy the conditions

$$V_1(\tau) - V_2(\tau) = 2 \frac{d^2}{d\tau^2} \log\left(\frac{F_1(\tau)}{F_2(\tau)}\right), \tag{18.3.11}$$

$$h_1 - h_2 = g \frac{d}{d\tau} \log\left(\frac{F_1(\tau)}{F_2(\tau)}\right)\Bigg|_{\tau=0} \tag{18.3.12}$$

(cf. Chadan and Sabatier, 1977).

In the Marchenko version (Marchenko, 1950) of the theory, the wave-functions $u_j(k, x)$ are required to become asymptotically equal at infinity, thus

$$u_1(k, x) - u_2(k, x) \to 0 \quad \text{as } x \to \infty, \tag{18.3.13}$$

uniformly in k. The potentials V_1, V_2 must approach each other closely enough so that the integral

$$\int \left|V_1(\tau) - V_2(\tau)\right| \tau \, d\tau, \tag{18.3.14}$$

converges at infinity. The Fredholm determinants

$$\Delta_j(\tau) = \det[1 - K_j]_\tau^\infty, \tag{18.3.15}$$

are defined on the infinite interval $[\tau, \infty]$, and

$$V_1(\tau) - V_2(\tau) = 2\frac{d^2}{d\tau^2} \log\left(\frac{\Delta_1(\tau)}{\Delta_2(\tau)}\right). \tag{18.3.16}$$

Equations (18.3.11) and (18.3.16) are important as they relate two sets of Fredholm determinants one defined on the finite interval $(0, \tau)$, and the other defined over the infinite interval (τ, ∞). What is crucial is the choice of the potentials. One of the potentials is imposed by the Fredholm determinant we want to study. The second comparison potential need not be the same near the origin and at large distances; it has to satisfy the boundary conditions (18.3.8) and (18.3.9) near the origin, and the convergence condition (18.3.14) at large distances. Also the choice should be simple enough so as to allow computation of the corresponding Fredholm determinants.

When the Gel'fand–Levitan and Marchenko formalisms are applied to the inverse scattering problem, it is customary to assume that the unknown wavefunctions $u_1(k, x)$ form a complete orthonormal set. Then Eq. (18.3.7) implies

$$K_1 = 0, \qquad F_1 = \Delta_1 = 1. \tag{18.3.17}$$

18.4 Application of the Gel'fand–Levitan Method

We apply the Gel'fand–Levitan formalism to the potentials

$$V_1(\tau) = W_\pm(\tau) = -2\frac{d^2}{d\tau^2} \log F_\pm(t) - 1, \quad \tau = \pi t, \tag{18.4.1}$$

with the comparison potential

$$V_2(\tau) = -1. \tag{18.4.2}$$

According to Eq. (18.2.6)

$$W_{\pm}(\tau) \sim -\frac{1}{4}\tau^{-2} \quad \text{as } \tau \to \infty. \tag{18.4.3}$$

We take $u_1(k, x)$ to be a complete orthonormal set of solutions of Eq. (18.3.6), so that Eq. (18.3.11) holds with

$$F_1(\tau) = 1, \qquad F_2(\tau) = F_{\pm}(\tau). \tag{18.4.4}$$

The wave-functions $u_2(k, x)$ must be cosines and sines of $((k^2 + 1)^{1/2}x)$ in order to satisfy Eq. (18.3.6). We have to normalize these wave-functions so that

$$K_2(x, y) = K_{\pm}(x, y), \tag{18.4.5}$$

with K_2 given by Eq. (18.3.7) and $K_{\pm}$ by Eq. (18.3.3). Thus we require

$$\int_0^\infty u_2(k, x)u_2(k, y)\, dk = \frac{2}{\pi}\int_1^\infty \begin{matrix}\cos\\ \sin\end{matrix} kx \begin{matrix}\cos\\ \sin\end{matrix} ky\, dk \tag{18.4.6}$$

which is satisfied by choosing

$$u_2(k, x) = s(k) \begin{matrix}\cos\\ \\ \sin\end{matrix} \left((k^2 + 1)^{1/2}x\right), \tag{18.4.7}$$

$$s(k) = \left(\frac{2}{\pi}\right)^{1/2}\left[\frac{k^2}{k^2 + 1}\right]^{1/4}. \tag{18.4.8}$$

We next have to determine the boundary conditions satisfied by $u_1(k, x)$ and $u_2(k, x)$ at $x = 0$. In the even case, when $V_1 = W_+$, Eqs. (18.3.8) and (18.3.9) hold with

$$q(k) = s(k), \quad g = 1, \quad h_2 = 0. \tag{18.4.9}$$

In this case Eq. (18.3.12) gives

$$h_1 = -\frac{d}{d\tau}\left[\log F_+(t)\right]_{\tau=0} = K_+(0, 0) = \frac{2}{\pi}, \quad \tau = \pi t. \tag{18.4.10}$$

Therefore the boundary conditions for u_1 are

$$u_1(k, 0) = s(k), \qquad u_1'(k, 0) = \frac{2}{\pi}\, s(k). \tag{18.4.11}$$

In the odd case, when $V_1 = W_-$, Eq. (18.3.12) gives

$$h_1 = h_2, \tag{18.4.12}$$

and the boundary conditions for u_1 are

$$u_1(k,0) = 0, \qquad u_1'(k,0) = (k^2+1)^{1/2}s(k). \tag{18.4.13}$$

We are now in a position to determine the behaviour of $u_1(k,x)$ as $x \to \infty$. Since the potential V_1 decreases according to Eq. (18.4.3) at infinity, and the $u_1(k,x)$ are an orthonormal system, we have

$$u_1(k,x) \sim \left(\frac{2}{\pi}\right)^{1/2} \begin{matrix}\cos\\ \sin\end{matrix}\big(kx + \eta(k)\big), \tag{18.4.14}$$

where the phase-shift $\eta(k)$ must be calculated separately for the even and odd cases. Following Jost (Jost, 1947), we define $J(k,x)$ to be the solution of Eq. (18.3.6) with potential V_1 and the asymptotic behaviour

$$J(k,x) \sim \exp(ikx) \quad \text{as } x \to \infty. \tag{18.4.15}$$

The functions

$$a(k) = J(k,0), \qquad b(k) = J'(k,0) \tag{18.4.16}$$

are the boundary values of a function analytic in the half-plane $(\operatorname{Im} k > 0)$, with the symmetry property

$$a(-k) = a^*(k), \qquad b(-k) = b^*(k), \quad k \text{ real}, \tag{18.4.17}$$

and the asymptotic behaviour

$$a(k) \sim 1, \quad b(k) \sim ik, \quad k \to \infty. \tag{18.4.18}$$

The Wronskian

$$J(-k,x)J'(k,x) - J'(-k,x)J(k,x) \tag{18.4.19}$$

is independent of x. Equating its value at $x = 0$ with its value at $x = \infty$, we find

$$a^*(k)b(k) - a(k)b^*(k) = 2ik. \tag{18.4.20}$$

The comparison of Eq. (18.4.14) with (18.4.15) implies

$$u_1(k,x) = \left(\frac{2}{\pi}\right)^{1/2} \begin{matrix}\mathrm{Re}\\ \mathrm{Im}\end{matrix}\left[J(k,x)\exp\bigl(i\eta(k)\bigr)\right]. \tag{18.4.21}$$

Consider first the even case. Then the boundary conditions (18.4.11) together with Eq. (18.4.21) imply

$$\mathrm{Re}\left[\exp\bigl(i\eta(k)\bigr)a(k)\right] = \left[\frac{k^2}{(k^2+1)}\right]^{1/4}. \tag{18.4.22}$$

$$\mathrm{Re}\left[\exp\bigl(i\eta(k)\bigr)\left(b(k) - \frac{2}{\pi}a(k)\right)\right] = 0. \tag{18.4.23}$$

The function

$$\varphi(k) = b(k) - \frac{2}{\pi}a(k) \tag{18.4.24}$$

is analytic in the upper half-plane and satisfies the same conditions (18.4.17), (18.4.18) as $b(k)$. According to Eq. (18.4.23)

$$\exp\bigl(i\eta(k)\bigr) = i\frac{\varphi^*(k)}{|\varphi(k)|}. \tag{18.4.25}$$

When we substitute Eq. (18.4.25) into (18.4.22) and make use of Eq. (18.4.20), we obtain

$$\left|\varphi(k)\right| = \bigl(k^2(k^2+1)\bigr)^{1/4}. \tag{18.4.26}$$

The only analytic function satisfying Eqs. (18.4.17), (18.4.18) and (18.4.26) is

$$\varphi(k) = \left[-k(k+i)\right]^{1/2}. \tag{18.4.27}$$

Putting Eq. (18.4.27) back into (18.4.25), we find

$$\exp\bigl(i\eta(k)\bigr) = \left[\frac{k-i}{k+i}\right]^{1/4}, \tag{18.4.28}$$

and so the phase-shift is given by

$$k\tan 2\eta(k) = -1. \tag{18.4.29}$$

In the odd case, the boundary conditions (18.4.13) imply

$$\mathrm{Im}\big[\exp\big(i\eta(k)\big)a(k)\big] = 0, \tag{18.4.30}$$

$$\mathrm{Im}\big[\exp\big(i\eta(k)\big)b(k)\big] = \big[k^2(k^2+1)\big]^{1/4}. \tag{18.4.31}$$

Equation (18.4.30) implies

$$\exp\big(i\eta(k)\big) = \frac{a^*(k)}{|a(k)|}, \tag{18.4.32}$$

and this with Eqs. (18.4.31) and (18.4.20) gives

$$\big|a(k)\big| = \left(\frac{k^2}{k^2+1}\right)^{1/4}. \tag{18.4.33}$$

The analytic function $a(k)$ is then

$$a(k) = \left(\frac{k}{k+i}\right)^{1/2}, \tag{18.4.34}$$

and Eq. (18.4.32) gives $\exp(i\eta(k)) = (\frac{k+i}{k-i})^{1/4}$, or

$$k\tan 2\eta(k) = +1. \tag{18.4.35}$$

Both cases are included in the formula

$$\eta(k) = \mp\frac{1}{2}\arctan(k^{-1}). \tag{18.4.36}$$

The potentials $W_\pm(\tau)$ are uniquely determined (Levinson, 1949) by the property that they give scattering states (18.4.14) with the phase-shifts (18.4.36), and no bound states.

It is a curious fact that the identity (6.4.41) can be written in the form

$$\left[\frac{d^2}{d\tau^2} - 1 - W_-(\tau)\right]u_B(\tau) = 0, \tag{18.4.37}$$

with

$$u_B(\tau) = \frac{F_+(t)}{F_-(t)} \sim 2^{1/2}e^{-\tau} \quad \text{as } \tau \to \infty. \tag{18.4.38}$$

Thus $u_B(\tau)$ is an acceptable bound-state wave-function with energy -1 in the potential $W_-(\tau)$. However u_B does not satisfy the boundary condition (18.4.13) for the odd case and is therefore irrelevant to the determination of W_-. Another alternative form of Eq. (6.4.41) is

$$\left[\frac{d^2}{d\tau^2} - 1 - W_+(\tau)\right]\left(u_B(\tau)\right)^{-1} = 0. \tag{18.4.39}$$

The wave-function $(u_B)^{-1}$ satisfies the correct boundary condition (18.4.11) for the even case at $\tau = 0$, but fails to converge at infinity, and is therefore also irrelevant to the determination of W_+.

18.5 Application of the Marchenko Method

We apply the Marchenko formalism to the potential $V_1(\tau)$ defined by Eq. (18.4.1). What should we choose for the comparison potential $V_2(\tau)$? It is possible to carry through the analysis with $V_2(\tau) = 0$. The calculations are then formally simple, but the integral (18.3.14) diverges in view of Eq. (18.4.3), and the results obtained are poorly convergent and of doubtful utility. The next most simple choice would be

$$V_2(\tau) = -\frac{1}{4}\tau^{-2}. \tag{18.5.1}$$

This makes the integral (18.3.14) converge. The wave-functions $u_2(k, x)$ are now Bessel functions of the quantity (kx), and the phase-shift produced by the potential $V_2(\tau)$ is $(\mp\pi/4)$ independent of k. The fact that this phase-shift agrees with the phase-shift (18.4.36) at $k = 0$ reflects the fact that V_1 and V_2 have the same behaviour at infinity. But the same argument carried one step further suggests a far better choice for V_2. The phase-shift (18.4.36) behaves like

$$\eta(k) \sim \mp\left(\frac{1}{4}\pi - \frac{1}{2}k\right), \tag{18.5.2}$$

with an error of order k^3 for small k. A Bessel function of the quantity $(k(x \mp 1/2))$ gives the phase-shift (18.5.2) for all k. We therefore choose for V_2 the potential

$$V_2(\tau) = -\frac{1}{4}\left(\tau \pm \frac{1}{2}\right)^{-2}, \tag{18.5.3}$$

with the expectation that this will make the difference $(V_1 - V_2)$ decrease much more rapidly as $\tau \to \infty$. The results of the calculation justify our expectation. The singularity of $V_2(\tau)$ at $\tau = 1/2$ (in the odd case) means that the analysis is valid only for $\tau > 1/2$. This is not a serious limitation, since our purpose is to study the behaviour of the potentials for large τ.

The wave-functions $u_2(k,x)$ are solutions of the equation

$$\left[\frac{d^2}{dx^2}+k^2+\frac{1}{4}\left(x\pm\frac{1}{2}\right)^{-2}\right]u_2(k,x)=0, \tag{18.5.4}$$

with asymptotic behaviour determined by Eq. (18.3.13). It is convenient to use the notations

$$j(z)=z^{1/2}J_0(z), \qquad y(z)=z^{1/2}Y_0(z), \tag{18.5.5}$$

$$h(z)=z^{1/2}H_0^1(z), \qquad k(z)=z^{1/2}K_0(z), \tag{18.5.6}$$

for the Bessel functions. We take for the wave-functions $u_1(k,x)$ the same complete orthonormal set that we studied in Section 18.4, with asymptotic behaviour given by Eqs. (18.4.14) and (18.4.36). Then Eq. (18.3.13) implies

$$u_2(k,x)=\alpha(k)j\left(k\left(x\pm\frac{1}{2}\right)\right)\pm\beta(k)y\left(k\left(x\pm\frac{1}{2}\right)\right), \tag{18.5.7}$$

with

$$\alpha(k)=\cos\left(\frac{\theta(k)}{2}\right), \qquad \beta(k)=\sin\left(\frac{\theta(k)}{2}\right), \tag{18.5.8}$$

$$\theta(k)=(k-\arctan k). \tag{18.5.9}$$

The Marchenko formula (18.3.16) becomes

$$V_1(\tau)=W_\pm(\tau)=-\frac{1}{4}\left(\tau\pm\frac{1}{2}\right)^{-2}-2\frac{d^2}{d\tau^2}\log\Delta_\pm(t), \tag{18.5.10}$$

$$\Delta_\pm(\tau)=\det[1-G_\pm]_\tau^\infty, \quad \tau=\pi t, \tag{18.5.11}$$

with the kernels $G_\pm$ defined by

$$G_\pm(x,y)=\delta(x-y)-\int_0^\infty u_2(k,x)u_2(k,y)\,dk. \tag{18.5.12}$$

Using the completeness relation

$$\delta(x-y)=\int_0^\infty j\left(k\left(x\pm\frac{1}{2}\right)\right)j\left(k\left(y\pm\frac{1}{2}\right)\right)dk. \tag{18.5.13}$$

we bring Eq. (18.5.12) to the form

$$G_{\pm}(x,y) = \frac{1}{2}\int_0^{\infty} \mathrm{Re}\left[\left(1-\exp(\mp i\theta(k))\right)h\left(k\left(x\pm\frac{1}{2}\right)\right)h\left(k\left(y\pm\frac{1}{2}\right)\right)\right]dk$$
$$= \frac{1}{4}\left[\int_0^{\infty} - \int_{-\infty}^{0}\right]\left(1-\exp(\mp i\theta(k))\right)h\left(k\left(x\pm\frac{1}{2}\right)\right)h\left(k\left(y\pm\frac{1}{2}\right)\right)dk. \tag{18.5.14}$$

The function

$$\exp(\mp i\theta(k)) = \exp(\mp ik)(1\pm ik)(k^2+1)^{-1/2} \tag{18.5.15}$$

is analytic in the upper half-plane with a cut from $(+i)$ to $(+i\infty)$. In the even case, the exponential growth of Eq. (18.5.15) is compensated by the exponential decrease of the Hankel functions in Eq. (18.5.14) for all positive x and y. In the odd case Eq. (18.5.15) decreases exponentially in the upper half-plane, but we must require $x, y > 1/2$ so that the term in Eq. (18.5.14) not involving (18.5.15) has exponential decrease. With this proviso, we may move the path of integration in both parts of Eq. (18.5.14) to the positive imaginary axis by writing $k = iz$. The terms involving Eq. (18.5.15) cancel along the cut, and we are left with

$$G_{\pm}(x,y) = \frac{2}{\pi^2}\left[\int_0^{\infty} dz - \int_0^{1} \exp(\pm\varphi(z))\,dz\right]k\left(z\left(x\pm\frac{1}{2}\right)\right)k\left(z\left(y\pm\frac{1}{2}\right)\right), \tag{18.5.16}$$

$$\varphi(z) = z - \operatorname{arc\,tanh} z. \tag{18.5.17}$$

The series expansion

$$\log\Delta_{\pm}(\tau) = \mathrm{tr}\left[\log(1-G_{\pm})\right]_{\tau}^{\infty} = -\sum_{1}^{\infty} n^{-1}\,\mathrm{tr}\left[(G_{\pm})^n\right]_{\tau}^{\infty} \tag{18.5.18}$$

converges absolutely for all positive τ in the even case, and at least for

$$\tau > \frac{1}{2} + (4\pi)^{-1} \tag{18.5.19}$$

in the odd case. The formula (18.5.10) with (18.5.16) and (18.5.18) provides a practical method for computing the potentials $W_{\pm}(\tau)$, either by using the series expansion or by finding numerically the eigenvalues of the kernels $G_{\pm}$.

The relations (18.4.1) and (18.5.10), connecting $W_\pm(\tau)$ with $F_\pm(t)$ and $\Delta_\pm(\tau)$, have the consequence that the quantity

$$\log F_\pm(t) + \frac{1}{4}\tau^2 + \frac{1}{8}\log\left|\tau \pm \frac{1}{2}\right| - \log\Delta_\pm(\tau), \quad \tau = \pi t, \tag{18.5.20}$$

is a linear function of t. But we know that as $t \to \infty$ the behaviour of $\log F_\pm(t)$ is governed by Eq. (18.2.6), while $\log\Delta_\pm(\tau)$ tends to zero. The asymptotic formula (18.2.6) can therefore be replaced by the identity

$$\log F_\pm(t) = -\frac{1}{4}\tau^2 \mp \frac{1}{2}\tau - \frac{1}{8}\log\left|\tau \pm \frac{1}{2}\right| \pm \frac{1}{4}\log 2 + \alpha + \log\Delta_\pm(\tau). \tag{18.5.21}$$

with

$$\alpha = \frac{1}{24}\log 2 + \frac{3}{2}\zeta'(-1), \tag{18.5.22}$$

This identity establishes the desired connection between the Fredholm determinants defined on $[0, \tau]$ and those defined on $[\tau, \infty]$.

18.6 Asymptotic Expansions

We wish to obtain the asymptotic expansion of $\log\Delta_\pm(\tau)$ in negative powers of $(\tau \pm 1/2)$. We can then use Eq. (18.5.21) to obtain the extension to negative powers of the asymptotic formula (18.2.6) for $\log F_\pm(t)$. The expansion of $\log\Delta_\pm(\tau)$ is derived from Eq. (18.5.18), the nth term of the sum giving contributions of order τ^{-3n} and higher. We expand the integrand of Eq. (18.5.16) in powers of z, so that

$$G_\pm(x, y) = \pm\frac{1}{3}G_3 \pm \frac{1}{5}G_5 - \frac{1}{18}G_6 \pm \frac{1}{7}G_7 + \cdots, \tag{18.6.1}$$

with the term

$$G_m = \frac{2}{\pi^2}\int_0^\infty z^m k\left(z\left(x \pm \frac{1}{2}\right)\right) k\left(z\left(y \pm \frac{1}{2}\right)\right) dz, \tag{18.6.2}$$

homogeneous of degree $(-m-1)$ in $(x \pm 1/2)$ and $(y \pm 1/2)$. The error involved in extending the integral from 1 to ∞ is of order $e^{-\tau}$ and is negligible in an asymptotic expansion. When the trace of $(G_\pm)^n$ is calculated, we obtain terms proportional to $(\tau \pm 1/2)^{-N}$ by forming products

$$G_{m_1}G_{m_2}\cdots G_{m_n}, \tag{18.6.3}$$

with

$$m_1 + m_2 + \cdots + m_n = N. \tag{18.6.4}$$

In Eq. (18.6.1) the odd terms carry the $\pm$ sign while the even terms do not. Equation (18.6.4) implies that terms with even N appear with the same coefficients in $\log \Delta_+$ and $\log \Delta_-$, while terms with odd N have equal and opposite coefficients. Therefore

$$\log \Delta_\pm(\tau) = \sum (\pm 1)^m a_m \left(\tau \pm \frac{1}{2}\right)^{-m}. \tag{18.6.5}$$

Moreover

$$a_1 = a_2 = a_4 = 0, \tag{18.6.6}$$

because no term of these orders can appear in any trace of $[G_\pm]^n$ according to Eq. (18.6.1). The formal relation

$$\log \Delta_\pm(-\tau) = \log \Delta_\mp(\tau) \tag{18.6.7}$$

holds for the asymptotic expansions (18.6.5), but cannot hold as an identity. On the contrary, if $\Delta_\pm(-\tau)$ is defined by Eq. (18.5.21), using the convention

$$F_\pm(-t) = \det[1 + K_\pm]_0^\tau, \tag{18.6.8}$$

which is the correct analytic continuation of Eq. (18.3.10), then Eq. (18.6.7) is definitely false. There is nothing paradoxical in this failure of Eq. (18.6.7), since the expansion (18.6.5) is not convergent.

The coefficients a_m in Eq. (18.6.5) can all be computed as traces of products of kernels (18.6.2). But we can find the first two non-vanishing coefficients a_3 and a_5 without calculating any integrals, by using the identity (6.4.42). It is convenient to work with the derivatives

$$L_\pm[\tau] = -\frac{d}{d\tau} \log F_\pm(t) = \frac{\tau}{2} \pm \frac{1}{2} + \frac{1}{8}\left(\tau \pm \frac{1}{2}\right)^{-1} - \frac{d}{d\tau} \log \Delta_\pm(\tau) \tag{18.6.9}$$

and their sum and difference

$$A = L_+ + L_-, \qquad B = L_+ - L_-. \tag{18.6.10}$$

The identity (6.4.42) then becomes

$$A' - B^2 = 0. \tag{18.6.11}$$

Equations (18.6.9) and (18.6.5) give to order τ^{-6}

$$\begin{aligned} A &= -\tau - \frac{1}{4}\tau\left(\tau^2 - \frac{1}{4}\right)^{-1} + 12a_3\tau^{-5}, \\ B &= -1 + \frac{1}{8}\left(\tau^2 - \frac{1}{4}\right)^{-1} - 6a_3\tau^{-4} - (15a_3 + 10a_5)\tau^{-6}. \end{aligned} \tag{18.6.12}$$

When we substitute Eq. (18.6.12) into (18.6.11), we find that all terms vanish to order τ^{-6} provided that

$$a_3 = -\frac{3}{256}, \qquad a_5 = -\frac{45}{2048}. \tag{18.6.13}$$

The vanishing of Eq. (18.6.11) provides a check of the consistency of our procedures. Unfortunately it is not possible to determine the coefficients beyond a_5 in this way, because Eq. (18.6.11) gives only one equation for each pair of unknowns (a_{2m}, a_{2m+1}).

Putting together Eqs. (18.6.5) and (18.6.13), we have the asymptotic expansion of $\log \Delta_\pm(\tau)$ in powers of τ^{-1},

$$\log \Delta_\pm(\tau) = \mp\frac{3}{256}\tau^{-3} + \frac{9}{512}\tau^{-4} \mp \frac{81}{2048}\tau^{-5} + \cdots. \tag{18.6.14}$$

The potentials which give rise to the phase-shifts (18.4.36) have the expansion

$$W_\pm(\tau) = -\frac{1}{4}\tau^{-2} \pm \frac{1}{4}\tau^{-3} - \frac{3}{16}\tau^{-4} \pm \frac{13}{32}\tau^{-5} - \frac{25}{32}\tau^{-6} \pm \frac{1239}{512}\tau^{-7} - \cdots. \tag{18.6.15}$$

Finally, the Fredholm determinants $F_\pm(t)$ have the asymptotic expansions

$$\begin{aligned} \log F_\pm(t) = &-\frac{1}{4}\pi^2 t^2 \mp \frac{1}{2}\pi t - \frac{1}{8}\log\left|\pi t \pm \frac{1}{2}\right| \pm \frac{1}{4}\log 2 + \alpha \\ &\mp \frac{3}{256}\left(\pi t \pm \frac{1}{2}\right)^{-3} \mp \frac{45}{2048}\left(\pi t \pm \frac{1}{2}\right)^{-5} + \cdots, \end{aligned} \tag{18.6.16}$$

with

$$\alpha = \frac{1}{24}\log 2 + \frac{3}{2}\zeta'(-1) \approx -0.219\,250\,583. \tag{18.6.17}$$

The numerical value of α is discussed in McCoy and Wu (1973), Appendix B. Dyson gave a few more terms in the above expansion, namely

$$a_6 = \frac{63}{4096}, \qquad a_7 = -\frac{3^2 \cdot 5^3 \cdot 7}{2^{16}}, \qquad a_8 = \frac{3^2 \cdot 5^3 \cdot 23}{2^{17}}. \tag{18.6.18}$$

The numerical precision of Eq. (18.6.16) is very good both for $F_+(t)$ and $F_-(t)$ beyond $t \geqslant 1$.

The coefficients a_n in these expansions can now be obtained with very little effort by using the non-linear differential equations satisfied by the functions A and B of Eq. (18.6.10), Eq. (18.6.11) being merely one of them. For details see Chapter 21. A table of values a_n for $n \leqslant 25$ is given in Appendix A.47.

Summary of Chapter 18

Three different methods of finding the asymptotic behaviour of $E_\beta(0; s)$ for large s are described. These complimentary methods gave for the first time the asymptotic series

$$\log F_\pm(t) = -\frac{1}{4}\pi^2 t^2 \mp \frac{1}{2}\pi t - \frac{1}{8}\log\left|\pi t \pm \frac{1}{2}\right| \pm \frac{1}{4}\log 2 + \alpha$$
$$+ \sum_{m=3}^{\infty} (\pm 1)^m a_m \left(\pi t \pm \frac{1}{2}\right)^{-m}, \tag{18.6.16}$$

with

$$\alpha = \frac{1}{24}\log 2 + \frac{3}{2}\zeta'(-1) \approx -0.219\,250\,583,$$
$$a_3 = -\frac{3}{256}, \qquad a_4 = 0, \qquad a_5 = -\frac{45}{2048}, \dots \tag{18.6.17}$$

and a laborious way to compute the higher terms.

The fourth and by far the fastest method to get these asymptotic series to almost any order in τ^{-k} is to use the non-linear differential equations satisfied by $F_\pm(t)$ and this will be discussed in Chapter 21. The only defect of this fourth method is that it does not give the value of the constant α.

The probability $E_\beta(0; s)$ that a randomly chosen interval of length $s = 2t$ contains no eigenvalue is given in terms of the $F_\pm(t)$ as

$$E_1(0; s) = F_+(t),$$
$$E_2(0; s) = F_+(t)F_-(t),$$

and

$$E_4\left(0; \frac{s}{2}\right) = \frac{1}{2}\big(F_+(t) + F_-(t)\big).$$

19

MATRIX ENSEMBLES AND CLASSICAL ORTHOGONAL POLYNOMIALS

One cannot but notice in Chapters 6, 7 and 8, for example, that the Gaussian ensembles are closely related to the Hermite polynomials. Orthogonal polynomials other than the Hermite polynomials have been extensively investigated Bateman (1953b), and one may think of replacing the Hermite polynomials by other polynomials, classical or not. We can define a matrix ensemble by giving the joint probability density function for its eigenvalues arbitrarily:

$$P(x_1, \ldots, x_N) = \prod_{i=1} w(x_i) \prod_{i<j} |x_i - x_j|^{\beta}, \tag{19.0.1}$$

where the function $w(x)$ can be chosen to suit the needs.

The choices $w(x) = \exp(-\beta x^2/2)$, $-\infty < x < \infty$, and $w(x) = 1$, $x = e^{i\theta}$, $0 \leqslant \theta \leqslant 2\pi$, correspond to the Gaussian and the circular ensembles, respectively. The choice

$$w(x) := w(a, b; x) = (1-x)^a (1+x)^b, \quad a, b > -1, \ -1 \leqslant x \leqslant 1, \tag{19.0.2}$$

gives an ensemble corresponding to the Jacobi polynomials. All other classical orthogonal polynomials (Gegenbauer, Tchebichev, Legendre, Laguerre, Hermite) can be obtained from these Jacobi polynomials by taking suitable values of a, b or by taking certain limits. Therefore, Nagao, Wadati, Ghosh and Pandey studied this most general case of the classical weights, Eq. (19.0.2), and thus deduced the correlation functions for the weights $w(x)$ corresponding to all the classical orthogonal polynomials. The expressions of the skew-orthogonal polynomials of the real type given by Ghosh and Pandey are specially elegant and we will follow them here. They are of course equivalent to the expressions derived earlier by Nagao and Wadati. Matrix ensembles corresponding to the weights $w(x) = \exp(-V(x))$, $V(x)$ a polynomial, will be briefly mentioned in Section 19.4.

The Jacobi polynomials $P_j^{(a,b)}(x)$, orthogonal over $(-1, 1)$ with the weight function $w(a, b; x) = (1 - x)^a (1 + x)^b$ have been known for a long time (cf. Bateman, 1953b, 10.8)

$$\int_{-1}^{1} w(a, b; x) P_j^{(a,b)}(x) P_k^{(a,b)}(x)\, dx = h_j^{(a,b)} \delta_{j,k}, \tag{19.0.3}$$

$$P_j^{(a,b)}(x) = k_j x^j + \text{lower powers of } x, \tag{19.0.4}$$

$$k_j = 2^{-j} \binom{2j + a + b}{j}, \tag{19.0.5}$$

$$h_j^{(a,b)} = \frac{2^{a+b+1}}{(2j + a + b + 1)} \frac{\Gamma(j + a + 1)\Gamma(j + b + 1)}{\Gamma(j + 1)\Gamma(j + a + b + 1)}. \tag{19.0.6}$$

19.1 Unitary Ensemble

When $\beta = 2$ we need the sum (cf. Section 5.7)

$$K_N(x, y) = w(a, b; x) \sum_{j=0}^{N-1} \left[h_j^{(a,b)}\right]^{-1} P_j^{(a,b)}(x) P_j^{(a,b)}(y) \tag{19.1.1}$$

$$= \frac{w(a, b; x)}{h_{N-1}^{(a,b)}} \frac{k_{N-1}}{k_N} \frac{P_N^{(a,b)}(x) P_{N-1}^{(a,b)}(y) - P_N^{(a,b)}(y) P_{N-1}^{(a,b)}(x)}{x - y}. \tag{19.1.2}$$

For large j and finite a, b with $x = \cos\theta$, $0 \leqslant \theta \leqslant \pi$, one has the asymptotic form (Szegö, 1939)

$$\left(\frac{w(a, b; x)}{h_j^{(a,b)}}\right)^{1/2} P_j^{(a,b)}(x)$$

$$= \sqrt{\frac{2}{\pi \sin\theta}} \cos\left[\left(j + \frac{a + b + 1}{2}\right)\theta - \left(a + \frac{1}{2}\right)\frac{\pi}{2}\right] + \mathrm{O}(j^{-1}), \tag{19.1.3}$$

so that taking $y = x + \Delta x = \cos(\theta + \Delta\theta)$, and neglecting terms of order N^{-1}, one has

$$K_N(x, y) = \frac{\sin(N\Delta\theta)}{\pi\,\Delta\theta\,\sin\theta} = \frac{\sin(N\Delta x(1-x^2)^{-1/2})}{\pi\,\Delta x}. \tag{19.1.4}$$

The level density is obtained by taking the limit $\Delta\theta \to 0$,

$$R_1(x) = \frac{N}{\pi\,\sin\theta} = \frac{N}{\pi\sqrt{1-x^2}}, \tag{19.1.5}$$

and the two-point function is

$$\lim_{N\to\infty} \frac{K_N(x, y)}{K_N(x, x)} = \frac{\sin(\pi r)}{\pi r}, \qquad r = (y - x)R_1(x), \tag{19.1.6}$$

independent of x and independent of a, b.

Results for other classical weights are deduced simply by giving particular values to a, b, except for Laguerre and Hermite where certain limits after appropriate scaling of the variable have to be taken while a or b or both do not remain finite. (See, e.g., Szegö, 1939, Eq. (5.3.4) and Section 5.6.) For example, in the Laguerre case $w(a; x) = x^a e^{-x}$, $a > -1$, $0 \leqslant x < \infty$, denote the associated Laguerre polynomials by $L_j^{(a)}(x)$, so that

$$\int_0^\infty w(a; x) L_j^{(a)}(x) L_k^{(a)}(x)\,dx = h_j^{(a)}\delta_{jk}, \tag{19.1.7}$$

$$L_j^{(a)}(x) = k_j x^j + \text{lower powers of } x, \tag{19.1.8}$$

$$k_j = (-1)^j/j!, \quad h_j^{(a)} = \Gamma(j + a + 1)/j!. \tag{19.1.9}$$

For large j and finite a with $x = (4j + 2a + 2)\cos^2\theta$, $0 \leqslant \theta \leqslant \pi/2$, one has

$$\left(\frac{w(a; x)}{h_j^{(a)}}\right)^{1/2} L_j^{(a)}(x) = \frac{(-1)^j}{\sqrt{\pi j \sin(2\theta)}} \sin\left[\left(j + \frac{a+1}{2}\right)(\sin(2\theta) - 2\theta) + \frac{3\pi}{4}\right]. \tag{19.1.10}$$

Level density is

$$R_1(x) = \frac{1}{2\pi}\sqrt{\frac{4N - x}{x}}, \qquad 0 \leqslant x \leqslant 4N, \tag{19.1.11}$$

and the two-point function is again $s(r) := \sin(\pi r)/(\pi r)$.

19.2 Orthogonal Ensemble

The usual Jacobi polynomials will be as in Eqs. (19.0.3)–(19.0.6). When $\beta = 1$ and the number of variables $N = 2m$ is even, we choose skew-orthogonal polynomials of the real type $R_j(x)$, not necessarily monic, but of precise degree j, as follows

$$R_{2j}(x) = P_{2j}^{(2a+1,2b+1)}(x), \tag{19.2.1}$$

$$R_{2j+1}(x) = -\big[w(a,b;x)\big]^{-1}\frac{d}{dx}\big[w(a+1,b+1;x)P_{2j}^{(2a+1,2b+1)}(x)\big], \tag{19.2.2}$$

$$r_j = h_{2j}^{(2a+1,2b+1)}, \tag{19.2.3}$$

so that for the $\psi_j(x)$ of Eq. (5.7.13)

$$\psi_j(x) := \int_{-1}^{1} \varepsilon(x-y)w(a,b;y)R_j(y)\,dy, \tag{19.2.4}$$

one has

$$\psi_{2j+1}(x) = -w(a+1,b+1;x)P_{2j}^{(2a+1,2b+1)}(x), \tag{19.2.5}$$

and for $j > 1$,

$$\begin{aligned}\psi_{2j}(x) = &-\frac{4j+2a+2b+1}{4j(j+a+b+1)}w(a+1,b+1;x)P_{2j-1}^{(2a+1,2b+1)}(x)\\ &+\frac{(j+a)(j+b)}{j(j+a+b+1)}\psi_{2j-2}(x),\end{aligned} \tag{19.2.6}$$

while

$$\psi_0(x) = \int_{-1}^{1} \varepsilon(x-y)w(a,b;y)\,dy. \tag{19.2.7}$$

To derive Eq. (19.2.6) one can use the relation

$$\begin{aligned}&\big[w(a,b;x)\big]^{-1}\frac{d}{dx}\big[w(a+1,b+1;x)P_j^{(2a+1,2b+1)}(x)\big]\\ &= \left[b-a-(b+a)x+(1-x^2)\frac{d}{dx}\right]P_j^{(2a+1,2b+1)}(x)\\ &= -\frac{(j+1)(j+2a+2b+3)}{(2j+2a+2b+3)}P_{j+1}^{(2a+1,2b+1)}(x)\\ &\quad+\frac{(j+2a+1)(j+2b+1)}{(2j+2a+2b+3)}P_{j-1}^{(2a+1,2b+1)}(x).\end{aligned} \tag{19.2.8}$$

The polynomials $R_j(x)$ satisfy the skew-orthogonality relations

$$\langle R_{2j}, R_{2k+1}\rangle_1 = h_{2j}^{(2a+1,2b+1)}\delta_{jk}, \tag{19.2.9}$$

$$\langle R_{2j}, R_{2k}\rangle_1 = \langle R_{2j+1}, R_{2k+1}\rangle_1 = 0. \tag{19.2.10}$$

Verification. One can convince oneself that $R_j(x)$ is a polynomial of degree j. It remains to verify their skew-orthogonality.

Without loss of generality, we can assume $j \leqslant k$ in the following verifications. By definition, Eq. (5.7.9),

$$\begin{aligned}\langle R_j, R_k\rangle_1 &= \int_{-1}^{1}\int_{-1}^{1} w(a,b;x)w(a,b;y)R_j(x)R_k(y)\varepsilon(y-x)\,dx\,dy\\ &= -\int_{-1}^{1} w(a,b;x)R_j(x)\psi_k(x)\,dx,\end{aligned} \tag{19.2.11}$$

so that from Eqs. (19.2.4)–(19.2.6) and the orthogonality of the Jacobi polynomials, Eq. (19.0.3), we get

$$\begin{aligned}\langle R_{2j}, R_{2k+1}\rangle_1 &= \int_{-1}^{1} w(2a+1,2b+1;x)P_{2j}^{(2a+1,2b+1)}(x)P_{2k}^{(2a+1,2b+1)}(x)\,dx\\ &= h_{2j}^{(2a+1,2b+1)}\delta_{jk},\end{aligned} \tag{19.2.12}$$

$$\begin{aligned}\langle R_{2j+1}, R_{2k+1}\rangle_1 &= \int_{-1}^{1} w(2a+1,2b+1;x)R_{2j+1}(x)P_{2k}^{(2a+1,2b+1)}(x)\,dx\\ &= 0,\end{aligned} \tag{19.2.13}$$

while using Eq. (19.2.5) several times and the orthogonality of the Jacobi polynomials, Eq. (19.0.3),

$$\langle R_{2j}, R_{2k}\rangle_1 = \text{const}\cdot\langle R_{2j}, R_{2j}\rangle_1 = 0. \tag{19.2.14}$$

The last step is by the anti-symmetry of the scalar product.

When the number of variables $N = 2m+1$ is odd, we choose skew-orthogonal polynomial of the real type $R_j(x)$, not necessarily monic, but of precise degree $j+1$ for $j < N-1$, as

$$R_{2j}(x) = P_{2j+1}^{(2a+1,2b+1)}(x) - \int_{-1}^{1} w(a,b;y)P_{2j+1}^{(2a+1,2b+1)}(y)\,dy, \tag{19.2.15}$$

$$R_{2j+1}(x) = -\left[w(a,b;x)\right]^{-1}\frac{d}{dx}\left[w(a+1,b+1,;x)P_{2j+1}^{(2a+1,2b+1)}(x)\right], \quad (19.2.16)$$

$$r_j = h_{2j+1}^{(2a+1,2b+1)}, \quad (19.2.17)$$

and

$$R_{2m}(x) = P_{2m}^{(2a+1,2b+1)}(x) \div \int_{-1}^{1} w(a,b;y)P_{2m}^{(2a+1,2b+1)}(y)\,dy. \quad (19.2.18)$$

These polynomials satisfy the skew-orthogonality relations and an extra relation

$$\langle R_{2j}, R_{2k+1}\rangle_1 = h_{2j+1}^{(2a+1,2b+2)}\delta_{jk}, \quad (19.2.19)$$

$$\langle R_{2j}, R_{2k}\rangle_1 = \langle R_{2j+1}, R_{2k+1}\rangle_1 = 0, \quad (19.2.20)$$

$$\int_{-1}^{1} w(a,b;x)R_j(x)\,dx = \delta_{j,2m}. \quad (19.2.21)$$

Verification. One can convince oneself that $R_j(x)$ is a polynomial of degree $j+1$ for $j < 2m$ and $R_{2m}(x)$ is a polynomial of degree $2m$. Verification of the skew-orthogonality is similar to the case when the number of variables was even, except that the degree of each polynomial is shifted by one. The constant added to the $R_{2j}(x)$ in Eq. (19.2.15) has no effect on the skew-orthogonality and makes it to satisfy the extra condition (19.2.21).

For the correlation functions one defines as usual the quaternion kernel, Eqs. (5.7.23)–(5.7.27), for $N = 2m$ or $N = 2m+1$,

$$K_{N1}(x,y) = \begin{bmatrix} S_N(x,y) & D_N(x,y) \\ J_N(x,y) & S_N(y,x) \end{bmatrix} + \begin{bmatrix} \alpha(x) & 0 \\ u(x)-u(y) & \alpha(y) \end{bmatrix}, \quad (19.2.22)$$

$$S_N(x,y) = \sum_{j=0}^{m-1} r_j^{-1}\left(\psi'_{2j+1}(x)\psi_{2j}(y) - \psi'_{2j}(x)\psi_{2j+1}(y)\right), \quad (19.2.23)$$

$$D_N(x,y) = \sum_{j=0}^{m-1} r_j^{-1}\left(\psi'_{2j+1}(x)\psi'_{2j}(y) - \psi'_{2j}(x)\psi'_{2j+1}(y)\right), \quad (19.2.24)$$

$$J_N(x,y) = I_N(x,y) - \varepsilon(x-y), \quad (19.2.25)$$

$$I_N(x,y) = \sum_{j=0}^{m-1} r_j^{-1}\left(\psi_{2j+1}(x)\psi_{2j}(y) - \psi_{2j}(x)\psi_{2j+1}(y)\right), \quad (19.2.26)$$

$$\alpha(x) = u(x) = 0, \quad \text{if } N = 2m, \quad \text{and} \quad (19.2.27)$$

$$\left.\begin{array}{l}\alpha(x) = w(a,b;x)R_{2m}(x) \\ u(x) = \int_{-1}^{1} \varepsilon(x-y)\alpha(y)\,dy\end{array}\right\}, \quad \text{if } N = 2m+1. \tag{19.2.28}$$

Nagao and Wadati (1992), give a Christoffel–Darboux like formula for the finite sums $S_N(x,y)$, $D_N(x,y)$ and $I_N(x,y)$. In the limit $N \to \infty$ the sums in the above equations can be replaced by integrals replacing ψ_j, ψ_j' by their large j approximations. As the final result one gets the level density, the limit of $S_N(x,x)$, as Eq. (19.1.5), and the 2-point kernel as

$$K_1(x,y) = \begin{bmatrix} s(r) & D(r) \\ J(r) & s(r) \end{bmatrix}, \quad r = (x-y)R_1(x), \tag{19.2.29}$$

with

$$s(r) = \frac{\sin(\pi r)}{\pi r}, \tag{19.2.30}$$

$$D(r) = \frac{d}{dr} s(r), \tag{19.2.31}$$

$$J(r) = I(r) - \varepsilon(r), \tag{19.2.32}$$

$$I(r) = \int_0^r s(t)\,dt, \tag{19.2.33}$$

independent of a, b.

For more details see either Nagao and Wadati (1992) or Ghosh and Pandey (2002).

Nagao and Wadati (1991a, 1992) expanded the skew-orthogonal polynomials $R_j(x)$ in terms of $P_j^{(a,b)}(x)$ in their first article, and in terms of $P_j^{(2a,2b)}(x)$, in their second, while Ghosh and Pandey did it in terms of $P_j^{(2a+1,2b+1)}(x)$. The coefficients in the first cases are complicated, while in the second case they simplify drastically. We presented here the simpler version.

For the Laguerre weight $w(a;x) = x^a \exp(-x)$, $a > -1$, $0 \leqslant x < \infty$, and the associated Laguerre polynomials $L_j^{(a)}(x)$, Eqs. (19.1.7)–(19.1.9), the skew-orthogonal polynomials of the real type are as follows.

For $N = 2m$ even,

$$R_{2j}(x) = L_{2j}^{(2a+1)}(2x), \tag{19.2.34}$$

$$R_{2j+1}(x) = \left[w(a;x)\right]^{-1} \frac{d}{dx}\left[w(a+1;x)L_{2j}^{(2a+1)}(2x)\right], \tag{19.2.35}$$

$$r_j = \frac{\Gamma(2j+2a+2)}{2^{2a+2}(2j)!}. \tag{19.2.36}$$

For $N = 2m + 1$ odd, if $j < m$,

$$R_{2j}(x) = L_{2j+1}^{(2a+1)}(2x) - \int_0^\infty w(a; y) L_{2j+1}^{(2a+1)}(2y)\, dy, \tag{19.2.37}$$

$$R_{2j+1}(x) = \left[w(a; x)\right]^{-1} \frac{d}{dx}\left[w(a+1; x) L_{2j+1}^{(2a+1)}(2x)\right], \tag{19.2.38}$$

$$r_j = \frac{\Gamma(2j + 2a + 3)}{2^{2a+3}(2j)!} \tag{19.2.39}$$

and

$$R_{2m}(x) = L_{2m}^{(2a+1)}(2x) \div \int_0^\infty w(a; y) L_{2m}^{(2a+1)}(y)\, dy. \tag{19.2.40}$$

Equation corresponding to (19.2.8) is now

$$\begin{aligned} &2\left[w(a; x)\right]^{-1} \frac{d}{dx}\left[w(a+1; x) L_j^{(2a+1)}(2x)\right] \\ &\quad = (j+1) L_{j+1}^{(2a+1)}(2x) - (j+a) L_{j-1}^{(2a+1)}(2x). \end{aligned} \tag{19.2.41}$$

The polynomials $R_j(x)$ (of degree j or $j + 1$ according as N is even or odd, for $j < N - 1$, the degree of $R_{N-1}(x)$ being $N - 1$) satisfy the skew-orthogonality relations

$$\langle R_{2j}, R_{2k+1}\rangle_1 = r_j \delta_{jk}, \tag{19.2.42}$$

$$\langle R_{2j}, R_{2k}\rangle_1 = \langle R_{2j+1}, R_{2k+1}\rangle_1 = 0, \tag{19.2.43}$$

and if N is odd, one has in addition

$$\int_0^\infty w(a; x) R_j(x)\, dx = \delta_{j,N-1}. \tag{19.2.44}$$

The verification of these facts proceeds as in the Jacobi case. The limit $N \to \infty$ can be taken similarly and one gets the level density (19.1.11) and the 2-point kernel (19.2.22).

19.3 Symplectic Ensemble

For the Jacobi weight, Eq. (19.0.2), an elegant form for the skew-orthogonal polynomials of the quaternion type has not been found. We report here Ghosh and Pandey's expressions for them, equivalent to the ones given earlier by Nagao and Wadati (1992) but a little simpler.

$$Q'_{2j+1}(x) = P_{2j}^{(a,b)}(x), \tag{19.3.1}$$

$$Q'_{2j}(x) = P_{2j-1}^{(a,b)}(x) + \eta_j Q'_{2j-2}(x), \tag{19.3.2}$$

$$\eta_j = \frac{(2j+a-1)(2j+b-1)(4j+a+b-5)}{(2j-1)(2j+a+b-1)(4j+a+b-1)}, \tag{19.3.3}$$

$$q_j = \frac{2h_{2j}^{(a,b)}}{4j+a+b-1}. \tag{19.3.4}$$

The $Q_j(x)$ or $P_j^{(a,b)}(x)$ for negative j is taken to be zero. Using the relation

$$\frac{d}{dx} P_j^{(a,b)}(x) = \frac{1}{2}(j+a+b+1) P_{j-1}^{(a+1,b+1)}(x), \tag{19.3.5}$$

and other relations connecting $P_j^{(a,b)}(x)$ to $P_k^{(\alpha,\beta)}(x)$ with $a-\alpha$, $b-\beta$ and $j-k$ taking values either 0 or ± 1 (Bateman, 1953b, 10.8.17, 10.8.34–37), one can express $Q_j(x)$ as a linear combination of the $P_k^{(a,b)}(x)$ with $k \leqslant j$. The expressions obtained are not very elegant and the verification that they satisfy skew-orthogonality relations needs some elementary, though heavy, algebraic work. The 2-point kernel $K_4(x, y)$ can then be constructed and the limit $N \to \infty$ can be taken. We will not do this here and refer to Ghosh and Pandey (2002) for details.

The final result is that the level density is again given by Eq. (19.1.5) and the 2-point kernel turns out to be

$$K_4(x, y) = \frac{1}{2}\begin{bmatrix} S_{2N}(x, y) & D_{2N}(x, y) \\ I_{2N}(x, y) & S_{2N}(y, x) \end{bmatrix}. \tag{19.3.6}$$

When $N \to \infty$, the functions S_{2N}, D_{2N} and I_{2N} appearing in the above equation become respectively $S(2r)$, $D(2r)$ and $I(2r)$.

For the Laguerre weight $w(a; x) = x^a e^{-x}$, $a > -1$, $0 \leqslant x < \infty$, the skew-orthogonal polynomials of the quaternion type are

$$Q'_{2j+1}(x) = L_{2j}^{(a)}(x), \tag{19.3.7}$$

$$Q'_{2j}(x) = -L_{2j-1}^{(a)}(x) - \frac{(2j+a-1)}{(2j-1)} Q'_{2j-2}(x), \tag{19.3.8}$$

$$q_j = h_{2j}^{(a)} = \frac{\Gamma(2j+a+1)}{(2j)!}, \tag{19.3.9}$$

or

$$Q_{2j+1}(x) = -L_{2j+1}^{(a)}(x) + L_{2j}^{(a)}(x), \tag{19.3.10}$$

$$Q_{2j}(x) = L_{2j}^{(a)}(x) - L_{2j-1}^{(a)}(x) - \frac{(2j+a-1)}{(2j-1)} Q_{2j-2}(x). \tag{19.3.11}$$

The 2-point kernel $K_4(x, y)$ is constructed as usual, Section 5.7, which in the limit of large N gives the level density, Eq. (19.1.11) and the 2-point kernel K_4, Eq. (19.3.6) with the functions S_{2N}, D_{2N} and I_{2N} replaced by the functions $s(2r)$, $D(2r)$ and $I(2r)$ respectively.

For more details see either Nagao and Wadati (1991a, 1992) or Ghosh and Pandey (2002).

19.4 Ensembles With Other Weights

Instead of the weights corresponding to the classical orthogonal polynomials, matrix ensembles with the probability density $P(A) \propto \exp(-\operatorname{tr} V(A))$, $V(x)$ a polynomial, have also been investigated (Brezin, Cicuta, Eynard, Forrester, ...). For the convergence of the integrals $V(x)$ should be a polynomials of even degree with the coefficient of the highest power positive. These ensembles are invariant under the change $A \to U^\dagger AU$, where the matrix U is complex unitary, real orthogonal or quaternion symplectic according as A is complex hermitian, real symmetric or quaternion self-dual. The resulting joint probability density of the eigenvalues is (19.0.1) with $w(x) = \exp(-V(x))$. The method of Chapter 5 based on orthogonal or skew-orthogonal polynomials gives again the level density and the correlation functions. The level density depends on $V(x)$; for example, its support may consist of a single interval as is the case with the Gaussian ensembles, or of several disconnected intervals. In case the level density is non-zero only in a single connected interval, it turns out surprisingly that in the limit of large matrices the properly scaled two level correlation functions, and hence the m-level correlation functions for any finite m, are identical to those of the Gaussian ones in all the three cases $\beta = 1, 2$ and 4. This property has been called "the universality". We will not pursue this line any more and refer the interested reader to the large existing literature.

19.5 Conclusion

One does not know any probability density for the matrix elements which will give rise to the probability density (19.0.1) for its eigenvalues; nor for any other weight corresponding to the classical orthogonal polynomials except for the Hermite and Laguerre weights.

The Laguerre weight arises as follows. As in Section 15.1 consider the ensemble of all $m \times n$, $m \leqslant n$, complex matrices A with the joint probability density $P(A) \propto$

$\exp(-(1/2)\operatorname{tr} AA^\dagger)$, i.e. the real and imaginary parts of the matrix elements are independent Gaussian random variables. Any such A can be written as (cf. Mehta, 1989MT, Section 4.8)

$$A = U^\dagger \Lambda V, \tag{19.5.1}$$

where U and V are unitary matrices and Λ is a diagonal matrix with real non-negative elements. The non-zero diagonal elements of Λ are known as the singular values of A; their squares are the (non-zero) eigenvalues of $AA^\dagger$ or of $A^\dagger A$ (they are the same). In some physical applications one may need the distribution of the eigenvalues not of A, but of $AA^\dagger$, the joint probability density of which can be derived from Eq. (19.5.1) to be (case $\beta = 2$)

$$K \exp\left(-\sum_1^m \lambda_i/2\right)|\Delta(\lambda)|^\beta \prod_i \lambda_i^{(n-m+1)\beta/2-1},$$

$$K = m!\, 2^{mn\beta/2}\big(\Gamma(\beta/2)\big)^{-m} \prod_{i=1}^m \Gamma\big((n-i+1)\beta/2\big)\Gamma\big((m-i+1)\beta/2\big). \tag{19.5.2}$$

For a derivation of this see, e.g., James (1964).

The conclusion is that all the local properties of the eigenvalues are the same as those of the corresponding Gaussian ensembles.

Summary of Chapter 19

Ensembles related to the classical orthogonal polynomials are considered. Gaussian ensembles are related to the Hermite polynomials. Instead of the Hermite ones if we choose other classical orthogonal polynomials, we get other ensembles. The most general of them including the others as special cases are related to the Jacobi weight (19.0.2).

The skew-orthogonal polynomials of the real type for the Jacobi weight $(1-x)^a(1+x)^b$ have much simpler expressions in terms of the Jacobi polynomials orthogonal with respect to the weight $(1-x)^{2a+1}(1+x)^{2b+1}$.

The eigenvalue density is different for different polynomials. However, in the limit of large matrices the m-point correlation functions for $m > 1$, and hence the spacing probability densities do not depend on this choice. They are the same as those of the corresponding Gaussian ensembles; they are "universal".

20

LEVEL SPACING FUNCTIONS $E_\beta(r, s)$; INTER-RELATIONS AND POWER SERIES EXPANSIONS

20.1 Three Sets of Spacing Functions; Their Inter-Relations

In earlier chapters we introduced $E_\beta(r, s)$, the probability that a randomly chosen interval of length s contains exactly r eigenvalues. Here, as usual, β is 1, 2 or 4 according as the ensemble considered is orthogonal, unitary or symplectic. In Chapters 6, 7 and 11 we computed these $E_\beta(r, s)$ in terms of spheroidal functions and their eigenvalues, and in Appendix A.13 gave a few terms in their power series expansions. Moreover, in Chapter 18 we dealt with the asymptotic expansion of $E_\beta(0, s)$. The $E_\beta(r, s)$, $s = 2t$, can be expressed as (cf. Eq. (6.4.35), Appendix A.7)

$$E_\beta(r, s) = \frac{1}{r!}\left(-\frac{\partial}{\partial z}\right)^r F_\beta(z, s)|_{z=1}, \tag{20.1.1}$$

$$F_\beta(z, s) = \prod_{j=0}^{\infty}(1 - z\lambda_j), \tag{20.1.2}$$

where $\lambda_j = \lambda_j(s)$ are the eigenvalues of an integral equation

$$\lambda f(x) = \int_{-s/2}^{s/2} K(\beta; x, y) f(y)\, dy, \tag{20.1.3}$$

with the kernels (cf. Appendix A.7)

$$K(2; x, y) = S(x, y), \tag{20.1.4}$$

$$K(1; x, y) = \begin{bmatrix} S(x, y) & D(x, y) \\ J(x, y) & S(x, y) \end{bmatrix}, \tag{20.1.5}$$

$$K(4; x, y) = \frac{1}{2}\begin{bmatrix} S(2x, 2y) & D(2x, 2y) \\ I(2x, 2y) & S(2x, 2y) \end{bmatrix}. \tag{20.1.6}$$

For $\beta = 2$, the kernel is simple, while for $\beta = 1$ and $\beta = 4$, the kernels are 2×2 matrices and the integral equation (20.1.3) is a set of two coupled equations.

For $\beta = 1$ and $\beta = 4$, the eigenvalues are doubly degenerate because the kernel is self-dual in the quaternion sense

$$K^T(\beta; y, x) = -\begin{bmatrix} 0 & 1 \\ -1 & 0 \end{bmatrix} K(\beta; x, y) \begin{bmatrix} 0 & 1 \\ -1 & 0 \end{bmatrix}. \tag{20.1.7}$$

In Eq. (20.1.2) the product is taken over distinct eigenvalues λ_j.

In the limit of very large matrices one has

$$S(x, y) = \frac{\sin \pi(x - y)}{\pi(x - y)}, \tag{20.1.8}$$

$$D(x, y) = \frac{\partial}{\partial x} S(x, y), \tag{20.1.9}$$

$$I(x, y) = \int_{-\infty}^{\infty} \varepsilon(x - \xi) S(\xi, y)\, d\xi, \tag{20.1.10}$$

$$J(x, y) = I(x, y) - \varepsilon(x - y) \tag{20.1.11}$$

and

$$\varepsilon(x) = \pm 1/2 \quad \text{according as } x > 0 \text{ or } x < 0. \tag{20.1.12}$$

In the case $\beta = 2$, the solutions of Eq. (20.1.3) are either even or odd. They are also the solutions of

$$\lambda f(x) = \int_{-s/2}^{s/2} S_{\pm}(x, y) f(y)\, dy, \tag{20.1.13}$$

with

$$S_{\pm}(x, y) = \frac{1}{2}\big[S(x, y) \pm S(x, -y)\big]. \tag{20.1.14}$$

We will use an even (odd) index to λ when it corresponds to an even (odd) solution, and write Eq. (20.1.2) for $\beta = 2$ as

$$F_2(z,s) = F_+(z,s)F_-(z,s), \tag{20.1.15}$$

$$F_+(z,s) = \prod_{i=0}^{\infty}(1 - z\lambda_{2i}), \tag{20.1.16}$$

$$F_-(z,s) = \prod_{i=0}^{\infty}(1 - z\lambda_{2i+1}). \tag{20.1.17}$$

In case $\beta = 1$ or $\beta = 4$, each component of any eigenfunction has a definite parity, and the parities of the two components of an eigenfunction are opposite.

If one sets

$$E_\pm(r,s) = \frac{1}{r!}\left(-\frac{\partial}{\partial z}\right)^r F_\pm(z,s)|_{z=1}, \tag{20.1.18}$$

or

$$F_\pm(z,s) = \sum_{r=0}^{\infty}(1-z)^r E_\pm(r,s), \tag{20.1.19}$$

then one knows that (Eqs. (7.5.27), (7.5.29) and (11.7.6))

$$E_\pm(r,s) = E_1(2r,s) + E_1(2r \mp 1, s), \quad r \geqslant 0, \tag{20.1.20}$$

$$E_4(r,s) = \frac{1}{2}\left[E_+(r,2s) + E_-(r,2s)\right], \quad r \geqslant 0 \tag{20.1.21}$$

($E_1(-1,s) \equiv 0$). We will present here still another proof of Eqs. (20.1.20) and (20.1.21).

Equation (20.1.8) and onwards have been written for very large matrices. They are the limit when $N \to \infty$ of the results for $N \times N$ matrices from the circular ensembles. For the Gaussian ensembles they are also the limit when $N \to \infty$ of the results for $N \times N$ matrices when one considers eigenvalues in the region of highest density (see, e.g., Chapters 6, 7, 8, and 11). For the case of finite $N \times N$ matrices Eq. (20.1.8) is replaced by (cf. Chapter 11)

$$S(\theta,\phi) = S(\theta - \phi) = \frac{1}{N}\sum_p e^{ip(\theta-\phi)} = \frac{\sin N(\theta-\phi)/2}{N\sin(\theta-\phi)/2}, \tag{20.1.22}$$

and in all other equations x, y and s are replaced respectively by θ, ϕ and 2α. Instead of an infinite number of eigenvalues we have a finite number N of them. For $\beta = 2$, the number of even (odd) eigenfunctions is $[(N+1)/2]$ ($[N/2]$). The even (odd) eigenfunctions are also the eigenfunctions of $S_+(\theta,\phi)$ ($S_-(\theta,\phi)$). In Eq. (20.1.22) p is summed over the values $-(N-1)/2, -(N-3)/2, \ldots, (N-3)/2, (N-1)/2$. Note that $S(\theta,\phi)$ depends only on the difference $\theta-\phi$.

It is somewhat convenient to argue when N is finite and later take the limit $N\to\infty$ while $s=N\alpha/\pi$, $x=N\theta/2\pi$, $y=N\phi/2\pi$ are kept finite.

20.2 Relation Between Odd and Even Solutions of Eq. (20.1.13)

The derivatives of the eigenfunctions of Eq. (20.1.3) for the three cases $\beta=1$, $\beta=2$ and $\beta=4$, can be expanded in terms of the eigenfunctions themselves for the case $\beta=2$. This can be seen directly from the integral equation for the finite N case, since the kernel is a sum of separable functions involving exponentials. Thus

$$f'_{2i+1}(\theta)=\sum_j c_{ij}f_{2j}(\theta),\quad f_{2i+1}(\theta)=\sum_j c_{ij}\int_0^\theta f_{2j}(\phi)\,d\phi, \tag{20.2.1}$$

$$f'_{2i}(\theta)=\sum_j d_{ij}f_{2j+1}(\theta),\quad f_{2i}(\theta)=f_{2i}(0)+\sum_j d_{ij}\int_0^\theta f_{2j+1}(\phi)\,d\phi. \tag{20.2.2}$$

The kernels $S_\pm(\theta,\phi)$ satisfy the obvious property

$$\frac{\partial}{\partial\theta}S_\pm(\theta,\phi)=-\frac{\partial}{\partial\phi}S_\mp(\theta,\phi), \tag{20.2.3}$$

and have the spectral representations

$$S_+(\theta,\phi)=\sum_i\lambda_{2i}f_{2i}(\theta)f_{2i}(\phi), \tag{20.2.4}$$

$$S_-(\theta,\phi)=\sum_i\lambda_{2i+1}f_{2i+1}(\theta)f_{2i+1}(\phi), \tag{20.2.5}$$

where $f_j(\theta)$ are normalized eigenfunctions of Eq. (20.1.13),

$$\int_{-\alpha}^{\alpha}f_i(\theta)f_j(\theta)\,d\theta=\delta_{ij}. \tag{20.2.6}$$

Differentiation of Eq. (20.1.13) and a partial integration gives

$$\begin{aligned}\lambda_{2i+1}f'_{2i+1}(\theta)&=\frac{\partial}{\partial\theta}\int_{-\alpha}^{\alpha}S_-(\theta,\phi)f_{2i+1}(\phi)\,d\phi\\&=-\int_{-\alpha}^{\alpha}\frac{\partial}{\partial\phi}S_+(\theta,\phi)f_{2i+1}(\phi)\,d\phi\\&=-2S_+(\theta,\alpha)f_{2i+1}(\alpha)+\int_{-\alpha}^{\alpha}S_+(\theta,\phi)f'_{2i+1}(\phi)\,d\phi.\end{aligned} \tag{20.2.7}$$

Or using Eqs. (20.2.1), (20.2.4) and (20.2.6),

$$\lambda_{2i+1} \sum_j c_{ij} f_{2j}(\theta) = -2 f_{2i+1}(\alpha) \sum_j \lambda_{2j} f_{2j}(\theta) f_{2j}(\alpha) + \sum_j c_{ij} \lambda_{2j} f_{2j}(\theta). \quad (20.2.8)$$

As $f_{2j}(\theta)$ are linearly independent even functions, one has

$$c_{ij} = \frac{-2\lambda_{2j}}{\lambda_{2i+1} - \lambda_{2j}} f_{2j}(\alpha) f_{2i+1}(\alpha), \quad (20.2.9)$$

and Eq. (20.2.1) can be written as

$$\begin{aligned} f_{2i+1}(\alpha) &= \sum_j c_{ij} \int_0^\alpha f_{2j}(\phi)\, d\phi \\ &= f_{2i+1}(\alpha) \sum_j \frac{-2\lambda_{2j}}{\lambda_{2i+1} - \lambda_{2j}} f_{2j}(\alpha) \int_0^\alpha f_{2j}(\phi)\, d\phi. \end{aligned} \quad (20.2.10)$$

Since $f_{2i+1}(\alpha) \neq 0$, from Eq. (20.2.9) one has

$$1 + \sum_j \frac{2\lambda_{2j}}{\lambda_{2i+1} - \lambda_{2j}} f_{2j}(\alpha) \int_0^\alpha f_{2j}(\phi)\, d\phi = 0, \quad (20.2.11)$$

for every i. Consider the rational function

$$G(z) = 1 + \sum_j \frac{z\lambda_{2j}}{1 - z\lambda_{2j}} f_{2j}(\alpha) \int_{-\alpha}^\alpha f_{2j}(\phi)\, d\phi. \quad (20.2.12)$$

This function has zeros at $z = 1/\lambda_{2i+1}$ from Eq. (20.2.11), has poles at $z = 1/\lambda_{2j}$, and is 1 at $z = 0$. Therefore

$$G(z) = \prod_j \frac{(1 - z\lambda_{2j+1})}{(1 - z\lambda_{2j})}, \quad (20.2.13)$$

i.e.,

$$\frac{F_-(z, \alpha)}{F_+(z, \alpha)} = 1 + \sum_j \frac{z\lambda_{2j}}{1 - z\lambda_{2j}} f_{2j}(\alpha) \int_{-\alpha}^\alpha f_{2j}(\phi)\, d\phi. \quad (20.2.14)$$

Starting with Eq. (20.2.2) one can similarly determine the coefficients d_{ij},

$$d_{ij} = \frac{-2\lambda_{2j+1}}{\lambda_{2i} - \lambda_{2j+1}} f_{2i}(\alpha) f_{2j+1}(\alpha), \quad (20.2.15)$$

and similarly,

$$f_{2i}(\alpha)\left[1+\sum_j \frac{2\lambda_{2j+1}}{\lambda_{2i}-\lambda_{2j+1}} f_{2j+1}(\alpha)\int_0^\alpha f_{2j+1}(\phi)\,d\phi\right] = f_{2i}(0). \tag{20.2.16}$$

But from Eqs. (20.1.13) and (20.2.2),

$$\begin{aligned}\lambda_{2i} f_{2i}(0) &= \int_{-\alpha}^{\alpha} S_+(0,\phi) f_{2i}(\phi)\,d\phi = \int_{-\alpha}^{\alpha} S(\theta) f_{2i}(\theta)\,d\theta \\ &= \int_{-\alpha}^{\alpha} S(\theta)\left[f_{2i}(0)+\sum_j d_{ij}\int_0^\theta f_{2j+1}(\phi)\,d\phi\right]d\theta, \end{aligned}\tag{20.2.17}$$

or, substituting for d_{ij} from Eq. (20.2.15),

$$\begin{aligned}&\big(\lambda_{2i}-2I(\alpha)\big) f_{2i}(0) \\ &= \sum_j \frac{-2\lambda_{2j+1}}{\lambda_{2i}-\lambda_{2j+1}} f_{2i}(\alpha) f_{2j+1}(\alpha)\int_{-\alpha}^{\alpha} d\theta\, S(\theta)\int_0^\theta d\phi\, f_{2j+1}(\phi), \end{aligned}\tag{20.2.18}$$

where

$$2I(\theta) = \int_{-\theta}^{\theta} S(\phi)\,d\phi = \int_{-\theta}^{\theta} S_+(\phi)\,d\phi = 2\int_0^\theta S(\phi)\,d\phi. \tag{20.2.19}$$

Now a partial integration gives

$$\begin{aligned}&\int_{-\alpha}^{\alpha} d\theta\, S(\theta)\int_0^\theta d\phi\, f_{2j+1}(\phi) \\ &= 2\int_0^\alpha d\theta\, S(\theta)\int_0^\alpha d\phi\, f_{2j+1}(\phi) - 2\int_0^\alpha d\theta\, f_{2j+1}(\theta)\int_0^\theta d\phi\, S(\phi) \\ &= \int_0^\alpha d\theta\, f_{2j+1}(\theta)\big[2I(\alpha)-2I(\theta)\big], \end{aligned}\tag{20.2.20}$$

so that from Eqs. (20.2.16) and (20.2.18), removing the common factor $f_{2i}(\alpha)$, and rearranging,

$$\lambda_{2i}-2I(\alpha)+\sum_j \frac{2\lambda_{2j+1}}{\lambda_{2i}-\lambda_{2j+1}} f_{2j+1}(\alpha)\int_0^\alpha d\theta\, f_{2j+1}(\theta)\big[\lambda_{2i}-2I(\theta)\big]\,d\theta = 0. \tag{20.2.21}$$

In other words, the function

$$1-2zI(\alpha)+\sum_j \frac{2z\lambda_{2j+1}}{1-z\lambda_{2j+1}} f_{2j+1}(\alpha)\int_0^\alpha d\theta\, f_{2j+1}(\theta)\big[1-2zI(\theta)\big] \qquad (20.2.22)$$

has zeros at $z=1/\lambda_{2i}$, has poles at $1/\lambda_{2j+1}$, and is 1 at $z=0$, so that it is equal to $F_+(z,\alpha)/F_-(z,\alpha)$, i.e.,

$$\frac{F_+(z,\alpha)}{F_-(z,\alpha)} = 1-2zI(\alpha) + \sum_j \frac{2z\lambda_{2j+1}}{1-z\lambda_{2j+1}} f_{2j+1}(\alpha)\int_0^\alpha d\theta\, f_{2j+1}(\theta)\big[1-2zI(\theta)\big]. \qquad (20.2.23)$$

Equation (20.2.14) is simpler than (20.2.23) since instead of (20.2.19) one has

$$\int_{-\theta}^{\theta} S_-(\phi)\, d\phi = 0. \qquad (20.2.24)$$

20.3 Relation Between $F_1(z,s)$ and $F_\pm(z,s)$

As for $\beta=2$ the even and odd solutions of Eq. (20.1.3) are also solutions of Eq. (20.1.13), similarly for $\beta=1$, the solutions of Eq. (20.1.3) are also solutions of

$$\mu\begin{bmatrix}\xi(\theta)\\ \eta(\theta)\end{bmatrix} = \int_{-\alpha}^{\alpha}\begin{bmatrix} S_\pm(\theta,\phi) & D_\mp(\theta,\phi)\\ J_\pm(\theta,\phi) & S_\mp(\theta,\phi)\end{bmatrix}\begin{bmatrix}\xi(\phi)\\ \eta(\phi)\end{bmatrix} d\phi, \qquad (20.3.1)$$

with

$$S_\pm(\theta,\phi) = \frac{1}{2}\big[S(\theta,\phi)\pm S(\theta,-\phi)\big], \qquad (20.3.2)$$

$$D_\pm(\theta,\phi) = \frac{1}{2}\big[D(\theta,\phi)\pm D(\theta,-\phi)\big], \qquad (20.3.3)$$

$$J_\pm(\theta,\phi) = \frac{1}{2}\big[J(\theta,\phi)\pm J(\theta,-\phi)\big]. \qquad (20.3.4)$$

Similar to Eq. (20.2.3), we have

$$\frac{\partial}{\partial\theta} D_\pm(\theta,\phi) = -\frac{\partial}{\partial\phi} D_\mp(\theta,\phi), \qquad (20.3.5)$$

$$\frac{\partial}{\partial\theta} J_\pm(\theta,\phi) = -\frac{\partial}{\partial\phi} J_\mp(\theta,\phi). \qquad (20.3.6)$$

The kernel $S_+(\theta,\phi)$ ($S_-(\theta,\phi)$) acting on any function $g(\phi)$ selects the even (odd) part of g and the result is an even (odd) function:

$$\int_{-\alpha}^{\alpha} S_\pm(\theta,\phi)g(\phi)\,d\phi = \int_{-\alpha}^{\alpha} S_\pm(\theta,\phi)\frac{1}{2}\big[g(\phi)\pm g(-\phi)\big]\,d\phi, \tag{20.3.7}$$

is an even (odd) function of θ. So that in the operator sense

$$S_+\circ S_- = S_-\circ S_+ = 0. \tag{20.3.8}$$

Similarly, $D_+(\theta,\phi)$ and $J_+(\theta,\phi)$ ($D_-(\theta,\phi)$ and $J_-(\theta,\phi)$) acting on any function $g(\phi)$ selects the even (odd) part of g and the result is an odd (even) function, so that

$$\begin{aligned} &D_+\circ J_+ = D_-\circ J_- = D_-\circ S_+ = D_+\circ S_- = J_+\circ S_- = J_-\circ S_+ = 0,\\ &J_+\circ D_+ = J_-\circ D_- = S_+\circ D_+ = S_-\circ D_- = S_+\circ J_+ = S_-\circ J_- = 0, \end{aligned} \tag{20.3.9}$$

i.e.,

$$\sigma_+\circ\sigma_- = \sigma_-\circ\sigma_+ = 0, \tag{20.3.10}$$

where

$$\sigma_\pm(\theta,\phi) = \begin{bmatrix} S_\pm(\theta,\phi) & D_\mp(\theta,\phi) \\ J_\pm(\theta,\phi) & S_\mp(\theta,\phi) \end{bmatrix}. \tag{20.3.11}$$

The kernels σ_+ and σ_- have the same set of eigenvalues, half of them being zero. In what follows we will be concerned with only the non-zero eigenvalues.

Equation (20.3.1) written in full reads for the upper sign, for example,

$$\mu\xi(\theta) = \int_{-\alpha}^{\alpha} \big[S_+(\theta,\phi)\xi(\phi) + D_-(\theta,\phi)\eta(\phi)\big]\,d\phi, \tag{20.3.12}$$

$$\mu\eta(\theta) = \int_{-\alpha}^{\alpha} \big[J_+(\theta,\phi)\xi(\phi) + S_-(\theta,\phi)\eta(\phi)\big]\,d\phi. \tag{20.3.13}$$

Differentiating Eq. (20.3.13) with respect to θ, and comparing with (20.3.12), one gets

$$\xi(\theta) = \frac{\mu}{\mu-1}\eta'(\theta). \tag{20.3.14}$$

Thus $\xi(\theta)$ and $\eta(\theta)$ have opposite parities, and $\xi(\theta)$ is proportional to the derivative of $\eta(\theta)$. With the upper (lower) sign in (20.3.1) $\xi(\theta)$ is even (odd) and $\eta(\theta)$ is odd (even).

Now from Eq. (20.2.3) and a partial integration,

$$\int_{-\alpha}^{\alpha} D_-(\theta,\phi)\eta(\phi)\,d\phi = \frac{\partial}{\partial\theta}\int_{-\alpha}^{\alpha} S_-(\theta,\phi)\eta(\phi)\,d\phi = -\int_{-\alpha}^{\alpha}\frac{\partial}{\partial\phi}S_+(\theta,\phi)\eta(\phi)\,d\phi$$

$$= -2S_+(\theta,\alpha)\eta(\alpha) + \int_{-\alpha}^{\alpha} S_+(\theta,\phi)\eta'(\phi)\,d\phi. \qquad (20.3.15)$$

For $\eta(\theta)$ an odd function, this gives with Eqs. (20.3.12) and (20.3.14)

$$\frac{\mu^2}{\mu-1}\eta'(\theta) = -2S_+(\theta,\alpha)\eta(\alpha) + \left(\frac{\mu}{\mu-1}+1\right)\int_{-\alpha}^{\alpha} S_+(\theta,\phi)\eta'(\phi)\,d\phi. \qquad (20.3.16)$$

Substituting the expansion of $\eta'(\theta)$ in terms of the $f_{2j}(\theta)$,

$$\eta'(\theta) = \sum_i c_i f_{2i}(\theta), \quad \eta(\theta) = \sum_i c_i \int_0^{\theta} f_{2i}(\phi)\,d\phi, \qquad (20.3.17)$$

in Eq. (20.3.16), and using (20.2.4),

$$\left(\frac{\mu^2}{\mu-1} - \frac{2\mu-1}{\mu-1}\lambda_{2i}\right)c_i + 2\lambda_{2i} f_{2i}(\alpha)\sum_j c_j \int_0^{\alpha} f_{2j}(\phi)\,d\phi = 0. \qquad (20.3.18)$$

So that the eigenvalues μ_i are the roots of the algebraic equation

$$\det\left[\left(\mu^2 - (2\mu-1)\lambda_{2i}\right)\delta_{ij} + 2(\mu-1)\lambda_{2i} f_{2i}(\alpha)\int_0^{\alpha} f_{2j}(\phi)\,d\phi\right] = 0. \qquad (20.3.19)$$

The function $F_1(z,\alpha)$ is therefore obtained by substituting $\mu = 1/z$ in the left-hand side of this equation and multiplying by an appropriate power of z to remove all its negative powers,

$$\begin{aligned} F_1(z,\alpha) &= \det\left[\left(1-(2z-z^2)\lambda_{2i}\right)\delta_{ij} + 2z(1-z)\lambda_{2i} f_{2i}(\alpha)\int_0^{\alpha} f_{2j}(\phi)\,d\phi\right] \\ &= \prod_i \left(1-(2z-z^2)\lambda_{2i}\right) \\ &\quad \times \left[1 + \sum_j \frac{2z(1-z)\lambda_{2j}}{1-(2z-z^2)\lambda_{2j}} f_{2j}(\alpha)\int_0^{\alpha} f_{2j}(\phi)\,d\phi\right]. \end{aligned} \qquad (20.3.20)$$

For the last equality note that

$$\begin{aligned} \det[a_i\delta_{ij} + b_i c_j] &= \det\begin{bmatrix} 1 & c_j \\ 0 & a_i\delta_{ij} + b_i c_j \end{bmatrix} = \det\begin{bmatrix} 1 & c_j \\ -b_i & a_i\delta_{ij} \end{bmatrix} \\ &= \prod_i a_i \left(1 + \sum_j \frac{b_j c_j}{a_j}\right). \end{aligned} \qquad (20.3.21)$$

Now

$$\prod_i \left(1-(2z-z^2)\lambda_{2i}\right) = F_+(2z-z^2,\alpha), \tag{20.3.22}$$

while using relation (20.2.14),

$$\begin{aligned}
1+2z(1-z)&\sum_i \frac{\lambda_{2i}}{1-(2z-z^2)\lambda_{2i}} f_{2i}(\alpha)\int_0^\alpha f_{2i}(\phi)\,d\phi \\
&= \frac{1}{2-z}\left(2-z+(1-z)\sum_i \frac{(2z-z^2)\lambda_{2i}}{1-(2z-z^2)\lambda_{2i}} f_{2i}(\alpha)\int_{-\alpha}^\alpha f_{2i}(\phi)\,d\phi\right) \\
&= \frac{1}{2-z}\left(1+(1-z)\frac{F_-(2z-z^2,\alpha)}{F_+(2z-z^2,\alpha)}\right),
\end{aligned} \tag{20.3.23}$$

so that finally,

$$(2-z)F_1(z,\alpha) = F_+(2z-z^2,\alpha)+(1-z)F_-(2z-z^2,\alpha). \tag{20.3.24}$$

This equation is equivalent to (20.1.20), since the left-hand side of Eq. (20.3.24) is

$$\begin{aligned}
(2-z)F_1(z,\alpha) &= \sum_r \left((1-z)^r E_1(r,\alpha)+(1-z)^{r+1}E_1(r,\alpha)\right) \\
&= \sum_r (1-z)^{2r}\left[E_1(2r,\alpha)+E_1(2r-1,\alpha)\right] \\
&\quad + \sum_r (1-z)^{2r+1}\left[E_1(2r,\alpha)+E_1(2r+1,\alpha)\right],
\end{aligned} \tag{20.3.25}$$

while on the right-hand side

$$\begin{aligned}
F_+(2z-z^2,\alpha) &= \sum_r (1-2z+z^2)^r E_+(r,\alpha) \\
&= \sum_r (1-z)^{2r} E_+(r,\alpha),
\end{aligned} \tag{20.3.26}$$

$$\begin{aligned}
(1-z)F_-(2z-z^2,\alpha) &= (1-z)\sum_r (1-2z+z^2)^r E_-(r,\alpha) \\
&= \sum_r (1-z)^{2r+1} E_-(r,\alpha).
\end{aligned} \tag{20.3.27}$$

Comparing the various powers of $(1-z)$, we get the equivalence of Eqs. (20.1.20) and (20.3.24).

One could have started with the lower sign in Eq. (20.3.1), and used Eq. (20.2.23) to arrive at the same result.

20.4 Relation Between $F_4(z,s)$ and $F_\pm(z,s)$

For the symplectic ensemble we can again separate $K(4;\theta,\phi)$ into even and odd parts,

$$K(4;\theta,\phi)=\sigma_+(\theta,\phi)+\sigma_-(\theta,\phi), \tag{20.4.1}$$

$$\sigma_\pm(\theta,\phi)=\frac{1}{2}\begin{bmatrix} S_\pm(2\theta,2\phi) & D_\mp(2\theta,2\phi) \\ I_\pm(2\theta,2\phi) & S_\mp(2\theta,2\phi) \end{bmatrix}, \tag{20.4.2}$$

where $S_\pm(\theta,\phi)$, $D_\pm(\theta,\phi)$ and $I_\pm(\theta,\phi)$ are given by Eqs. (20.3.2), (20.3.3) and a similar equation

$$I_\pm(\theta,\phi)=\frac{1}{2}\big[I(\theta,\phi)\pm I(\theta,-\phi)\big]. \tag{20.4.3}$$

The eigenvalues of the integral equation (20.1.3) are again also the eigenvalues of an integral equation with the kernel either $\sigma_+(\theta,\phi)$ or $\sigma_-(\theta,\phi)$ and the components of the eigenfunctions have definite opposite parities. It is convenient to take 2θ, 2ϕ as new variables and write the integral equation (20.1.3) as

$$\mu\begin{bmatrix}\xi(\theta)\\ \eta(\theta)\end{bmatrix}=\frac{1}{2}\int_{-2\alpha}^{2\alpha}\begin{bmatrix} S_\pm(\theta,\phi) & D_\mp(\theta,\phi) \\ I_\pm(\theta,\phi) & S_\mp(\theta,\phi) \end{bmatrix}\begin{bmatrix}\xi(\phi)\\ \eta(\phi)\end{bmatrix}d\phi. \tag{20.4.4}$$

Following Section 20.3 we find now $\xi(\theta)=\eta'(\theta)$. The arguments proceed as in Section 20.3, equations corresponding to (20.3.18) and (20.3.19) are now

$$(\mu-\lambda_{2i})c_i+\lambda_{2i}f_{2i}(2\alpha)\sum_j c_j\int_0^{2\alpha}f_{2j}(\phi)\,d\phi=0, \tag{20.4.5}$$

$$\det\left[(\mu-\lambda_{2i})\delta_{ij}+\lambda_{2i}f_{2i}(2\alpha)\int_0^{2\alpha}f_{2j}(\phi)\,d\phi\right]$$
$$=\prod_i(\mu-\lambda_{2i})\left[1+\sum_j\frac{\lambda_{2j}}{1-\lambda_{2j}}f_{2j}(2\alpha)\int_0^{2\alpha}f_{2j}(\phi)\,d\phi\right]=0 \tag{20.4.6}$$

so that with Eq. (20.2.14)

$$F_4(z,\alpha)=\prod_i(1-z\lambda_{2i})\left[1+\sum_j\frac{z\lambda_{2j}}{1-z\lambda_{2j}}f_{2j}(2\alpha)\int_0^{2\alpha}f_{2j}(\phi)\,d\phi\right]$$
$$=F_+(z,2\alpha)\left[\frac{1}{2}+\frac{1}{2}\frac{F_-(z,2\alpha)}{F_+(z,2\alpha)}\right]. \tag{20.4.7}$$

The final result is

$$F_4(z,\alpha) = \frac{1}{2}\big[F_+(z,2\alpha) + F_-(z,2\alpha)\big], \tag{20.4.8}$$

which is Eq. (20.1.21).

20.5 Power Series Expansions of $E_\beta(r,s)$

Dietz and Haake derived the general term in the power series expansions of $E_\beta(0,s)$. Their method adapted to the case of $E_\beta(r,s)$ is presented here.

Consider the $N \times N$ determinants,

$$f(z,s) = \det\left[\delta_{jk} - z\frac{\sin(j-k)\pi s/N}{(j-k)\pi}\right], \tag{20.5.1}$$

$$f_\pm(z,s) = \det\left[\delta_{jk} - \frac{z}{2}\left(\frac{\sin(j-k)\pi s/N}{(j-k)\pi} \pm \frac{\sin(j+k)\pi s/N}{(j+k)\pi}\right)\right]. \tag{20.5.2}$$

In the limit $N \to \infty$, they become (see Chapters 6 and 11) the functions $F_2(z,s)$, $F_\pm(z,s)$ of Eqs. (20.1.15)–(20.1.17) giving the spacing functions $E_\beta(r,s)$ through Eqs. (20.1.1), (20.1.18), (20.1.20), (20.1.21).

Following Dietz and Haake we expand $f(z,s)$ in powers of s and then take the limit $N \to \infty$. This will give us the coefficient of s^m for any m. The same will be true for $f_\pm(z,s)$, if we restrict ourselves to the even or odd contributions. The expansions of $E_2(r,s)$ and $E_\pm(r,s)$ will then be obtained by differentiation with respect to z. The power series of $E_1(r,s)$ and $E_4(r,s)$ can then be deduced from those of $E_\pm(r,s)$ using Eqs. (20.1.20) and (20.1.21).

To get the power series

$$f(z,s) = \sum_{m=0}^{\infty} f_m(z)\cdot s^m, \tag{20.5.3}$$

let us write the $N \times N$ determinant as a sum of products of N elements. In each element we insert the power series of $\sin x$ in powers of $x = \pi s(j-k)/N$. Finally, we use the binomial expansion of $(j-k)^{2n}$ to get

$$f(z,s) = \sum_P (-1)^P \prod_{k=1}^{N}\left\{\delta_{k.Pk} + \frac{z}{\pi}\sum_{m=0}^{\infty}(-1)^{m+1}\frac{(\pi s/N)^{2m+1}}{(2m+1)!}\right.$$
$$\left.\times\left(\sum_{q=0}^{2m}\binom{2m}{q}(-k)^{2m-q}(Pk)^q\right)\right\}, \tag{20.5.4}$$

where P is a permutation of the indices $1, 2, \ldots, N$ and $(-1)^P$ its parity.

The zeroth order coefficient is evidently $f_0(z) = 1$, as it should. For the coefficient $f_m(z)$, consider the case when n out of the N factors in the sum (20.5.4) are of non-vanishing order in s. Let these orders be $2m_1 + 1, 2m_2 + 1, \ldots, 2m_n + 1$. The remaining $N - n$ factors in that term are of the type $\delta_{k,Pk}$. The contribution of such terms with fixed $m_1, m_2, \ldots, m_n$ is

$$T\big(z, s, n, \{m_i\}\big) = \frac{z^n}{\pi^n n!} \sum_{k_1=1}^{N} \cdots \sum_{k_n=1}^{N} (-1)^{m_1+m_2+\cdots+m_n+n}$$
$$\times \sum_{q_1=0}^{2m_1} \cdots \sum_{q_n=0}^{2m_n} \prod_{j=1}^{n} \frac{(\pi s/N)^{2m_j+1}}{(2m_j+1)!} \binom{2m_j}{q_j} (-k_j)^{2m_j-q_j}$$
$$\times \sum_{\overline{P}} (-1)^{\overline{P}} (\overline{P}k_1)^{q_1} \cdots (\overline{P}k_n)^{q_n}. \tag{20.5.5}$$

The $n!$ permutations $\overline{P}$ to be summed over in (20.5.5) act on the n indices in the n summation variables $k_1, \ldots, k_n$; they are the remaining ones of the original $N!$ permutations P in Eq. (20.5.4), if one takes account of the factors $\delta_{k,Pk}$. The sum over $n!$ permutations $\overline{P}$ gives an $n \times n$ determinant $\det[k_i^{q_j}]$ and Eq. (20.5.5) becomes

$$T\big(z, s, n, \{m_i\}\big) = \frac{z^n}{\pi^n n!} \prod_{j=1}^{n} \frac{(-1)^{m_j+1}(\pi s/N)^{2m_j+1}}{(2m_j+1)!}$$
$$\times \sum_{q_1=0}^{2m_1} \cdots \sum_{q_n=0}^{2m_n} (-1)^{q_1+\cdots+q_n} \binom{2m_1}{q_1} \cdots \binom{2m_n}{q_n}$$
$$\times \sum_{k_1=1}^{N} \cdots \sum_{k_n=1}^{N} \det\big[k_i^{2m_i-q_i+q_j}\big]. \tag{20.5.6}$$

To get $f_m(z)$ one has to sum $T(z, s, n, \{m_i\})$ over all $m_1, \ldots, m_n$ from 0 to ∞, and over n from 1 to ∞. Whenever any $q_i = q_j$, or any $2m_i - q_i = 2m_j - q_j$, the $n \times n$ determinant in (20.5.6) is zero. Thus $T(z, s, n, \{m_i\})$ is non-zero only when (i) all the q_i are different from one another, and (ii) all the $2m_i - q_i$ are different from one another. This gives a constraint on the summation index n which contribute to the final power m of s,

$$m = \sum_{i=1}^{n} (2m_i + 1) = n + 2\sum_{i=1}^{n} m_i. \tag{20.5.7}$$

The resulting constraint is $n^2 \leqslant m$, and can be established by recursion. Also Eq. (20.5.7) tells us that $m-n$ must be an even integer. Thus

$$f_m(z) = \left(\frac{\pi}{N}\right)^m \sum_{n=1}^{\infty} (-1)^{(m+n)/2} \frac{z^n}{\pi^n n!} \sum_{m_1=0}^{\infty} \cdots \sum_{m_n=0}^{\infty} \delta\left(\frac{m-n}{2}, \sum_{i=1}^{n} m_i\right)$$
$$\times \left(\prod_{j=1}^{n} \frac{1}{(2m_j+1)!}\right) \sum_{q_1=0}^{2m_1} \cdots \sum_{q_n=0}^{2m_n} \left(\prod_{j=1}^{n} (-1)^{q_j} \binom{2m_j}{q_j}\right)$$
$$\times \sum_{k_1=0}^{N} \cdots \sum_{k_n=0}^{N} \det\left[k_i^{2m_i-q_i+q_j}\right]; \tag{20.5.8}$$

here and elsewhere $\delta(j,k)$ is equal to one if $j=k$ and zero if $j \neq k$.

When $N \to \infty$, the sums over the k_i can be carried out as

$$\sum_{k=1}^{N} k^x = \frac{N^{x+1}}{x+1}\big(1 + \mathrm{O}(1/N)\big), \quad \text{for } x > 0. \tag{20.5.9}$$

The sum over the k_i then gives the leading term

$$\sum_{k_1=0}^{N} \cdots \sum_{k_n=0}^{N} \det\left[k_i^{2m_i-q_i+q_j}\right] = N^m \det\left[\frac{1}{2m_i - q_i + q_j + 1}\right]. \tag{20.5.10}$$

Therefore for $m \geqslant 1$, one has

$$F_m(z) := \lim_{N\to\infty} f_m(z)$$
$$= \sum_{n=1}^{m \geqslant n^2} \frac{(-1)^{(m+n)/2} \pi^{m-n} z^n}{n!} \sum_{m_1=0}^{\infty} \cdots \sum_{m_n=0}^{\infty} \delta\left(\frac{m-n}{2}, \sum_{i=1}^{n} m_i\right)$$
$$\times \sum_{q_1=0}^{2m_1} \cdots \sum_{q_n=0}^{2m_n} \det\left[\frac{1}{2m_i - q_i + q_j + 1}\right]$$
$$\times \left(\prod_{j=1}^{n} \frac{(-1)^{q_j}}{(2m_j+1)!} \binom{2m_j}{q_j}\right). \tag{20.5.11}$$

Considering only the even or odd contributions one gets

$$\begin{aligned} F_m^{(+)}(z) &:= \lim_{N\to\infty} f_m^{(+)}(z) \\ &= \sum_{n=1}^{m\geqslant n(2n-1)} \left(\frac{\pi}{2}\right)^{m-n} \frac{(-1)^{(m+n)/2} z^n}{n!} \sum_{m_1=0}^{\infty} \cdots \sum_{m_n=0}^{\infty} \delta\left(\frac{m-n}{2}, \sum_{i=1}^{n} m_i\right) \\ &\quad \times \sum_{q_1=0}^{m_1} \cdots \sum_{q_n=0}^{m_n} \det\left[\frac{1}{2m_i - 2q_i + 2q_j + 1}\right] \\ &\quad \times \left(\prod_{j=1}^{n} \frac{1}{(2m_j+1)!} \binom{2m_j}{2q_j}\right), \end{aligned} \tag{20.5.12}$$

and

$$\begin{aligned} F_m^{(-)}(z) &:= \lim_{N\to\infty} f_m^{(-)}(z) \\ &= \sum_{n=1}^{m\geqslant n(2n+1)} \left(\frac{\pi}{2}\right)^{m-n} \frac{(-1)^{(m+3n)/2} z^n}{n!} \sum_{m_1=0}^{\infty} \cdots \sum_{m_n=0}^{\infty} \delta\left(\frac{m-n}{2}, \sum_{i=1}^{n} m_i\right) \\ &\quad \times \sum_{q_1=0}^{m_1-1} \cdots \sum_{q_n=0}^{m_n-1} \det\left[\frac{1}{2m_i - 2q_i + 2q_j + 1}\right] \\ &\quad \times \left(\prod_{j=1}^{n} \frac{1}{(2m_j+1)!} \binom{2m_j}{2q_j+1}\right). \end{aligned} \tag{20.5.13}$$

Note that the constraints on n are quite severe. For example, in Eq. (20.5.11), the smallest even (odd) value of m for which two different values of n are allowed is 16 (25); (The value $n = 1$ contributes only to $m = 1$;) and in Eqs. (20.5.12) and (20.5.13) these smallest values of m are respectively 28 (15) and 36 (21).

We can write Eq. (20.5.11) more conveniently as

$$\begin{aligned} F(z,t) &= \sum_{m=0}^{\infty} F_m(z) s^m \\ &= 1 + \sum_{n=1}^{\infty} \frac{(zs)^n}{n!} \sum_{j=n(n-1)/2}^{\infty} (-1)^j a_{jn}^{(2)} (\pi s)^{2j}, \end{aligned} \tag{20.5.14}$$

or taking account of Eq. (20.1.1),

$$E_2(r,s) = \delta(r,0) + (-1)^r \sum_{n=1}^{\infty} \binom{n}{r} \frac{s^n}{n!} \sum_{j=n(n-1)/2}^{\infty} (-1)^j a_{jn}^{(2)} (\pi s)^{2j}, \tag{20.5.15}$$

where

$$a_{jn}^{(2)} = (-1)^n \sum_{m_1=0}^{\infty} \cdots \sum_{m_n=0}^{\infty} \delta\left(j, \sum_{i=1}^{n} m_i\right) \cdot \left\{\prod_{k=1}^{n} \frac{1}{(2m_k+1)!}\right\}$$
$$\times \sum_{q_1=0}^{2m_1} \cdots \sum_{q_n=0}^{2m_n} \det\left[\frac{1}{2m_i - q_i + q_k + 1}\right] \cdot \left\{\prod_{k=1}^{n} (-1)^{q_k} \binom{2m_k}{q_k}\right\}. \tag{20.5.16}$$

Similarly, Eqs. (20.5.12), (20.5.13), (20.1.18) and (20.1.21) can be written as

$$F_\pm(z,t) = \sum_{m=0}^{\infty} F_m^{(\pm)}(z) s^m = 1 + \sum_{n=1}^{\infty} \frac{(zs)^n}{n!} \sum_{j=n_\pm}^{\infty} (-1)^j a_{jn}^{(\pm)} (\pi s)^{2j}, \tag{20.5.17}$$

$$E_\pm(r,s) = \delta(r,0) + (-1)^r \sum_{n=1}^{\infty} \binom{n}{r} \frac{s^n}{n!} \sum_{j=n_\pm}^{\infty} (-1)^j a_{jn}^{(\pm)} (\pi s)^{2j}, \tag{20.5.18}$$

$$E_4(r,s) = \delta(r,0) + (-1)^r \sum_{n=1}^{\infty} \binom{n}{r} \frac{s^n}{n!} \sum_{j=n(n-1)}^{\infty} (-1)^j a_{jn}^{(4)} (\pi s)^{2j}, \tag{20.5.19}$$

where $n_+ = n(n-1)$, $n_- = n^2$,

$$a_{jn}^{(+)} = (-1)^n 2^{-2j} \sum_{m_1=0}^{\infty} \cdots \sum_{m_n=0}^{\infty} \delta\left(j, \sum_{i=1}^{n} m_i\right) \cdot \left\{\prod_{k=1}^{n} \frac{1}{(2m_k+1)!}\right\}$$
$$\times \sum_{q_1=0}^{m_1} \cdots \sum_{q_n=0}^{m_n} \det\left[\frac{1}{2m_i - 2q_i + 2q_k + 1}\right] \cdot \left\{\prod_{k=1}^{n} \binom{2m_k}{2q_k}\right\}, \tag{20.5.20}$$

$$a_{jn}^{(-)} = 2^{-2j} \sum_{m_1=0}^{\infty} \cdots \sum_{m_n=0}^{\infty} \delta\left(j, \sum_{i=1}^{n} m_i\right) \cdot \left\{\prod_{k=1}^{n} \frac{1}{(2m_k+1)!}\right\}$$
$$\times \sum_{q_1=0}^{m_1-1} \cdots \sum_{q_n=0}^{m_n-1} \det\left[\frac{1}{2m_i - 2q_i + 2q_k + 1}\right] \cdot \left\{\prod_{k=1}^{n} \binom{2m_k}{2q_k+1}\right\}, \tag{20.5.21}$$

and

$$a_{jn}^{(4)} = 2^{n+2j-1}\left(a_{jn}^{(+)} + a_{jn}^{(-)}\right). \tag{20.5.22}$$

For $n = 1$, the sums appearing in Eqs. (20.5.16), (20.5.20) and (20.5.21) are simple,

$$a_{j1}^{(\pm)} = -\frac{1}{2}\delta(j,0) \mp \left[2(2j+1)(2j+1)!\right]^{-1}, \tag{20.5.23}$$

$$a_{j1}^{(2)} = a_{j1}^{(4)} = -\delta(j,0); \tag{20.5.24}$$

while for $n = 2$, we have (see Appendix A.42)

$$a_{j2}^{(\pm)} = \frac{1}{(2j+2)!}\left(\pm\frac{j+1}{2j+1} - \frac{2^{2j}}{(j+1)(2j+1)}\right) + \frac{1}{(2j+2)!}\sum_{i=0}^{j}\frac{1}{(2i+1)(2j-2i+1)}\left\{\frac{1}{4}\binom{2j+2}{2i+1} \mp 1\right\}, \tag{20.5.25}$$

$$a_{j2}^{(2)} = -2^{2j+1}\left[(j+1)(2j+1)(2j+2)!\right]^{-1}, \tag{20.5.26}$$

$$a_{j2}^{(4)} = \frac{2^{2j}}{(2j+2)!}\left(-\frac{2^{2j+2}}{(j+1)(2j+1)} + \sum_{i=0}^{j}\frac{1}{(2i+1)(2j-2i+1)}\binom{2j+2}{2i+1}\right); \tag{20.5.27}$$

and for $n = 3$,

$$a_{j3}^{(2)} = \frac{2^{2j+1}\cdot 3}{(j+1)(2j+1)(2j+2)!}\left(1 - \frac{4(2j+1)}{(2j+3)^2}\sum_{\nu=1}^{2j+2}\frac{1}{\nu}\right). \tag{20.5.28}$$

The general expressions for $a_{jn}^{(\pm)}$ or for $a_{jn}^{(2)}$ when $n \geqslant 3$ are not as simple. The values of $a_{j2}^{(\pm)}$, $a_{j3}^{(\pm)}$, $a_{j3}^{(2)}$ and $a_{j4}^{(2)}$, for relatively low values of j can be calculated with some effort. A small list of them is given in Appendix A.43.

Summary of Chapter 20

New proofs of the known Eqs. (20.1.20) and (20.1.21) are given. They along with Eqs. (20.1.1), (20.1.15)–(20.1.18) relate the spacing functions $E_\beta(r,s)$ for $\beta = 1$, 2 and 4. Equations (20.2.14) and (20.2.24) relating odd and even spheroidal functions are recovered on the way. Equations (20.1.20) and (20.1.21) are useful to derive the power series expansions and the asymptotic behaviour of the spacing functions $E_\beta(r,s)$.

21

FREDHOLM DETERMINANTS AND PAINLEVÉ EQUATIONS

21.1 Introduction

Around 1980 Jimbo et al. (1980) discovered that the logarithmic derivative of the spacing function $E_2(0, s)$ satisfies a second order non-linear differential equation, and they even used it to compute a few more terms in the power series expansion of $E_2(0, s)$. However, this discovery was not widely appreciated for more than ten years. The reason perhaps was that the length of the paper with such an abundant notation and complicated logic discouraged the others not familiar with the mathematical language (cf. Appendix A.44.)

Actually related to the Fredholm determinant of the sine kernel $K(x, y) = \sin((x - y)\pi)/((x - y)\pi)$ one can introduce one complex and two real functions which satisfy five relations involving their first derivatives. Elimination of all but one of these functions leads to a second order non-linear differential equation satisfied by that function; the differential equation discovered by Jimbo et al. is one of them. All of these equations are the so called Painlevé equations.

Consider the integral equation

$$\lambda g(x) = \int_{-t}^{t} K(x, y) g(y)\, dy, \tag{21.1.1}$$

$$K(x, y) = \frac{\sin(x - y)\pi}{(x - y)\pi}, \tag{21.1.2}$$

whose solutions can always be chosen to be either even, $g(-x) = g(x)$, or odd, $g(-x) = -g(x)$. The set of even solutions $g_{2n}(x)$, labeled by even subscripts, consists of all the solutions of an integral equation obtained from (21.1.1) when $K(x, y)$ there is replaced by

$$K_{+}(x, y) = \frac{1}{2}\{K(x, y) + K(-x, y)\}. \tag{21.1.3}$$

Similarly the set of odd solutions $g_{2n+1}(x)$ of (21.1.1), labeled by odd subscripts, consists of all the solutions of an integral equation with the kernel

$$K_{-}(x, y) = \frac{1}{2}\{K(x, y) - K(-x, y)\}. \tag{21.1.4}$$

Let us denote the Fredholm determinants corresponding to $K(x, y)$, $K_{+}(x, y)$ and $K_{-}(x, y)$, by $F(z, t)$, $F_{+}(z, t)$ and $F_{-}(z, t)$, respectively

$$F(z, t) := F_{+}(z, t)F_{-}(z, t) = \det[1 - zK_t] = \prod_{n=0}^{\infty}(1 - z\lambda_n) \tag{21.1.5}$$

$$= \sum_{n=0}^{\infty}\frac{(-z)^n}{n!}\int_{-t}^{t} dx_1 \cdots dx_n \det\big[K(x_i, x_j)\big]_{i,j=1,\dots,n}, \tag{21.1.6}$$

and similar expressions for $F_{+}(z, t)$ and $F_{-}(z, t)$. In Eq. (21.1.5) K_t is the integral operator with kernel K on the interval $(-t, t)$. For convenience we will not indicate the dependence on the real variable z in what follows.

Let us define the functions A, B and S as follows.

$$A(t) = -\frac{1}{2}\frac{d}{dt}\big[\log F_{+}(z, t) + \log F_{-}(z, t)\big], \tag{21.1.7}$$

$$B(t) = -\frac{1}{2}\frac{d}{dt}\big[\log F_{+}(z, t) - \log F_{-}(z, t)\big], \tag{21.1.8}$$

and

$$S(t) = e^{i\pi t} + \sum_{n=1}^{\infty} z^n \int_{-t}^{t} dx\, e^{i\pi x} K^{(n)}(x, t)$$

$$= e^{i\pi t} + \sum_{n=1}^{\infty} z^n \int_{-t}^{t} dx\, e^{-i\pi x} K^{(n)}(x, -t), \tag{21.1.9}$$

where the nth power of $K(x, y) = \sin(x-y)\pi/(x-y)\pi$, Eq. (21.1.2), is defined by

$$K^{(1)}(x, y) = K(x, y),$$
$$K^{(n)}(x, y) = \int_{-t}^{t} d\xi\, K(x, \xi) K^{(n-1)}(\xi, y), \quad n > 1. \tag{21.1.10}$$

Among these three functions A, B and S we have the following five relations

$$\frac{dA}{dt} = 2B^2, \tag{21.1.11}$$
$$\frac{d}{dt}(tA) = z|S|^2, \tag{21.1.12}$$
$$\frac{d}{dt}(tB) = z \operatorname{Re} S^2, \tag{21.1.13}$$
$$2\pi t B = z \operatorname{Im} S^2, \tag{21.1.14}$$
$$\frac{dS}{dt} = i\pi S + 2S^* B, \tag{21.1.15}$$
$$= i\pi S - \frac{iz}{2\pi t} S^* \left(S^2 - S^{*2}\right), \tag{21.1.16}$$

where "Re" ("Im") means "the real (imaginary) part of", $i = \sqrt{-1}$ and S^* is the complex conjugate of S.

We will also prove (cf. Eqs. (20.2.14) and (20.2.23)) that

$$\frac{F_-(z,t)}{F_+(z,t)} = 1 + \sum_{n=0}^{\infty} \frac{z\lambda_{2n}}{1 - z\lambda_{2n}} g_{2n}(t) \int_{-t}^{t} g_{2n}(x)\, dx \div \int_{-t}^{t} g_{2n}^2(x)\, dx, \tag{21.1.17}$$

$$\frac{F_+(z,t)}{F_-(z,t)} = 1 - 2zI(t) + \sum_{n=0}^{\infty} \frac{2z\lambda_{2n+1}}{1 - z\lambda_{2n+1}} g_{2n+1}(t) \int_0^t g_{2n+1}(x)\left[1 - 2I(x)\right] dx$$
$$\div \int_{-t}^{t} g_{2n+1}^2(x)\, dx, \tag{21.1.18}$$

$$I(x) = \int_0^x \frac{\sin(\pi y)}{\pi y}\, dy. \tag{21.1.19}$$

The "sine kernel" $K(x, y) = \sin(x-y)\pi/(x-y)\pi$ is the first important one of a class of kernels of the form $[\varphi(x)\psi'(y) - \varphi'(x)\psi(y)]/(x-y)$ when the functions $\varphi(x)$ and $\psi(x)$ satisfy certain differential equations of the form

$$m(x)\varphi'(x) = M(x)\varphi(x) + N(x)\psi(x),$$

$$m(x)\psi'(x) = -P(x)\varphi(x) - M(x)\psi(x), \tag{21.1.20}$$

with m, M, N and P polynomials. The Fredholm determinant of such kernels over a domain J, the union of a finite number of disjoint intervals $J = \bigcup_{i=1}^{m}(a_{2i-1}, a_{2i})$, as a function of the end points a_j satisfies a set of coupled partial differential equations (cf. Tracy and Widom, 1994b).

Here we will consider only the sine kernel in the simplest case of one connected interval $(-t, t)$.

21.2 Proof of Eqs. (21.1.11)–(21.1.17)

The proof consists in expanding everything in powers of z and comparing coefficients of z^n. The variable t appears at two places, in the limits of integration and in the integrand. When there is a derivative under the integral sign, we follow a method first used by Gaudin; replace $(\partial/\partial x)K(x, y)$ by $-(\partial/\partial y)K(x, y)$ and integrate by parts so as to shift the derivative by one step. Repeating this several times, the derivative is moved from one end to the other where it combines with another derivative, if any. All integrated terms are collected on the way. This way of power series expansions, though a little cumbersome, may be easy to follow. There are more elegant ways of presenting them in terms of projectors in the Hilbert space using Dirac's compact notation. Fortunately at least two such "translations" exist in print (Dyson, 1995; Tracy and Widom, 1994b, 1994c) and still another one due to Mahoux is given in Appendix A.16. Readers familiar with formal mathematical manipulations may prefer to replace the rest of this section by Appendix A.16. Also the original Japanese version (Jimbo et al., 1980) probably contains more ideas still to be exploited.

As the eigenvalues λ_j of the integral equation (21.1.1) are known to be all real and $1 \geqslant \lambda_0 \geqslant \lambda_1 \geqslant \cdots \geqslant 0$, the series expansions in powers of z used in the following are absolutely convergent for $|z| < 1/\lambda_0 \leqslant 1$ and the heuristic proofs given here can perfectly be justified. In what follows z is a real variable with $0 \leqslant z \leqslant 1$.

From the Fredholm theory of integral equations (Goursat, 1956), one can write symbolically,

$$\begin{aligned}\log F(z, t) &= \log\det[1 - zK_t] \\ &= \operatorname{Tr}\log[1 - zK_t] \qquad (21.2.1) \\ &= -\sum_{n=1}^{\infty} \frac{z^n}{n} \operatorname{Tr}\big[K_t^{(n)}\big] \\ &= -\sum_{n=1}^{\infty} \frac{z^n}{n} \int_{-t}^{t} dx_1 \cdots dx_n\, K(x_1, x_2)K(x_2, x_3) \cdots K(x_n, x_1), \qquad (21.2.2)\end{aligned}$$

$$(K_t f)(x) = \int_{-t}^{t} dy\, K(x, y) f(y). \tag{21.2.3}$$

For $F_{\pm}(z, t)$, this same Eq. (21.2.1) holds provided we replace $K(x, y)$ by $K_{\pm}(x, y)$. Differentiating this equation with respect to t, we have

$$\begin{aligned}\frac{\partial}{\partial t} \log F(z, t) &= -zK(t, t) - zK(-t, -t) \\ &\quad - \sum_{n=1}^{\infty} z^{n+1} \int_{-t}^{t} dx_1 \cdots dx_n \big\{ K(t, x_1) K(x_1, x_2) \cdots K(x_n, t) \\ &\quad + K(-t, x_1) K(x_1, x_2) \cdots K(x_n, -t) \big\} \\ &= -2zK(t, t) - 2 \sum_{n=1}^{\infty} z^{n+1} \int_{-t}^{t} dx_1 \cdots dx_n\, K(t, x_1) K(x_1, x_2) \cdots K(x_n, t).\end{aligned} \tag{21.2.4}$$

To derive the last equality, we have used the fact that

$$K(x, y) = K(-x, -y). \tag{21.2.5}$$

Similarly we have

$$\begin{aligned}\frac{\partial}{\partial t} \log F_{\pm}(z, t) &= -2zK_{\pm}(t, t) - 2 \sum_{n=1}^{\infty} z^{n+1} \int_{-t}^{t} dx_1 \cdots dx_n \\ &\quad \times K_{\pm}(t, x_1) K_{\pm}(x_1, x_2) \cdots K_{\pm}(x_{n-1}, x_n) K_{\pm}(x_n, t) \\ &= -2zK_{\pm}(t, t) - 2 \sum_{n=1}^{\infty} z^{n+1} \int_{-t}^{t} dx_1 \cdots dx_n \\ &\quad \times K(t, x_1) K(x_1, x_2) \cdots K(x_{n-1}, x_n) K_{\pm}(x_n, t).\end{aligned} \tag{21.2.6}$$

Thus $A(t)$ and $B(t)$ defined in Eqs. (21.1.7) and (21.1.8) can be written as follows

$$\begin{aligned}A(t) &= zK(t, t) + \sum_{n=1}^{\infty} z^{n+1} \int_{-t}^{t} dx_1 \cdots dx_n\, K(t, x_1) K(x_1, x_2) \cdots K(x_{n-1}, x_n) K(x_n, t) \\ &= \big(zK(1 - zK)^{-1}\big)(t, t),\end{aligned} \tag{21.2.7}$$

and

$$\begin{aligned} B(t) &= zK(t,-t) \\ &\quad + \sum_{n=1}^{\infty} z^{n+1} \int_{-t}^{t} dx_1 \cdots dx_n\, K(t,x_1)K(x_1,x_2)\cdots K(x_{n-1},x_n)K(x_n,-t) \\ &= \big(zK(1-zK)^{-1}\big)(t,-t) \\ &= \big(zK(1-zK)^{-1}\big)(-t,t). \end{aligned} \tag{21.2.8}$$

Differentiating $A(t)$ we get

$$\frac{dA(t)}{dt} = A_1 + A_2 + A_3 + A_4, \tag{21.2.9}$$

where

$$A_1 = \sum_{n=1}^{\infty} z^{n+1} \int_{-t}^{t} dx_1 \cdots dx_n \frac{\partial K(t,x_1)}{\partial t} K(x_1,x_2)\cdots K(x_{n-1},x_n)K(x_n,t), \tag{21.2.10}$$

$$A_2 = \sum_{n=1}^{\infty} z^{n+1} \int_{-t}^{t} dx_1 \cdots dx_n\, K(t,x_1)K(x_1,x_2)\cdots K(x_{n-1},x_n)\frac{\partial K(x_n,t)}{\partial t}, \tag{21.2.11}$$

$$\begin{aligned} A_3 &= \sum_{n=1}^{\infty} z^{n+1} \sum_{j=1}^{n} \int_{-t}^{t} dx_1 \cdots dx_{j-1}\, dx_{j+1} \cdots dx_n \\ &\quad \times K(t,x_1)\cdots K(x_{j-1},t)K(t,x_{j+1})\cdots K(x_n,t), \end{aligned} \tag{21.2.12}$$

$$\begin{aligned} A_4 &= \sum_{n=1}^{\infty} z^{n+1} \sum_{j=1}^{n} \int_{-t}^{t} dx_1 \cdots dx_{j-1}\, dx_{j+1} \cdots dx_n \\ &\quad \times K(t,x_1)\cdots K(x_{j-1},-t)K(-t,x_{j+1})\cdots K(x_n,t). \end{aligned} \tag{21.2.13}$$

Now as $K(x,y)$ depends only on the difference $x-y$, we can, following Gaudin, replace $\partial K(x_n,t)/\partial t$ in A_2 by $-\partial K(x_n,t)/\partial x_n$. Integration by parts with respect to x_n pushes the derivative one step left. In the integral

$$\int_{-t}^{t} dx_1 \cdots dx_n\, K(t,x_1)K(x_1,x_2)\cdots \frac{\partial K(x_{n-1},x_n)}{\partial x_n} K(x_n,t) \tag{21.2.14}$$

replace $\partial K(x_{n-1},x_n)/\partial x_n$ by $-\partial K(x_{n-1},x_n)/\partial x_{n-1}$ and integrate by parts over x_{n-1}. And so on, till the partial derivative is pushed to the extreme left so as to be recognized as A_1. All the integrated terms reconstitute A_3 and A_4. Thus these step by step integrations give

$$A_2 = -A_1 - A_3 + A_4. \tag{21.2.15}$$

Also one can convince oneself from Eqs. (21.2.8) and (21.2.13) that

$$A_4 = B^2(t). \tag{21.2.16}$$

With Eqs. (21.2.9), (21.2.15) and (21.2.16) the proof of Eq. (21.1.11) is complete. Note that we have used only the property that $K(x, y)$ is an even function of $x - y$.

To arrive at Eqs. (21.1.12)–(21.1.14) we need to use the explicit form (21.1.2) of $K(x, y)$. With the notation,

$$k(x, y) = \frac{\sin(x-y)\pi t}{(x-y)\pi} = \frac{1}{2}\int_{-t}^{t} d\xi \exp i\pi(x-y)\xi \tag{21.2.17}$$

we get from Eq. (21.2.7) by a change of variables $x_j \to t x_j$,

$$\begin{aligned} tA(t) &= zt + \sum_{n=1}^{\infty} z^{n+1} \int_{-1}^{1} dx_1 \cdots dx_n\, k(1, x_1)k(x_1, x_2)\cdots k(x_{n-1}, x_n)k(x_n, 1) \\ &= zt + \sum_{n=1}^{\infty} \left(\frac{z}{2}\right)^{n+1} \int_{-1}^{1} dx_1 \cdots dx_n \int_{-t}^{t} dy_1 \cdots dy_{n+1} \\ &\quad \times \exp i\pi(y_1 - y_1x_1 + y_2x_1 - y_2x_2 + \cdots + y_nx_{n-1} - y_nx_n + y_{n+1}x_n - y_{n+1}). \end{aligned} \tag{21.2.18}$$

Similarly

$$\begin{aligned} tB(t) &= z\frac{\sin(2\pi t)}{2\pi} + \sum_{n=1}^{\infty} z^{n+1} \int_{-1}^{1} dx_1 \cdots dx_n \\ &\quad \times k(1, x_1)k(x_1, x_2)\cdots k(x_{n-1}, x_n)k(x_n, -1) \\ &= z\frac{\sin(2\pi t)}{2\pi} + \sum_{n=1}^{\infty} \left(\frac{z}{2}\right)^{n+1} \int_{-1}^{1} dx_1 \cdots dx_n \int_{-t}^{t} dy_1 \cdots dy_{n+1} \\ &\quad \times \exp i\pi(y_1 - y_1x_1 + y_2x_1 - y_2x_2 + \cdots + y_nx_{n-1} - y_nx_n + y_{n+1}x_n + y_{n+1}). \end{aligned} \tag{21.2.19}$$

A differentiation with respect to t gives

$$\frac{d}{dt}(tA(t)) = X_1 + X_2, \tag{21.2.20}$$

$$X_1 = \frac{z}{2} + \sum_{n=1}^{\infty} \left(\frac{z}{2}\right)^{n+1} \sum_{j=1}^{n+1} \int_{-1}^{1} dx_1 \cdots dx_n \int_{-t}^{t} dy_1 \cdots dy_{j-1} \, dy_{j+1} \cdots dy_{n+1}$$
$$\times \exp i\pi (y_1 - y_1 x_1 + \cdots - y_{j-1} x_{j-1} + t x_{j-1} - t x_j$$
$$+ y_{j+1} x_j - \cdots + y_{n+1} x_n - y_{n+1}), \tag{21.2.21}$$

$$X_2 = \frac{z}{2} + \sum_{n=1}^{\infty} \left(\frac{z}{2}\right)^{n+1} \sum_{j=1}^{n+1} \int_{-1}^{1} dx_1 \cdots dx_n \int_{-t}^{t} dy_1 \cdots dy_{j-1} \, dy_{j+1} \cdots dy_{n+1}$$
$$\times \exp i\pi (y_1 - y_1 x_1 + \cdots - y_{j-1} x_{j-1} - t x_{j-1} + t x_j$$
$$+ y_{j+1} x_j - \cdots + y_{n+1} x_n - y_{n+1}). \tag{21.2.22}$$

Now one can convince oneself that

$$X_1 = \frac{z}{2} S S^* = X_2, \tag{21.2.23}$$

with $S = S(t)$ given by Eq. (21.1.9), and S^* is its complex conjugate.

Similar manipulations will give

$$\frac{d}{dt}(t B(t)) = B_1 + B_2, \tag{21.2.24}$$

$$B_1 = \frac{z}{2} e^{2i\pi t} + \sum_{n=1}^{\infty} \left(\frac{z}{2}\right)^{n+1} \sum_{j=1}^{n+1} \int_{-1}^{1} dx_1 \cdots dx_n \int_{-t}^{t} dy_1 \cdots dy_{j-1} \, dy_{j+1} \cdots dy_{n+1}$$
$$\times \exp i\pi (y_1 - y_1 x_1 + \cdots - y_{j-1} x_{j-1} + t x_{j-1} - t x_j + y_{j+1} x_j - \cdots$$
$$+ y_{n+1} x_n + y_{n+1})$$
$$= \frac{z}{2} S^2 \tag{21.2.25}$$

and

$$B_2 = \frac{z}{2} e^{-2i\pi t} + \sum_{n=1}^{\infty} \left(\frac{z}{2}\right)^{n+1} \sum_{j=1}^{n+1} \int_{-1}^{1} dx_1 \cdots dx_n \int_{-t}^{t} dy_1 \cdots dy_{j-1} \, dy_{j+1} \cdots dy_{n+1}$$
$$\times \exp i\pi (y_1 - y_1 x_1 + \cdots - y_{j-1} x_{j-1} - t x_{j-1} + t x_j + y_{j+1} x_j - \cdots$$
$$+ y_{n+1} x_n + y_{n+1})$$
$$= \frac{z}{2} S^{*2}. \tag{21.2.26}$$

Equations (21.2.20) and (21.2.23) give Eq. (21.1.12), while Eqs. (21.2.24)–(21.2.26) give (21.1.13).

To get Eq. (21.1.14) we will calculate the imaginary part of $S^2(t)$. For this purpose let us first integrate over all the variables y_j in $S(t)$ to write Eq. (21.1.9) in the form

$$S(t) = e^{i\pi t} + \sum_{n=1}^{\infty} z^n \int_{-1}^{1} dx_1 \cdots dx_n \, e^{i\pi t x_1} k(x_1, x_2) k(x_2, x_3) \cdots k(x_{n-1}, x_n) k(x_n, 1) \tag{21.2.27}$$

where $k(x, y)$ is given by Eq. (21.2.17). Changing the sign of each of the integration variables, we can write this equation also as

$$S(t) = e^{i\pi t} + \sum_{n=1}^{\infty} z^n \int_{-1}^{1} d\xi_1 \cdots d\xi_n \, e^{-i\pi t \xi_1} k(\xi_1, \xi_2) k(\xi_2, \xi_3) \cdots k(\xi_{n-1}, \xi_n) k(\xi_n, -1). \tag{21.2.28}$$

Multiplying the two expressions (21.2.27) and (21.2.28) we see that the coefficient of z^n in $S^2(t)$ is

$$\begin{aligned}
&\int_{-1}^{1} d\xi_1 \cdots d\xi_n \, e^{i\pi t(1-\xi_1)} k(\xi_1, \xi_2) \cdots k(\xi_{n-1}, \xi_n) k(\xi_n, -1) \\
&\quad + \int_{-1}^{1} dx_1 \cdots dx_n \, e^{i\pi t(1+x_1)} k(x_1, x_2) \cdots k(x_{n-1}, x_n) k(x_n, 1) \\
&\quad + \sum_{j=2}^{n-1} \int_{-1}^{1} dx_1 \cdots dx_j \, d\xi_1 \cdots d\xi_{n-j} \, e^{i\pi(x_1-\xi_1)} k(x_1, x_2) \cdots k(x_{j-1}, x_j) k(x_j, 1) \\
&\quad \times k(\xi_1, \xi_2) \cdots k(\xi_{n-j-1}, \xi_{n-j}) k(\xi_{n-j}, -1).
\end{aligned} \tag{21.2.29}$$

Its imaginary part is therefore

$$\begin{aligned}
&\int_{-1}^{1} dx_1 \cdots dx_n \, k(1, x_1) k(x_1, x_2) \cdots k(x_{n-1}, x_n) k(x_n, -1) \\
&\quad \times \pi \left(1 - x_1 + x_n + 1 + \sum_{j=2}^{n-1} (x_{j-1} - x_j) \right) \\
&\quad = 2\pi \int_{-1}^{1} dx_1 \cdots dx_n \, k(1, x_1) k(x_1, x_2) k(x_1, x_2) \cdots k(x_{n-1}, x_n) k(x_n, -1).
\end{aligned} \tag{21.2.30}$$

But on integrating over all the variables y_j in Eq. (21.2.19) (the first equality there) one sees that the coefficient of z^{n+1} in $tB(t)$ is

$$\int_{-1}^{1} dx_1 \cdots dx_n\, k(1, x_1)k(x_1, x_2)k(x_1, x_2) \cdots k(x_{n-1}, x_n)k(x_n, -1). \quad (21.2.31)$$

Thus

$$2\pi t B(t) = z \operatorname{Im} S^2(t). \quad (21.2.32)$$

This is Eq. (21.1.14). Note that unlike Eq. (21.1.11), this one is valid only for the particular kernel (21.1.2).

To differentiate $S(t)$, it is convenient first to write it in the form (Eqs. (21.1.9) and (21.2.27))

$$\begin{aligned} S(t) &= e^{i\pi t} + \sum_{n=1}^{\infty} z^n \int_{-1}^{1} dx_1 \cdots dx_n\, e^{i\pi t x_1} k(x_1, x_2) \cdots k(x_{n-1}, x_n)k(x_n, 1) \\ &= e^{i\pi t} + \sum_{n=1}^{\infty} z^n \int_{-t}^{t} dx_1 \cdots dx_n\, e^{i\pi x_1} K(x_1, x_2) \cdots K(x_{n-1}, x_n)K(x_n, t), \end{aligned} \quad (21.2.33)$$

so that

$$\frac{dS}{dt} = i\pi e^{i\pi t} + S_1 + S_2 + S_3, \quad (21.2.34)$$

$$S_1 = \sum_{n=1}^{\infty} z^n \int_{-t}^{t} dx_1 \cdots dx_n\, e^{i\pi x_1} K(x_1, x_2) \cdots K(x_{n-1}, x_n) \frac{\partial K(x_n, t)}{\partial t}, \quad (21.2.35)$$

$$S_2 = \sum_{n=1}^{\infty} z^n \sum_{j=1}^{n} \int_{-t}^{t} dx_1 \cdots dx_n\, e^{i\pi x_1} K(x_1, x_2) \cdots K(x_{j-1}, t)K(t, x_{j+1}) \cdots K(x_{n-1}, x_n)K(x_n, t), \quad (21.2.36)$$

$$S_3 = \sum_{n=1}^{\infty} z^n \sum_{j=1}^{n} \int_{-t}^{t} dx_1 \cdots dx_n\, e^{i\pi x_1} K(x_1, x_2) \cdots K(x_{j-1}, -t)K(-t, x_{j+1}) \cdots K(x_{n-1}, x_n)K(x_n, t). \quad (21.2.37)$$

For S_1 we follow the method of displacing the derivative; replace $\partial K(x_n, t)/\partial t$ by $-\partial K(x_n, t)/\partial x_n$ and integrate by parts on x_n, then in the integral

$$\int_{-t}^{t} dx_1 \cdots dx_n \, e^{i\pi x_1} K(x_1, x_2) \cdots \frac{\partial K(x_{n-1}, x_n)}{\partial x_n} K(x_n, t) \tag{21.2.38}$$

replace $\partial K(x_{n-1}, x_n)/\partial x_n$ by $-\partial K(x_{n-1}, x_n)/\partial x_{n-1}$ and integrate by parts over x_{n-1}, and so on, till the differentiation sign is pushed to the extreme left, where it finally disappears. Thus

$$S_1 = -S_2 + S_3 + i\pi \sum_{n=1}^{\infty} z^n \int_{-t}^{t} dx_1 \cdots dx_n \, e^{i\pi x_1} K(x_1, x_2) \cdots K(x_n, t), \tag{21.2.39}$$

and from Eqs. (21.2.33), (21.2.34) and (21.2.39),

$$\frac{dS}{dt} = i\pi S + 2S_3. \tag{21.2.40}$$

One can convince oneself from Eqs. (21.2.8), (21.2.33) and (21.2.37) that

$$S_3 = S^* \cdot B. \tag{21.2.41}$$

Equations (21.2.40), (21.2.41) and (21.1.14) imply Eqs. (21.1.15) and (21.1.16).

Provided the summation and integration can be interchanged, relation (21.1.17) can be written as

$$\frac{F_-}{F_+} = 1 + \int_{-t}^{t} \big(zK(1 - zK)^{-1}\big)(t, x)\, dx. \tag{21.2.42}$$

One has only to expand the "resolvent" in terms of the normalized eigenfunctions

$$\big(zK(1 - zK)^{-1}\big)(t, x) = \sum_{n=0}^{\infty} \frac{z\lambda_n}{1 - z\lambda_n} g_n(t) g_n(x) \div \int_{-t}^{t} g_n^2(y)\, dy \tag{21.2.43}$$

and note that the odd functions contribute nothing on integration. We will prove relation (21.1.17) in the form (21.2.42).

Let us calculate the logarithmic derivatives with respect to t of the two sides of (21.2.42).

$$\left(\frac{F_+}{F_-}\right) \frac{d}{dt} \left(\frac{F_-}{F_+}\right) = 2B(t) = 2\big(zK(1 - zK)^{-1}\big)(t, -t) \tag{21.2.44}$$

from Eqs. (21.1.8) and (21.2.8). Also the derivative of the right-hand side of Eq. (21.2.42) is

$$\frac{d}{dt}\left\{1+\sum_{n=1}^{\infty} z^n \int_{-t}^{t} dx_1 \cdots dx_n\, K(t,x_1)\cdots K(x_{n-1},x_n)\right\}$$
$$=\sum_{n=1}^{\infty} z^n \sum_{j=1}^{n} \int_{-t}^{t} dx_1 \cdots dx_{j-1}\, dx_{j+1} \cdots dx_n$$
$$\times \{K(t,x_1)\cdots K(x_{j-1},t)K(t,x_{j+1})\cdots K(x_{n-1},x_n)$$
$$+K(t,x_1)\cdots K(x_{j-1},-t)K(-t,x_{j+1})\cdots K(x_{n-1},x_n)\}$$
$$+\sum_{n=1}^{\infty} z^n \int dx_1 \cdots dx_n \frac{\partial K(t,x_1)}{\partial t} K(x_1,x_2)\cdots K(x_{n-1},x_n). \quad (21.2.45)$$

In the last line of the above equation one can again shift the partial derivation to the extreme right by successively replacing $\partial K(x,y)/\partial x$ by $-\partial K(x,y)/\partial y$ and integrating by parts. The expression in the last line of (21.2.45) is therefore

$$\sum_{n=1}^{\infty} z^n \sum_{j=1}^{n} \int_{-t}^{t} dx_1 \cdots dx_{j-1}\, dx_{j+1} \cdots dx_n$$
$$\times \big(-K(t,x_1)\cdots K(x_{j-1},t)K(t,x_{j+1})\cdots K(x_{n-1},x_n)$$
$$+K(t,x_1)\cdots K(x_{j-1},-t)K(-t,x_{j+1})\cdots K(x_{n-1},x_n)\big). \quad (21.2.46)$$

Also

$$\sum_{n=1}^{\infty} z^n \sum_{j=1}^{n} \int_{-t}^{t} dx_1 \cdots dx_{j-1}\, dx_{j+1} \cdots dx_n$$
$$\times K(t,x_1)\cdots K(x_{j-1},-t)K(-t,x_{j+1})\cdots K(x_{n-1},x_n)$$
$$=\left[zK(1-zK)^{-1}\right](t,-t)\left\{1+\int_{-t}^{t} dx\left[zK(1-zK)^{-1}\right](-t,x)\right\}, \quad (21.2.47)$$

and

$$\int_{-t}^{t} dx\left[zK(1-zK)^{-1}\right](-t,x)=\int_{-t}^{t} dx\left[zK(1-zK)^{-1}\right](t,-x)$$
$$=\int_{-t}^{t} dx\left[zK(1-zK)^{-1}\right](t,x). \quad (21.2.48)$$

Putting together Eqs. (21.2.44)–(21.2.48) we see that the logarithmic derivatives of the two sides of (21.2.42) are equal. In addition (21.2.42) is obviously valid for $t = 0$. Thus (21.2.42) (or Eq. (21.1.17)) is valid for all t.

Just as Eq. (21.1.17) is equivalent to Eq. (21.2.42), so is the Eq. (21.1.18) to

$$\frac{F_+(z,t)}{F_-(z,t)} = 1 - 2zI(t) + 4\int_{-t}^{t} zK(1-zK)^{-1}(t,x)\big[\varepsilon(x) - I(x)\big]\,dx \quad (21.2.49)$$

with $\varepsilon(x) = (1/2)\,\mathrm{sign}(x)$. Proof of this last equation can be supplied along the same lines.

21.3 Differential Equations for the Functions *A*, *B* and *S*

From Eqs. (21.1.11) and (21.1.12) one has

$$A = z|S|^2 - 2tB^2. \quad (21.3.1)$$

A careful examination of Eqs. (21.1.11)–(21.1.14), and (21.3.1) shows that any one of the three functions S, A and B completely determines the other two. Also we may deduce a differential equation for any of the three functions A, B or S by eliminating the other two. Thus from Eq. (21.1.16) we have

$$i\frac{dS}{dt} + \pi S = \frac{z}{2\pi t} S^*\left(S^2 - S^{*2}\right). \quad (21.1.16)$$

From Eqs. (21.1.12)–(21.1.14) we have

$$\left(t\frac{dA}{dt} + A\right)^2 = \left(t\frac{dB}{dt} + B\right)^2 + (2\pi t B)^2. \quad (21.3.2)$$

Also from Eq. (21.1.11) with one differentiation we have

$$\frac{t}{2}\frac{d^2A}{dt^2} + \frac{dA}{dt} = 2B\left(t\frac{dB}{dt} + B\right). \quad (21.3.3)$$

Eliminating $t\,dB/dt + B$ between Eqs. (21.3.2) and (21.3.3) and replacing $2B^2$ with dA/dt, Eq. (21.1.11), we get

$$\left(\frac{t}{2}\frac{d^2A}{dt^2} + \frac{dA}{dt}\right)^2 + 4\pi^2 t^2\left(\frac{dA}{dt}\right)^2 = 2\frac{dA}{dt}\left(t\frac{dA}{dt} + A\right)^2. \quad (21.3.4)$$

(Jimbo et al., 1980, were the first to derive this equation for the particular value $z = 1$.)

On the other hand, if we differentiate Eq. (21.3.2), use Eq. (21.3.3), and cancel out the non-zero factor $t\,dB/dt + B$, we get

$$4B\left(t\frac{dA}{dt}+A\right)=t\frac{d^2B}{dt^2}+2\frac{dB}{dt}+4\pi^2 tB. \tag{21.3.5}$$

Squaring this and using Eq. (21.3.2) again to eliminate, this time, $t\,dA/dt + A$, we get

$$\left(\frac{t}{2}\frac{d^2B}{dt^2}+\frac{dB}{dt}+2\pi^2 tB\right)^2=4B^2\left\{\left(t\frac{dB}{dt}+B\right)^2+4\pi^2t^2B^2\right\}. \tag{21.3.6}$$

This last equation could have been obtained also as follows. Multiply Eq. (21.1.15) throughout by S and write it as a differential equation for S^2. Separate the real and imaginary parts of this equation and eliminate $\operatorname{Re} S^2$; one thus gets a differential equation for $\operatorname{Im} S^2$, i.e., for $B(t)$, Eq. (21.1.14).

Note that if S is a solution of Eq. (21.1.16), then iS, $-S$ and $-iS$ are others. This four-fold degeneracy is removed by the initial condition $S(0) = 1$. Changing S into iS changes the sign of B and interchanges F_+ and F_-.

From any of the differential equations (21.1.16), (21.3.4) or (21.3.6) and the appropriate initial conditions one can work out the power series expansions of S, A, B, F_+ and F_-. For example, from Eq. (21.1.16) and $S(0) = 1$, one gets successively,

$$\begin{aligned}S(z,t) &= 1+2zt+\left(4z^2-\frac{\pi^2}{2}\right)t^2+\left(8z^3-\frac{7\pi^2}{9}z\right)t^3+\cdots\\ &\quad+i\left(\pi t-\frac{\pi^3}{6}t^3+\frac{2\pi^3}{9}t^4+\frac{\pi^5}{120}t^5+\cdots\right),\end{aligned} \tag{21.3.7}$$

$$\begin{aligned}B(z,t) &= z\left[1+2zt+\left(4z^2-\frac{2\pi^2}{3}\right)t^2+\left(8z^3-\frac{8\pi^2}{9}z\right)t^3\right.\\ &\quad\left.+\left(16z^4-\frac{20\pi^2}{9}z^2+\frac{2\pi^4}{15}\right)t^4+\left(32z^5-\frac{16\pi^2}{3}z^3+\frac{92\pi^4}{675}z\right)t^5+\cdots\right],\end{aligned} \tag{21.3.8}$$

$$\begin{aligned}A(z,t) &= z\left[1+2zt+4z^2t^2+\left(8z^3-\frac{8\pi^2}{9}z\right)t^3\right.\\ &\quad\left.+\left(16z^4-\frac{20\pi^2}{9}z^2\right)t^4+\left(32z^5-\frac{16\pi^2}{3}z^3+\frac{64\pi^4}{225}z\right)t^5+\cdots\right].\end{aligned} \tag{21.3.9}$$

From these series, Eqs. (21.1.7)–(21.1.8) and the initial conditions $F_\pm(z,0) = 1$, one can compute the power series for $F_\pm(z,t)$, recovering those obtained in Section 20.5.

To get rid of as many clumsy factors π as possible in the next section, we decide from now on to shift from the variables t and z to the new variables

$$\tau = \pi t \quad \text{and} \quad \zeta = \frac{2z}{\pi}. \tag{21.3.10}$$

We also define new functions

$$\mathcal{A}(\zeta,\tau) = \frac{2}{\pi}A(z,t) \quad \text{and} \quad \mathcal{B}(\zeta,\tau) = \frac{2}{\pi}B(z,t), \tag{21.3.11}$$

and rename $\mathcal{F}_\pm(\zeta,\tau)$ and $\mathcal{S}(\zeta,\tau)$ the functions $F_\pm(z,t)$ and $S(z,t)$. Suppressing the dependence on z or ζ we will also write them as $\mathcal{A}(\tau)$, $\mathcal{B}(\tau)$, $\mathcal{F}_\pm(\tau)$ or $\mathcal{S}(\tau)$ for short.

With these definitions, Eqs. (21.1.7), (21.1.8), (21.1.11), (21.1.13), (21.1.14), (21.1.16), (21.3.1), (21.3.4) and (21.3.6) become

$$\mathcal{A}(\tau) = -\frac{d}{d\tau}\{\log\mathcal{F}_+(\tau) + \log\mathcal{F}_-(\tau)\}, \tag{21.3.12}$$

$$\mathcal{B}(\tau) = -\frac{d}{d\tau}\{\log\mathcal{F}_+(\tau) - \log\mathcal{F}_-(\tau)\}, \tag{21.3.13}$$

$$\frac{d}{d\tau}\mathcal{A}(\tau) = \mathcal{B}^2(\tau), \tag{21.3.14}$$

$$\frac{d}{d\tau}\{\tau\mathcal{B}(\tau)\} = \zeta\,\mathrm{Re}\{\mathcal{S}(\tau)^2\}, \tag{21.3.15}$$

$$2\tau\mathcal{B}(\tau) = \zeta\,\mathrm{Im}\{\mathcal{S}(\tau)^2\}, \tag{21.3.16}$$

$$\frac{d\mathcal{S}}{d\tau} = i\mathcal{S} + \frac{\zeta}{\tau}\mathcal{S}^*\left(\frac{\mathcal{S}^2 - \mathcal{S}^{*2}}{4i}\right), \tag{21.3.17}$$

$$\mathcal{A}(\tau) = \zeta|\mathcal{S}(\tau)|^2 - \tau\mathcal{B}^2(\tau), \tag{21.3.18}$$

$$\left(\tau\frac{d^2\mathcal{A}}{d\tau^2} + 2\frac{d\mathcal{A}}{d\tau}\right)^2 + 16\tau^2\left(\frac{d\mathcal{A}}{d\tau}\right)^2 - 4\frac{d\mathcal{A}}{d\tau}\left(\tau\frac{d\mathcal{A}}{d\tau} + \mathcal{A}\right)^2 = 0, \tag{21.3.19}$$

$$\left(\tau\frac{d^2\mathcal{B}}{d\tau^2} + 2\frac{d\mathcal{B}}{d\tau} + 4\tau\mathcal{B}\right)^2 - 4\mathcal{B}^2\left\{\left(\tau\frac{d\mathcal{B}}{d\tau} + \mathcal{B}\right)^2 + 4\tau^2\mathcal{B}^2\right\} = 0, \tag{21.3.20}$$

etc.

21.4 Asymptotic Expansions for Large Positive τ

Any one of the functions $\mathcal{S}(\zeta,\tau)$, $\mathcal{A}(\zeta,\tau)$ or $\mathcal{B}(\zeta,\tau)$ is completely determined by the differential equation it satisfies, and initial conditions at $\tau = 0$. In particular, its asymptotic behaviour at $\tau = \infty$ is also completely determined. But to derive this asymptotic behaviour one has to connect solutions at $\tau = 0$ and $\tau = \infty$, which is difficult. To avoid this one may postulate a particular series expansion for $\mathcal{S}(\zeta, \tau)$ around $\tau = \infty$ (see Eq. (21.4.5) below), and find that the resulting series for $\log \mathcal{F}_{\pm}$ coincide with the asymptotic expansions found in Chapter 18, for $\zeta = 2/\pi$. Then, one can expand $\mathcal{S}(\zeta, \tau)$ around $\zeta = 2/\pi$.

More precisely, if one writes the Taylor expansions in ζ, around $\zeta = 2/\pi$, of the functions $\mathcal{S}$, $\mathcal{A}$, $\mathcal{B}$ and $\mathcal{F}_{\pm}$ as follows

$$\mathcal{S}(\zeta, \tau) = 2\sqrt{\frac{\tau}{\zeta}} \sum_{n=0}^{\infty} \left[\alpha \left(\zeta - \frac{2}{\pi} \right) \frac{e^{2\tau}}{\sqrt{\tau}} \right]^n \mathcal{S}_n(\tau), \tag{21.4.1}$$

$$\mathcal{A}(\zeta, \tau) = 2 \sum_{n=0}^{\infty} \left[\alpha \left(\zeta - \frac{2}{\pi} \right) \frac{e^{2\tau}}{\sqrt{\tau}} \right]^n \mathcal{A}_n(\tau), \tag{21.4.2}$$

$$\mathcal{B}(\zeta, \tau) = 2 \sum_{n=0}^{\infty} \left[\alpha \left(\zeta - \frac{2}{\pi} \right) \frac{e^{2\tau}}{\sqrt{\tau}} \right]^n \mathcal{B}_n(\tau), \tag{21.4.3}$$

$$\mathcal{F}_{\pm}(\zeta, \tau) = - \sum_{n=0}^{\infty} \left[\alpha \left(\zeta - \frac{2}{\pi} \right) \frac{e^{2\tau}}{\sqrt{\tau}} \right]^n \mathcal{F}_{\pm n}(\tau), \tag{21.4.4}$$

where α is a constant, then, for τ positive and $\gg 1$, the coefficient functions $\mathcal{S}_n(\tau)$, $\mathcal{A}_n(\tau)$, $\mathcal{B}_n(\tau)$ and $\mathcal{F}_{\pm n}(\tau)$ have asymptotic expansions in powers of τ^{-1}. A few terms of these expansions for low values of n are listed in Appendix A.46.

To derive these expansions, let us look for a solution of Eq. (21.3.17) of the form

$$\tau^{\gamma} \sum_{j=0}^{\infty} s_j \tau^{-j}, \tag{21.4.5}$$

where γ and the s_j are complex constants. It turns out that, up to the four-fold degeneracy mentioned earlier in Section 21.3, the solution is unique and given by the right-hand side of Eq. (A.46.1). From this particular solution, using successively Eqs. (21.3.16), (21.3.18), (21.3.12) and (21.3.13), we get the corresponding functions $\mathcal{A}$, $\mathcal{B}$ and $\log \mathcal{F}_{\pm}$. The first two of them are given by Eqs. (A.46.6) and (A.46.11), while

$$\log \mathcal{F}_{\pm} = -\frac{\tau^2}{4} \mp \frac{\tau}{2} - \frac{1}{8} \log\left(\tau \pm \frac{1}{2} \right) + c_{\pm} + \sum_{n=3}^{\infty} (\mp 1)^n a_n \left(\tau \pm \frac{1}{2} \right)^{-n}, \tag{21.4.6}$$

with undetermined constants $c_{\pm}$. The coefficients a_n can be computed recursively. A few of them are listed in Appendix A.47.

The above asymptotic series for $\log \mathcal{F}_\pm$ coincides with that found in Chapter 18, Eq. (18.6.16). Therefore we know that (A.46.1) is indeed the asymptotic expansion of $\mathcal{S}(2/\pi, \tau)$.

Next, let us expand $\mathcal{S}(\zeta, \tau)$ as follows

$$\mathcal{S}(\zeta, \tau) = \frac{1}{\sqrt{\zeta}} \sum_{n=0}^{\infty} \left(\zeta - \frac{2}{\pi}\right)^n \left[R_n(\tau) + i I_n(\tau)\right], \tag{21.4.7}$$

where R_n and I_n are real functions of τ, and let us define the two component column

$$\phi_n(\tau) = \begin{pmatrix} R_n(\tau) \\ I_n(\tau) \end{pmatrix}. \tag{21.4.8}$$

Then, Eq. (21.3.17) implies that $\phi_n(\tau)$ satisfies for $n \geqslant 1$ the linear differential equation

$$\frac{d\phi_n}{d\tau} = \mathcal{M}\phi_n + \psi_n, \tag{21.4.9}$$

where $\mathcal{M}$ is the 2×2 matrix

$$\begin{aligned} \mathcal{M}(\tau) &= \begin{pmatrix} 2R_0 I_0/\tau & R_0^2/\tau - 1 \\ 1 - I_0^2/\tau & -2R_0 I_0/\tau \end{pmatrix} \\ &= \begin{pmatrix} 2 - \dfrac{1}{4\tau^2} - \dfrac{13}{64\tau^4} + \cdots & \dfrac{1}{2\tau} + \dfrac{1}{16\tau^3} - \dfrac{1}{16\tau^4} + \cdots \\ \dfrac{1}{2\tau} + \dfrac{1}{16\tau^3} + \dfrac{1}{16\tau^4} + \cdots & -2 + \dfrac{1}{4\tau^2} + \dfrac{13}{64\tau^4} + \cdots \end{pmatrix}, \end{aligned} \tag{21.4.10}$$

and where the inhomogeneous term ψ_n is a function of $\phi_0, \phi_1, \ldots, \phi_{n-1}$.

For $n = 1$, the equation is homogeneous, $\psi_1 = 0$. Two linearly independent particular solutions are

$$\phi_1^+(\tau) = e^{2\tau} \begin{pmatrix} \operatorname{Re} \mathcal{S}_1(\tau) \\ \operatorname{Im} \mathcal{S}_1(\tau) \end{pmatrix}, \tag{21.4.11}$$

$$\phi_1^-(\tau) = e^{-2\tau} \begin{pmatrix} \operatorname{Im} \mathcal{S}_1(-\tau) \\ \operatorname{Re} \mathcal{S}_1(-\tau) \end{pmatrix}, \tag{21.4.12}$$

where $\mathcal{S}_1(\tau)$ is given by Eq. (A.46.2). The second one has a sub-dominant behaviour at infinity as compared with the first one. So we write

$$\phi_1(\tau) = 2\alpha\phi_1^+(\tau), \tag{21.4.13}$$

where α is a constant which, once more, has to be determined by using the initial condition at the origin, $\mathcal{S}_1(0) = 1$. Although the problem is linear, it is complicated by the fact that the equation involves the solution $\mathcal{S}_0$ of a non linear (Painlevé) equation.

For $n = 2$, one has

$$\psi_2 = \frac{1}{\tau} \begin{pmatrix} 2R_0 R_1 I_1 + I_0 R_1^2 \\ -2I_0 R_1 I_1 - R_0 I_1^2 \end{pmatrix}. \tag{21.4.14}$$

This inhomogeneous term increases at infinity like $e^{4\tau}$. Consequently, any solution of Eq. (21.4.9) for $n = 2$ also increases at infinity like $e^{4\tau}$. Since the solutions of the homogeneous equation increase at most like $e^{2\tau}$, the dominant terms in the asymptotic expansion of ϕ_2 are independent of the initial condition at the origin. Since we are interested only in the exponentially dominant terms, no new unknown constant appears here, and one obtains

$$\phi_2(\tau) = 2\alpha^2 \frac{e^{4\tau}}{\sqrt{\tau}} \begin{pmatrix} \operatorname{Re} \mathcal{S}_2(\tau) \\ \operatorname{Im} \mathcal{S}_2(\tau) \end{pmatrix}, \tag{21.4.15}$$

where $\mathcal{S}_2(\tau)$ is given by Eq. (A.46.3).

The same phenomenon happens at each further step $n = 3, 4, \ldots$ of the procedure. This readily leads to the expansion (21.4.1), and the expressions given in Eqs. (A.46.2) to (A.46.5).

The calculation of the asymptotic expansions (21.4.2)–(21.4.4) and those in Appendix A.46 is then straightforward. Note that all the integration constants that appear when calculating $\mathcal{F}_{\pm n}$, for $n \geqslant 1$, with the use of Eqs. (21.3.12) and (21.3.13), are irrelevant sub-dominant terms.

To find the unknown constant α we proceed as follows. It is known (see, for example, Eq. (18.1.14)) that

$$1 - \lambda_j \approx \frac{\pi^{j+1}}{j!} 2^{2j+3/2} s^{j+1/2} e^{-\pi s}, \tag{21.4.16}$$

for j finite and $t \gg 1$. Substituting it in the dominant terms

$$\frac{E_+(n,s)}{E_+(0,s)} \approx \prod_{j=0}^{n-1} \frac{\lambda_{2j}}{1-\lambda_{2j}}, \qquad \frac{E_-(n,s)}{E_-(0,s)} \approx \prod_{j=0}^{n-1} \frac{\lambda_{2j+1}}{1-\lambda_{2j+1}} \tag{21.4.17}$$

gives

$$\frac{E_+(n,s)}{E_+(0,s)} \approx \left(\prod_{j=0}^{n-1} (2j)! \right) \pi^{-n^2} 2^{-n(2n-1/2)} s^{-n(2n-1)/2} e^{\pi n s}, \tag{21.4.18}$$

$$\frac{E_-(n,s)}{E_-(0,s)} \approx \left(\prod_{j=0}^{n-1}(2j+1)!\right)\pi^{-n(n+1)}2^{-n(2n+3/2)}s^{-n(2n+1)/2}e^{\pi ns}. \tag{21.4.19}$$

The expansions (A.46.16)–(A.46.23) and Eq. (21.4.7) will all agree with the above if we choose the constant α as

$$\alpha = \frac{\sqrt{\pi}}{8}. \tag{21.4.20}$$

From the expansions of $\mathcal{S}_j(\tau)$ listed in Appendix A.46, we observe that the first terms are given by

$$\begin{aligned} S_j(\tau) &= 1 + \frac{7j-4}{(16\tau)} + \frac{7(7j^2+12j-16)}{2!(16\tau)^2} + \frac{(7j)^3+48(7j)^2+6536j-6720}{3!(16\tau)^3} + \cdots \\ &\quad + \frac{i}{(8\tau)^j}\left(1 + \frac{19j-12}{(16\tau)} + \frac{(19j)^2+252j-400}{2!(16\tau)^2} + \cdots\right). \end{aligned} \tag{21.4.21}$$

This formula gives the correct coefficient of τ^{-k} up to $k=j$ in the real part, and up to $k=2j$ in the imaginary part.

We observe also that

$$\mathcal{A}_j(\tau) + \mathcal{B}_j(\tau) = 2\,\mathrm{Re}\,\mathcal{S}_j(\tau) + \mathrm{O}(\tau^{-2j}), \tag{21.4.22}$$

$$\mathcal{A}_j(\tau) - \mathcal{B}_j(\tau) = 2\,\mathrm{Im}\,\mathcal{S}_j(\tau) + \mathrm{O}(\tau^{-3j}). \tag{21.4.23}$$

We checked Eq. (21.4.22) for $j \leqslant 8$ and Eq. (21.4.23) for $j \leqslant 5$.

21.5 Fifth and Third Painlevé Transcendents

The second order non-linear differential equations satisfied by the functions A, B and S are Painlevé equations (see Appendix A.45)

We will be concerned with the third and the fifth of these transcendents (P3 and P5), solutions of

$$\frac{d^2y}{d\tau^2} = \frac{1}{y}\left(\frac{dy}{d\tau}\right)^2 - \frac{1}{\tau}\frac{dy}{d\tau} + \frac{\alpha y^2+\beta}{\tau} + \gamma y^3 + \frac{\delta}{y} \tag{21.5.1}$$

and

$$\frac{d^2y}{d\tau^2} = \left(\frac{1}{2y} + \frac{1}{y-1}\right)\left(\frac{dy}{d\tau}\right)^2 - \frac{1}{\tau}\frac{dy}{d\tau} + \frac{(y-1)^2}{\tau^2}\left(\alpha y + \frac{\beta}{y}\right) + \gamma\frac{y}{\tau} + \delta\frac{y(y+1)}{y-1} \tag{21.5.2}$$

respectively, where α, β, γ and δ are complex constants.

We will show that

(1) The real part of $\mathcal{S}(\tau)$ can be expressed in terms of a P5, Eq. (21.5.2), with $\alpha = -\beta = 1/32$, $\gamma = 0$, $\delta = -2$; and near $\tau = 0$,

$$y_{r5}(\tau) = 1 + 2\left(\frac{\tau}{\zeta}\right)^{1/2} + \frac{2\tau}{\zeta} + 2(1-\zeta^2)\left(\frac{\tau}{\zeta}\right)^{3/2} + \mathrm{O}(\tau^2). \quad (21.5.3)$$

It can also be expressed in terms of a P3, Eq. (21.5.1), with $\alpha = -\gamma = \delta = -1$, $\beta = 3$; and near $\tau = 0$,

$$y_{r3}(\tau) = -\frac{1}{\zeta} + \left(3 - \frac{1}{\zeta^2}\right)\tau + \left(-\frac{1}{\zeta^3} + \frac{3}{\zeta} - 2\zeta\right) + \mathrm{O}(\tau^3). \quad (21.5.4)$$

(2) The imaginary part of $\mathcal{S}(\tau)$ is expressed in terms of a P5 with $\alpha = -\beta = 1/32$, $\gamma = 0$, $\delta = -2$; and near $\tau = 0$

$$y_{i5}(\tau) = -1 - 2(\zeta\tau)^{1/2} - 2\zeta\tau - 4(\zeta\tau)^{3/2} - 6\zeta^2\tau^2 + \mathrm{O}(\tau^{5/2}). \quad (21.5.5)$$

It can also be expressed in terms of a P3, Eq. (21.5.1), with $\alpha = -\gamma = \delta = -1$, $\beta = 3$; and near $\tau = 0$,

$$y_{i3}(\tau) = \frac{\tau}{3} + \frac{\tau^3}{45} - \frac{\zeta\tau^4}{27} + \mathrm{O}(\tau^5). \quad (21.5.6)$$

Any one of the above two P5's or two P3's can be simply expressed in terms of the other.

(3) The function $\mathcal{A}(\tau)$ is expressed in terms of a P5, Eq. (21.5.2), with either $\alpha = 1/2$, $\beta = \gamma = 0$, $\delta = 8$; and near $\tau = 0$,

$$y_{a51} = -\zeta\tau - \zeta^2\tau^2 - \left(\zeta^3 - \frac{4}{3}\zeta\right)\tau^3 + \mathrm{O}(\tau^4), \quad (21.5.7)$$

or with $\alpha = \beta = 0$, $\gamma = -4i$, $\delta = 8$; and near $\tau = 0$,

$$\begin{aligned} y_{a52} &= 1 - 4i\tau + 4(i\zeta - 2)\tau^2 + 16\left(\zeta - \frac{2i}{3}\right)\tau^3 + \mathrm{O}(\tau^4) \\ &= \exp\left[-4i\left(\tau - \zeta\tau^2 + \mathrm{O}(\tau^4)\right)\right]. \end{aligned} \quad (21.5.8)$$

If we set $y_{a52} = \exp(-2i\rho)$, then $\rho(\tau)$ is real for real τ, satisfies the non-linear differential equation

$$\left[\frac{d^2\rho}{d\tau^2} + \frac{1}{\tau}\left(\frac{d\rho}{d\tau} - 2\right)\right] = \left[\left(\frac{d\rho}{d\tau}\right)^2 - 4\right]\cot\rho; \tag{21.5.9}$$

and near $\tau = 0$ has the expansion

$$\rho(\tau) = 2\tau\left[1 - \zeta\tau + O(\tau^3)\right]. \tag{21.5.10}$$

Taking $x = 2\tau$ as the new variable, Eq. (21.5.9) goes over to

$$\frac{d^2\rho}{dx^2} = \left[\left(\frac{d\rho}{dx}\right)^2 - 1\right]\cot\rho + \frac{1}{x}\left(1 - \frac{d\rho}{dx}\right)^2, \tag{21.5.11}$$

with

$$\rho(x) = x - \frac{\zeta}{2}x^2 + O(x^4), \tag{21.5.12}$$

a problem studied in detail by Suleimanov (see in Its and Novokshenov (1986)). In particular he gives the behaviour of $\rho(x)$ when $\zeta < 2/\pi$, $\zeta = 2/\pi$ and $\zeta > 2/\pi$.

It can also be expressed in terms of a P3, Eq. (21.5.1), with $\alpha = \beta = \gamma = -\delta = 1$; and near $\tau = 0$,

$$y_{a31} = -\tau + \zeta\tau^2 - \frac{\tau^3}{3} + O(\tau^4); \tag{21.5.13}$$

or with another choice

$$y_{a32} = -\frac{1}{\tau} - \zeta - \left(\zeta^2 - \frac{1}{3}\right)\tau + O(\tau^2). \tag{21.5.14}$$

(4) The function $\mathcal{B}(\tau)$ (which is almost the imaginary part of $\mathcal{S}^2(\tau)$) is expressed in terms of a P5, Eq. (21.5.2), with $\alpha = \delta = 0$, $\beta = -1/8$, $\gamma = 1$; and near $\tau = 0$,

$$y_{b1} = 1 - \frac{1}{\zeta}(2\tau)^{1/2} + 4\tau - \left(\zeta + \frac{1}{\zeta}\right)(2\tau)^{3/2} - 4\left(\zeta^2 - \frac{20}{9}\right)\tau^2 + O(\tau^{5/2}). \tag{21.5.15}$$

It is also expressed in terms of a P3, Eq. (21.5.1) with $\alpha = \beta = -\gamma = \delta = -1$; and near $\tau = 0$,

$$y_{b2} = \frac{1}{\tau} + \zeta + \left(\zeta^2 - \frac{1}{3}\right)\tau + O(\tau^2). \tag{21.5.16}$$

To start with, consider Eq. (21.3.17). Let us prove that its general solution is a homographic function of a P5. Writing $\mathcal{S}=(R+iI)/\sqrt{\zeta}$, where R and I are real, and separating the real and imaginary parts, we get

$$R=\frac{I'\tau}{\tau-I^2},\quad I=\frac{R'\tau}{R^2-\tau}. \tag{21.5.17}$$

This pair of equations expresses R in terms of I and I', and I in terms of R and R'. In the study of Painlevé equations one usually exhibits such a pair of equations, as we will see.

Elimination of I gives

$$R''=\frac{R}{R^2-\tau}R'^2-\left(\frac{R^2}{R^2-\tau}\right)\frac{R'}{\tau}+\frac{R}{\tau}(R^2-\tau). \tag{21.5.18}$$

Elimination of R gives the same equation for I. Eq. (21.5.18) has singular points at $\pm\sqrt{\tau}$, and ∞. To displace them to 0, 1 and ∞ we set $y_{r5}=(R+\sqrt{\tau})/(R-\sqrt{\tau})$, and $y_{i5}=(I+\sqrt{\tau})/(I-\sqrt{\tau})$, then $y_r(\tau)$ and $y_i(\tau)$ satisfy the fifth Painlevé equation (21.5.2) with the parameters $\alpha=-\beta=1/32$, $\gamma=0$ and $\delta=-2$. Near the origin $\tau=0$, $\mathcal{S}(\tau)$ is given by Eqs. (21.3.7), (21.3.10), which implies that for $\tau\ll 1$, $y_{r5}(\tau)$ and $y_{i5}(\tau)$ are given by Eqs. (21.5.3) and (21.5.5).

Consider the couple of equations

$$f=\frac{\sqrt{\tau}}{R^2-\tau}\left(R'-\frac{R}{\tau}\right),\quad R=-\frac{\sqrt{\tau}}{f^2-1}\left(f'+\frac{3}{2}\frac{f}{\tau}\right). \tag{21.5.19}$$

Eliminating f one gets Eq. (21.5.18), while eliminating R one gets

$$f''=\frac{f}{f^2-1}f'^2-\frac{f'}{\tau}-\frac{9}{4\tau^2}\frac{f}{f^2-1}+f(f^2-1). \tag{21.5.20}$$

Set $f=\sqrt{u/(u-1)}$, or $u=f^2/(f^2-1)$, so that

$$u''=\frac{3u-1}{2u(u-1)}u'^2-\frac{u'}{\tau}+\frac{9u}{2\tau^2}(u-1)^2-2u. \tag{21.5.21}$$

This is again almost a P5 with $\delta=0$, and becomes a standard P5 on taking τ^2 as the new independent variable.

Consider now the couple of equations

$$\frac{u+1}{u-1}=\omega^2-\omega'-\frac{2\omega}{\tau},\quad \omega=\frac{2\tau u}{\tau u'-3u(u-1)}. \tag{21.5.22}$$

Eliminating ω, one gets (21.5.21), while eliminating u one gets a P3 for ω with $\alpha = -\gamma = \delta = -1, \beta = 3$.

We now turn to Eq. (21.3.19). Its solution is also a homographic function of a P5. To see this, note that it is almost homogeneous in $\mathcal{A}$. Therefore let (Cosgrove and Scoufis, 1993),

$$\mathcal{A} = e^{-w}, \qquad u = \tau w', \tag{21.5.23}$$

and for convenience let (Gromak and Lukashevich, 1990)

$$v = u - 1 - \frac{\tau u'}{u}. \tag{21.5.24}$$

Then

$$\mathcal{A}' = -u\mathcal{A}/\tau, \qquad (\tau\mathcal{A})' = \tau\mathcal{A}' + \mathcal{A} = -(u-1)\mathcal{A}, \tag{21.5.25}$$

$$(\tau\mathcal{A})'' = \tau\mathcal{A}'' + 2\mathcal{A}' = -\{u'\mathcal{A} + (u-1)\mathcal{A}'\} = \frac{uv}{\tau}\mathcal{A}. \tag{21.5.26}$$

Disregarding the non-zero factors, Eq. (21.3.19) now reads

$$v^2 + 16\tau^2 + 4\tau\frac{(u-1)^2}{u}\mathcal{A} = 0. \tag{21.5.27}$$

The derivative of this last equation gives

$$vv' + 16\tau + 2\tau\frac{(u-1)^2}{u}\left(-\frac{u'}{u} + \frac{2u'}{u-1} + \frac{1}{\tau} - \frac{u}{\tau}\right)\mathcal{A} = 0. \tag{21.5.28}$$

Eliminating $\mathcal{A}$ between the last two equations, we get

$$vv' + 16\tau - \frac{1}{2}(v^2 + 16\tau^2)\left(-\frac{u'}{u} + \frac{2u'}{u-1} - \frac{u-1}{\tau}\right) = 0. \tag{21.5.29}$$

Using Eq. (21.5.24), one finds by elementary algebraic manipulations that v factorizes in the left-hand side of Eq. (21.5.29). The solution $v = 0$ gives $u = 1/(1 - c\tau)$, with c an integration constant. However, this cannot correspond to the value of $\mathcal{A}(\tau)$ at $\tau = 0$. The other possibility is

$$v' = v\left(-\frac{u'}{2u} + \frac{u'}{u-1} - \frac{u-1}{2\tau}\right) - 8\tau\frac{u+1}{u-1}, \tag{21.5.30}$$

or using Eq. (21.5.24)

$$\frac{u+1}{u-1} = 2\frac{v - \tau v'}{v^2 + 16\tau^2}. \tag{21.5.31}$$

Equation (21.5.24) expresses v in terms of u and u', while Eq. (21.5.31) expresses u in terms of v and v'.

On writing v in terms of $u \equiv y_{a51}$, one obtains the Painlevé equation of the fifth kind, Eq. (21.5.2), with the parameters $\alpha = 1/2$, $\beta = \gamma = 0$ and $\delta = 8$. And on writing u in terms of v, one obtains

$$v'' = \left(\frac{v}{v^2 + 16\tau^2}\right)v'^2 - \left(\frac{v^2 - 16\tau^2}{v^2 + 16\tau^2}\right)\frac{v'}{\tau} - \frac{16v}{v^2 + 16\tau^2} + \frac{v+2}{4\tau^2}(v^2 + 16\tau^2). \tag{21.5.32}$$

Setting $v = 4i\tau\xi$ we get

$$\xi'' = \frac{\xi}{\xi^2 - 1}\xi'^2 - \frac{\xi'}{\tau} + \frac{2i}{\tau}(\xi^2 - 1) - 4\xi(\xi^2 - 1). \tag{21.5.33}$$

This last equation has singular points at -1, 1 and ∞. As before, one sets $\xi = (y_{a52} + 1)/(y_{a52} - 1)$, then y_{a52} satisfies the fifth Painlevé equation (21.5.2) with the parameters $\alpha = \beta = 0$, $\gamma = -4i$ and $\delta = 8$. This equation has been investigated by Jimbo et al. (1980) and by McCoy and Tang (1986).

Near $\tau = 0$, $\mathcal{A}$ is given by Eqs. (21.3.9), (21.3.10), which implies that y_{a51} and y_{a52} are given by Eqs. (21.5.7) and (21.5.8).

In Eq. (21.5.33) if one writes $\xi = (y^2 - 1)/(2iy)$, then one gets

$$y'' = \frac{y'^2}{y} - \frac{y'}{\tau} + \frac{1}{\tau}(y^2 + 1) + y^3 - \frac{1}{y}, \tag{21.5.34}$$

a P3 with $\alpha = \beta = \gamma = -\delta = 1$. Expressing y in terms of ξ,

$$y_{\pm} = i\xi \pm \sqrt{1 + (i\xi)^2}, \tag{21.5.35}$$

one gets Eq. (21.5.13) or (21.5.14).

Finally, consider Eq. (21.3.20). We write it as

$$b'' + 4b = \frac{b}{\tau}\sqrt{b'^2 + 4b^2}, \quad b \equiv 2\tau\mathcal{B}. \tag{21.5.36}$$

To see that this is an algebraic transform of a Painlevé transcendent, we follow Bureau (1972), and consider the pair of equations

$$b = \frac{\tau\psi'}{\psi - 2\tau}, \quad 2\psi = b' + \sqrt{b'^2 + 4b^2}. \tag{21.5.37}$$

This pair is like Eqs. (21.5.17). If we eliminate ψ, we get Eq. (21.5.36); and if we eliminate b, we get

$$\psi'' = \left(\frac{1}{\psi} + \frac{1}{\psi - 2\tau}\right)\frac{\psi'^2}{2} - \left(\frac{\psi}{\psi - 2\tau}\right)\frac{\psi'}{\tau} + \frac{\psi}{\tau}(\psi - 2\tau). \tag{21.5.38}$$

To put this last equation in the canonical form, let

$$\varphi = \frac{\psi - 2\tau}{\psi}, \quad \psi = -\frac{2\tau}{\varphi - 1}, \tag{21.5.39}$$

so that

$$\varphi'' = \left(\frac{1}{2\varphi} + \frac{1}{\varphi - 1}\right)\varphi'^2 - \frac{\varphi'}{\tau} - \frac{(\varphi - 1)^2}{2\tau^2\varphi} + 2\varphi. \tag{21.5.40}$$

This is almost the fifth Painlevé equation. To put it in standard form, take $\tau^2/2$ as the new independent variable. Equation (21.5.40) transforms into the fifth Painlevé equation (21.5.2) with $\alpha = \delta = 0$, $\beta = -1/8$ and $\gamma = 1$.

As $\delta = 0$, we can transform it into a third Painlevé equation (21.5.1). To do it, we follow Gromak and Lukashevich (1990), and consider the pair of equations

$$\varphi = \frac{\eta' + \eta^2 - 1}{\eta' + \eta^2 + 1}, \quad \eta = \frac{2\tau\varphi}{\tau\varphi' - \varphi + 1}. \tag{21.5.41}$$

Eliminating η, we get Eq. (21.5.40) for φ; and eliminating φ, we get for η the third Painlevé equation in its standard form (21.5.1), with $\alpha = \beta = -\gamma = \delta = -1$.

Some one parameter families of solutions of the Painlevé equations of this section are known, but none of them satisfy our conditions at $\tau = 0$.

21.6 Solution of Eq. (21.3.6) for Large *t*

By a comparison with the continuum model, Dyson showed (1995) that for $s \gg 1$ and $|1 - z| \ll 1$, $B(s)$ is nearly a Jacobian elliptic function. We can recover this result as follows. Let us write Eq. (21.3.6) as

$$\left(B'' + 4\pi^2 B + \frac{2B'}{t}\right)^2 = 16B^2\left\{\left(B' + \frac{B}{t}\right)^2 + 4\pi^2 B^2\right\}. \tag{21.6.1}$$

If we ignore the terms in $1/t$, then

$$\frac{B'' + 4\pi^2 B}{\sqrt{B'^2 + 4\pi^2 B^2}} = \pm 4B, \tag{21.6.2}$$

or

$$d\left(\sqrt{B'^2 + 4\pi^2 B^2}\right) = \pm d\left(2B^2\right). \tag{21.6.3}$$

On integration this gives

$$B'^2 + 4\pi^2 B^2 = 4\left(B^2 + c\right)^2, \tag{21.6.4}$$

with c an integration constant. As B' equals the square root of a fourth degree polynomial in B, B is a Jacobian elliptic function. Let us look for a solution of the form $a\,\mathrm{sn}(u, \lambda)$, where sn is the elliptic sine function, a and λ are constants, and u is a function of t to be determined. Then

$$\begin{aligned} B'^2 &= a^2 u'^2 \left(1 - \mathrm{sn}^2(u, \lambda)\right)\left(1 - \lambda^2 \mathrm{sn}^2(u, \lambda)\right) \\ &= a^2 u'^2 \left(1 - \frac{B^2}{a^2}\right)\left(1 - \lambda^2 \frac{B^2}{a^2}\right), \end{aligned} \tag{21.6.5}$$

or

$$\begin{aligned} B'^2 + 4\pi^2 B^2 - 4\left(B^2 + c\right)^2 &= a^2 u'^2 - 4c^2 + B^2\left(4\pi^2 - 8c - u'^2\left(1 + \lambda^2\right)\right) \\ &\quad + B^4\left(\frac{\lambda^2 u'^2}{a^2} - 4\right). \end{aligned} \tag{21.6.6}$$

Choose a, u and λ so that the coefficients on the right-hand side of Eq. (21.6.6) are zero; This gives

$$u' = \frac{2c}{a} = \frac{2a}{\lambda}, \quad u'^2\left(1 + \lambda^2\right) = 4\left(\pi^2 - 2c\right), \tag{21.6.7}$$

or

$$u' = \frac{2\pi}{1 + \lambda}, \qquad a = \sqrt{c\lambda} = \frac{\pi\lambda}{1 + \lambda}. \tag{21.6.8}$$

Hence

$$B(t) = \frac{\pi\lambda}{1 + \lambda}\,\mathrm{sn}\left(\frac{2\pi t}{1 + \lambda} + b, \lambda\right) \tag{21.6.9}$$

is the general solution of Eq. (21.6.2); b and λ are integration constants. This gives the solution of Eq. (21.6.1) to the dominant order.

To calculate the corrections in $1/t$, $1/t^2$, etc. the procedure would be to substitute

$$B = \frac{\pi\lambda}{1+\lambda} \operatorname{sn}\left(\frac{2\pi t}{1+\lambda} + b, \lambda\right) + \frac{f_1(t)}{t} + \frac{f_2(t)}{t^2} + \cdots \tag{21.6.10}$$

and compare coefficients of various negative powers of t to find equations satisfied by $f_1(t)$, $f_2(t)$, etc. Of course, this cannot give us the constants λ and b in terms of the initial conditions $B(0) = z$ and $B'(0) = 2z^2$, since Eq. (21.6.10) is not valid near $t = 0$.

To see that $b = 0$ and λ varies slowly with t, one needs a better method (see, e.g., Dyson, 1995).

Summary of Chapter 21

Differential relations among the Fredholm determinants of the even and odd parts of the "sine kernel"

$$K(x, y) = \frac{\sin \pi(x - y)}{\pi(x - y)}$$

and an auxiliary function $S(z, t)$ are studied. This leads to second order non linear differential equations, the so called Painlevé equations, for functions related to the spacing functions and allows either their power series expansions near $t = 0$ or their asymptotic expansions for large t.

22

MOMENTS OF THE CHARACTERISTIC POLYNOMIAL IN THE THREE ENSEMBLES OF RANDOM MATRICES

Integer moments of the characteristic polynomial of a random matrix taken from any of the three ensembles, orthogonal, unitary or symplectic, can be obtained either as a determinant or a Pfaffian or as a sum of determinants. For Gaussian ensembles comparing the two expressions of the same moment one gets two remarkable identities, one between an $n \times n$ determinant and an $m \times m$ determinant and another between the Pfaffian of a $2n \times 2n$ anti-symmetric matrix and a sum of $m \times m$ determinants.

22.1 Introduction

As in Chapter 5 consider a non-negative function $w(x)$ with all its moments finite,

$$\int x^m w(x)\, dx < \infty, \quad m = 0, 1, \ldots . \tag{22.1.1}$$

With this weight function $w(x)$ let us define, as in Chapter 5, three scalar products, one symmetric and two anti-symmetric, as

$$\langle f, g\rangle_2 := \int f(x)g(x)w(x)\,dx = \langle g, f\rangle_2, \tag{22.1.2}$$

$$\langle f, g\rangle_4 := \int [f(x)g'(x) - f'(x)g(x)]w(x)\,dx = -\langle g, f\rangle_4, \tag{22.1.3}$$

$$\langle f, g\rangle_1 := \iint f(x)g(y)\,\mathrm{sign}(y-x)\sqrt{w(x)w(y)}\,dx\,dy = -\langle g, f\rangle_1, \tag{22.1.4}$$

and introduce polynomials C_n, Q_n and R_n of degree n, satisfying the orthogonality relations

$$\langle C_n, C_m\rangle_2 = c_n\delta_{n,m}, \tag{22.1.5}$$

$$\langle Q_{2n}, Q_{2m}\rangle_4 = \langle Q_{2n+1}, Q_{2m+1}\rangle_4 = 0, \quad \langle Q_{2n}, Q_{2m+1}\rangle_4 = q_n\delta_{n,m}, \tag{22.1.6}$$

$$\langle R_{2n}, R_{2m}\rangle_1 = \langle R_{2n+1}, R_{2m+1}\rangle_1 = 0, \quad \langle R_{2n}, R_{2m+1}\rangle_1 = r_n\delta_{n,m}. \tag{22.1.7}$$

We will take these polynomials to be monic, i.e., the coefficient of the highest power will be taken to be one. As we said in Section 5.10, the above conditions determine completely the $C_n(x)$, $Q_{2n}(x)$ and $R_{2n}(x)$, while $Q_{2n+1}(x)$ can be replaced by $Q_{2n+1}(x) + a\,Q_{2n}(x)$ with an arbitrary constant a. Similarly, $R_{2n+1}(x)$ can be replaced by $R_{2n+1}(x) + aR_{2n}(x)$ with arbitrary a. We will choose these constants so that the coefficient of x^{2n} in $Q_{2n+1}(x)$ and in $R_{2n+1}(x)$ is zero, thus fixing them also completely.

The subscript $\beta = 1$, 2 or 4 is used to remind that it was convenient to use these polynomials to express the correlation functions of the eigenvalues of real symmetric, complex hermitian and quaternion self-dual random matrices respectively (cf. Chapters 6, 7, 8). The polynomials $C_j(x)$ are said to be orthogonal while the polynomials $Q_j(x)$ and $R_j(x)$ are said to be skew-orthogonal of the quaternion type and the real type respectively.

In what follows n and m are non-negative integers. It can be checked that all the given formulas remain valid on replacing a non-existing sum by 0, a non-existing product, integral or determinant by 1 and forgetting non-existing rows or columns in a matrix.

It is known (Brézin and Hikami, 2000) that

$$I_2(n, m; x) := \int \cdots \int \Delta_n^2(\mathbf{y}) \prod_{j=1}^{n} [(x - y_j)^m w(y_j)\,dy_j] \tag{22.1.8}$$

can be expressed as an $m \times m$ determinant. We will see below that similar integrals

$$I_\beta(n, m; x) := \int \cdots \int |\Delta_n(\mathbf{y})|^\beta \prod_{j=1}^{n} \{(x - y_j)^m [w(y_j)]^{1/(1+\delta_{\beta,1})}\,dy_j\} \tag{22.1.9}$$

for $\beta = 1$ or 4, can as well be expressed as a sum of $m \times m$ determinants. They are the mth moments of the characteristic polynomial of an $n \times n$ random matrix A

$$\langle \det(xI - A)^m \rangle = \frac{I_\beta(n, m; x)}{I_\beta(n, 0; x)}, \tag{22.1.10}$$

the parameter β taking the values 1, 2 or 4 according as A is real symmetric, complex hermitian or quaternion self-dual and the probability density of the eigenvalues $\mathbf{y}$ being $|\Delta_n(\mathbf{y})|^\beta \prod_{j=1}^n [w(y_j)]^{1/(1+\delta_{\beta,1})}$. The special case $w(y) = e^{-ay^2}$ arises when the probability density of the random matrix is invariant under any change of basis and the algebraically independent parameters specifying the matrix elements are also statistically independent.

Instead of the real symmetric, complex hermitian and quaternion self-dual matrices some authors considered the corresponding ensembles of unitary matrices. The mth moment of the characteristic polynomial in $x = \exp(i\alpha)$

$$\int_0^{2\pi} \cdots \int_0^{2\pi} \prod_{1 \leqslant j < k \leqslant n} |e^{i\theta_j} - e^{i\theta_k}|^\beta \prod_{j=1}^n \left[|e^{i\alpha} - e^{i\theta_j}|^m \, d\theta_j \right] \tag{22.1.11}$$

is a constant independent of α which was evaluated using Selberg's integral (cf. Chapter 17). In the limit of large n with fixed m, this constant has a close relation with the moments of the absolute values of the Riemann zeta function $\zeta(z)$ or of the L-functions on the critical line $\mathrm{Re}\, z = 1/2$. Similarly, the moments of the derivative with respect to x of the characteristic polynomial are related to the moments of the absolute values of the derivative $\zeta'(z)$ on the critical line.

In case of the real symmetric, complex hermitian and quaternion self-dual matrices, which are the only ones we consider here, the mth moment of the characteristic polynomial is a polynomial of order mn, and we do not know whether their zeros are related with the Riemann zeta function.

22.2 Calculation of $I_\beta(n, m; x)$

The case $\beta = 2$ is the simplest involving the better known determinants, while the other two cases $\beta = 1$ and 4 are similar involving the less familiar Pfaffians.

The $I_\beta(n, m; x)$ can be expressed in two different forms:

(i) a determinant or a Pfaffian of a matrix of size which depends on n; i.e.,

- an $n \times n$ determinant in case $\beta = 2$;
- a Pfaffian of a $2n \times 2n$ anti-symmetric matrix in case $\beta = 4$;
- a Pfaffian of a $n \times n$ (respectively $(n+1) \times (n+1)$) anti-symmetric matrix in case $\beta = 1$ for n even (respectively odd);

(ii) a sum of determinants of matrices of size which depends on m; i.e.,

- an $m \times m$ determinant in case $\beta = 2$;
- a sum of $m \times m$ determinants in case $\beta = 4$ for m even;
- a sum of $m \times m$ (respectively $(m+1) \times (m+1)$) determinants in case $\beta = 1$ for n even (respectively odd).

The usefulness of the expressions above, depending on the form of the weight function $w(x)$, may be limited. When the weight is Gaussian, two of the forms supplying alternate expressions for $I_2(n, m; x)$ as $n \times n$ and $m \times m$ determinants lead us to an interesting known identity. Similarly, the two expressions of $I_4(n, m; x)$ lead us to an identity between a Pfaffian of an $2n \times 2n$ anti-symmetric matrix and a sum of $m \times m$ determinants.

22.2.1 $I_\beta(n, m; x)$ as a determinant or a Pfaffian of a matrix of size depending on n. Choose $P_j(x)$ and $\pi_j(x)$ any monic polynomials of degree j.

The case $\beta = 2$. Write $\Delta_n(\mathbf{y})$ as an $n \times n$ Vandermonde determinant

$$\begin{aligned}\Delta_n(\mathbf{y}) &= \det\left[y_j^{k-1}\right]_{j,k=1,\ldots,n} = \det\left[P_{k-1}(y_j)\right]_{j,k=1,\ldots,n} \\ &= \det\left[\pi_{k-1}(y_j)\right]_{j,k=1,\ldots,n}.\end{aligned} \tag{22.2.1}$$

Then as in Chapter 5

$$\begin{aligned}&\int \cdots \int \Delta_n^2(\mathbf{y}) \prod_{j=1}^{n} [f(y_j)\, dy_j] \\ &\quad = \int \cdots \int \det\left[P_{k-1}(y_j)\right] \det\left[\pi_{k-1}(y_j)\right] \prod_{j=1}^{n} [f(y_j)\, dy_j] \\ &\quad = n! \int \cdots \int \det\left[\pi_{k-1}(y_j)\right] \prod_{j=1}^{n} [P_{j-1}(y_j) f(y_j)\, dy_j] \\ &\quad = n! \int \cdots \int \det\left[P_{j-1}(y_j)\pi_{k-1}(y_j) f(y_j)\right] \prod_{j=1}^{n} dy_j \\ &\quad = n! \det[\phi_{2;\, j,k}]_{j,k=0,\ldots,n-1},\end{aligned} \tag{22.2.2}$$

where

$$\phi_{2;\, j,k} := \int P_j(y) \pi_k(y) f(y)\, dy. \tag{22.2.3}$$

In the first three lines of the above equation the indices j, k take the values from 1 to n. In the second line we have replaced the first determinant by a single term which is allowed by the symmetry of the integrand in the y_j; while the last line is obtained by integrating over the y_j each of them occurring only in one column.

Setting $f(y) = (x-y)^m w(y)$, we get a first form of $I_2(n, m; x)$ in terms of an $n \times n$ determinant

$$I_2(n, m; x) = n! \det\left[\phi_{2;j,k}(x)\right]_{j,k=0,\dots,n-1}, \tag{22.2.4}$$

where

$$\phi_{2;j,k}(x) := \int P_j(y)\pi_k(y)(x-y)^m w(y)\,dy. \tag{22.2.5}$$

The case $\beta = 4$. Write $\Delta_n^4(\mathbf{y})$ as a $2n \times 2n$ determinant (see, e.g., Mehta, 1989MT, Section 7.1)

$$\begin{aligned}\Delta_n^4(\mathbf{y}) &= \det\left[y_j^k \quad k y_j^{k-1}\right]_{j=1,\dots,n;\ k=0,\dots,2n-1} \\ &= \det\left[P_k(y_j) \quad P_k'(y_j)\right]_{j=1,\dots,n;\ k=0,\dots,2n-1}.\end{aligned} \tag{22.2.6}$$

Each variable occurs in two columns, a column of monic polynomials and a column of its derivatives. Expand the determinant and integrate to see that (see, e.g., Appendix A.17)

$$\int \cdots \int \Delta_n^4(\mathbf{y}) \prod_{j=1}^{n} f(y_j)\,dy_j = \sum \pm \phi_{4;j_1,j_2}\phi_{4;j_3,j_4}\cdots\phi_{4;j_{2n-1},j_{2n}} \tag{22.2.7}$$

$$= n!\,\mathrm{pf}[\phi_{4;j,k}]_{j,k=0,\dots,2n-1}, \tag{22.2.8}$$

where

$$\phi_{4;j,k} := \int \left[P_j(y)P_k'(y) - P_j'(y)P_k(y)\right] f(y)\,dy = -\phi_{4;k,j}. \tag{22.2.9}$$

In Eq. (22.2.7) the sum is over all permutations $\left(\begin{smallmatrix} 0 & 1 & \dots & 2n-1 \\ j_1 & j_2 & \dots & j_{2n}\end{smallmatrix}\right)$ of the $2n$ indices $(0, 1, \dots, 2n-1)$ with $j_1 < j_2, \dots, j_{2n-1} < j_{2n}$, the sign being plus or minus according as this permutation is even or odd. In Eq. (22.2.8) *pf* means the Pfaffian. Setting $f(y) = (x-y)^m w(y)$ as before, we get a first form of $I_4(n, m; x)$ in terms of the Pfaffian of a $2n \times 2n$ anti-symmetric matrix

$$I_4(n, m; x) = n!\,\mathrm{pf}\left[\phi_{4;j,k}(x)\right]_{j,k=0,\dots,2n-1}, \tag{22.2.10}$$

where

$$\phi_{4;j,k}(x) := \int \big[P_j(y)P_k'(y) - P_j'(y)P_k(y)\big](x-y)^m w(y)\,dy. \tag{22.2.11}$$

The case $\beta = 1$. The difficulty of the absolute value of $\Delta_n(\mathbf{y})$ can be overcome by ordering the variables. Writing

$$\Delta_n(\mathbf{y}) \prod_{j=1}^{n} f(y_j) = \det\big[P_{k-1}(y_j)f(y_j)\big]_{j,k=1,\dots,n}, \tag{22.2.12}$$

with $f(y) = (x-y)^m \sqrt{w(y)}$ and using a result of de Bruijn (1955) one gets a first form of $I_1(n,m;x)$ in terms of the Pfaffian of an $n \times n$ (respectively $(n+1) \times (n+1)$) anti-symmetric matrix for n even (respectively odd)

$$I_1(n,m;x) = \begin{cases} n!\,\mathrm{pf}\big[\phi_{1;j,k}(x)\big]_{j,k=0,\dots,n-1}, & n \text{ even}, \\ n!\,\mathrm{pf}\begin{bmatrix} \phi_{1;j,k}(x) & \alpha_j(x) \\ -\alpha_k(x) & 0 \end{bmatrix}_{j,k=0,\dots,n-1}, & n \text{ odd}, \end{cases} \tag{22.2.13}$$

where

$$\phi_{1;j,k}(x) := \iint P_j(z)P_k(y)(x-z)^m(x-y)^m \sqrt{w(z)w(y)}\,\mathrm{sign}(y-z)\,dz\,dy, \tag{22.2.14}$$

and

$$\alpha_j(x) := \int P_j(y)(x-y)^m \sqrt{w(y)}\,dy. \tag{22.2.15}$$

Another way to get the same result is (cf. Chapter 7) to order the variables, integrate over the alternate ones, remove the ordering over the remaining alternate variables and observe that the result is given by Eqs. (22.2.13)–(22.2.15).

An alternative of the expressions above for $\beta = 1$, *2 and 4*. If one replaces the monic polynomials $P_j(y)$ and $\pi_j(y)$ by $P_j(y-x)$ and $\pi_j(y-x)$, Eqs. (22.2.4), (22.2.10) and (22.2.13) become

$$I_2(n,m;x) = (-1)^{mn} n!\,\det\big[\varphi_{2;j,k}(x)\big]_{j,k=0,\dots,n-1}, \tag{22.2.16}$$

$$I_4(n,m;x) = (-1)^{mn} n!\,\mathrm{pf}\big[\varphi_{4;j,k}(x)\big]_{j,k=0,\dots,2n-1}, \tag{22.2.17}$$

$$I_1(n,m;x) = \begin{cases} n!\,\mathrm{pf}\big[\varphi_{1;j,k}(x)\big]_{j,k=0,\dots,n-1}, & n \text{ even}, \\ (-1)^m n!\,\mathrm{pf}\begin{bmatrix} \varphi_{1;j,k}(x) & \theta_j(x) \\ -\theta_k(x) & 0 \end{bmatrix}_{j,k=0,\dots,n-1}, & n \text{ odd}, \end{cases} \tag{22.2.18}$$

where

$$\varphi_{2;j,k}(x) := \int P_j(y-x)\pi_k(y-x)(y-x)^m w(y)\,dy, \tag{22.2.19}$$

$$\varphi_{4;j,k}(x) := \int \big[P_j(y-x)P_k'(y-x) - P_j'(y-x)P_k(y-x)\big] \times (y-x)^m w(y)\,dy, \tag{22.2.20}$$

$$\varphi_{1;j,k}(x) := \iint P_j(z-x)P_k(y-x)\big[(z-x)(y-x)\big]^m \times \sqrt{w(z)w(y)}\,\mathrm{sign}(y-z)\,dz\,dy, \tag{22.2.21}$$

$$\theta_j(x) := \int P_j(y-x)(y-x)^m \sqrt{w(y)}\,dy. \tag{22.2.22}$$

22.2.2 $I_\beta(n, m; x)$ as determinants of size depending on m

The case $\beta = 2$. For $I_2(n, m; x)$, Eq. (22.1.8), let us write the integrand as the product of two determinants: $\Delta_n(\mathbf{y})$ given in Eq. (22.2.1) and

$$\begin{aligned}\Delta_n(\mathbf{y}) \prod_{j=1}^{n}(x-y_j)^m &= b \det\left[y_j^{k-1} \quad \left(\frac{d}{dx}\right)^l x^{k-1} \right]_{\substack{j=1,\dots,n\\ k=1,\dots,n+m\\ l=0,\dots,m-1}} \\ &= b \det\big[P_{k-1}(y_j) \quad P_{k-1}^{(l)}(x)\big]_{\substack{j=1,\dots,n\\ k=1,\dots,n+m\\ l=0,\dots,m-1}},\end{aligned} \tag{22.2.23}$$

where

$$b = \left(\prod_{l=0}^{m-1} l!\right)^{-1}. \tag{22.2.24}$$

$P_k(x)$ is any monic polynomial of degree k and $P_k^{(l)}(x)$ is its lth derivative. Expression (22.2.23) is an $(m+n) \times (m+n)$ determinant the last m columns of which are $P_k(x)$ and its successive derivatives (see, e.g., Mehta, 1989MT, Section 7.1). As the integrand in Eq. (22.1.8) is symmetric in the y_j, we can replace the first $n \times n$ determinant by a single term and multiply the result by $n!$

$$I_2(n, m; x) = bn! \int \cdots \int \det\big[P_{k-1}(y_j) \quad P_{k-1}^{(l)}(x)\big]_{\substack{j=1,\dots,n\\ k=1,\dots,n+m\\ l=0,\dots,m-1}}$$

$$\times \prod_{j=1}^{n} [P_{j-1}(y_j) w(y_j)\, dy_j],$$

$$= bn! \int \cdots \int \det\big[P_{j-1}(y_j) P_{k-1}(y_j) w(y_j),\ P_{k-1}^{(l)}(x)\big]_{\substack{j=1,\dots,n\\ k=1,\dots,n+m\\ l=0,\dots,m-1}}$$

$$\times \prod_{j=1}^{n} dy_j. \tag{22.2.25}$$

As each variable y_j occurs only in one column, we can integrate over them independently

$$I_2(n,m;x) = bn! \det\left[\int P_{j-1}(y) P_{k-1}(y) w(y)\, dy \quad P_{k-1}^{(l)}(x)\right]_{\substack{j=1,\dots,n\\ k=1,\dots,n+m\\ l=0,\dots,m-1}}. \tag{22.2.26}$$

If we choose the polynomials $P_j(x)$ to be the orthogonal polynomials $C_j(x)$, Eq. (22.1.5), then one gets a second form for $I_2(n,m;x)$

$$I_2(n,m;x) = bn! c_0 \cdots c_{n-1} \det\big[C_{n+k}^{(l)}(x)\big]_{k,l=0,\dots,m-1}, \tag{22.2.27}$$

an $m \times m$ determinant whose first column consists of $C_{n+k}(x)$, $k = 0, 1, \dots, m-1$, and the other $m-1$ columns are the successive derivatives of the first column. Therefore from Eq. (22.1.10)

$$\langle \det(xI - A)^m \rangle = b \det\big[C_{n+k}^{(l)}(x)\big]_{k,l=0,\dots,m-1}. \tag{22.2.28}$$

The case $\beta = 4$. This method applies only for even moments. For $I_4(n, 2m; x)$, Eq. (22.1.9), one can write the integrand as a single determinant (see, e.g., Mehta, 1989MT, Section 7.1.2)

$$\Delta_n^4(\mathbf{y}) \prod_{j=1}^{n} (x-y)^{2m} = b \det\left[y_j^k \quad k y_j^{k-1} \quad \left(\frac{d}{dx}\right)^l x^k\right]_{\substack{j=1,\dots,n\\ k=0,\dots,n+m-1\\ l=0,\dots,m-1}}$$

$$= b \det\big[P_k(y_j) \quad P_k'(y_j) \quad P_k^{(l)}(x)\big]_{\substack{j=1,\dots,n\\ k=0,\dots,n+m-1\\ l=0,\dots,m-1}}, \tag{22.2.29}$$

where b is given by Eq. (22.2.24). Each variable y_j occurs in two columns. Expanding the determinant and integrating one sees that the result is a sum of products of the form

$$\pm b\, a_{s_1,s_2} \cdots a_{s_{2n-1},s_{2n}} \det\big[P_{j_k}^{(l-1)}(x)\big]_{k,l=1,\dots,m}, \tag{22.2.30}$$

with

$$a_{j,k} := \int \left[P_j(y)P_k'(y) - P_j'(y)P_k(y)\right]w(y)\,dy = \langle P_j, P_k\rangle_4, \tag{22.2.31}$$

the indices $s_1, \ldots, s_{2n}, j_1, \ldots, j_m$ are all distinct, chosen from $0, 1, \ldots, 2n+m-1$ and the sign is plus or minus according as the permutation

$$\begin{pmatrix} 0 & 1 & \ldots & 2n-1 & 2n & \ldots & 2n+m-1 \\ s_1 & s_2 & \ldots & s_{2n} & j_1 & \ldots & j_m \end{pmatrix} \tag{22.2.32}$$

is even or odd. If we choose the polynomials $P_j(x)$ to be the skew-orthogonal polynomials $Q_j(x)$, Eq. (22.1.6), then all the $a_{s,j}$ except $a_{2s,2s+1} = q_s$ will be zero and we get a second form of $I_4(n, 2m; x)$ in terms of a sum of $m \times m$ determinants

$$I_4(n, 2m; x) = bn! \sum_{(s)} q_{s_1} \cdots q_{s_n} \det\left[Q_{j_k}^{(l-1)}(x)\right]_{k,l=1,\ldots,m}, \tag{22.2.33}$$

where the sum is over all choices of $s_1 < \cdots < s_n$ such that $2s_1, 2s_1+1, \ldots, 2s_n, 2s_n + 1, j_1, \ldots, j_m$ are all the indices from 0 to $2n+m-1$ and moreover $j_1 < \cdots < j_m$.

The case $\beta = 1$. For $I_1(n, m; x)$, Eq. (22.1.9), the absolute value sign of $\Delta_n(\mathbf{y})$ is the main difficulty. Ordering the variables and using the same method as above, in Section 22.2.1, one has

$$\begin{aligned} I_1(n, m; x) &= \int \cdots \int \left|\Delta_n(\mathbf{y})\right| \prod_{j=1}^{n} \left[(x - y_j)^m \sqrt{w(y_j)}\,dy_j\right] \\ &= n! \int \cdots \int_{y_1 \leqslant \cdots \leqslant y_n} \Delta_n(\mathbf{y}) \prod_{j=1}^{n} \left[(x - y_j)^m \sqrt{w(y_j)}\,dy_j\right] \\ &= bn! \int \cdots \int_{y_1 \leqslant \cdots \leqslant y_n} \det\left[P_k(y_j), P_k^{(l)}(x)\right]_{\substack{j=1,\ldots,n \\ k=0,\ldots,n+m-1 \\ l=0,\ldots,m-1}} \\ &\quad \times \prod_{j=1}^{n} \left[\sqrt{w(y_j)}\,dy_j\right], \end{aligned} \tag{22.2.34}$$

with b as in Eq. (22.2.24). Integrating successively over the alternate variables $y_1, y_3, y_5, \ldots$ and then removing the restriction over the remaining variables $y_2, y_4, \ldots$

as indicated at the end of Section 22.2.1 above, one gets for even n

$$I_1(n,m;x) = b\frac{n!}{[n/2]!}\int\cdots\int \det\big[G_k(y_{2j}) \quad P_k(y_{2j}) \quad P_k^{(l)}(x)\big]_{\substack{j=1,\ldots,[n/2]\\ k=0,\ldots,n+m-1\\ l=0,\ldots,m-1}}$$

$$\times \prod_{j=1}^{[n/2]}\big[\sqrt{w(y_{2j})}\,dy_{2j}\big], \tag{22.2.35}$$

with

$$G_k(y_2) := \int^{y_2} P_k(y_1)\sqrt{w(y_1)}\,dy_1. \tag{22.2.36}$$

In case n is odd, one more column of the numbers $g_k := \int P_k(y)\sqrt{w(y)}\,dy$ appears in the determinant just after the column of $P_k(y_{n-1})$.

The present situation is exactly as in the case $\beta = 4$ with each variable occurring in two columns. Expanding the determinant one sees that the result contains the expressions

$$\begin{aligned}
&\int\big[G_j(y)P_k(y) - G_k(y)P_j(y)\big]\sqrt{w(y)}\,dy \\
&= \iint P_j(x)P_k(y)\,\mathrm{sign}(y-x)\sqrt{w(x)w(y)}\,dx\,dy \\
&= \langle P_j, P_k\rangle_1.
\end{aligned} \tag{22.2.37}$$

If we choose $P_j(x) = R_j(x)$, Eq. (22.1.7), then writing the result separately for even and odd n for clarity, one has

$$I_1(2n,m;x) = b(2n)!\sum_{(s)} r_{s_1}\cdots r_{s_n}\det\big[R_{j_k}^{(l-1)}(x)\big]_{k,l=1,\ldots,m}, \tag{22.2.38}$$

where the sum is taken over all choices of $s_1 < \cdots < s_n$ such that $2s_1, 2s_1+1, \ldots, 2s_n, 2s_n+1, j_1, \ldots, j_m$ are all the indices from 0 to $2n+m-1$ and moreover $j_1 < \cdots < j_m$ and

$$I_1(2n+1,m;x) = b(2n+1)!\sum_{(s)} r_{s_1}\cdots r_{s_n}\det\big[g_{j_k} \quad R_{j_k}^{(l-1)}(x)\big]_{k=1,\ldots,m+1;\ l=1,\ldots,m}, \tag{22.2.39}$$

where now the sum is taken over all choices of $s_1 < \cdots < s_n$ such that $2s_1, 2s_1+1, \ldots, 2s_n, 2s_n+1, j_1, \ldots, j_m, j_{m+1}$ are all the indices from 0 to $2n+m$ and moreover $j_1 < \cdots < j_{m+1}$.

Note that Eqs. (5.11.1), (5.11.2) and (5.11.4) are particular cases ($m = 1$) of Eqs. (22.2.27), (22.2.33) and (22.2.38) respectively. If we shift the index of the last row in the right-hand side of Eq. (22.2.29) for $m = 1$ from n to $n + 1$, i.e., replace the last row

$$\left[P_n(y_j),\, P_n'(y_j),\, P_n(x)\right],$$

by the row

$$\left[P_{n+1}(y_j),\, P_{n+1}'(y_j),\, P_{n+1}(x)\right],$$

we get the integrand of Eq. (5.11.3). Following the procedure which leads to Eq. (22.2.33) we get Eq. (5.11.3). Similarly, if we shift the index of the last row from n to $n + 1$ in the right-hand side of Eq. (22.2.34) for $m = 1$ and follow the procedure leading to Eq. (22.2.38) we get Eq. (5.11.5).

22.3 Special Case of the Gaussian Weight

These formulas for $I_\beta(n, m; x)$ will have little use if one does not know the polynomials $C_n(x)$, $Q_n(x)$ or $R_n(x)$. Fortunately one knows them for all classical weights. As an example we give them here for the Gaussian weight $w(x) = e^{-x^2}$ over $[-\infty, \infty]$ in terms of Hermite polynomials $H_n(x) := e^{x^2}(-d/dx)^n e^{-x^2}$. Their verification is straightforward. One has

$$C_n(x) = 2^{-n} H_n(x), \tag{22.3.1}$$

$$c_n = 2^{-n} n! \sqrt{\pi}, \tag{22.3.2}$$

$$Q_{2n}(x) = \sum_{j=0}^{n} 2^{-2j} \frac{n!}{j!} H_{2j}(x), \qquad Q_{2n+1}(x) = 2^{-2n-1} H_{2n+1}(x), \tag{22.3.3}$$

$$q_n = 2^{-2n}(2n + 1)! \sqrt{\pi}, \tag{22.3.4}$$

$$R_{2n}(x) = 2^{-2n} H_{2n}(x), \qquad R_{2n+1}(x) = 2^{-2n}\left[x H_{2n}(x) - H_{2n}'(x)\right], \tag{22.3.5}$$

$$r_n = 2^{1-2n}(2n)! \sqrt{\pi}. \tag{22.3.6}$$

In this case some results can be given other forms. In particular, using the recurrence relation

$$2x H_n(x) = H_{n+1}(x) + H_n'(x), \tag{22.3.7}$$

or

$$x C_n(x) = C_{n+1}(x) + \frac{1}{2} C_n'(x), \tag{22.3.8}$$

one can replace the $C_{n+k}^{(l)}$ in the determinant (22.2.27) by $C_{n+k+l}(x)$ so that

$$I_2(n,m;x) = bn!c_0\cdots c_{n-1}(-2)^{m(m-1)/2}\det\big[C_{n+j+k}(x)\big]_{j,k=0,1,\ldots,m-1}$$
$$= \pi^{n/2}\frac{2^{-n(n-1)/2}\prod_{j=0}^{n} j!}{(-2)^{-m(m-1)/2}\prod_{j=0}^{m-1} j!}\det\big[C_{n+j+k}(x)\big]_{j,k=0,1,\ldots,m-1}. \tag{22.3.9}$$

Also for any non-negative integer j (Bateman, 1953b, Chapter 10.13, (31))

$$\int_{-\infty}^{\infty}(y-x)^j e^{-y^2}\,dy = \sqrt{\pi}\,i^j\,C_j(ix), \tag{22.3.10}$$

with $i=\sqrt{-1}$. Therefore, choosing $P_j(y-x)=\pi_j(y-x)=(y-x)^j$ in Eqs. (22.2.19) and (22.2.20), we get

$$\varphi_{2;\,j,k}(x) = \sqrt{\pi}\,i^{m+j+k}\,C_{m+j+k}(ix), \tag{22.3.11}$$

$$\varphi_{4;\,j,k}(x) = \sqrt{\pi}\,i^{m+j+k-1}(k-j)C_{m+j+k-1}(ix), \tag{22.3.12}$$

and Eqs. (22.2.16) and (22.2.17) then give

$$I_2(n,m;x) = (-i)^{mn}n!\pi^{n/2}(-1)^{n(n-1)/2}$$
$$\times\det\big[C_{m+j+k}(ix)\big]_{j,k=0,\ldots,n-1}, \tag{22.3.13}$$

$$I_4(n,m;x) = (-i)^{mn}n!\pi^{n/2}\,\mathrm{pf}\big[(k-j)C_{m+j+k-1}(ix)\big]_{j,k=0,\ldots,2n-1}. \tag{22.3.14}$$

Equations (22.3.9) and (22.3.13) give a relation almost symmetric in n and m. Writing them again

$$\frac{I_2(n,m;x)}{I_2(m,n;ix)} = (-i)^{mn}\frac{\pi^{n/2}2^{-n(n-1)/2}\prod_{j=0}^{n} j!}{\pi^{m/2}2^{-m(m-1)/2}\prod_{j=0}^{m} j!}, \tag{22.3.15}$$

or equivalently

$$\frac{\det[C_{n+j+k}(x)]_{j,k=0,\ldots,m-1}}{\det[C_{m+j+k}(ix)]_{j,k=0,\ldots,n-1}} = (-i)^{mn}\frac{(-2)^{n(n-1)/2}\prod_{j=0}^{m-1} j!}{(-2)^{m(m-1)/2}\prod_{j=0}^{n-1} j!}. \tag{22.3.16}$$

Equations (22.2.33) and (22.3.14) give another identity relating the Pfaffian

$$\mathrm{pf}\big[(k-j)C_{2m+j+k-1}(ix)\big]_{j,k=0,\ldots,2n-1},$$

to a sum of $m \times m$ determinants.

A more general identity

$$\frac{\int_{-\infty}^{\infty}\cdots\int_{-\infty}^{\infty}|\Delta_n(\mathbf{y})|^{2\gamma}\prod_{j=1}^{n}[(x-y_j)^m e^{-ay_j^2}\,dy_j]}{\int_{-\infty}^{\infty}\cdots\int_{-\infty}^{\infty}|\Delta_m(\mathbf{y})|^{2/\gamma}\prod_{j=1}^{m}[(ix-y_j)^n e^{-ay_j^2/\gamma}\,dy_j]} = K(\gamma,a,n,m), \quad (22.3.17)$$

with $K(\gamma,a,n,m)$ a constant, is probably true (see, e.g., Forrester and Witte (2001)). In particular

$$\frac{I_4(n,m;x)}{I_1(m,n;ix)} = (-i)^{mn}\frac{\pi^{n/2}2^{-n^2}\prod_{j=0}^{n}(2j)!}{2^{3m/2}\prod_{j=0}^{m}\Gamma(1+j/2)}. \quad (22.3.18)$$

22.4 Average Value of $\prod_{i=1}^{m}\det(x_i I - A)\prod_{j=1}^{\ell}\det(z_j I - A)^{-1}$ for the Unitary Ensemble

In Section 22.2.2 above we described how to express the average of $\det(xI - A)^m$ for the unitary ensemble, $\beta = 2$, as a single $m \times m$ determinant. One can repeat the same procedure to get the average of $\prod_{i=1}^{m}\det(x_i I - A)$,

$$M(m,\mathbf{x}) := \left\langle\prod_{i=1}^{m}\det(x_i I - A)\right\rangle \quad (22.4.1)$$

$$= K_n\int\cdots\int\prod_{j=1}^{n}dy_j\,w(y_j)\Delta_n^2(y)\prod_{i=1}^{m}(x_i - y_j) \quad (22.4.2)$$

$$= \frac{K_n}{\Delta_m(x)}\int\cdots\int\prod_{j=1}^{n}dy_j\,w(y_j)\det\bigl[\pi_r(y_j)\bigr]$$
$$\times\det\bigl[P_{k-1}(y_j) \quad P_{k-1}(x_i)\bigr] \quad (22.4.3)$$

$$= \frac{K_n}{\Delta_m(x)}n!\int\cdots\int\prod_{j=1}^{n}dy_j\,w(y_j)\pi_{j-1}(y_j)$$
$$\times\det\bigl[P_{k-1}(y_j) \quad P_{k-1}(x_i)\bigr] \quad (22.4.4)$$

$$= \frac{K_n}{\Delta_m(x)}n!\int\cdots\int\prod_{j=1}^{n}dy_j\,w(y_j)$$
$$\times\det\bigl[\pi_{j-1}(y_j)P_{k-1}(y_j) \quad P_{k-1}(x_i)\bigr] \quad (22.4.5)$$

$$= \frac{K_n}{\Delta_m(x)}n!\det\bigl[\phi_{j,k} \quad P_{k-1}(x_i)\bigr], \quad (22.4.6)$$

with

$$\phi_{j,k} = \int dy\, w(y)\pi_{j-1}(y)P_{k-1}(y). \tag{22.4.7}$$

In the above K_n is the normalization constant

$$K_n = \int \cdots \int \prod_{j=1}^{n} dy_j\, w(y_j)\Delta_n^2(y), \tag{22.4.8}$$

$\pi_j(x)$ and $P_j(x)$ are any monic polynomials; the indices take the values as $r = 0, \ldots, n-1$; $i = 1, \ldots, m$; $j = 1, \ldots, n$; $k = 1, \ldots, m+n$; so that the determinants in Eq. (22.4.3) are of order n and $n+m$ respectively while the determinant in Eqs. (22.4.4), (22.4.5) and (22.4.6) is of order $n+m$. The step from Eq. (22.4.3) to (22.4.4) is possible because of the symmetry of the integrand in the n variables y_j while that from (22.4.5) to (22.4.6) is possible because each variable occurs in only one column.

The monic polynomials $\pi_j(x)$ and $P_j(x)$ being arbitrary, if we choose them as $\pi_j(x) = P_j(x) = C_j(x)$, orthogonal with the weight $w(x)$

$$\int dy\, w(y)C_j(y)C_k(y) = c_j\delta_{j,k}, \tag{22.4.9}$$

then $\phi_{j,k} = c_{j-1}\delta_{j,k}$ and Eq. (22.4.6) simplifies to

$$M(m, \mathbf{x}) = \frac{K_n}{\Delta_m(x)} \prod_{j=0}^{n-1} c_j \det\left[C_{n+i-1}(x_j)\right]_{i,j=1,\ldots,m} \tag{22.4.10}$$

$$= \left[\Delta_m(x)\right]^{-1} \det\left[C_{n+i-1}(x_j)\right]_{i,j=1,\ldots,m}, \tag{22.4.11}$$

since K_n is just the inverse of $\prod_{j=0}^{n-1} c_j$.

Even more remarkable is the fact that in this particular case $\beta = 2$, not only the positive moments, but a mixture of positive and negative moments of characteristic polynomials can be expressed as a single determinant. If $\ell < n$, $\operatorname{Im} z_k \neq 0$ for $k = 1, \ldots, \ell$, then

$$M(m, \ell, \mathbf{x}, \mathbf{z}) := \left\langle \prod_{i=1}^{m} \det(x_i I - A) \Big/ \prod_{k=1}^{\ell} \det(z_k I - A) \right\rangle \tag{22.4.12}$$

$$= K_n \int \cdots \int \prod_{j=1}^{n} dy_j\, w(y_j)\Delta_n^2(y) \prod_{i=1}^{m}(x_i - y_j) \Big/ \prod_{k=1}^{\ell}(z_k - y_j) \tag{22.4.13}$$

$$= \left[\Delta_m(x)\Delta_\ell(z)\right]^{-1}\prod_{r=1}^{\ell} c_{n-r}^{-1}$$

$$\times \det\left[\tilde{C}_{n-\ell+r}(z_k),\ C_{n-\ell+r}(x_i)\right]_{\substack{k=1,\ldots,\ell;\ i=1,\ldots,m,\\ r=0,\ldots,\ell+m-1}} \qquad (22.4.14)$$

where as before $C_j(x)$ are the monic orthogonal polynomials with the weight $w(x)$, Eq. (22.4.9), and $\tilde{C}_j(z)$ are their Cauchy transforms

$$\tilde{C}_j(z) = \int dy\, w(y)\frac{C_j(y)}{z-y}, \qquad (22.4.15)$$

these later are well defined since $\operatorname{Im} z \neq 0$.

Here we will give the proof of Eq. (22.4.14) only for the simplest case $\ell = 1$, indicate the changes necessary for $\ell > 1$ and refer to Fyodorov and Strahov (2003) for further details.

In case $\ell = 1$, a decomposition in partial fractions gives

$$\left[\det(zI-A)\right]^{-1} = \prod_{j=1}^{n}\frac{1}{z-y_j}$$

$$= \sum_{j=1}^{n}\frac{a_j}{z-y_j}, \qquad (22.4.16)$$

$$a_j = \prod_{k(\neq j)}(y_j-y_k)^{-1}. \qquad (22.4.17)$$

From the symmetry of the integrand on all the variables y_j,

$$\int\cdots\int\prod_{j=1}^{n} dy_j\, w(y_j)\Delta_n(y)\det\left[C_{k-1}(y_j)\quad C_{k-1}(x_i)\right]\prod_{j=1}^{n}\frac{1}{z-y_j}$$

$$= \int\cdots\int\prod_{j=1}^{n} dy_j\, w(y_j)\Delta_n(y)\det\left[C_{k-1}(y_j)\quad C_{k-1}(x_i)\right]\sum_{j=1}^{n}\frac{a_j}{z-y_j} \qquad (22.4.18)$$

$$= n\int\cdots\int\prod_{j=1}^{n} dy_j\, w(y_j)\Delta_n(y)\det\left[C_{k-1}(y_j)\quad C_{k-1}(x_i)\right]\frac{a_n}{z-y_n} \qquad (22.4.19)$$

$$= n \int \cdots \int \prod_{j=1}^{n} dy_j \, w(y_j) \Delta_{n-1}(y_1, \ldots, y_{n-1})(z - y_n)^{-1}$$

$$\times \det\bigl[C_{k-1}(y_j) \quad C_{k-1}(x_i)\bigr] \tag{22.4.20}$$

$$= n! \int \cdots \int \prod_{j=1}^{n} dy_j \, w(y_j) C_{j-1}(y_j)(z - y_n)^{-1}$$

$$\times \det\bigl[C_{k-1}(y_j) \quad C_{k-1}(x_i)\bigr] \tag{22.4.21}$$

$$= n! \int \cdots \int \prod_{j=1}^{n} dy_j \, w(y_j)(z - y_n)^{-1}$$

$$\times \det\bigl[C_{j-1}(y_j) C_{k-1}(y_j) \quad C_{k-1}(x_i)\bigr] \tag{22.4.22}$$

$$= n! c_0 c_1 \cdots c_{n-2} \det\bigl[\tilde{C}_{n+r-1} \quad C_{n+r-1}(x_i)\bigr]_{r=0,\ldots,m-1}, \tag{22.4.23}$$

which is Eq. (22.4.14) for $\ell = 1$ if we put in the normalization constant.

When $\ell > 1$, decomposition into partial fractions is more involved. For example,

$$\prod_{j=1}^{n} \frac{1}{(z_1 - y_j)(z_2 - y_j)} = \frac{1}{z_1 - z_2} \sum_{1 \leqslant j < k \leqslant n} \frac{a_{j,k}}{(z_1 - y_j)(z_2 - y_k)}, \tag{22.4.24}$$

$$a_{j,k} = \prod_{\ell(\neq j,k)} \bigl[(y_j - y_\ell)(y_k - y_\ell)\bigr]^{-1}. \tag{22.4.25}$$

Taking the symmetry into account one can replace the sum in Eq. (22.4.24) by a single term corresponding to $j = n - 1$, $k = n$ and multiply the result by $n(n - 1)$. The factor $a_{n-1,n}$ reduces $\Delta_n(y)$ to $\Delta_{n-2}(y_1, \ldots, y_{n-2})$, the denominators $z_1 - y_{n-1}$ and $z_2 - y_n$ go in the columns $C_{k-1}(y_{n-1})$ and $C_{k-1}(y_n)$, while the first $n - 2$ columns on integration will contain only one non-zero element each, due to orthogonality of the polynomials $C_j(y)$.

For further details see Fyodorov and Strahov (2003).

Summary of Chapter 22

The mth moment of the characteristic polynomial of an $n \times n$ random matrix is expressed:

- as an $n \times n$, Eqs. (22.2.4) and (22.2.16), or as an $m \times m$, Eq. (22.2.27), determinant for the unitary ensemble ($\beta = 2$);

- as a Pfaffian of a $2n \times 2n$ anti-symmetric matrix, Eqs. (22.2.10) and (22.2.17), or a sum of $m \times m$ determinants in case m is even, Eq. (22.2.33) for the symplectic ensemble ($\beta = 4$);

and

- as a Pfaffian of an $n \times n$ (respectively $(n+1) \times (n+1)$) anti-symmetric matrix, Eqs. (22.2.13) and (22.2.18), or a sum of $m \times m$ (respectively $(m+1) \times (m+1)$) determinants, Eq. (22.2.38) (respectively (22.2.39)) for the orthogonal ensemble ($\beta = 1$) for n even (respectively odd).

In the Gaussian case $w(y) = e^{-y^2}$ this leads to the remarkable identity almost symmetric in m and n, Eq. (22.3.15),

$$I_2(n, m; x) = \text{const} \cdot I_2(m, n; ix), \tag{22.4.1}$$

or, Eq. (22.3.16),

$$\det\big[H_{m+j+k}(x)\big]_{j,k=0,\ldots,n-1} = \text{const} \cdot \det\big[H_{n+j+k}(ix)\big]_{j,k=0,\ldots,m-1}, \tag{22.4.2}$$

and another identity expressing $\text{pf}[(k-j)H_{2m+j+k}(ix)]_{j,k=0,\ldots,2n-1}$, Eq. (22.3.14), as a sum of $m \times m$ determinants, Eq. (22.2.33). Also probably the remarkable general identity (22.3.17) and in particular (22.3.18) holds.

In case of the unitary ensemble, $\beta = 2$, the average value of the ratio

$$\prod_{i=1}^{m} \det(x_i I - A) \prod_{j=1}^{\ell} \det(z_j I - A)^{-1},$$

for $m \geqslant 0$ and $\ell < n$ is expressed as a single $(m+\ell) \times (m+\ell)$ determinant, of which the first ℓ columns contain Cauchy transforms at z_j of the monic orthogonal polynomials $C_j(x)$ with the weight $w(x)$ and the last m columns contain those polynomials themselves at x_j.

23

HERMITIAN MATRICES COUPLED IN A CHAIN

In the study of planar diagrams of quantum field theory, of 2-d gravity and in some other applications one encounters integrals of the form

$$\int \exp\left[-\sum_j \operatorname{tr} V_j(A_j) + \sum_{i<j} c_{ij} \operatorname{tr} A_i A_j \right] \prod_j dA_j, \tag{23.0.1}$$

where $V_j(x)$ is a polynomial in x of even degree, $A_1, A_2, \ldots$ are $n \times n$ matrices and c_{ij} are constants. The coefficient of the highest power of x in every polynomial $V_j(x)$ is positive and the constants c_{ij} are such that the integral (23.0.1) converges. This is an ensemble of several coupled matrices.

Till now we have studied the simplest case of one matrix; there were no cross terms containing the c_{ij}. The case of coupled matrices may be represented by a graph where each matrix is represented by a point, and two points representing matrices A_i and A_j are joined by a line marked c_{ij} if the coupling factor $\exp[c_{ij} \operatorname{tr}(A_i A_j)]$ is present in the probability density. When several matrices are coupled the probability density for the eigenvalues is known only in the case where these matrices are complex hermitian and the graph has a tree structure, i.e., does not have a closed path.

In this chapter we will consider the simplest case of a tree, i.e., that of a chain of p complex hermitian $n \times n$ matrices with the joint probability density for their elements

$$F(A_1, \ldots, A_p) \propto \exp\left[-\operatorname{tr}\left\{\frac{1}{2}V_1(A_1) + V_2(A_2) + \cdots + V_{p-1}(A_{p-1}) + \frac{1}{2}V_p(A_p)\right\}\right] \times \exp\left[\operatorname{tr}\{c_1 A_1 A_2 + c_2 A_2 A_3 + \cdots + c_{p-1} A_{p-1} A_p\}\right], \tag{23.0.2}$$

where $V_j(x)$ are real polynomials of even order with positive coefficients of their highest powers and c_j are real constants. For each j, the eigenvalues of the matrix A_j are real and will be denoted by $\mathbf{x_j} := \{x_{j,1}, x_{j,2}, \ldots, x_{j,n}\}$. From Appendix A.5 one can deduce that

$$\int \exp\left[c\operatorname{tr}(U_1 A_1 U_1^{-1} \cdot U_2 A_2 U_2^{-1})\right] dU_1\, dU_2 \propto \left[\Delta(\mathbf{x}_1)\Delta(\mathbf{x}_2)\right]^{-1} \det[e^{cx_{1,i}x_{2,k}}]. \tag{23.0.3}$$

The probability density for the eigenvalues of all the p matrices resulting from Eq. (23.0.2) is thus

$$\begin{aligned} F(\mathbf{x_1}; \ldots; \mathbf{x_p}) &= C\exp\left[-\sum_{r=1}^{n}\left\{\frac{1}{2}V_1(x_{1,r}) + V_2(x_{2,r}) + \cdots + V_{p-1}(x_{p-1,r}) + \frac{1}{2}V_p(x_{p,r})\right\}\right] \\ &\quad \times \prod_{1 \leqslant r < s \leqslant n} (x_{1,r} - x_{1,s})(x_{p,r} - x_{p,s}) \\ &\quad \times \det[e^{c_1 x_{1,r} x_{2,s}}]\det[e^{c_2 x_{2,r} x_{3,s}}] \cdots \det[e^{c_{p-1} x_{p-1,r} x_{p,s}}] \qquad (23.0.4) \\ &= C\left[\prod_{1 \leqslant r < s \leqslant n} (x_{1,r} - x_{1,s})(x_{p,r} - x_{p,s})\right]\left[\prod_{k=1}^{p-1} \det\left[w_k(x_{k,r}, x_{k+1,s})\right]_{r,s=1,\ldots,n}\right], \qquad (23.0.5) \end{aligned}$$

where

$$w_k(\xi, \eta) := \exp\left[-\frac{1}{2}V_k(\xi) - \frac{1}{2}V_{k+1}(\eta) + c_k \xi\eta\right], \tag{23.0.6}$$

and C is a normalization constant such that the integral of F over the np variables $x_{i,r}$ is 1. We will be interested in the correlation functions

$$R_{k_1,\ldots,k_p}(x_{1,1}, \ldots, x_{1,k_1}; \ldots; x_{p,1}, \ldots, x_{p,k_p}) := \int F(\mathbf{x_1}; \ldots; \mathbf{x_p}) \prod_{j=1}^{p} \frac{n!}{(n-k_j)!} \prod_{r_j=k_j+1}^{n} dx_{j,r_j}. \tag{23.0.7}$$

Loosely speaking, this is the density of ordered sets of k_j eigenvalues of A_j within small intervals around $x_{j,1}, \ldots, x_{j,k_j}$ for $j = 1, 2, \ldots, p$. Here and in what follows, all the integrals are taken over $-\infty$ to ∞ unless explicitly stated otherwise.

The general correlation function (23.0.7) can be written as a single $m \times m$ determinant with $m = k_1 + k_2 + \cdots + k_p$. The result is given in Section 23.1 and the proof in Section 23.2. This result is a generalization of that in Chapter 6 for a single hermitian matrix, (the case $p = 1$), according to which the correlation function of k eigenvalues is given by a $k \times k$ determinant of the form:

$$R_k(x_1, \ldots, x_k) = \det\bigl[K(x_i, x_j)\bigr]_{i,j=1,k}, \tag{23.0.8}$$

the kernel $K(x, y)$ depending on the polynomial $V(x)$.

In Sections 23.3 and 23.4 we present an investigation of the spacing functions $E(m_1, I_1; m_2, I_2; \ldots; m_p, I_p)$, the probability that exactly m_j eigenvalues of A_j lie in the interval I_j for $j = 1, \ldots, p$. Finally in Section 23.5 we will see that the zeros of the bi-orthogonal polynomials involved are real, simple and lie in the expected intervals.

23.1 General Correlation Function

To express our result we need some notations. Recall that a polynomial is called monic when the coefficient of the highest power is one. With a monic polynomial $P_j(\xi)$ of degree j let us write

$$P_{1,j}(\xi) := P_j(\xi), \tag{23.1.1}$$

and recursively,

$$P_{i,j}(\xi) := \int P_{i-1,j}(\eta) w_{i-1}(\eta, \xi)\, d\eta, \quad 2 \leqslant i \leqslant p \tag{23.1.2}$$

$$:= \int P_j(\eta) U_{Li}(\eta, \xi)\, d\eta, \tag{23.1.3}$$

$$U_{Li}(\eta, \xi) := (w_1 * w_2 * \cdots * w_{i-1})(\eta, \xi). \tag{23.1.4}$$

Similarly, with a monic polynomial $Q_j(\xi)$ of degree j we will write

$$Q_{p,j}(\xi) := Q_j(\xi), \tag{23.1.5}$$

$$Q_{i,j}(\xi) := \int w_i(\xi, \eta) Q_{i+1,j}(\eta)\, d\eta, \quad 1 \leqslant i \leqslant p-1 \tag{23.1.6}$$

$$:= \int U_{Ri}(\xi, \eta) Q_j(\eta)\, d\eta, \tag{23.1.7}$$

$$U_{Ri}(\xi, \eta) := (w_i * w_{i+1} * \cdots * w_{p-1})(\xi, \eta). \tag{23.1.8}$$

With arbitrary monic polynomials $P_j(\xi)$ and $Q_j(\xi)$, $j = 0, 1, 2, \ldots$, we can write the product of differences as $n \times n$ determinants

$$\prod_{1 \leqslant i < j \leqslant n} (\xi_j - \xi_i) = \det[\xi_i^{j-1}]_n = \det[P_{j-1}(\xi_i)]_n = \det[Q_{j-1}(\xi_i)]_n. \tag{23.1.9}$$

As usual, the first equality is known as the Vandermonde determinant, while the later equalities are obtained by the observation that a determinant is not changed if we add to any of its rows a linear combination of its other rows, and in particular, if we add to its jth row an arbitrary linear combination of the rows $1, 2, \ldots, j-1$. The idea is to replace the powers ξ^j by arbitrary monic polynomials $P_j(\xi)$ and choose these polynomials in a convenient way to facilitate later computations. If the polynomials V_j and the constants c_j are such that the moment matrix $[M_{i,j}]$, $i, j = 0, 1, \ldots, n$, is non-singular for every n, where

$$M_{i,j} := \int \xi^i (w_1 * w_2 * \cdots * w_{p-1})(\xi, \eta) \eta^j \, d\xi \, d\eta, \tag{23.1.10}$$

and for $k > 1$,

$$(w_{i_1} * w_{i_2} * \cdots * w_{i_k})(\xi, \eta) := \int w_{i_1}(\xi, \xi_1) w_{i_2}(\xi_1, \xi_2) \cdots w_{i_k}(\xi_{k-1}, \eta) \, d\xi_1 \cdots d\xi_{k-1}, \tag{23.1.11}$$

then it is always possible to choose the polynomials $P_j(\xi)$ and $Q_j(\xi)$ such that

$$\int P_j(\xi)(w_1 * w_2 * \cdots * w_{p-1})(\xi, \eta) Q_k(\eta) \, d\xi \, d\eta = h_j \delta_{j,k}, \tag{23.1.12}$$

i.e., they are bi-orthogonal with the weight $(w_1 * w_2 * \cdots * w_{p-1})(x, y)$. This means that the functions $P_{i,j}(\xi)$ and $Q_{i,j}(\xi)$, which are not necessarily polynomials, are orthogonal

$$\int P_{i,j}(\xi) Q_{i,k}(\xi) \, d\xi = h_j \delta_{j,k}, \tag{23.1.13}$$

for $i = 1, 2, \ldots, p$ and $j, k = 0, 1, 2, \ldots$.

Now define

$$K_{i,j}(\xi, \eta) := H_{i,j}(\xi, \eta) - E_{i,j}(\xi, \eta), \tag{23.1.14}$$

where

$$H_{i,j}(\xi, \eta) := \sum_{\ell=0}^{n-1} \frac{1}{h_\ell} Q_{i,\ell}(\xi) P_{j,\ell}(\eta), \tag{23.1.15}$$

$$E_{i,j}(\xi,\eta) := \begin{cases} 0, & \text{if } i \geqslant j, \quad (23.1.16a)\\ w_i(\xi,\eta), & \text{if } i = j-1, \quad (23.1.16b)\\ (w_i * w_{i+1} * \cdots * w_{j-1})(\xi,\eta), & \text{if } i < j-1. \quad (23.1.16c)\end{cases}$$

Theorem 23.1.1. *The correlation function* (23.0.7) *is equal to*

$$R_{k_1,\ldots,k_p}(x_{1,1},\ldots,x_{1,k_1};\ldots;x_{p,1},\ldots,x_{p,k_p}) \\ = \det[K_{i,j}(x_{i,r},x_{j,s})]_{i,j=1,\ldots,p;r=1,\ldots,k_i;s=1,\ldots,k_j}. \tag{23.1.17}$$

This determinant has $m = k_1 + \cdots + k_p$ rows and m columns; the first k_1 rows and k_1 columns are labeled by the pair of indices $(1,r)$, $r = 1,\ldots,k_1$; the next k_2 rows and k_2 columns are labeled by the pair of indices $(2,r)$, $r = 1,\ldots,k_2$, and so on. Each variable $x_{i,r}$ appears in exactly one row and one column, this row and column crossing at the main diagonal. If all the eigenvalues of a matrix A_j are not observed (are integrated out), then no row or column corresponding to them appears in Eq. (23.1.17).

23.2 Proof of Theorem 23.1.1

The theorem is proved by recurrence. We will first prove that Eq. (23.1.17) holds for the initial case $k_1 = n,\ldots,k_p = n$. Next we will prove that if $R_{k_1,\ldots,k_\ell,\ldots,k_p}$ has the form (23.1.17) then $R_{k_1,\ldots,k_\ell-1,\ldots,k_p}$ obtained by integrating out one of the $x_{\ell,t}$, has the same form. Thus the theorem is a consequence of the following two lemmas:

Lemma 23.2.1. *The $np \times np$ determinant* $\det[K_{i,j}(x_{i,r},x_{j,s})]$, $i,j = 1,\ldots,p$; $r,s = 1,\ldots,n$, *is, apart from a constant, equal to the probability density* $F(\mathbf{x_1};\ldots;\mathbf{x_p})$, *Eq.* (23.0.4),

$$\det[K_{i,j}(x_{i,r},x_{j,s})]_{\substack{i,j=1,\ldots,p\\ r,s=1,\ldots,n}} = \left(\prod_{\ell=0}^{n-1} h_\ell^{-1}\right) C^{-1} F(\mathbf{x_1};\ldots;\mathbf{x_p}). \tag{23.2.1}$$

Lemma 23.2.2. *Using the convolution $*$ defined in* (23.1.11):

$$(f*g)(\xi,\eta) := \int f(\xi,\zeta)g(\zeta,\eta)\,d\zeta, \tag{23.2.2}$$

let's assume that the p^2 functions $K_{i,j}(x,y)$, $i,j = 1,\ldots,p$, are such that:

$$K_{i,j} * K_{j,k} = \begin{cases} K_{i,k}, & \text{if } i \geqslant j \geqslant k,\\ -K_{i,k}, & \text{if } i < j < k,\\ 0, & \text{otherwise.}\end{cases} \tag{23.2.3}$$

Then the integral of the $m \times m$ determinant $\det[K_{i,j}(x_{i,r}, x_{j,s})]$ $(i, j = 1, \ldots, p;\ r = 1, \ldots, k_i;\ s = 1, \ldots, k_j;\ k_1 \geqslant 0, \ldots, k_p \geqslant 0;\ m = k_1 + k_2 + \cdots + k_p)$, *over $x_{\ell,t}$ is proportional to the $(m-1) \times (m-1)$ determinant obtained from it by removing the row and the column containing the variable $x_{\ell,t}$. The constant of proportionality is $\alpha_\ell - k_\ell + 1$, with*

$$\alpha_\ell = \int K_{\ell,\ell}(x, x)\, dx. \tag{23.2.4}$$

We recall here a result of Section 5.1, Theorem 5.1.4.

Let the function $K(x, y)$ be such that $K * K = K$, then

$$\int \det[K(x_i, x_j)]_{i,j=1,\ldots,k}\, dx_k = (\alpha - k + 1) \det[K(x_i, x_j)]_{i,j=1,\ldots,k-1}, \tag{23.2.5}$$

with

$$\alpha = \int K(x, x)\, dx. \tag{23.2.6}$$

Our Lemma 2 above is a generalization of this one when the matrix elements $K(x_i, x_j)$ are replaced by $k_i \times k_j$ matrices $K_{i,j}(x_{i,r}, x_{j,s})$.

Proof of Lemma 23.2.1. Consider the $np \times np$ matrix $[H_{i,j}(x_{i,r}, x_{j,s})]$, Eq. (23.1.15), the rows of which are denoted by the pair of indices (i, r) and the columns by (j, s); $i, j = 1, \ldots, p$; $r, s = 1, \ldots, n$. This matrix can be written as the product of two rectangular matrices $[Q_{i,\ell}(x_{i,r})]$ and $[P_{j,\ell}(x_{j,s})/h_\ell]$ respectively of sizes $np \times n$ and $n \times np$, with $\ell = 0, 1, \ldots, n-1$. The rows of the first matrix $[Q_{i,\ell}(x_{i,r})]$ are numbered by the pair (i, r) and its columns by ℓ. For $[P_{j,\ell}(x_{j,s})/h_\ell]$ the rows are numbered by ℓ and the columns by (j, s). Cutting the matrix $[H_{i,j}(x_{i,r}, x_{j,s})]$ into $n \times n$ blocks, we can write,

$$\begin{aligned} H &= \begin{bmatrix} \overline{Q}_1\overline{P}_1 & \overline{Q}_1\overline{P}_2 & \ldots & \overline{Q}_1\overline{P}_p \\ \overline{Q}_2\overline{P}_1 & \overline{Q}_2\overline{P}_2 & \ldots & \overline{Q}_2\overline{P}_p \\ \ldots & \ldots & \ldots & \ldots \\ \overline{Q}_p\overline{P}_1 & \overline{Q}_p\overline{P}_2 & \ldots & \overline{Q}_p\overline{P}_p \end{bmatrix} \\ &= \begin{bmatrix} \overline{Q}_1 \\ \overline{Q}_2 \\ \vdots \\ \overline{Q}_p \end{bmatrix}_{np\times n} [\overline{P}_1 \quad \overline{P}_2 \quad \ldots \quad \overline{P}_p]_{n\times np}, \end{aligned} \tag{23.2.7}$$

where $[\overline{Q}_i]_{r\ell} := [Q_{i\ell}(x_{ir})]$ and $[\overline{P}_j]_{\ell s} := [P_{j\ell}(x_{j,s})/h_\ell]$ are $n \times n$ matrices. Equation (23.2.7) implies that the rank of $[H_{i,j}(x_{i,r}, x_{j,s})]$ is at most n. The rows of $\overline{P}_1$ and the columns of $\overline{Q}_p$ contain distinct monic polynomials, their ranks are therefore n, thus the rank of $[H_{i,j}(x_{i,r}, x_{j,s})]$ is n, and the last $n(p-1)$ columns can be linearly expressed in terms of the first n columns. In view of Eqs. (23.1.14) and (23.1.16a) the first n columns of $[H_{i,j}(x_{i,r}, x_{j,s})]$ are identical with the first n columns of $[K_{i,j}(x_{i,r}, x_{j,s})]$. The determinant of the later is therefore not changed if we subtract from its last $n(p-1)$ columns the corresponding $n(p-1)$ columns of the former. Thus

$$\det[K_{i,j}(x_{i,r}, x_{j,s})] = \det[\,H_{i,1}(x_{i,r}, x_{1,s}) \quad -E_{i,j}(x_{i,r}, x_{j,s})\,]_{\substack{i=1,\dots,p;\,j=2,\dots,p \\ r,s=1,2,\dots,n}}$$

$$= \det\begin{bmatrix} H_{1,1} & -E_{1,2} & -E_{1,3} & \dots & -E_{1,p} \\ H_{2,1} & 0 & -E_{2,3} & \dots & -E_{2,p} \\ \vdots & \vdots & \vdots & \ddots & \vdots \\ H_{p-1,1} & 0 & 0 & \dots & -E_{p-1,p} \\ H_{p,1} & 0 & 0 & \dots & 0 \end{bmatrix}. \tag{23.2.8}$$

From Eq. (23.1.16a) the last n rows of this matrix corresponding to $i = p$ have non-zero elements only in the first n columns; also the matrix $[E_{i,j}(x_{i,r}, x_{j,s})]$ is block triangular, $E_{i,j}(\xi, \eta)$ being zero for $i \geqslant j$. Therefore,

$$\begin{aligned} &\det[K_{i,j}(x_{i,r}, x_{j,s})] \\ &\quad = \det[H_{p,1}(x_{p,r}, x_{1,s})]\,\det[E_{i,j}(x_{i,r}, x_{j,s})]_{\substack{i=1,\dots,p-1;\ j=2,\dots,p \\ r,s=1,2,\dots,n}} \\ &\quad = \det[H_{p,1}(x_{p,r}, x_{1,s})] \prod_{j=2}^{p} \det[E_{j-1,j}(x_{j-1,r}, x_{j,s})] \\ &\quad = \left(\prod_{\ell=0}^{n-1} h_\ell^{-1}\right) \det[Q_{p,\ell}(x_{p,r})]\,\det[P_{1,\ell}(x_{1,s})] \prod_{j=1}^{p-1} \det[w_j(x_{j,r}, x_{j+1,s})], \end{aligned} \tag{23.2.9}$$

and from Eqs. (23.0.5), (23.1.9) one gets Eq. (23.2.1). This ends the proof. □

The learned reader will have recognized that the above $E_{i,j}$'s play the same role as the ε in Chapter 10 in the case of a single matrix of the circular orthogonal ensemble.

Proof of Lemma 23.2.2. We want to integrate the $m \times m$ determinant $\det[K_{i,j}(x_{i,r}, x_{j,s})]$ over $x_{\ell,t}$. We can write the expansion of the determinant as a sum over $m!$ permutations, writing these permutations as a product of mutually exclusive cycles. The variable $x_{\ell,t}$ occurs in the row and the column labeled by (ℓ, t) (recall that rows and columns are labeled by a pair of indices). If (ℓ, t) forms a cycle by itself, then by Eq. (23.2.4) integration over $x_{\ell,t}$ gives a factor α_ℓ, and its coefficient is just the expansion of the $(m-1) \times (m-1)$ determinant obtained by removing the row and the column containing $x_{\ell,t}$. If (ℓ, t) occurs in a longer cycle, say in the permutation $\sigma = ((i,r), (\ell,t), (j,s), \ldots)(\ldots)\ldots$, then from Eq. (23.2.3) integration over $x_{\ell,t}$ decreases the length of the cycle containing (ℓ, t) by one, giving the permutation $\sigma' = ((i,r), (j,s), \ldots)(\ldots)\ldots$, and multiplies the corresponding term by a factor $+1$ if $i \geqslant \ell \geqslant j$, by -1 if $i < \ell < j$ and by 0 otherwise. So the question is, given the permutation σ', in how many ways can one insert (ℓ, t) in any of its cycles with the algebraic weights $+1$, -1 and 0 to get a permutation σ, or equivalently, how many properly weighted permutations σ give the same σ' on removing (ℓ, t). Fortunately, it turns out that this number is independent of σ', so that the sum over the $(m-1)!$ permutations σ' gives back the $(m-1) \times (m-1)$ determinant obtained by removing the row and the column containing $x_{\ell,t}$.

Let us represent the cycles of permutations by a graph. Since the discussion of the weight $+1$, -1 or 0 depends only on the first index, only this first index $i, j, \ldots$ of each pair of indices will be plotted against the place number where they occur. For example, the cycle $((4,2), (2,6), (3,6), (1,5), (2,4), (2,2))$ is represented on Figure 23.1, where points at successive heights 4, 2, 3, 1, 2, 2 are joined successively by line segments or "sides". Note that we identify the 7th point and the first one. Permutation σ' is thus represented by a certain number of closed directed polygons corresponding to its mutually exclusive cycles. Addition of ℓt in one of the cycles of σ' amounts to the addition of a point at a height ℓ in the corresponding polygon. If this added point lies on a non-ascending side, then the weight multiplying the corresponding σ is $+1$, if it lies on an ascending side, the weight is -1, and this weight is zero otherwise. In other words, each

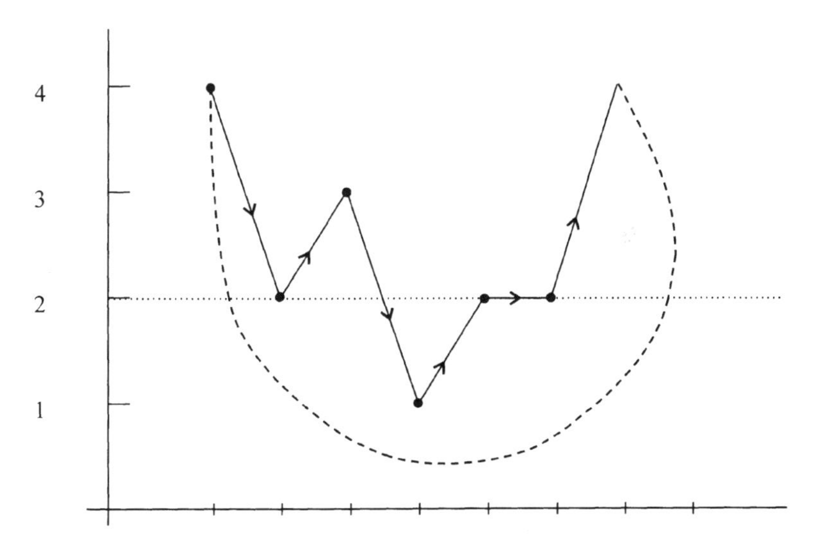

Figure 23.1. The permutation $\sigma' = (4, 2, 3, 1, 2, 2)$; $\ell = 2$.

downward crossing of the line at height ℓ, with or without stops, contributes a factor $+1$, each point on this height contributes a factor $+1$ and each upward crossing, with or without stops, contributes a factor -1. The graph of σ' consisting of closed loops, the number of upward crossings is equal to the number of downward crossings at any height, and the corresponding contributions cancel out. The algebraic sum of all such coefficients is thus seen to be the number of points at height ℓ in the graph of σ', i.e., it is $k_\ell - 1$. Also the permutations σ and σ' have opposite signs, since only one of their cycle lengths differ by unity. Thus

$$\int \det[K_{i,j}(x_{i,r}, x_{j,s})]_{m\times m}\, dx_{\ell,t} = (\alpha_\ell - k_\ell + 1) \det[K_{i,j}(x_{i,r}, x_{j,s})]_{(m-1)\times(m-1)}, \tag{23.2.10}$$

where the integrand on the left-hand side is an $m \times m$ determinant, $i, j = 1, \ldots, p$; $r = 1, \ldots, k_i$; $s = 1, \ldots, k_j$; $m = k_1 + k_2 + \cdots + k_p$ and the result on the right-hand side is the $(m-1) \times (m-1)$ determinant obtained from the integrand by removing the row and the column containing the variable $x_{\ell,t}$. This ends the proof. □

Using Eqs. (23.1.13)–(23.1.16), one verifies that the $H_{i,j}$ and $E_{i,j}$ satisfy the following relations

$$H_{i,j} * H_{j,k} = H_{i,k}, \tag{23.2.11}$$

$$H_{i,j} * E_{j,k} = \begin{cases} H_{i,k}, & \text{if } j < k, \\ 0, & \text{if } j \geqslant k, \end{cases} \tag{23.2.12}$$

$$E_{i,j} * H_{j,k} = \begin{cases} H_{i,k}, & \text{if } i < j, \\ 0, & \text{if } i \geqslant j, \end{cases} \tag{23.2.13}$$

$$E_{i,j} * E_{j,k} = \begin{cases} E_{i,k}, & \text{if } i < j < k, \\ 0, & \text{if either } i \geqslant j \text{ or } j \geqslant k. \end{cases} \tag{23.2.14}$$

This implies for the $K_{i,j} = H_{i,j} - E_{i,j}$, Eq. (23.1.14), the relations (23.2.3). Also

$$\alpha_\ell = \int K_{\ell,\ell}(\xi, \xi)\, d\xi = n. \tag{23.2.15}$$

Using Lemma 1 once and Lemma 2 several times one gets the normalization constant C,

$$C = (n!)^{-p} \prod_{\ell=0}^{n-1} h_\ell^{-1}. \tag{23.2.16}$$

Again with repeated use of Lemma 23.2.2 and from the definition, Eq. (23.0.7), of the correlation function $R_{k_1,\ldots,k_p}(x_{1,1}, \ldots, x_{1,k_1}; \ldots; x_{p,1}, \ldots, x_{p,k_p})$, one gets Eq. (23.1.17).

23.3 Spacing Functions

Let us recall here a few facts concerning the case of one matrix. For an $n \times n$ complex Hermitian matrix A with matrix elements probability density $\exp[-\operatorname{tr} V(A)]$, the probability density of its eigenvalues $\mathbf{x} := \{x_1, x_2, \ldots, x_n\}$ is (cf. Chapter 3)

$$F(\mathbf{x}) \propto \exp\left[-\sum_{j=1}^{n} V(x_j)\right] \prod_{1 \leqslant i < j \leqslant n} (x_j - x_i)^2, \tag{23.3.1a}$$

$$\propto \det[K(x_i, x_j)]_{i,j=1,\ldots,n}, \tag{23.3.1b}$$

where $V(x)$ is a real polynomial of even order, the coefficient of the highest power being positive; $K(x, y)$ is defined by

$$K(x, y) := \exp\left[-\frac{1}{2}V(x) - \frac{1}{2}V(y)\right] \sum_{i=0}^{n-1} \frac{1}{h_i} P_i(x) P_i(y), \tag{23.3.2}$$

$P_i(x)$ is a real polynomial of degree i and the polynomials are chosen orthogonal with the weight $\exp[-V(x)]$,

$$\int P_i(x) P_j(x) \exp[-V(x)]\, dx = h_i \delta_{i,j}. \tag{23.3.3}$$

Here and in what follows, all the integrals are taken from $-\infty$ to $+\infty$, unless explicitly stated otherwise.

The correlation function $R_k(x_1, \ldots, x_k)$, i.e., the density of ordered sets of k eigenvalues within small intervals around $x_1, \ldots, x_k$, ignoring the other eigenvalues, is

$$R_k(x_1, \ldots, x_k) := \frac{n!}{(n-k)!} \int F(\mathbf{x})\, dx_{k+1} \cdots dx_n$$

$$= \det[K(x_i, x_j)]_{i,j=1,\ldots,k}. \tag{23.3.4}$$

The spacing function $E(k, I)$, i.e., the probability that a chosen domain I contains exactly k eigenvalues $(0 \leqslant k \leqslant n)$, is

$$E(k, I) := \frac{n!}{k!(n-k)!} \int F(\mathbf{x}) \left[\prod_{j=1}^{k} \chi(x_j)\right] \left[\prod_{j=k+1}^{n} [1 - \chi(x_j)]\right] dx_1 \cdots dx_n$$

$$= \frac{1}{k!} \left(\frac{d}{dz}\right)^k R(z, I)|_{z=-1}, \tag{23.3.5}$$

where $\chi(x)$ is the characteristic function of the domain I,

$$\chi(x) := \begin{cases} 1, & \text{if } x \in I, \\ 0, & \text{otherwise,} \end{cases} \tag{23.3.6}$$

and $R(z, I)$ is the generating function of the integrals over I of the correlation functions $R_k(x_1, \ldots, x_k)$,

$$R(z, I) := \int F(\mathbf{x}) \prod_{j=1}^{n} [1 + z\chi(x_j)]\, dx_j = \sum_{k=0}^{n} \frac{\rho_k}{k!} z^k, \tag{23.3.7}$$

$$\rho_k = \begin{cases} 1, & k = 0, \\ \displaystyle\int R_k(x_1, \ldots, x_k) \prod_{j=1}^{k} \chi(x_j)\, dx_j, & \text{otherwise.} \end{cases} \tag{23.3.8}$$

The $R(z, I)$ of Eq. (23.3.7) can be expressed as a determinant

$$R(z, I) = \det[G_{ij}]_{i,j=0,\ldots,n-1}, \tag{23.3.9}$$

where, using the orthogonality, Eq. (23.3.3), of polynomials $P_i(x)$ and splitting the constant and linear terms in z, G_{ij} reads

$$G_{ij} = \frac{1}{h_i} \int P_i(x) P_j(x) \exp[-V(x)][1 + z\chi(x)]\, dx = \delta_{ij} + \overline{G}_{ij}. \tag{23.3.10}$$

Finally, $R(z, I)$ can also be written as the Fredholm determinant

$$R(z, I) = \prod_{k=1}^{n} [1 + \lambda_k(z, I)] \tag{23.3.11}$$

of the integral equation

$$\int N(x, y) f(y)\, dy = \lambda f(x), \tag{23.3.12}$$

where remarkably the kernel $N(x, y)$ is simply $zK(x, y)\chi(y)$ with $K(x, y)$ of Eq. (23.3.2). The $\lambda_i(z, I)$ are the eigenvalues of the above equation and also of the matrix $[\overline{G}_{ij}]$.

Here we will consider the spacing function $E(k_1, I_1; \ldots; k_p, I_p)$, i.e., the probability that the domain I_j contains exactly k_j eigenvalues of the matrix A_j for $j = 1, \ldots, p$, $0 \leqslant k_j \leqslant n$. The domains I_j may have overlaps. As in the one matrix case one has evidently

$$E(k_1, I_1; \ldots; k_p, I_p) = \frac{1}{k_1!}\left(\frac{\partial}{\partial z_1}\right)^{k_1} \cdots \frac{1}{k_p!}\left(\frac{\partial}{\partial z_p}\right)^{k_p} \times R(z_1, I_1; \ldots; z_p, I_p)|_{z_1=\cdots=z_p=-1}, \tag{23.3.13}$$

with the generating function

$$R(z_1, I_1; \ldots; z_p, I_p) = \int F(\mathbf{x_1}; \ldots; \mathbf{x_p}) \prod_{j=1}^{p} \prod_{r=1}^{n} [1 + z_j \chi_j(x_{jr})]\, dx_{jr} \tag{23.3.14}$$

$$= \sum_{k_1=0}^{n} \cdots \sum_{k_p=0}^{n} \frac{\rho_{k_1,\ldots,k_p}}{k_1! \cdots k_p!} z_1^{k_1} \cdots z_p^{k_p}, \tag{23.3.15}$$

$\rho_{k_1,\ldots,k_p} = 1$, if $k_1 = \cdots = k_p = 0$ and

$$\rho_{k_1,\ldots,k_p} = \prod_{j=1}^{p} \left[\int_{I_j} \prod_{r=1}^{k_j} dx_{jr} \right] R_{k_1,\ldots,k_p}(x_{11}, \ldots, x_{1k_1}; \ldots; x_{p1}, \ldots, x_{pk_p}), \tag{23.3.16}$$

otherwise, $\chi_j(x)$ being the characteristic function of the domain I_j, Eq. (23.3.6).

The function $R(z_1, I_1; \ldots; z_p, I_p)$ will be expressed as a $n \times n$ determinant. It will also be written as a Fredholm determinant, the kernel of which will now depend on the variables $z_1, \ldots, z_p$ and the domains $I_1, \ldots, I_p$ in a more involved way than in the one matrix case. In particular, it does not have the remarkable form mentioned after Eq. (23.3.12).

23.4 The Generating Function $R(z_1, I_1; \ldots; z_p, I_p)$

The expression of F, Eq. (23.0.5), contains a product of determinants. As the product of differences

$$\Delta(\mathbf{x_1}) = \prod_{1 \leqslant r < s \leqslant n} (x_{1s} - x_{1r}), \tag{23.4.1}$$

and $\det[w_1(x_{1r}, x_{2s})]$ are completely antisymmetric and other factors in the integrand of Eq. (23.3.14) are completely symmetric in the variables $x_{11}, \ldots, x_{1n}$, one can replace $\det[w_1(x_{1r}, x_{2s})]$ under the integral sign in Eq. (23.3.14) by a single term, say the diagonal one, and multiply by $n!$. This single term is invariant under a permutation of the variables x_{1r} and simultaneously the same permutation on the variables x_{2r}.

Therefore, after integration over the x_{1r}, $r = 1, \ldots, n$, the integrand, excluding the factor $\det[w_2(x_{2r}, x_{3s})]$, is completely antisymmetric in the variables $x_{21}, \ldots, x_{2n}$ and so one can replace the second determinant $\det[w_2(x_{2r}, x_{3s})]$ by a single term, say the diagonal one, and multiply the result by $n!$. In this way, under the integral sign one can replace successively each of the $p-1$ determinants $\det[w_k(x_{kr}, x_{k+1s})]$ by a single term multiplying the result each time by $n!$

$$R(z_1, I_1; \ldots; z_p, I_p) = (n!)^{p-1} C \int \Delta(\mathbf{x_1}) \Delta(\mathbf{x_p}) \left[\prod_{j=1}^{p-1} \prod_{r=1}^{n} w_j(x_{jr}, x_{j+1r}) \right] \times \left[\prod_{j=1}^{p} \prod_{r=1}^{n} [1 + z_j \chi_j(x_{jr})] \, dx_{jr} \right]. \tag{23.4.2}$$

Recall that a polynomial is called monic when the coefficient of the highest power is one. Also recall that the product of differences $\Delta(\mathbf{x})$ can be written as a determinant

$$\Delta(\mathbf{x}) = \det[x_i^{j-1}] = \det[P_{j-1}(x_i)] = \det[Q_{j-1}(x_i)], \tag{23.4.3}$$

where $P_j(x)$ and $Q_j(x)$ are arbitrary monic polynomials of degree j. As usual, we will choose these polynomials real and bi-orthogonal, Eq. (23.1.12)

$$\int P_j(x)(w_1 * w_2 * \cdots * w_{p-1})(x, y) Q_k(y) \, dx \, dy = h_j \delta_{jk}, \tag{23.4.4}$$

with the obvious notation

$$(f * g)(x, y) = \int f(x, \xi) g(\xi, y) \, d\xi. \tag{23.4.5}$$

The normalization constant C is (Eq. (23.2.16)),

$$C = (n!)^{-p} \prod_{i=0}^{n-1} h_i^{-1}. \tag{23.4.6}$$

Now expand the determinant as a sum over $n!$ permutations $(i) := \binom{0 \; \ldots \; n-1}{i_1 \; \ldots \; i_n}$, $\sigma(i)$ being its sign,

$$\det[P_{s-1}(x_{1r})] = \sum_{(i)} \sigma(i) P_{i_1}(x_{11}) P_{i_2}(x_{12}) \cdots P_{i_n}(x_{1n}). \tag{23.4.7}$$

Doing the same thing for $\det[Q_{s-1}(x_{pr})]$ and integrating over all the np variables x_{jr}; $j = 1, \ldots, p$; $r = 1, \ldots, n$ in Eq. (23.4.2), one gets

$$R(z_1, I_1; \ldots; z_p, I_p) = \frac{1}{n!} \sum_{(i)} \sum_{(j)} \sigma(i)\sigma(j) G_{i_1 j_1} G_{i_2 j_2} \cdots G_{i_n j_n}$$
$$= \det[G_{ij}]_{i,j=0,\ldots,n-1}, \tag{23.4.8}$$

where

$$G_{ij} = \frac{1}{h_i} \int P_i(x_1) \left[\prod_{k=1}^{p-1} w_k(x_k, x_{k+1}) \right] Q_j(x_p) \left[\prod_{k=1}^{p} [1 + z_k \chi_k(x_k)] \, dx_k \right]. \tag{23.4.9}$$

When all the z_k vanish, G_{ij} is equal to δ_{ij} as a consequence of the bi-orthogonality, Eq. (23.4.4), of the polynomials $P_i(x)$ and $Q_i(x)$. Let us define $\overline{G}_{ij}$ as follows

$$\overline{G}_{ij} := G_{ij} - \delta_{ij}, \tag{23.4.10}$$

so that

$$\overline{G}_{ij} = \frac{1}{h_i} \int P_i(x_1) \left[\prod_{k=1}^{p-1} w_k(x_k, x_{k+1}) \right] Q_j(x_p) \left[\prod_{k=1}^{p} [1 + z_k \chi_k(x_k)] - 1 \right] \left[\prod_{k=1}^{p} dx_k \right]. \tag{23.4.11}$$

Any $n \times n$ determinant is the product of its n eigenvalues and therefore one has

$$R(z_1, I_1; \ldots; z_p, I_p) = \prod_{k=1}^{n} [1 + \lambda_k(z_1, I_1; \ldots; z_p, I_p)], \tag{23.4.12}$$

where the $\lambda_k(z_1, I_1; \ldots; z_p, I_p)$ are the n roots (not necessarily distinct, either real or pairwise complex conjugates, since $\overline{G}_{ij}$ is real) of the algebraic equation in λ

$$\det[\overline{G}_{ij} - \lambda \delta_{ij}] = 0. \tag{23.4.13}$$

One can always write a Fredholm integral equation with a separable kernel whose eigenvalues are identical to these (cf. Mehta and Shukla (1994)) for the case of $p = 2$ matrices). Indeed, for any eigenvalue λ the system of linear equations

$$\sum_{j=0}^{n-1} \overline{G}_{ij} \xi_j = \lambda \xi_i \tag{23.4.14}$$

has at least one solution ξ_i, $i = 0, \ldots, n-1$, not all zero. Multiplying both sides of the above equation by $Q_i(x)$, summing over i and using Eq. (23.4.11) gives the Fredholm equation

$$\int N(x, x_p) f(x_p)\, dx_p = \lambda f(x), \tag{23.4.15}$$

where

$$f(x) := \sum_{i=0}^{n-1} \xi_i Q_i(x), \tag{23.4.16}$$

$$N(x, x_p) := \sum_{i=0}^{n-1} \frac{1}{h_i} Q_i(x) \int P_i(x_1) \left[\prod_{k=1}^{p-1} w_k(x_k, x_{k+1}) \right] \times \left[\prod_{k=1}^{p} [1 + z_k \chi_k(x_k)] - 1 \right] \left[\prod_{k=1}^{p-1} dx_k \right]. \tag{23.4.17}$$

Hence if λ is an eigenvalue of the matrix $[\overline{G}_{ij}]$, it is also an eigenvalue of the integral equation (23.4.15). Conversely, since the kernel $N(x, x_p)$ is a sum of separable ones and since the polynomials $Q_i(x)$ for $i = 0, \ldots, n-1$ are linearly independent, if λ and $f(x)$ are, respectively an eigenvalue and an eigenfunction of this integral equation, then $f(x)$ is necessarily of the form

$$f(x) = \sum_{i=0}^{n-1} \xi_i Q_i(x), \tag{23.4.18}$$

and the ξ_i, $i = 0, \ldots, n-1$, not all zero, satisfy Eq. (23.4.14). Therefore λ is a root of Eq. (23.4.13).

When one considers the eigenvalues of a single matrix anywhere in the chain, disregarding those of the other matrices, everything works as if one is dealing with the one matrix case and formulas (23.3.2), (23.3.5), (23.3.7) and (23.3.11) are valid with minor replacements. Similarly, when one considers properties of the eigenvalues of k $(1 \leqslant k \leqslant p)$ matrices situated anywhere in the chain, not necessarily consecutive, everything works as if one is dealing with a chain of only k matrices; the presence of other matrices modifying only the couplings.

To say something more about the general case seems difficult.

When $V_j(x) = a_j x^2$, $j = 1, \ldots, p$, then the polynomials $P_j(x)$ and $Q_j(x)$ are Hermite polynomials $P_j(x) = H_j(\alpha x)$, $Q_j(x) = H_j(\beta x)$, the constants α and β depending on the parameters a_j and the couplings c_j. In this particular case the calculation can be pushed to the end (see Appendix A.49).

23.5 The Zeros of the Bi-Orthogonal Polynomials

Though not very useful, we record here as Theorem 23.5.4 a curious fact about the real zeros of the functions $P_{i,j}(x)$ and $Q_{i,j}(x)$. We will follow closely the arguments of Ercolani and McLaughlin (2001).

Theorem 23.5.1. *Let $W(x,y)$ be a positive weight function with all its moments $M_{i,j} = \iint W(x,y)x^i y^j\,dx\,dy$ finite, the determinant of the moments $\det[M_{i,j}]_{i,j=0,\ldots,n} \neq 0$ for $n \geqslant 0$, and $\det[W(x_i,y_j)]_{i,j=1,\ldots,n} > 0$ for $x_1 < x_2 < \cdots < x_n$, $y_1 < y_2 < \cdots < y_n$. Then unique monic polynomials $p_j(x)$ and $q_j(x)$ exist which are bi-orthogonal with the weight $W(x,y)$*

$$\iint W(x,y)p_j(x)q_k(y)\,dx\,dy = h_j\delta_{j,k} \tag{23.5.1}$$

and the zeros of $p_j(x)$ and $q_j(x)$ are real, simple and lie in the respective x and y supports of $W(x,y)$.

For the existence of the bi-orthogonal polynomials see Chapter 5, Sections 5.10 and 5.11. The proof of the remaining statement results from the following two lemmas.

Lemma 23.5.2. *For any monic polynomial $p_j(x)$ of degree j, the function $P_j(x) = \int W(x,y)p_j(y)\,dy$ has at most j real distinct zeros in the x support of $W(x,y)$. Similarly, for any monic polynomial $q_j(x)$ of degree j, the function $Q_j(y) = \int q_j(x)W(x,y)\,dx$ has at most j real distinct zeros in the y support of $W(x,y)$.*

Proof. Let, if possible, $z_1, z_2, \ldots, z_m$, $m > j$, be the distinct real zeros of $P_j(x)$ in the x support of $W(x,y)$. Since

$$P_j(x) = \sum_{k=0}^{j} a_k T_k(x), \quad \text{with } T_k(x) = \int W(x,y)y^k\,dy \tag{23.5.2}$$

and

$$P_j(z_\ell) = 0, \quad \ell = 1,2,\ldots,m,\ m > j, \tag{23.5.3}$$

one has

$$0 = \det\begin{bmatrix} T_0(z_1) & T_1(z_1) & \ldots & T_j(z_1) \\ \vdots & \vdots & & \vdots \\ T_0(z_{j+1}) & T_1(z_{j+1}) & \ldots & T_j(z_{j+1}) \end{bmatrix}$$

$$= \int \cdots \int \det \begin{bmatrix} W(z_1, y_1) & y_2 W(z_1, y_2) & \ldots & y_{j+1}^j W(z_1, y_{j+1}) \\ \vdots & \vdots & \ldots & \vdots \\ W(z_{j+1}, y_1) & y_2 W(z_{j+1}, y_2) & \ldots & y_{j+1}^j W(z_{j+1}, y_{j+1}) \end{bmatrix} dy_1 dy_2 \cdots dy_{j+1}$$

$$= \int \cdots \int y_2 y_3^2 \cdots y_{j+1}^j \det[W(z_i, y_k)]_{i,k=1,\ldots,j+1} dy_1 \cdots dy_{j+1}$$

$$= \frac{1}{(j+1)!} \int \cdots \int \det[y_i^{k-1}]_{i,k=1,\ldots,j+1} \det[W(z_i, y_k)]_{i,k=1,\ldots,j+1} dy_1 \cdots dy_{j+1}. \tag{23.5.4}$$

But for $z_1 < z_2 < \cdots < z_{j+1}$ the integrand in the last line above is non-negative and not always zero, so that the integral can not be zero; a contradiction. Thus m can not be greater than j.

The proof for the function $Q_j(y)$ is similar. □

Lemma 23.5.3. *Let $p_j(x)$ and $q_j(x)$ be monic polynomials of order j bi-orthogonal with the weight $W(x, y)$.*

$$\iint W(x, y) p_j(x) q_k(y)\, dx\, dy = h_j \delta_{j,k}. \tag{23.5.5}$$

Then each of $p_j(x)$ and $P_j(x) = \int W(x, y) p_j(y)\, dy$ has at least j distinct real zeros of odd multiplicity in the x support of $W(x, y)$. Similarly, $q_j(y)$ and $Q_j(y) = \int q_j(x) W(x, y)\, dx$ each has at least j distinct real zeros of odd multiplicity in the y support of $W(x, y)$.

Proof. Let, if possible, $z_1 < z_2 < \cdots < z_m$, $m < j$, be the only real zeros of $P_j(x)$ of odd multiplicity in the x support of $W(x, y)$. Set $R(x) = \prod_{k=1}^m (x - z_k)$. Then as $m < j$, one has

$$\int P_j(x) R(x)\, dx = 0. \tag{23.5.6}$$

But $P_j(x)$ and $R(x)$ change sign simultaneously as x passes through the values $z_1, z_2, \ldots, z_m$ and at no other value in the x support of $W(x, y)$. So the product $P_j(x) R(x)$ never changes sign, in contradiction to Eq. (23.5.6). Therefore $P_j(x)$ has at least j real distinct zeros of odd multiplicity in the x support of $W(x, y)$.

To prove that $p_j(x)$ has at least j real distinct zeros in the x support of $W(x, y)$, let, if possible, $z_1, z_2, \ldots, z_m$, $m < j$, be the only such zeros. Set

$$R(x) = \det \begin{bmatrix} P_0(x) & P_1(x) & \ldots & P_m(x) \\ P_0(z_1) & P_1(z_1) & \ldots & P_m(z_1) \\ \vdots & \vdots & & \vdots \\ P_0(z_m) & P_1(z_m) & \ldots & P_m(z_m) \end{bmatrix}. \tag{23.5.7}$$

Since $m < j$, bi-orthogonality gives

$$\int p_j(x)R(x)\,dx = 0. \tag{23.5.8}$$

But

$$\begin{aligned}
R(x) &= \int \det\begin{bmatrix} W(x,\xi_0) & W(x,\xi_1)p_1(\xi_1) & \dots & W(x,\xi_m)p_m(\xi_m) \\ W(z_1,\xi_0) & W(z_1,\xi_1)p_1(\xi_1) & \dots & W(z_1,\xi_m)p_m(\xi_m) \\ \vdots & \vdots & & \vdots \\ W(z_m,\xi_0) & W(z_m,\xi_1)p_1(\xi_1) & \dots & W(z_m,\xi_m)p_m(\xi_m) \end{bmatrix} d\xi_0\,d\xi_1\cdots d\xi_m \\
&= \int \det\begin{bmatrix} W(x,\xi_0) & W(x,\xi_1)\xi_1 & \dots & W(x,\xi_m)\xi_m^m \\ W(z_1,\xi_0) & W(z_1,\xi_1)\xi_1 & \dots & W(z_1,\xi_m)\xi_m^m \\ \vdots & \vdots & & \vdots \\ W(z_m,\xi_0) & W(z_m,\xi_1)\xi_1 & \dots & W(z_m,\xi_m)\xi_m^m \end{bmatrix} d\xi_0\,d\xi_1\cdots d\xi_m \\
&= \frac{1}{(m+1)!}\int \det\begin{bmatrix} W(x,\xi_0) & W(x,\xi_1) & \dots & W(x,\xi_m) \\ W(z_1,\xi_0) & W(z_1,\xi_1) & \dots & W(z_1,\xi_m) \\ \vdots & \vdots & & \vdots \\ W(z_m,\xi_0) & W(z_m,\xi_1) & \dots & W(z_m,\xi_m) \end{bmatrix} \\
&\quad \times \prod_{0\leqslant r<s\leqslant m} (\xi_s - \xi_r)\,d\xi_0\,d\xi_1\cdots d\xi_m
\end{aligned} \tag{23.5.9}$$

which says that $z_1, z_2, \dots, z_m$ are the only real distinct zeros of $R(x)$ in the x support of $W(x,y)$ and therefore $p_j(x)R(x)$ has a constant sign, in contradiction to (23.5.8).

The proof for $q_j(y)$ and $Q_j(y)$ is similar. □

Now we come back to the functions $P_{i,j}(x)$ and $Q_{i,j}(x)$ of Section 23.1. Let $w_i(x,y)$ be given by Eq. (23.0.6).

Theorem 23.5.4. *If $P_j(x)$ and $Q_j(x)$ are bi-orthogonal polynomials satisfying Eq. (23.1.12), then $P_{i,j}(x)$ and $Q_{i,j}(x)$ each have exactly j distinct real zeros of odd multiplicity. In particular, the zeros of the bi-orthogonal polynomials $P_j(x) \equiv P_{1,j}(x)$ and $Q_j(x) \equiv Q_{p,j}(x)$ are real and simple.*

The proof of this is the result of a number of lemmas which follow.

Lemma 23.5.5. *For $x_1 < x_2 < \dots < x_n$, $y_1 < y_2 < \dots < y_n$,*

$$\det[w_i(x_j, y_k)]_{j,k=1,\dots,n} > 0. \tag{23.5.10}$$

This can be seen as follows. Let $X = [x_i \delta_{i,j}]$ and $Y = [y_i \delta_{i,j}]$ be two $n \times n$ diagonal matrices with diagonal elements $x_1, \ldots, x_n$ and $y_1, \ldots, y_n$ respectively. Then the integral of $\exp[-c \operatorname{tr} U X U^{-1} Y]$ over the $n \times n$ unitary matrices U is given by (cf. Appendix A.5)

$$K \frac{\det[\exp(-c x_i y_i)]_{i,j=1,\ldots,n}}{\Delta(\mathbf{x})\Delta(\mathbf{y})}, \tag{23.5.11}$$

where K is a positive constant depending on c and n. Hence

$$\exp\left[-\frac{1}{2}\sum_{j=1}^{n}(V_i(x_j) + V_{i+1}(y_j))\right] \int dU\, e^{-c_i \operatorname{tr} U X U^{-1} Y} = K \frac{\det[w_i(x_j, y_k)]_{j,k=1,\ldots,n}}{\Delta(\mathbf{x})\Delta(\mathbf{y})}. \tag{23.5.12}$$

The left-hand side is evidently positive while on the right-hand side the denominator is positive since $x_1 < x_2 < \cdots < x_n$ and $y_1 < y_2 < \cdots < y_n$. From this Eq. (23.5.10) follows.

Lemma 23.5.6. *For* $x_1 < x_2 < \cdots < x_n$, $y_1 < y_2 < \cdots < y_n$,

$$\det[(w_{i_1} * w_{i_2} * \cdots * w_{i_\ell})(x_j, y_k)]_{j,k=1,\ldots,n} > 0. \tag{23.5.13}$$

Proof. Binet–Cauchy formula tells us that (cf. Mehta (1989MT), Section 3.7)

$$\det[(w_{i_1} * w_{i_2})(x_j, y_k)]_{j,k=1,\ldots,n}$$

is equal to

$$\int_{\xi_1<\xi_2<\cdots<\xi_n} \det[w_{i_1}(x_j, \xi_k)]_{j,k=1,\ldots,n} \cdot \det[w_{i_2}(\xi_j, y_k)]_{j,k=1,\ldots,n}\, d\xi_1 \cdots d\xi_n. \tag{23.5.14}$$

By Lemma 23.5.5 the integrand is every where positive, so Lemma 23.5.6 is proved for the case $\ell = 2$. The proof is now completed by induction on ℓ, using again the Binet–Cauchy formula. □

Lemma 23.5.7. *For any monic polynomial $P_j(x)$ of degree j, $P_{i,j}(x)$, $1 \leqslant i \leqslant p$, has at most j distinct real zeros. Similarly, for any monic polynomial $Q_j(x)$ of degree j, $Q_{i,j}(x)$, $1 \leqslant i \leqslant p$, has at most j distinct real zeros.*

Proof. In the proof of Lemma 23.5.2 replace $P_j(x)$, $T_k(x)$ and $W(z_\ell, y_k)$ respectively by $P_{i,j}(x)$, $T_{i,k}(x)$ and $U_{Li}(z_\ell, y_k)$.

The proof for $Q_{i,j}(x)$ is similar. □

Lemma 23.5.8. *Let the real constants $c_1, \dots, c_{p-1}$, none of them being zero, be such that*

$$M_{i,j} := \int x^i U_{Lp}(x,y) y^j \, dx\, dy \equiv \int x^i (w_1 * w_2 * \cdots * w_{p-1})(x,y) y^j \, dx\, dy \tag{23.5.15}$$

exist for all $i, j \geqslant 0$. Then

$$D_n := \det[M_{i,j}]_{i,j=0,\dots,n} \neq 0, \tag{23.5.16}$$

for any $n \geqslant 0$.

Proof. Let, if possible, $D_n = 0$ for some n. Then $\sum_{j=0}^n M_{i,j} q_j = 0$, q_j not all zero, and

$$\int x^i U_{Lp}(x,y) \sum_{j=0}^{n} q_j y^j \, dx\, dy = 0, \quad i = 0, 1, \dots, n, \tag{23.5.17}$$

or

$$\int p_i(x) U_{Lp}(x,y) \sum_{j=0}^{n} q_j y^j \, dx\, dy = 0, \tag{23.5.18}$$

for any polynomial $p_i(x)$ of degree $i \leqslant n$. But

$$\int U_{Lp}(x,y) \sum_{j=0}^{n} q_j y^j \, dy, \tag{23.5.19}$$

has at most n distinct real zeros (Lemma 23.5.7). So one can choose $p_i(x)$ such that

$$p_i(x) \int U_{Lp}(x,y) \sum_{j=0}^{n} q_j y^j \, dy > 0, \tag{23.5.20}$$

in contradiction to Eq. (23.5.18). So $D_n \neq 0$ and bi-orthogonal polynomials $P_j(x)$, $Q_j(x)$ exist. □

Lemma 23.5.9. *Let $P_j(x)$, $Q_j(x)$ be the bi-orthogonal polynomials, Eq.* (23.1.12); *or with the definitions* (23.1.2) *and* (23.1.6)

$$\int P_{i,j}(x) Q_{i,k}(x) \, dx = h_j \delta_{j,k}, \quad 1 \leqslant i \leqslant p. \tag{23.5.21}$$

Then $P_{i,j}(x)$ has at least j real distinct zeros of odd multiplicity. So does have $Q_{i,j}(x)$.

Proof. Let, if possible, $z_1 < z_2 < \cdots < z_m$, $m < j$, be the only real zeros of $P_{i,j}(x)$ of odd multiplicity. Set

$$R(x) = \det \begin{bmatrix} Q_{i,0}(x) & Q_{i,1}(x) & \dots & Q_{i,m}(x) \\ Q_{i,0}(z_1) & Q_{i,1}(z_1) & \dots & Q_{i,m}(z_1) \\ \vdots & \vdots & & \vdots \\ Q_{i,0}(z_m) & Q_{i,1}(z_m) & \dots & Q_{i,m}(z_m) \end{bmatrix} \tag{23.5.22}$$

$$= \int U_{Ri}(x,\xi) \sum_{k=0}^{m} \alpha_k \xi^k \, d\xi, \tag{23.5.23}$$

with some constants α_k depending on $z_1, \dots, z_m$.

Since $m < j$, the bi-orthogonality gives

$$\int P_{i,j}(x) R(x)\, dx = 0. \tag{23.5.24}$$

However, $R(x)$ can also be written as

$$R(x) = \int \det \begin{bmatrix} U_{Ri}(x,\xi_0) & U_{Ri}(x,\xi_1)\xi_1 & \dots & U_{Ri}(x,\xi_m)\xi_m^m \\ U_{Ri}(z_1,\xi_0) & U_{Ri}(z_1,\xi_1)\xi_1 & \dots & U_{Ri}(z_1,\xi_m)\xi_m^m \\ \vdots & \vdots & & \vdots \\ U_{Ri}(z_m,\xi_0) & U_{Ri}(z_m,\xi_1)\xi_1 & \dots & U_{Ri}(z_m,\xi_m)\xi_m^m \end{bmatrix} d\xi_0\, d\xi_1 \cdots d\xi_m$$

$$= \int \det \begin{bmatrix} U_{Ri}(x,\xi_0) & U_{Ri}(x,\xi_1) & \dots & U_{Ri}(x,\xi_m) \\ U_{Ri}(z_1,\xi_0) & U_{Ri}(z_1,\xi_1) & \dots & U_{Ri}(z_1,\xi_m) \\ \vdots & \vdots & & \vdots \\ U_{Ri}(z_m,\xi_0) & U_{Ri}(z_m,\xi_1) & \dots & U_{Ri}(z_m,\xi_m) \end{bmatrix} \xi_1 \xi_2^2 \cdots \xi_m^m \, d\xi_0\, d\xi_1 \cdots d\xi_m$$

$$= \frac{1}{(m+1)!} \int \det \begin{bmatrix} U_{Ri}(x,\xi_0) & U_{Ri}(x,\xi_1) & \dots & U_{Ri}(x,\xi_m) \\ U_{Ri}(z_1,\xi_0) & U_{Ri}(z_1,\xi_1) & \dots & U_{Ri}(z_1,\xi_m) \\ \vdots & \vdots & & \vdots \\ U_{Ri}(z_m,\xi_0) & U_{Ri}(z_m,\xi_1) & \dots & U_{Ri}(z_m,\xi_m) \end{bmatrix}$$

$$\times \prod_{0 \leqslant r < s \leqslant m} (\xi_s - \xi_r)\, d\xi_0\, d\xi_1 \cdots d\xi_m$$

$$= \int_{\xi_0 < \xi_1 < \cdots < \xi_m} \det \begin{bmatrix} U_{Ri}(x,\xi_0) & U_{Ri}(x,\xi_1) & \dots & U_{Ri}(x,\xi_m) \\ U_{Ri}(z_1,\xi_0) & U_{Ri}(z_1,\xi_1) & \dots & U_{Ri}(z_1,\xi_m) \\ \vdots & \vdots & & \vdots \\ U_{Ri}(z_m,\xi_0) & U_{Ri}(z_m,\xi_1) & \dots & U_{Ri}(z_m,\xi_m) \end{bmatrix}$$

$$\times \prod_{0 \leqslant r < s \leqslant m} (\xi_s - \xi_r)\, d\xi_0\, d\xi_1 \cdots d\xi_m. \tag{23.5.25}$$

Thus $R(x)$ is represented as an integral whose integrand has a fixed sign determined by the relative ordering of the numbers $x, z_1, z_2, \ldots, z_m$ (Lemma 23.5.6). It thus follows that $R(x)$ changes sign when x passes through any of the points z_k, $k = 1, \ldots, m$, and at no other value of x. In other words, $z_1, \ldots, z_m$ are the only real zeros of $R(x)$ having an odd multiplicity. And therefore $P_{i,j}(x)R(x)$ has a constant sign, so that

$$\int P_{i,j}(x)R(x)\,dx \neq 0, \tag{23.5.26}$$

in contradiction to (23.5.24).

The proof for $Q_{i,j}(x)$ is similar. □

As a consequence we have the integral representations of $P_{i,j}(x)$ for $i > 1$ and of $Q_{i,j}(x)$ for $i < p$ involving their respective zeros

$$P_{i,j}(x) \propto \int \det \begin{bmatrix} U_{Li}(\xi_0, x) & U_{Li}(\xi_1, x) & \ldots & U_{Li}(\xi_j, x) \\ U_{Li}(\xi_0, z_1) & U_{Li}(\xi_1, z_1) & \ldots & U_{Li}(\xi_j, z_1) \\ \vdots & \vdots & & \vdots \\ U_{Li}(\xi_0, z_j) & U_{Li}(\xi_1, z_j) & \ldots & U_{Li}(\xi_j, z_j) \end{bmatrix} \times \prod_{0 \leqslant r < s \leqslant j} (\xi_s - \xi_r)\, d\xi_0\, d\xi_1 \cdots d\xi_j, \tag{23.5.27}$$

$$Q_{i,j}(x) \propto \int \det \begin{bmatrix} U_{Ri}(x, \xi_0) & U_{Ri}(x, \xi_1) & \ldots & U_{Ri}(x, \xi_j) \\ U_{Ri}(z_1, \xi_0) & U_{Ri}(z_1, \xi_1) & \ldots & U_{Ri}(z_1, \xi_j) \\ \vdots & \vdots & & \vdots \\ U_{Ri}(z_j, \xi_0) & U_{Ri}(z_j, \xi_1) & \ldots & U_{Ri}(z_j, \xi_j) \end{bmatrix} \times \prod_{0 \leqslant r < s \leqslant j} (\xi_s - \xi_r)\, d\xi_0\, d\xi_1 \cdots d\xi_j. \tag{23.5.28}$$

Lemmas 23.5.7 and 23.5.9 tell us that if $P_j(x)$ and $Q_j(x)$ are bi-orthogonal polynomials satisfying Eq. (23.1.12), then $P_{i,j}(x)$ and $Q_{i,j}(x)$ each have exactly j distinct real zeros of odd multiplicity. In particular, the zeros of the bi-orthogonal polynomials $P_j(x) \equiv P_{1,j}(x)$ and $Q_j(x) \equiv Q_{p,j}(x)$ are real and simple.

With a little more effort one can perhaps show that all the real zeros of $P_{i,j}(x)$ and of $Q_{i,j}(x)$ are simple. Other zeros, if any, must be complex. Whether the zeros of $P_j(x)$ ($Q_j(x)$) interlace for successive j, remains an open question.

Summary of Chapter 23

For a linear chain of coupled hermitian matrices with the probability density

$$F(A_1, \ldots, A_p) \propto \exp\left[-\operatorname{tr}\left\{\frac{1}{2}V_1(A_1) + V_2(A_2) + \cdots + V_{p-1}(A_{p-1}) + \frac{1}{2}V_p(A_p)\right\}\right] \times \exp\left[\operatorname{tr}\{c_1 A_1 A_2 + c_2 A_2 A_3 + \cdots + c_{p-1} A_{p-1} A_p\}\right], \qquad (23.0.2)$$

we expressed the most general correlation function as a single determinant, Eq. (23.1.17). The spacing functions are partial derivatives of a generating function, Eq. (23.3.13). And this later is expressed as a certain Fredholm determinant with a kernel which is complicated in general. The zeros of the bi-orthogonal polynomials involved are shown to be real and simple. Whether these zeros interlace for successive orders of the polynomials, as they do for orthogonal polynomials, and as numerical evidence suggests, remains an open question.

If the weight $w(x, y)$ is such that the moments $M_{i,j} = \int w(x, y)x^i y^j \, dx \, dy$ exist for all $i, j \geqslant 0$ and the moment matrix $[M_{i,j}]_{i,j=0,\ldots,n}$ is non-singular for $n \geqslant 0$, then bi-orthogonal polynomials exist. Moreover, if $\det[w(x_i, y_j)]_{i,j=1,\ldots,n} > 0$ for $x_1 < x_2 < \cdots < x_n$, $y_1 < y_2 < \cdots < y_n$, then their zeros are real, simple and lie in the respective supports of the weight $w(x, y)$. The same is true for a weight which is a convolution of several such weights.

24

GAUSSIAN ENSEMBLES. EDGE OF THE SPECTRUM

The density of nuclear levels increases steeply almost like an exponential in the experimentally observed energy range. On the other hand, the eigenvalue density for the Gaussian ensembles is a semicircle in the first approximation:

$$\sigma(x) \approx \pi^{-1}(2N - x^2)^{1/2}. \tag{24.0.1}$$

Therefore one might think that near the lower end, $x = -(2N)^{1/2}$, this density looks like the actual rise in nuclear level density. Although the deviations must be small compared with the dominant behavior (24.0.1), the tail might still contain an infinite number of eigenvalues. For example, $\sigma(x)$ may be proportional to $N^{1/3}$ in a region extending to $N^{1/6}$, so that the number of eigenvalues not accounted for by (24.0.1) will be proportional to $N^{1/2}$, increasing rapidly with the dimension number N of the matrices in the set. If this were the case, we should expect that the correlation functions and the spacing distribution in the tail part would be nearer the actual situation. However, a somewhat laborious calculation (Bronk, 1964a, 1965) shows that this is not true. Near $x = \pm(2N)^{1/2}$ the eigenvalue density $\sigma_N(x)$ is $\sim N^{1/6}$ in a region of extent $\sim N^{-1/6}$, so that the total number of eigenvalues in the tail part remains finite and amounts to only a few, even when $N \to \infty$. What can one say about the distribution of these few eigenvalues, their spacings, ...?

24.1 Level Density Near the Inflection Point

For the Gaussian unitary ensemble the eigenvalue density $\sigma_N(x)$ is the sum (cf. (6.2.10) and Appendix A.9)

$$\sigma_N(x) = \sum_{j=0}^{N-1} \varphi_j^2(x) \equiv \sum_{j=0}^{N-1} \{2^j j! \pi^{1/2}\}^{-1} e^{-x^2} \{H_j(x)\}^2 \tag{24.1.1}$$

$$= (N/2)^{1/2} [\varphi_N'(x)\varphi_{N-1}(x) - \varphi_N(x)\varphi_{N-1}'(x)], \tag{24.1.2}$$

where φ' is the derivative of φ. To have any similarity with the exponential the function $\sigma_N(x)$ must be convex from below, and we will be interested only in that region. Let us therefore determine the inflection point x_0 such that $\sigma_N(x)$ is convex from below for all $x > x_0$. Differentiating (24.1.2) and substituting from the differential equation

$$\varphi_j''(x) + (2j + 1 - x^2)\varphi_j(x) = 0, \tag{24.1.3}$$

satisfied by the harmonic oscillator function $\varphi_j(x)$, we get

$$\sigma_N'(x) = -(2N)^{1/2} \varphi_N(x)\varphi_{N-1}(x).$$

Differentiating once more, we obtain

$$\sigma_N''(x) = -(2N)^{1/2} [\varphi_N'(x)\varphi_{N-1}(x) + \varphi_N(x)\varphi_{N-1}'(x)]. \tag{24.1.4}$$

We are interested in the location of the largest zero of $\sigma_N''(x)$. For $x \geqslant (2N)^{1/2}$, $\varphi_N(x)$ and $\varphi_{N-1}(x)$ are both positive and decreasing so that $\varphi_N'(x)$ and $\varphi_{N-1}'(x)$ are both negative. Because the outermost maxima of $\varphi_j(x)$ move out with the increase of j, $\varphi_{N-1}'(x)$ is negative and $\varphi_N(x)$ is positive when $\varphi_N'(x)$ first becomes zero. As we decrease x across the value $(2N)^{1/2}$, $\varphi_N(x)$ will attain its maximum value and then decrease to zero, whereas $\varphi_{N-1}(x)$ will always remain positive. Thus $\sigma_N''(x)$ changes sign as x varies from the largest zero of $\varphi_N'(x)$ to the largest zero of $\varphi_N(x)$ and therefore must vanish somewhere in between. These largest zeros lie very near each other and their location is known (Szegö, 1939):

$$x_0 \approx (2N)^{1/2} - 1.856(2N)^{-1/6}. \tag{24.1.5}$$

We are interested in estimating the number of eigenvalues larger than x_0:

$$\int_{x_0}^{\infty} \sigma_N(x)\, dx. \tag{24.1.6}$$

To estimate $\varphi_j(x)$ near the transition point put

$$x = (2j+1)^{1/2} + 2^{-1/3}(2j+1)^{-1/6}t, \tag{24.1.7}$$

so that the differential equation (24.1.3) for $\varphi_j(x)$ is transformed to

$$\frac{d^2}{dt^2}\tilde{\phi}_j(t) = t\tilde{\phi}_j(t), \quad t \cdot j^{-2/3} \ll 1 \ll j; \tag{24.1.8}$$

$$\varphi_j(x) = \tilde{\phi}_j(t). \tag{24.1.9}$$

This is Airy's equation. One should choose the solution that goes to zero for $t \to \infty$.

Realizing that the only terms contributing appreciably to the summation (24.1.1) are those with a large j we set (c_1, c_2 being constants, not always the same)

$$j = N - 1 - \mu, \tag{24.1.10}$$

$$(2N)^{1/2} + y = (2j+1)^{1/2} + 2^{-1/3}(2j+1)^{-1/6}t, \tag{24.1.11}$$

expand in powers of μ/N, and keep only the dominant terms to get

$$\{\tilde{\phi}_{N-\mu-1}(t)\}^2 \approx c_1 N^{-1/6} \exp\left[-c_2 N^{1/6}\left(y + \frac{\mu}{(2N)^{1/2}}\right)\right]. \tag{24.1.12}$$

Replacing the summation in Eq. (24.1.1) by an integration

$$\begin{aligned}\sigma_N(x) &\approx \text{const} \cdot N^{-1/6} \exp\{-c_2 N^{1/6} y\} \int_0^N \exp(-c_2\, N^{-1/3}\mu)\, d\mu \\ &\approx \text{const} \cdot N^{1/6} \exp(-c_2 N^{1/6} y).\end{aligned} \tag{24.1.13}$$

Thus the eigenvalue density in the tail varies as $N^{1/6}$ extending over a region of order $N^{-1/6}$ beyond the "semi-circle" point; and the average number of eigenvalues in the tail beyond that point is finite and of the order of 1.

The eigenvalue density (7.2.32) for the Gaussian orthogonal ensemble has one more term. To get it in the tail of the semicircle we need an estimation of

$$I = m^{1/2} \int_{x_0}^{\infty} dx \left[\varphi_{2m-1}(x) \int_0^x \varphi_{2m}(y)\, dy\right], \tag{24.1.14}$$

where $N = 2m$ and x_0 has the same meaning as before. This can be estimated with some effort to be a small constant independent of m. For details see Bronk (1965).

For the Gaussian symplectic ensemble, the level density is (Eq. (8.2.2), case $n = 1$),

$$\sigma_{N4}(x) = \frac{1}{\sqrt{2}} \sum_{0}^{2N} \varphi_j^2\left(x\sqrt{2}\right).$$

Its estimation in the tail of the semi-circle follows from what we have already said before.

24.2 Spacing Functions

For the spacing functions, as in Chapter 6, one should take the limit of $K_N(x, y) = \sum_{j=0}^{N-1} \varphi_j(x)\varphi_j(y)$ for x and y around $\sqrt{2N}$ and large N. Now writing more carefully Eqs. (24.1.7) and (24.1.8) one has

$$\begin{aligned} K(x, y) &= \lim_{N\to\infty} \frac{1}{\sqrt{2}N^{1/6}} K_N\left(\sqrt{2N} + \frac{x}{\sqrt{2}N^{1/6}}, \sqrt{2N} + \frac{y}{\sqrt{2}N^{1/6}}\right) \\ &= \frac{Ai(x)Ai'(y) - Ai'(x)Ai(y)}{x - y} \end{aligned} \tag{24.2.1}$$

where $Ai(x)$ is the Airy function satisfying

$$Ai''(x) = xAi(x) \tag{24.2.2}$$

and derivatives are indicated by primes.

As in Chapter 6, or Appendix A.7, the probability $E(r, (s, \infty))$ that the interval (s, ∞) contains exactly r eigenvalues for the Gaussian unitary ensemble is given by the rth derivative of the Fredholm determinant at $z = 1$,

$$E\big(r, (s, \infty)\big) = \frac{1}{r!}\left(-\frac{\partial}{\partial z}\right)^r \det[1 - zK]|_{z=1}, \tag{24.2.3}$$

where now $K(x, y)$, the so called "Airy kernel", is given by Eq. (24.2.1) above, $s \leqslant x, y < \infty$.

As the "sine kernel", Eq. (6.3.13), is the "square" of another symmetric kernel, Eq. (6.3.16), so is the Airy kernel. In fact from Eqs. (24.2.1) and (24.2.2)

$$\begin{aligned} \left(\frac{\partial}{\partial x} + \frac{\partial}{\partial y}\right) K(x, y) &= \frac{Ai(x)Ai''(y) - Ai''(x)Ai(y)}{x - y} \\ &= -Ai(x)Ai(y), \end{aligned} \tag{24.2.4}$$

and also

$$\left(\frac{\partial}{\partial x}+\frac{\partial}{\partial y}\right)\int_0^\infty Ai(x+z)Ai(y+z)\,dz=\int_0^\infty \frac{\partial}{\partial z}\big[Ai(x+z)Ai(y+z)\big]\,dz$$
$$=-Ai(x)Ai(y), \tag{24.2.5}$$

so that $(\partial/\partial x+\partial/\partial y)$ of

$$K(x,y)-\int_0^\infty Ai(x+z)Ai(y+z)\,dz, \tag{24.2.6}$$

is zero and so the later is a function of $x-y$, say $f(x-y)$. But this $f(x-y)$ is zero for x and y both going to ∞ with any fixed value of $x-y$. Hence expression (24.2.6) is identically zero. Thus

$$K(x,y)=\int_0^\infty Ai(x+z)Ai(y+z)\,dz$$
$$=\int_s^\infty Ai(x+z-s)Ai(z+y-s)\,dz, \tag{24.2.7}$$

i.e., the kernel $K(x,y)$ is the square of the kernel $Ai(x+y-s)$ over the interval (s,∞).

As the sine kernel commutes with a second order linear differential operator, Eq. (6.3.22), so does the Airy kernel with the operator

$$L:=\frac{d}{dx}(x-s)\frac{d}{dx}-x(x-s), \tag{24.2.8}$$

for functions vanishing at s and at ∞. It is simpler to verify that L commutes with the kernel $Ai(x+y-s)$, "square root" of the Airy kernel $K(x,y)$, Eq. (24.2.1).

In the case of the sine kernel the corresponding differential equation was extensively studied and the solutions were known as the spheroidal functions. In the case of the Airy kernel the differential equation

$$\left[\frac{d}{dx}(x-s)\frac{d}{dx}-x(x-s)-\lambda\right]f(x)=0, \tag{24.2.9}$$

did not receive much attention and its solutions are not known.

The sine kernel $K_S(x,y)=\sin\pi(x-y)/\pi(x-y)$ satisfies the equation

$$\int_{-\infty}^\infty K_S(x,y)K_S(y,z)\,dy=K_S(x,z).$$

Airy kernel does satisfy the same equation (cf. Eq. (A.10.8)).

24.3 Differential Equations; Painlevé

As the functions related to the Fredholm determinant corresponding to the sine kernel satisfy second order non-linear differential equations, so do the functions related to the Fredholm determinant corresponding to the Airy kernel, which we now take up. Here we will consider only the case of a single connected interval (s, ∞) and will indicate the old and cumbersome method of expanding everything in powers of the auxiliary variable z. For the union of a finite number of disjoint intervals and a neat way to derive the results using a compact notation one may refer to Appendix A.16 or consult Tracy and Widom (1994b).

From Eq. (24.2.3) we can symbolically write the "resolvent" (suppressing the dependence on z for convenience) as (see, e.g., Goursat, 1956)

$$\begin{aligned} R(s) &:= \frac{d}{ds} \log\det[1 - zK] \\ &= \frac{d}{ds} \operatorname{tr}\log[1 - zK] \\ &= -\sum_{n=1}^{\infty} \frac{z^n}{n} \frac{d}{ds} \int_s^{\infty} K^{(n)}(x, x)\, dx \\ &= \sum_{n=0}^{\infty} z^{n+1} K^{(n+1)}(s, s) \end{aligned} \tag{24.3.1}$$

where by convention

$$K^{(0)}(x, y) = \delta(x - y), \tag{24.3.2}$$

$$K^{(1)}(x, y) = K(x, y) \equiv \frac{Ai(x)Ai'(y) - Ai'(x)Ai(y)}{x - y}, \tag{24.3.3}$$

$$K^{(n)}(x, y) = \int_s^{\infty} K(x, z) K^{(n-1)}(z, y)\, dz, \quad n > 1, \tag{24.3.4}$$

$$K(x, x) = \lim_{\varepsilon \to 0} K(x + \varepsilon, x) = \left[Ai'(x)\right]^2 - x\left[Ai(x)\right]^2. \tag{24.3.5}$$

The variable s appears in $R(s)$ in two ways, as the lower limits of integration and explicitly in the first and last integrand. So differentiating $R(s)$ with respect to s, we get terms obtained by replacing the intermediate variables by s,

$$-\sum_{n=1}^{\infty} z^{n+1} \sum_{j=1}^{n} K^{(j)}(s, s) K^{(n+1-j)}(s, s), \quad 1 \leqslant j \leqslant n, \tag{24.3.6}$$

and (apart from $dK(s,s)/ds = -Ai^2(s)$) the two terms

$$R_1 = \sum_{n=1}^{\infty} z^{n+1} \int_s^{\infty} \frac{\partial K(s,x)}{\partial s} K^{(n)}(x,s)\, dx \tag{24.3.7}$$

and

$$R_2 = \sum_{n=1}^{\infty} z^{n+1} \int_s^{\infty} K^{(n)}(s,x) \frac{\partial K(x,s)}{\partial s}. \tag{24.3.8}$$

We now use a procedure originally due to Gaudin. In R_1 replace

$$\partial K(s,x)/\partial s \quad \text{by} \quad -\partial K(s,x)/\partial x - Ai(s)Ai(x)$$

and integrate the first term by parts so as to push the partial derivative one step to the right. Then replace

$$\partial K(x_1,x_2)/\partial x_1 \quad \text{by} \quad -\partial K(x_1,x_2)/\partial x_2 - Ai(x_1)Ai(x_2)$$

and integrate the first term by parts so as to push the partial derivative one more step to the right. Repeating this process again and again we move the partial derivative to the extreme right where it combines with R_2. Thus

$$R'(s) = -zq^2(s), \tag{24.3.9}$$

with

$$\begin{aligned} q(s) &= Ai(s) + \sum_{n=1}^{\infty} z^n \int_s^{\infty} Ai(x) K^{(n)}(x,s)\, dx \\ &= Ai(s) + \int_s^{\infty} dx\, Ai(x) \left[zK(1-zK)^{-1} \right](x,s) \\ &= Ai(s) + \int_s^{\infty} dx \left[zK(1-zK)^{-1} \right](s,x) Ai(x) \\ &= \int_s^{\infty} dx \left[(1-zK)^{-1} \right](s,x) Ai(x) \\ &= \int_s^{\infty} dx\, Ai(x) \left[(1-zK)^{-1} \right](x,s). \end{aligned} \tag{24.3.10}$$

Differentiating $q(s)$ and using again the same procedure to push the partial derivative to the other end we get

$$q'(s) = p(s) - q(s)u(s), \tag{24.3.11}$$

$$\begin{aligned} p(s) &= Ai'(s) + \int_s^\infty dx \left[zK(1-zK)^{-1}\right](s,x)Ai'(x), \\ &= \int_s^\infty dx \left[(1-zK)^{-1}\right](s,x)Ai'(x) \\ &= \int_s^\infty dx\, Ai'(x)\left[(1-zK)^{-1}\right](x,s), \end{aligned} \tag{24.3.12}$$

$$u(s) = z\int_s^\infty dx \int_s^\infty dy\, Ai(x)\left[(1-zK)^{-1}\right](x,y)Ai(y). \tag{24.3.13}$$

Differentiating $p(s)$ and $u(s)$ and pushing the partial derivative from one end to the other we get

$$p'(s) = w(s) - q(s)v(s), \tag{24.3.14}$$

$$u'(s) = -zq^2(s), \tag{24.3.15}$$

$$\begin{aligned} w(s) &= \int_s^\infty dx \int_s^\infty dy\, Ai(x)\left[(1-zK)^{-1}\right](x,y)Ai''(y) \\ &= \int_s^\infty dx \int_s^\infty dy\, Ai(x)\left[(1-zK)^{-1}\right](x,y)yAi(y), \end{aligned} \tag{24.3.16}$$

$$\begin{aligned} v(s) &= z\int_s^\infty dx \int_s^\infty dy\, Ai(x)\left[(1-zK)^{-1}\right](x,y)Ai'(y) \\ &= \int_s^\infty dx \int_s^\infty dy\, Ai'(x)\left[(1-zK)^{-1}\right](x,y)Ai(y). \end{aligned} \tag{24.3.17}$$

Using the relation

$$K(x,y)y = xK(x,y) - Ai(x)Ai'(y) + Ai'(x)Ai(y), \tag{24.3.18}$$

we can move the factor y under the integral sign of $w(s)$ from the right end to the left end;

$$w(s) = sq(s) + p(s)u(s) - q(s)v(s) \tag{24.3.19}$$

so that

$$p'(s) = sq(s) + p(s)u(s) - 2q(s)v(s). \tag{24.3.20}$$

Here the chain closes; no new quantities need be introduced.

In $v(s)$ there is a derivative at one end which we can push from end to end getting one more relation

$$2v(s) = u^2(s) - zq^2(s). \tag{24.3.21}$$

From Eqs. (24.3.9) and (24.3.15) with the boundary condition $R(\infty) = u(\infty) = 0$ one sees that $R(s) = u(s)$.

Differentiating once more Eq. (24.3.11) and using Eqs. (24.3.11), (24.3.20), (24.3.15) and (24.3.21),

$$\frac{d^2q}{ds^2} = sq + 2zq^3. \tag{24.3.22}$$

This is the Painlevé-2 equation in its standard form for $y = \sqrt{z}q(s)$. (Cf. Appendix A.45.) The boundary condition is $q(s) \equiv q(z,s) \approx Ai(s)$ as $s \to \infty$, $Ai(x)$ being the Airy function.

Multiplying this last equation through out by $2q'(s)$ and integrating one gets

$$q'^2 = sq^2 + zq^4 + \int_s^\infty dx\, q^2(x) = sq^2 + zq^4 - R,$$

so that

$$R(s) = sq^2(s) + zq^4(s) - q'^2(s). \tag{24.3.23}$$

Also differentiating Eq. (24.3.9), squaring and using Eq. (24.3.23) again

$$(R'')^2 = (2zqq')^2 = 4z^2q^2\left(sq^2 + zq^4 - R\right) = -4R'\left(-sR' + R'^2 - zR\right). \tag{24.3.24}$$

This is the equation satisfied by the "resolvent" $R(s)$.

From Eqs. (24.3.9) and (24.3.22) one has the simple expression for the Fredholm determinant in terms of a Painlevé-2 transcendent

$$\det[1 - zK] = \exp\left(-z\int_s^\infty (x-s)q^2(z,x)\,dx\right). \tag{24.3.25}$$

One can numerically integrate Eq. (24.3.22) and then Eq. (24.3.25) to get for example the probability $E(0,s)$ or $E(1,s)$ that the interval (s,∞) contains no eigenvalue or exactly one eigenvalue, or the probability densities $\tilde{F}(0,s)$ and $\tilde{F}(1,s)$ from their derivatives; cf. Figures 24.1 and 24.2.

We considered here only the Gaussian unitary ensemble. The limiting kernels in the other two cases $\beta = 1$ and $\beta = 4$ are 2×2 matrices, cf. Chapter 20, giving rise to coupled equations. Tracy and Widom have treated these cases as well. We content ourselves to reproduce here their curves for the spacing functions $p_\beta(s) \equiv p_\beta(0,(s,\infty)) = d^2E_\beta(0,(s,\infty))/ds^2$, Figure 24.3.

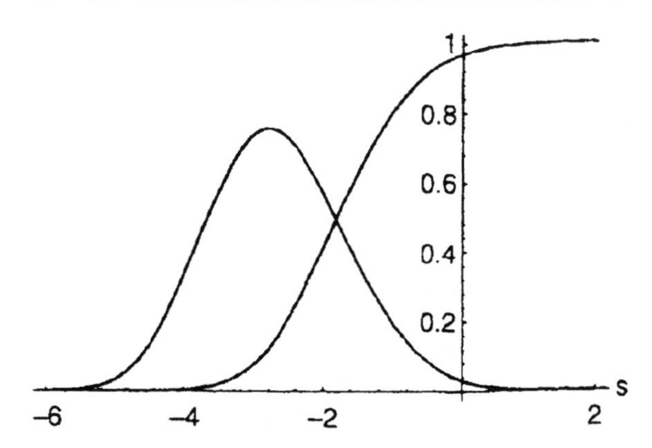

Figure 24.1. The probabilities $E(0, s)$ and $E(1, s)$. Of course, $E(0, s) \to 1$ as $s \to \infty$. From Tracy and Widom (1994b). Reprinted by permission.

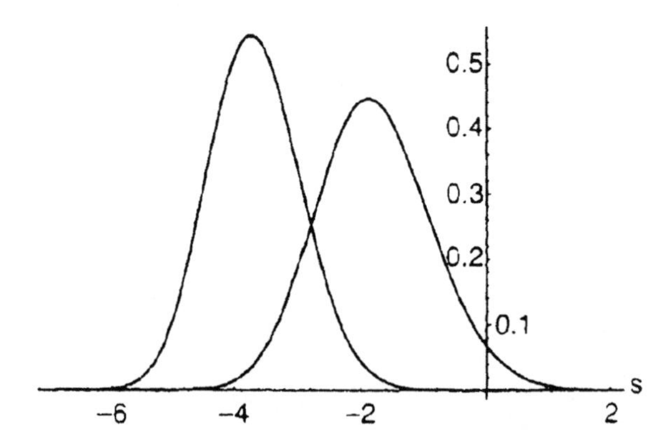

Figure 24.2. The probability densities $\tilde{F}(0, s)$ and $\tilde{F}(1, s)$; the right and left curves respectively. From Tracy and Widom (1994b). Reprinted by permission.

Summary to Chapter 24

For the Gaussian ensembles the dominant part of the eigenvalue density is a "semi-circle". How good is this estimate in the tail part? It is found that the decrease is quite sharp; in the tail part the eigenvalue density $\sigma_N(x)$ is $N^{1/6}$ extending to $N^{-1/6}$ beyond the "semi-circle" limit point, so that the total number of eigenvalues lying outside the "semi-circle" remains finite when $N \to \infty$.

The probability that the interval (s, ∞) contains exactly r eigenvalues is obtained from a Fredholm determinant with the so called "Airy" kernel, Eq. (24.2.1). The Airy kernel is the square of another kernel, Eq. (24.2.7), and commutes with a second or-

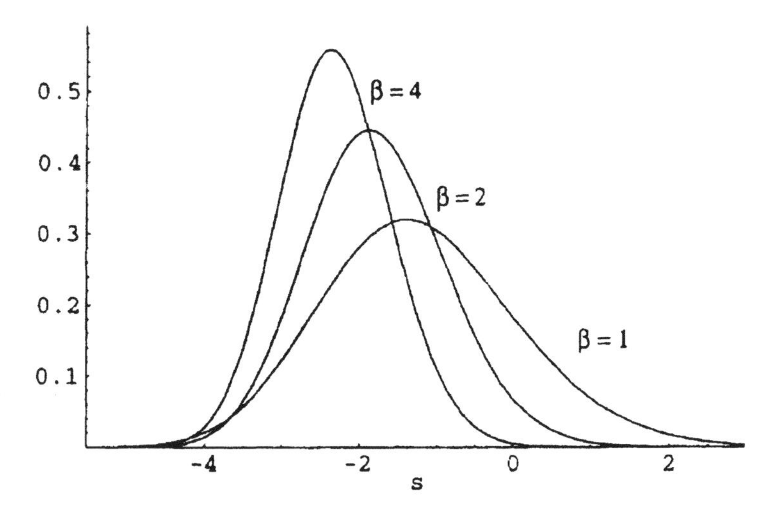

Figure 24.3. The probability densities $p_\beta(s)$ for $\beta = 1, 2$ and 4. From Tracy and Widom (1996). Reprinted by permission.

der linear differential operator, Eq. (24.2.8). The Fredholm determinant itself can be expressed in terms of a Painlevé-2 transcendent, Eq. (24.3.25).

Probably the same situation prevails for a much larger class of matrix ensembles.

25

RANDOM PERMUTATIONS, CIRCULAR UNITARY ENSEMBLE (CUE) AND GAUSSIAN UNITARY ENSEMBLE (GUE)

25.1 Longest Increasing Subsequences in Random Permutations

If $\sigma = \begin{pmatrix} 1 & 2 & \ldots & k \\ \sigma(1) & \sigma(2) & \ldots & \sigma(k) \end{pmatrix}$ is a permutation of $(1, 2, \ldots, k)$, then σ has an increasing (respectively decreasing) subsequence of length n if there exist indices $1 \leqslant i_1 < i_2 < \cdots < i_n \leqslant k$ such that $\sigma(i_1) < \sigma(i_2) < \cdots < \sigma(i_n)$ (respectively $\sigma(i_1) > \sigma(i_2) > \cdots > \sigma(i_n)$). A famous old result of Erdös and Szekeres (1938) states that every permutation of $(1, \ldots, k)$ contains either an increasing or a decreasing subsequence of length $\geqslant \sqrt{k}$. Let $L(k)$ be the length of the longest increasing subsequence of a permutation of $(1, 2, \ldots, k)$. Ulam (1961) studied numerically the distribution of $L(k)$ for random permutations and conjectured that its average value is $c\sqrt{k}$ with c a constant of the order of 2. Hammersley (1972) proved that the average value of $L(k)$ is indeed $c\sqrt{k}$ and $\pi/2 \leqslant c \leqslant e$. Later authors narrowed down the upper and lower bounds on c and finally established that $c = 2$ (Vershik and Kerov (1977), Logan and Shepp (1977), Pipel (1990), and references given there).

How $L(k)$ are distributed?

25.2 Random Permutations and the Circular Unitary Ensemble

Let $I(k, n)$ be the number of permutations of $(1, 2, \ldots, k)$ in which the length of the longest increasing subsequence is less than or equal to n. By combinatorial arguments Gessel (1990) showed that the generating function

$$G_n(z) = \sum_{k=0}^{\infty} I(k, n) \frac{z^{2k}}{(k!)^2}, \tag{25.2.1}$$

can be expressed as the $n \times n$ determinant

$$G_n(z) = \det[I_{p-q}(2z)]_{p,q=0,1,\ldots,n-1}, \tag{25.2.2}$$

where $I_p(z)$ is the modified Bessel function

$$I_p(z) = \frac{1}{2\pi} \int_0^{2\pi} e^{z \cos\theta + ip\theta}\, d\theta. \tag{25.2.3}$$

Now if in Chapter 5, Section 5.3 we set $x = e^{i\theta}$, $f(x) = \exp(2z\cos\theta)$, $w(x)\,dx = d\theta$, $C_p(x) = e^{ip\theta}$, $\bar{C}_p(x) = e^{-ip\theta}$, $c_p = 2\pi$, then Eq. (5.3.8) takes the form

$$\Psi_{p,q}\left(e^{2z\cos\theta}\right) = I_{p-q}(2z), \tag{25.2.4}$$

and Section 5.3 gives

$$\left\langle \prod_{j=1}^{n} \exp(2z\cos\theta_j) \right\rangle := \frac{1}{(2\pi)^n} \frac{1}{n!} \int_0^{2\pi} \cdots \int_0^{2\pi} \prod_{j=1}^{n} \exp(2z\cos\theta_j) \times \prod_{1 \leqslant p < q \leqslant n} \left|e^{i\theta_p} - e^{i\theta_q}\right|^2 d\theta_1 \cdots d\theta_n \tag{25.2.5}$$

$$= \det[I_{p-q}(2z)]_{p,q=0,1,\ldots,n-1}, \tag{25.2.6}$$

so that

$$G_n(z) = \left\langle \prod_{j=1}^{n} \exp(2z\cos\theta_j) \right\rangle = \left\langle \exp\left(2z \sum_{j=1}^{n} \cos\theta_j\right) \right\rangle \tag{25.2.7}$$

$$= \sum_{k=0}^{\infty} \frac{z^{2k}}{(2k)!} \left\langle \left(2 \sum_{j=1}^{n} \cos\theta_j\right)^{2k} \right\rangle, \tag{25.2.8}$$

since the odd terms in the expansion of the exponential are zero by symmetry. Moreover,

$$\left\langle \left(2\sum_{j=1}^{n}\cos\theta_j\right)^{2k} \right\rangle = \left\langle \left[\sum_{j=1}^{n}(e^{i\theta_j}+e^{-i\theta_j})\right]^{2k} \right\rangle$$

$$= \sum_{\ell=0}^{2k}\binom{2k}{\ell}\left\langle \left(\sum_{j=1}^{n}e^{i\theta_j}\right)^{\ell}\left(\sum_{j=1}^{n}e^{-i\theta_j}\right)^{2k-\ell} \right\rangle \tag{25.2.9}$$

$$= \frac{(2k)!}{(k!)^2}\left\langle \left|\sum_{j=1}^{n}e^{i\theta_j}\right|^{2k} \right\rangle, \tag{25.2.10}$$

since the only non-zero contribution comes from the term when the powers of the two complex conjugate quantities are equal, i.e., from the term $\ell = 2k-\ell$.

Putting together Eqs. (25.2.7)–(25.2.10), one has

$$G_n(z) = \sum_{k=0}^{\infty}\frac{z^{2k}}{(k!)^2}\left\langle \left|\sum_{j=1}^{n}e^{i\theta_j}\right|^{2k} \right\rangle. \tag{25.2.11}$$

Thus Gessel's result expresses $I(k,n)$ as the average value of $|\operatorname{tr} U|^{2k}$ for matrices U taken from the $n \times n$ circular unitary ensemble (CUE) (cf. Chapters 10, 11):

$$I(k,n) = \left\langle \left|\sum_{j=1}^{n}e^{i\theta_j}\right|^{2k} \right\rangle$$

$$= \frac{1}{(2\pi)^n}\frac{1}{n!}\int_0^{2\pi}\cdots\int_0^{2\pi}\left|\sum_{j=1}^{n}\exp(i\theta_j)\right|^{2k}$$

$$\times \prod_{1\leqslant p<q\leqslant n}|e^{i\theta_p}-e^{i\theta_q}|^2\,d\theta_1\cdots d\theta_n, \tag{25.2.12}$$

and the question is reduced to the calculation of such averages. This interpretation gives in particular the value of the integral (25.2.12) for $k \leqslant n$

$$\frac{1}{(2\pi)^n}\frac{1}{n!}\int_0^{2\pi}\cdots\int_0^{2\pi}\left|\sum_{j=1}^{n}\exp(i\theta_j)\right|^{2k}\prod_{1\leqslant p<q\leqslant n}|e^{i\theta_p}-e^{i\theta_q}|^2\,d\theta_1\cdots d\theta_n = k!,$$

$$0 \leqslant k \leqslant n. \tag{25.2.13}$$

For small values of k or of n, see Appendix A.51.

For n finite the asymptotic behavior of $I(k,n)$ follows from the results of Regev (1981)

$$I(k,n) \approx n^{2k+n^2/2}(2\pi)^{-(n-1)/2}(2k)^{-(n^2-1)/2}\prod_{j=1}^{n-1} j!, \quad k \to \infty. \tag{25.2.14}$$

Odlyzko et al. studied $I(k,n)$ numerically for $k \leqslant 120$ preparing tables of $L(k,n) = I(k,n) - I(k,n-1)$, i.e., the number of permutations of $(1,\ldots,k)$ in which the length of the longest increasing subsequence is exactly n. The graph of $\log L(k,n)$ versus n seems to follow a smooth curve rising steeply, almost vertically, at the origin, attaining a maximum around $2\sqrt{k}$ and falling linearly at $n = k$ (cf. Tables 25.1, 25.2 and 25.3). Can one derive an equation of this curve?

25.3 Robinson–Schensted Correspondence

In the theory of permutations, the notion of a standard Young tableau is important. If integers $\lambda_1,\ldots,\lambda_j$ are such that $\lambda_1 \geqslant \lambda_2 \geqslant \cdots \geqslant \lambda_j > 0$, and $k = \lambda_1 + \cdots + \lambda_j$, then $\lambda \equiv \{\lambda_1,\ldots,\lambda_j\}$ is called a partition of k and is usually represented by a (Young) diagram with λ_1 consecutive boxes in the first row, λ_2 consecutive boxes in the second row, and so on, all these sequences of boxes starting at the same vertical line at the left. For example, the partition $\{4, 2, 2\}$ is represented by

(25.3.1)

When the boxes of a Young diagram are filled in with integers 1 to k (each box receiving one integer) such that the integers appearing in each row and in each column are in strictly increasing order, then it is called a standard Young tableau. For example,

$$P: \begin{array}{|c|c|c|c|} \hline 1 & 2 & 4 & 6 \\ \hline 3 & 5 \\ \cline{1-2} 7 & 8 \\ \cline{1-2} \end{array} \qquad Q: \begin{array}{|c|c|c|c|} \hline 1 & 3 & 4 & 8 \\ \hline 2 & 5 \\ \cline{1-2} 6 & 7 \\ \cline{1-2} \end{array} \tag{25.3.2}$$

are two standard Young tableaux having the same shape $(4, 2, 2)$. The number of standard Young tableaux having a given shape is known to be given by explicit formulas in terms of the so called hook lengths of the shape and does not concern us here.

Robinson and Schensted found a one to one correspondence between a permutation and a pair of standard Young tableaux of the same shape. Given the permutation of $(1,\ldots,k)$ two standard Young tableaux P and Q of the same shape are constructed step by step according to a set of specified rules. This construction is best explained by

Table 25.1. Enumeration of permutations of $1, \ldots, k$ with $k = 30$ by values of the length of the longest increasing subsequence, $L(k, n) = I(k, n) - I(k, n-1)$, where $I(k, n)$ denotes the number of permutations of $(1, \ldots, k)$ having the length of the longest increasing subsequence less than or equal to n. $L(30, 1) = 1$ (there is 1 permutation with $n = 1$), $L(30, 2) = 3814986502092303$, etc. From Odlyzko, private communication

1:1
2:3814986502092303
3:122238896672891001069665
4:1790036582998939530743648877
5:449044243619862872721423598179
6:10236819433951393776243660748875
7:50241067877038219983230124657600
8:86511371455863277882723853476200
9:70971582765623356071324810857700
10:33700117351593715495661064101700
11:10447178628714722178634866396630
12:2277900847905046253535807880680
13:366440157064983378222220318530
14:44912755712412555783652789980
15:4289203871330156652985437480
16:324301002215082697285357800
17:19633107355949074371195000
18:959064229546178387532600
19:37982369568044622191625
20:1222055891584247185425
21:31925927141978856309
22:675007128155925069
23:11475430101232224
24:155228816648544
25:1644397829384
26:13319151176
27:79490741
28:328861
29:841
30:1

an example. Consider the permutation $\binom{1\,2\,3\,4\,5\,6\,7\,8}{3\,4\,1\,2\,7\,6\,8\,5}$. We start the tableaux P and Q by one box each, with 3 in the box of P and 1 in the box of Q corresponding to the first column entries in the permutation. Thus

$$P: \boxed{3} \qquad Q: \boxed{1} \tag{25.3.3}$$

Next we look at the second entry, 4; $4 > 3$, so we add a new box to the first row of P and a new box to the first row of Q, write 4 in the new box of P and 2 in the new box of Q; thus

$$P: \boxed{3}\boxed{4} \qquad Q: \boxed{1}\boxed{2} \tag{25.3.4}$$

Table 25.2. The same as Table 25.1 but for $k = 60$. From Odlyzko, private communication

1:1
2:15838509645961200426867727790388 95
3:35358010112347692425762860373008396032460841074812 9
4:17080691328825216538079811628828842602913045806045692424793199
5:17524302825007966090501884321361592986082556954968188486776569054170 1
6:93361519849307080211439112179568136778191621646404527876278830055347609 01
7:15180807338873516832021030140438444665815147021460591742801378406314408952231
8:22334944749484956902431105687452229832621595022835516892738911050997037646392 03
9:60002895752771099779779088462943847999099581712023250349374731986619450937660387
10:4681044407221266448128396321775561872819533309163225124590262917955291900841400 03
11:145532705437438575698254535186430657953648186790100201042377606224074097806267840 5
12:2259251055120372007733214696091079754018818083465717757461536975882962682765500625
13:2062265432178679983886852088922462401452557170316484374161761008379074310593517320
14:1243711511999821270591207565082889798761871176715300197918122808539228337822802740
15:537394830317050100339379519887032754646946119740464857911956705681737098244483360
16:1759231034235535719477619062786762451289739501001292331195633463264641048552768 60
17:45368617608497201905530039854748875664926717869975676357175086535901817870722812
18:94796038560305031579556851460637006725863578662081880425477385096396428886412 24
19:16387590091101218239825060044873038382415495507286625077753642305030635749753 16
20:2381888588175596539077577668910401316360693594922221344031478022546538547717 28
21:2948048767004722380392174789010902355631916600101345128103891298417031075312 0
22:31391980388285282862037109314559062852767089404609771281564468608856509273 10
23:29003056393202200211822060244775357501163113213893082887752963597079436270 0
24:23413655153323993212944806641432538187723487636946542215904509952446312690
25:16613935428055130717424170679898226596863648667569205918866364296631786 80
26:10414590036322014483046686657102315241819980099734155323043893524910725 8
27:5792237419925383613451898590561388009628831263062366370911868157401964
28:2868694673819196818222224697604922550547765331891831543188637523122 30
29:1269187185548182862659345802512560685759613114378273890357382598179 6
30:5029680586285726622771915665753736264156249152521549549726650084 52
31:1789460770029970329595261721561400979949464928384663161853006027 6
32:572672495594979262825338794738392020356281425452234267523956756
33:165113872207955676046858694485444444896127174737028045397681196
34:42944777682837494500889195658890287626247199702145964821255 6
35:10085939380850856133146998090323665735800777575608380654076
36:2140470590462530162888888773190221688888803440263783963 80
37:41065535476551473417108441729099099335532576644951114 05
38:71234520883705192260893127544474396900805731744717605
39:11171129510737042891640603020127531847122825028439 05
40:158314683653240272182995233071207313289009719945 05
41:20260744481586451879256098891357005163892522457 9
42:233911381147268850227765477830679405985356349 9
43:24328028687991328153614089674352694270581559
44:227535457430697745412499435864265131254799
45:19094646784190651978021317588969392507 54
46:14338575949172527832964867110076216498
47:9602477649339128451235480278680173 8
48:57119423894186917577984943763285 8
49:300310105323461983624329498843 8
50:13871858035569655993122428198
51:5588228944519012585653798 2
52:19453894588019188516414 2
53:57849946841676837554 7
54:14476874826014624 67
55:2988846947868807
56:4953109533951
57:6329639181
58:5851621
59:3481
60:1

Table 25.3. The same as Table 25.1 but for $k = 90$. From Odlyzko, private communication

1:1

2:100013460080035478192939925053654186436246108995079 9

3:30503479645734269022870556950168828935101816680540919718170405054372279536786 14

4:118720940615909838431874868305198637118991272677934156670142623293999556129117188371351145457116 2

5:14479568760433244909259595688013998553185125019866177284483581572438769347220611952148689495776965369232381 74

6:578236721230054528958680213726846053928350621919051236198983645292162909567710270650787563235072496452187593768135090

7:981140031051541984961596528108087435654648882396061958133346874054316541957856478885113807464215531200764354464863669335110

8:342899800591645286251655695405502880028247868134604578443990465591612257410463815299665338744145095209870438811690638965946581 30

9:702555008513453885267194407798034272168947939965398716616950982733565254691059145789854953154395425075396894632742257444557822 29540

10:174685006339488010696117831865613033708391087165376398336762935741443991431323167116876198111636696463495208583626128802011504301 78380

11:886489684016239977018269997410836052728425609969615046413177151882564775606228635596506675865063901134708370740974989928857442021333970

12:133664425780385910664168208685900311879595466611200489899823676282432371452870861715900189298281084391356938356502973019979720523883 52470

13:787984778094902094881849235229695144603802746240706779955409325008449445674125554431297739270027592951545412528346282853080355139867872 34

14:223081540789791057206182282458347292824074360038617889328789788117520282109119928523687714138206840589463701960276723839614642808292177846

15:354938619969288886735285871207667486495436734413523693523957198265241300201465357702005268418945906530527903531267701648811116882933051680

16:358017221938882581790960985701813475004381810514628358518594906556356349492924722460371742912180085845528159222988031747841438661172896440

17:250655302488123359102054888110242384601359832080075145983715483970871920980090862377475321001844038170028172942242636444186990485833650960

18:130196674695618727509003213182526326905586759529544529157604710154367542770050470719022644664679781575880259576861055424922552185453636840

19:526535415809512950218896313341395621352460252059859032023808671814463107925854956560371021000202585866478368469609566491800623949394054 30

20:171699454892496446192589554779158592801544159076666519516154977976031506235848814344033137706209346117890337536521905917891094452139353 30

21:463287938214215649649956420031205342245271167291309042872920995865051165517270409005950741876214496867624231064810516813102599729187909 4

22:105480906906341658053171055242503989833607223232501667278870180714858011735197427244803151881039581842906999132780091227956628765758782 6

23:205754863868200788258368010141444543172036127291122997617330575587403320260171761719166236567242977725027604220575236394781200518574435

24:348061293507153525173299245934331598628168739263039068732612514863123474549813384878412141558515121615101530721995958916055810259951 01

25:515696231614764237605964560483694085458693812855462025510507740869570297724632001722048505429429041472083832240142564287654812616619

26:674746773291706641541501901690487506198082434756130258384929989973597064435545132363664175602608806184769419189223271716663741673101

27:785081999909464845792515411094873085532513560017248637285673450856254735383089341433380708555979898043839714616261529300844237546 96

28:817144100616284875397161967169010076330759436221001469847342454850455346471900932011703324777203252481657898230771137034102341455 2

29:764759619527925875328195453841813434910968239175713526267040170518147894895245418218058957987515588776558012262827798123936134504

30:646462914505019760130071200602764036261513481492624013978981737155261605219824528747699052307740126006995281800298140145938940 72

31:495530426860316291114394573109419606332340413096004366739794038266128336295635802108129853906402641598525751141945138126316296 0

32:345640757500481383309039280118258961232393016067226348418529947633138876745998662757339309471513277517867051109373663096136400

33:220070332446065833463113003729509850265660690268416217315005253555275891670288741917350096249276617014212018159355554662986 40

34:128260212533212454310135128298313690847868911350551175787427485058679870131438699726148509319269964775510640997386446349288 0

35:685970434369960928325070098318124270160615390420110310409102701927486165175397984383851468752065833594455779600150160561 20

36:337429920080136337436920643993541547543603853067336515075216516459560774531876224624619702563903540645303266682056127472 8

37:152972636683601436789225515466107381364404188483368133550131098561417471099496481914282397634591910611884422981306703256

38:640323074496645321168797386400968656318757600270770550335106358961693528574259934955360192949820028236251571176099066 4

39:247895082951400282259151070276293172763662457683006026755823550895161752965434760154891656951948260146882954882041967

40:888956299151841312343900148227481898965139183186624496874310189744861904084689236427774220574951536243641666805453 5

41:295689234974242832494073010907751238375353476588398080612720646488714639902346295991925984951247226190477545079895

42:913431192119989662008780854727815005156417566405969382587280696850022983112348861938959126266159724923673489295

43:262356741023638621960459241636528332404362763436187298799626509580056961584100723921520856518579510375649010570

44:701338273262021180406490608439485687664058299652640747344454224342578835381542242470446265354089302130896177 0

45:174654532787921164833827802754142011707155401693334063351644799913737484052534569588877942001839020268081770

46:405513414576446821607654267450643226876249734513717172529502795688214722489829502518365189283877430236033 0

47:878455763787009339332285721734881941573703006105020897870262711704773987996228280254196539507716038355 80

48:177665160672780262107900067986258628036230507315133830546165083816644102065254898967314968061564363558 0

49:335656023971626871834274493837349382604004108579439990695744843009890027400959696494194532739835863 80

50:592657043430392616528107236781393086422224247308394571411201512542940516845016873567505709019139980

51:978367974176270636110015417622160233246459723065280083857202913066188312446296122091665572253467 0

52:151053337165402662595753716037701718650532740100756604535996027223018560852441073630509137652270

53:218169645369291107515341999713650395191138439810070906398649408185242591022201677299191450507 0

54:294827541050211829049703195445685492319613744631121021342461767877671819138674609640358930 70

55:372814238943906162162652703070492425887677831343236490129454084794141571215897613044678695

56:441139053816240497287846954826587670856051572549541748987389356914138583028992644088965 5

57:488417547921937266123965981327546333802375468065239136917472646440596544465514450782 55

58:505917110633761822572084391349359565323480986417512238709616830004670499386770489055

59:490166684276109092778798671835009041078363697106186204871787554751417330140415018 0

60:444069930697544044830419834426788283757362621411226492611513736991027755656461 80

61:376036526188147968010494812120216918711222255178795662097559647151894522679268

62:297486059141915603160503478100288667769627387653573580010079682067733867062 8

63:219737521697410340592157302096941404749657015492809359780699952234742250 08

64:151439458645631421656113123883188419803818592123965056490363615020790128

65:973008563071099118790139506711973666406008508931715595819129223522768

66:582274176398312719312165023225460568154009789272476609429876477128 0

67:324191574334121019594633465572410350678182382922188898918174027 80

68:167726861640295922577686591572074363062795220519912500362914780

69:805228952494192801504844315783671654299366672350492752357980

(continued on next page)

Table 25.3. (Continued)

70:3581442053191044718645644649859768798492759249365893412380
71:14730926916044630890759250917325129656345776940079800845
72:55916545091632769874075315311859320804778380934550845
73:195420340013708454241955126584589610185183208237845
74:627127865914234260752165889096491598463906019045
75:1842336146468696169806210012765510230989332570
76:4937170478379622848647534638856335667750330
77:12020272672486567674375039277300137844730
78:26461613882365295949357259482612587930
79:52379652048029632774477431104534180
80:92614360586028292816142891390180
81:145113199312105623546393842420
82:199538720798846062539888820
83:237894262849311200884270
84:242144218879304680270
85:206193769623472270
86:142850993609326
87:77320519321
88:30662281
89:7921
90:1

Next is 1, which is < 4 and < 3, so we replace 3 by 1 and displace 3 to the second row in P, adding a box in Q in the same place where the new box in P appears inscribing 3 in it, thus

$$P: \begin{array}{|c|c|}\hline 1 & 4 \\ \hline 3 & \\ \cline{1-1}\end{array} \qquad Q: \begin{array}{|c|c|}\hline 1 & 2 \\ \hline 3 & \\ \cline{1-1}\end{array} \tag{25.3.5}$$

Next comes 2, which is > 1 and < 4, so we replace 4 by 2 in P and look for a place for 4 in the second row of P. As $4 > 3$ we place it to the right of 3, at the same time adding a box to Q in the same place where the new box in P appears. Thus

$$P: \begin{array}{|c|c|}\hline 1 & 2 \\ \hline 3 & 4 \\ \hline\end{array} \qquad Q: \begin{array}{|c|c|}\hline 1 & 2 \\ \hline 3 & 4 \\ \hline\end{array} \tag{25.3.6}$$

Next, $7 > 2$, so it goes in the first row next to 2, and Q gets an additional box, thus

$$P: \begin{array}{|c|c|c|}\hline 1 & 2 & 7 \\ \hline 3 & 4 \\ \cline{1-2}\end{array} \qquad Q: \begin{array}{|c|c|c|}\hline 1 & 2 & 5 \\ \hline 3 & 4 \\ \cline{1-2}\end{array} \tag{25.3.7}$$

Next, $6 > 2$ and < 7, so 7 is replaced by 6, and 7 being > 4 sits next to it in P, Q getting a box in the same place,

$$P: \begin{array}{|c|c|c|}\hline 1 & 2 & 6 \\ \hline 3 & 4 & 7 \\ \hline\end{array} \qquad Q: \begin{array}{|c|c|c|}\hline 1 & 2 & 5 \\ \hline 3 & 4 & 6 \\ \hline\end{array} \tag{25.3.8}$$

and so on.

$$P: \begin{array}{|c|c|c|c|}\hline 1 & 2 & 6 & 8 \\ \hline 3 & 4 & 7 \\ \cline{1-3} \end{array} \qquad Q: \begin{array}{|c|c|c|c|}\hline 1 & 2 & 5 & 7 \\ \hline 3 & 4 & 6 \\ \cline{1-3} \end{array} \tag{25.3.9}$$

and finally

$$P: \begin{array}{|c|c|c|c|}\hline 1 & 2 & 5 & 8 \\ \hline 3 & 4 & 6 \\ \cline{1-3} 7 \\ \cline{1-1} \end{array} \qquad Q: \begin{array}{|c|c|c|c|}\hline 1 & 2 & 5 & 7 \\ \hline 3 & 4 & 6 \\ \cline{1-3} 8 \\ \cline{1-1} \end{array} \tag{25.3.10}$$

The tableaux are now complete.

At each step one box is added and the tableaux P and Q keep the same shape, remain standard and by the construction algorithm are uniquely determined by the given permutation. Also any two standard tableaux of the same shape determine by the inverse process a unique permutation. For example, the pair of tableaux (25.3.2) correspond to the permutation $\binom{1\,2\,3\,4\,5\,6\,7\,8}{7\,1\,3\,8\,5\,2\,4\,6}$.

This one to one correspondence between a permutation and a pair of (standard Young) tableaux of the same shape has one property, among others, that the length of the first row (respectively column) of either P or Q is the length of the longest increasing (resp. decreasing) subsequence in the permutation.

25.4 Random Permutations and GUE

Baik et al. (2003) showed that for random permutations the distribution of the length of the first row (or column) of P (or of Q) is identical to the distribution of the properly centered and scaled largest eigenvalue of matrices taken from the Gaussian unitary ensemble (GUE); that the distribution of the length of the second row (or column) of P (or of Q) is identical to the distribution of the properly centered and scaled second largest eigenvalue of matrices taken from the GUE; ....

Summary of Chapter 25

The average value of $|\operatorname{tr} U|^{2k}$ for unitary $n \times n$ matrices (U is said to belong to the circular unitary ensemble or CUE) is the number of permutations of $(1, 2, \ldots, k)$ having a longest increasing subsequence of length less than or equal to n. The distribution of the length of the longest increasing subsequence in random permutations (properly centered and scaled) is the same as that of the (properly centered and scaled) largest eigenvalue of random Gaussian hermitian matrices H (H is said to belong to the Gaussian unitary ensemble or GUE).

26

PROBABILITY DENSITIES OF THE DETERMINANTS; GAUSSIAN ENSEMBLES

The question about the distribution of the determinant was asked once by (Wigner, 1965b, p. 459) who gave the mean value of the logarithm of the $n \times n$ determinant $[\delta_{jk} + \varepsilon_{jk}]$ in the limit $n \to \infty$ with ε_{jk} ($\varepsilon_{jk} = \varepsilon_{kj}$, if real and $\varepsilon_{jk} = \varepsilon_{kj}^*$, if complex) of order $1/n$.

In Section 15.4 we studied the probability density of the determinant of a Gaussian random real, complex or quaternion matrix with no symmetry. For this we calculated the Mellin transforms of the determinants using the method of singular values. Now we will study the same for matrices from the three Gaussian ensembles, orthogonal, unitary, and symplectic, i.e., the matrices we will consider now are real symmetric, complex hermitian or quaternion self-dual. For this we will use the method of eigenvalues.

26.1 Introduction

The Mellin transform explores a function only for the real positive variable, so one needs to consider both the even and odd parts of the function to be studied. As usual our method consists in calculating explicitly the Mellin transforms of the even and odd parts of the probability density of the determinant $y = x_1 \cdots x_n$, where $x_1, \ldots, x_n$ are the eigenvalues of the random matrix. For this calculation we use the method explained in Chapter 5 (Sections 5.3–5.5) to compute the expectation value of any function of eigenvalues of the form $\prod_{j=1}^{n} \phi(x_j)$. We then use the inverse Mellin transform to get the even and odd parts of the probability density of the determinant (PDD). Integer moments of this probability density are also deduced.

For the Gaussian ensembles the probability density of the matrix elements will be taken to be

$$F(A) \propto \exp\left(-\frac{\beta}{2} \operatorname{tr} A^2\right), \tag{26.1.1}$$

so that the joint probability density of the eigenvalues $\mathbf{x} := \{x_1, x_2, \ldots, x_n\}$ will be (cf. Chapter 3)

$$F_\beta(\mathbf{x}) := C_{n\beta} \exp\left(-\frac{\beta}{2} \sum_{j=1}^{n} x_j^2\right) |\Delta(\mathbf{x})|^\beta, \tag{26.1.2}$$

$$\Delta(\mathbf{x}) := \prod_{1 \leqslant j < k \leqslant n} (x_k - x_j), \tag{26.1.3}$$

$$C_{n\beta} := (2\pi)^{-n/2} \beta^{n(n-1)\beta/4+n/2} \prod_{j=1}^{n} \frac{\Gamma(1+\beta/2)}{\Gamma(1+j\beta/2)}. \tag{26.1.4}$$

If one takes $F(A) \propto \exp(-a \operatorname{tr} A^2)$, the dependence of the various quantities on the scale factor a can be deduced by a dimensional argument. This dependence is indicated in Table 26.1.

The order of the matrices being n, the PDD will be denoted by $g_n(\beta, y)$,

$$g_n(\beta, y) := \int_{-\infty}^{\infty} F_\beta(\mathbf{x}) \delta(y - x_1 x_2 \cdots x_n)\, dx_1\, dx_2 \cdots dx_n, \tag{26.1.5}$$

for $\beta = 1, 2$ or 4. The Mellin transforms of the even and odd parts of the PDD, i.e., of

$$g_n^{\pm}(\beta, y) := \frac{1}{2}\left[g_n(\beta, y) \pm g_n(\beta, -y)\right] \tag{26.1.6}$$

will be denoted by $\mathcal{M}_n^{\pm}(\beta, s)$

$$\mathcal{M}_n^{\pm}(\beta, s) := \int_0^\infty y^{s-1} g_n^{\pm}(\beta, y)\, dy, \tag{26.1.7}$$

Table 26.1.

$C_{n\beta}(a) = a^{n/2+\beta n(n-1)/4} C_{n\beta}(1)$
$g_n(\beta, y; a) = a^{n/2} g_n(\beta, a^{n/2} y; 1)$
$\mathcal{M}_n^{\pm}(\beta, s; a) = a^{n(1-s)/2} \mathcal{M}_n^{\pm}(\beta, s; 1)$
$M_\beta(n, q; a) = a^{-qn/2} M_\beta(n, q; 1)$

so that

$$\mathcal{M}_n^+(\beta, s) := \int_0^\infty y^{s-1} g_n^+(\beta, y)\, dy \tag{26.1.8}$$

$$= \frac{1}{2}\int_{-\infty}^{\infty} \cdots \int_{-\infty}^{\infty} F_\beta(\mathbf{x}) |x_1 \cdots x_n|^{s-1}\, dx_1 \cdots dx_n \tag{26.1.9}$$

$$= \frac{1}{2}\langle |x_1 \cdots x_n|^{s-1}\rangle, \tag{26.1.10}$$

and similarly

$$\mathcal{M}_n^-(\beta, s) := \int_0^\infty y^{s-1} g_n^-(\beta, y)\, dy \tag{26.1.11}$$

$$= \frac{1}{2}\int_{-\infty}^{\infty} \cdots \int_{-\infty}^{\infty} F_\beta(\mathbf{x})\, \mathrm{sign}(x_1 \cdots x_n) |x_1 \cdots x_n|^{s-1}\, dx_1 \cdots dx_n \tag{26.1.12}$$

$$= \frac{1}{2}\langle \mathrm{sign}(x_1 \cdots x_n) |x_1 \cdots x_n|^{s-1}\rangle. \tag{26.1.13}$$

Thus we can write the above equations together as

$$\mathcal{M}_n^\pm(\beta, s) := \int_0^\infty y^{s-1} g_n^\pm(\beta, y)\, dy \tag{26.1.14}$$

$$= \frac{1}{2}\int_{-\infty}^{\infty} F_\beta(\mathbf{x}) |x_1 \cdots x_n|^{s-1} \varepsilon^\pm(x_1 \cdots x_n)\, dx_1 \cdots dx_n, \tag{26.1.15}$$

with $\varepsilon^+(x) = 1$ and $\varepsilon^-(x) = \mathrm{sign}(x)$.

The moments of the probability density $g_n(\beta, y)$ can be expressed in terms of the Mellin transforms $\mathcal{M}_n^\pm(\beta, s)$ as follows. For $q = 0, 1, \ldots$

$$\begin{aligned} M_\beta(n, q) &:= \int_{-\infty}^\infty g_n(\beta, y) y^q\, dy \\ &= \int_0^\infty \big[g_n(\beta, y) + (-1)^q g_n(\beta, -y)\big] y^q\, dy \\ &= \big[1 + (-1)^q\big] \int_0^\infty g_n^+(\beta, y) y^q\, dy + \big[1 - (-1)^q\big] \int_0^\infty g_n^-(\beta, y) y^q\, dy \\ &= \big[1 + (-1)^q\big] \mathcal{M}_n^+(\beta, q+1) + \big[1 - (-1)^q\big] \mathcal{M}_n^-(\beta, q+1). \end{aligned} \tag{26.1.16}$$

Normalization implies that

$$M_\beta(n,0) = 2\mathcal{M}_n^+(\beta,1) = 1, \tag{26.1.17}$$

and the symmetry implies that

$$M_\beta(2m+1, 2p+1) = 0 = \mathcal{M}_{2m+1}^-(\beta, s), \tag{26.1.18}$$

while Eq. (17.8.5) says that

$$M_\beta(2m,1) = (-1)^m \frac{(2m)!}{2^{2m} m!}. \tag{26.1.19}$$

The last three equations may serve as a check on the calculations.

The value of $g_n(\beta, y)$ and its derivatives at $y = 0$ can be obtained from its Mellin transform $\mathcal{M}_n(\beta, s)$ by

$$g_n(\beta,0) = \lim_{s\to 0}[s\,\mathcal{M}_n(\beta,s)] = \lim_{s\to 0}[s\,\mathcal{M}_n^+(\beta,s)], \tag{26.1.20}$$

$$g_n'(\beta,0) = -\lim_{s\to 0}[s(s-1)\,\mathcal{M}_n(\beta,s-1)], \tag{26.1.21}$$

etc.

An integration over x_n in Eq. (26.1.5) gives

$$g_n(\beta,y) = C_{n\beta}\int_{-\infty}^{\infty}\cdots\int_{-\infty}^{\infty}\exp\left[-(\beta/2)\left(\sum_{j=1}^{n-1}x_j^2 + \frac{y^2}{(x_1\cdots x_{n-1})^2}\right)\right]\frac{1}{|x_1\cdots x_{n-1}|}$$
$$\times\prod_{j=1}^{n-1}\left|x_j - \frac{y}{x_1\cdots x_{n-1}}\right|^\beta \prod_{1\leqslant i<j\leqslant n-1}|x_i - x_j|^\beta\,dx_1\cdots dx_{n-1}. \tag{26.1.22}$$

The integrand in the above equation is the product of an exponential term by a polynomial in $y, x_1, \ldots, x_{n-1}$ and divided by $|x_1\cdots x_{n-1}|^{n\beta-\beta+1}$. When any x_j goes either to infinity or to zero the integrand has a decreasing Gaussian factor. It follows that the integral (26.1.22) is convergent. Furthermore, the integrand is clearly a continuous function of y. Therefore, from Eq. (26.1.22) the probability density $g_n(\beta, y)$ is a continuous bounded function for any real y. From Eq. (26.1.6) the even and odd parts $g_n^\pm(\beta, y)$ of $g_n(\beta, y)$ are also bounded and continuous. Their Mellin transforms $\mathcal{M}_n^\pm(\beta, s)$ are analytic in the right-half complex s-plane $\mathrm{Re}\, s > 0$. Thus they are uniquely determined by the inverse Mellin transform of $\mathcal{M}_n^\pm(\beta, s)$ (Widder, 1971, 5.7 Corollary 7.3a).

Expansions of $g_n(\beta, y)$ near $y = 0$ and $y = \infty$ can be obtained from its Mellin transform, it can be numerically calculated for small values of n and represented graphically.

26.2 Gaussian Unitary Ensemble

If $\Phi(\mathbf{x}) = \prod_{j=1}^{n} \phi(x_j)$, then the average value of $\Phi(\mathbf{x})$ from Chapter 5, Section 5.3 is

$$\begin{aligned}\langle \Phi(\mathbf{x}) \rangle &:= \int_{-\infty}^{\infty} \cdots \int_{-\infty}^{\infty} F_2(\mathbf{x}) \Phi(\mathbf{x})\, dx_1 \cdots dx_n \\ &= C_{n2} n! \det[\Phi_{i,j}(2)]_{i,j=0,\dots,n-1},\end{aligned} \tag{26.2.1}$$

where

$$\Phi_{i,j}(2) := \int_{-\infty}^{\infty} P_i(x) Q_j(x) e^{-x^2} \phi(x)\, dx \tag{26.2.2}$$

and $P_i(x)$ and $Q_i(x)$ are any monic polynomials of degree i. Recall that a polynomial is called monic if the coefficient of its highest power is one.

26.2.1 Mellin Transform of the PDD. From Eq. (26.1.10) the Mellin transform of the even part of $g_n(2, y)$ is

$$\begin{aligned}\mathcal{M}_n^+(2, s) &= \frac{1}{2} \langle |x_1 \cdots x_n|^{s-1} \rangle \\ &= \frac{1}{2} C_{n2} n! \det[\Phi_{i,j}^+(2)]_{i,j=0,\dots,n-1},\end{aligned} \tag{26.2.3}$$

where with $\phi(x) = |x|^{s-1}$ in Eq. (26.2.2), one has

$$\Phi_{i,j}^+(2) := \int_{-\infty}^{\infty} P_i(x) Q_j(x) e^{-x^2} |x|^{s-1}\, dx. \tag{26.2.4}$$

We can choose $P_i(x)$ and $Q_i(x)$ as any monic polynomials of degree i. They can be chosen to make the matrix $[\Phi_{i,j}^+(2)]$ diagonal. But let us take $P_i(x) = Q_i(x) = x^i$. Then

$$\begin{aligned}\Phi_{i,j}^+(2) &:= \int_{-\infty}^{\infty} x^{i+j} |x|^{s-1} e^{-x^2}\, dx, \quad \operatorname{Re} s > 0 \\ &= \begin{cases} \Gamma[(s+i+j)/2], & i+j \text{ even}, \\ 0, & i+j \text{ odd}. \end{cases}\end{aligned} \tag{26.2.5}$$

The alternate elements of the $n \times n$ determinant $[\Phi_{i,j}^+(2)]$ being zero, we can rearrange its rows and columns so as to collect the zero elements separate from the non-zero elements. Thus

$$\begin{aligned}&\det[\Phi_{i,j}^+(2)]_{i,j=0,\dots,n-1} \\ &\quad = \det[\Phi_{2i,2j}^+(2)]_{i,j=0,\dots,[(n-1)/2]} \det[\Phi_{2i+1,2j+1}^+(2)]_{i,j=0,\dots,[(n-2)/2]},\end{aligned} \tag{26.2.6}$$

where $[x]$ denotes the largest integer less than or equal to x. It is straightforward to evaluate the determinants $[\Phi^{+}_{2i,2j}(2)]$ and $[\Phi^{+}_{2i+1,2j+1}(2)]$ (cf. Appendix A.18, Eq. (A.18.7)).

$$\det\left[\Gamma\left(\frac{s}{2}+i+j\right)\right]_{i,j=0,\dots,[(n-1)/2]} = \prod_{j=0}^{[(n-1)/2]}\left[j!\Gamma\left(\frac{s}{2}+j\right)\right], \tag{26.2.7}$$

$$\det\left[\Gamma\left(\frac{s}{2}+i+j+1\right)\right]_{i,j=0,\dots,[n/2]} = \prod_{j=0}^{[n/2]}\left[j!\Gamma\left(\frac{s}{2}+j+1\right)\right]. \tag{26.2.8}$$

Putting in the constants, we get

$$\mathcal{M}^{+}_{n}(2,s) = \frac{1}{2}\prod_{j=1}^{n}\frac{\Gamma(s/2+b^{+}_{j}(2))}{\Gamma(1/2+b^{+}_{j}(2))}, \quad \operatorname{Re} s > 0, \tag{26.2.9}$$

where

$$b^{+}_{j}(2) := [j/2], \quad j = 1, 2, \dots. \tag{26.2.10}$$

Similarly the Mellin transform of the odd part of $g_n(2, y)$ is

$$\begin{aligned}\mathcal{M}^{-}_{n}(2,s) &= \frac{1}{2}\langle \operatorname{sign}(x_1\cdots x_n)|x_1\cdots x_n|^{s-1}\rangle \\ &= \frac{1}{2}C_{n2}n!\det[\Phi^{-}_{i,j}(2)]_{i,j=0,\dots,n-1},\end{aligned} \tag{26.2.11}$$

where with $\phi(x) = \operatorname{sign}(x)|x|^{s-1}$ and $P_i(x) = Q_i(x) = x^i$ in Eq. (26.2.2), one has

$$\begin{aligned}\Phi^{-}_{i,j}(2) &:= \int_{-\infty}^{\infty} x^{i+j}\operatorname{sign}(x)|x|^{s-1}e^{-x^2}\,dx, \quad \operatorname{Re} s > 0 \\ &= \begin{cases}\Gamma[(s+i+j)/2], & i+j \text{ odd}, \\ 0, & i+j \text{ even}.\end{cases}\end{aligned} \tag{26.2.12}$$

In the $n\times n$ determinant $[\Phi^{-}_{i,j}(2)]$, the zero and non-zero elements can again be separated and the two resulting determinants computed. One has

$$\det[\Phi^{-}_{i,j}(2)]_{i,j=0,\dots,n-1} = \begin{cases}(-1)^{n/2}\displaystyle\prod_{j=0}^{n/2-1}\left[j!\Gamma\left(\frac{s+1}{2}+j\right)\right]^2, & n \text{ even}, \\ 0, & n \text{ odd}.\end{cases} \tag{26.2.13}$$

So that the Mellin transform of the odd part of $g_n(2, y)$ is zero for n odd and

$$\mathcal{M}_n^-(2, s) = (-1)^{n/2}\frac{1}{2}\prod_{j=1}^{n}\frac{\Gamma(s/2 + b_j^-(2))}{\Gamma(1/2 + b_j^+(2))}, \quad \operatorname{Re} s > 0, \tag{26.2.14}$$

for n even, where

$$b_j^-(2) := [(j-1)/2] + 1/2, \quad j = 1, 2, \ldots. \tag{26.2.15}$$

26.2.2 Inverse Mellin Transforms. Looking at the tables of integral transforms one finds that the Mellin transform of a Meijer G-function is the ratio of products of gamma functions. In particular (Bateman, 1954, 6.9, (14))

$$\int_0^\infty x^{s-1} G_{0,n}^{n,0}(x \mid b_1, \ldots, b_n)\, dx = \prod_{j=1}^{n}\Gamma(s + b_j). \tag{26.2.16}$$

By a change of x into y^2 and s into $s/2$, this can be written as

$$2\int_0^\infty y^{s-1} G_{0,n}^{n,0}(y^2 \mid b_1, \ldots, b_n)\, dy = \prod_{j=1}^{n}\Gamma\left(\frac{s}{2} + b_j\right). \tag{26.2.17}$$

Comparing this last equation with Eqs. (26.2.9) and (26.2.14), one gets for $y \geqslant 0$

$$g_n^+(2, y) = \text{const}\cdot G_{0,n}^{n,0}(y^2 \mid b_1^+(2), \ldots, b_n^+(2)) \tag{26.2.18}$$

$$g_n^-(2, y) = \begin{cases} (-1)^{n/2}\text{const}\cdot G_{0,n}^{n,0}(y^2 \mid b_1^-(2), \ldots, b_n^-(2)), & n \text{ even}, \\ 0, & n \text{ odd}. \end{cases} \tag{26.2.19}$$

Taking into account the symmetry properties of $g_n^\pm(2, y)$, one obtains for all real y the following results

$$g_{2m+1}(2, y) = N_{2m+1} G_{0,2m+1}^{2m+1,0}(y^2 \mid 0, 1, 1, 2, 2, \ldots, m, m), \tag{26.2.20}$$

$$g_{2m}(2, y) = N_{2m}\left[G_{0,2m}^{2m,0}(y^2 \mid 0, 1, 1, 2, 2, \ldots, m-1, m-1, m)\right.$$
$$\left. + (-1)^m \operatorname{sign}(y) G_{0,2m}^{2m,0}\left(y^2 \,\middle|\, \frac{1}{2}, \frac{1}{2}, \frac{3}{2}, \frac{3}{2}, \ldots, m - \frac{1}{2}, m - \frac{1}{2}\right)\right], \tag{26.2.21}$$

$$N_{2m+1} = 2^{m(2m+1)}\pi^{-(m+1/2)} \prod_{j=0}^{m} \left[\frac{j!}{(2j)!}\right]^2, \tag{26.2.22}$$

$$N_{2m} = 2^{2m^2}\pi^{-m}\frac{(2m)!}{m!} \prod_{j=0}^{m} \left[\frac{j!}{(2j)!}\right]^2, \tag{26.2.23}$$

$$g_{2m+1}(2,0) = 2^{m(2m+1)}\pi^{-(m+1/2)}(m!)^{-2} \prod_{j=0}^{m} \frac{(j!)^4}{((2j)!)^2}, \tag{26.2.24}$$

$$g_{2m}(2,0) = 2^{2m^2}\pi^{-m}\frac{m}{(2m)!} \prod_{j=0}^{m-1} \frac{j!^4}{((2j)!)^2}. \tag{26.2.25}$$

Here $G^{n,0}_{0,n}$ is a Meijer G-function (Luke, 1969, Chapter 5; Bateman, 1953a, 5.3–5.6). The constants N_n and $g_n(2,0)$ were fixed from Eqs. (26.1.17), (26.2.3), (26.2.9), (26.2.10) and (26.2.13). For details see, e.g., Mehta and Normand (1998).

At $y = 0$, $g_n(2, y)$ and its first derivative are continuous and one has

$$g'_{2m+1}(2,0) = 0, \tag{26.2.26}$$

$$g'_{2m}(2,0) = (-1)^m \times \infty. \tag{26.2.27}$$

The non-zero moments of the probability density $g_n(2, y)$ are found to be (cf. Eqs. (26.1.16), (26.2.9) and (26.2.14))

$$M_2(2m+1, 2p) = \frac{\Gamma(p+1/2)}{\Gamma(1/2)} \prod_{j=1}^{m} \left[\frac{\Gamma(p+j+1/2)}{\Gamma(j+1/2)}\right]^2, \tag{26.2.28}$$

$$M_2(2m, 2p+1) = (-1)^m \frac{\Gamma(1/2)}{\Gamma(m+1/2)} \prod_{j=0}^{m-1} \left[\frac{\Gamma(p+j+3/2)}{\Gamma(j+1/2)}\right]^2, \tag{26.2.29}$$

$$M_2(2m, 2p) = \frac{\Gamma(p+m+1/2)\Gamma(1/2)}{\Gamma(p+1/2)\Gamma(m+1/2)} \prod_{j=0}^{m-1} \left[\frac{\Gamma(p+j+1/2)}{\Gamma(j+1/2)}\right]^2. \tag{26.2.30}$$

The behaviour of $g_n(y)$ near $y = 0$ and for large $|y|$ can be determined from that of the Meijer G-functions (cf. Appendix A.52).

For small values of n one may compute $g_n(2, y)$ numerically (Mehta and Normand, 1998); they are plotted in Figure 26.1 for $n = 1, 2, 3$ and 4.

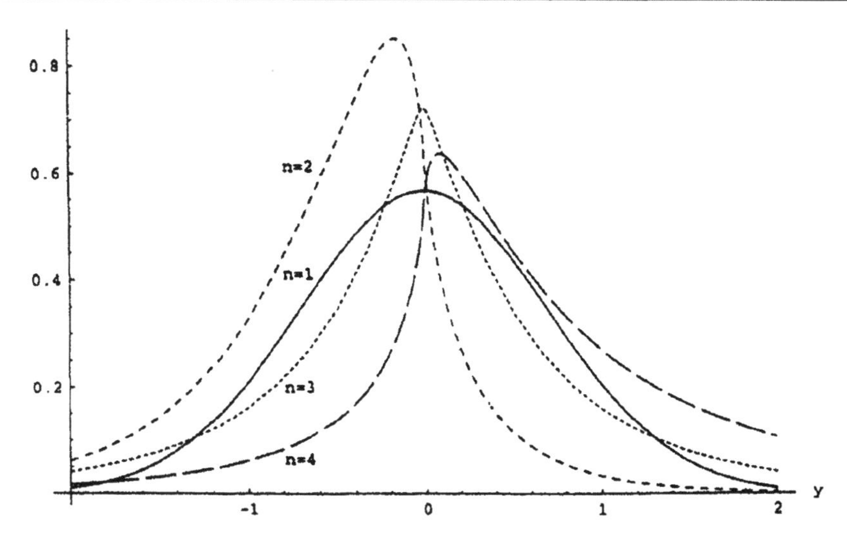

Figure 26.1. The probability densities $g_n(2, y)$ for $n = 1, 2, 3$ and 4. From Mehta and Normand (1998).

26.3 Gaussian Symplectic Ensemble

The probability density for the eigenvalues of a matrix taken from the Gaussian symplectic ensemble is (Chapter 3)

$$F_4(\mathbf{x}) = C_{n4} \exp\left(-2\sum_{j=1}^{n} x_j^2\right) |\Delta(\mathbf{x})|^4. \tag{26.3.1}$$

The Mellin transforms of the even and odd parts of the probability density $g_n(4, y)$ of its determinant $y = x_1 \cdots x_n$ can again be considered

$$\mathcal{M}_n^{\pm}(4, s) = \frac{1}{2}\int_{-\infty}^{\infty} F_4(\mathbf{x})|x_1 \cdots x_n|^{s-1} \varepsilon^{\pm}(x_1 \cdots x_n)\, dx_1 \cdots dx_n, \tag{26.3.2}$$

with $\varepsilon^+(x) = 1$ and $\varepsilon^-(x) = \text{sign}(x)$. So that from Section 5.4 one has

$$\mathcal{M}_n^{\pm}(4, s) = \frac{1}{2} C_{n4} n! \,\text{pf}[\Phi_{ij}^{\pm}(4)]_{i,j=0,1,\ldots,2n-1}, \tag{26.3.3}$$

where the $P_i(x)$ are arbitrary monic polynomials and

$$\Phi_{ij}^{\pm}(4) := \int_{-\infty}^{\infty} e^{-2x^2} |x|^{s-1} \varepsilon^{\pm}(x) \big[P_i(x) P_j'(x) - P_i'(x) P_j(x)\big]\, dx. \tag{26.3.4}$$

Choosing $P_i(x) = x^i$, one sees that

$$\Phi^+_{ij}(4) = \begin{cases} 2^{-(i+j+s-1)/2}(j-i)\Gamma((i+j+s-1)/2), & i+j \text{ odd}, \\ 0, & i+j \text{ even}, \end{cases}$$

$$\mathrm{Re}\, s > 0, \tag{26.3.5}$$

$$\Phi^-_{ij}(4) = \begin{cases} 2^{-(i+j+s-1)/2}(j-i)\Gamma((i+j+s-1)/2), & i+j \text{ even}, \\ 0, & i+j \text{ odd}, \end{cases}$$

$$\mathrm{Re}\, s > 0. \tag{26.3.6}$$

The alternate elements of the matrices $[\Phi^\pm_{ij}(4)]_{i,j=0,1,\ldots,2n-1}$ are zero and they can be collected together by a rearrangement of the rows and columns without changing the Pfaffian, except possibly for the sign which can be fixed at the end. Thus one finds that

$$\mathcal{M}^+_n(4,s) = \frac{1}{2} C_{n4} n! \det\left[2^{-(i+j+s/2)}(2j-2i+1)\Gamma\left(i+j+\frac{s}{2}\right)\right]_{i,j=0,\ldots,n-1}, \tag{26.3.7}$$

$$\begin{aligned}\mathcal{M}^-_n(4,s) = {} & \frac{1}{2}(-1)^{n/2} C_{n4} n! \\ & \times \mathrm{pf}\left[2^{-(i+j+(s-1)/2)}(2j-2i)\Gamma\left(i+j+\frac{s-1}{2}\right)\right]_{i,j=0,1,\ldots,n-1} \\ & \times \mathrm{pf}\left[2^{-(i+j+(s+1)/2)}(2j-2i)\Gamma\left(i+j+\frac{s}{2}\right)\right]_{i,j=0,1,\ldots,n-1}, \quad n \text{ even},\end{aligned} \tag{26.3.8}$$

$$\mathcal{M}^-_n(4,s) = 0, \quad n \text{ odd}, \tag{26.3.9}$$

so that from Appendix A.18 for the even part

$$\begin{aligned}\mathcal{M}^+_n(4,s) = {} & \frac{1}{2} C_{n4} n! 2^{-n(n-1)-ns/2} 2^n \prod_{i=0}^{n-1} i!\Gamma\left(\frac{s}{2}+i\right) \\ & \times \sum_{k=0}^{n} (-1)^k \binom{n}{k} \left(\frac{s-1}{4}\right)_k \left(\frac{s+1}{4}\right)_{n-k} \\ = {} & \frac{1}{2} 2^{n(1-s)/2} \prod_{i=0}^{n-1} \frac{\Gamma(i+s/2)}{\Gamma(i+3/2)} \sum_{k=0}^{n} (-1)^k \binom{n}{k} \left(\frac{s-1}{4}\right)_k \left(\frac{s+1}{4}\right)_{n-k},\end{aligned} \tag{26.3.10}$$

with the Pochhammer symbol

$$(a)_0 = 1, \quad (a)_k = \Gamma(a+k)/\Gamma(a). \tag{26.3.11}$$

For the odd part, $\mathcal{M}^-_{2m+1}(4,s) = 0$, and if $n = 2m$ is even, one has

$$\begin{aligned} \mathcal{M}^-_n(4,s) &= \frac{1}{2}(-1)^{n/2} C_{n4} n! 2^{-n(n-1)+n-ns/2} \\ &\quad \times \operatorname{pf}\left[(j-i)\Gamma\left(j+i+\frac{s-1}{2}\right)\right]_{i,j=0,\ldots,n-1} \\ &\quad \times \operatorname{pf}\left[(j-i)\Gamma\left(j+i+\frac{s+1}{2}\right)\right]_{i,j=0,\ldots,n-1}. \end{aligned} \tag{26.3.12}$$

Putting in the value of C_{n4} from Eq. (26.1.4) and that of the Pfaffians from (A.18.21) this simplifies to

$$\mathcal{M}^-_{2m}(4,s) = \frac{1}{2}(-1)^m \frac{(2m)!}{2^{ms} m!} \prod_{j=1}^{2m} \frac{\Gamma(j+(s-1)/2)}{\Gamma(j+1/2)!}. \tag{26.3.13}$$

Thus $g^+_n(4,y)$, the inverse Mellin transform of $\mathcal{M}^+_n(4,s)$, is a linear combination of Meijer G-functions, cumbersome to write; while $g^-_n(4,y)$ for $n = 2m$ even is the Meijer G-function

$$g^-_{2m}(4,y) = (-1)^m 2^{2m^2+m-1} \pi^{-m} \frac{(2m)!}{m!} \prod_{j=1}^{2m} \frac{j!}{(2j)!} G^{2m,0}_{0,2m}\left(2^m y^2 \,\middle|\, \frac{1}{2}, \frac{3}{2}, \ldots, \frac{4m-1}{2}\right), \tag{26.3.14}$$

and $g^-_n(4,y)$ for $n = 2m+1$ odd is zero.

From Eqs. (26.1.20) and (26.3.10) we deduce the value of $g_n(4,0)$ as

$$g_1(4,0) = \sqrt{\frac{2}{\pi}}, \tag{26.3.15}$$

and for $n > 1$,

$$\begin{aligned} g_n(4,0) &= \lim_{s\to 0}[s\,\mathcal{M}_n(4,s)] = \lim_{s\to 0}[s\,\mathcal{M}^+_n(4,s)] \\ &= \frac{2^{(n+2)/2}}{\sqrt{\pi}} \prod_{j=1}^{n-1} \frac{\Gamma(j)}{\Gamma(j+3/2)} \sum_{k=0}^{n} (-1)^k \binom{n}{k} \left(-\frac{1}{4}\right)_k \left(\frac{1}{4}\right)_{n-k}. \end{aligned} \tag{26.3.16}$$

Non-zero integer moments of $g_n(4, y)$ are

$$M_4(2m, 2p+1) = (-1)^m \frac{(2m)!}{2^{2m(p+1)} m!} \prod_{j=1}^{2m} \frac{\Gamma(p+j+1/2)}{\Gamma(j+1/2)}, \tag{26.3.17}$$

$$M_4(n, 2p) = 2^{-np} \prod_{j=0}^{n-1} \frac{\Gamma(j+p+1/2)}{\Gamma(j+3/2)} \sum_{k=0}^{n} (-1)^k \binom{n}{k} \left(\frac{p}{2}\right)_k \left(\frac{p+1}{2}\right)_{n-k}. \tag{26.3.18}$$

26.4 Gaussian Orthogonal Ensemble

The probability density for the eigenvalues of a matrix taken from the Gaussian orthogonal ensemble is (Chapter 3)

$$F_1(\mathbf{x}) = C_{n1} \exp\left(-\frac{1}{2}\sum_{j=1}^{n} x_j^2\right) |\Delta(\mathbf{x})|. \tag{26.4.1}$$

As in Section 26.1 above and Section 5.5 the Mellin transform of the even and odd parts of the PDD is a Pfaffian of an anti-symmetric matrix of order n or $n+1$ whichever is even, with elements

$$\Phi_{i,j}^{\pm}(1) = \int_{-\infty<x\leqslant y<\infty} dx\, dy\, e^{-(x^2+y^2)/2} [P_i(x)P_j(y) - P_j(x)P_i(y)] |xy|^{s-1} \varepsilon^{\pm}(xy), \tag{26.4.2}$$

$$a_i^{\pm} = \int_{-\infty}^{\infty} dx\, e^{-x^2/2} P_i(x) |x|^{s-1} \varepsilon^{\pm}(x), \tag{26.4.3}$$

$$\varepsilon^{+}(x) = 1, \quad \varepsilon^{-}(x) = \text{sign}(x), \tag{26.4.4}$$

and arbitrary monic polynomials $P_i(x)$. Choosing $P_i(x) = x^i$, one sees that $\Phi_{i,j}^{\pm}(1) = 0$ if $i+j$ is even, $a_{2i+1}^{+} = a_{2i}^{-} = 0$, and the Pfaffian simplifies to a determinant. Thus when n is even, $n = 2m$,

$$\mathcal{M}_n^{\pm}(1, s) = \frac{1}{2} C_{n1} n! \det[\Phi_{2i,2j+1}^{\pm}(1)]_{i,j=0,1,\ldots,m-1}, \tag{26.4.5}$$

and when n is odd, $n = 2m+1$,

$$\mathcal{M}_n^{\pm}(1, s) = \frac{1}{2} C_{n1} n! \det[\Phi_{2i,2j+1}^{\pm}(1) \quad a_{2i}^{\pm}]_{i=0,1,\ldots,m;\, j=0,1,\ldots,m-1}, \tag{26.4.6}$$

where

$$\Phi^{\pm}_{2i,2j+1}(1) = 2\int_{-\infty<x\leqslant y<\infty} dx\,dy\,e^{-(x^2+y^2)/2}|xy|^{s-1}x^{2i}y^{2j+1}\varepsilon^{\pm}(xy), \quad (26.4.7)$$

$$a^{\pm}_{2i} = \int_{-\infty}^{\infty} dx\,e^{-x^2/2}x^{2i}|x|^{s-1}\varepsilon^{\pm}(x). \quad (26.4.8)$$

One immediately notices that $a^{-}_{2i} = 0$, and so $\mathcal{M}^{-}_{n}(1,s) = 0$ for n odd. Now splitting the domain of integration $-\infty < x \leqslant y < \infty$ in three parts $-\infty < x \leqslant y \leqslant 0$, $-\infty < x \leqslant 0 \leqslant y < \infty$, and $0 \leqslant x \leqslant y < \infty$, changing variables from x, y to $-y$, $-x$ in the first part and rearranging one gets

$$\Phi^{\pm}_{2i,2j+1}(1) = 2\psi_{i,j} \pm \frac{1}{2}\eta_{2i}\eta_{2j+1}, \quad (26.4.9)$$

$$\psi_{i,j} = \int_{0\leqslant x\leqslant y<\infty} dx\,dy\,e^{-(1/2)(x^2+y^2)}(xy)^{s-1}(x^{2i}y^{2j+1} - x^{2j+1}y^{2i}), \quad (26.4.10)$$

$$\eta_j = 2^{(j+s)/2}\Gamma\left(\frac{j+s}{2}\right). \quad (26.4.11)$$

An integration by parts gives

$$\psi_{i+1,j} = (2i+s)\psi_{i,j} - \Gamma\left(i+j+s+\frac{1}{2}\right), \quad (26.4.12)$$

$$\psi_{i,j+1} = (2j+s+1)\psi_{i,j} + \Gamma\left(i+j+s+\frac{1}{2}\right). \quad (26.4.13)$$

Iteration of these equations give

$$\psi_{i,j} = \frac{\Gamma(i+s/2)}{\Gamma(s/2)}2^{i}\left[\psi_{0,j} - \Gamma\left(\frac{s}{2}\right)\sum_{k=0}^{i-1}\frac{\Gamma(k+j+s+1/2)}{2^{k+1}\Gamma(k+1+s/2)}\right] \quad (26.4.14)$$

$$= \frac{\Gamma(j+(s+1)/2)}{\Gamma((s+1)/2)}2^{j}\left[\psi_{i,0} + \Gamma\left(\frac{s+1}{2}\right)\sum_{k=0}^{j-1}\frac{\Gamma\, k+i+s+1/2)}{2^{k+1}\Gamma(k+1+(s+1)/2)}\right]. \quad (26.4.15)$$

Also

$$\eta_{i+2} = (i+s)\eta_i. \quad (26.4.16)$$

26.5 Gaussian Orthogonal Ensemble. Case $n = 2m + 1$ Odd

For $n = 2m + 1$ odd, Eq. (26.4.6), subtracting a suitable multiple of the ith row from the $(i + 1)$st for $i = m - 1, m - 2, \ldots, 0$, the last column can be replaced by zeros except for one element a_0^+. Thus for $n = 2m + 1$ odd, Eqs. (26.4.6) and (26.4.12) give

$$\mathcal{M}_n^+(1, s) = \frac{1}{2} C_{n1} n! a_0^+ 2^m \det\left[\Gamma\left(i + j + s + \frac{1}{2}\right)\right]_{i,j=0,1,\ldots,m-1}. \tag{26.5.1}$$

Substituting the values of a_0^+ and $\det[\Gamma(i + j + s + 1/2)]$ from Eqs. (26.4.8) and (A.18.7) and fixing the constant from the condition $\mathcal{M}_n(1, 1) = \mathcal{M}_n^+(1, 1) = 1/2$ (or substituting the values of the constants), one has

$$\mathcal{M}_{2m+1}^+(1, s) = \frac{1}{2} \Gamma\left(\frac{s}{2}\right) 2^{s/2} (2\pi)^{-1/2} \prod_{i=0}^{m-1} \frac{\Gamma(i + s + 1/2)}{\Gamma(i + 3/2)}. \tag{26.5.2}$$

Making use of the identity $\sqrt{\pi}\,\Gamma(2z) = 2^{2z-1}\Gamma(z)\Gamma(z + 1/2)$, one can write the last equation as

$$\mathcal{M}_{2m+1}^+(1, s) = \frac{1}{2} 2^{(2m+1)(s-1)/2} \Gamma\left(\frac{s}{2}\right) \pi^{-1/2} \prod_{j=1}^{2m} \frac{\Gamma(s/2 + b_j^+(1))}{\Gamma(1/2 + b_j^+(1))}, \tag{26.5.3}$$

with

$$b_j^+(1) = \frac{1}{2}\left[\frac{j}{2}\right] + \frac{1}{4}, \quad j \geqslant 1. \tag{26.5.4}$$

The inverse Mellin transform of $\mathcal{M}_n^+(1, s)$ for $n = 2m + 1$ odd is the Meijer G-function

$$\begin{aligned} g_{2m+1}(1, y) &= g_{2m+1}^+(1, y) \\ &= N \cdot G_{0,2m+1}^{2m+1,0}(2^{-2m-1} y^2 \mid 0, b_1^+(1), \ldots, b_{2m}^+(1)), \end{aligned} \tag{26.5.5}$$

$$N = \pi^{m-1/2} 2^{-2m^2-2m+1/2} \prod_{j=1}^{m} \frac{(2j)!}{j!}. \tag{26.5.6}$$

For example,

$$g_1(1, y) = \frac{1}{\sqrt{2\pi}} G_{(0,1)}^{(1,0)}\left(\frac{y^2}{2} \middle| 0\right) = \frac{1}{\sqrt{2\pi}} \exp(-y^2/2), \tag{26.5.7}$$

$$g_3(1, y) = \pi^{-3/2} G_{(0,3)}^{(3,0)}\left(\frac{y^2}{8} \middle| 0, \frac{1}{4}, \frac{3}{4}\right), \tag{26.5.8}$$

$$G^{(3,0)}_{(0,3)}\left(\frac{y^2}{8}\,\middle|\,0, \frac{1}{4}, \frac{3}{4}\right) = \sum_{j=0}^{\infty} \frac{(-y^2/8)^j}{j!}\left[\Gamma\left(\frac{1}{4}-j\right)\Gamma\left(\frac{3}{4}-j\right)\right.$$
$$+\Gamma\left(-\frac{1}{4}-j\right)\Gamma\left(\frac{1}{2}-j\right)(y^2/8)^{1/4}$$
$$\left.+\Gamma\left(-\frac{3}{4}-j\right)\Gamma\left(-\frac{1}{2}-j\right)(y^2/8)^{3/4}\right]$$
$$= \pi\sqrt{2} - 4\sqrt{\pi}\,\Gamma(3/4)8^{-1/4}y^{1/2}$$
$$+\frac{2}{3}\sqrt{\pi}\,\Gamma(1/4)8^{-3/4}y^{3/2} + O(y^2). \tag{26.5.9}$$

The value at $y = 0$ is

$$g_{2m+1}(1,0) = \lim_{s\to 0}[s\mathcal{M}^{+}_{2m+1}(1,s)] = \frac{1}{\sqrt{2\pi}}\frac{2^{2m}m!}{(2m)!}. \tag{26.5.10}$$

Integer moments of $g_{2m+1}(1, y)$ are

$$M_1(2m+1, 2p) = 2\mathcal{M}^{+}_{2m+1}(1, 2p+1)$$
$$= 2^p\Gamma(p+1/2)\pi^{-1/2}\prod_{j=1}^{m}\frac{\Gamma(2p+j+1/2)}{\Gamma(j+1/2)}$$
$$= \frac{(2p)!}{2^p p!}\prod_{j=1}^{m}\frac{\Gamma(2p+j+1/2)}{\Gamma(j+1/2)} = \frac{(2p)!}{2^p p!}\prod_{j=1}^{m}\prod_{k=1}^{2p}(j+k-1/2)$$
$$= \frac{(2p)!}{2^p p!}\prod_{k=1}^{2p}\frac{\Gamma(m+k+1/2)}{\Gamma(k+1/2)} \tag{26.5.11}$$

and $M_1(2m+1, 2p+1) = 0$.

26.6 Gaussian Orthogonal Ensemble. Case $n = 2m$ Even

This case is more complicated. For $n = 2m$ even, we have from Eq. (26.4.9)

$$\det\left[\Phi^{\pm}_{2i,2j+1}(1)\right]_{i,j=0,1,\ldots,m-1} = \det\left[2\psi_{i,j} \pm \frac{1}{2}\eta_{2i}\eta_{2j+1}\right]_{i,j=0,\ldots,m-1} \tag{26.6.1}$$
$$= \det\begin{bmatrix} 1 & \eta_{2j+1} \\ 0 & 2\psi_{i,j} \pm \frac{1}{2}\eta_{2i}\eta_{2j+1}\end{bmatrix}_{i,j=0,1,\ldots,m-1}$$

$$= 2^m \det \begin{bmatrix} 1 & \frac{1}{2}\eta_{2j+1} \\ \mp\frac{1}{2}\eta_{2i} & \psi_{i,j} \end{bmatrix}_{i,j=0,\dots,m-1} . \tag{26.6.2}$$

The determinant in the last equation can almost be triangulated by using either Eq. (26.4.12) or (26.4.13). No suitable compact form of $\mathcal{M}_{2m}^{\pm}(1,s)$ is known whose inverse Mellin transform can be readily written down for all m.

For the particular case $n = 2$, one finds

$$\mathcal{M}_2^{\pm}(1,s) = \frac{1}{4\sqrt{\pi}}\Phi_{0,1}^{\pm} = \frac{1}{2\sqrt{\pi}}\left[\psi_{0,0} \pm \sqrt{\frac{\pi}{2}}\Gamma(s)\right]. \tag{26.6.3}$$

But the PDD for the 2×2 matrices from the Gaussian orthogonal ensemble can be computed directly:

$$g_2(1,y) := \frac{1}{4\sqrt{\pi}}\int_{-\infty}^{\infty} dx_1\, dx_2\, |x_1 - x_2| e^{-(x_1^2+x_2^2)/2}\delta(y - x_1x_2). \tag{26.6.4}$$

A change of variables

$$u_1 = x_1 + x_2, \quad u_2 = x_1x_2, \quad \frac{\partial(u_1,u_2)}{\partial(x_1,x_2)} = x_1 - x_2, \tag{26.6.5}$$

gives

$$g_2(1,y) = \frac{2}{4\sqrt{\pi}}\int du_1\, du_2\, e^{-u_1^2/2}e^{u_2}\delta(y - u_2), \tag{26.6.6}$$

and one should be careful about the limits of integration. A factor 2 appears on the right-hand side of the above equation because u_1, u_2 determine x_1, x_2 only up to a permutation. From $u_1^2 - 4u_2 = (x_1 - x_2)^2 \geqslant 0$, one sees that if $y \leqslant 0$, u_1 varies freely from $-\infty$ to $+\infty$ and if $y \geqslant 0$, the interval $(-2\sqrt{u_2}, +2\sqrt{u_2})$ is excluded for u_1. Thus

$$g_2(1,y) = \begin{cases} \dfrac{1}{2\sqrt{\pi}}e^y \displaystyle\int_{-\infty}^{\infty} e^{-u_1^2/2}\, du_1 = \frac{1}{\sqrt{2}}e^y, & y \leqslant 0, \\ \dfrac{2}{2\sqrt{\pi}}e^y \displaystyle\int_{2\sqrt{y}}^{\infty} e^{-u_1^2/2}\, du_1 = \frac{1}{\sqrt{2}}e^y \operatorname{erfc}\left(\sqrt{2y}\right), & y \geqslant 0 \end{cases} \tag{26.6.7}$$

with the error function

$$\operatorname{erfc}(x) := \frac{2}{\sqrt{\pi}}\int_x^{\infty} e^{-t^2}\, dt. \tag{26.6.8}$$

From the last equation one gets the integer moments in a straightforward way

$$\begin{aligned} M_1(2,q) &= \int_{-\infty}^{\infty} g_2(1,y) y^q \, dy \\ &= \frac{1}{\sqrt{2}} \int_{-\infty}^{0} e^y y^q \, dy + \frac{1}{\sqrt{2}} \int_0^{\infty} e^y y^q \operatorname{erfc}(\sqrt{2y}) \, dy. \end{aligned} \tag{26.6.9}$$

The first integral is trivial, while for the second an integration by parts gives

$$\begin{aligned} \int_0^{\infty} e^y y^q \operatorname{erfc}(\sqrt{2y}) \, dy &= \Big[\operatorname{erfc}(\sqrt{2y})(-1)^q q! e^y e_q(y) \Big]_0^{\infty} \\ &\quad + (-1)^q q! \sqrt{\frac{2}{\pi}} \int_0^{\infty} e^{-y} y^{-1/2} e_q(y) \, dy, \end{aligned} \tag{26.6.10}$$

where $e_q(y)$ denotes the finite sum of $q+1$ terms

$$e_q(y) = \sum_{r=0}^{q} (-1)^r \frac{y^r}{r!}. \tag{26.6.11}$$

Hence

$$\begin{aligned} \int_0^{\infty} e^y y^q \operatorname{erfc}(\sqrt{2y}) \, dy &= (-1)^q q! \left[-1 + \sqrt{\frac{2}{\pi}} \sum_{r=0}^{q} (-1)^r \frac{\Gamma(r+1/2)}{r!} \right] \\ &= (-1)^q q! \left[-1 + \sqrt{2} \sum_{r=0}^{q} (-1)^r \frac{(2r)!}{2^{2r}(r!)^2} \right], \end{aligned} \tag{26.6.12}$$

and therefore

$$M_1(2,q) = (-1)^q \, q! T_q, \tag{26.6.13}$$

$$T_q := \sum_{r=0}^{q} (-1)^r \frac{(2r)!}{2^{2r}(r!)^2}. \tag{26.6.14}$$

Delannay and Le Caër (2000) give the integer moments of the PDD for all even n as finite hypergeometric series

$$M_1(2m, 2p) = \frac{2^p}{\sqrt{\pi}} \frac{\Gamma(m+1/2)\Gamma(p+1/2)}{\Gamma(m+p+1/2)}$$

$$\times \prod_{j=1}^{2p} \frac{\Gamma(j+m+1/2)}{\Gamma(j+1/2)} F\left(p+\frac{1}{2}, -p; m+p+\frac{1}{2}; \frac{1}{2}\right), \tag{26.6.15}$$

$$M_1(2m, 2p+1) = \frac{2^p}{\sqrt{\pi}} \frac{\Gamma(m+1/2)\Gamma(p+3/2)}{\Gamma(m+p+3/2)}$$

$$\times \prod_{j=1}^{2p+1} \frac{\Gamma(j+m+1/2)}{\Gamma(j+1/2)} F\left(p+\frac{3}{2}, -p; m+p+\frac{3}{2}; \frac{1}{2}\right). \tag{26.6.16}$$

Summary of Chapter 26

The probability density of the determinant (PDD) of a random matrix taken from the three Gaussian ensembles GOE, GUE and GSE is investigated by first calculating their Mellin transforms using the joint probability density of their eigenvalues. The inverse Mellin transforms then give the PDDs themselves.

For the Gaussian unitary ensemble the PDD is a Meijer G-function or a linear combination of two Meijer G-functions depending on whether n, the size of the matrix, is odd or even. For the Gaussian orthogonal ensemble the PDD is a Meijer G-function if n, the size of the matrix, is odd. When n is even, the PDD is more complicated. The PDD of a 2×2 matrix can be explicitly calculated (cf. Eq. (26.6.7)). For the Gaussian symplectic ensemble the PDD is in all the cases a complicated combination of several Meijer G-functions. The integer moments of the PDD are known in all the cases, Eqs. (26.1.18), (26.2.28)–(26.2.30), (26.3.17)–(26.3.18), (26.5.11), and (26.6.15)–(26.6.16).

27

RESTRICTED TRACE ENSEMBLES

As mentioned towards the end of Chapter 2, Gaussian ensembles are unsatisfactory because the various matrix elements $H_{ij}^{(\lambda)}$ are not equally weighted. Apart from Dyson's method, efforts have been made to equalize this weighting in a straightforward manner. For example, by diagonalizing on a computer a large number of random matrices (Porter and Rosenzweig, 1960a), the elements of which can be made to conform to a given probability law, we can learn a lot about their eigenvalue distributions. Such knowledge, although useful, is purely empirical, and we restrict ourselves in this chapter to only those cases in which these empirical findings can be put on a firmer footing.

27.1 Fixed Trace Ensemble; Equivalence of Moments

When working with large but finite dimensional Hermitian matrices, we cannot allow the elements to grow indefinitely because then one would be unable to normalize. Gaussian ensembles overcome this difficulty by giving exponentially vanishing weights to large values of matrix elements. Another method will be to apply a cut-off. Proceeding from the analogy of a fixed energy in classical statistical mechanics, Rosenzweig defines (1963) his "fixed trace" ensemble by the requirement that the trace of H^2 be fixed to a number r^2 with no other constraint. The number r is called the strength of the ensemble. The joint probability density function for the matrix elements of H is therefore given by

$$P_r(H) = K_r^{-1} \delta\left[\frac{1}{r^2} \operatorname{tr} H^2 - 1\right], \tag{27.1.1}$$

with

$$K_r = \int_{-\infty}^{\infty} \cdots \int_{-\infty}^{\infty} \delta\left(\frac{1}{r^2} \operatorname{tr} H^2 - 1\right) \prod_{\lambda} \prod_{i \leqslant j} dH_{ij}^{(\lambda)}. \tag{27.1.2}$$

This probability density function is invariant under a change of basis

$$H' = W^R H W, \quad W^R W = 1, \tag{27.1.3}$$

where W is an orthogonal, unitary, or symplectic matrix according to the three possibilities noted in Chapter 2, and W^R is the transpose, Hermitian conjugate or the dual of W. This is evident from the fact that under such a transformation the volume element $dH = \prod_{\lambda} \prod_{i \leqslant j} dH_{ij}^{(\lambda)}$ and the quantity $\operatorname{tr} H^2$ are invariant.

The important thing to be noted about these ensembles is their moment equivalence with Gaussian ensembles of large dimensions. More precisely, if we choose the constant a in (2.6.18) to give

$$\langle \operatorname{tr} H^2 \rangle_G \approx K_G^{-1} \int_{-\infty}^{\infty} \operatorname{tr} H^2 e^{-a \operatorname{tr} H^2} dH = r^2, \tag{27.1.4}$$

then for any fixed value of the sum

$$s = \sum_{\lambda} \sum_{i \leqslant j} \eta_{ij}^{(\lambda)}, \quad \eta_{ij}^{(\lambda)} \geqslant 0, \tag{27.1.5}$$

the ratio of the moments

$$M_r(N, \eta) = \left\langle \prod_{\lambda} \prod_{i \leqslant j} (H_{ij}^{(\lambda)})^{\eta_{ij}^{(\lambda)}} \right\rangle_r \tag{27.1.6}$$

and

$$M_G(N, \eta) = \left\langle \prod_{\lambda} \prod_{i \leqslant j} (H_{ij}^{(\lambda)})^{\eta_{ij}^{(\lambda)}} \right\rangle_G \tag{27.1.7}$$

tends to unity as the number of dimensions N tends to infinity. The subscripts r and G denote that the average is taken in the fixed trace and Gaussian ensembles, respectively.

Notice the analogy with the assumption

$$\langle E \rangle_{\text{grand canonical}} = E_{\text{canonical}}, \tag{27.1.8}$$

made in classical statistical mechanics to prove the equivalence there.

From (3.3.8)–(3.3.10) we get, with a little manipulation,

$$K_G = a^{-N/2-\beta N(N-1)/4} C_{N\beta}, \tag{27.1.9}$$

so that a partial differentiation with respect to a gives

$$\langle \operatorname{tr} H^2 \rangle_G = (N/2 + \beta N(N-1)4) a^{-1}. \tag{27.1.10}$$

Therefore we make the choice

$$a = (2r^2)^{-1} N[1 + \beta(N-1)/2]. \tag{27.1.11}$$

Next, to calculate $M_r(N, \eta)$, put

$$H_{ij}^{(\lambda)} = a^{1/2} r h_{ij}^{(\lambda)} \xi^{-1/2}, \tag{27.1.12}$$

where ξ is a parameter. This gives

$$\begin{aligned} & M_r(N, \eta) \left(\frac{\xi}{ar^2} \right)^{N/2+\beta N(N-1)/4+s/2} \\ & = K_r^{-1} \int_{-\infty}^{\infty} \cdots \int_{-\infty}^{\infty} \delta\left(\frac{a}{\xi} \operatorname{tr} h^2 - 1 \right) \prod_{\lambda} \prod_{i \leqslant j} \left[\left(h_{ij}^{(\lambda)} \right)^{\eta_{ij}^{(\lambda)}} dh_{ij}^{(\lambda)} \right]. \end{aligned} \tag{27.1.13}$$

Multiplying both sides by $e^{-\xi}$ and integrating (first!) on ξ from 0 to ∞, we get

$$\begin{aligned} & M_r(N, \eta) \Gamma(L + s/2) L^{-L-s/2} \\ & = K_r^{-1} \int_{-\infty}^{\infty} \cdots \int_{-\infty}^{\infty} e^{-a \operatorname{tr} h^2} \prod_{\lambda} \prod_{i \leqslant j} \left[\left(h_{ij}^{(\lambda)} \right)^{\eta_{ij}^{(\lambda)}} dh_{ij}^{(\lambda)} \right], \end{aligned} \tag{27.1.14}$$

where we have put

$$L = ar^2 = N/2 + \beta N(N-1)/4. \tag{27.1.15}$$

Or

$$M_r(N, \eta) = \frac{L^{L+s/2}}{\Gamma(L + s/2)} \frac{K_G}{K_r} M_G(N, \eta). \tag{27.1.16}$$

Setting $\eta_{ij}^{(\lambda)} = 0$ in the above and using the normalization condition $M_r(N, 0) = M_G(N, 0) = 1$, we get the ratio of the constants K_G and K_r. Substituting this ratio we then obtain

$$M_r(N, \eta) = \frac{L^{s/2} \Gamma(L)}{\Gamma(L + s/2)} M_G(N, \eta). \tag{27.1.17}$$

As $N \to \infty$, $L \to \infty$, and we can use Stirling's formula for the gamma functions

$$\Gamma(x) = x^{x-1/2} e^{-x} (2\pi)^{1/2} [1 + O(1/x)],$$

to prove the asymptotic equality of all the finite moments $s \ll N$.

It is not very clear whether this moment equivalence implies that all local statistical properties of the eigenvalues in the two sets of ensembles are identical. This is so because these local properties of eigenvalues may not be expressible only in terms of finite moments of the matrix elements.

27.2 Probability Density of the Determinant

As in Section 26.1 one may calculate the Mellin transforms of the even and odd parts of the probability density of the determinant (PDD) for matrices in the fixed trance ensembles

$$M_F^{\pm}(\beta, s, r) = \int_D \prod_{i=1}^{N} [|x_i|^{s-1} \varepsilon^{\pm}(x_i)] |\Delta(x)|^{\beta} \prod_{i=1}^{N} dx_i, \tag{27.2.1}$$

where the domain D is $\sum_{i=1}^{N} x_i^2 = r^2$, $\Delta(x)$ is the product of differences

$$\Delta(x) = \prod_{1 \leqslant i < j \leqslant N} (x_j - x_i), \tag{27.2.2}$$

$\varepsilon^+(x) = 1$, $\varepsilon_-(x) = \text{sign}(x)$ and $\beta = 1$, 2 or 4 according as the matrices are real symmetric, complex Hermitian or quaternion self-dual.

By a change of scale, $x_i = r y_i$, one has

$$M_F^{\pm}(\beta, s, r) = r^{Ns+N(N-1)\beta/2} M_F^{\pm}(\beta, s, 1). \tag{27.2.3}$$

As observed by Le Caër and Delannay (2003) these $M_F^{\pm}$ in turn are related to the Mellin transforms of the PDD of matrices from the corresponding Gaussian ensembles

$$M_G^{\pm}(\beta, s) = \int_{-\infty}^{\infty} \cdots \int_{-\infty}^{\infty} \prod_{i=1}^{N} [|x_i|^{s-1} \varepsilon^{\pm}(x_i)] |\Delta(x)|^{\beta} \prod_{i=1}^{N} \exp(-x_i^2)\, dx_i. \tag{27.2.4}$$

In fact, a change of variables from Cartesian to polar coordinates in (27.2.4) shows that

$$\begin{aligned} M_G^{\pm}(\beta, s) &= M_F^{\pm}(\beta, s, 1) \int_0^{\infty} e^{-r^2} r^{Ns+N(N-1)\beta/2-1}\, dr \\ &= \frac{1}{2} \Gamma\left(\frac{1}{2} Ns + \frac{1}{4} N(N-1)\beta\right) M_F^{\pm}(\beta, s, 1). \end{aligned} \tag{27.2.5}$$

Therefore

$$M_F^{\pm}(\beta, s, r) = \frac{2r^{Ns+N(N-1)\beta/2}}{\Gamma(\frac{1}{2}Ns + \frac{1}{4}N(N-1)\beta)} M_G^{\pm}(s, \beta). \tag{27.2.6}$$

Hence if one knows the Mellin transform of the probability density of the determinant (PDD) for matrices in the Gaussian ensembles, one deduces the same for matrices in the corresponding fixed trace ensembles.

Thus from Eqs. (26.2.9), (26.2.14), (27.2.6) and

$$\Gamma(Nz) = (2\pi)^{(1-N)/2} N^{Nz-1/2} \prod_{j=1}^{N} \Gamma\left(z + \frac{j-1}{2}\right), \tag{27.2.7}$$

we get for $\beta = 2$,

$$M_F^{+}(2, s, r) = \text{const} \cdot r^{Ns} N^{-s/2} \prod_{j=1}^{N} \frac{\Gamma(s/2 + b_j^{+}(2))}{\Gamma(s/2 + a_j(2))}, \tag{27.2.8}$$

$$b_j^{+}(2) = \left[\frac{j}{2}\right], \quad a_j(2) = \frac{N-1}{2} + \frac{j-1}{N}, \quad 1 \leqslant j \leqslant N, \tag{27.2.9}$$

$M_F^{-}(2, s, r) = 0$ for N odd, and

$$M_F^{-}(2, s, r) = \text{const} \cdot (-1)^{N/2} r^{Ns} N^{-s/2} \prod_{j=1}^{N} \frac{\Gamma(s/2 + b_j^{-}(2))}{\Gamma(s/2 + a_j(2))}, \tag{27.2.10}$$

for N even. So that the PDD for fixed trace complex Hermitian matrices is a Meijer G-function for N odd and a linear combination of two Meijer G-functions for N even,

$$P_{F(2m+1)}(2, y) = \text{const} \cdot G_{2m+1,2m+1}^{2m+1,0}\left(\frac{r^{4m+2}y^2}{2m+1} \middle| \begin{matrix} 0, 1, 1, \ldots, m, m \\ a_1, a_2, \ldots, a_{2m+1} \end{matrix}\right), \tag{27.2.11}$$

$$\begin{aligned} P_{F(2m)}(2, y) = \text{const} \cdot \Bigg[& G_{2m,2m}^{2m,0}\left(\frac{r^{4m}y^2}{2m} \middle| \begin{matrix} 0, 1, 1, \ldots, m-1, m-1, m \\ a_1, a_2, \ldots, a_{2m} \end{matrix}\right) \\ & + (-1)^m G_{2m,2m}^{2m,0}\left(\frac{r^{2m}y^2}{2m} \middle| \begin{matrix} \frac{1}{2}, \frac{1}{2}, \frac{3}{2}, \frac{3}{2}, \ldots, m-\frac{1}{2}, m-\frac{1}{2} \\ a_1, a_2, \ldots, a_{2m} \end{matrix}\right)\Bigg], \end{aligned} \tag{27.2.12}$$

$$a_j \equiv a_j(2) = \frac{N-1}{2} + \frac{j-1}{N}. \tag{27.2.13}$$

Similarly, for $\beta = 1$ and $N = 2m + 1$ odd we get the PDD for fixed trace real symmetric matrices from Eqs. (26.5.5) and (27.2.6) as

$$P_{F(2m+1)}(1, r) = \text{const} \cdot G^{2m+1,0}_{2m+1,2m+1}\left(\frac{r^{4m+2}y^2}{2m+1}\,\middle|\,\begin{matrix}0, \frac{1}{4}, \frac{1}{4}, \frac{3}{4}, \frac{3}{4}, \ldots, \frac{2m-1}{4}, \frac{2m-1}{4} \\ \frac{1}{2}a_1, \frac{1}{2}a_2, \ldots, \frac{1}{2}a_{2m+1}\end{matrix}\right). \tag{27.2.14}$$

Except for the level density (see the next section), which is a semi-circle, one does not know how to get, for example, the two level correlation function for the fixed trace ensembles.

27.3 Bounded Trace Ensembles

Instead of keeping the trace constant, we might require it to be bounded (Bronk, 1964a). We would then obtain a bounded trace ensemble defined by the joint probability density function

$$P_B(H) = \begin{cases} \text{constant}, & \text{if } \operatorname{tr} H^2 \leqslant r^2, \\ 0, & \text{if } \operatorname{tr} H^2 > r^2. \end{cases} \tag{27.3.1}$$

This probability density is again invariant under a change of basis, Eq. (27.1.3).

The density of eigenvalues for this ensemble can easily be found in the existing literature. A theorem of Stieltjes (1914) states that if there are N unit masses located at the variable points $x_1, x_2, \ldots, x_N$ in the interval $[-\infty, \infty]$ such that their moment of inertia is bounded by

$$\sum_{i=1}^{N} x_i^2 \leqslant N(N-1)/2, \tag{27.3.2}$$

the unique maximum of the function

$$V(x_1, \ldots, x_N) = \prod_{1 \leqslant i < j \leqslant N} |x_i - x_j|^\beta \tag{27.3.3}$$

will be obtained when the x_i are the zeros of the Hermite polynomial

$$H_N(x) = e^{x^2}\left(-\frac{d}{dx}\right)^N e^{-x^2}. \tag{27.3.4}$$

Thus, making the usual assumption of classical statistical mechanics that the actual eigenvalue density makes the logarithm of $V(x_1, \ldots, x_N)$ a maximum, we get this result:

The eigenvalue density for the bounded trace ensemble is identical to the density of zeros of Hermite-like polynomial

$$e^{N(N-1)x^2/2r^2}\left(-\frac{d}{dx}\right)^N e^{-N(N-1)x^2/2r^2}, \tag{27.3.5}$$

and for large N is given by

$$\sigma(x) \cong \begin{cases} \dfrac{N^2}{2\pi r^2}\left(\dfrac{4r^2}{N} - x^2\right)^{1/2}, & \text{if } |x| < 2rN^{-1/2}, \\ 0, & \text{if } |x| > 2rN^{-1/2}. \end{cases} \tag{27.3.6}$$

To work out the eigenvalue spacing distribution is much more difficult.

Summary of Chapter 27

Two types of ensembles (of real symmetric, complex Hermitian or quaternion self-dual matrices) are considered.

Fixed trace; in which the trace of H^2 is kept constant with no other restriction on the matrix elements of H. These ensembles are shown to be equivalent to the Gaussian ensembles as far as finite moments of the matrix elements are concerned. The probability density of their determinants is deduced from that of the corresponding Gaussian ensembles.

Bounded trace; in which the trace of H^2 is less than or equal to a given constant. For these ensembles the eigenvalue density is a "semi circle", just as for the Gaussian ensembles.

APPENDICES

A.1 Numerical Evidence in Favor of Conjectures 1.2.1 and 1.2.2

In Section 1.2 we referred to the extended numerical experience about random matrices. Here are some details about what we know.

The first large scale use of a computer to study random matrices is due to Porter and Rosenzweig (1960a, 1960b). In the late fifties they generated and diagonalized a few tens of thousands of real symmetric matrices with random elements. In particular, they studied the following probability densities for the matrix elements H_{ij}, $i \leqslant j$;

(1)
$$P(H_{ij}) = \begin{cases} \dfrac{1}{2}, & \text{if } |H_{ij}| \leqslant 1, \\ 0, & \text{otherwise;} \end{cases}$$

(2)
$$P(H_{ij}) = \frac{1}{2}\big(\delta(H_{ij}+1) + \delta(H_{ij}-1)\big),$$

(3)
$$P(H_{ij}) = (2\pi)^{-1/2} \exp\big(-H_{ij}^2/2\big).$$

They found that (for $N \times N$ matrices) the eigenvalue density always converged to the "semi-circle" law

$$\sigma(x) = \begin{cases} (1/\pi)\sqrt{2N - x^2}, & \text{if } |x^2| \leqslant 2N, \\ 0, & \text{otherwise,} \end{cases} \tag{A.1.1}$$

and the spacing probability density for consecutive eigenvalues converged always to the same curve which is nearly the "Wigner surmise"

$$p_W(s) = (\pi s/2)\exp\left(-\pi s^2/4\right), \quad s = S/\langle S\rangle. \tag{A.1.2}$$

The convergence was rapid, it was reached for matrices of order $\approx 20 \times 20$.

The Gaussian probability density, case (3) above, received later an analytical treatment, which covers a large part of this book. This is one of the few cases where we know for certain all correlation and cluster functions.

Brody, French and coworkers in the sixties generated matrices of a different random nature. They considered a finite number of shell model states (in nuclear physics), which were partially filled with nucleons. The two nucleon forces were taken random having some prescribed probability density. The shell model Hamiltonians so generated were thus matrices the elements of which depended on a small number of random variables; most of the matrix elements were *not* statistically independent. Diagonalizing a large number of such matrices, they found that the eigenvalue density is not a "semi-circle", but a Gaussian centered at the origin. However, the surprising thing is that the spacing probability density for consecutive eigenvalues tends always to the same curve which is nearly the "Wigner surmise", Eq. (A.1.2).

Similar computations have since been carried out for quantum chaotic systems, billiards of various shapes and boundary conditions. The eigenvalue density is again something else. However, the spacing probability density for consecutive eigenvalues in the absence of the magnetic field is the same old curve approximated by the "Wigner surmise", Eq. (A.1.2); in the presence of a strong magnetic field this curve is different and corresponds to that derived for the Gaussian unitary ensemble, Chapter 6.

Thus empirical evidence tells us that if the matrices are real symmetric and their $N(N+1)/2$ linearly independent elements depend on independent random variables almost as numerous, then the eigenvalue density tends to the "semi-circle" law, Eq. (A.1.1), as the order $N \to \infty$. If the matrix elements depend on a smaller number, say about N, of independent random variables, then the eigenvalue density is something else and *not* a "semi-circle". But the spacing probability density for consecutive eigenvalues is always the same for large N. Chaotic systems in a magnetic field correspond to matrices which are not real, and have a different spacing probability density, that of the Gaussian unitary ensemble.

A.2 The Probability Density of the Spacings Resulting from a Random Superposition of *n* Unrelated Sequences of Energy Levels

In Section 1.4 we said that for a simple sequence of levels (i.e., levels having the same spin and parity) the probability density for the nearest neighbor spacings is nicely approximated by the Wigner surmise. For a mixed sequence this probability density results from a random superposition of its constituent simple sequences as follows.

Let ρ_i be the level density in the ith sequence and $p_i(\rho_i S)\rho_i\, dS$, the probability that a spacing in the ith sequence will have a value between S and $S + dS$. Because $p_i(x)$

is normalized and the level density is the inverse of the mean spacing, we have

$$\int_0^\infty p_i(x)\,dx = 1, \qquad \int_0^\infty x p_i(x)\,dx = 1. \tag{A.2.1}$$

Let $\Psi_i(x)$ and $E_i(x)$ be defined by

$$\Psi_i(x) = \int_0^x p_i(y)\,dy = 1 - \int_x^\infty p_i(y)\,dy = 1 - \int_0^\infty p_i(x+y)\,dy, \tag{A.2.2}$$

and

$$\begin{aligned} E_i(x) &= \int_x^\infty [1 - \Psi_i(y)]\,dy = \int_x^\infty \left[\int_0^\infty p_i(y+z)\,dz\right] dy \\ &= \int_0^\infty \int_0^\infty p_i(x+y+z)\,dy\,dz, \end{aligned} \tag{A.2.3}$$

so that $\Psi_i(\rho_i S)$ is the probability that a spacing in the ith sequence is less than or equal to S and $E_i(\rho_i S)$ is the probability that a given interval of length S will not contain any of the levels belonging to the sequence i.

Consider the system resulting from the superposition of n sequences. The total density is

$$\rho = \sum_i \rho_i. \tag{A.2.4}$$

Let $P(\rho S)\rho\,dS$ be the probability that a spacing will lie between S and $S + dS$. Analogously to (A.2.2) and (A.2.3), we introduce the functions $\Psi(x)$ and $E(x)$ by

$$\Psi(x) = \int_0^x P(y)\,dy = 1 - \int_0^\infty P(x+y)\,dy \tag{A.2.5}$$

and

$$E(x) = \int_x^\infty [1 - \Psi(y)]\,dy = \int_0^\infty \int_0^\infty P(x+y+z)\,dy\,dz. \tag{A.2.6}$$

From the observation that $E(\rho S)$ is the probability that a given interval of length S will not contain any of the levels and the randomness of the superposition we have

$$E(\rho S) = \prod_i E_i(\rho_i S). \tag{A.2.7}$$

Introducing the fractional densities

$$f_i = \frac{\rho_i}{\rho}, \quad \sum_i f_i = 1 \tag{A.2.8}$$

and the variable $x = \rho S$, we have

$$E(x) = \prod_i E_i(f_i x). \tag{A.2.9}$$

By differentiating (A.2.9) twice, we obtain

$$P(x) = \frac{d^2E}{dx^2} = E(x)\left\{\sum_i f_i^2 \frac{p_i(f_i x)}{E_i(f_i x)} + \left[\sum_i f_i \frac{1-\Psi_i(f_i x)}{E_i(f_i x)}\right]^2 - \sum_i \left(f_i \frac{1-\Psi_i(f_i x)}{E_i(f_i x)}\right)^2\right\}, \tag{A.2.10}$$

which was the purpose of this appendix.

We now consider three special cases.

(1) Let the levels in each of the sequences be independent of one another so that $p_i(x) = e^{-x}$. In this case

$$1 - \Psi_i(x) = E_i(x) = e^{-x}, \tag{A.2.11}$$

and (A.2.10) yields $P(x) = e^{-x}$, which verifies the obvious fact that the random superposition of sequences of independent random levels produces a sequence of independent random levels.

(2) Let all fractional densities be equal to $1/n$ and take the limit as n goes to ∞. Let $x = ny$ so that

$$P(ny) = [E(y)]^n \left\{\frac{1}{n}\frac{p(y)}{E(y)} + \left(1 - \frac{1}{n}\right)\left[\frac{1-\Psi(y)}{E(y)}\right]^2\right\}. \tag{A.2.12}$$

From (A.2.1), (A.2.2), and (A.2.3) we have

$$\Psi(0) = 0, \quad E(0) = 1, \quad E'(0) = -1. \tag{A.2.13}$$

Therefore, taking the limit as $n \to \infty$, $y \to 0$, whereas $ny = x$ is fixed, we see that the terms in the square brackets tend to 1, while from

$$E(y) \approx E(0) + yE'(0) + \cdots = 1 - y + \cdots,$$

keeping only the first term,

$$\left[E(y)\right]^n \approx (1-y)^n = \left(1-\frac{x}{n}\right)^n \to e^{-x}. \tag{A.2.14}$$

This is a verification of the heuristic reasoning that if the number n of the sequences to be superimposed is large, a level belonging to a sequence will almost certainly be followed by a level of another sequence and these two levels will be independent, whatever $p(x)$ may be.

(3) For the "Wigner surmise"

$$p(x) = \frac{\pi}{2} x \exp\left(-\frac{\pi}{4}x^2\right), \tag{A.2.15}$$

we have

$$1 - \Psi(x) = \exp\left(-\frac{\pi}{4}x^2\right), \quad E(x) = 1 - \frac{2}{\sqrt{\pi}} \int_0^{x\sqrt{\pi}/2} e^{-y^2}\, dy. \tag{A.2.16}$$

Because $p(0) = 0$, we have

$$P(0) = 1 - \sum_{i=1}^{n} f_i^2, \tag{A.2.17}$$

and, in particular, $P(0) \neq 0$.

(4) For the Gaussian ensembles the functions $\Psi(x)$ and $E(x)$ are tabulated in Appendix A.15.

A.3 Some Properties of Hermitian, Unitary, Symmetric or Self-Dual Matrices

For completeness we collect here a few properties of matrices which we need in Chapters 2, 3 and 9. A proof of them can be found in any standard text on matrix theory, for example, Mehta (1989MT).

A.3.1 A matrix S can be diagonalized by a unitary matrix (i.e., given S, one can find a U with $U^\dagger S U = D$, U unitary, D diagonal), if and only if $SS^\dagger = S^\dagger S$.

Thus Hermitian and unitary matrices can be diagonalized by a unitary matrix. The diagonal elements of D are the eigenvalues of S; they are real for Hermitian S and of the form $\exp(i\theta_j)$, θ_j real, for unitary S.

A.3.2 Any unitary matrix S can be written as $S = \exp(iH)$, where H is Hermitian. Moreover, if S is symmetric, then H is symmetric; if S is self-dual, then H is self-dual.

For if $S = U^\dagger \exp(i\theta)U$, with U unitary and θ diagonal and real, then with $H = U^\dagger \theta U$ one has $S = \exp(iH)$. The symmetry or self-duality of S evidently implies that of H.

A.3.3 Any unitary symmetric matrix S can be written, in many ways, as $S = UU^T$, where U is unitary. Any unitary self-dual matrix S can be written, in many ways, as $S = UU^R$, where U is unitary. The transpose (respectively dual) of U is denoted by U^T (respectively U^R).

For if $S = \exp(iH)$, then write $U = \exp(iH/2)R$, where R is any real orthogonal (respectively quaternion real symplectic) matrix; for example, $R = I$, the unit matrix.

A.3.4 A symmetric Hermitian or a symmetric unitary matrix can be diagonalized by a real orthogonal matrix. A self-dual Hermitian or a self-dual unitary matrix can be diagonalized by a quaternion real symplectic matrix.

For if $A = U^\dagger DU = A^T$ with U unitary and D diagonal, then $U = U^*$, and U is also real, i.e., U is real orthogonal. Similarly, if $A = U^\dagger DU = A^R$, with U unitary and D diagonal, then U is quaternion real, i.e., U is quaternion real symplectic.

A.3.5 Any anti-symmetric Hermitian or anti-symmetric unitary matrix S can be reduced to the block diagonal form with a real orthogonal matrix; i.e., given an anti-symmetric Hermitian or anti-symmetric unitary matrix S, one can find an R with $R^T SR = D$, R real orthogonal and D block diagonal, $R^T R = 1$, $D_{2j-1,2j} = -D_{2j,2j-1} = \mu_j$, all other elements of D being zero. When S is Hermitian, μ_j are real and when S is unitary, μ_j have the form $\exp(i\theta_j)$.

A.3.6 Any anti-self-dual Hermitian or anti-self-dual unitary matrix can be reduced to the block diagonal form with a quaternion real symplectic matrix.

A.4 Counting the Dimensions of $T_{\beta G}$ and $T'_{\beta G}$ (Chapter 3) and of $T_{\beta C}$ and $T'_{\beta C}$ (Chapter 10)

When we require that two of the eigenvalues be equal, we drop a number of parameters needed to specify a certain two-dimensional subspace, the subspace of these two equal eigenvalues. However, this degenerate eigenvalue is itself one real parameter. Thus, if $f(N, \beta)$ is the number of independent real parameters needed to specify a particular matrix from the ensemble $E_{\beta G}$, the number needed to specify a matrix from the ensemble $E_{\beta G}$ with two equal eigenvalues is

$$f(N, \beta) - f(2, \beta) + 1. \tag{A.4.1}$$

In other words if the number of dimensions of the space $T_{\beta G}$ is $f(N, \beta)$, that of the space $T'_{\beta G}$ is $f(N, \beta) - f(2, \beta) + 1$.

Now to specify a matrix from any of the ensembles $E_{\beta G}$ we need specify only the matrix elements H_{ij} with $i \leqslant j$. The diagonal elements are real and therefore require N real parameters for their specification. The off-diagonal elements H_{ij} with $i < j$ are $N(N-1)/2$ in number and they need $N(N-1)\beta/2$ real parameters. Thus

$$f(N, \beta) = N + \frac{1}{2}N(N-1)\beta. \tag{A.4.2}$$

By inserting $\beta = 1, 2,$ or 4 into (A.4.2) and (A.4.1) we get the dimensions of $T_{\beta G}$ and $T'_{\beta G}$.

To count the dimensions of the spaces $T_{\beta C}$ and $T'_{\beta C}$ we must find the corresponding numbers $f(N, \beta)$, and for this purpose it is sufficient to consider matrices in the neighborhood of unity. Let us then have

$$S = 1 + iA.$$

Where A is infinitesimal. Since S is unitary,

$$S^{\dagger}S \equiv (1 - iA^{\dagger})(1 + iA) = 1,$$

or, up to terms linear in A,

$$A = A^{\dagger}; \tag{A.4.3}$$

A is then Hermitian. If, in addition, S is symmetric (self-dual), then A is symmetric (self-dual). Thus the number of independent real parameters needed to specify a symmetric unitary, self-dual unitary, or unitary matrix S is the same as that needed to specify a symmetric Hermitian, self-dual Hermitian, or Hermitian matrix A, respectively. Thus the dimensions of $T_{\beta C}$ and $T_{\beta G}$ are equal, and hence also those of $T'_{\beta C}$ and $T'_{\beta G}$.

A.5 An Integral Over The Unitary Group, Eq. (3.5.1) or Eq. (14.3.1)

We want to show that

$$\int \exp\left(-\frac{1}{2t}\operatorname{tr}(A - UBU^{-1})^2\right) dU$$
$$= c\big(\Delta(\vec{a})\Delta(\vec{b})\big)^{-1} \det\left[\exp\left(-\frac{1}{2t}(a_j - b_k)^2\right)\right], \tag{A.5.1}$$

where the integral is taken over the group of $n \times n$ unitary matrices U. Here A is an $n \times n$ Hermitian matrix with eigenvalues $a_1, \ldots, a_n$ and $\Delta(\vec{a})$ is the product of their differences, $\Delta(\vec{a}) = \prod_{1\leqslant j<k\leqslant n}(a_j - a_k)$; similarly the Hermitian matrix B has eigenvalues $b_1, \ldots, b_n$ and $\Delta(\vec{b})$ is the product of their differences. The determinant on the right hand side is $n \times n$ with $j, k = 1, \ldots, n$; and the constant c,

$$c = t^{n(n-1)/2} \prod_{j=1}^{n} j!,$$

is neither needed nor evaluated here.

Instead of the above we will derive the equivalent formula

$$\begin{aligned}
&\int \exp\left(-\frac{1}{2t}\operatorname{tr}(A-B)^2\right) f(\vec{b})\, dB \\
&= \int \exp\left(-\frac{1}{2t}\operatorname{tr}(A-B)^2\right) f(\vec{b})\Delta^2(\vec{b})\, d\vec{b}\, dU \\
&= c' \cdot n! \int \exp\left(-\frac{1}{2t}\sum_{j=1}^{n}(a_j - b_j)^2\right) f(\vec{b}) \frac{\Delta(\vec{b})}{\Delta(\vec{a})}\, d\vec{b} \qquad \text{(A.5.2)} \\
&= c' \int \left(\Delta(\vec{a})\Delta(\vec{b})\right)^{-1} \det\left[\exp\left(-\frac{1}{2t}(a_j - b_k)^2\right)\right] f(\vec{b})\Delta^2(\vec{b})\, d\vec{b}, \qquad \text{(A.5.3)}
\end{aligned}$$

valid for an arbitrary symmetric function $f(\vec{b})$ of the eigenvalues of B. Here

$$dB = \prod_j dB_{jj} \prod_{j<k} d(\operatorname{Re} B_{jk})\, d(\operatorname{Im} B_{jk}), \qquad \text{(A.5.4)}$$

$$d\vec{b} = \prod_{j=1}^{n} db_j, \qquad \text{(A.5.5)}$$

all the integrals are taken from $-\infty$ to ∞, and c' is another constant. The second equality (A.5.3) is easier since under the integral sign one can replace

$$n!\Delta(b) \exp\left(-\frac{1}{2t}\sum_j (a_j - b_j)^2\right),$$

by the sum over all permutations P of the indices $1, 2, \ldots, n$, as

$$\begin{aligned}
&\Delta(b) \sum_P (-1)^P \exp\left(-\frac{1}{2t}\sum_j (a_j - b_{Pj})^2\right) \\
&= \Delta(b) \det\left[\exp\left(-\frac{1}{2t}\sum_j (a_j - b_k)^2\right)\right]_{j,k=1,\ldots,n}.
\end{aligned}$$

To see the first equality (A.5.2), let us recall that the partial differential equation (p.d.Eq.)

$$\frac{\partial F}{\partial t} = \frac{1}{2} \sum_j D_j \frac{\partial^2 F}{\partial x_j^2} \tag{A.5.6}$$

with the initial condition $F(\vec{x}; 0) = f(\vec{x})$, $\vec{x} \equiv (x_1, \ldots, x_n)$, has the unique solution (cf. Morse and Feshbach, 1953)

$$F(\vec{x}; t) = \int K(\vec{x}, \vec{y}; t) f(\vec{y}) \, d\vec{y}, \quad d\vec{y} = dy_1 \cdots dy_n, \tag{A.5.7}$$

where

$$K(\vec{x}, \vec{y}; t) = \left(\prod_i (2\pi D_i t)^{-1/2} \right) \cdot \exp\left(- \sum_j (x_j - y_j)^2 / 2D_j t \right). \tag{A.5.8}$$

If we take

$$K(A, B; t) = 2^{-n/2} (\pi t)^{-n^2/2} \exp\left(- \operatorname{tr}(A - B)^2 / 2t\right), \tag{A.5.9}$$

$$\begin{aligned} \operatorname{tr}(A - B)^2 = {} & \sum_j (A_{jj} - B_{jj})^2 \\ & + 2 \sum_{j<k} \left((\operatorname{Re} A_{jk} - \operatorname{Re} B_{jk})^2 + (\operatorname{Im} A_{jk} - \operatorname{Im} B_{jk})^2 \right), \end{aligned} \tag{A.5.10}$$

then

$$F(A; t) = \int K(A, B; t) f(B) \, dB, \tag{A.5.11}$$

satisfies the p.d.Eq.

$$\frac{\partial F}{\partial t} = \frac{1}{2} \nabla_A^2 F, \tag{A.5.12}$$

$$\nabla_A^2 = \sum_j \frac{\partial^2}{\partial A_{jj}^2} + \frac{1}{2} \sum_{j<k} \left(\frac{\partial^2}{\partial (\operatorname{Re} A_{jk})^2} + \frac{\partial^2}{\partial (\operatorname{Im} A_{jk})^2} \right), \tag{A.5.13}$$

with the initial condition $F(A; 0) = f(A)$.

Hermitian matrices A and B can be diagonalized by unitary matrices,

$$A = U_A a U_A^{-1}, \quad B = U_B b U_B^{-1}, \quad a = [a_j \delta_{jk}], \quad b = [b_j \delta_{jk}].$$

Changing variables as in Section 3.3, we get

$$\nabla_A^2 = (\Delta(\vec{a}))^{-2} \sum_j \frac{\partial}{\partial a_j} (\Delta(\vec{a}))^2 \frac{\partial}{\partial a_j} + \nabla_{U_A}^2.$$

Now

$$\begin{aligned}
&\frac{1}{\Delta(\vec{a})} \sum_j \frac{\partial^2}{\partial a_j^2} \Delta(\vec{a}) f - (\Delta(\vec{a}))^{-2} \sum_j \frac{\partial}{\partial a_j} (\Delta(\vec{a}))^2 \frac{\partial f}{\partial a_j} \\
&= \frac{1}{\Delta(\vec{a})} \sum_j \frac{\partial}{\partial a_j} \left(\Delta(\vec{a}) \sum_k{}' (a_j - a_k)^{-1} \right) f \\
&= \sum_j \left(\left(\sum_k{}' \frac{1}{a_j - a_k} \right)^2 - \sum_k{}' \left(\frac{1}{a_j - a_k} \right)^2 \right) f \\
&= \sum_j \sum_{k \neq p}{}' (a_j - a_k)^{-1} (a_j - a_p)^{-1} f,
\end{aligned}$$

where a prime on the summation means that the sum does not contain any singular term. Now the last sum is equal to $P(\vec{a}) f / \Delta(\vec{a})$, where $P(\vec{a})$ is some polynomial in $a_1, \ldots, a_n$ of degree at least two less than that of $\Delta(\vec{a})$. But $P(\vec{a})/\Delta(\vec{a})$ is symmetric in $a_1, \ldots, a_n$, so that $P(\vec{a})$ is anti-symmetric in $a_1, \ldots, a_n$ and hence divisible by $\Delta(\vec{a})$. The conclusion is that $P(\vec{a})$ is identically zero. Hence

$$\nabla_A^2 = (\Delta(\vec{a}))^{-1} \sum_j \frac{\partial^2}{\partial a_j^2} \Delta(\vec{a}) + \nabla_{U_A}^2, \tag{A.5.14}$$

$$\begin{aligned}
&F(\vec{a}, U_A; t) \\
&\quad = \text{const} \cdot \int \exp\left(-\frac{1}{2t} \operatorname{tr}(a - U B U^{-1})^2 \right) f(\vec{b}, U_B) \Delta^2(\vec{b})\, d\vec{b}\, dU_B,
\end{aligned} \tag{A.5.15}$$

where $U = U_A^{-1} U_B$, and $\Delta^2(\vec{b})$ comes from the Jacobian (cf. Section 3.3). If $f(\vec{b}, U_B) = f(\vec{b})$ is independent of U_B, then so is $F(A)$ of U_A, and

$$F(\vec{a}; t) = \text{const} \cdot \int \exp\left(-\frac{1}{2t} \operatorname{tr}(a - U B U^{-1})^2 \right) f(\vec{b}) \Delta^2(\vec{b})\, d\vec{b}\, dU, \tag{A.5.16}$$

$$F(\vec{a}; 0) = f(\vec{a}), \tag{A.5.17}$$

satisfies the p.d.Eq.

$$\frac{\partial}{\partial t}\big(\Delta(\vec{a})F(\vec{a};t)\big)=\frac{1}{2}\sum_j\frac{\partial^2}{\partial a_j^2}\big(\Delta(\vec{a})F(\vec{a};t)\big), \tag{A.5.18}$$

with the initial condition $\Delta(\vec{a})F(\vec{a};0)=\Delta(\vec{a})f(\vec{a})$. Therefore $\Delta(\vec{a})F(\vec{a};t)$ is given by

$$\Delta(\vec{a})F(\vec{a};t)=(2\pi t)^{-n/2}\int\exp\left(-\frac{1}{2t}\sum_{j=1}^{n}(a_j-b_j)^2\right)\Delta(\vec{b})f(\vec{b})\,d\vec{b}. \tag{A.5.19}$$

Comparing (A.5.16) and (A.5.19) we get

$$\begin{aligned}&\int\exp\left(-\frac{1}{2t}\operatorname{tr}\left(a-UBU^{-1}\right)^2\right)f(\vec{b})\Delta^2(\vec{b})\,d\vec{b}\,dU\\&=\text{const}\cdot\int\exp\left(-\frac{1}{2t}\sum_j(a_j-b_j)^2\right)\frac{\Delta(\vec{b})}{\Delta(\vec{a})}f(\vec{b})\,d\vec{b}\end{aligned} \tag{A.5.20}$$

which is (A.5.2).

For more details see Mehta (1981, 1989MT) and references given there.

A.6 The Minimum Value of W, Proof of Eq. (4.1.6)

To get the minimum value of the potential energy W we present Stieltjes' (1914) ingenious arguments. The existence of a minimum is clear. Let the points $x_1,\ldots,x_N$ make

$$W=\frac{1}{2}\sum_1^N x_j^2-\sum_{\ell\leqslant i<j\leqslant N}\ln|x_i-x_j| \tag{A.6.1}$$

a minimum; then

$$0=-\frac{\partial W}{\partial x_j}\equiv -x_j+\sum_{i(\neq j)}\frac{1}{x_j-x_i}. \tag{A.6.2}$$

Consider the polynomial

$$g(x)=(x-x_1)(x-x_2)\cdots(x-x_N), \tag{A.6.3}$$

which has $x_1,x_2,\ldots,x_N$ as its zeros. Logarithmic differentiation gives

$$\sum_{i(\neq j)}\frac{1}{x-x_i}=\frac{g'(x)}{g(x)}-\frac{1}{x-x_j}=\frac{(x-x_j)g'(x)-g(x)}{(x-x_j)g(x)}. \tag{A.6.4}$$

Near $x = x_j$ the left-hand side is continuous, while the right-hand side takes the indeterminate form 0/0. Taking the limit $x \to x_j$ one gets from d'Hopital's rule on two differentiations

$$\sum_{i(\neq j)} \frac{1}{x_j - x_i} = \lim_{x \to x_j} \frac{(x - x_j)g'(x) - g(x)}{(x - x_j)g(x)}$$

$$= \lim_{x \to x_j} \frac{(x - x_j)g''(x)}{(x - x_j)g'(x) + g(x)} = \frac{g''(x_j)}{2g'(x_j)}, \tag{A.6.5}$$

so that (A.6.2) can be written as

$$g''(x_j) - 2x_j g'(x_j) = 0. \tag{A.6.6}$$

This means that the polynomial

$$g''(x) - 2xg'(x),$$

of order N, has its zeros at $x_1, \ldots, x_N$ and therefore it must be proportional to $g(x)$. Comparing the coefficients of x^N, we see that

$$g''(x) - 2xg'(x) + 2Ng(x) = 0. \tag{A.6.7}$$

The polynomial solution of this differential equation is uniquely determined to be the Hermite polynomial of order N:

$$H_N(x) = N! \sum_{m=0}^{[N/2]} \frac{(-1)^m (2x)^{N-2m}}{m!(N-2m)!}. \tag{A.6.8}$$

The discriminant of $H_N(x)$ (Szegö, 1959) is

$$\prod_{1 \leqslant i < j \leqslant N} (x_i - x_j)^2 = 2^{-N(N-1)/2} \prod_{j=1}^{N} j^j, \tag{A.6.9}$$

and from (A.6.8) we get

$$\sum_1^N x_j^2 = N(N-1)/2. \tag{A.6.10}$$

Thus the minimum value of W is

$$W_0 = \frac{1}{4} N(N-1)(1 + \ln 2) - \frac{1}{2} \sum_1^N j \ln j. \tag{A.6.11}$$

A.7 Relation Between R_n, T_n and $E(n; 2\theta)$. Equivalence of Eqs. (6.1.3) and (6.1.4)

With $R_n(x_1, \ldots, x_n)$ and $T_n(x_1, \ldots, x_n)$ defined by (6.1.2) and (6.1.3), let us write

$$r_0 = 1, \tag{A.7.1}$$

$$r_n = \int_{-\infty}^{\infty} \cdots \int_{-\infty}^{\infty} R_n(x_1, \ldots, x_n) \prod_1^n \big(a(x_i)\, dx_i\big), \quad n \geqslant 1, \tag{A.7.2}$$

$$t_n = \int_{-\infty}^{\infty} \cdots \int_{-\infty}^{\infty} T_n(x_1, \ldots, x_n) \prod_1^n \big(a(x_i)\, dx_i\big), \quad n \geqslant 1, \tag{A.7.3}$$

and the generating functions

$$R(z) = \sum_{n=0}^{\infty} \frac{r_n}{n!} z^n \equiv 1 + \sum_{n=1}^{\infty} \frac{r_n}{n!} z^n, \tag{A.7.4}$$

$$T(z) = \sum_{n=1}^{\infty} (-1)^{n-1} \frac{t_n}{n!} z^n, \tag{A.7.5}$$

where the function $a(x)$ is arbitrary. The numbers r_n and t_n may be related by the use of (6.1.3). In fact,

$$\begin{aligned} t_n &= \sum_G (-1)^{n-m} (m-1)! \frac{n!}{G_1! \cdots G_m!} \frac{1}{m!} r_{G_1} r_{G_2} \cdots r_{G_m} \\ &= \sum_G (-1)^{n-m} \frac{n!}{m} \prod_{j=1}^m \frac{r_{G_j}}{G_j!}, \end{aligned} \tag{A.7.6}$$

the sum being taken over all partitions G of n:

$$n = G_1 + \cdots + G_m, \quad G_j \geqslant 1,\ m \geqslant 1. \tag{A.7.7}$$

We now relate the functions $T(z)$ and $R(z)$:

$$\begin{aligned} T(z) &= \sum_{n=1}^{\infty} \frac{(-1)^{n-1}}{n!} t_n z^n \\ &= \sum_{n=1}^{\infty} \sum_{G(n)} \frac{(-1)^{m-1}}{m} \prod_{j=1}^m \frac{r_{G_j}}{G_j!} z^{G_j}. \end{aligned} \tag{A.7.8}$$

As we are summing finally over all integers n, the restriction

$$\sum_{j=1}^{m} G_j = n \tag{A.7.9}$$

may be removed; we then have

$$\begin{aligned} T(z) &= \sum_{m, G_1, \ldots, G_m = 1}^{\infty} \frac{(-1)^{m-1}}{m} \prod_{j=1}^{m} \frac{r_{G_j}}{G_j!} z^{G_j} \\ &= \sum_{m=1}^{\infty} \frac{(-1)^{m-1}}{m} \left(\sum_{\ell=1}^{\infty} \frac{r_\ell}{\ell!} z^\ell \right)^m = \ln R(z); \end{aligned} \tag{A.7.10}$$

or

$$R(z) = \exp\bigl(T(z)\bigr). \tag{A.7.11}$$

Expanding the exponential and equating coefficients of z^n on both sides, we get

$$\begin{aligned} \frac{r_n}{n!} &= \text{coefficient of } z^n \text{ in } \sum_{m=1}^{n} \frac{1}{m!} \left(\sum_{\ell=1}^{n} \frac{(-1)^{\ell-1}}{\ell!} t_\ell z^\ell \right)^m \\ &= \sum_G \frac{1}{m!} \prod_{j=1}^{m} \left(\frac{(-1)^{G_j - 1}}{G_j!} t_{G_j} \right); \quad \sum_{j=1}^{m} G_j = n, \end{aligned} \tag{A.7.12}$$

or

$$r_n = \sum_G \frac{(-1)^{n-m}}{m!} \frac{n!}{G_1! G_2! \cdots G_m!} t_{G_1} t_{G_2} \cdots t_{G_m}. \tag{A.7.13}$$

In view of the arbitrariness of $a(x)$, this relation is identical to (6.1.4).

If we take $a(x)$ as the characteristic function of the interval $(-\theta, \theta)$:

$$a(x) = \begin{cases} 1, & \text{if } |x| \leqslant \theta, \\ 0, & \text{otherwise,} \end{cases} \tag{A.7.14}$$

then the probability that the interval $(-\theta, \theta)$ contains exactly n levels can be written as

$$\begin{aligned} E(n; 2\theta) = {} & \frac{N!}{n!(N-n)!} \int_{-\infty}^{\infty} \cdots \int_{-\infty}^{\infty} dx_1 \cdots dx_N \prod_{i=1}^{n} a(x_i) \\ & \times \prod_{j=n+1}^{N} \bigl(1 - a(x_j)\bigr) P_N(x_1, \ldots, x_N). \end{aligned} \tag{A.7.15}$$

Expanding the product $\prod(1-a(x_j))$ and regrouping similar terms we get

$$E(n;2\theta) = \frac{1}{n!}\sum_{j=n}^{N}\frac{(-1)^{j-n}}{(j-n)!}\int_{-\infty}^{\infty}\cdots\int_{-\infty}^{\infty}R_j(x_1,\ldots,x_j)\prod_{i=1}^{j}a(x_i)\,dx_i$$

$$= \frac{(-1)^n}{n!}\sum_{j=n}^{N}\frac{(-1)^j}{(j-n)!}r_j$$

$$= \frac{(-1)^n}{n!}\left(\frac{d}{dz}\right)^n\sum_{j=0}^{N}\frac{(-1)^j}{j!}r_j z^j|_{z=1}. \tag{A.7.16}$$

In terms of the generating functions this becomes

$$E(n;2\theta) = \frac{1}{n!}\left(-\frac{d}{dz}\right)^n R(-z)|_{z=1}$$

$$= \frac{1}{n!}\left(-\frac{d}{dz}\right)^n \exp\big(T(-z)\big)\big|_{z=1}. \tag{A.7.17}$$

Substituting the expression for T_n, Eq. (6.2.9),

$$T_{n2}(x_1,\ldots,x_n) = \sum K_N(x_1,x_2)K_N(x_2,x_3)\cdots K_N(x_n,x_1) \tag{A.7.18}$$

in (A.7.3), we get

$$t_{n2} = (n-1)!\int_{-\theta}^{\theta}\cdots\int_{-\theta}^{\theta}K_N(x_1,x_2)K_N(x_2,x_3)\cdots K_N(x_n,x_1)\,dx_1\cdots dx_n$$

$$= (n-1)!\sum_{j=0}^{N-1}\lambda_j^n, \tag{A.7.19}$$

so that

$$T_2(z) = \sum_{j=1}^{\infty}\frac{(-1)^{j-1}}{j}z^j\sum_i \lambda_i^j = \sum_i \ln(1+z\lambda_i). \tag{A.7.20}$$

Hence from (A.7.17)

$$E(n;2\theta) = \frac{1}{n!}\left(-\frac{d}{dz}\right)^n\prod_i(1-z\lambda_i)|_{z=1}. \tag{A.7.21}$$

Taking the limit $N \to \infty$, we can replace $K_N(x, y)$ by $\mathcal{K}(\xi, \eta)$ and $E(n; 2\theta)$ by $E_2(n; s)$.

Had we taken the expression of T_n for orthogonal, symplectic or the non-invariant Gaussian ensemble of Chapter 14, we would find the corresponding expression for $E(n; s)$. For example, for the non-invariant Gaussian ensemble of Chapter 14

$$T_n(x_1, \ldots, x_n) = \frac{1}{2} \operatorname{tr} \sum \phi(x_1, x_2)\phi(x_2, x_3) \cdots \phi(x_n, x_1), \tag{A.7.22}$$

where $\phi(x, y)$ is the 2×2 matrix, Eq. (14.1.14). The integral equation

$$\int_{-\theta}^{\theta} \phi(x, y) F(y)\, dy = \lambda F(x) \tag{A.7.23}$$

has N distinct eigenvalues λ_j, each occurring twice, so that t_n is again given by (A.7.19), and $E(n; 2\theta)$ by (A.7.21). In the limit $N \to \infty$ one replaces ϕ by σ, Eq. (14.1.34),

$$\sigma(x, y) = \begin{bmatrix} \dfrac{\sin(x-y)\pi}{(x-y)\pi} & D(x-y) \\ J(x-y) & \dfrac{\sin(x-y)\pi}{(x-y)\pi} \end{bmatrix}, \tag{A.7.24}$$

$$D(r) = -\frac{1}{\pi} \int_0^{\pi} t \sin(rt) \exp(2\rho^2 t^2)\, dt, \tag{A.7.25}$$

$$J(r) = -\frac{1}{\pi} \int_{\pi}^{\infty} \frac{\sin(rt)}{t} \exp(-2\rho^2 t^2)\, dt. \tag{A.7.26}$$

Equation (A.7.21) is now

$$E(n; s) = \frac{1}{n!} \left(-\frac{d}{dz} \right)^n \prod_{i=0}^{\infty} (1 - z\lambda_i) \Big|_{z=1}, \tag{A.7.27}$$

with λ_i the distinct eigenvalues of

$$\int_{-t}^{t} \sigma(x, y) F(y)\, dy = \lambda F(x). \tag{A.7.28}$$

A hierarchical relation between R_n and R_{n+1} may be obtained by considering α^2 in Chapter 14 as the "time" variable, constructing a Brownian motion model as in Chapter 9 and solving the partial differential equation so obtained. For more details see French et al. (1988), p. 229, Eq. (72).

A.8 Relation Between $E(n; s)$, $\tilde{F}(n; s)$ and $p(n; s)$: Eqs. (6.1.16), (6.1.17) and (6.1.18)

Let $E(0; x)$ be the probability that an interval of length x is empty of eigenvalues. Then $E(0; x + \delta x)$ is the probability that the interval $\delta x + x$ is empty, and $E(0; x) - E(0; x + \delta x)$ is the probability that the interval x is empty and δx is not empty. The probability that δx will contain more than one eigenvalue is of second or higher order in δx. Therefore, taking the limit $\delta x \to 0$ and keeping only the first order terms, we get $-(dE(0; x)/dx)\delta x$ as the probability that x is empty and δx contains one eigenvalue. This is equivalent to $F(0; x) = -dE(0; x)/dx$. By a similar argument we obtain $p(0; x) = -dF(0; x)/dx$.

$$\delta x \qquad\qquad x$$

Let $E(n; x)$ be the probability that the interval x contains exactly n eigenvalues. Increasing x by δx in $E(n; x)$ and subtracting it from $E(n; x)$ we find that either one of the eigenvalues in x moves in δx or a new eigenvalue appears in δx. Therefore,

$$-\frac{dE(n; x)}{dx}\delta x = -\operatorname{prob}\begin{pmatrix} x & n-1 \\ \delta x & 1 \end{pmatrix} + \operatorname{prob}\begin{pmatrix} x & n \\ \delta x & 1 \end{pmatrix},$$

where $\operatorname{prob}\left(\begin{smallmatrix} x & r \\ \delta x & s \end{smallmatrix}\right)$ means the probability that x and δx contain r and s eigenvalues respectively. This is equivalent to Eq. (6.1.16b). A similar argument gives Eq. (6.1.16c).

The inverse relations, Eqs. (6.1.17) and (6.1.18), are easy consequences of Eqs. (6.1.16) and can be derived step by step starting at $n = 0$.

The quantity $B_1'(t)$ or $E_1'(s)$ of Section 11.4 is the probability that the interval $s = 2t$ contains either zero or one eigenvalue, i.e., $E_1'(x) = E_1(0; x) + E_1(1; x)$. So that

$$\frac{d^2}{dx^2}\left(E_1(0; x) + E_1'(x)\right) = \frac{d^2}{dx^2}\left(2E_1(0; x) + E_1(1; x)\right) = p(1; x),$$

i.e., $p(1; x)\delta x$ is the probability that the interval between two eigenvalues containing exactly one eigenvalue inside it has a length between x and $x + \delta x$. In other words it is the probability density for the next nearest neighbor spacings.

A.9 The Limit of $\sum_0^{N-1} \varphi_j^2(x)$. Eq. (6.2.11)

The dominant term in $\sum_0^{N-1} \varphi_j^2(x)$ may be obtained with ease by a physical argument.

The $\varphi_j(x)$ is the normalized oscillator function, so that $\varphi_j^2(x)\, dx$ gives the probability that an oscillator in the jth state is found in the interval $(x, x + dx)$. Consider N oscillators, one each in the states $0, 1, \ldots, N-1$, so that when N is large, $\sum_0^{N-1} \varphi_j^2(x)$

is the density of the particles at the point x. The particles are fermions and the temperature is zero, for there is not more than one particle in each state and all states up to a certain energy (Fermi energy) are filled. The Fermi momentum corresponding to this maximum energy can be obtained from the differential equation satisfied by $\varphi_{N-1}(x)$.

$$\hbar^2 \frac{d^2}{dx^2}\varphi_{N-1}(x) + \hbar^2(2N-1-x^2)\varphi_{N-1}(x) = 0,$$

so that

$$-p_F^2 + (2N-1-x^2)\hbar^2 = 0.$$

Because our oscillators are one-dimensional, their density is given by

$$\sigma(x) \approx \frac{1}{(2\pi\hbar)} \int_{-p_F}^{p_F} dp = \frac{1}{2\pi\hbar} 2p_F, \quad p \leqslant p_F.$$

From the last two equations we get

$$\sigma(x) = \begin{cases} \frac{1}{\pi}(2N-x^2)^{1/2}, & x^2 \leqslant 2N, \\ 0, & x^2 \geqslant 2N. \end{cases}$$

This is the dominant term. Terms of the next lower order cannot be obtained from physical arguments alone. To get further information about $\sigma(x)$ we may write from the formula of Christoffel–Darboux (Bateman, 1953b)

$$\sum_0^{N-1} \varphi_j^2(x) = N\varphi_N^2(x) - \left[N(N+1)\right]^{1/2} \varphi_{N-1}(x)\varphi_{N+1}(x),$$

and use the known (Erdelyi, 1960) asymptotic behavior of the functions $\varphi_{N-1}(x)$, $\varphi_N(x)$, and $\varphi_{N+1}(x)$ for the various intervals of x.

The level density $\sigma_N(x)$ for the Gaussian orthogonal ensemble contains two extra terms, namely, $\sqrt{N/2}\varphi_{N-1}(x)\int_{-\infty}^{\infty} \varepsilon(x-t)\varphi_N(t)\,dt$ and $\varphi_{2m}(x)/\int_{-\infty}^{\infty}\varphi_{2m}(t)\,dt$. In the limit $N \to \infty$, their contribution is negligible.

A.10 The Limits of $\sum_0^{N-1} \varphi_j(x)\varphi_j(y)$, etc. Sections 6.2, 7.2 and 8.2

The Christoffel–Darboux formula gives (Bateman, 1953b)

$$\sum_0^{N-1} \varphi_j(x)\varphi_j(y) = \left(\frac{1}{2}N\right)^{1/2} \left[\frac{\varphi_N(x)\varphi_{N-1}(y) - \varphi_N(y)\varphi_{N-1}(x)}{x-y}\right]. \qquad \text{(A.10.1)}$$

Let $N = 2m$, $2m^{1/2}x = \pi t\xi$, $2m^{1/2}y = \pi t\eta$ and let us take the limit $m \to \infty$, $x \to 0$, $y \to 0$; whereas ξ and η are finite. Using the formula (Bateman, 1953b)

$$\lim(-1)^m m^{1/4}\varphi_{2m}(x) = \pi^{-1/2}\cos(\pi t\xi),$$
$$\lim(-1)^m m^{1/4}\varphi_{2m+1}(x) = \pi^{-1/2}\sin(\pi t\xi),$$

we get

$$\lim \sum_0^{2m-1} \varphi_j(x)\varphi_j(y) = \frac{2m^{1/2}}{\pi}\frac{\sin(\pi\xi t)\cos(\pi\eta t) - \cos(\pi\xi t)\sin(\pi\eta t)}{\pi\xi t - \pi\eta t}$$
$$= \frac{2\sqrt{m}}{\pi}\frac{\sin(\xi-\eta)\pi t}{(\xi-\eta)\pi t}, \tag{A.10.2}$$

which is (6.2.14). To obtain the limit of

$$K_m(x, y) = \sum_0^{m-1} \varphi_{2j}(x)\varphi_{2j}(y)$$

we observe that $K_m(x, y)$ is the even part in x of (A.10.2).

$$K_m(x, y) = \frac{1}{2}\sum_0^{2m-1} \left[\varphi_j(x)\varphi_j(y) + \varphi_j(-x)\varphi_j(y)\right].$$

Therefore

$$\lim K_m(x, y) = \frac{1}{2}\frac{2\sqrt{m}}{\pi}\left[\frac{\sin(\xi-\eta)\pi}{(\xi-\eta)\pi} + \frac{\sin(\xi+\eta)\pi}{(\xi+\eta)\pi}\right] \tag{A.10.3}$$
$$= \frac{2\sqrt{m}}{\pi}\mathcal{K}_+(\xi, \eta), \quad \text{say.} \tag{A.10.4}$$

Also

$$\lim \sum_0^{m-1} \varphi_{2j}(x)\varphi'_{2j}(y) = \lim \frac{\partial}{\partial y}\sum_0^{m-1} \varphi_{2j}(x)\varphi_{2j}(y)$$
$$= \frac{2\sqrt{m}}{\pi}\frac{\partial}{\partial\eta}\lim K_m(x, y)$$
$$= \left(\frac{2\sqrt{m}}{\pi}\right)^2 \frac{\partial}{\partial\eta}\mathcal{K}_+(\xi, \eta), \tag{A.10.5}$$

and

$$\lim \sum_0^{m-1} \varphi_{2j}(x) \int_0^y \varphi_{2j}(z)\, dz = \lim \int_0^y K_m(x,z)\, dz = \int_0^\eta \overline{\mathcal{K}}(\xi,\zeta)\, d\zeta. \tag{A.10.6}$$

By writing similar equations for

$$\lim \sum_0^{m-1} \varphi'_{2j}(x)\varphi_{2j}(y) \quad \text{and} \quad \lim \sum_0^{m-1} \varphi_{2j}(y) \int_0^x \varphi_{2j}(z)\, dz,$$

and combining, we obtain (7.2.36)–(7.2.38).

Let us make an observation here. From the orthonormality of the $\varphi_j(x)$

$$\int_{-\infty}^{\infty} \varphi_j(x)\varphi_k(x)\, dx = \delta_{jk}$$

one knows that $K_N(x,y) = \sum_{j=0}^{N-1} \varphi_j(x)\varphi_j(y)$ satisfies

$$\int_{-\infty}^{\infty} K_N(x,y)K_N(y,z)\, dy = K_N(x,z).$$

Setting $x = \pi u/\sqrt{2N}$, $y = \pi v/\sqrt{2N}$, $z = \pi w/\sqrt{2N}$ and taking the limit $N \to \infty$, one gets

$$\int_{-\infty}^{\infty} \frac{\sin\pi(u-v)}{\pi(u-v)} \frac{\sin\pi(v-w)}{\pi(v-w)}\, dv = \frac{\sin\pi(u-w)}{\pi(u-w)}. \tag{A.10.7}$$

Similarly, from

$$\begin{aligned} K_{\text{Airy}}(x,y) &= \lim_{N\to\infty} \frac{1}{\sqrt{2}N^{1/6}} K_N\left(\sqrt{2N} + \frac{x}{\sqrt{2}\,N^{1/6}}, \sqrt{2N} + \frac{y}{\sqrt{2}\,N^{1/6}}\right) \\ &= \frac{Ai(x)Ai'(y) - Ai'(x)Ai(y)}{x-y}, \end{aligned}$$

where $Ai(x)$ is the Airy function satisfying

$$Ai''(x) = xAi(x),$$

and primes denote derivatives, one deduces that

$$\int_{-\infty}^{\infty} K_{\text{Airy}}(x,y)K_{\text{Airy}}(y,z)\, dy = K_{\text{Airy}}(x,z). \tag{A.10.8}$$

A.11 The Fourier Transforms of the Two-Point Cluster Functions

The functions $s(r)$, $(d/dr)(s(r))$, and $\int_0^r s(z)\,dz$ are given only for positive values of r by (6.2.14), (7.2.37) and (7.2.38); for negative values of r they are defined by the statement that they are even functions of r. Therefore writing $F(f(x))$ for the Fourier transform of $f(x)$,

$$
\begin{aligned}
F\big(s(r)\big) &= \int_{-\infty}^{\infty} e^{2\pi ikr} s(r)\,dr = 2\int_0^{\infty} \cos(2\pi|k|r)\frac{\sin\pi r}{\pi r}\,dr \\
&= \int_0^{\infty} \big(\sin(2|k|+1)\pi r - \sin(2|k|-1)\pi r\big)\frac{dr}{\pi r} \\
&= \frac{1}{2}\big(\text{sign}(2|k|+1) - \text{sign}(2|k|-1)\big) \\
&= \begin{cases} 1, & \text{if } |k| < 1/2, \\ 1/2, & \text{if } |k| = 1/2, \\ 0, & \text{if } |k| > 1/2. \end{cases}
\end{aligned} \tag{A.11.1}
$$

$$
\begin{aligned}
F\big[s^2(r)\big] &\equiv \int_{-\infty}^{\infty} e^{2\pi ikr} s^2(r)\,dr = 2\int_0^{\infty} \cos(2\pi|k|r)\frac{\sin^2\pi r}{\pi^2 r^2}\,dr \\
&= \int_0^{\infty} \cos(2\pi|k|r)\frac{1-\cos 2\pi r}{\pi^2 r^2}\,dr \\
&= \int_0^{\infty} (2\pi^2 r^2)^{-1}\big[2\cos(2\pi|k|r) - \cos(2|k|+2)\pi r - \cos(2|k|-2)\pi r\big]\,dr.
\end{aligned}
$$

Now

$$
\begin{aligned}
\int_0^{\infty} r^{-2}(\cos ar - \cos br)\,dr &= \int_a^b d\lambda \int_0^{\infty} r^{-1}\sin\lambda r\,dr \\
&= \frac{\pi}{2}\int_a^b d\lambda\,\text{sign}\,\lambda = \frac{\pi}{2}\big(|b| - |a|\big),
\end{aligned}
$$

so that

$$
\begin{aligned}
F\big[s^2(r)\big] &= (2\pi^2)^{-1}\frac{\pi}{2}2\pi\big[(|k|+1) + \big|(|k|-1)\big| - 2|k|\big] \\
&= \frac{1}{2}\big[(1-|k|) + \big|(1-|k|)\big|\big] \\
&= \begin{cases} 1-|k|, & \text{if } |k| < 1, \\ 0, & \text{if } |k| > 1. \end{cases}
\end{aligned} \tag{A.11.2}
$$

By partial integration we have

$$F\left[\frac{d}{dr}s(r)\right] = 2\int_0^\infty \cos(2\pi|k|r)\frac{d}{dr}\frac{\sin\pi r}{\pi r}\,dr$$
$$= 2\left\{-1 + |k|\int_0^\infty r^{-1}\left[\cos(2|k|-1)\pi r - \cos(2|k|+1)\pi r\right]dr\right\}$$
$$= 2\left(-1 + |k|\int_{(2|k|-1)\pi}^{(2|k|+1)\pi} d\lambda \int_0^\infty dr\sin\lambda r\right).$$

Now

$$\int_0^\infty dr\sin\lambda r = \lim_{\alpha\to 0}\int_0^\infty e^{-\alpha r}\sin\lambda r\,dr = \lim_{\alpha\to 0}\frac{\lambda}{\alpha^2+\lambda^2} = \frac{1}{\lambda},$$

so that

$$F\left(\frac{d}{dr}s(r)\right) = 2\left(-1 + |k|\ln\frac{(2|k|+1)}{(2|k|-1)}\right). \tag{A.11.3}$$

Also by partial integration

$$F\left\{\left[\int_0^r s(z)\,dz\right]\left[\frac{d}{dr}s(r)\right]\right\}$$
$$= -F\left[s^2(r)\right] + 4\pi|k|\int_0^\infty dr\sin(2\pi|k|r)\frac{\sin\pi r}{\pi r}\left[\int_0^r s(z)\,dz\right],$$

and

$$2\int_0^\infty r^{-1}\sin(2\pi|k|r)\sin\pi r\left(\int_0^r\frac{\sin\pi x}{x}\,dx\right)dr$$
$$= \int_{(2|k|-1)\pi}^{(2|k|+1)\pi} dz\int_0^\infty dr\sin zr\left(\int_0^r\frac{\sin\pi x}{x}\,dx\right)$$
$$= \int_{(2|k|-1)\pi}^{(2|k|+1)\pi} dz\int_0^\infty dr\frac{\sin\pi r}{r}\frac{\cos zr}{z}$$
$$= \int_{(2|k|-1)\pi}^{(2|k|+1)\pi} dz\,\frac{1}{2z}\left(\mathrm{sign}(\pi+z) + \mathrm{sign}(\pi-z)\right)$$
$$= \begin{cases} -\frac{1}{2}\ln|(2|k|-1)|, & \text{if } |k|<1,\\ 0, & \text{if } |k|>1,\end{cases}$$

so that

$$F\left\{\left[\int_0^r s(z)\,dz\right]\left[\frac{d}{dr}s(r)\right]\right\} = \begin{cases} -1+|k|-|k|\ln|(2|k|-1)|, & \text{if } |k|<1, \\ 0, & \text{if } |k|>1. \end{cases} \tag{A.11.4}$$

Combining Eqs. (A.11.1)–(A.11.4), we get all the Fourier transforms quoted in Chapters 6, 7 and 8.

A.12 Some Applications of Gram's Formula

We have used Gram's result repeatedly, e.g., in Sections 6.3, 7.6, 11.2 and 13.2. It may be stated as follows (cf. Mehta, 1989MT).

Let v_i, $i = 1, 2, \ldots, m$, be m vectors and let $v_{i\nu}$, $\nu = 1, 2, \ldots, n$, be their components along some basis. Form the scalar products

$$b_{ij} = (v_i, v_j) = \sum_{\nu=1}^{n} v_{i\nu} v_{j\nu}^*, \quad i, j = 1, 2, \ldots, m;$$

then

$$\det\begin{bmatrix} b_{11} & b_{12} & \ldots & b_{1m} \\ b_{21} & b_{22} & \ldots & b_{2m} \\ \ldots & \ldots & \ldots & \ldots \\ b_{m1} & b_{m2} & \ldots & b_{mm} \end{bmatrix} = \frac{1}{m!} \sum_{\nu_1,\nu_2,\ldots,\nu_m} \left| \det\begin{bmatrix} v_{1\nu_1} & \ldots & v_{1\nu_m} \\ \ldots & \ldots & \ldots \\ v_{m\nu_m} & \ldots & v_{m\nu_m} \end{bmatrix} \right|^2, \tag{A.12.1}$$

where on the right-hand side we sum over all possible ways of choosing $\nu_1, \nu_2, \ldots, \nu_m$ among $1, 2, \ldots, n$. The summation over ν may be finite, denumerable, or continuously infinite, the summation sign being replaced by an integration over a suitable measure.

If we make the correspondence

$$v_{i\nu} \to \varphi_{i-1}(x), \quad \sum_{\nu} \to \int_{\text{out}} dx = \left(\int_{-\infty}^{-\theta} + \int_{\theta}^{\infty}\right) dx,$$

the scalar products become

$$\begin{aligned} b_{ij} &= \sum_{\nu} v_{i\nu} v_{j\nu}^* \to \left(\int_{-\infty}^{-\theta} + \int_{\theta}^{\infty}\right) \varphi_{i-1}(x)\varphi_{j-1}(x)\,dx \\ &= \delta_{ij} - \int_{-\theta}^{\theta} \varphi_{i-1}(x)\varphi_{j-1}(x)\,dx = g_{ij}, \end{aligned}$$

and Eq. (A.12.1) yields (6.3.3). If we make the correspondence

$$v_{i\nu} \to \varphi_{2i-2}(x), \quad \sum_{\nu} \to \int_{\theta}^{\infty} dx,$$

then the scalar products become

$$b_{ij} = \sum_{\nu} v_{i\nu} v_{j\nu}^{*} \to \int_{\theta}^{\infty} \varphi_{2i-2}(x)\varphi_{2j-2}(x)\, dx,$$

and Eq. (A.12.1) gives Eq. (7.6.2). If we make the correspondence

$$v_{i\nu} \to \sqrt{2}\, e^{-(1/2)y^2} \frac{y^{2i+1/2}}{(y^2+\theta^2)^{1/4}}, \quad \sum_{\nu} \to \int_{0}^{\infty} dy,$$

the scalar products become

$$b_{ij} = \sum_{\nu} v_{i\nu} v_{j\nu}^{*} \to \int_{0}^{\infty} dy\, 2e^{-y^2} \frac{y^{2i+2j+1}}{(y^2+\theta^2)^{1/2}} = \eta_{i+j},$$

and Eq. (A.12.1) yields (7.6.9). If we make the correspondence

$$v_{j\nu} \to (2\pi)^{-1/2} e^{ip\theta}, \quad \sum_{\nu} \to \int_{\alpha}^{2\pi-\alpha} d\theta,$$

the scalar products become

$$b_{jk} = (2\pi)^{-1} \int_{\alpha}^{2\pi-\alpha} d\theta\, e^{i(p-q)\theta},$$

and we get (11.2.3).

A.13 Power Series Expansions of Eigenvalues, of Spheroidal Functions and of Various Probabilities

The first few terms in the power series expansions of the functions $f_j(x)$, Eq. (6.3.17), λ_j, Eq. (6.3.18), $E_\beta(n;s)$, Eqs. (6.3.27), (6.4.30), (7.3.19), (7.4.13), (7.5.21), and $p_\beta(n;s)$, Eqs. (6.4.32), (6.4.33), (7.4.14), (7.4.18), (7.4.19) are reproduced here.

$$\lambda_0 = s - \frac{\pi^2}{36}s^3 + \frac{23\pi^4}{32400}s^5 - \frac{79\pi^6}{5715360}s^7 + O(s^9),$$

$$\lambda_1 = \frac{\pi^2}{36}s^3 - \frac{\pi^4}{1200}s^5 + \frac{41\pi^6}{2940000}s^7 + O(s^9),$$
$$\lambda_2 = \frac{\pi^4}{8100}s^5 - \frac{\pi^6}{2857680}s^7 + O(s^9),$$
$$\lambda_3 = \frac{\pi^6}{4410000}s^7 + O(s^9),$$
$$\lambda_j = O(s^{2j+1}),$$

$$f_0(x) = \left(\frac{1}{2}\right)^{1/2}\left(\left(1 - \frac{\pi^4 s^4}{12960}\right) + \left(-\frac{\pi^2 s^2}{36} + \frac{\pi^4 s^4}{4536}\right)P_2(x) + \frac{\pi^4 s^4}{8400}P_4(x) + O(s^6)\right),$$
$$f_1(x) = \left(\frac{3}{2}\right)^{1/2}\left(\left(1 - \frac{3\pi^4 s^4}{140000}\right)P_1(x) + \left(-\frac{\pi^2 s^2}{100} - \frac{\pi^4 s^4}{45000}\right)P_3(x) + \frac{\pi^4 s^4}{35280}P_5(x) + O(s^6)\right),$$
$$f_2(x) = \left(\frac{5}{2}\right)^{1/2}\left(\left(\frac{\pi^2 s^2}{180} - \frac{\pi^4 s^4}{22680}\right) + \left(1 - \frac{545\pi^4 s^4}{6223392}\right)P_2(x) + \left(-\frac{3\pi^2 s^2}{490} - \frac{23\pi^4 s^4}{384160}\right)P_4(x) + O(s^6)\right),$$
$$f_j(x) = \left(\frac{2j+1}{2}\right)^{1/2} P_j(x) + O(s^2),$$

where $P_j(x)$ are the Legendre polynomials,

$$E_1(0;s) = 1 - s + \frac{\pi^2 s^3}{36} - \frac{\pi^4 s^5}{1200} + \frac{\pi^4 s^6}{8100} + \frac{\pi^6 s^7}{70560} - \frac{\pi^6 s^8}{264600} - \frac{\pi^8 s^9}{6531840} + O(s^{10}),$$
$$E_1(1;s) = s - \frac{\pi^2 s^3}{18} + \frac{\pi^4 s^5}{600} - \frac{\pi^4 s^6}{8100} - \frac{\pi^6 s^7}{35280} + \frac{\pi^6 s^8}{264600} + O(s^9),$$
$$E_1(2;s) = \frac{\pi^2 s^3}{36} - \frac{\pi^4 s^5}{1200} - \frac{\pi^4 s^6}{8100} + O(s^7),$$

$$E_2(0;s) = 1 - s + \frac{\pi^2 s^4}{36} - \frac{\pi^4 s^6}{675} + \frac{\pi^6 s^8}{17640} - \frac{\pi^6 s^9}{291600} - \frac{\pi^8 s^{10}}{637875} + O(s^{11}),$$
$$E_2(1;s) = s - \frac{\pi^2 s^4}{18} + \frac{2\pi^4 s^6}{675} - \frac{\pi^6 s^8}{8820} + \cdots,$$
$$E_2(2;s) = \frac{\pi^2 s^4}{36} - \frac{\pi^4 s^6}{675} + \cdots,$$

$$E_4(0;s) = E_1(0;2s) + \frac{1}{2}E_1(1;2s) = 1 - s + \frac{8\pi^4 s^6}{2025} - \frac{16\pi^6 s^8}{33075} + \cdots,$$

$$E_4(n;s) = E_1(2n;2s) + \frac{1}{2}E_1(2n-1;2s) + \frac{1}{2}E_1(2n+1;2s), \quad n \geqslant 1,$$

$$p_1(0;s) = \frac{\pi^2 s}{6} - \frac{\pi^4 s^3}{60} + \frac{\pi^4 s^4}{270} + \frac{\pi^6 s^5}{1680} - \frac{\pi^6 s^6}{4725} - \frac{\pi^8 s^7}{90720} + \cdots,$$

$$p_1(1;s) = \frac{\pi^4 s^4}{270} - \frac{\pi^6 s^6}{4725} + \cdots,$$

$$p_1(2;s) = \frac{\pi^8 s^8}{1764000} + \cdots,$$

$$p_2(0;s) = \frac{\pi^2 s^2}{3} - \frac{2\pi^4 s^4}{45} + \frac{\pi^6 s^6}{315} - \frac{\pi^6 s^7}{4050} - \frac{2\pi^8 s^8}{14175} + \cdots,$$

$$p_2(1;s) = \frac{\pi^6 s^7}{4050} + \cdots,$$

$$p_2(2;s) = \frac{\pi^{12} s^{14}}{5358150000} + \cdots,$$

$$p_4(0;s) = \frac{16\pi^4 s^4}{135} - \frac{128\pi^6 s^6}{4725} + \cdots,$$

$$B_1(t;y) = 1 - \frac{\pi^2}{24}(t^2 + y^2) + \frac{\pi^4}{1920}(t^4 + 6t^2y^2 + y^4) + \frac{\pi^4 t^3}{5400}(t^2 - 5y^2) + \cdots,$$

$$B_2(t;y) = 1 - \frac{2}{9}\pi^2 t(3y^2 + t^2) + \frac{4}{225}\pi^4 t(5y^4 + 10y^2t^2 + t^4) + \cdots,$$

$$\mathcal{B}_1(x_1, x_2) = 1 - \frac{\pi^2}{12}\left(x_1^2 + x_2^2\right) + \frac{\pi^4}{240}\left(x_1^4 + x_2^4\right) - \frac{\pi^4}{1350}\left(x_1^5 - 5x_1^2x_2^2(x_1 + x_2) + x_2^5\right) + \cdots,$$

$$\mathcal{B}_2(x_1, x_2) = 1 - \frac{\pi^2}{9}\left(x_1^3 + x_2^3\right) + \frac{2\pi^4}{225}\left(x_1^5 + x_2^5\right) + \cdots,$$

$$\mathcal{P}_1(x_1, x_2) = \frac{\pi^4}{45}x_1x_2(x_1 + x_2) + \cdots.$$

From the method of Chapter 20 or 21 one can get more terms without much effort. For example, Dietz computed the power series expansion of $E_\beta(0, s)$ up to s^{40} to get a good Padé approximation for small as well as large s. Here we almost copy from Dietz and Haake (1990).

$$F_+(z,s) = 1 - z\left(s + \sum_{n=1}^{\infty} \frac{(-\pi^2)^n \cdot s^{2n+1}}{2\cdot(2n+1)\cdot(2n+1)!}\right) + z^2\left(\frac{\pi^4\cdot s^6}{2^2\cdot 3^4\cdot 5^2} - \frac{\pi^6\cdot s^8}{2^3\cdot 3^3\cdot 5^2\cdot 7^2}\right.$$
$$+ \frac{29\cdot\pi^8\cdot s^{10}}{2^6\cdot 3^5\cdot 5^4\cdot 7^2} - \frac{47\cdot\pi^{10}\cdot s^{12}}{2^7\cdot 3^6\cdot 5^3\cdot 7^2\cdot 11^2} + \frac{19\cdot 487\cdot\pi^{12}\cdot s^{14}}{2^{10}\cdot 3^5\cdot 5^3\cdot 7^4\cdot 11^2\cdot 13^2}$$
$$- \frac{307\cdot\pi^{14}\cdot s^{16}}{2^{11}\cdot 3^6\cdot 5^4\cdot 7^3\cdot 11^2\cdot 13^2} + \frac{1797581\cdot\pi^{16}\cdot s^{18}}{2^{13}\cdot 3^{12}\cdot 5^4\cdot 7^3\cdot 11^2\cdot 13^2\cdot 17^2}$$
$$- \frac{4495537\cdot\pi^{18}\cdot s^{20}}{2^{14}\cdot 3^{10}\cdot 5^5\cdot 7^3\cdot 11^2\cdot 13^2\cdot 17^2\cdot 19^2}$$
$$+ \frac{28943983\cdot\pi^{20}\cdot s^{22}}{2^{18}\cdot 3^{11}\cdot 5^4\cdot 7^3\cdot 11^4\cdot 13^2\cdot 17^2\cdot 19^2}$$
$$\left. - \frac{171859117\cdot\pi^{22}\cdot s^{24}}{2^{19}\cdot 3^{13}\cdot 5^3\cdot 7^4\cdot 11^3\cdot 13^2\cdot 17^2\cdot 19^2\cdot 23^2} + \cdots\right)$$
$$+ z^3\left(-\frac{\pi^{12}\cdot s^{15}}{2^4\cdot 3^{10}\cdot 5^6\cdot 7^4} + \frac{\pi^{14}\cdot s^{17}}{2^6\cdot 3^9\cdot 5^5\cdot 7^4\cdot 11^2} - \frac{\pi^{16}\cdot s^{19}}{2^3\cdot 3^{10}\cdot 5^5\cdot 7^4\cdot 11^2\cdot 13^2}\right.$$
$$\left. + \frac{9059\cdot\pi^{18}\cdot s^{21}}{2^{10}\cdot 3^{13}\cdot 5^6\cdot 7^6\cdot 11^2\cdot 13^2} - \frac{492083\cdot\pi^{20}\cdot s^{23}}{2^{12}\cdot 3^{12}\cdot 5^6\cdot 7^5\cdot 11^3\cdot 13^2\cdot 17^2} + \cdots\right)$$
$$+ \cdots + z^r \mathrm{O}\left(s^{r(2r-1)}\right),$$

$$F_-(z,s) = 1 + z\sum_{n=1}^{\infty} \frac{(-\pi^2)^n \cdot s^{2n+1}}{2\cdot(2n+1)\cdot(2n+1)!} + z^2\left(\frac{\pi^8\cdot s^{10}}{2^6\cdot 3^4\cdot 5^4\cdot 7^2} - \frac{\pi^{10}\cdot s^{12}}{2^7\cdot 3^8\cdot 5^3\cdot 7^2}\right.$$
$$+ \frac{251\cdot\pi^{12}\cdot s^{14}}{2^{10}\cdot 3^7\cdot 5^3\cdot 7^4\cdot 11^2} - \frac{1973\cdot\pi^{14}\cdot s^{16}}{2^{11}\cdot 3^8\cdot 5^4\cdot 7^3\cdot 11^2\cdot 13^2}$$
$$\left. + \frac{1777\cdot\pi^{16}\cdot s^{18}}{2^{13}\cdot 3^{11}\cdot 5^4\cdot 7^3\cdot 11^2\cdot 13^2} - \frac{19441\cdot\pi^{18}\cdot s^{20}\cdot z^2}{2^{14}\cdot 3^9\cdot 5^6\cdot 7^3\cdot 11^2\cdot 13^2\cdot 17^2} + \cdots\right)$$
$$+ z^3\left(-\frac{\pi^{18}\cdot s^{21}}{2^{10}\cdot 3^{14}\cdot 5^6\cdot 7^6\cdot 11^2} + \cdots\right) + \cdots + z^r \mathrm{O}\left(s^{r(2r+1)}\right),$$

$$F_2(z,s) = 1 - z\cdot s - z^2\sum_{n=1}^{\infty} \frac{(-4\pi^2)^n \cdot s^{2n+2}}{(n+1)\cdot(2n+1)\cdot(2n+2)!}$$

$$+z^3\left(-\frac{\pi^6 \cdot s^9}{2^4 \cdot 3^6 \cdot 5^2} + \frac{\pi^8 \cdot s^{11}}{2 \cdot 3^4 \cdot 5^4 \cdot 7^2} - \frac{\pi^{10} \cdot s^{13}}{2^2 \cdot 3^8 \cdot 5^3 \cdot 7^2}\right.$$
$$+\frac{2293 \cdot \pi^{12} \cdot s^{15}}{3^9 \cdot 5^5 \cdot 7^4 \cdot 11^2} - \frac{3581 \cdot \pi^{14} \cdot s^{17}}{2^6 \cdot 3^8 \cdot 5^4 \cdot 7^3 \cdot 11^2 \cdot 13^2}$$
$$\left.+\frac{71 \cdot \pi^{16} \cdot s^{19}}{2^2 \cdot 3^{11} \cdot 5^4 \cdot 7^3 \cdot 11^2 \cdot 13^2} - \frac{94789 \cdot \pi^{18} \cdot s^{21}}{2^3 \cdot 3^{11} \cdot 5^5 \cdot 7^5 \cdot 11^2 \cdot 13^2 \cdot 17^2} + \cdots\right)$$
$$+z^4\left(\frac{\pi^{12} \cdot s^{16}}{2^8 \cdot 3^8 \cdot 5^6 \cdot 7^2} - \frac{\pi^{14} \cdot s^{18}}{2^4 \cdot 3^{11} \cdot 5^5 \cdot 7^4} + \frac{83 \cdot \pi^{16} \cdot s^{20}}{2^7 \cdot 3^9 \cdot 5^7 \cdot 7^4 \cdot 11^2} - \cdots\right)$$
$$+\cdots + z^r O\left(s^{r^2}\right).$$

A.14 Numerical Tables of $\lambda_j(s)$, $b_j(s)$ and $E_\beta(n;s)$ for $\beta = 1, 2$ and 4

Table A.14.1. Eigenvalues λ_j of the Integral Equation (6.3.14), for $0 \leqslant j \leqslant 8$, $0 \leqslant s \leqslant 5$

s	$\lambda_0(s)$	$\lambda_1(s)$	$\lambda_2(s)$	$\lambda_3(s)$	$\lambda_4(s)$	$\lambda_5(s)$	$\lambda_6(s)$	$\lambda_7(s)$	$\lambda_8(s)$
0.127	0.12676	0.00056							
0.255	0.25019	0.00444	0.00001						
0.382	0.36724	0.01463	0.00010						
0.509	0.47534	0.03355	0.00041						
0.637	0.57258	0.06279	0.00124	0.00001					
0.891	0.73072	0.15395	0.00650	0.00010					
1.146	0.84107	0.28243	0.02186	0.00056	0.00001				
1.401	0.91141	0.43134	0.05555	0.00222	0.00004				
1.655	0.95288	0.57972	0.11544	0.00697	0.00020				
1.910	0.97583	0.70996	0.20514	0.01820	0.00071	0.00002			
2.165	0.98793	0.81233	0.32102	0.04102	0.00214	0.00007			
2.419	0.99409	0.88538	0.45226	0.08154	0.56681	0.00022	0.00001		
2.674	0.99715	0.93336	0.58414	0.14510	0.01339	0.00065	0.00002		
2.928	0.99864	0.96278	0.70298	0.23361	0.02866	0.00172	0.00007		
3.183	0.99936	0.97987	0.79992	0.34356	0.05602	0.00418	0.00019	0.00001	
3.438	0.99970	0.98937	0.87226	0.46601	0.10050	0.00932	0.00052	0.00002	
3.692	0.99986	0.99450	0.92218	0.58890	0.16613	0.01950	0.00128	0.00006	
3.947	0.99994	0.99719	0.95442	0.70071	0.25388	0.03781	0.00295	0.00015	0.00001
4.202	0.99997	0.99859	0.97415	0.79357	0.36019	0.06844	0.00639	0.00038	0.00002
4.456	0.99999	0.99929	0.98571	0.86457	0.47705	0.11572	0.01306	0.00091	0.00005
4.711	1.00000	0.99965	0.99226	0.91499	0.59392	0.18282	0.02520	0.00202	0.00012
4.966		0.99983	0.99588	0.94863	0.70068	0.27007	0.04598	0.04286	0.00028
5.093		0.99988	0.99701	0.96055	0.74790	0.32028	0.06078	0.00613	0.00042

Table A.14.2. Numerical Values of $b_j(s) \equiv f_{2j}(1) \int_{-1}^{1} f_{2j}(x)\,dx / \int_{-1}^{1} f_{2j}^2(x)\,dx$, for $0 \leqslant j \leqslant 3$, $0 \leqslant s \leqslant 5$. The $f_j(x)$ are the spheroidal functions, solutions of Eq. (6.3.14)

s	b_0	b_1	b_2	b_3	b_4
0.000	1.00000				
0.127	0.99556	0.00444			
0.255	0.98230	0.01768	0.00002		
0.382	0.96040	0.03949	0.00011		
0.509	0.93021	0.06946	0.00033		
0.637	0.89225	0.10694	0.00081		
0.891	0.79630	0.20060	0.00309	0.00002	
1.146	0.68148	0.31011	0.00835	0.00007	
1.401	0.55958	0.42183	0.01836	0.00023	
1.655	0.44242	0.52179	0.03516	0.00063	0.00001
1.910	0.33878	0.59892	0.06082	0.00147	0.00002
2.165	0.25291	0.64676	0.09721	0.00308	0.00005
2.419	0.18520	0.66324	0.14552	0.00592	0.00011
2.674	0.13369	0.64973	0.20569	0.01064	0.00025
2.928	0.09549	0.61021	0.27574	0.01803	0.00051
3.183	0.06766	0.55075	0.35146	0.02912	0.00099
3.438	0.04765	0.47883	0.42662	0.04505	0.00181
3.692	0.03339	0.40225	0.49401	0.06709	0.00317
3.947	0.02331	0.32783	0.54689	0.09649	0.00532
4.202	0.01622	0.26039	0.58023	0.13423	0.00862
4.456	0.01126	0.20250	0.59141	0.18074	0.01353
4.711	0.00779	0.15483	0.58028	0.23549	0.02065
4.966	0.00539	0.11679	0.54896	0.29657	0.03067
5.093	0.00447	0.10101	0.52694	0.32847	0.03703

Table A.14.3. The Probabilities $E_1(n; s)$ of Having n Eigenvalues of a Real Symmetric Matrix in an Interval s for $0 \leqslant n \leqslant 7$, $0 \leqslant s \leqslant 5$

s	$E_1(0;s)$	$E_1(1;s)$	$E_1(2;s)$	$E_1(3;s)$	$E_1(4;s)$	$E_1(5;s)$	$E_1(6;s)$	$E_1(7;s)$
0.000	1.00000							
0.127	0.87324	0.12620	0.00056					
0.255	0.74980	0.24576	0.00444					
0.382	0.63270	0.35267	0.01460	0.00004				
0.509	0.52445	0.44200	0.03336	0.00019				
0.573	0.47425	0.47893	0.04644	0.00039				
0.637	0.42689	0.51031	0.06209	0.00071				
0.891	0.26753	0.57844	0.14928	0.00473	0.00001			
1.146	0.15546	0.56171	0.26444	0.01823	0.00016			
1.401	0.08367	0.48373	0.38194	0.04971	0.00096			
1.655	0.04167	0.37567	0.47249	0.10612	0.00402	0.00002		
1.910	0.01920	0.26555	0.51451	0.18780	0.01279	0.00014		
2.165	0.00818	0.17178	0.50144	0.28523	0.03269	0.00068		
2.419	0.00322	0.10204	0.44206	0.38032	0.06982	0.00254	0.00002	
2.674	0.00117	0.05577	0.35500	0.45211	0.12814	0.00773	0.00009	
2.928	0.00039	0.02808	0.26096	0.48437	0.20603	0.01978	0.00039	

(continued on next page)

Table A.14.3. (*Continued*)

s	$E_1(0;s)$	$E_1(1;s)$	$E_1(2;s)$	$E_1(3;s)$	$E_1(4;s)$	$E_1(5;s)$	$E_1(6;s)$	$E_1(7;s)$
3.183	0.00012	0.01304	0.17617	0.47130	0.29444	0.04352	0.00140	0.00001
3.438	0.00003	0.00559	0.10946	0.41885	0.37802	0.08372	0.00428	0.00005
3.692	0.00001	0.00221	0.06270	0.34136	0.43954	0.14273	0.01126	0.00020
3.947		0.00081	0.03314	0.25590	0.46572	0.21791	0.02581	0.00071
4.202		0.00027	0.01618	0.17684	0.45179	0.30039	0.05228	0.00223
4.456		0.00008	0.00730	0.11282	0.40273	0.37638	0.09453	0.00607
4.711		0.00002	0.00304	0.06652	0.33077	0.43087	0.15385	0.01458
4.966		0.00001	0.00117	0.03628	0.25083	0.45252	0.22692	0.03117
5.093			0.00071	0.02602	0.21209	0.44948	0.26607	0.04373

Table A.14.4. The Probabilities $E_2(n;s)$ of Having n Eigenvalues of a Random Hermitian Matrix in an Interval s for $0 \leqslant n \leqslant 7$, $0 \leqslant s \leqslant 5$

s	$E_2(0;s)$	$E_2(1;s)$	$E_2(2;s)$	$E_2(3;s)$	$E_2(4;s)$	$E_2(5;s)$	$E_2(6;s)$	$E_2(7;s)$
0.000	1.00000							
0.127	0.87275	0.12718	0.00007					
0.255	0.74647	0.25242	0.00111					
0.382	0.62344	0.37115	0.00541					
0.509	0.50685	0.47700	0.01614	0.00001				
0.637	0.40008	0.56327	0.03661	0.00004				
0.891	0.22632	0.65684	0.11610	0.00074				
1.146	0.11149	0.63644	0.24674	0.00533				
1.401	0.04747	0.52729	0.40249	0.02270	0.00005			
1.655	0.01739	0.37809	0.53689	0.06717	0.00046			
1.910	0.00547	0.23561	0.60522	0.15101	0.00270			
2.165	0.00147	0.12759	0.58709	0.27268	0.01114	0.00002		
2.419	0.00034	0.05991	0.49506	0.40982	0.03468	0.00019		
2.674	0.00007	0.02433	0.36473	0.52458	0.08518	0.00111		
2.928	0.00001	0.00852	0.23518	0.58041	0.17104	0.00482	0.00001	
3.183		0.00257	0.13262	0.56033	0.28814	0.01627	0.00007	
3.438		0.00066	0.06528	0.47461	0.41494	0.04411	0.00040	
3.692		0.00015	0.02796	0.35367	0.51765	0.09868	0.00188	
3.947		0.00003	0.01040	0.23199	0.56466	0.18595	0.00695	0.00002
4.202			0.00335	0.13381	0.54181	0.29997	0.02093	0.00013
4.456			0.00093	0.06772	0.45897	0.41949	0.05222	0.00066
4.711			0.00022	0.02999	0.34379	0.51330	0.10998	0.00271
4.966			0.00005	0.01159	0.22769	0.55320	0.19836	0.00908
5.093			0.00002	0.00684	0.17687	0.54838	0.25230	0.01549

Table A.14.5. Probabilities $E_4(n;s)$ of Having n Eigenvalues of a Random Self-Dual Hermitian Matrix in an Interval s for $0 \leqslant n \leqslant 7, 0 \leqslant s \leqslant 6$

s	$E_4(0;s)$	$E_4(1;s)$	$E_4(2;s)$	$E_4(3;s)$	$E_4(4;s)$	$E_4(5;s)$	$E_4(6;s)$	$E_4(7;s)$
0.000	1.00000							
0.127	0.87268	0.12732						
0.255	0.74545	0.25445	0.00010					
0.382	0.61903	0.37996	0.00100					
0.509	0.49565	0.49940	0.00495					
0.637	0.37938	0.60461	0.01600					
0.891	0.18829	0.73217	0.07950	0.00003				
1.146	0.07208	0.71061	0.21663	0.00068				
1.401	0.02067	0.56458	0.40825	0.00650				
1.655	0.00438	0.36923	0.59322	0.03317	0.00001			
1.910	0.00068	0.19603	0.69624	0.10687	0.00019			
2.165	0.00008	0.08238	0.67242	0.24321	0.00191			
2.419	0.00001	0.02672	0.53883	0.42302	0.01143			
2.674		0.00656	0.35802	0.59051	0.04488	0.00004		
2.928		0.00120	0.19485	0.67862	0.12487	0.00045		
3.183		0.00016	0.08520	0.64929	0.26207	0.00328		
3.438		0.00002	0.02922	0.51973	0.43513	0.01590	0.00001	
3.692			0.00772	0.34691	0.59068	0.05459	0.00009	
3.947			0.00155	0.19083	0.66723	0.13961	0.00078	
4.202			0.00023	0.08513	0.63202	0.27789	0.00472	
4.456			0.00003	0.03021	0.50335	0.44626	0.02015	0.00001
4.711				0.00832	0.33602	0.59211	0.06339	0.00016
4.966				0.00176	0.18533	0.65888	0.15285	0.00117
5.220				0.00019	0.07777	0.62056	0.29517	0.00630
5.475				0.00004	0.03048	0.48899	0.45617	0.02430
5.730					0.00856	0.32485	0.59459	0.07176
5.984					0.00190	0.18029	0.65110	0.16510
6.239					0.00033	0.08162	0.60330	0.30687
6.366					0.00012	0.05087	0.54700	0.38653

A.15 Numerical Values of $E_\beta(0;s)$, $\Psi_\beta(s)$ and $p_\beta(0;s)$ for $\beta = 1, 2$ and $s \leqslant 3.7$

Table A.15.1.

s	$E_1(0;s)$	$E_2(0;s)$	$\Psi_1(s)$	$\Psi_2(s)$	$p_1(0;s)$	$p_2(0;s)$
0.000	1.00000	1.00000				
0.064	0.93641	0.93634	0.00333	0.00028	0.10431	0.01326
0.127	0.87324	0.87275	0.01323	0.00223	0.20620	0.05221
0.191	0.81090	0.80937	0.02948	0.00742	0.30346	0.11438
0.255	0.74980	0.74647	0.05173	0.01721	0.39414	0.19592
0.318	0.69028	0.68436	0.07949	0.03267	0.47657	0.29186
0.382	0.63270	0.62344	0.11220	0.05455	0.54944	0.39649
0.446	0.57734	0.56413	0.14922	0.08321	0.61178	0.50384
0.509	0.52445	0.50685	0.18986	0.11864	0.66295	0.60804

(continued on next page)

Table A.15.1. *(Continued)*

s	$E_1(0; s)$	$E_2(0; s)$	$\Psi_1(s)$	$\Psi_2(s)$	$p_1(0; s)$	$p_2(0; s)$
0.573	0.47425	0.45204	0.23339	0.16045	0.70266	0.70370
0.637	0.42689	0.40008	0.27908	0.20796	0.73095	0.78628
0.764	0.34112	0.30596	0.37411	0.31607	0.75477	0.89937
0.891	0.26753	0.22632	0.46963	0.43360	0.73965	0.93351
1.019	0.20589	0.16171	0.56114	0.55059	0.69350	0.89288
1.146	0.15546	0.11149	0.64529	0.65853	0.62542	0.79483
1.273	0.11515	0.07411	0.71986	0.75156	0.54452	0.66302
1.401	0.08367	0.04747	0.78376	0.82693	0.45891	0.52079
1.528	0.05962	0.02929	0.83680	0.88453	0.37514	0.38659
1.655	0.04167	0.01739	0.87956	0.92621	0.29791	0.27194
1.783	0.02856	0.00994	0.91307	0.95482	0.23010	0.18166
1.910	0.01920	0.00547	0.93862	0.97350	0.17304	0.11543
2.037	0.01265	0.00289	0.95759	0.98510	0.12679	0.06986
2.165	0.00818	0.00147	0.97133	0.99197	0.09058	0.04032
2.292	0.00518	0.00072	0.98103	0.99585	0.06314	0.02221
2.419	0.00322	0.00034	0.98772	0.99794	0.04295	0.01168
2.546	0.00196	0.00015	0.99221	0.99902	0.02853	0.00588
2.674	0.00117	0.00007	0.99517	0.99955	0.01851	0.00283
2.801	0.00069	0.00003	0.99706	0.99980	0.01173	0.00130
2.928	0.00039	0.00001	0.99825	0.99992	0.00727	0.00057
3.056	0.00022		0.99898	0.99997	0.00440	0.00024
3.183	0.00012		0.99942	0.99999	0.00261	0.00010
3.310	0.00007		0.99968	1.00000	0.00151	0.00004
3.438	0.00003		0.99982		0.00085	0.00001
3.565	0.00002		0.99991		0.00047	
3.692	0.00001		0.99995		0.00026	

A.16 Proof of Eqs. (21.1.11)–(21.1.16), (24.3.11), (24.3.15) and (24.3.20) Using a Compact Notation

Expanding the quantities in powers of z and comparing coefficients to prove various relations in Chapters 21 and 24 is somewhat cumbersome. The same thing can be achieved using compact notations which avoid series expansions and replace tedious integrations by parts by simple commutation relations, thus making those manipulations more transparent and hence perhaps easier to understand. Two good transcriptions exist in print, that of Dyson (1995) and of Tracy and Widom (1994b, 1994c). Here we reproduce still another one kindly written by Mahoux and assimilating the essentials of the above two.

The contents of this appendix is pure algebra. No attempt is made to give a full mathematical definition of the various operators introduced (domains, possible self-adjointness, ...).

A.16.1 Preliminaries. Using Dirac notation of *bras* and *kets*, we write $\langle x|f\rangle$ any element $f(x)$ of the space $L^2(\mathbb{R}, dx)$. Here, the *localized states* $|x\rangle$ satisfy the orthogonality and completion relations

$$\langle x|x'\rangle = \delta(x - x'), \tag{A.16.1}$$

$$\int_{-\infty}^{\infty} dx\, |x\rangle\langle x| = 1. \tag{A.16.2}$$

We denote by X and Π the *position* and *parity* operators

$$X|x\rangle := x|x\rangle, \tag{A.16.3}$$

$$\Pi|x\rangle := |-x\rangle, \tag{A.16.4}$$

and by D the derivation operator such that, for any $f(x) \in L^2(\mathbb{R}, dx)$,

$$\langle x|D|f\rangle := \frac{d}{dx}\langle x|f\rangle, \tag{A.16.5}$$

where the derivative in the right-hand side has to be understood in the sense of distributions. D is antihermitian, $D^\dagger = -D$, thus

$$\langle f|D|x\rangle = -\frac{d}{dx}\langle f|x\rangle. \tag{A.16.6}$$

Let $\mathcal{F}$ be the *Fourier transform* operator with kernel

$$\langle x|\mathcal{F}|y\rangle := \frac{1}{\sqrt{2}}\exp(i\pi xy). \tag{A.16.7}$$

$\mathcal{F}$ is unitary, $\mathcal{F}^{-1} = \mathcal{F}^\dagger$, and it enjoys the obvious properties:

$$\mathcal{F}D = -i\pi\, X\mathcal{F}, \tag{A.16.8}$$

$$\text{if } \langle x|f\rangle \text{ is real, then } \langle x|\mathcal{F}|f\rangle = \langle f|\mathcal{F}|x\rangle. \tag{A.16.9}$$

We also define, for t positive, a *dilatation* operator $\mathbb{k}_t$ such that

$$\mathbb{k}_t|x\rangle := t^{1/2}|tx\rangle. \tag{A.16.10}$$

One can check that $\mathbb{k}_t$ satisfies the relations

$$\mathbb{k}_t^\dagger = \mathbb{k}_t^{-1} = \mathbb{k}_{1/t} = \mathcal{F}\mathbb{k}_t\mathcal{F}^\dagger = \mathcal{F}^\dagger\mathbb{k}_t\mathcal{F}. \tag{A.16.11}$$

A.16.2 The Sine Kernel: Definitions and Properties. We now define a collection of operators $L^2(\mathbb{R}, dx) \to L^2(\mathbb{R}, dx)$:

(a) For t positive, the projector

$$P_t := \int_{-t}^{t} dx\, |x\rangle\langle x|, \tag{A.16.12}$$

the kernel of which is $\langle x|P_t|y\rangle = \delta(x-y)\chi_t(x)$, where $\chi_t(x)$ is the characteristic function of the interval $[-t, t]$ (equal to 1 on this interval and 0 outside). With the definition (A.16.10) of $\Bbbk_t$, one can check that

$$P_t = \Bbbk_t P_1 \Bbbk_t^{-1}. \tag{A.16.13}$$

(b) The Fourier transforms $\mathbb{K}$ and $\mathbb{L}_t$ of respectively the projectors P_1 and P_t:

$$\mathbb{K} := \mathcal{F}^{-1} P_1 \mathcal{F}, \qquad \mathbb{L}_t := \mathcal{F} P_t \mathcal{F}^{-1}. \tag{A.16.14}$$

$\mathbb{K}$ and $\mathbb{L}_t$ are themselves projectors ($\mathbb{K}^\dagger = \mathbb{K} = \mathbb{K}^2$ and $\mathbb{L}_t^\dagger = \mathbb{L}_t = \mathbb{L}_t^2$) and their kernels, real and symmetric, write

$$\langle x|\mathbb{K}|y\rangle = \frac{\sin \pi(x-y)}{\pi(x-y)}, \qquad \langle x|\mathbb{L}_t|y\rangle = \frac{\sin \pi t(x-y)}{\pi(x-y)}. \tag{A.16.15}$$

From these expressions, one immediately derives the relation

$$\mathbb{K} = \Bbbk_t \mathbb{L}_t \Bbbk_t^{-1}. \tag{A.16.16}$$

With these definitions, the *sine kernel* $K(x, y)$ of Eqs. (21.1.1) and (21.1.2) is the kernel of the t-dependent operator $K = P_t \mathbb{K} P_t$. As for the Fredholm determinant $F(z, t)$ of Eq. (21.1.5) and the functions $F_\pm(z, t)$, they write

$$F(z, t) := \det(1 - z\mathbb{K}P_t) = \exp \operatorname{Tr} \ln(1 - z\mathbb{K}P_t),$$

$$\frac{F_+(z, t)}{F_-(z, t)} = \exp \operatorname{Tr} \Pi \ln(1 - z\mathbb{K}P_t).$$

(c) For z complex, three operators $\mathcal{K}(z, t)$, $\mathcal{L}(z, t)$ and $\mathcal{N}(z, t)$:

$$\mathcal{K}(z, t) := z\mathbb{K}(1 - zP_t\mathbb{K})^{-1} = (1 - z\mathbb{K}P_t)^{-1} z\mathbb{K}, \tag{A.16.17}$$

$$\mathcal{L}(z, t) := z\mathbb{L}_t(1 - zP_1\mathbb{L}_t)^{-1} = (1 - z\mathbb{L}_t P_1)^{-1} z\mathbb{L}_t, \tag{A.16.18}$$

$$\mathcal{N}(z, t) := \mathcal{F}(1 - zP_t\mathbb{K})^{-1} = (1 - z\mathbb{L}_t P_1)^{-1}\mathcal{F}. \tag{A.16.19}$$

Since $\mathbb{K}$ and P_t are projectors, one has, for any $f \in L^2(\mathbb{R}, dx)$:

$$\|\mathbb{K}P_t|f\rangle\| \leqslant \|P_t|f\rangle\| \leqslant \|f\|.$$

Thus, for any f and g in $L^2(\mathbb{R}, dx)$, $\langle f|\mathcal{K}(z,t)|g\rangle$ is an analytic function of the complex variable z in the circle $|z| < 1$. The same is true of course for the operators $\mathcal{L}(z,t)$ and $\mathcal{N}(z,t)$. From now on, we restrict z to real values so that $\mathcal{K}(z,t)$ and $\mathcal{L}(z,t)$ become Hermitian.

Note that $\mathcal{F}$, $\Bbbk_t$, P_t, P_1, $\mathbb{K}$, $\mathbb{L}_t$, $\mathcal{K}(z,t)$, $\mathcal{L}(z,t)$ and $\mathcal{N}(z,t)$, commute with Π (parity invariance), and that, as a consequence of Eqs. (A.16.13) and (A.16.16):

$$\mathcal{K}(z,t) = \Bbbk_t \mathcal{L}(z,t) \Bbbk_t^{-1}. \tag{A.16.20}$$

With these definitions, the functions $A(z,t)$, $B(z,t)$ and $S(z,t)$, introduced in Eqs. (21.1.7) to (21.1.9), write

$$A(z,t) = \langle t|\mathcal{K}(z,t)|t\rangle, \tag{A.16.21}$$

$$B(z,t) = \langle t|\mathcal{K}(z,t)|-t\rangle = \langle -t|\mathcal{K}(z,t)|t\rangle, \tag{A.16.22}$$

$$S(z,t) = \sqrt{2}\,\langle 1|\mathcal{N}(z,t)|t\rangle = \sqrt{2}\,\langle t|\mathcal{N}^{\dagger}(z,t)|-1\rangle, \tag{A.16.23}$$

where $|t\rangle$, $|-t\rangle$, $|1\rangle$ and $|-1\rangle$ are $|x\rangle$ states localized at $x = t, -t, 1$ and -1, respectively. The second expression of $S(z,t)$ in (A.16.23) results from the property (A.16.9) of $\mathcal{F}$ and the fact that $\langle x|(1 - zP_t\mathbb{K})^{-1}|t\rangle$ is real.

Equation (A.16.20) provides us with new expressions of $A(z,t)$ and $B(z,t)$:

$$tA(z,t) = \langle 1|\mathcal{L}(z,t)|1\rangle, \tag{A.16.24}$$

$$tB(z,t) = \langle 1|\mathcal{L}(z,t)|-1\rangle. \tag{A.16.25}$$

The Sine Kernel: Differential Equations. We now prove that $A(z,t)$, $B(z,t)$ and $S(z,t)$ satisfy the following system of coupled equations (to simplify the writing, we drop the z and t dependences):

$$\partial_t A = 2\,B^2, \tag{A.16.26}$$

$$\partial_t(tA) = z|S|^2, \tag{A.16.27}$$

$$\partial_t(tB) = z\Re\big(S^2\big), \tag{A.16.28}$$

$$2\pi t B = z\Im\big(S^2\big), \tag{A.16.29}$$

$$\partial_t S = i\pi S + 2S^* B. \tag{A.16.30}$$

Here, ∂_t stands for the derivation $\partial/\partial t$ with respect to t.

We first derive the following relations:

$$[D, \mathbb{K}] = [D, \mathbb{L}_t] = 0, \tag{A.16.31}$$

$$[D, P_t] = -|t\rangle\langle t| + |-t\rangle\langle -t|, \tag{A.16.32}$$

$$\partial_t P_t = |t\rangle\langle t| + |-t\rangle\langle -t|. \tag{A.16.33}$$

Equations (A.16.31) express the translational invariance of $\mathbb{K}$ and $\mathbb{L}_t$. The proof of Eq. (A.16.33) is elementary. As for Eq. (A.16.32), its proof goes as follows. From Eqs. (A.16.5), (A.16.6) and (A.16.12), we get successively

$$\langle x|[D, P_t]|y\rangle = (\partial_x + \partial_y)\langle x|P_t|y\rangle, \tag{A.16.34}$$

$$= (\partial_x + \partial_y)\delta(x-y)\chi_t(x), \tag{A.16.35}$$

$$= \delta(x-y)\big(\delta(x+t) - \delta(x-t)\big). \tag{A.16.36}$$

Indeed, Eq. (A.16.34) results from Eqs. (A.16.5) and (A.16.6), Eq. (A.16.35) from the definition (A.16.12) of P_t, and Eq. (A.16.36) from the derivation of $\chi_t(x)$. Then, one checks that the last expression (A.16.36) is equivalent to Eq. (A.16.32).

As consequences of Eqs. (A.16.31) to (A.16.33), we derive the three following relations:

$$[D, \mathcal{K}] = \mathcal{K}\big(-|t\rangle\langle t| + |-t\rangle\langle -t|\big)\mathcal{K}, \tag{A.16.37}$$

$$\partial_t \mathcal{K} = \mathcal{K}\big(|t\rangle\langle t| + |-t\rangle\langle -t|\big)\mathcal{K}, \tag{A.16.38}$$

$$\partial_t \mathcal{L} = z\mathcal{N}\big(|t\rangle\langle t| + |-t\rangle\langle -t|\big)\mathcal{N}^\dagger. \tag{A.16.39}$$

Using the identity $[A, B^{-1}] = -B^{-1}[A, B]B^{-1}$, where A and B are operators (B invertible), we obtain:

$$\begin{aligned}
\big[D, (1 - z\mathbb{K}P_t)^{-1} z\mathbb{K}\big] &= \big[D, (1 - z\mathbb{K}P_t)^{-1}\big] z\mathbb{K}, \\
&= (1 - z\mathbb{K}P_t)^{-1}[D, z\mathbb{K}P_t](1 - z\mathbb{K}P_t)^{-1} z\mathbb{K}, \\
&= \mathcal{K}[D, P_t]\mathcal{K},
\end{aligned}$$

which proves Eq. (A.16.37). The proof of Eq. (A.16.38) is similar.

Furthermore, from the definition (A.16.18) of $\mathcal{L}$:

$$\begin{aligned}
\partial_t \mathcal{L} &= (1 - z\mathbb{L}_t P_1)^{-1}(z\partial_t \mathbb{L}_t) P_1 (1 - z\mathbb{L}_t P_1)^{-1} z\mathbb{L}_t + (1 - z\mathbb{L}_t P_1)^{-1}(z\partial_t \mathbb{L}_t), \\
&= (1 - z\mathbb{L}_t P_1)^{-1}(z\partial_t \mathbb{L}_t)\big((1 - zP_1\mathbb{L}_t)^{-1} zP_1\mathbb{L}_t + 1\big), \\
&= z(1 - z\mathbb{L}_t P_1)^{-1}(\partial_t \mathbb{L}_t)(1 - zP_1\mathbb{L}_t)^{-1}, \\
&= z(1 - z\mathbb{L}_t P_1)^{-1}\mathcal{F}(\partial_t P_t)\mathcal{F}^\dagger(1 - zP_1\mathbb{L}_t)^{-1},
\end{aligned}$$

which proves Eq. (A.16.39).

From Eqs. (A.16.5), (A.16.6), (A.16.21), (A.16.34) and (A.16.35), we obtain

$$\partial_t A = \langle t|[D,\mathcal{K}] + \partial_t \mathcal{K}|t\rangle,$$
$$= 2\langle t|\mathcal{K}|-t\rangle\langle -t|\mathcal{K}|t\rangle,$$

which proves Eq. (A.16.26). Next, from Eqs. (A.16.24) and (A.16.39), we get

$$\partial_t (tA) = z\langle 1|\mathcal{N}\big(|t\rangle\langle t| + |-t\rangle\langle -t|\big)\mathcal{N}^\dagger|1\rangle,$$
$$= 2\langle 1|\mathcal{N}|t\rangle\langle t|\mathcal{N}^\dagger|1\rangle.$$

The last expression is obtained by using the relation $\langle 1|\mathcal{N}|-t\rangle = \langle t|\mathcal{N}^\dagger|1\rangle$. It proves Eq. (A.16.27). Similarly

$$\partial_t (tB) = z\langle 1|\mathcal{N}\big(|t\rangle\langle t| + |-t\rangle\langle -t|\big)\mathcal{N}^\dagger|-1\rangle,$$
$$= z\big(\langle 1|\mathcal{N}|t\rangle^2 + \langle t|\mathcal{N}^\dagger|1\rangle^2\big),$$

which proves Eq. (A.16.28).

Next, from Eq. (A.16.23), one has

$$\partial_t S = \sqrt{2}\,\langle 1|(\partial_t \mathcal{N}) - \mathcal{N}D|t\rangle. \tag{A.16.40}$$

Let us show the relation

$$\partial_t \mathcal{N} - \mathcal{N}D = i\pi X\mathcal{N} + 2\mathcal{N}|-t\rangle\langle -t|\mathcal{K}, \tag{A.16.41}$$

which, inserted in Eq. (A.16.40), proves Eq. (A.16.30). From the definition of $\mathcal{N}$:

$$\partial_t \mathcal{N} - \mathcal{N}D = \mathcal{F}\big(\partial_t (1 - zP_t\mathbb{K})^{-1} - (1 - zP_t\mathbb{K})^{-1}D\big),$$
$$= \mathcal{F}\big(\partial_t (1 - zP_t\mathbb{K})^{-1} + \big[D, (1 - zP_t\mathbb{K})^{-1}\big] - D(1 - zP_t\mathbb{K})^{-1}\big),$$
$$= \mathcal{F}(1 - zP_t\mathbb{K})^{-1}\big(\partial_t P_t + [D, P_t]\big)z\mathbb{K}(1 - zP_t\mathbb{K})^{-1} + i\pi X\mathcal{N}.$$

In this last expression, use has been made of Eq. (A.16.8). It proves Eq. (A.16.41) and thus Eq. (A.16.30).

Finally, from Eqs. (A.16.3) and (A.16.22), one has

$$\langle t|[X,\mathcal{K}]|-t\rangle = 2tB, \tag{A.16.42}$$

and from the definition (A.16.17) of $\mathcal{K}$

$$[X,\mathcal{K}] = (1 - z\mathbb{K}P_t)^{-1}[X, z\mathbb{K}](1 - zP_t\mathbb{K})^{-1}. \tag{A.16.43}$$

Then, one can check that both sides of the following equality have the same kernel $i \sin \pi(x-y)$:

$$i\pi[X, \mathbb{K}] = \mathcal{F}^\dagger\big(|-1\rangle\langle -1| - |1\rangle\langle 1|\big)\mathcal{F}. \tag{A.16.44}$$

These last three equations (A.16.42) to (A.16.44), with the first expression (A.16.19) of $\mathcal{N}$, lead to

$$\begin{aligned} 2\pi t B &= \frac{z}{i}\langle t|\mathcal{N}^\dagger\big(|-1\rangle\langle -1| - |1\rangle\langle 1|\big)\mathcal{N}|-t\rangle, \\ &= \frac{z}{i}\langle t|\mathcal{N}^\dagger\big(|-1\rangle\langle 1| - |1\rangle\langle -1|\big)\mathcal{N}|t\rangle. \end{aligned}$$

To write the last right-hand side, use has been made of the parity invariance of $\mathcal{N}$. With the definition (A.16.23) of S, it proves Eq. (A.16.29).

A.16.3 The Airy Kernel. In order to stress the parallelism between the two constructions with the sine and Airy kernels, we depart here from the definitions of Tracy and Widom (1994b, 1994c). In particular, we do not include a factor $\sqrt{\lambda}$ in the definition of q_j and p_j, and keep writing z instead of λ. So, what they call q_j and p_j coincide in our notation with $\sqrt{z}\,q_j$ and $\sqrt{z}\,p_j$, with $\lambda = z$. Then, the variable z appears in Eqs. (A.16.59) to (A.16.64) as in Eqs. (A.16.27) to (A.16.29).

Note also that, since indeed this section is independent of the previous one, we reuse here several symbols to name new objects.

The Airy Kernel: Definitions and Properties. (a) Let J be the set of disjoint intervals on $\mathbb{R}$

$$J = \bigcup_{j=1}^{m} [a_{2j-1}, a_{2j}], \tag{A.16.45}$$

with $a_{2j-1} < a_{2j}$. We call P_J the projector

$$P_J := \int_J dx\, |x\rangle\langle x|. \tag{A.16.46}$$

In the particular case when J coincides with the positive real axis, we call this projector P_0:

$$P_0 := \int_0^\infty dx\, |x\rangle\langle x|. \tag{A.16.47}$$

(b) Let $\langle x|Ai\rangle = Ai(x)$ be the Airy function. Note that, like the states $|x\rangle$, it does not belong to $L^2(\mathbb{R}, dx)$. The differential equation it satisfies writes, in our notations,

$$D^2|Ai\rangle = X|Ai\rangle. \tag{A.16.48}$$

Define the operator $\mathcal{A}$ with the real and symmetric kernel

$$\langle x|\mathcal{A}|y\rangle := Ai(x+y). \tag{A.16.49}$$

$\mathcal{A}$ is indeed a square root of the operator unity (Hastings and McLeod (1980)) and, since it is Hermitian, it is unitary:

$$\mathcal{A} = \mathcal{A}^{-1} = \mathcal{A}^{\dagger}. \tag{A.16.50}$$

We now define the *Airy transform* $\mathbb{K}$ of the projector P_0 (compare with Eqs. (A.16.14)):

$$\mathbb{K} := \mathcal{A} P_0 \mathcal{A}^{-1}. \tag{A.16.51}$$

$\mathbb{K}$ is itself a projector ($\mathbb{K}^{\dagger} = \mathbb{K} = \mathbb{K}^2$), and its kernel, real and symmetric, writes (see Eq. (24.2.7)):

$$\langle x|\mathbb{K}|y\rangle = \int_0^{\infty} dz\, Ai(x+z)Ai(y+z), \tag{A.16.52}$$

$$= \frac{Ai(x)Ai'(y) - Ai'(x)Ai(y)}{x-y}. \tag{A.16.53}$$

The first expression (A.16.52) is nothing but the definition (A.16.51). The second one (A.16.53) entails

$$[\mathbb{K}, X] = \big\{|Ai\rangle\langle Ai|, D\big\}, \tag{A.16.54}$$

where $\{\ ,\ \}$ denotes an anticommutator.

With these definitions, the *Airy kernel* $K(x, y)$ of Tracy and Widom is the kernel of the z and J-dependent operator $K = zP_J\mathbb{K}P_J$, and their kernel $R(x, y) = \langle x|(1 - z\mathbb{K}P_J)^{-1}z\mathbb{K}|y\rangle$ is to be compared with the kernel of the operator $\mathcal{K}(z, t)$ in the previous section.

(c) Then, their functions q_j, p_j, u and v write (up to factors $\sqrt{z}$ for the first two):

$$q_j(z, \boldsymbol{a}) := \langle a_j|(1 - z\mathbb{K}P_J)^{-1}|Ai\rangle = \langle Ai|(1 - zP_J\mathbb{K})^{-1}|a_j\rangle, \tag{A.16.55}$$

$$p_j(z, \boldsymbol{a}) := \langle a_j|(1 - z\mathbb{K}P_J)^{-1}D|Ai\rangle = -\langle Ai|D(1 - zP_J\mathbb{K})^{-1}|a_j\rangle, \tag{A.16.56}$$

$$u(z, \boldsymbol{a}) := \langle Ai|zP_J(1 - z\mathbb{K}P_J)^{-1}|Ai\rangle, \tag{A.16.57}$$

$$v(z, \boldsymbol{a}) := \langle Ai|zP_J(1 - z\mathbb{K}P_J)^{-1}D|Ai\rangle, \tag{A.16.58}$$

where we use the notation $\boldsymbol{a} = \{a_1, a_2, \ldots, a_{2m}\}$.

The Airy Kernel: Differential Equations. We now prove that the above functions satisfy the following system of coupled equations (once again, to simplify the writing, we drop the z and $\boldsymbol{a}$ dependences):

$$\frac{\partial q_j}{\partial a_k} = (-)^k z \frac{q_j p_k - p_j q_k}{a_j - a_k} q_k \quad (k \neq j), \tag{A.16.59}$$

$$\frac{\partial p_j}{\partial a_k} = (-)^k z \frac{q_j p_k - p_j q_k}{a_j - a_k} p_k \quad (k \neq j), \tag{A.16.60}$$

$$\frac{\partial q_j}{\partial a_j} = -z \sum_{k \neq j} (-)^k \frac{q_j p_k - p_j q_k}{a_j - a_k} q_k + p_j - q_j u, \tag{A.16.61}$$

$$\frac{\partial p_j}{\partial a_j} = -z \sum_{k \neq j} (-)^k \frac{q_j p_k - p_j q_k}{a_j - a_k} p_k + a_j q_j + p_j u - 2 q_j v, \tag{A.16.62}$$

$$\frac{\partial u}{\partial a_k} = (-)^k z q_k^2, \tag{A.16.63}$$

$$\frac{\partial v}{\partial a_k} = (-)^k z q_k p_k. \tag{A.16.64}$$

Let $\boldsymbol{d}_a$ be the exterior differentiation with respect to the a_k's:

$$\boldsymbol{d}_a := \sum_{k=1}^{2m} da_k \frac{\partial}{\partial a_k}. \tag{A.16.65}$$

We first derive the following relations:

$$[D, \mathbb{K}] = -|Ai\rangle\langle Ai|, \tag{A.16.66}$$

$$[D, P_J] = -\sum_{k=1}^{2m} (-)^k \, |a_k\rangle\langle a_k|, \tag{A.16.67}$$

$$\boldsymbol{d}_a P_J = \sum_{k=1}^{2m} (-)^k |a_k\rangle\langle a_k| \, da_k. \tag{A.16.68}$$

From Eq. (A.16.53), and the differential equation satisfied by $Ai(x)$, we get

$$\begin{aligned} \langle x|[D, \mathbb{K}]|y\rangle &= (\partial_x + \partial_y) \frac{Ai(x)Ai'(y) - Ai'(x)Ai(y)}{x - y} \\ &= -Ai(x)Ai(y). \end{aligned}$$

This is Eq. (A.16.66). The proofs of Eqs. (A.16.67) and (A.16.68) parallel those of Eqs. (A.16.32) and (A.16.33).

Next, we show that, for $j \neq k$,

$$\langle a_j|(1-z\mathbb{K}P_J)^{-1}z\mathbb{K}|a_k\rangle = z\frac{q_j p_k - p_j q_k}{a_j - a_k}. \tag{A.16.69}$$

One has first

$$\begin{aligned}
&\left[X, (1-z\mathbb{K}P_J)^{-1}z\mathbb{K}\right] \\
&\quad = (1-z\mathbb{K}P_J)^{-1}[X, z\mathbb{K}]P_J(1-z\mathbb{K}P_J)^{-1}z\mathbb{K} + (1-z\mathbb{K}P_J)^{-1}[X, z\mathbb{K}], \\
&\quad = (1-z\mathbb{K}P_J)^{-1}[X, z\mathbb{K}](1-zP_J\mathbb{K})^{-1}.
\end{aligned}$$

Taking the matrix element $\langle a_j|\ldots|a_k\rangle$ of both sides, we obtain, with the expression (A.16.54) of the commutator $[X, \mathbb{K}]$:

$$\begin{aligned}
&(a_j - a_k)\langle a_j|(1-z\mathbb{K}P_J)^{-1}z\mathbb{K}|a_k\rangle \\
&\quad = -z\langle a_j|(1-z\mathbb{K}P_J)^{-1}D|Ai\rangle\langle Ai|(1-zP_J\mathbb{K})^{-1}|a_k\rangle \\
&\qquad - z\langle a_j|(1-z\mathbb{K}P_J)^{-1}|Ai\rangle\langle Ai|D(1-zP_J\mathbb{K})^{-1}|a_k\rangle.
\end{aligned}$$

This is Eq. (A.16.69).

Let us now calculate $\boldsymbol{d}_a q_j$:

$$\boldsymbol{d}_a q_j = \langle a_j|(D\,da_j + \boldsymbol{d}_a)(1-z\mathbb{K}P_J)^{-1}|Ai\rangle.$$

Commuting D with $(1-z\mathbb{K}P_J)^{-1}$, we get

$$\begin{aligned}
&(D\,da_j + \boldsymbol{d}_a)(1-z\mathbb{K}P_J)^{-1} \\
&\quad = (1-z\mathbb{K}P_J)^{-1}D\,da_j \\
&\qquad + (1-z\mathbb{K}P_J)^{-1}\big([D, z\mathbb{K}P_J]\,da_j + z\mathbb{K}(\boldsymbol{d}_a P_J)\big)(1-z\mathbb{K}P_J)^{-1},
\end{aligned}$$

and from Eqs. (A.16.66) to (A.16.68)

$$[D, z\mathbb{K}P_J]\,da_j + z\mathbb{K}(\boldsymbol{d}_a P_J) = -|Ai\rangle\langle Ai|zP_J\,da_j + z\mathbb{K}\sum_{k\neq j}(-)^k|a_k\rangle\langle a_k|(da_k - da_j).$$

From these last three equations, we obtain

$$\begin{aligned}
\boldsymbol{d}_a q_j &= \sum_{k\neq j}(-)^k\langle a_j|(1-z\mathbb{K}P_J)^{-1}z\mathbb{K}|a_k\rangle\langle a_k|(1-z\mathbb{K}P_J)^{-1}|Ai\rangle(da_k - da_j) \\
&\quad + \big(-\langle a_j|(1-z\mathbb{K}P_J)^{-1}|Ai\rangle\langle Ai|zP_J(1-z\mathbb{K}P_J)^{-1}|Ai\rangle \\
&\quad + \langle a_j|(1-z\mathbb{K}P_J)^{-1}D|Ai\rangle\big)\,da_j.
\end{aligned}$$

Thanks to Eq. (A.16.69), the right-hand side can be expressed in terms of the functions q_k, p_k, u and v:

$$\boldsymbol{d}_a q_j = \Big(-\sum_{k\neq j}(-)^k z\frac{q_j p_k - p_j q_k}{a_j - a_k} q_k - q_j u + p_j\Big)\, da_j + \sum_{k\neq j}(-)^k z\frac{q_j p_k - p_j q_k}{a_j - a_k} q_k\, da_k.$$

This is nothing but Eqs. (A.16.59) and (A.16.61).

Similarly:

$$\begin{aligned}\boldsymbol{d}_a p_j &= \langle a_j|(D\,da_j + \boldsymbol{d}_a)(1 - z\mathbb{K}P_J)^{-1}D|Ai\rangle,\\ &= \sum_{k\neq j}(-)^k\langle a_j|(1 - z\mathbb{K}P_J)^{-1}z\mathbb{K}|a_k\rangle\langle a_k|(1 - z\mathbb{K}P_J)^{-1}D|Ai\rangle(da_k - da_j)\\ &\quad + \big(-\langle a_j|(1 - z\mathbb{K}P_J)^{-1}|Ai\rangle\langle Ai|zP_J(1 - z\mathbb{K}P_J)^{-1}D|Ai\rangle\\ &\quad + \langle a_j|(1 - z\mathbb{K}P_J)^{-1}D^2|Ai\rangle\big)\, da_j.\end{aligned}$$

The only quantity which remains to calculate here is $\langle a_j|(1 - z\mathbb{K}P_J)^{-1}D^2|Ai\rangle$ which, thanks to Eq. (A.16.48), writes $\langle a_j|(1 - z\mathbb{K}P_J)^{-1}X|Ai\rangle$. Commuting X with $(1 - z\mathbb{K}P_J)^{-1}$, we get, with the help of Eq. (A.16.54),

$$(1 - z\mathbb{K}P_J)^{-1}X = X(1 - z\mathbb{K}P_J)^{-1} + (1 - z\mathbb{K}P_J)^{-1}\big\{|Ai\rangle\langle Ai|, D\big\}zP_J(1 - z\mathbb{K}P_J)^{-1},$$

and we end up with the following result

$$\begin{aligned}\boldsymbol{d}_a p_j = &\Big(-\sum_{k\neq j}(-)^k z\frac{q_j p_k - p_j q_k}{a_j - a_k} p_k + a_j q_j + p_j u - 2q_j v\Big)\, da_j\\ &+ \sum_{k\neq j}(-)^k z\frac{q_j p_k - p_j q_k}{a_j - a_k} p_k\, da_k,\end{aligned}$$

which is nothing but Eqs. (A.16.60) and (A.16.62).

Finally:

$$\begin{aligned}\boldsymbol{d}_a u &= \langle Ai|\boldsymbol{d}_a(1 - zP_J\mathbb{K})^{-1}zP_J|Ai\rangle,\\ \boldsymbol{d}_a v &= \langle Ai|\boldsymbol{d}_a(1 - zP_J\mathbb{K})^{-1}zP_J D|Ai\rangle,\end{aligned}$$

and

$$\begin{aligned}\boldsymbol{d}_a(1 - zP_J\mathbb{K})^{-1}zP_J &= (1 - zP_J\mathbb{K})^{-1}(\boldsymbol{d}_a P_J)z\mathbb{K}(1 - zP_J\mathbb{K})^{-1}zP_J\\ &\quad + (1 - zP_J\mathbb{K})^{-1}z(\boldsymbol{d}_a P_J),\\ &= (1 - zP_J\mathbb{K})^{-1}z(\boldsymbol{d}_a P_J)(1 - z\mathbb{K}P_J)^{-1}.\end{aligned}$$

With the expression (A.16.68) of $\boldsymbol{d}_a P_J$, we get

$$\boldsymbol{d}_a u = z\sum_{k=1}^{2m}(-)^k \langle Ai|(1 - zP_J\mathbb{K})^{-1}|a_k\rangle\langle a_k|(1 - z\mathbb{K}P_J)^{-1}|Ai\rangle,$$

$$\boldsymbol{d}_a v = z\sum_{k=1}^{2m}(-)^k \langle Ai|(1 - zP_J\mathbb{K})^{-1}|a_k\rangle\langle a_k|(1 - z\mathbb{K}P_J)^{-1}D|Ai\rangle.$$

These are Eqs. (A.16.63) and (A.16.64).

The reader will convince himself that when the set J reduces to the unique interval $(s, +\infty)$, the differential equations (A.16.59) to (A.16.63) boil down to Eqs. (24.3.11), (24.3.20) and (24.3.15).

A.17 Use of Pfaffians in Some Multiple Integrals. Proof of Eqs. (7.3.6), (14.3.6)–(14.3.8) and (A.25.6)

Consider an anti-symmetric matrix $A = [a_{ij}]$, $i, j = 1, 2, \ldots, n$, $a_{ij} = -a_{ji}$. If its order n is odd, then from

$$\det A = \det A^T = (-1)^n \det A$$

one sees that its determinant is zero. On the other hand if n is even, it is known that its determinant is a perfect square (see, e.g., Mehta, 1989MT); the square root is

$$\left(\det[a_{ij}]\right)^{1/2} = \frac{1}{(n/2)!}\sum \pm a_{i_1 i_2} a_{i_3 i_4} \cdots a_{i_{n-1} i_n}, \tag{A.17.1}$$

where the summation is extended over all permutations $(i_1, i_2, \ldots, i_n)$ of $(1, 2, \ldots, n)$, with the restrictions $i_1 < i_2, i_3 < i_4, \ldots, i_{n-1} < i_n$, and the sign is plus or minus depending on whether the permutation

$$\begin{pmatrix} 1 & 2 & \ldots & n \\ i_1 & i_2 & \ldots & i_n \end{pmatrix}$$

is even or odd. The expression (A.17.1) is known as the "Pfaffian".

Now consider evaluating the integrals

$$I_1 = \int \cdots \int \det\big[A_i(x_j) \quad B_i(x_j)\big]\, d\mu(x_1) \cdots d\mu(x_m), \tag{A.17.2}$$

and

$$I_2 = \int \cdots \int \det\begin{bmatrix} \alpha_0 & \beta(x_j) & \gamma(x_j) \\ \alpha_i & A_i(x_j) & B_i(x_j) \end{bmatrix} d\mu(x_1) \cdots d\mu(x_m), \tag{A.17.3}$$

$i = 1, \ldots, 2m$, $j = 1, \ldots, m$. The integrands in I_1 and I_2 are determinants of orders $2m \times 2m$ and $(2m+1) \times (2m+1)$, respectively; each variable occurs in two columns and $d\mu(x)$ is a suitable measure. Define

$$\rho_j = \int \big(\beta(x) B_j(x) - A_j(x)\gamma(x)\big)\, d\mu(x), \tag{A.17.4}$$

$$a_{ij} = \int \big(A_i(x) B_j(x) - A_j(x) B_i(x)\big)\, d\mu(x), \tag{A.17.5}$$

so that

$$a_{ij} = -a_{ji}. \tag{A.17.6}$$

Expanding the integrand in (A.17.2) and integrating independently over all the variables we make the following observations:

(1) The integral I_1 is a sum of terms, each being a product of m numbers a_{ij};
(2) The indices of the various a_{ij} occurring in any of the above terms are all different. In totality they are all the indices from 1 to $2m$.
(3) We may restrict i to be less than j in each of a_{ij} occurring in I_1; for if i is not less than j we may replace a_{ij} by $-a_{ji}$.
(4) The coefficient of the term $a_{i_1 i_2} a_{i_3 i_4} \cdots a_{i_{2m-1} i_{2m}}$ in I_1 is $+1$ or -1, depending on whether the permutation

$$\begin{pmatrix} i_1 & i_2 & \ldots & i_{2m} \\ 1 & 2 & \ldots & 2m \end{pmatrix}$$

is even or odd.

From these observations and (A.17.1) we conclude that

$$I_1 = m!\big(\det[a_{ij}]_{i,j=1,2,\ldots,2m}\big)^{1/2}. \tag{A.17.7}$$

Similarly, the integral I_2 is

$$I_2 = m!\,\mathrm{pf}\, M = m!(\det M)^{1/2}, \tag{A.17.8}$$

where M is an $(2m+2) \times (2m+2)$ anti-symmetric matrix with elements

$$\begin{aligned} &M_{1j} = -M_{j1} = \alpha_{j-2}, && j = 2, \ldots, 2m+2, \\ &M_{2j} = -M_{j2} = \rho_{j-2}, && j = 3, \ldots, 2m+2, \\ &M_{jk} = a_{j-2,k-2}, && 3 \leqslant j, k \leqslant 2m+2, \\ &M_{jj} = 0, && j = 1, \ldots, 2m+2. \end{aligned} \tag{A.17.9}$$

Equations (A.17.7) and (A.17.8) are general results. If $a_{ij} = 0$, for $i + j$ even, then the matrix $[a_{ij}]$ has a checker board structure, every alternate element being zero; and there are simplifications. In this case I_1 is an $m \times m$ determinant and I_2 is a sum of two $(m+1) \times (m+1)$ determinants.

$$I_1 = m! \det[a_{2i-1,2j}]_{i,j=1,\ldots,m}, \tag{A.17.10}$$

and

$$I_2 = m! \det\begin{bmatrix} 0 & \rho_{2j} \\ \alpha_{2i-1} & a_{2i-1,2j} \end{bmatrix} + m! \det\begin{bmatrix} \alpha_0 & -\alpha_{2j} \\ \rho_{2i-1} & a_{2i-1,2j} \end{bmatrix}. \tag{A.17.11}$$

The integrand in Eq. (7.3.5) is also a determinant where each variable occurs in two columns. By parity argument

$$a_{2i+1,2j+1} = \int_{\text{out}} \big(F_{2i}(x)\varphi_{2j}(x) - F_{2j}(x)\varphi_{2i}(x)\big)\, dx = 0, \tag{A.17.12}$$

$$a_{2i,2j} = \int_{\text{out}} \big(F_{2i-1}(x)\varphi'_{2j-2}(x) - F_{2j-1}(x)\varphi'_{2i-2}(x)\big)\, dx = 0, \tag{A.17.13}$$

while

$$\begin{aligned} a_{2i+1,2j+2} &= \int_{\text{out}} \big(F_{2i}(x)\varphi'_{2j}(x) - F_{2j+1}(x)\varphi_{2i}(x)\big)\, dx \\ &= \int_{y<x,\ \text{out}} \int \big(\varphi_{2i}(y)\varphi'_{2j}(x) - \varphi'_{2j}(y)\varphi_{2i}(x)\big)\, dx \\ &= -2 \int_{\text{out}} \varphi_{2i}(x)\varphi_{2j}(x)\, dx = -2\left(\delta_{ij} - \int_{-\theta}^{\theta} \varphi_{2i}(x)\varphi_{2j}(x)\, dx\right) \\ &= -2\bar{g}_{ij}. \end{aligned} \tag{A.17.14}$$

From Eqs. (A.17.2), (A.17.10) and (A.17.14) we get Eq. (7.3.6). Similarly, from (A.17.11) we get (7.5.12).

A.18 Calculation of Certain Determinants

A.18.1 Consider the determinant $[\delta_{ij} - u_i v_j]$.

To evaluate such a determinant it is convenient to border it with an extra row and an extra column. Thus

$$\det\begin{bmatrix} 1 - u_1 v_1 & -u_1 v_2 & \ldots & -u_1 v_n \\ -u_2 v_1 & 1 - u_2 v_2 & \ldots & -u_2 v_n \\ \ldots & \ldots & \ldots & \ldots \\ -u_n v_1 & -u_n v_2 & \ldots & 1 - u_n v_n \end{bmatrix}$$

$$= \det \begin{bmatrix} 1 & 0 & 0 & \dots & 0 \\ u_1 & 1-u_1v_1 & -u_1v_2 & \dots & -u_1v_n \\ u_2 & -u_2v_1 & 1-u_2v_2 & \dots & -u_1v_n \\ \dots & \dots & \dots & \dots & \dots \\ u_n & -u_nv_1 & -u_nv_2 & \dots & 1-u_nv_n \end{bmatrix}$$

$$= \det \begin{bmatrix} 1 & v_1 & v_2 & \dots & v_n \\ u_1 & 1 & 0 & \dots & 0 \\ u_2 & 0 & 1 & \dots & 0 \\ \dots & \dots & \dots & \dots & \dots \\ u_n & 0 & 0 & \dots & 1 \end{bmatrix} = 1 - \sum_{j=1}^{n} u_j v_j. \tag{A.18.1}$$

The cofactor b_{ij} of the element (i, j) in $\det[\delta_{ij} - u_i v_j]$ is easy to calculate by the same bordering

$$b_{ij} = \delta_{ij}\left(1 - \sum_{k=1}^{n} u_k v_k\right) + u_j v_i. \tag{A.18.2}$$

A.18.2 To calculate a determinant of the form $\det[A + u_i - u_j - \text{sign}(i-j)]$, $i, j = 1, \dots, n$, we border it again with two rows and two columns; thus

$$\det\big[A + u_i - u_j - \text{sign}(i-j)\big]$$

$$= \det \begin{bmatrix} 1 & 0 & 0 \\ A & 1 & -A + u_j \\ -A - u_i & 0 & A + u_i - u_j - \text{sign}(i-j) \end{bmatrix}$$

$$= \det \begin{bmatrix} 1 & 0 & 1 \\ A & 1 & u_j \\ -u_i & 1 & -\text{sign}(i-j) \end{bmatrix} = \det \begin{bmatrix} 0 & 1 & 1 \\ -1 & A & u_j \\ -1 & -u_i & -\text{sign}(i-j) \end{bmatrix}$$

$$= \det \begin{bmatrix} 0 & 1 & 1 \\ -1 & 0 & u_j \\ -1 & -u_i & -\text{sign}(i-j) \end{bmatrix} + A \det \begin{bmatrix} 0 & 1 \\ -1 & -\text{sign}(i-j) \end{bmatrix}. \tag{A.18.3}$$

The last two determinants are of anti-symmetric matrices; their orders differ by one. If n is odd, the first determinant is zero and if n is even, the second one is zero. The non-zero determinant is the square of its Pfaffian. Thus

$$\det\big[A + u_i - u_j - \text{sign}(i-j)\big]$$

$$= \begin{cases} A, & \text{if } n \text{ is odd}, \quad \text{(A.18.4)} \\ (1 + u_1 - u_2 + \dots + u_{n-1} - u_n)^2, & \text{if } n \text{ is even}. \quad \text{(A.18.5)} \end{cases}$$

A.18.3 Consider the determinant of the $n \times n$ matrix $[\Gamma(b + i + j]_{i,j=0,\dots,n-1}$, with b some complex number, not zero or a negative integer. This determinant, denoted by

$D_{1,0}(n;b)$, is not changed if one replaces the row $R_i = \Gamma(b+i+j)$ by

$$R_i - (b+i-1)R_{i-1} = \Gamma(b+i+j) - (b+i-1)\Gamma(b+i+j-1) = j\Gamma(b+i+j-1).$$

Doing this successively for $i = n-1, n-2, \ldots, 1$, reduces all the elements but one in the first column to zero. Taking the factors j out, one therefore has

$$\begin{aligned} D_{1,0}(n;b) &:= \det\left[\Gamma(b+i+j)\right]_{i,j=0,1,\ldots,n-1} \\ &= \Gamma(b)\det\left[j\Gamma(b+i+j-1)\right]_{i,j=1,\ldots,n-1} \\ &= (n-1)!\Gamma(b)\det\left[\Gamma(b+i+j+1)\right]_{i,j=0,1,\ldots,n-2} \\ &= (n-1)!\Gamma(b)D_{1,0}(n-1;b+1). \end{aligned} \tag{A.18.6}$$

Iteration of the last equation gives

$$D_{1,0}(n;b) = \prod_{j=0}^{n-1} j!\,\Gamma(b+j). \tag{A.18.7}$$

All the replacements above could have been done in one step. Replace R_i by

$$\begin{aligned} R_i + \sum_{k=0}^{i-1}(-1)^{k+i}\binom{i}{k}\frac{\Gamma(b+i)}{\Gamma(b+k)}R_k &= \Gamma(b+i)\sum_{k=0}^{i}(-1)^{k+i}\binom{i}{k}\frac{\Gamma(b+j+k)}{\Gamma(b+k)} \\ &= \frac{j!}{(j-i)!}\Gamma(b+j). \end{aligned} \tag{A.18.8}$$

The last line results from the identity

$$\sum_{k=0}^{i}(-1)^{k+i}\binom{i}{k}\frac{\Gamma(b+j+k)}{\Gamma(b+k)} = \frac{j!}{(j-i)!}\frac{\Gamma(b+j)}{\Gamma(b+i)}, \tag{A.18.9}$$

which can be proved either by recurrence on i or by differentiating j times $(x-1)^i \times x^{b+j-1}$ and setting $x=1$;

$$\left(\frac{d}{dx}\right)^j (x-1)^i x^{b+j-1}|_{x=1} = \sum_{k=0}^{i}(-1)^{k+i}\binom{i}{k}\left(\frac{d}{dx}\right)^j x^{b+j+k-1}|_{x=1}, \tag{A.18.10}$$

which is also

$$= \sum_{k=0}^{j} \binom{j}{k} \left(\frac{d}{dx}\right)^{k} (x-1)^{i} \left(\frac{d}{dx}\right)^{j-k} x^{b+j-1}|_{x=1}$$

$$= \binom{j}{i} \left(\frac{d}{dx}\right)^{i} (x-1)^{i} \left(\frac{d}{dx}\right)^{j-i} x^{b+j-1}. \tag{A.18.11}$$

This makes the determinant triangular with diagonal elements $j!\Gamma(b+j)$ and we get back Eq. (A.18.7).

A.18.4 Consider the determinant of the $n \times n$ matrix $[1/\Gamma(c+i+j]_{i,j=0,\ldots,n-1}$, with c some complex number, not zero or a negative integer. This determinant, denoted by $D_{0,1}(n;c)$, is not changed if one replaces the row $R_i = 1/\Gamma(c+i+j)$ by

$$R_i - \frac{1}{(c+i-1)} R_{i-1} = -\frac{j}{(c+i-1)} \frac{1}{\Gamma(c+i+j)}.$$

Doing this successively for $i = n-1, n-2, \ldots, 1$, reduces all the elements but one in the first column to zero. Taking the factors $-j/(c+i-1)$ out, one therefore has

$$D_{0,1}(n;c) = \left[\Gamma(c)\right]^{-1} (n-1)! \prod_{i=0}^{n-2} \frac{-1}{c+i} \det\left[1/\Gamma(c+i+j)\right]_{i,j=1,\ldots,n-1}$$

$$= (-1)^{n-1} \frac{(n-1)!}{\Gamma(c+n-1)} D_{0,1}(n-1;c+2). \tag{A.18.12}$$

Iteration of the last equation gives

$$D_{0,1}(c;n) = \prod_{j=0}^{n-1} (-1)^{j} j! \left[\Gamma(c+n+j-1)\right]^{-1}. \tag{A.18.13}$$

All the replacements above could have been done in one step making the determinant triangular. Replace R_i by

$$R_i + \sum_{k=0}^{i-1} (-1)^{k+i} \binom{i}{k} \frac{\Gamma(c+i+k-1)}{\Gamma(c+2i-1)} R_k$$

$$= (-1)^{i} \frac{1}{\Gamma(c+2i-1)} \sum_{k=0}^{i} (-1)^{k} \binom{i}{k} \frac{\Gamma(c+i+k-1)}{\Gamma(c+j+k)}$$

$$= (-1)^{i} \frac{j!}{(j-i)!} \frac{\Gamma(c+i-1)}{\Gamma(c+2i-1)\Gamma(c+i+j)}. \tag{A.18.14}$$

The last line above results from the identity

$$\sum_{k=0}^{i}(-1)^k\binom{i}{k}\frac{\Gamma(c+k+i-1)}{\Gamma(c+k+j)}=\frac{j!}{(j-i)!}\frac{\Gamma(c+i-1)}{\Gamma(c+i+j)},\tag{A.18.15}$$

which can be proved by recurrence on i.

The determinant $D_{0,1}(n;c)$ is then the product of the diagonal elements $(-1)^j j!\Gamma(c+j-1)/[\Gamma(c+2j-1)\Gamma(c+2j)]$ in (A.18.14), giving back (A.18.13).

A.18.5 Consider the determinant of the $n\times n$ matrix $[\Gamma(b+i+j)/\Gamma(c+i+j)]_{i,j=0,1,\ldots,n-1}$ with some complex numbers b and c, not zero or negative integers. This determinant, denoted by $D_{1,1}(n;b,c)$, is not changed if we replace the row R_i by the linear combination $R_i-[(b+i-1)/(c+i-1)]R_{i-1}$,

$$\begin{aligned}
R_i &- \frac{b+i-1}{c+i-1}R_{i-1}\\
&= \frac{\Gamma(b+i+j)}{\Gamma(c+i+j)}-\frac{(b+i-1)\Gamma(b+i+j-1)}{(c+i-1)\Gamma(c+i+j-1)}\\
&= \frac{\Gamma(b+i+j-1)}{\Gamma(c+i+j)}\frac{1}{(c+i-1)}\\
&\quad\times\left[(b+i+j-1)(c+i-1)-(b+i-1)(c+i+j-1)\right]\\
&= \frac{\Gamma(b+i+j-1)}{\Gamma(c+i+j)}\frac{j(c-b)}{(c+i-1)}.
\end{aligned}$$

This operation replaces the first column by zeros except for one element. Taking the common factors $j(c-b)/(c+i-1)$ out, we get

$$\begin{aligned}
D_{1,1}(n;b,c) &:= \det\left[\frac{\Gamma(b+i+j)}{\Gamma(c+i+j)}\right]_{i,j=0,1,\ldots,n-1}\\
&= (n-1)!(c-b)^{n-1}\frac{\Gamma(b)}{\Gamma(c+n-1)}D_{1,1}(n-1;b+1,c+2).
\end{aligned}\tag{A.18.16}$$

An iteration of the last equation gives

$$\begin{aligned}
D_{1,1}(n;b,c) &= \prod_{j=0}^{n-1}(n-1-j)!(c-b+j)^{n-j-1}\frac{\Gamma(b+j)}{\Gamma(c+n+j-1)}\\
&= \prod_{j=0}^{n-1}j!(c-b+j)^{n-j-1}\frac{\Gamma(b+j)}{\Gamma(c+n+j-1)}.
\end{aligned}\tag{A.18.17}$$

As in Appendices A.18.3 and A.18.4 above, the triangulation of $D_{1,1}(n;b,c)$ could have been performed in one step. Replace the row R_i by

$$\sum_{k=0}^{i}(-1)^{i-k}\binom{i}{k}\frac{\Gamma(b+i)}{\Gamma(b+k)}\frac{\Gamma(c+i+k-1)}{\Gamma(c+2i-1)}R_k,$$

and use the identity

$$\sum_{k=0}^{i}(-1)^{i-k}\binom{i}{k}\frac{\Gamma(b+k+j)}{\Gamma(b+k)}\frac{\Gamma(c+i+k-1)}{\Gamma(c+k+j)}$$
$$=\frac{j!}{(j-i)!}\frac{\Gamma(c-b+i)}{\Gamma(c-b)}\frac{\Gamma(b+j)}{\Gamma(b+i)}\frac{\Gamma(c+i-1)}{\Gamma(c+i+j)}, \tag{A.18.18}$$

which can be proved by recurrence on i.

The product of the diagonal elements in (A.18.18) gives back $D_{1,1}(n;b,c)$, Eq. (A.18.17).

A.18.6 Consider the Pfaffian of the $2n \times 2n$ anti-symmetric matrix

$$M := \left[P_{1,0}(n;b)\right] \equiv \left[(j-i)\Gamma(b+i+j)\right], \quad i,j=0,1,\ldots,2n-1,$$

with b a complex number not zero or negative integer. To calculate this Pfaffian we introduce the $2n \times 2n$ matrix A with elements

$$\begin{aligned} A_{i,i} &= 1, \quad i=0,1,\ldots,2n-1, \\ A_{i,i-1} &= -\frac{i(b+i-1)}{i-1}, \quad i=2,3,\ldots,2n-1, \\ A_{i,j} &= 0, \quad \text{otherwise}. \end{aligned} \tag{A.18.19}$$

Then $\det A = 1$, and $\operatorname{pf}[AMA^T] = \operatorname{pf} M \cdot \det A = \operatorname{pf} M$. Now

$$\begin{aligned} \left(AMA^T\right)_{i,j} &= \sum_{k,l} A_{i,k}A_{j,l}M_{k,l} \\ &= M_{i,j} - \frac{i(b+i-1)}{i-1}M_{i-1,j} - \frac{j(b+j-1)}{j-1}M_{i,j-1} \\ &\quad + \frac{ij(b+i-1)(b+j-1)}{(i-1)(j-1)}M_{i-1,j-1}. \end{aligned}$$

Substituting the expressions for the matrix elements, after some elementary algebra this simplifies to

$$\left(AMA^T\right)_{0,1} = -\left(AMA^T\right)_{1,0} = \Gamma(b+1),$$

$$\left(AMA^T\right)_{0,j} = -\left(AMA^T\right)_{j,0} = 0, \quad \text{if either } j = 0, \text{ or } j \geqslant 2,$$

$$\left(AMA^T\right)_{i,j} = ij(j-i)\Gamma(b+i+j-2), \quad i, j \geqslant 1,$$

so that

$$\begin{aligned} \operatorname{pf} P_{1,0}(n; b) &:= \operatorname{pf}\left[(j-i)\Gamma(b+i+j)\right]_{i,j=0,1,\ldots,2n-1} \\ &= (2n-1)!\Gamma(b+1)\operatorname{pf} P_{1,0}(n-1; b+2), \end{aligned} \tag{A.18.20}$$

which by iteration gives

$$\operatorname{pf} P_{1,0}(n; b) = \prod_{j=0}^{n-1}(2j+1)!\Gamma(b+2j+1). \tag{A.18.21}$$

A.18.7 Consider the Pfaffian of the $2n \times 2n$ anti-symmetric matrix $P_{1,1}(n; b, c) := M$, with elements $M_{i,j} = (j-i)\Gamma(b+i+j)/\Gamma(c+i+j)$, $i, j = 0, 1, \ldots, 2n-1$, with some complex numbers b and c, not zero or negative integers.

To calculate this Pfaffian we introduce the $2n \times 2n$ matrix A with elements

$$\begin{aligned} &A_{i,i} = 1, \quad i = 0, 1, \ldots, 2n-1, \\ &A_{i,i-1} = -\frac{i(b+i-1)}{(i-1)(c+i-1)}, \quad i = 2, 3, \ldots, 2n-1, \\ &A_{i,j} = 0, \quad \text{otherwise.} \end{aligned} \tag{A.18.22}$$

Then $\det A = 1$, and $\operatorname{pf}[AMA^T] = \operatorname{pf} M \cdot \det A = \operatorname{pf} M$. Now

$$\begin{aligned} \left(AMA^T\right)_{i,j} &= \sum_{k,l} A_{i,k}A_{j,l}M_{k,l} \\ &= M_{i,j} - \frac{i(b+i-1)}{(i-1)(c+i-1)}M_{i-1,j} - \frac{j(b+j-1)}{(j-1)(c+j-1)}M_{i,j-1} \\ &\quad + \frac{ij(b+i-1)(b+j-1)}{(i-1)(j-1)(c+i-1)(c+j-1)}M_{i-1,j-1}. \end{aligned}$$

Substituting the expressions of $M_{i,j}$, after some elementary but heavy algebra this simplifies to

$$(j-i)\frac{(i+1)(j+1)}{(c+i)(c+j)}(c-b)(c-b+1)\frac{\Gamma(b+i+j)}{\Gamma(c+i+j+2)}.$$

Taking the common factors out, this gives

$$\begin{aligned}
\operatorname{pf} P_{1,1}(n;b,c) &= \operatorname{pf}\left[(j-i)\frac{\Gamma(b+i+j)}{\Gamma(c+i+j)}\right]_{i,j=0,1,\ldots,2n-1} \\
&= \frac{\Gamma(b+1)}{\Gamma(c+1)}\left[(c-b)(c-b+1)\right]^{n-1}\prod_{j=1}^{2n-2}\frac{j+1}{c+j} \\
&\quad \times \operatorname{pf} P_{1,1}(n-1;b+2,c+4) \\
&= \left[(c-b)(c-b+1)\right]^{n-1}(2n-1)!\frac{\Gamma(b+1)}{\Gamma(c+2n-1)} \\
&\quad \times \operatorname{pf} P_{1,1}(n-1;b+2,c+4).
\end{aligned} \tag{A.18.23}$$

Iteration of the last equation gives

$$\begin{aligned}
\operatorname{pf} P_{1,1}(n;b,c) &= \prod_{j=0}^{n-1}\left[(c-b+2j)(c-b+2j+1)\right]^{n-j-1} \\
&\quad \times \frac{(2j+1)!\Gamma(b+2j+1)}{\Gamma(c+2n+2j-1)}.
\end{aligned} \tag{A.18.24}$$

A.18.8 Consider the determinant of the $n \times n$ matrix

$$M_n(a,b) = \left[(a+j-i)\Gamma(b+j+i)\right]_{i,j=0,1,\ldots,n-1}, \tag{A.18.25}$$

or its Pfaffian when $a = 0$, (b neither zero nor a negative integer). The result is

$$M_n(a,b) = \det\left[(a+j-i)\Gamma(b+j+i)\right]_{i,j=0,1,\ldots,n-1} \tag{A.18.26}$$

$$= D_n \prod_{i=0}^{n-1} i!\Gamma(b+i), \tag{A.18.27}$$

$$\begin{aligned}
\sqrt{M_{2n}(0,b)} &= \operatorname{Pf}\left[(j-i)\Gamma(b+j+i)\right]_{i,j=0,1,\ldots,2n-1} \\
&= \prod_{i=0}^{n-1}(2i+1)!\Gamma(b+2i+1),
\end{aligned} \tag{A.18.28}$$

where

$$D_n = \det\left[a\delta_{i,j} - \delta_{i,j+1} + j(b+i)\delta_{i,j-1}\right] \tag{A.18.29}$$

$$= \text{coeff. of } (z^n/n!) \text{ in } (1-z)^{-(b+a)/2}(1+z)^{-(b-a)/2} \tag{A.18.30}$$

$$= \sum_{k=0}^{n} (-1)^k \binom{n}{k} \left(\frac{b-a}{2}\right)_k \left(\frac{b+a}{2}\right)_{n-k}, \tag{A.18.31}$$

with the Pochhammer's symbol

$$(a)_0 = 1, \quad (a)_n = \Gamma(a+n)/\Gamma(a). \tag{A.18.32}$$

The expression for D_n simplifies when a is a small integer or when $a = b$.

We need the following lemma.

Lemma. *For j a non-negative integer and A and B complex numbers one has the identity*

$$F(j, A, B) \equiv \sum_{k=0}^{j} (-1)^k \binom{j}{k} (A+j-k)_k (B)_{j-k} = (B-A-j+1)_j. \tag{A.18.33}$$

Proof. The lemma is trivial for $j = 0$ and easy to verify for $j = 1$. Suppose that it is true for some positive integer j. Then for the next integer

$$F(j+1, A, B) = \sum_{k=0}^{j+1} (-1)^k \binom{j+1}{k} (A+j+1-k)_k (B)_{j+1-k}$$

$$= \sum_{k=0}^{j+1} (-1)^k \left[\binom{j}{k} + \binom{j}{k-1}\right] (A+j+1-k)_k (B)_{j+1-k}$$

$$= B \sum_{k} (-1)^k \binom{j}{k} (A+1+j-k)_k (B+1)_{j-k}$$

$$- (A+j) \sum_{k} (-1)^{k-1} \binom{j}{k-1} (A+j-k+1)_{k-1} (B)_{j-k+1}$$

$$= BF(j, A+1, B+1) - (A+j)\, F(j, A, B)$$

$$= (B-A-j)F(j, A, B) = (B-A-j)_{j+1}. \tag{A.18.34}$$

And so it is true for every positive integer j. Note that the binomial coefficient $\binom{j}{k}$ is zero if the integer k is either negative or is greater than j. Actually, $k! = \Gamma(k+1)$ can be replaced by ∞ whenever k is a negative integer. □

Corollary.

$$\sum_{k=0}^{j}(-1)^k\binom{j}{k}(j-k)(A+j-k)_k(B)_{j-k} = jB(B-A-j+2)_{j-1}. \quad \text{(A.18.35)}$$

Proof.

$$\begin{aligned}\text{Left-hand side} &= jB\sum_{k=0}^{j-1}(-1)^k\binom{j-1}{k}(A+j-k)_k(B+1)_{j-1-k} \\ &= jBF(j-1, A+1, B+1) = jB(B-A-j+2)_{j-1}. \square \quad \text{(A.18.36)}\end{aligned}$$

Determinant. The determinant of a matrix is not changed if we add to any of its row (column) an arbitrary linear combination of the other rows (columns). Replacing the jth column $C_j = (a+j-i)\Gamma(b+j+i)$ by

$$C'_j = \sum_{k=0}^{j}(-1)^k\binom{j}{k}\frac{\Gamma(b+j)}{\Gamma(b+j-k)}C_{j-k} \quad \text{(A.18.37)}$$

$$\begin{aligned} &= (a-i)\sum_{k=0}^{j}(-1)^k\binom{j}{k}(b+j-k)_k\Gamma(b+i+j-k) \\ &\quad + \sum_{k=0}^{j}(-1)^k\binom{j}{k}(j-k)(b+j-k)_k\Gamma(b+i+j-k) \\ &= \Gamma(b+i)\big[(a-i)F(j,b,b+i) + j(b+i)F(j-1,b+1,b+i+1)\big] \\ &= \Gamma(b+i)\big[(a-i)(i-j+1)_j + j(b+i)(i-j+2)_{j-1}\big] \\ &= \Gamma(b+i)\left[(a-i)\frac{i!}{(i-j)!} + j(b+i)\frac{i!}{(i-j+1)!}\right], \quad \text{(A.18.38)}\end{aligned}$$

where in the third line above we have used the lemma and its corollary. Taking out the factors $\Gamma(b+i)$ one has

$$M_n(a,b) = \left[\prod_{i=0}^{n-1}\Gamma(b+i)\right]\det\left[(a-i)\frac{i!}{(i-j)!} + j(b+i)\frac{i!}{(i-j+1)!}\right]. \quad \text{(A.18.39)}$$

Now replace the row

$$R_i = \left[(a-i)\frac{i!}{(i-j)!} + j(b+i)\frac{i!}{(i-j+1)!}\right], \quad \text{(A.18.40)}$$

by the linear combination

$$R_i' = \sum_{k=0}^{i} (-1)^k \binom{i}{k} R_{i-k} \tag{A.18.41}$$

$$= \sum_{k=0}^{i} (-1)^k \binom{i}{k} \left[(a-i+k)\frac{(i-k)!}{(i-k-j)!} + j(b+i-k)\frac{(i-k)!}{(i-k-j+1)!} \right]$$

$$= i! \sum_{k=0}^{i} (-1)^k \left[\frac{a-i}{k!(i-j-k)!} + \frac{1}{(k-1)!(i-j-k)!} \right.$$

$$\left. + \frac{j(b+i)}{k!(i-k-j+1)!} - \frac{j}{(k-1)!(i-k-j+1)!} \right]$$

$$= i! \left[\frac{(a-i)}{(i-j)!} \sum_{k=0}^{i} (-1)^k \binom{i-j}{k} - \frac{1}{(i-j-1)!} \sum_{k=0}^{i} (-1)^{k-1} \binom{i-j-1}{k-1} \right.$$

$$\left. + \frac{j(b+i)}{(i-j+1)!} \sum_{k=0}^{i} (-1)^k \binom{i-j+1}{k} + \frac{j}{(i-j)!} \sum_{k=0}^{i} (-1)^{k-1} \binom{i-j}{k-1} \right]$$

$$= i! \left[(a-i)\delta_{i,j} - \delta_{i,j+1} + j(b+i)\delta_{i,j-1} + j\delta_{i,j} \right]$$

$$= i! \left[a\delta_{i,j} - \delta_{i,j+1} + j(b+i)\delta_{i,j-1} \right]. \tag{A.18.42}$$

Thus

$$M_n(a,b) = D_n \prod_{i=0}^{n-1} i!\Gamma(b+i), \tag{A.18.43}$$

where

$$D_n = \det\left[a\delta_{i,j} - \delta_{i,j+1} + j(b+i)\delta_{i,j-1} \right]. \tag{A.18.44}$$

Expanding by the last row and last column, one gets the recurrence relation

$$D_{n+1} = aD_n + n(b+n-1)D_{n-1}, \tag{A.18.45}$$

$$D_0 = 1, \quad D_1 = a, \quad D_2 = a^2 + b, \quad \ldots. \tag{A.18.46}$$

To find D_n introduce the generating function

$$f(z) = \sum_{n=0}^{\infty} \frac{z^n}{n!} D_n. \tag{A.18.47}$$

Multiplying Eq. (A.18.45) on both sides by $z^n/n!$ and summing over n from 0 to ∞, one has

$$f'(z) = af(z) + bzf(z) + z^2 f'(z), \tag{A.18.48}$$

since

$$\sum_{n=0}^{\infty} D_{n+1} \frac{z^n}{n!} = \frac{d}{dz} \sum_{n=0}^{\infty} D_n \frac{z^n}{n!}, \tag{A.18.49}$$

and

$$\begin{aligned}\sum_{n=0}^{\infty} n(b+n-1) D_{n-1} \frac{z^n}{n!} &= bz \sum_{n=1}^{\infty} D_{n-1} \frac{z^{n-1}}{(n-1)!} + z^2 \sum_{n=2}^{\infty} D_{n-1} \frac{z^{n-2}}{(n-2)!} \\ &= bzf(z) + z^2 f'(z),\end{aligned} \tag{A.18.50}$$

hence

$$(1-z^2) f'(z) = (a+bz) f(z), \tag{A.18.51}$$

or

$$f(z) = (1-z)^{-(b+a)/2} (1+z)^{-(b-a)/2}. \tag{A.18.52}$$

This gives Eqs. (A.18.27) and (A.18.31).

If $a = 0$,

$$f(z) = (1-z^2)^{-b/2} = \sum_{n=0}^{\infty} \frac{z^{2n}}{n!} \left(\frac{b}{2}\right)_n, \tag{A.18.53}$$

so that the determinant (A.18.26) of an $n \times n$ matrix is zero when n is odd, as it should, and when n is even it is

$$\begin{aligned} M_{2n}(0,b) &= \frac{(2n)!}{n!} \left(\frac{b}{2}\right)_n \prod_{i=0}^{2n-1} i!\Gamma(b+i) \\ &= \left[\prod_{i=0}^{n-1} (2i+1)!\Gamma(b+2i+1)\right]^2. \end{aligned} \tag{A.18.54}$$

The Pfaffian is the square root of this and its sign can be fixed by looking at one of the terms.

When $a = b$,

$$f(z) = (1-z)^{-a} \quad \text{and} \quad D_n = (a)_n. \tag{A.18.55}$$

When a is a small integer, expression (A.18.31) for D_n simplifies. For example, when $a = 1$,

$$f(z) = (1+z)(1-z^2)^{-(b+1)/2} = (1+z)\sum_{n=0}^{\infty} \frac{z^{2n}}{n!}\left(\frac{b+1}{2}\right)_n, \tag{A.18.56}$$

and

$$D_{2n} = \frac{(2n)!}{n!}\left(\frac{b+1}{2}\right)_n, \tag{A.18.57}$$

$$D_{2n+1} = \frac{(2n+1)!}{n!}\left(\frac{b+1}{2}\right)_n. \tag{A.18.58}$$

Or when $a = 2$,

$$f(z) = (1+z)^2(1-z^2)^{-(b+2)/2} = (1+z)^2\sum_{n=0}^{\infty} \frac{z^{2n}}{n!}\left(\frac{b+2}{2}\right)_n, \tag{A.18.59}$$

and

$$D_{2n} = \frac{(2n)!}{n!}\left(\frac{b+2}{2}\right)_n + \frac{(2n)!}{(n-1)!}\left(\frac{b+2}{2}\right)_{n-1}, \tag{A.18.60}$$

$$D_{2n+1} = 2\cdot\frac{(2n+1)!}{n!}\left(\frac{b+2}{2}\right)_n, \tag{A.18.61}$$

etc.

A.19 Power-Series Expansion of $I_m(\theta)$, Eq. (7.6.12)

Expanding the integrand in (7.6.10) in powers of θ and integrating, we get

$$\eta_i(\theta) = \xi_{2i} - \frac{1}{2}\theta^2\xi_{2i-2} + \frac{3}{8}\theta^4\xi_{2i-4} + \cdots, \tag{A.19.1}$$

where

$$\xi_{2i} = 2\int_0^{\infty} e^{-y^2} y^{2i}\, dy = \Gamma\left(i+\frac{1}{2}\right).$$

Taking terms only up to θ^4, we put the expansion (A.19.1) in the determinant (7.6.9). We see that in writing the determinant as a sum of several terms many of them vanish because two rows are proportional. Thus we may write

$$I_m(\theta) = \text{const} \cdot \left(a_0 - \frac{1}{2}\theta^2 a_1 + \frac{3}{8}\theta^4 (a_2 - a_3) + \cdots \right), \tag{A.19.2}$$

where

$$a_0 = \det[\xi_{2i+2j}]_{i,j=1,\ldots,m-1},$$

$$a_1 = \det \begin{bmatrix} \xi_2 & \xi_4 & \cdots & \xi_{2m-2} \\ \xi_6 & \xi_8 & \cdots & \xi_{2m+2} \\ \cdots & \cdots & \cdots & \cdots \\ \xi_{2m} & \xi_{2m+2} & \cdots & \xi_{4m-4} \end{bmatrix}, \quad a_2 = \det \begin{bmatrix} \xi_0 & \xi_2 & \cdots & \xi_{2m-4} \\ \xi_6 & \xi_8 & \cdots & \xi_{2m+2} \\ \cdots & \cdots & \cdots & \cdots \\ \xi_{2m} & \xi_{2m+2} & \cdots & \xi_{4m-4} \end{bmatrix},$$

$$a_3 = \det \begin{bmatrix} \xi_2 & \xi_4 & \cdots & \xi_{2m-2} \\ \xi_4 & \xi_6 & \cdots & \xi_{2m} \\ \xi_8 & \xi_{10} & \cdots & \xi_{2m+4} \\ \cdots & \cdots & \cdots & \cdots \\ \xi_{2m} & \xi_{2m+2} & \cdots & \xi_{4m-4} \end{bmatrix}.$$

The evaluation of determinants whose elements are gamma functions is almost as easy as those whose elements are the successive factorials. Taking out all the common factors, one may reduce these determinants to the triangular form by simple operations (cf. Appendix A.18). Thus we obtain

$$a_1 = \frac{4}{3}(m-1)a_0, \qquad a_2 = \frac{8}{3}m(m-1)a_0$$

and

$$a_3 = \frac{8}{45}(m-1)(m-2)a_0.$$

Putting these values in (A.19.2) we get (7.6.12).

A.20 Proof of the Inequalities (7.6.15)

Let $u_i = \theta^2/y_i^2$ so that $u_i \geqslant 0$. The second inequality in (7.6.15)

$$\prod_1^n (1+u_i)^{-1/2} \leqslant 1$$

is immediate, for each factor in the product lies between 0 and 1.

The first inequality can be proved by induction. Suppose that

$$1-\frac{1}{2}\sum_{i=1}^{r}u_i \leqslant \prod_{i=1}^{r}(1+u_i)^{-1/2}$$

is true for $1 \leqslant r \leqslant n$ and let us prove then that

$$1-\frac{1}{2}\sum_{i=1}^{n+1}u_i \leqslant \prod_{i=1}^{n+1}(1+u_i)^{-1/2}.$$

Let $\sum_1^{n+1} u_i \leqslant 2$, for otherwise the left-hand side will be negative and therefore smaller than the right-hand side, which is positive. Thus

$$\left(1-\frac{1}{2}\sum_{1}^{r}u_i\right)\left(1-\frac{1}{2}\sum_{r+1}^{n+1}u_i\right) \leqslant \prod_{1}^{r}(1+u_i)^{-1/2}\prod_{r+1}^{n+1}(1+u_i)^{-1/2}, \quad r \leqslant n,$$

for both quantities in the product on both sides are positive. Therefore we have

$$\prod_{i=1}^{n+1}(1+u_i)^{-1/2} \geqslant 1-\frac{1}{2}\sum_{i=1}^{n+1}u_i+\left(\frac{1}{2}\sum_{1}^{r}u_i\right)\left(\frac{1}{2}\sum_{r+1}^{n+1}u_i\right) \geqslant 1-\frac{1}{2}\sum_{i=1}^{n+1}u_i.$$

Also it is easy to verify that

$$1-\frac{1}{2}u_1 \leqslant (1+u_1)^{-1/2},$$

and the proof is complete.

A.21 Proof of Eqs. (10.1.11) and (10.2.11)

Let dM and dM' be connected by a similarity transformation

$$dM' = A\,dM\,A^{-1}, \tag{A.21.1}$$

where A is non singular. We now show that the Jacobian

$$J = \frac{\partial(dM'_{ij})}{\partial(dM_{ij})} \tag{A.21.2}$$

is unity.

Considering the various matrix elements dM'_{ij} as components of a single vector (and similarly for dM_{ij}), we can write (A.21.1) as

$$dM'_{ij} = \sum_{k,\ell} A_{ik} A^{-1T}_{j\ell}\, dM_{k\ell}$$

or

$$dM' = (A \times A^{-1T})\, dM, \tag{A.21.3}$$

where the direct or the Kronecker product $(A \times B)$ is defined by the equation

$$(A \times B)_{ij,k\ell} = A_{ik} B_{j\ell}.$$

The Jacobian (A.21.2) is thus seen to be equal to the determinant of $(A \times A^{-1T})$.

Now it can be verified that if P and Q are matrices of the order $(n \times n)$, whereas R and S are of the order $(m \times m)$,

$$(P \times R) \cdot (Q \times S) = (P \cdot Q) \times (R \cdot S), \tag{A.21.4}$$

where a dot means ordinary matrix multiplication. From (A.21.4) we obtain

$$(R \times P) = (R \times 1_n) \cdot (1_m \times P),$$

where 1_r is the $(r \times r)$ unit matrix. Taking determinants on both sides of this equation we have

$$\det(R \times P) = (\det R)^n (\det P)^m. \tag{A.21.5}$$

Thus

$$J = \frac{\partial(dM'_{ij})}{\partial(dM_{ij})} = \det(A \times A^{-1T}) = \left[(\det A)(\det A^{-1})\right]^N = 1,$$

which establishes the result we wanted.

A.22 Proof of the Inequality (12.1.5)

Consider all $N(N-1)/2$ chords that join the N points $A_1, A_2, \ldots, A_N$ which lie on the unit circle. For definiteness let the angle variables of these points be in increasing order. Moreover, in the subsequent argument, whenever the index of any point exceeds N we subtract a multiple of N so that it is one of the numbers $1, 2, \ldots, N$. We want to maximize the product of the lengths of all the chords.

Let us divide the set of chords into a number of classes. In the first class we put the chords $A_1A_2, A_2A_3, \ldots, A_{N-1}A_N, A_NA_1$. In the second one we put

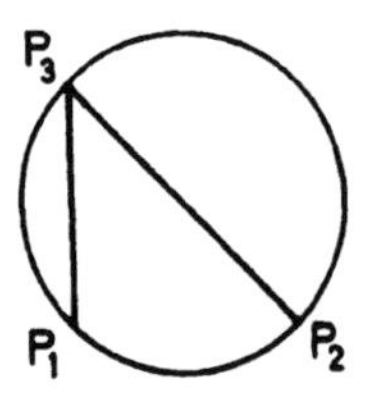

Figure A.22.1. When P_1 and P_2 are fixed and P_3 varies over the upper arc of the circle, then the maximum of the product of P_1P_3 and P_2P_3 is attained when $P_1P_3 = P_2P_3$.

$A_1A_3, A_3A_5, A_5A_7, \ldots$ and so on until we get back to A_1. If A_2 is left out, we construct a separate class of $A_2A_4, A_4A_6, \ldots$ until we get back to A_2. In the next class we put the chords $A_1A_4, A_4A_7, A_7A_10, \ldots$ until A_1 is repeated. If A_2 is left out, we construct a separate class $A_2A_5, A_5A_8, \ldots$. Similarly for A_3. And so on, until all the chords are exhausted. We maximize the product of the lengths of the chords belonging to a particular class. It is conceivable that the maximization conditions are different for different classes. If this occurs, we will really be in trouble.

Notice that if the points P_1, P_2 are fixed, but the point P_3 varies over the upper arc, the product of the chords P_1P_3, P_2P_3 is maximum when the two chords are equal. (See Figure A.22.1.) From this it follows that the product of the chords belonging to any one class is maximum when the points $A_1, A_2, \ldots, A_N$ lie at the vertices of a regular polygon. This condition is the same for any of the classes and the above mentioned trouble does not arise.

A.23 Good's Proof of Eq. (12.1.16)

For non-negative integers $a_1, a_2, \ldots, a_n$ let us write

$$F(\vec{x};\vec{a}) = \prod_{i\neq j=1}^{n} \left(1 - \frac{x_j}{x_i}\right)^{a_j},$$

and

$$M(\vec{a}) = \frac{(a_1 + \cdots + a_n)!}{a_1! \cdots a_n!}.$$

When $F(\vec{x};\vec{a})$ is expanded in positive and negative powers of $x_1, x_2, \ldots, x_n$, let the constant term be $G(\vec{a})$. We will show that $G(\vec{a}) = M(\vec{a})$.

Consider the polynomial

$$P(x) = \sum_{j=1}^{n} \prod_{i=1}^{n}{}' \frac{(x - x_i)}{(x_j - x_i)},$$

where the prime on the product means that the term $i = j$ is not taken. This polynomial, of degree $n-1$, takes the value 1 at $x = x_1, x = x_2, \ldots, x = x_n$; therefore it is identically equal to 1. Put $x = 0$ to get

$$\sum_{j=1}^{n} \prod_{i=1}^{n}{}' \left(1 - \frac{x_j}{x_i}\right)^{-1} = 1.$$

Multiplying $F(\vec{x}; \vec{a})$ by this function, we see that if $a_j > 0$, for every $j = 1, 2, \ldots, n$, then

$$F(\vec{x}; \vec{a}) = \sum_{j=1}^{n} F(\vec{x}; a_1, \ldots, a_{j-1}, a_j - 1, a_{j+1}, \ldots, a_n),$$

so that

$$G(\vec{a}) = \sum_{j=1}^{n} G(a_1, \ldots, a_{j-1}, a_j - 1, a_{j+1}, \ldots, a_n). \tag{A.23.1}$$

If some $a_j = 0$, then x_j occurs only in negative powers in $F(\vec{x}; \vec{a})$, and $G(\vec{a})$ is then equal to the constant term in

$$F(x_1, \ldots, x_{j-1}, x_{j+1}, \ldots, x_n; a_1, \ldots, a_{j-1}, a_{j+1}, \ldots, a_n);$$

that is

$$G(\vec{a}) = G(a_1, \ldots, a_{j-1}, a_{j+1}, \ldots, a_n), \quad \text{if } a_j = 0. \tag{A.23.2}$$

Also obviously

$$G(\vec{0}) = 1. \tag{A.23.3}$$

The recursive Eqs. (A.23.1)–(A.23.3) determine uniquely $G(\vec{a})$. But $M(\vec{a})$ satisfies these three equations. Therefore $G(\vec{a}) = M(\vec{a})$.

A.24 Some Recurrence Relations and Integrals Used in Chapter 14

In this appendix we collect some recurrence relations and identities often used in Chapter 14 and in later appendices.

A.24.1 The harmonic oscillator functions

$$\begin{aligned}\varphi_j(x) &= \left(2^j j! \sqrt{\pi}\right)^{-1/2} \exp(x^2/2) \left(-\frac{d}{dx}\right)^j \exp(-x^2) \\ &= \left(2^j j! \sqrt{\pi}\right)^{-1/2} \exp(-x^2/2) H_j(x),\end{aligned}$$

with $H_j(x)$, the Hermite polynomials, obey the orthonormality

$$\int_{-\infty}^{\infty} \varphi_i(x)\varphi_j(x)\,dx = \delta_{ij}, \tag{A.24.1}$$

and the recurrence relations

$$\sqrt{2}\varphi_j'(x) = \sqrt{j}\,\varphi_{j-1}(x) - \sqrt{j+1}\,\varphi_{j+1}(x), \tag{A.24.2}$$

$$\sqrt{2}x\varphi_j(x) = \sqrt{j}\,\varphi_{j-1}(x) + \sqrt{j+1}\,\varphi_{j+1}(x). \tag{A.24.3}$$

These equalities follow from those for Hermite polynomials (Bateman, 1953b; Szegö, 1939),

$$H_j'(x) = 2jH_{j-1}(x),$$
$$2xH_j(x) = H_{j+1}(x) + 2jH_{j-1}(x).$$

A.24.2 The functions $\psi_j(x)$ and $A_j(x)$ defined by (14.1.27) and (14.1.28) satisfy the equations

$$\sqrt{2}\,\psi_j(x) = \left[(1+\alpha^2)/(1-\alpha^2)\right]^j \times \left[\sqrt{j}\,(1-\alpha^2)\varphi_{j-1}(x) - \sqrt{j+1}\,(1+\alpha^2)\varphi_{j+1}(x)\right], \tag{A.24.4}$$

$$\sqrt{2}\,\varphi_j(x) = \left[(1+\alpha^2)/(1-\alpha^2)\right]^j \times \left[\sqrt{j}\,(1-\alpha^2)A_{j-1}(x) - \sqrt{j+1}\,(1+\alpha^2)A_{j+1}(x)\right]. \tag{A.24.5}$$

These relations can be derived from (A.24.2) and (A.24.3).

A.24.3 The orthonormality relations involving $\psi_j(x)$, $A_j(x)$ and $\varphi_j(x)$ are

$$\int_{-\infty}^{\infty} \varphi_i(x)\varphi_j(x)\,dx = \delta_{ij}, \tag{A.24.1}$$

$$\int_{-\infty}^{\infty} \psi_i(x)A_j(x)\,dx = -\delta_{ij}, \tag{A.24.6}$$

$$\int_{-\infty}^{\infty} \psi_i(x)\varphi_j(x)\,dx = \int_{-\infty}^{\infty} A_i(x)\varphi_j(x)\,dx = 0, \quad i+j \text{ even}. \tag{A.24.7}$$

The second is obtained by partial integration and the third by a parity argument. Note that $\varphi_j(x)$ has the parity of j, while $\psi_j(x)$ and $A_j(x)$ have parities opposite to that of j.

A.24.4 The recurrence relations

$$
\begin{aligned}
&\psi_{2j}(x)A_{2j}(y)\\
&\quad= -\varphi_{2j+1}(x)\varphi_{2j+1}(y) + (1-\alpha^2)\big\{\big[(1+\alpha^2)/(1-\alpha^2)\big]^{2j}\sqrt{j}\,\varphi_{2j-1}(x)A_{2j}(y)\\
&\qquad - \big[(1+\alpha^2)/(1-\alpha^2)\big]^{2j+2}\sqrt{j+1}\,\varphi_{2j+1}(x)A_{2j+2}(y)\big\}, \qquad \text{(A.24.8)}\\
&\psi_{2j+1}(x)A_{2j+1}(y)\\
&\quad= -\varphi_{2j}(x)\varphi_{2j}(y) + (1-\alpha^2)\big\{\big[(1+\alpha^2)/(1-\alpha^2)\big]^{2j}\sqrt{j}\,\varphi_{2j}(x)A_{2j-1}(y)\\
&\qquad - \big[(1+\alpha^2)/(1-\alpha^2)\big]^{2j+2}\sqrt{j+1}\,\varphi_{2j+2}(x)A_{2j+1}(y)\big\}, \qquad \text{(A.24.9)}\\
&\varphi_{2j}(x)A_{2j}(y) + A_{2j+1}(x)\varphi_{2j+1}(y)\\
&\quad= (1-\alpha^2)\big\{\big[(1+\alpha^2)/(1-\alpha^2)\big]^{2j}\sqrt{j}\,A_{2j-1}(x)A_{2j}(y)\\
&\qquad - \big[(1+\alpha^2)/(1-\alpha^2)\big]^{2j+2}\sqrt{j+1}\,A_{2j+1}(x)A_{2j+2}(y)\big\}, \qquad \text{(A.24.10)}\\
&\psi_{2j}(x)\varphi_{2j}(y) + \varphi_{2j+1}(x)\psi_{2j+1}(y)\\
&\quad= (1-\alpha^2)\big\{\big[(1+\alpha^2)/(1-\alpha^2)\big]^{2j}\sqrt{j}\,\varphi_{2j-1}(x)\varphi_{2j}(y)\\
&\qquad - \big[(1+\alpha^2)/(1-\alpha^2)\big]^{2j+2}\sqrt{j+1}\,\varphi_{2j+1}(x)\varphi_{2j+2}(y)\big\}, \qquad \text{(A.24.11)}
\end{aligned}
$$

can be derived from (A.24.4) and (A.24.5). These relations are useful to verify the equivalence of expressions (14.1.16) and (14.1.17), of (14.1.18) and (14.1.19) and of (14.1.21) and (14.1.22).

A.24.5 We need the integral

$$
\begin{aligned}
&\int_{-\infty}^{\infty} \varphi_{2j}(y)\exp\left(-\frac{1}{2}\alpha^2 y^2\right) dy\\
&\quad= \big[2\pi/(1+\alpha^2)\big]^{1/2}\frac{(2j)!}{j!}\left(2^{2j}(2j)!\sqrt{\pi}\right)^{-1/2}\big[(1-\alpha^2)/(1+\alpha^2)\big]^{j}. \qquad \text{(A.24.12)}
\end{aligned}
$$

To prove this equation we use the generating function

$$
\begin{aligned}
\exp(-y^2+2yz-z^2) &= \sum_{j=0}^{\infty}\frac{z^j}{j!}\left(-\frac{d}{dy}\right)^j \exp(-y^2)\\
&= \sum_{j=0}^{\infty}\frac{z^j}{j!}\exp\left(-\frac{1}{2}y^2\right)\left(2^j j!\sqrt{\pi}\right)^{1/2}\varphi_j(y), \qquad \text{(A.24.13)}
\end{aligned}
$$

to write

$$\sum_{j=0}^{\infty} \frac{z^j}{j!}(2^j j!\sqrt{\pi})^{1/2} \int_{-\infty}^{\infty} \varphi_j(y) \exp\left(-\frac{1}{2}\alpha^2 y^2\right) dy$$
$$= \int_{-\infty}^{\infty} \exp\left[-\frac{1}{2}(1+\alpha^2)y^2 + 2yz - z^2\right] dy$$
$$= \exp\left[z^2\left(\frac{1-\alpha^2}{1+\alpha^2}\right)\right] \int_{-\infty}^{\infty} \exp\left[-\frac{1+\alpha^2}{2}\left(y - \frac{2}{1+\alpha^2}z\right)^2\right] dy$$
$$= \left(\frac{2\pi}{1+\alpha^2}\right)^{1/2} \sum_{j=0}^{\infty} \frac{z^{2j}}{j!}\left(\frac{1-\alpha^2}{1+\alpha^2}\right)^j.$$

Equating coefficients of z^{2j} on both sides we get (A.24.12).

A.24.6 We need the convolution integrals

$$\int_{-\infty}^{\infty} g(x, y)\psi_j(y)\, dy = \varphi_j(x), \tag{A.24.14}$$

$$\int_{-\infty}^{\infty} g(x, y)\varphi_j(y)\, dy = A_j(x), \tag{A.24.15}$$

with $g(x, y)$ given by Eq. (14.1.23). By partial integration (A.24.14) is equivalent to

$$\int_{-\infty}^{\infty} \exp\left[-\frac{1}{2}\alpha^2(x^2+y^2) - \left(\frac{1-\alpha^4}{4\alpha^2}\right)(x-y)^2\right]\varphi_j(y)\, dy$$
$$= \frac{2\alpha\sqrt{\pi}}{1+\alpha^2}\left(\frac{1-\alpha^2}{1+\alpha^2}\right)^j \varphi_j(x). \tag{A.24.16}$$

To prove this we expand both sides of the identity

$$\int_{-\infty}^{\infty} \exp\left[2yz - z^2 - \frac{1}{2}(1+\alpha^2)y^2 - \left(\frac{1-\alpha^4}{4\alpha^2}\right)(x-y)^2\right] dy$$
$$= \frac{2\alpha\sqrt{\pi}}{1+\alpha^2} \exp\left[-\left(\frac{1-\alpha^2}{1+\alpha^2}\right)^2 z^2 + 2\left(\frac{1-\alpha^2}{1+\alpha^2}\right)zx - \frac{1}{2}(1-\alpha^2)x^2\right] \tag{A.24.17}$$

in powers of z and use the generating function (A.24.13).

Equation (A.24.15) can be rewritten as

$$\left(\frac{1+\alpha^2}{1-\alpha^2}\right)^{j+1/2}\int_{-\infty}^{\infty} f(x-y)\exp\left(-\frac{1}{2}\alpha^2y^2\right)\varphi_j(y)\,dy$$
$$=\int_{-\infty}^{\infty}\exp\left(\frac{1}{2}\alpha^2y^2\right)\varphi_j(y)\,\mathrm{sign}(x-y)\,dy.$$

This last equation is true, since the derivatives of both sides are equal, Eq. (A.24.16), and the equality holds at one point $x=\infty$. In fact at $x=\infty$, $f(x-y)=\mathrm{sign}(x-y)=1$, and we have to make sure that

$$\left(\frac{1+\alpha^2}{1-\alpha^2}\right)^{j+1/2}\int_{-\infty}^{\infty}\exp\left(-\frac{1}{2}\alpha^2y^2\right)\varphi_j(y)\,dy$$
$$=\int_{-\infty}^{\infty}\exp\left(\frac{1}{2}\alpha^2y^2\right)\varphi_j(y)\,dy. \tag{A.24.18}$$

For j odd, both sides are zero and the equality is evident. For j even, we use Eq. (A.24.12) and the one obtained by changing the sign of α^2 in it. Taking ratios we get (A.24.18).

A.24.7 Recurrence and Orthonormality Relations Between $\boldsymbol{\Psi}_j(x)$, $A_j(x)$ and $\varphi_j(x)$. The function $\boldsymbol{\Psi}_j(x)$ and $\mathbf{A}_j(x)$, Eqs. (14.2.20) and (14.2.21), are obtained from $\psi_j(x)$ and $A_j(x)$ Eqs. (14.1.27) and (14.1.28), by changing the sign of α^2. So from (A.24.4)–(A.24.7) we get

$$\sqrt{2}\,\boldsymbol{\Psi}_j(x)=\left[(1-\alpha^2)/(1+\alpha^2)\right]^j$$
$$\times\left[\sqrt{j}\,(1+\alpha^2)\varphi_{j-1}(x)-\sqrt{j+1}\,(1-\alpha^2)\varphi_{j+1}(x)\right], \tag{A.24.19}$$

$$\sqrt{2}\,\varphi_j(x)=\left[(1-\alpha^2)/(1+\alpha^2)\right]^j$$
$$\times\left[\sqrt{j}\,(1+\alpha^2)\mathbf{A}_{j-1}(x)-\sqrt{j+1}\,(1-\alpha^2)\mathbf{A}_{j+1}(x)\right], \tag{A.24.20}$$

$$\int_{-\infty}^{\infty}\boldsymbol{\Psi}_i(x)\mathbf{A}_j(x)\,dx=-\delta_{ij}, \tag{A.24.21}$$

$$\int_{-\infty}^{\infty}\boldsymbol{\Psi}_i(x)\varphi_j(x)\,dx=\int_{-\infty}^{\infty}\mathbf{A}_i(x)\varphi_j(x)\,dx=0,\quad i+j \text{ even}. \tag{A.24.22}$$

A.24.8 We also need the convolution integrals

$$\int_{-\infty}^{\infty}\mathbf{g}(x,y)\varphi_j(y)\,dy=\boldsymbol{\Psi}_j(x), \tag{A.24.23}$$

$$\int_{-\infty}^{\infty} \mathbf{g}(x, y)\mathbf{A}_j(y)\, dy = \varphi_j(x). \tag{A.24.24}$$

Equation (A.24.23) can be established from the definitions of $\mathbf{g}(x, y)$, $\boldsymbol{\Psi}_j(x)$, Eqs. (14.2.18), (14.2.20), and using Eqs. (A.24.16), (A.24.2) and (A.24.3). For Eq. (A.24.24) we have from the definitions of $\mathbf{g}(x, y)$ and $\mathbf{A}_j(x)$, Eqs. (14.2.18), (14.2.21)

$$\begin{aligned}
&\int_{-\infty}^{\infty} \mathbf{g}(x, y)\mathbf{A}_j(y)\, dy \\
&\quad = -\frac{(1+\alpha^2)(1-\alpha^4)}{4\alpha^3\sqrt{\pi}} \left(\frac{1+\alpha^2}{1-\alpha^2}\right)^j \iint_{-\infty}^{\infty} dy\, dz \\
&\qquad \times \varepsilon(y-z)\varphi_j(z) \exp\left[-\frac{1}{2}\alpha^2(x^2+z^2)\right](x-y)\exp\left(-\frac{1-\alpha^4}{4\alpha^2}(x-y)^2\right),
\end{aligned}$$

with $2\varepsilon(x) = \text{sign}(x)$, Eq. (14.1.29). Changing the variable y to $t = x - y$, one can integrate over t,

$$\frac{1-\alpha^4}{-2\alpha^2} \int_{-\infty}^{\infty} \varepsilon(x-z-t)t \exp\left(-\frac{1-\alpha^4}{4\alpha^2}t^2\right) dt = \exp\left(-\frac{1-\alpha^4}{4\alpha^2}(x-z)^2\right).$$

Now using Eq. (A.24.16) one gets (A.24.24).

A.25 Normalization Integral, Eq. (14.1.11)

The constant in (14.1.11) being fixed by the condition

$$\int_{-\infty}^{\infty} \cdots \int_{-\infty}^{\infty} p(x_1, \ldots, x_N)\, dx_1 \cdots dx_N = 1, \tag{A.25.1}$$

with $p(x_1, \ldots, x_N)$ given by (14.1.7), we shall evaluate the integral

$$C_N^{-1} = \int_{-\infty}^{\infty} \cdots \int_{-\infty}^{\infty} dx_1 \cdots dx_N \exp\left(-\frac{1+\alpha^2}{2}\sum x_i^2\right)\Delta(x)\,\text{Pf}[F_{ij}], \tag{A.25.2}$$

with

$$\Delta(x) = \prod_{i<j}(x_i - x_j), \tag{14.1.8}$$

and F_{ij} defined by Eqs. (14.1.9) and (14.1.9′).

The Pfaffian and $\Delta(x)$ are alternating functions of the variables $x_1, \ldots, x_N$. The $\Delta(x)$ can be written as a determinant

$$\exp\Big(-\sum x_i^2/2\Big)\Delta(x) = \prod_0^{N-1} \big(2^{-j} j! \sqrt{\pi}\big)^{1/2} \det\big[\varphi_{i-1}(x_j)\big]_{i,j=1,\ldots,N}, \tag{A.25.3}$$

where $\varphi_i(x)$ is the normalized "oscillator function" given by (14.1.26). The number of terms in the Pfaffian is $(2m)!/(2^m m!)$ with $N = 2m$ or $N = 2m - 1$.

It is convenient to discuss separately the cases N even and N odd. Let us first take the case $N = 2m$ even. From the symmetry and what we said above, one has

$$C_N^{-1} = \prod_0^{N-1} \big(2^{-j} j! \sqrt{\pi}\big)^{1/2} \int_{-\infty}^{\infty} \cdots \int_{-\infty}^{\infty} dx_1 \cdots dx_N \exp\Big(-\frac{1}{2}\alpha^2 \sum x_i^2\Big)$$
$$\times \operatorname{Pf}\big[f(x_i - x_j)\big] \det\big[\varphi_{i-1}(x_j)\big] \tag{A.25.4}$$
$$= \prod_0^{N-1} \big(2^{-j} j! \sqrt{\pi}\big)^{1/2} \frac{(2m)!}{2^m m!} \int_{-\infty}^{\infty} \cdots \int_{-\infty}^{\infty} dx_1 \cdots dx_N \exp\Big(-\frac{1}{2}\alpha^2 \sum x_i^2\Big)$$
$$\times \prod_{j=1}^{m} f(x_{2j} - x_{2j-1}) \det\big[\varphi_{i-1}(x_j)\big]. \tag{A.25.5}$$

For the last integral, we have from the theory of Pfaffians (Mehta, 1989MT or Appendix A.17)

$$C_N^{-1} \prod_0^{N-1} \big(2^{-j} j! \sqrt{\pi}\big)^{-1/2} \left(\frac{(2m)!}{2^m m!}\right)^{-1} = m! 2^m \operatorname{Pf}[a_{ij}]_{i,j=0,1,\ldots,2m-1}, \tag{A.25.6}$$

where for $i, j = 0, 1, \ldots, 2m - 1$,

$$a_{ij} = \frac{1}{2} \int_{-\infty}^{\infty} \int_{-\infty}^{\infty} \exp\Big[-\frac{1}{2}\alpha^2 (x^2 + y^2)\Big] f(y - x)\big[\varphi_i(x)\varphi_j(y) - \varphi_j(x)\varphi_i(y)\big]\, dx\, dy$$
$$= \int_{-\infty}^{\infty} \int_{-\infty}^{\infty} \exp\Big[-\frac{1}{2}\alpha^2 (x^2 + y^2)\Big] f(y - x)\varphi_i(x)\varphi_j(y)\, dx\, dy. \tag{A.25.7}$$

From parity one sees that $a_{ij} = 0$ if $i + j$ is even. Thus the Pfaffian reduces to a determinant

$$\operatorname{Pf}[a_{ij}]_{i,j=0,1,\ldots,2m-1} = \det[a_{2i,2j+1}]_{i,j=0,1,\ldots,m-1}. \tag{A.25.8}$$

The determinant is not changed if we add to any row (column) a constant multiple of another row (column). To choose convenient multiples, we observe that using the identity (A.24.4)

$$\sqrt{2}\frac{d}{dy}\left[\varphi_j(y)\exp\left(-\frac{1}{2}\alpha^2y^2\right)\right]$$
$$=\exp\left(-\frac{1}{2}\alpha^2y^2\right)\left[\sqrt{j}\,(1-\alpha^2)\varphi_{j-1}(y)-\sqrt{j+1}\,(1+\alpha^2)\varphi_{j+1}(y)\right], \quad \text{(A.25.9)}$$

an integration by parts in Eq. (A.25.7) gives

$$(1/\sqrt{2})\left[\sqrt{j}\,(1-\alpha^2)a_{i,j-1}-\sqrt{j+1}\,(1+\alpha^2)a_{i,j+1}\right]$$
$$=-\frac{2}{\sqrt{\pi}}\left(\frac{1-\alpha^4}{4\alpha^2}\right)^{1/2}\int_{-\infty}^{\infty}\int_{-\infty}^{\infty}\exp\left(-\frac{1}{2}\alpha^2(x^2+y^2)-\frac{1-\alpha^4}{4\alpha^2}(x-y)^2\right)$$
$$\times\,\varphi_i(x)\varphi_j(y)\,dx\,dy$$
$$=-2\left[(1-\alpha^2)/(1+\alpha^2)\right]^{j+1/2}\delta_{ij}. \quad \text{(A.25.10)}$$

In the last step we made use of Eqs. (A.24.16) and (A.24.1). Thus

$$a_{2i,2j+1}-\left(\frac{2j}{2j+1}\right)^{1/2}\left(\frac{1-\alpha^2}{1+\alpha^2}\right)a_{2i,2j-1}$$
$$=\frac{2}{1+\alpha^2}\left(\frac{2}{2j+1}\right)^{1/2}\left(\frac{1-\alpha^2}{1+\alpha^2}\right)^{2j+1/2}\delta_{ij}$$
$$=b_{ij}, \quad \text{say,} \quad \text{(A.25.11)}$$

and

$$\det[a_{2i,2j+1}]=\det[b_{ij}]=\prod_{j=0}^{m-1}\left[\frac{2}{1+\alpha^2}\left(\frac{2}{2j+1}\right)^{1/2}\left(\frac{1-\alpha^2}{1+\alpha^2}\right)^{2j+1/2}\right]. \quad \text{(A.25.12)}$$

From (A.25.6), (A.25.8) and (A.25.12) we finally have for $N=2m$,

$$C_N^{-1}=\prod_0^{N-1}(2^{-j}j!\sqrt{\pi})^{1/2}N!\prod_0^{m-1}\left[\frac{2}{1+\alpha^2}\left(\frac{2}{2j+1}\right)^{1/2}\left(\frac{1-\alpha^2}{1+\alpha^2}\right)^{2j+1/2}\right], \quad \text{(A.25.13)}$$

which is Eq. (14.1.11), since

$$\prod_0^{N-1}(\sqrt{\pi}\,j!)^{1/2} = \prod_0^{m-1}\left(\pi(2j)!(2j+1)!\right)^{1/2} = \prod_0^{m-1}\left[(2j+1)^{1/2}\pi^{1/2}\Gamma(2j+1)\right]$$

$$= \prod_0^{m-1}\left[(2j+1)^{1/2}2^{2j}\Gamma(j+1/2)\Gamma(j+1)\right]$$

$$= 2^{m(m-1)}\prod_0^{m-1}(2j+1)^{1/2}\prod_1^{2m}\left(\frac{2}{j}\Gamma(1+j/2)\right). \tag{A.25.14}$$

When $N = 2m - 1$ is odd, we have similarly

$$C_N^{-1}\prod_0^{N-1}\left(2^{-j}j!\sqrt{\pi}\right)^{-1/2}$$

$$= \frac{(2m)!}{2^m m!}\int_{-\infty}^{\infty}\cdots\int_{-\infty}^{\infty}dx_1\cdots dx_N\,\exp\left(-\frac{1}{2}\alpha^2\sum x_i^2\right)$$

$$\times\prod_{j=1}^{m-1}f(x_{2j}-x_{2j-1})\det\left[\varphi_{i-1}(x_j)\right]$$

$$= \frac{(2m)!}{2^m m!}2^{m-1}(m-1)!\,\mathrm{Pf}[a_{ij}]_{i,j=0,1,\ldots,2m-1}, \tag{A.25.15}$$

where now for $i, j = 0, 1, \ldots, 2m-2$, a_{ij} is given by (A.25.7), and

$$a_{i,2m-1} = -a_{2m-1,i}$$

$$= \int_{-\infty}^{\infty}\exp\left(-\frac{1}{2}\alpha^2x^2\right)\varphi_i(x)\,dx, \quad i = 0, 1, \ldots, 2m-2, \tag{A.25.16}$$

$$a_{2m-1,2m-1} = 0. \tag{A.25.17}$$

Again we have $a_{ij} = 0$, if $i + j$ is even, so that

$$\mathrm{Pf}[a_{ij}]_{i,j=0,1,\ldots,2m-1} = \det[a_{2i,2j+1}]_{i,j=0,1,\ldots,m-1}. \tag{A.25.18}$$

We can again, without changing the value of the determinant, replace $a_{2i,2j+1}$ by b_{ij}, where for $j = 0, 1, \ldots, m-2$ and $i = 0, 1, \ldots, m-1$ the b_{ij} is given by Eq. (A.25.11), while for $i = 0, 1, \ldots, m-1$

$$b_{i,m-1} = \int_{-\infty}^{\infty} \exp\left(-\frac{1}{2}\alpha^2 x^2\right) \varphi_{2i}(x)\,dx$$
$$= \left(\frac{2\pi}{1+\alpha^2}\right)^{1/2} \left(2^{2i}(2i)!\sqrt{\pi}\right)^{-1/2} \frac{(2i)!}{i!} \left(\frac{1-\alpha^2}{1+\alpha^2}\right)^i \qquad \text{(A.25.19)}$$

(Eq. (A.24.12)).

Collecting the results for $N = 2m - 1$, one has

$$C_N^{-1} = \prod_0^{N-1} \left(2^{-j} j! \sqrt{\pi}\right)^{1/2} N! \prod_0^{m-2} \left[\frac{2}{1+\alpha^2} \left(\frac{2}{2j+1}\right)^{1/2} \left(\frac{1-\alpha^2}{1+\alpha^2}\right)^{2j+1/2}\right]$$
$$\times \left(\frac{2\pi}{1+\alpha^2}\right)^{1/2} \left(2^{2m-2}(2m-2)!\sqrt{\pi}\right)^{-1/2} \frac{(2m-2)!}{(m-1)!} \left(\frac{1-\alpha^2}{1+\alpha^2}\right)^{m-1}, \qquad \text{(A.25.20)}$$

which on similar manipulations can be seen to be Eq. (14.1.11) with $N = 2m - 1$.

A.26 Another Normalization Integral, Eq. (14.2.9)

The constant in Eq. (14.2.9) being fixed by the normalization condition

$$\int_{-\infty}^{\infty} \cdots \int_{-\infty}^{\infty} \mathbf{p}(x_1, \ldots, x_{2N})\,dx_1 \cdots dx_{2N} = 1,$$

with $\mathbf{p}(x_1, \ldots, x_{2N})$ given by (14.2.7), we need the integral

$$\mathbf{C}_N^{-1} = \int_{-\infty}^{\infty} \cdots \int_{-\infty}^{\infty} dx_1 \cdots dx_{2N} \exp\left(-\frac{1+\alpha^2}{2} \sum_1^{2N} x_i^2\right) \Delta(x)\,\mathrm{Pf}\left[\mathbf{F}(x_i - x_j)\right],$$

$$\mathbf{F}(x) = x \exp\left[-(1-\alpha^4)x^2/4\alpha^2\right].$$

We follow the reasoning of Appendix A.25 for the case of an even number of variables and write

$$\mathbf{C}_N^{-1} \prod_0^{2N-1} \left(2^{-j} j! \sqrt{\pi}\right)^{-1/2} = (2N)!\,\mathrm{Pf}[a_{ij}]_{i,j=0,1,\ldots,2N-1}$$
$$= (2N)!\det[a_{2i,2j+1}]_{i,j=0,1,N-1},$$

where

$$a_{ij} = \iint_{-\infty}^{\infty} \mathbf{F}(y-x)\exp\left[-\frac{1}{2}\alpha^2(x^2+y^2)\right]\varphi_i(x)\varphi_j(y)\,dx\,dy$$
$$= \iint_{-\infty}^{\infty} (y-x)\varphi_i(x)\varphi_j(y)\exp\left(-\frac{1}{2}\alpha^2(x^2+y^2) - \frac{1-\alpha^4}{4\alpha^2}(x-y)^2\right)dx\,dy.$$

Now from (A.24.3), (A.24.16) and (A.24.1) one gets after a little algebra

$$a_{2i,2j+1} = \frac{2\alpha^3\sqrt{2\pi}}{(1+\alpha^2)^2}\left(\frac{1-\alpha^2}{1+\alpha^2}\right)^{2i}\left(\sqrt{2i+1}\,\delta_{ij} + \sqrt{2i}\,\delta_{i,j+1}\right).$$

Thus

$$\mathbf{C}_N^{-1} = \prod_0^{2N-1}\left(2^{-j}j!\sqrt{\pi}\right)^{1/2}(2N)!\prod_0^{N-1}\left[\frac{2\alpha^3\sqrt{2\pi}}{(1+\alpha^2)^2}\left(\frac{1-\alpha^2}{1+\alpha^2}\right)^{2j}\sqrt{2j+1}\right],$$

which is the same as Eq. (14.2.9).

A.27 Joint Probability Density as a Determinant of a Self-Dual Quaternion Matrix. Section 14.4, Eqs. (14.4.2) and (14.4.5)

Equation (14.4.2) will be verified separately for the cases N even and N odd. The manipulations are similar to those of Chapter 5.

When N is even, we observe that the $2N \times 2N$ matrix

$$G = \begin{bmatrix} S_N(x_i,x_j) & D_N(x_i,x_j) \\ I_N(x_i,x_j) & S_N(x_j,x_i) \end{bmatrix}_{i,j=1,\dots,N}, \tag{A.27.1}$$

can be written as a product of two rectangular matrices of orders $2N \times N$ and $N \times 2N$ respectively

$$G = G_1G_2,$$
$$G_1 = \begin{bmatrix} \varphi_{2k}(x_i) & -\psi_{2k}(x_i) \\ -A_{2k}(x_i) & \varphi_{2k}(x_i) \end{bmatrix}_{i=1,2,\dots,N;\ k=0,1,\dots,N/2-1},$$
$$G_2 = \begin{bmatrix} \varphi_{2k}(x_j) & \psi_{2k}(x_j) \\ A_{2k}(x_j) & \varphi_{2k}(x_j) \end{bmatrix}_{k=0,1,\dots,N/2-1;\ j=1,2,\dots,N}. \tag{A.27.2}$$

The rank of G_1 or of G_2 is at most N; it is in fact N, since the $\varphi_{2k}(x)$ is an orthonormal sequence. The rank of G, the product of G_1 and G_2, is therefore at most N (see, e.g., Mehta, 1989MT); in fact we know that it is N, since its first N rows are linearly independent. Therefore, the last N rows of G are linear combinations of its first N rows and *vice versa*. The ordinary determinant of the $2N \times 2N$ matrix

$$[\phi(x_i, x_j)] = \begin{bmatrix} S_N(x_i, x_j) & D_N(x_i, x_j) \\ J_N(x_i, x_j) & S_N(x_j, x_i) \end{bmatrix}_{i,j=1,\dots,N}$$

is therefore not changed if we subtract from its last N rows the corresponding rows of G:

$$\begin{aligned} \det[\phi(x_i, x_j)] &= \det \begin{bmatrix} S_N(x_i, x_j) & D_N(x_i, x_j) \\ g(x_i, x_j) & 0 \end{bmatrix} \\ &= (-1)^N \det[g(x_i, x_j)] \det[D_N(x_i, x_j)]. \end{aligned} \tag{A.27.3}$$

But

$$\det[g(x_i, x_j)] = 2^{-N} \left(\frac{1+\alpha^2}{1-\alpha^2} \right)^{N/2} \exp\left(-\alpha^2 \sum x_i^2\right) \det[f(x_i - x_j)], \tag{A.27.4}$$

and

$$\begin{aligned} \det[D_N(x_i, x_j)] &= \det[\varphi_{2k}(x_i) \psi_{2k}(x_i)] \begin{bmatrix} \psi_{2k}(x_j) \\ -\varphi_{2k}(x_j) \end{bmatrix} \\ &= \left(\det[\varphi_{2k}(x_i) \psi_{2k}(x_i)]\right)^2. \end{aligned} \tag{A.27.5}$$

Now ψ_{2k} being a linear combination of φ_{2k-1} and φ_{2k+1}, Eq. (A.24.4), we can replace the columns $\psi_0, \psi_2, \dots, \psi_{2N-2}$ successively by $\varphi_1, \varphi_3, \dots, \varphi_{2N-1}$ in the determinant. The result is

$$\begin{aligned} \det[D_N(x_i, x_j)] &\propto \{\det[\varphi_{j-1}(x_i)]\}^2 \\ &\propto \exp\left(-\sum x_i^2\right) \{\det[x_i^{j-1}]\}^2 \\ &\propto \exp\left(-\sum x_i^2\right) [\Delta(x)]^2. \end{aligned} \tag{A.27.6}$$

From Eqs. (A.27.3), (A.27.4) and (A.27.6) we get

$$\det[\phi(x_i, x_j)] \propto \exp\left(-(1+\alpha^2) \sum_1^N x_i^2\right) \det[f(x_i - x_j)] [\Delta(x)]^2,$$

which in view of Eq. (14.1.7) is

$$\det[\phi(x_i, x_j)] \propto [p(x_1, \dots, x_N)]^2. \tag{A.27.7}$$

(Note that here the matrix is written in its partitioned form, its elements being ordinary numbers, not quaternions, and the determinant is the usual one.)

When $N = 2m + 1$ odd, the method used above fails at one point; the $N \times N$ matrix

$$D_N(x_i, x_j) = \begin{bmatrix} \varphi_{2k+1}(x_i) & -\psi_{2k+1}(x_i) \end{bmatrix} \begin{bmatrix} \psi_{2k+1}(x_j) \\ \varphi_{2k+1}(x_j) \end{bmatrix} \tag{A.27.8}$$

of rank $N - 1$ has a zero determinant, while its multiplying factor becomes infinite due to the extra terms ξ_N and μ_N.

Instead of G consider the $2N \times 2N$ matrix

$$G_\delta = \begin{bmatrix} \varphi_{2k+1}(x_i) & -\psi_{2k+1}(x_i) & \delta\varphi_{2m}(x_i) \\ -A_{2k+1} & \varphi_{2k+1}(x_i) & (c\delta)^{-1}\exp(-\alpha^2 x_i^2/2) \end{bmatrix} \times \begin{bmatrix} \varphi_{2k+1}(x_j) & \psi_{2k+1}(x_j) \\ A_{2k+1}(x_j) & \varphi_{2k+1}(x_j) \\ (c\delta)^{-1}\exp(-\alpha^2 x_j^2/2) & \delta\varphi_{2m}(x_j) \end{bmatrix}, \tag{A.27.9}$$

where δ is arbitrary and

$$c = \int_{-\infty}^{\infty} \varphi_{2m}(t)\exp(-\alpha^2 t^2/2)\, dt. \tag{A.27.10}$$

Thus

$$G_\delta = \begin{bmatrix} S_N(x_i, x_j) + \xi_N(x_i, x_j) & D_N(x_i, x_j) + \delta^2\varphi_{2m}(x_i)\varphi_{2m}(x_j) \\ I_N(x_i, x_j) + (c\delta)^{-2}\exp[-\alpha^2(x_i^2 + x_j^2)/2] & S_N(x_j, x_i) + \xi_N(x_j, x_i) \end{bmatrix}, \tag{A.27.11}$$

with ξ_N, S_N, I_N and D_N given by Eqs. (14.1.15), (14.1.17), (14.1.22) and (14.1.19). The rank of G_δ is N. The determinant of the $2N \times 2N$ matrix

$$[\phi_\delta(x_i, x_j)] = \begin{bmatrix} S_N(x_i, x_j) + \xi_N(x_i, x_j) & D_N(x_i, x_j) + \delta^2\varphi_{2m}(x_i)\varphi_{2m}(x_j) \\ J_N(x_i, x_j) & S_N(x_j, x_i) + \xi_N(x_j, x_i) \end{bmatrix}, \tag{A.27.12}$$

is not changed if we subtract from its last N rows the corresponding rows of G_δ; the resulting determinant factorises,

$$\det[\phi_\delta(x_i, x_j)] = \det[D_N(x_i, x_j) + \delta^2\varphi_{2m}(x_i)\varphi_{2m}(x_j)](-1)^N \det[g(x_i, x_j) - (c\delta)^{-2}\exp\{-\alpha^2(x_i^2 + x_j^2)/2\} + \mu(x_i, x_j) - \mu(x_j, x_i)]. \tag{A.27.13}$$

The first factor is

$$\delta^2\{\det[\varphi_1(x_i), \psi_1(x_i);\ \varphi_3(x_i), \psi_3(x_i);\ \ldots;\ \varphi_{2m-1}(x_i), \psi_{2m-1}(x_i);\ \varphi_{2m}(x_i)]\}^2.$$

We can replace the last but one column by a linear combination of the last two columns. Choosing this combination properly, Eq. (A.24.4), ψ_{2m-1} can be replaced by φ_{2m-2}. Then the column ψ_{2m-3} can be replaced by the column φ_{2m-4}, and so on. Thus the column ψ_{2k+1} is replaced by φ_{2k} for $k = 0, 1, \ldots, m-1$, the determinant being multiplied by a constant. Thus the first factor in (A.27.13) is proportional to

$$\delta^2\{\det[\varphi_{j-1}(x_i)]_{i,j=1,2,\ldots,2m+1}\}^2 \propto \delta^2 \exp\Big(-\sum x_i^2\Big)[\Delta(x)]^2. \tag{A.27.14}$$

The second factor in Eq. (A.27.13) is proportional to

$$\exp\Big(-\alpha^2\sum x_i^2\Big)\det[f(x_i - x_j) - k\delta^{-2} + k'(h(x_i) - h(x_j))], \tag{A.27.15}$$

where

$$h(x) = \int_{-\infty}^{\infty} \exp(\alpha^2 t^2/2)\varepsilon(x-t)\varphi_{2m}(t)\,dt, \tag{A.27.16}$$

$2\varepsilon(x) = \operatorname{sign} x$, Eq. (14.1.29), and k and k' are certain constants. But

$$\begin{aligned}
&\det[f(x_i - x_j) - k\delta^{-2} + k'(h(x_i) - h(x_j))] \\
&= \det\begin{bmatrix} f(x_i - x_j) - k\delta^{-2} + k'(h(x_i) - h(x_j)) & 0 & k\delta^{-2} - k'h(x_i) \\ k\delta^{-2} + k'h(x_j) & 1 & -k\delta^{-2} \\ 0 & 0 & 1 \end{bmatrix} \qquad \text{(A.27.17)} \\
&= \det\begin{bmatrix} f(x_i - x_j) & -k'h(x_i) & 1 \\ k'h(x_j) & -k\delta^{-2} & 1 \\ -1 & -1 & 0 \end{bmatrix} \\
&= -k\delta^{-2}\det\begin{bmatrix} f(x_i - x_j) & 1 \\ -1 & 0 \end{bmatrix} + \det\begin{bmatrix} f(x_i - x_j) & -k'h(x_i) & 1 \\ k'h(x_j) & 0 & 1 \\ -1 & -1 & 0 \end{bmatrix}. \qquad \text{(A.27.18)}
\end{aligned}$$

The second determinant in Eq. (A.27.18) above, being that of an anti-symmetric matrix of odd order $N+2$, is zero.

Thus the second factor of Eq. (A.27.13) is proportional to

$$\delta^{-2}\exp\Big(-\alpha^2\sum x_i^2\Big)\det\begin{bmatrix} f(x_i - x_j) & 1 \\ -1 & 0 \end{bmatrix}. \tag{A.27.19}$$

From Eqs. (A.27.13), (A.27.14) and (A.27.19) on taking the limit $\delta \to 0$, we get for N odd,

$$\det[\phi(x_i, x_j)] \propto \left[\exp\left(-(1+\alpha^2)\sum x_i^2/2\right)\Delta(x)\right]^2 \det[F_{ij}]$$
$$\propto [p(x_1, \ldots, x_N)]^2, \tag{A.27.20}$$

in view of Eq. (14.1.7). The value (A.27.10) of the constant c is not required for this appendix; it will be needed in Appendix A.28.

Thus whether N is even or odd, the determinant of the $2N \times 2N$ matrix $[\phi(x_i, x_j)]$ is proportional to $\{p(x_1, \ldots, x_N)\}^2$, Eq. (A.27.7) or (A.27.20). Therefore, from Eq. (5.1.15), considering $\phi(x_i, x_j)$ as a quaternion element of an $N \times N$ self-dual quaternion matrix

$$\det[\phi(x_i, x_j)] \propto p(x_1, \ldots, x_N). \tag{A.27.21}$$

The constant of proportionality is fixed by the normalization, Theorem 5.1.4 applied N times and the fact that

$$\int_{-\infty}^{\infty} \phi(x, x)\, dx = \int_{-\infty}^{\infty} [S_N(x, x) + \xi_N(x, x)]\, dx = N. \tag{A.27.22}$$

Arguments similar to the even N case given above show that the ordinary $2N \times 2N$ determinant

$$\det[\mathbf{\Phi}(x_i, x_j)] \propto [\mathbf{p}(x_1, \ldots, x_{2N})]^2, \tag{A.27.23}$$

which is Eq. (14.4.5) except for the constant. This constant is again determined by normalization, Theorem 5.1.4, and the fact that

$$\int_{-\infty}^{\infty} \mathbf{\Phi}(x, x)\, dx = \int_{-\infty}^{\infty} \mathbf{S}_N(x, x)\, dx = 2N. \tag{A.27.24}$$

A.28 Verification of Eq. (14.4.3)

We will verify the equation

$$\int_{-\infty}^{\infty} \phi(x, y)\phi(y, z)\, dy = \phi(x, z) + \tau\phi(x, z) - \phi(x, z)\tau, \tag{14.4.3}$$

where ϕ is defined by Eq. (14.1.14) and

$$\tau = \frac{1}{2}\begin{bmatrix} 1 & 0 \\ 0 & -1 \end{bmatrix}. \tag{A.28.1}$$

We will also verify the same equation when ϕ is replaced by $\mathbf{\Phi}$, Eq. (14.2.10), and τ by $-\tau$.

Writing for brevity $F^\dagger(x, y) = F(y, x)$ and

$$F * G \equiv \int_{-\infty}^{\infty} F(x, y) G(y, z)\, dy, \tag{A.28.2}$$

Eq. (14.4.3) above is

$$\begin{aligned} \phi * \phi &\equiv \begin{bmatrix} S_N + \xi_N & D_N \\ J_N & S_N^\dagger + \xi_N^\dagger \end{bmatrix} * \begin{bmatrix} S_N + \xi_N & D_N \\ J_N & S_N^\dagger + \xi_N^\dagger \end{bmatrix} \\ &= \begin{bmatrix} S_N + \xi_N & 2D_N \\ 0 & S_N^\dagger + \xi_N^\dagger \end{bmatrix}, \end{aligned} \tag{A.28.3}$$

where S_N, ξ_N, D_N and J_N are defined by (14.1.17), (14.1.15), (14.1.19) and (14.1.20). This amounts to verifying that

$$\begin{aligned} (S_N + \xi_N) * (S_N + \xi_N) &= S_N + \xi_N, \\ D_N * J_N = 0 &= J_N * (S_N + \xi_N), \\ (S_N + \xi_N) * D_N &= D_N, \end{aligned} \tag{A.28.4}$$

and, as a consequence,

$$\begin{aligned} (S_N + \xi_N)^\dagger * (S_N + \xi_N)^\dagger &= (S_N + \xi_N)^\dagger, \\ J_N * D_N = 0 &= (S_N + \xi_N)^\dagger * J_N, \\ D_N * (S_N + \xi_N)^\dagger &= D_N. \end{aligned} \tag{A.28.5}$$

For the verification of (A.28.4) we will repeatedly use results of orthonormality and convolution integrals of Appendices A.24, A.24.3 and A.24.6. Thus from the expressions (14.1.17), (14.1.19) and (14.1.22) and the equations of Appendix A.24.3, we have

$$\begin{aligned} S_N * S_N &= S_N, \qquad & S_N * D_N &= D_N, \\ D_N * I_N &= S_N, \qquad & I_N * S_N &= I_N, \end{aligned} \tag{A.28.6}$$

while from Appendix A.24.6, we have

$$g * S_N = -I_N, \qquad D_N * g = -S_N. \tag{A.28.7}$$

For N odd we also need

$$\xi_N * \xi_N = \xi_N, \qquad \mu_N^\dagger * \xi_N = \mu_N^\dagger, \tag{A.28.8}$$

$$g * \xi_N = \mu_N^\dagger, \tag{A.28.9}$$

and

$$S_N * \xi_N, \qquad \xi_N * S_N, \qquad \xi_N * D_N,$$
$$\mu_N * \xi_N, \qquad (\mu_N - \mu_N^\dagger) * S_N, \qquad D_N * (\mu_N - \mu_N^\dagger). \tag{A.28.10}$$

Those in (A.28.8) are easily verified, and (A.28.9) is a consequence of (A.24.15). For (A.28.10) we need to know

$$\int_{-\infty}^{\infty} \varphi_{2j+1}(y)\varphi_{2m}(y)\,dy, \qquad \int_{-\infty}^{\infty} \psi_{2j+1}(y)A_{2m}(y)\,dy,$$
$$\int_{-\infty}^{\infty} \varphi_{2j+1}(y)\exp(-\alpha^2 y^2/2)\,dy, \qquad \int_{-\infty}^{\infty} A_{2m}(y)\varphi_{2m}(y)\,dy, \tag{A.28.11}$$

and

$$\int_{-\infty}^{\infty} A_{2j+1}(y)\varphi_{2m}(y)\,dy, \qquad \int_{-\infty}^{\infty} \varphi_{2j+1}(y)A_{2m}(y)\,dy, \tag{A.28.12}$$

$$\int_{-\infty}^{\infty} \exp(-\alpha^2 y^2/2)\psi_{2j+1}(y)\,dy. \tag{A.28.13}$$

Integrals (A.28.11) are zero by parity. For (A.28.12) we note that A_{2j+1} is a linear combination of φ_{2j} and A_{2j-1}, Eq. (A.24.5), and hence of $\varphi_{2j}, \varphi_{2j-2}, \ldots, \varphi_2, \varphi_0$. Similarly φ_{2j+1} is a linear combination of $\psi_{2j}, \psi_{2j-2}, \ldots, \psi_2, \psi_0$ by Eq. (A.24.4). Thus A_{2j+1} is orthogonal to φ_{2m} and φ_{2j+1} is orthogonal to A_{2m}, for $j < m$ (see Appendix A.24.3). The integrand in (A.28.13) is a perfect derivative, Eq. (14.1.27), of a quantity vanishing at both ends.

Therefore all the integrals in (A.28.11)–(A.28.13) are zero and so are (A.28.10).

Thus we have verified Eq. (14.4.3) in all the cases for $\phi(x, y)$ given by Eq. (14.1.14).

The verification of (14.4.3) with ϕ replaced by Φ, Eq. (14.2.10), and τ replaced by $-\tau$ is similar. From the expressions (14.2.11), (14.2.13), (14.2.16) and (14.2.18) together with (A.24.1), (A.24.21)–(A.24.24) we have

$$\mathbf{S}_N * \mathbf{S}_N = \mathbf{S}_N, \quad \mathbf{S}_N * \mathbf{D}_N = \mathbf{D}_N, \quad \mathbf{I}_N * \mathbf{S}_N = \mathbf{I}_N,$$
$$\mathbf{S}_N * g = -\mathbf{D}_N, \quad \mathbf{D}_N * \mathbf{I}_N = \mathbf{S}_N, \quad g * \mathbf{I}_N = \mathbf{S}_N,$$

or

$$\Phi * \Phi \equiv \begin{bmatrix} \mathbf{S}_N & \mathbf{K}_N \\ \mathbf{I}_N & \mathbf{S}_N^\dagger \end{bmatrix} * \begin{bmatrix} \mathbf{S}_N & \mathbf{K}_N \\ \mathbf{I}_N & \mathbf{S}_N^\dagger \end{bmatrix} = \begin{bmatrix} \mathbf{S}_N & 0 \\ 2\mathbf{I}_N & \mathbf{S}_N^\dagger \end{bmatrix} = \Phi + \Phi\tau - \tau\Phi,$$

with τ given by (A.28.1).

A.29 The Limits of $J_N(x, y)$ and $D_N(x, y)$ as $N \to \infty$. Asymptotic Forms of $J(r; \rho)$ and $D(r; \rho)$. Sections 14.1 and 14.2

The limits of $I_N(x, y)$ and $D_N(x, y)$ for $N \to \infty$ can be obtained by taking the Fourier transform of Eq. (A.28.7). Writing $F(f(x))$ for the Fourier transform of $f(x)$, as in Appendix A.11, we have (Bateman, 1954, vol. 1), for $b > 0$,

$$\begin{aligned} F\big(\operatorname{erf}(bx)\big) &= 2i \int_0^\infty \sin(2\pi kx)\operatorname{erf}(bx)\,dx \\ &= \frac{i}{\pi k}\exp(-\pi^2 k^2/b^2), \end{aligned} \tag{A.29.1}$$

and

$$\begin{aligned} F\left(\frac{\sin(bx)}{x}\right) &= 2\int_0^\infty \cos(2\pi kx)\frac{\sin(bx)}{x}\,dx \\ &= \begin{cases} \pi, & \text{if } 2\pi|k| < b, \\ \pi/2, & \text{if } 2\pi|k| = b, \\ 0, & \text{if } 2\pi|k| > b, \end{cases} \end{aligned} \tag{A.29.2}$$

which along with Eq. (A.28.7) give for the Fourier transform of I_N

$$F(I_N) \approx \frac{i}{2\pi k}\exp(-8\pi^2\rho^2 k^2)\cdot \begin{cases} 1, & \text{if } |k| < 1/2, \\ 1/2, & \text{if } |k| = 1/2, \\ 0, & \text{if } |k| > 1/2. \end{cases} \tag{A.29.3}$$

An inverse Fourier transform then gives

$$\begin{aligned} I_N(x, y) &\approx -\frac{1}{\pi}\int_0^r dt \int_0^\pi dk\cos(kt)\exp(-2\rho^2 k^2) \\ &= -\frac{1}{\pi}\int_0^\pi dk\,\frac{\sin(kr)}{k}\exp(-2\rho^2 k^2), \end{aligned} \tag{A.29.4}$$

so that

$$J_N(x, y) \approx J(r; \rho) = -\frac{1}{\pi}\int_0^r dt \int_\pi^\infty dk\cos(kt)\exp(-2\rho^2 k^2). \tag{A.29.5}$$

Similarly from the second equality in Eq. (A.28.7) we have

$$D_N(x, y) \approx \big(R_1(x)\big)^2 D(r; \rho), \tag{A.29.6}$$

$$\begin{aligned} D(r; \rho) &= \frac{d}{dr}\left(\frac{1}{\pi}\int_0^{\pi} dk \cos(kr)\exp(2\rho^2 k^2)\right) \\ &= -\frac{1}{\pi}\int_0^{\pi} dk\, k \sin(kr)\exp(2\rho^2 k^2). \end{aligned} \tag{A.29.7}$$

Integration gives

$$\begin{aligned} &\int_{\pi}^{\infty} \frac{\sin(kr)}{k}\exp(-2\rho^2 k^2)\, dk \\ &= \exp(-r^2/8\rho^2)\,\mathrm{Im}\int_{\pi}^{\infty} \exp\big(-2\rho^2\big(k - ir/(4\rho^2)\big)^2\big)\, dk/k \\ &= \exp(-r^2/(8\rho^2))\,\mathrm{Im}\left(\exp\left(-2\rho^2\left(k - \frac{ir}{4\rho^2}\right)^2\right)\cdot\left(-\frac{1}{4\rho^2 k(k - ir/(4\rho^2))} + \cdots\right)\right)_{\pi}^{\infty} \\ &= \exp(-2\rho^2\pi^2)\,\mathrm{Im}\, e^{i\pi r}\left(\frac{-1}{\pi(4\pi\rho^2 - ir)} + \cdots\right) \\ &= \frac{1}{\pi}\exp(-2\rho^2\pi^2)\left(\frac{r\cos(\pi r) + 4\pi\rho^2\sin(\pi r)}{16\pi^2\rho^4 + r^2} + \cdots\right). \end{aligned} \tag{A.29.8}$$

Similarly

$$\int_0^{\pi} k\sin(kr)\exp(2\rho^2 k^2)\, dk = \pi\exp(2\pi^2\rho^2)\left(\frac{r\cos(\pi r) - 4\pi\rho^2\sin(\pi r)}{16\pi^2\rho^4 + r^2} + \cdots\right). \tag{A.29.9}$$

A.30 Evaluation of the Integral (15.1.9) for Complex Matrices

In the exposition given here and in Appendix A.32 we follow the method of Ginibre (1965). Appendix A.33 gives an alternative simpler proof due to Dyson. We start with the proposition:

Any complex nonsingular $N \times N$ matrix X can be expressed in one and only one way as

$$X = UYV, \tag{A.30.1}$$

where U is a unitary matrix, Y is a triangular matrix with all diagonal elements equal to unity, $y_{ij} = 0$, $i > j$; $y_{ii} = 1$, and V is a diagonal matrix with real positive diagonal elements.

Proof. Given $X=[x_{ij}]$, we solve the homogeneous linear equations in u_{rj}:

$$\begin{aligned}\sum_{j=1}^{N} u_{rj}x_{ji} &= 0, \quad i<r,\\ \sum_{j=1}^{N} u_{rj}u_{ij}^{*} &= 0, \quad i>r,\end{aligned} \tag{A.30.2}$$

successively for $r=N, N-1, \ldots, 1$. Because the number of unknowns is always one greater than the number of equations, the u_{rj} for a fixed r are not all zero. We may normalize them to satisfy

$$\sum_{j=1}^{N} u_{rj}u_{rj}^{*} = 1, \tag{A.30.3}$$

without disturbing the equalities (A.30.2). Thus we have found a unitary matrix $U_1 = [u_{rj}]$ such that $Y_1 = U_1 X$ is triangular $(Y_1)_{ij} = 0, i>j$. Because X is nonsingular, all diagonal elements of Y_1 are different from zero. Writing the diagonal elements of Y_1 in the polar form, $(Y_1)_{jj} = v_j \exp(i\theta_j)$, we construct two diagonal matrices, one unitary, U_2, with diagonal elements $(U_2)_{jj} = \exp(i\theta_j)$, and the other positive definite, V, with diagonal elements $V_{jj} = v_j$. Putting $Y = U_2^{\dagger} Y_1 V^{-1}$ and $U = U_1^{\dagger} U_2$, we see that $X = UYV$, where U, Y, and V have the properties required in the proposition. Next let $X = UYV = U'Y'V'$; then $U'^{\dagger}U = Y'V'V^{-1}Y^{-1}$ is unitary (left-hand side) as well as triangular with real positive diagonal elements (right-hand side). Thus $U'^{\dagger}U$ and $Y'V'V^{-1}Y^{-1}$ are unit matrices. A comparison of the diagonal elements on the two sides of $YV = Y'V'$ now gives $V = V'$. The decomposition $X = UYV$ is therefore unique. □

Using the fact that $S = XEX^{-1}$, $U^{\dagger}U = 1$, and $EV = VE$, we can write

$$\operatorname{tr}(S^{\dagger}S) = \operatorname{tr}\big[E^{\dagger}X^{\dagger}XE(X^{\dagger}X)^{-1}\big] = \operatorname{tr}\big[E^{\dagger}Y^{\dagger}YE(Y^{\dagger}Y)^{-1}\big], \tag{A.30.4}$$

$$dA = X^{-1}\,dX = V^{-1}Y^{-1}(U^{-1}\,dU)YV + V^{-1}Y^{-1}\,dYV + V^{-1}\,dV. \tag{A.30.5}$$

The volume element $\prod_{i\neq j} dA_{ij}^{(0)}\,dA_{ij}^{(1)}$ needed in (15.1.9) is the quotient of the volume element $\prod_{ij} dA_{ij}^{(0)}\,dA_{ij}^{(1)}$ by that of the set of all complex diagonal matrices. We put aside the quantities that do not depend on the eigenvalues, for they give only multiplicative constants. All of these constants can be adjusted in the final normalization. From (A.30.5) and the structure of Y and V we see that

$$\prod_{i,j} dA_{ij}^{(0)}\,dA_{ij}^{(1)} = \prod_{i<j}(Y^{-1}\,dY)_{ij}^{(0)}(Y^{-1}\,dY)_{ij}^{(1)}a, \tag{A.30.6}$$

where a depends only on U and V. We replace $\prod_{i\neq j} dA_{ij}^{(0)}\, dA_{ij}^{(1)}$ in (15.1.9) with $\prod_{i<j}(Y^{-1}\,dY)_{ij}^{(0)}(Y^{-1}\,dY)_{ij}^{(1)}$ and calculate

$$\int \exp[-\operatorname{tr}(E^{\dagger}HEH^{-1})]\prod_{i<j}(Y^{-1}\,dY)_{ij}^{(0)}(Y^{-1}\,dY)_{ij}^{(1)}, \tag{A.30.7}$$

where

$$H = Y^{\dagger}Y. \tag{A.30.8}$$

The matrix H is Hermitian. Any of its upper left diagonal block of size n is obtained from the upper left diagonal block of Y of the same size: $H_n = Y_n^{\dagger}Y_n$. Therefore, for every n, $\det H_n = 1$, and the diagonal elements H_{nn} are successively and uniquely determined once the off-diagonal elements are given. Thus we need $N(N-1)$ real parameters to specify H, the same number needed to specify Y. We can further convince ourselves that the correspondence of Y and H is one to one. However, we do not need this last result, for we are omitting the constants anyway. Now we make a change of variables. First, because $\det Y = 1$,

$$\prod_{i<j}(Y^{-1}\,dY)_{ij}^{(0)}(Y^{-1}\,dY)_{ij}^{(1)} = \prod_{i<j} dY_{ij}^{(0)}\,dY_{ij}^{(1)}. \tag{A.30.9}$$

Next, we take $H_{ij}^{(0)}$, $H_{ij}^{(1)}$ for $i<j$ as independent variables. The superscripts (0) and (1) denote, as always, the real and the imaginary parts. From

$$H_{ij} = Y_{ij} + \sum_{k<i} Y_{ki}^{*}Y_{kj}, \quad i<j,$$

one can easily calculate the Jacobian of the transformation from Y to H, it being unity. The integral (15.1.9) is

$$J = C\int \exp[-\operatorname{tr}(E^{\dagger}HEH^{-1})]\prod_{i<j} dH_{ij}^{(0)}\,dH_{ij}^{(1)}, \tag{A.30.10}$$

where C is a constant.

The integration over H is done in N steps. At every step we integrate over the variables of the last column and thus decrease by one the size of the matrix, whose structure remains the same. For this we need the recursion relation (A.30.17) derived below.

Let $H' = Y_n^{\dagger}Y_n$, $E' = [z_i\delta_{ij}]_{i,j=1,2,\ldots,n}$, be the relevant matrices of order n and H, E be those obtained from H', E' by removing the last row and last column. Let the Greek indices run from 1 to n, and the Latin indices from 1 to $n-1$. Let $\Delta'_{\alpha\beta}$ be the cofactor of

$H'_{\alpha\beta}$ in H' and Δ_{ij}, the cofactor of H_{ij} in H. Let $g_i = H'_{in}$. Because $\det H' = \det H = 1$, we have

$$\Delta'_{\alpha\beta} = (H'^{-1})_{\beta\alpha}, \quad \Delta_{ij} = (H^{-1})_{ji}. \tag{A.30.11}$$

Expanding $\det H'$, Δ'_{in}, Δ'_{ij} by the last row and last column, we have

$$1 = H'_{nn} - \sum_{i,j} g_i^* g_j \Delta_{ji}, \tag{A.30.12}$$

$$\Delta_{in} = -\sum_{\ell} \Delta_{i\ell} g_\ell^*, \tag{A.30.13}$$

$$\Delta'_{ij} = H'_{nn}\Delta_{ij} - \sum_{\ell,k} g_k^* g_\ell \Delta_{ij}^{\ell k}, \tag{A.30.14}$$

where $\Delta_{ij}^{\ell k}$ is the cofactor obtained from H by removing the ith and ℓth rows and the jth and kth columns. Sylvester's theorem (cf. Mehta, 1989MT) expresses $\Delta_{ij}^{\ell k}$ in terms of Δ_{rs}

$$\Delta_{ij}^{\ell k} = \Delta_{ij}\Delta_{\ell k} - \Delta_{ik}\Delta_{\ell j}. \tag{A.30.15}$$

In writing (A.30.15), we have replaced $\det H$ by unity on the left-hand side. Let

$$\phi_n = \mathrm{tr}(E'^{\dagger} H' E' H'^{-1}) = \sum_{\alpha,\beta} z_\alpha^* z_\beta H'_{\alpha\beta}\Delta'_{\alpha\beta}. \tag{A.30.16}$$

Separating the last row and last column and making use of (A.30.11) to (A.30.15), we get, after some simplification,

$$\phi_n = |z_n|^2 + \phi_{n-1} + \langle g^* | H^{-1}(E^\dagger - z_n^*) H (E - z_n) H^{-1} | g \rangle, \tag{A.30.17}$$

where

$$\langle g^* | B | g \rangle = \sum_{i,j} g_i^* B_{ij} g_j. \tag{A.30.18}$$

Substituting (A.30.17) for $n = N$ in (A.30.10), we get

$$J = Ce^{-|z_N|^2} \int e^{-\phi_{N-1}} \prod_{1 \leqslant i < j \leqslant N-1} dH_{ij}^{(0)}\, dH_{ij}^{(0)} \times \int \exp[-\langle g^* | H^{-1}(E^\dagger - z_N^*) H (E - z_N) H^{-1} | g\rangle] \prod_{1 \leqslant i \leqslant N-1} dg_i^{(0)}\, dg_i^{(1)}. \tag{A.30.19}$$

The last integral is immediate and gives

$$\pi^{N-1}\{\det[H^{-1}(E^{\dagger}-z_N^*)H(E-z_N)H^{-1}]\}^{-1}=\pi^{N-1}\prod_{i=1}^{N-1}|z_i-z_N|^{-2}. \quad (A.30.20)$$

The process can be repeated N times and we finally get

$$J=C\exp\left(-\sum_1^N|z_i|^2\right)\prod_{1\leqslant i<j\leqslant N}|z_i-z_j|^{-2}, \quad (A.30.21)$$

where C is a new constant.

A.31 A Few Remarks About the Eigenvalues of a Quaternion Real Matrix and its Diagonalization

A quaternion-real matrix S is one whose elements are real quaternions (cf. Chapter 2 or Section 5.1). If we replace the elements of an $N \times N$ quaternion matrix S by their 2×2 matrix representation (2.4.3), we get a $2N \times 2N$ matrix $\Theta(S)$ with complex elements. A real quaternion is represented by a 2×2 matrix of the form

$$\begin{bmatrix} a & -b^* \\ b & a^* \end{bmatrix},$$

so that the matrix $\Theta(S)$ has the form

$$\Theta(S)=\begin{bmatrix} a_{ij} & -b_{ij}^* \\ b_{ij} & a_{ij}^* \end{bmatrix}. \quad (A.31.1)$$

The matrix $\Theta(S)$ has $2N$ (complex or c-) eigenvalues and at least one (complex or c-) eigenvector belonging to each distinct eigenvalue. If $[x_i, y_i]^T$ is a c-eigenvector of $\Theta(S)$ belonging to the c-eigenvalue α,

$$\sum_j\begin{bmatrix} a_{ij} & -b_{ij}^* \\ b_{ij} & a_{ij}^* \end{bmatrix}\begin{bmatrix} x_j \\ y_j \end{bmatrix}=\alpha\begin{bmatrix} x_i \\ y_i \end{bmatrix}, \quad (A.31.2)$$

$$\alpha=\alpha_0+i\alpha_1; \quad \alpha_0,\alpha_1 \text{ real}, \quad (A.31.3)$$

that is,

$$\sum_j(a_{ij}x_j-b_{ij}^*y_j)=\alpha x_i,$$

$$\sum_j(b_{ij}x_j+a_{ij}^*y_j)=\alpha y_i. \quad (A.31.2')$$

Then, taking the complex conjugate of these equations and changing the order in which they are written, we see that

$$\sum_j \begin{bmatrix} a_{ij} & -b_{ij}^* \\ b_{ij} & a_{ij}^* \end{bmatrix} \begin{bmatrix} -y_j^* \\ x_j^* \end{bmatrix} = \alpha^* \begin{bmatrix} -y_i^* \\ x_i^* \end{bmatrix}, \tag{A.31.4}$$

that is $[-y_i^*, x_i^*]^T$ is another c-eigenvector belonging to the c-eigenvalue α^*. We now write (A.31.2) and (A.31.4) together:

$$\sum_j \begin{bmatrix} a_{ij} & -b_{ij}^* \\ b_{ij} & a_{ij}^* \end{bmatrix} \begin{bmatrix} x_j & -y_j^* \\ y_j & x_j^* \end{bmatrix} = \begin{bmatrix} x_i & -y_i^* \\ y_i & x_i^* \end{bmatrix} \begin{bmatrix} \alpha & 0 \\ 0 & \alpha^* \end{bmatrix}, \tag{A.31.5}$$

or, in the quaternion notation,

$$Sx = x\alpha, \qquad \alpha = \alpha_0 + \alpha_1 e_1. \tag{A.31.6}$$

We see that the eigenvectors and eigenvalues of S are real quaternions. Moreover, the (quaternion or q-) eigenvalue does not contain the e_2 and e_3 parts that give rise to off-diagonal terms in its 2×2 matrix representation. Thus the quaternion α in (A.31.6) may be identified with the complex number α in (A.31.3).

We say that two quaternions λ_1 and λ_2 are essentially distinct if the equation $\lambda_1 \mu = \mu \lambda_2$ implies $\mu = 0$. If $x_1, x_2, \ldots, x_r$ are q-eigenvectors belonging to the essentially distinct q-eigenvalues $\lambda_1, \lambda_2, \ldots, \lambda_r$, then $x_1, x_2, \ldots, x_r$ are right linearly independent; that is the right linear (vector) equation

$$x_1 c_1 + x_2 c_2 + \cdots + x_r c_r = 0 \tag{A.31.7}$$

implies

$$c_1 = c_2 = \cdots = c_r = 0.$$

A proof can be supplied by induction.

However, the right linear independence of a set of q-vectors does not necessarily lead to their left linear independence, as may be seen from the following example. The vectors

$$x_1 = \begin{bmatrix} 1 \\ e_1 \end{bmatrix}, \qquad x_2 = \begin{bmatrix} e_2 \\ -e_3 \end{bmatrix}, \tag{A.31.8}$$

are right linearly independent, but they are left linearly dependent. Thus we are still far from the diagonalization of S by purely quaternion means.

However, we may again use the intermediatory of the matrix $\Theta(S)$. If all the c-eigenvalues $z_1, z_1^*, \ldots, z_N, z_N^*$ of $\Theta(S)$ are distinct, none of them being real, the complex matrix X whose columns are the eigenvectors of $\Theta(S)$ belonging to these c-eigenvalues, is nonsingular and

$$X^{-1}\Theta(S)X = E, \tag{A.31.9}$$

where E is diagonal with diagonal elements $z_1, z_1^*, \ldots, z_N, z_N^*$. It is easy to be convinced that when X^{-1} is re-expressed as an $N \times N$ quaternion matrix all its elements will be real quaternions. Therefore, if all the N (quaternion) eigenvalues of S are essentially distinct, a quaternion real matrix X exists such that

$$S = XEX^{-1}, \tag{A.31.10}$$

where E is diagonal and quaternion real.

A.32 Evaluation of the Integral Corresponding to (15.2.9)

As in Appendix A.30, we decompose the $2N \times 2N$ matrix X into the unique product $X = UYV$, where U is unitary, Y is triangular with unit diagonal elements $Y_{ij} = 0$, $i > j$; $Y_{ii} = 1$, and V is diagonal with real positive elements. Moreover, because X has the form

$$\begin{bmatrix} a_{ij} & -b_{ij}^* \\ b_{ij} & a_{ij}^* \end{bmatrix}, \tag{A.32.1}$$

the U, Y and V all have the same form. In particular, $Y_{2i-1,2i} = -Y_{2i,2i-1}^* = 0$, and $V_{2i-1,2i-1} = V_{2i,2i}$.

In fact any matrix A having the form (A.32.1) satisfies the relation $ZA = A^*Z$, where Z is given by (2.4.1). And conversely, if $ZA = A^*Z$, then A has the form (A.32.1). From $ZX = X^*Z$ one sees that

$$U^T ZU = Y^* VZV^{-1}Y^{-1} \tag{A.32.2}$$

is unitary and antisymmetric (the left-hand side) and has nonzero elements only in the 2×2 blocks along the principal diagonal (comparison of elements on the two sides). One also has V_{ii} real and positive. Thus $U^T ZU = Z$. Substituting this in (A.32.2), we get $Y^{*-1}ZY = VZV^{-1} = Z$. Thus U, Y and V all have the form of (A.32.1).

If we let $H = Y^\dagger Y$, then, because $H^\dagger = H$ and Y has the form (A.32.1), $H_{2i-1,2i-1} = H_{2i,2i} = h_i$ and $H_{2i-1,2i} = H_{2i,2i-1} = 0$. Moreover, h_i is completely determined by the condition $\det H_{2i-1} = \det Y_{2i-1}^\dagger Y_{2i-1} = 1$. Thus we may consider H_{ij}, $i < j$ and $(i, j) \neq (2k-1, 2k)$ as independent complex variables. As in Appendix A.30, we change from the volume element

$$\prod_{i<j} \prod_{\lambda=0}^{1} dA_{2i-1,2j}^{(\lambda)} \, dA_{2i,2j}^{(\lambda)} \quad \text{to} \quad \prod dY_{ij}^{(0)} \, dY_{ij}^{(1)},$$

and finally to $\prod dH_{ij}^{(0)}\, dH_{ij}^{(1)}$, where the product $\prod$ over the elements of $\mathrm{d}Y$ or $\mathrm{d}H$ are taken over all $i < j$ except the pairs $(i, j) = (2k - 1, 2k)$.

To calculate the integral

$$\int \exp[-(1/2)\,\mathrm{tr}(E^{\dagger} H E H^{-1})] \prod dH_{ij}^{(0)}\, dH_{ij}^{(1)}, \tag{A.32.3}$$

a recurrence relation similar to (A.30.17) is needed. Let H', E' denote matrices of order $2n$; H'', E'', their upper left diagonal blocks of order $(2n - 1)$, and H, E, their upper left diagonal blocks of order $2n - 2$. The cofactors of H', H'', and H are denoted, respectively, by Δ', Δ'', and Δ. Let

$$g = (g_i), \quad g' = (g_i'), \tag{A.32.4}$$

where

$$\begin{aligned} &g_i = H'_{i,2n-1}, \qquad g_i' = H'_{i,2n}, \quad i = 1, 2, \ldots, 2n - 2, \\ &g_{2n-1} = g_{2n} = g'_{2n-1} = g'_{2n} = 0. \end{aligned} \tag{A.32.5}$$

Then

$$g'_{2i} = g^*_{2i-1}, \qquad g'_{2i-1} = -g^*_{2i},$$

or

$$g' = -Zg^*, \tag{A.32.6}$$

where Z is given by (2.4.1). Using the facts

$$H'_{2n-1,2n-1} = H'_{2n,2n} = h_n, \qquad H'_{2n-1,2n} = H'_{2n,2n-1} = 0, \tag{A.32.7}$$

and

$$\det H' = \det H'' = \det H = 1, \tag{A.32.8}$$

we get by expanding according to the last row and last column

$$1 = h_n - \sum_{i,j} g_i^* g_j \Delta_{ji}, \tag{A.32.9}$$

$$\Delta''_{2n-1,\ell} = \sum_k g_k \Delta_{k\ell}, \tag{A.32.10}$$

$$\begin{aligned} \Delta''_{ij} &= h_n \Delta_{ij} - \sum_{k,\ell} g_k^* g_\ell \Delta_{ij}^{\ell k} \\ &= \Delta_{ij} + \sum_{k,\ell} g_k^* g_\ell \Delta_{ik} \Delta_{\ell j}, \end{aligned} \tag{A.32.11}$$

where in the last step of (A.32.11) we have used (A.32.9) and (A.30.15). The matrices H', H'', and H are positive definite and so are their inverses. In particular,

$$\sum_{i,j} f_i^* \Delta_{ij} f_j > 0, \tag{A.32.12}$$

if not all f_i are zero. By equating to zero the expansion of a determinant whose first $2n-1$ rows are identical with those of H' and whose last row is identical with the last but one of H', we get

$$\begin{aligned} 0 &= \sum_i g_i^* \Delta'_{2n,i} + h_n \Delta'_{2n,2n-1} \\ &= \sum_{i,j} g_i^* g'_j \Delta''_{ji} + h_n \sum_j g'_j \Delta''_{j,2n-1}, \end{aligned} \tag{A.32.13}$$

which on making use of (A.32.9), (A.32.10), and (A.32.11) gives

$$0 = \sum_{j,p} g'_j g_p^* \Delta_{jp} \left(1 + \sum_{i,q} g_i^* \Delta_{qi} g_q\right). \tag{A.32.14}$$

In view of (A.32.12), this is equivalent to

$$0 = \sum_{j,p} g'_j \Delta_{jp} g_p^* = \left(\sum_{j,p} g_j'^* \Delta_{pj} g_p\right)^*, \tag{A.32.15}$$

or

$$\sum_j \Delta''_{j,2n-1} g'_j = (H''^{-1} g')_{2n-1} = 0. \tag{A.32.15$'$}$$

Next we put

$$\phi_n = \frac{1}{2} \operatorname{tr}(E'^{\dagger} H' E' H'^{-1}), \tag{A.32.16}$$

and apply (A.30.17) twice to get

$$\phi_n = |z_n|^2 + \phi_{n-1} + \frac{1}{2}\langle g^* | U | g\rangle + \frac{1}{2}\langle g'^* | V | g'\rangle. \tag{A.32.17}$$

The notation is that of (A.30.18)

$$\langle f^* | B | f\rangle = \sum_{i,j} f_i^* B_{ij} f_i, \tag{A.32.18}$$

where

$$U = H^{-1}(E^{\dagger} - z_n^*)H(E - z_n)H^{-1}, \tag{A.32.19}$$

$$V = H''^{-1}(E''^{\dagger} - z_n)H''(E'' - z_n^*)H''^{-1}. \tag{A.32.20}$$

From (A.32.15) we see that V is essentially equal to U.

$$\langle g'^*|V|g'\rangle = \langle g'^*|U|g'\rangle. \tag{A.32.21}$$

Last, from (A.32.6), $U^{\dagger} = U$, and $UZ = ZU^*$, Z given by (2.4.1), we have

$$\langle g'^*|U|g'\rangle = \langle g^*|U|g\rangle. \tag{A.32.22}$$

Collecting (A.32.17), (A.32.21) and (A.32.22), the recurrence relation becomes

$$\phi_n = |z_n|^2 + \phi_{n-1} + \langle g^*|U|g\rangle, \tag{A.32.23}$$

where U is given by (A.32.19).

The rest of the integration is identical to that in Appendix A.30.

A.33 Another Proof of Eqs. (15.1.10) and (15.2.10)

A shorter proof of Eqs. (15.1.10) and (15.2.10) due to Dyson is as follows.

Consider first the case when S is a matrix with complex elements. Given S, one can always find a unitary matrix U such that $U^{\dagger}SU = T$ is triangular, i.e., $T_{jk} = 0$ if $j > k$ (cf. Mehta, 1989MT). The diagonal elements $T_{jj} = z_j$ of T are the eigenvalues of S. The matrix S has N^2 complex elements, so that the number of real parameters entering S is $2N^2$. The number of real parameters entering U is N^2 and that entering T is $N(N+1)$. So the number of real parameters in U and T seem to exceed those in S by N. Actually, S does not determine U and T uniquely. If we replace U by UV where V is any unitary diagonal matrix, then $(UV)^{\dagger}S(UV) = V^{\dagger}TV$ remains triangular. We will use this freedom below to impose N conditions on the variations of U.

Differentiating $S = UTU^{\dagger}$, we get

$$\begin{aligned} dS &= U\big(dT + U^{\dagger}\,dU\,T - TU^{\dagger}\,dU\big)U^{\dagger} \\ &= U\,dA\,U^{\dagger}, \end{aligned} \tag{A.33.1}$$

$$\begin{aligned} dA &= dT + U^{\dagger}\,dU\,T - TU^{\dagger}\,dU \\ &= dT + i(dH\,T - T\,dH), \end{aligned} \tag{A.33.2}$$

where dT is triangular and $dH = -iU^\dagger dU$ is Hermitian. Let us restrict the variations in U so that

$$dH_{jj} = -i(U^\dagger dU)_{jj} = 0, \quad j = 1, 2, \ldots, N. \tag{A.33.3}$$

In terms of its components Eq. (A.33.2) reads

$$dA_{jk} = dT_{jk} + i(T_{kk} - T_{jj})\, dH_{jk} + i \sum_{\ell<k} dH_{j\ell}\, T_{\ell k} - i \sum_{j<\ell} T_{j\ell}\, dH_{\ell k}. \tag{A.33.4}$$

Let us order the indices (j, k), $1 \leqslant j,\ k \leqslant N$, so that (j_1, k_1) precedes (j_2, k_2) if either $j_1 > j_2$ or $j_1 = j_2$ and $k_1 < k_2$. If $j \leqslant k$ we take $T_{j,k}$ and if $j > k$ we take $H_{j,k}$ as the variable. In other words the variables $A_{j,k}$, $(T_{j,k}, H_{j,k})$ are arranged as

$$A_{N,1}, A_{N,2}, \ldots, A_{N,N}, A_{N-1,1}, \ldots, A_{N-1,N}, \ldots, A_{1,1}, \ldots, A_{1,N};$$

$$H_{N,1}, H_{N,2}, \ldots, H_{N,N-1}, T_{N,N}, H_{N-1,1}, \ldots, H_{N-1,N-2};$$

$$T_{N-1,N-1}, T_{N-1,N}, H_{N-2,1}, \ldots, T_{2,N}, T_{1,1}, T_{1,2}, \ldots, T_{1,N}.$$

With this ordering, the elements above the main diagonal in the Jacobian matrix $\partial(A_{jk})/\partial(H_{jk}, T_{jk})$ are all zero, and the diagonal elements are either $i(T_{kk} - T_{jj})$ or 1. Therefore,

$$\begin{aligned} \mu(dS) &= \prod_{j,k} dS_{jk}\, dS^*_{jk} = \prod_{j,k} dA_{jk}\, dA^*_{jk} \\ &= \prod_{j<k} |T_{jj} - T_{kk}|^2 dH_{jk}\, dH^*_{jk} \prod_{j\leqslant k} dT_{jk}\, dT^*_{jk} \\ &= \prod_{j<k} |z_j - z_k|^2 \prod_j dz_j\, dz^*_j \prod_{j<k} dT_{jk}\, dT^*_{jk}\, dH_{jk}\, dH^*_{jk}, \end{aligned} \tag{A.33.5}$$

where $z_j = T_{jj}$ are the eigenvalues of S. Also

$$\operatorname{tr}(S^\dagger S) = \operatorname{tr}(T^\dagger T) = \sum_j |z_j|^2 + \sum_{j<k} |T_{jk}|^2, \tag{A.33.6}$$

$$\int \exp\Bigl(-\sum_{j<k} |T_{jk}|^2\Bigr) \prod_{j<k} dT_{jk}\, dT^*_{jk} = \pi^{N(N+1)/2}, \tag{A.33.7}$$

and

$$\int \prod_{j<k} dH_{jk}\, dH^*_{jk} = \Omega_U \Big/ \int \prod_j dH_{jj} = \Omega_U/(2\pi)^N, \tag{A.33.8}$$

where Ω_U is the volume of the unitary group. Collecting the results one gets

$$\int \exp(-\operatorname{tr}(S^\dagger S))\mu(dS) = \int P_c(z_1, \ldots, z_N) \prod_j dz_j\, dz_j^*, \tag{A.33.9}$$

with

$$P_c(z_1, \ldots, z_N) = \text{const} \cdot \prod_{j<k} |z_j - z_k|^2 \exp\Bigl(-\sum_j |z_j|^2\Bigr). \tag{A.33.10}$$

When S has quaternion real elements, then U is symplectic, T is triangular quaternion real, diagonal elements of T do not contain the e_2 or e_3 quaternion units, the eigenvalues of S are $x_j \pm iy_j$ if $T_{jj} = x_j + e_1 y_j$, idH is anti-self-dual and the proof given above works with minor modifications.

When S is real and has only real eigenvalues, all distinct, then U is real orthogonal, T is real triangular, dH is real symmetric and the proof given above is again valid. If S is real and has some complex eigenvalues, then $S = UTU^\dagger$ with U real orthogonal is not possible and the proof given here fails.

A.34 Proof of Eq. (15.2.38)

Let us put

$$\psi(x, y) = (2\pi)^{-1/2} \sum_{k=0}^{\infty} I_{k+1/2}(x)\bigl[y^{k+1/2} - y^{-(k+1/2)}\bigr]. \tag{A.34.1}$$

Differentiating (A.34.1) with respect to x and using the relation

$$I_\nu'(x) = \frac{1}{2}\bigl[I_{\nu+1}(x) + I_{\nu-1}(x)\bigr], \tag{A.34.2}$$

we have

$$\begin{aligned}
\frac{\partial \psi}{\partial x} &= (2\pi)^{-1/2} \frac{1}{2} \sum_{k=0}^{\infty} \bigl[I_{k+3/2}(x) + I_{k-1/2}(x)\bigr]\bigl[y^{k+1/2} - y^{-(k+1/2)}\bigr] \\
&= (2\pi)^{-1/2} \frac{1}{2} \sum_{k=0}^{\infty} I_{k+1/2}(x)\bigl[y^{k+1/2} - y^{-(k+1/2)}\bigr](y + y^{-1}) \\
&\quad + (2\pi)^{-1/2} \frac{1}{2} (y^{1/2} - y^{-1/2})\bigl(I_{1/2}(x) + I_{-1/2}(x)\bigr) \\
&= \frac{1}{2}(y + y^{-1})\psi(x, y) + (2\pi)^{-1/2} \frac{1}{2}(y^{1/2} - y^{-1/2}) \left(\frac{2}{\pi x}\right)^{1/2} e^x,
\end{aligned} \tag{A.34.3}$$

where we have used the fact that

$$I_{1/2}(x) + I_{-1/2}(x) = \left(\frac{2}{\pi x}\right)^{1/2} (\sinh x + \cosh x). \tag{A.34.4}$$

The differential Eq. (A.34.3) can be immediately solved to give us

$$\psi(x, y) = \frac{1}{2\pi} \exp\left[\frac{1}{2}(y + y^{-1})x\right](y^{1/2} - y^{-1/2}) \int_0^x \exp\left[x' - \frac{1}{2}(y + y^{-1})x'\right]\frac{dx'}{\sqrt{x'}}. \tag{A.34.5}$$

Changing the integration variable from x' to $t = 1 - x'/x$, we get

$$\psi(x, y) = \frac{1}{2\pi} e^x (y^{1/2} - y^{-1/2})\sqrt{x} \int_0^1 \exp\left[\frac{1}{2}(y^{1/2} - y^{-1/2})^2 xt\right]\frac{dt}{\sqrt{1-t}}. \tag{A.34.6}$$

Finally putting $x = zz^*$ and $y = z^*/z$ we have

$$\psi(x, y) = \phi(z, z^*), \tag{A.34.7}$$

and (A.34.6) gives (15.2.38).

A.35 Partial Triangulation of a Matrix

Let S be a real $n \times n$ random matrix, the elements of S being independent identically distributed Gaussian variables

$$P(S)\, d\mu(S) = \exp(-\operatorname{tr} S^T S) \prod_{i,j=1}^{n} dS_{i,j}. \tag{A.35.1}$$

Let Q be a real orthogonal matrix such that

$$Q^T S Q = \begin{bmatrix} A & B \\ 0 & C \end{bmatrix}, \tag{A.35.2}$$

with A, B and C real block matrices of orders $n_1 \times n_1$, $n_1 \times n_2$ and $n_2 \times n_2$ respectively ($n_1 + n_2 = n$), while 0 is an $n_2 \times n_1$ zero matrix. In integrals containing S we want to change variables from those of S to those of A, B and C and therefore need the Jacobian.

The number of independent real parameters entering S, A, B and C are respectively n^2, n_1^2, $n_1 n_2$ and n_2^2. Those entering Q are $f(n) \equiv n(n-1)/2$, but all of them can not be independent. In fact additional orthogonal transformations of dimensions n_1 and n_2 corresponding to the square matrices A and C leave the block triangular form (A.35.2) unchanged. So given A, B and C the number of independent real parameters entering Q is $f(n) - f(n_1) - f(n_2) = n_1 n_2$, and one has to choose them so that the calculation of the Jacobian is convenient.

From

$$S = QRQ^T, \quad R = \begin{bmatrix} A & B \\ 0 & C \end{bmatrix}, \tag{A.35.3}$$

one has

$$\operatorname{tr} S^T S = \operatorname{tr} R^T R = \operatorname{tr}(A^T A + B^T B + C^T C) \tag{A.35.4}$$

$$\begin{aligned} dS &= Q[dR + Q^T dQ\, R + R\, dQ^T\, Q]Q^T \\ &= Q[dR + dH\, R - R\, dH]Q^T \\ &= Q\, dM\, Q^T, \end{aligned} \tag{A.35.5}$$

where $dH = Q^T dQ = -dQ^T Q$ is a real anti-symmetric matrix and $dM = dR + dH\, R - R\, dH$. From the block structure of R, $dR_{i,j} = 0$ if $n_1 < i \leqslant n$ and $1 \leqslant j \leqslant n_1$. We will choose the independent parameters in Q such that the variations $dH_{i,j} \neq 0$ for $n_1 < i \leqslant n$ and $1 \leqslant j \leqslant n_1$.

From (A.35.5) we have $\mu(dS) = \mu(dM)$. Also

$$\frac{\partial M_{i,j}}{\partial R_{k,\ell}} = \delta_{i,k}\delta_{j,\ell} \quad \text{if either } k \leqslant n_1 \text{ or } k > n_1 \text{ and } \ell > n_1; \tag{A.35.6}$$

$$\frac{\partial M_{i,j}}{\partial H_{k,\ell}} = \delta_{i,k} R_{\ell,j} - R_{i,k}\delta_{j,\ell} \quad \text{if } k > n_1 \text{ and } \ell \leqslant n_1. \tag{A.35.7}$$

The determinant of the Jacobian matrix is therefore

$$\det\left[\frac{\partial^2 M}{\partial R \partial H}\right] = \det[I_{n_2} \times A - C \times I_{n_1}]. \tag{A.35.8}$$

If u is an eigenvector of A with the eigenvalue a and v an eigenvector of C with the eigenvalue c, then $v \times u$ is an eigenvector of $I_{n_2} \times A - C \times I_{n_1}$ with the eigenvalue $a - c$. Multiple eigenvalues of A or of C can be ignored since they occur in a space of measure zero. Thus if $a_1, a_2, \ldots, a_{n_1}$ are the eigenvalues of A and $c_1, c_2, \ldots, c_{n_2}$ are those of C, then $a_i - c_j$, $i = 1, \ldots, n_1$, $j = 1, \ldots, n_2$, are all the eigenvalues of $I_{n_2} \times A - C \times I_{n_1}$ and

$$\det[I_{n_2} \times A - C \times I_{n_1}] = \prod_{i=1}^{n_1} \prod_{j=1}^{n_2} (a_i - c_j). \tag{A.35.9}$$

With minor changes in the above reasoning one arrives at the following conclusion. If S is an $n \times n$ complex matrix, U a complex unitary $n \times n$ matrix such that

$$U^{\dagger} S U = \begin{bmatrix} A & B \\ 0 & C \end{bmatrix}, \tag{A.35.10}$$

with A, B, C complex block matrices of dimensions $n_1 \times n_1$, $n_1 \times n_2$ and $n_2 \times n_2$ respectively, then the Jacobian of the transformation from the $2n^2$ variables of S to those of A, B, C and U will be

$$\prod_{i=1}^{n_1} \prod_{j=1}^{n_2} (\lambda_i - \mu_j), \tag{A.35.11}$$

where $\lambda_1, \ldots, \lambda_{n_1}$ are the (complex) eigenvalues of A and $\mu_1, \ldots, \mu_{n_2}$ are those of C.

A.36 Average Number of Real Eigenvalues of a Real Gaussian Random Matrix

The calculation involves three steps.

(1) Suppose that the $n \times n$ matrix A has a real eigenvalue λ. Then there exists a real orthogonal matrix Q, in fact many, such that $Q^T A Q = \left[\begin{pmatrix} B & 0 \\ w & \lambda \end{pmatrix}\right]$, where 0 is a column of $n-1$ zeros, w is a $1 \times (n-1)$ row and B is an $(n-1) \times (n-1)$ real matrix. For this it suffices to take the (real) eigenvector corresponding to λ as the last column of Q. Change the variables from the n^2 elements of A to the set $\lambda, w_1, \ldots, w_{n-1}$, the $(n-1)^2$ elements of B and some $n-1$ parameters p_j characterizing Q. This replaces $\operatorname{tr} A^T A$ by

$$\lambda^2 + \sum_{j=1}^{n-1} w_j^2 + \operatorname{tr} B^T B, \tag{A.36.1}$$

and $\mu(dA) = \prod_{i,j=1}^{n} dA_{i,j}$ by

$$|\det J|\, d\lambda \prod_{i,j=1}^{n-1} dB_{i,j} \prod_{i=1}^{n-1} dp_i \prod_{j=1}^{n-1} dw_j, \tag{A.36.2}$$

where J is the Jacobian. A convenient choice of the parameters p_j in Q gives (cf. Appendix A.35)

$$\det J = \det(B - \lambda I), \tag{A.36.3}$$

where I is the $(n-1)\times(n-1)$ unit matrix. Thus

$$\exp(-\operatorname{tr} A^T A)\prod_{i,j=1}^{n} dA_{i,j} = \exp\left(-\lambda^2 - \sum_{j=1}^{n-1} w_j^2 - \operatorname{tr} B^T B\right)\left|\det(B-\lambda I)\right| \times d\lambda \prod_{i,j=1}^{n-1} dB_{i,j}\prod_{i=1}^{n-1} dp_i \prod_{j=1}^{n-1} dw_j. \tag{A.36.4}$$

(2) The probability density $\rho(\lambda)$ of the real eigenvalues λ of A is the integral of (A.36.4) over all the variables except λ

$$\rho(\lambda) \propto \int \exp\left[-\lambda^2 - \sum_{j=1}^{n-1} w_j^2 - \operatorname{tr} B^T B\right]\left|\det(B-\lambda I)\right| \prod_{i,j=1}^{n-1} dB_{i,j}\prod_{i=1}^{n-1} dp_i \prod_{j=1}^{n-1} dw_j$$

$$\propto \int \exp[-\lambda^2 - \operatorname{tr} B^T B]\left|\det(B-\lambda I)\right| \prod_{i,j=1}^{n-1} dB_{i,j}. \tag{A.36.5}$$

Here the integral over the $B_{i,j}$ is difficult due to the presence of the absolute value sign of the determinant. We do not understand how this integral is done. Referring to Muirhead's book *Aspects of multivariate statistical theory*, John Wesley, New York, 1982, pp. 237, 447, Edelman et al. (1994) give the result as

$$\pi^{-n^2/2}\int_{-\infty}^{\infty}\left|\det(A-\lambda I)\right|\exp\left(-\sum_{i,j=1}^{n} A_{i,j}^2\right)\prod_{i,j=1}^{n} dA_{i,j}$$

$$= \frac{\Gamma((n+1)/2))}{\sqrt{\pi}\,\Gamma(n)}\left[e^{\lambda^2}\int_{2\lambda^2}^{\infty} e^{-t}t^{n-1}\,dt + 2^{n-1}|\lambda|^n \int_0^{\lambda^2} e^{-t}t^{(n-2)/2}\,dt\right]. \tag{A.36.6}$$

(3) To get the average number E_n of real eigenvalues of A, one finally integrates the probability density $\rho(\lambda)$ over λ from $-\infty$ to ∞. The final answer is quoted in Eq. (15.3.2).

A.37 Probability Density of the Eigenvalues of a Real Random Matrix When k of Its Eigenvalues Are Real

We will take

$$P(A)\mu(dA) = \pi^{-n^2/2}\exp[-\operatorname{tr}(A^\dagger A)]\prod_{i,j} dA_{i,j}, \tag{A.37.1}$$

so that the elements of the real n by n matrix A are independent random Gaussian (normal) variables with variance one. Suppose that A has k real eigenvalues $\lambda_1, \ldots, \lambda_k$ and $m = (n-k)/2$ pairs of complex conjugate eigenvalues $x_1 \pm iy_1, \ldots, x_m \pm iy_m$. We want to find the joint probability density of the real variables $\lambda_1, \ldots, \lambda_k, x_1, y_1, \ldots, x_m, y_m$.

Lemma A.37.1. *Let X be an $m \times n$ (real) matrix. Define the linear operator*

$$\Omega(X) = XB - CX, \tag{A.37.2}$$

where B and C are given square matrices of dimensions n and m respectively. If b_j are the eigenvalues of B and c_j are those of C then the eigenvalues of Ω are $b_i - c_j$.

The operator Ω can be represented as a Kronecker product

$$\Omega = I \times B^T - C \times I. \tag{A.37.3}$$

If v_i is an eigenvector of B^T with eigenvalue b_i and w_j is an eigenvector of C with eigenvalue c_j, then $w_j \times v_i^T$ is an eigenvector of Ω with the eigenvalue $b_i - c_j$.

Lemma A.37.2. *Any real $n \times n$ matrix A having k real and m pairs of complex conjugate eigenvalues ($n = k + 2m$) is orthogonally similar to an upper quasi-triangular matrix R; i.e., $A = QRQ^T$ where*

$$R = \begin{bmatrix} \lambda_1 & \cdots & R_{1,k} & R_{1,k+1} & \cdots & R_{1,k+m} \\ \vdots & \ddots & \vdots & \vdots & \cdots & \vdots \\ 0 & \cdots & \lambda_k & R_{k,k+1} & \cdots & R_{k,k+m} \\ 0 & \cdots & 0 & Z_{k+1} & \cdots & R_{k+1,k+m} \\ \vdots & \cdots & \vdots & \vdots & \ddots & \vdots \\ 0 & \cdots & 0 & 0 & \cdots & Z_m \end{bmatrix}, \tag{A.37.4}$$

R is an $n \times n$ matrix, $n = k + 2m$, with blocks

$$R_{i,j} \text{ of size } \begin{cases} 1 \text{ by } 1 & \text{if } i \leqslant k,\ j \leqslant k, \\ 1 \text{ by } 2 & \text{if } i \leqslant k,\ j > k, \\ 2 \text{ by } 1 & \text{if } i > k,\ j \leqslant k, \\ 2 \text{ by } 2 & \text{if } i > k,\ j > k. \end{cases} \tag{A.37.5}$$

Here for $j \leqslant k$, $\lambda_j = R_{j,j}$ are the real eigenvalues of A and for $j > k$, $R_{j,j} = Z_j$ is the 2×2 block

$$Z_j = \begin{bmatrix} x_j & b_j \\ -c_j & x_j \end{bmatrix}, \quad b_j c_j > 0,\ b_j \geqslant c_j, \tag{A.37.6}$$

so that the complex eigenvalues of A are $x_j \pm iy_j$ where $y_j = \sqrt{b_j c_j}$. Finally $R_{i,j}$ is a zero block if $i > j$.

Proof. If λ_1 is a real eigenvalue of A, the corresponding eigenvector q_1 can be chosen real, $Aq_1 = \lambda_1 q_1$. Construct a real orthogonal matrix $Q_1 = [q_1, \alpha_1]$ whose first column is q_1. Transforming A by Q_1

$$Q_1^T A Q_1 \equiv \begin{bmatrix} q_1^T \\ \alpha_1^T \end{bmatrix} A [q_1 \quad \alpha_1] = \begin{bmatrix} q_1^T A q_1 & q_1^T A \alpha_1 \\ \alpha_1^T A q_1 & \alpha_1^T A \alpha_1 \end{bmatrix} = \begin{bmatrix} \lambda_1 & q_1^T A \alpha_1 \\ 0 & \alpha_1^T A \alpha_1 \end{bmatrix}, \quad \text{(A.37.7)}$$

the first column has zeros below the main diagonal. The process can be repeated for all the real eigenvalues $\lambda_2, \ldots, \lambda_k$ of A

$$Q_r^T A Q_r = \begin{bmatrix} \lambda_1 & \cdots & * & * \\ \vdots & \ddots & \vdots & \vdots \\ 0 & \cdots & \lambda_k & * \\ 0 & \cdots & 0 & A_1 \end{bmatrix}, \quad \text{(A.37.8)}$$

where a star denotes possible non-zero elements.

Now the real $(n-k) \times (n-k)$ matrix A_1 has only complex eigenvalues. For the pair of complex conjugate eigenvalues $x_1 \pm iy_1$ let $R_1 \pm iS_1$ be the corresponding eigenvectors, $A(R \pm iS) = (x_1 \pm iy_1)(R_1 \pm iS)$. The real columns R_1 and S_1 cannot be linearly dependent, since otherwise y_1 has to be zero, and the eigenvalue real. Construct a real orthogonal matrix Q_2 whose first two columns are linear combinations of R_1 and S_1 and transform A_1 by Q_2,

$$Q_2^T A Q_2 \equiv \begin{bmatrix} R_1^T \\ R_1^T + \alpha S_1^T \\ \beta^T \end{bmatrix} A_1 [R_1 \quad R_1 + \alpha S_1 \quad \beta] = \begin{bmatrix} * & * & * \\ * & * & * \\ 0 & 0 & A_2 \end{bmatrix}, \quad \text{(A.37.9)}$$

thus getting a 2×2 non-zero block in the upper left-hand corner with the eigenvalues $x_1 \pm iy_1$ and zero columns below it. This process can be repeated and one gets the quasi-triangular structure (A.37.4). It remains to verify that the 2×2 diagonal blocks are orthogonally similar to Z_j in (A.37.6). For this one can use the following result.

Any $n \times n$ real matrix A is orthogonally similar to a matrix with equal diagonal elements, i.e., there exists a real orthogonal matrix Q such that the diagonal elements of $Q^T A Q$ are all equal to tr A/n. (For a proof see, for example, Mehta (1989MT), Section 11.7.2.)

The block structure of R indicated above is convenient, it allows to handle simultaneously all the possibilities; only thing one has to remember is that the index i refers to a block of size one or two according as $i \leqslant k$ or $i > k$ respectively.

For an $n \times n$ real matrix A having k real eigenvalues λ_j and $m = (n-k)/2$ pairs of complex conjugate eigenvalues $x_j \pm iy_j$, the n^2 independent parameters are found in the new variables of (A.37.4) as follows. Q has $n(n-1)/2$ parameters; Λ has k; Z has $3m = 3(n-k)/2$; and R^U has $k(k-1)/2 + 2km + 4m(m-1)/2 = (n^2 - 2n + k)/2$ parameters. Q is the $n \times n$ real orthogonal matrix; Λ is the set of real eigenvalues $\lambda_1, \dots, \lambda_k$; Z is the set of 2×2 diagonal blocks $Z_1, \dots, Z_m$ and R^U denotes the upper quasi-triangular part of R.

To make the change of variables one has to compute the Jacobian. Since multiple eigenvalues form a set of measure zero, one may disregard them. Given A, the matrices Q and R are almost unique; they will be unique if one imposes some (arbitrary) extra requirements such as $\lambda_1 \geqslant \cdots \geqslant \lambda_k$, $x_1 \geqslant \cdots \geqslant x_m$ and the first row, say, of Q is positive. These requirements give a constant factor in the Jacobian. □

Lemma A.37.3. *The Jacobian of the transformation $A = QRQ^T$ is*

$$\prod_{i,j} dA_{i,j} \propto \Delta_0 \prod_{i=1}^{m} (b_i - c_i)\, dH\, d\Lambda\, dZ\, dR^U, \tag{A.37.10}$$

$$\Delta_0 = \prod_{1 \leqslant \ell < j \leqslant k+m} \left|\lambda(R_{\ell,\ell}) - \lambda(R_{j,j})\right|, \tag{A.37.11}$$

$$dH = \prod_{1 \leqslant i < j \leqslant n} \left(Q^T dQ\right)_{i,j} = - \prod_{1 \leqslant i < j \leqslant n} \left(dQ^T Q\right)_{i,j}, \tag{A.37.12}$$

$$d\Lambda = d\lambda_1 \cdots d\lambda_k, \tag{A.37.13}$$

$$dZ = \prod_{j=1}^{m} dx_j\, db_j\, dc_j, \tag{A.37.14}$$

$$dR^U = \prod_{i<j} dR_{i,j}. \tag{A.37.15}$$

Here Δ_0 denotes the product of differences of an eigenvalue of $R_{\ell,\ell}$ and an eigenvalue of $R_{j,j}$. Thus if $j \leqslant k$, $R_{j,j}$ has only one eigenvalue λ_j; whereas if $j > k$ then $R_{j,j}$ has two eigenvalues $x_j \pm iy_j$. The difference of every distinct pair of eigenvalues of A appear as a factor in Δ_0 except for the pair of complex conjugates. More explicitly Δ_0 contains the following factors:

$$\prod_{1 \leqslant i < j \leqslant k} |\lambda_i - \lambda_j|, \tag{A.37.16}$$

$$\prod_{i=1}^{k} \prod_{j=1}^{m} \{(\lambda_i - x_j)^2 + y_j^2\}, \tag{A.37.17}$$

$$\prod_{1 \leqslant i < j \leqslant m} [\{(x_i - x_j)^2 + (y_i - y_j)^2\}], \tag{A.37.18}$$

and

$$\prod_{1\leqslant i<j\leqslant m} \left[\{(x_i - x_j)^2 + (y_i + y_j)^2\}\right]. \tag{A.37.19}$$

Proof. We proceed as in Appendix A.33. From $A = QRQ^T$, we have

$$\begin{aligned}
\operatorname{tr} AA^T &= \operatorname{tr} RR^T = \sum_{i=1}^{k} \lambda_i^2 + 2\sum_{i=1}^{m} (x_i^2 + y_i^2) + \operatorname{tr} R^U R^{UT}, \\
dA &= Q(dR + Q^T dQ\, R + R\, dQ^T Q) Q^T \\
&= Q\, dM\, Q^T, \\
dM &= dR + Q^T dQ\, R - RQ^T dQ \\
&= dR + dH\, R - R\, dH,
\end{aligned} \tag{A.37.20}$$

where dR is quasi-triangular and dH is real anti-symmetric.

It is evident that $\mu(dA) = \mu(dM)$, since orthogonal matrices do not contribute to the Jacobian. Writing explicitly the matrix elements, for $i > j$ we have $dR_{i,j} = 0$, and

$$dM_{i,j} = dH_{i,j}\, R_{j,j} - R_{i,i}\, dH_{i,j} + \sum_{k<\ell<j} dH_{i,\ell}\, R_{\ell,j} - \sum_{k<i<\ell} R_{i,\ell}\, dH_{\ell,j}. \tag{A.37.21}$$

The differentials in the above have either one or two rows and one or two columns. Let us order the indices (i, j), $1 \leqslant i, j \leqslant k + m$, so that (i_1, j_1) precedes (i_2, j_2) if either $i_1 > i_2$ or $i_1 = i_2$ and $j_1 < j_2$. The Jacobian matrix is quasi-triangular and from Lemma A.37.1 one has

$$\begin{aligned}
\prod_{i>j} \mu(dM_{i,j}) &= \prod_{i>j} \mu(dH_{i,j}\, R_{j,j} - R_{i,i}\, dH_{i,j}) \\
&= \Delta_0 \prod_{i>j} \mu(dH_{i,j}).
\end{aligned} \tag{A.37.22}$$

For $i = j \leqslant k$,

$$dM_{i,i} = d\lambda_i + \cdots, \tag{A.37.23}$$

where ... denotes terms in $dH_{i,j}$ with $i \neq j$, terms not contributing to the Jacobian. For $i = j > k$, since dH is a 2×2 anti-symmetric matrix,

$$\begin{aligned}
dM_{j,j} &= dZ_j + dH_{j,j}\, Z_j - Z_j\, dH_{j,j} \\
&= \begin{bmatrix} dx_j + (b_j - c_j)\, dh_j & db_j \\ -dc_j & dx_j + (c_j - b_j)\, dh_j \end{bmatrix},
\end{aligned} \tag{A.37.24}$$

and

$$\frac{\partial(M_{j,j})}{\partial(x_j, b_j, c_j, h_j)} = \det\begin{bmatrix} 1 & 0 & 0 & b_j - c_j \\ 0 & 1 & 0 & 0 \\ 0 & 0 & -1 & 0 \\ 1 & 0 & 0 & c_j - b_j \end{bmatrix} = 2(b_j - c_j). \tag{A.37.25}$$

Finally, for $i < j$, $dM_{i,j} = dR^U_{i,j} +$ terms not contributing to the Jacobian; so that

$$\prod_{i<j} \mu(dM_{i,j}) = \mu(dR^U). \tag{A.37.26}$$

This completes the proof of Lemma A.37.3.

To get the joint probability density of the eigenvalues one still has to change the variables x_j, b_j and c_j in Z_j to the conjugate pairs of eigenvalues $x_j \pm iy_j$ and integrate over all the extra variables.

Given the 2 by 2 matrix

$$\begin{bmatrix} x & b \\ -c & x \end{bmatrix}, \quad bc > 0,\ b \geqslant c,$$

let $\xi = b - c$. Then the Jacobian of the change of variables from b, c to ξ, $y = \sqrt{bc}$ is

$$\frac{\partial(\xi, y)}{\partial(b, c)} = \det\begin{bmatrix} 1 & c/(2y) \\ -1 & b/(2y) \end{bmatrix} = \frac{b+c}{2y} = \sqrt{\frac{\xi^2 + 4y^2}{2y}}. \tag{A.37.27}$$

Now one needs the integral

$$\int_0^\infty d\xi \, \frac{\xi}{\sqrt{\xi^2 + 4y^2}} \exp(-\xi^2)\, d\xi,$$

which by a change of the variable $\eta = \sqrt{(\xi^2 + 4y^2)}$ is seen to be proportional to the error function $\operatorname{erfc}(2y)$. Thus the joint probability density of the eigenvalues $\lambda_1, \ldots, \lambda_k$; $x_1 \pm iy_1, \ldots, x_m \pm iy_m$ is

$$C_{n,k} \exp\left[-\sum_{j=1}^{k} \lambda_j^2 + 2\sum_{j=1}^{m} (y_j^2 - x_j^2) \right] \prod_{j=1}^{m} [y_j \operatorname{erfc}(2y_j)]. \tag{A.37.28}$$

To get the correct value of the constant $C_{n,k}$ one has to keep track of the constants at every step of change of variables making sure that the correspondence is one to one. For this we refer to Edelman's original paper (Edelman, 1997) quoting here only the result

$$C_{n,k}^{-1} = 2^{-2m} k! m! \prod_{j=1}^{n} \Gamma(j/2). \tag{A.37.29}$$

To get the probability $P_{n,k}$ that there are exactly k real (and hence $m = (n-k)/2$ pairs of complex conjugate) eigenvalues, one should integrate the joint probability density (A.37.28) over all the variables, over the λ_j and the x_j from $-\infty$ to $+\infty$ and over the y_j from 0 to ∞. This can be done with some effort when either k is small or m is small. For general values of k and m, no simplification is known.

A.38 Variance of the Number Statistic. Section 16.1

The variance of the number statistic is most easily calculated via the two level correlation or cluster function. However, we will compute it also via the n-level spacing functions; it will serve as an illustration and as a check of the consistency.

If $R_1(x)$ and $R_2(x, y)$ are the one and two level correlation functions, then

$$\langle n^2 \rangle = \int_{-L}^{L} \int_{-L}^{L} \big(R_1(x)\delta(x-y) + R_2(x, y)\big)\, dx\, dy. \tag{A.38.1}$$

Measuring the distances in terms of the local mean spacing $D = R_1^{-1}(x)$, one has

$$\begin{aligned}(\delta n)^2 \equiv \langle n^2 \rangle - \langle n \rangle^2 &= \int_{-s/2}^{s/2} \int_{-s/2}^{s/2} \big(\delta(x-y) - Y(x-y)\big)\, dx\, dy \\ &= s - 2\int_0^s (s-\xi) Y(\xi)\, d\xi, \end{aligned} \tag{A.38.2}$$

where $2L = sD$, and $Y(x-y)$ is the two level cluster function. Substituting the expressions of Y from Chapters 6, 7 and 8, we get the variance of n for the various ensembles.

Unitary: $\beta = 2$,

$$\begin{aligned} Y(\xi) &= \left(\frac{\sin(\pi\xi)}{\pi\xi}\right)^2, \\ (\delta n)_2^2 &= s - \frac{2}{\pi^2} \int_0^{\pi s} (\pi s - \xi) \left(\frac{\sin\xi}{\xi}\right)^2 d\xi \\ &= \frac{2s}{\pi} \int_{\pi s}^{\infty} \left(\frac{\sin\xi}{\xi}\right)^2 d\xi + \frac{1}{\pi^2} \int_0^{2\pi s} \frac{1-\cos\xi}{\xi}\, d\xi \\ &= \frac{1}{\pi^2}\big(\ln(2\pi s) + \gamma + 1\big) + O(s^{-1}), \end{aligned} \tag{A.38.3}$$

where γ is Euler's constant, $\gamma = 0.5772\ldots$.

Orthogonal: $\beta = 1$,

$$Y(\xi) = \left(\frac{\sin(\pi\xi)}{\pi\xi}\right)^2 + \frac{d}{d\xi}\left(\frac{\sin(\pi\xi)}{\pi\xi}\right) \cdot \int_\xi^\infty \frac{\sin(\pi t)}{\pi t}\, dt,$$

$$\begin{aligned}
(\delta n)_1^2 &= s - \frac{2}{\pi^2}\int_0^{\pi s} (\pi s - \xi)\left(\left(\frac{\sin\xi}{\xi}\right)^2 + \frac{d}{d\xi}\left(\frac{\sin\xi}{\xi}\right)\int_\xi^\infty \frac{\sin t}{t}\, dt\right) d\xi \\
&= \frac{4s}{\pi}\int_{\pi s}^\infty \left(\frac{\sin\xi}{\xi}\right)^2 d\xi + \frac{2}{\pi^2}\int_0^{2\pi s} \frac{1-\cos\xi}{\xi}\, d\xi - \frac{1}{4} + \frac{1}{\pi^2}\left(\int_{\pi s}^\infty \frac{\sin t}{t}\, dt\right)^2 \\
&= \frac{2}{\pi^2}\left(\ln(2\pi s) + \gamma + 1 - \frac{\pi^2}{8}\right) + \mathrm{O}(s^{-1}).
\end{aligned} \tag{A.38.4}$$

Symplectic: $\beta = 4$,

$$Y(\xi) = \left(\frac{\sin(2\pi\xi)}{2\pi\xi}\right)^2 - \int_0^\xi \frac{\sin(2\pi t)}{2\pi t}\, dt \cdot \frac{d}{d\xi}\frac{\sin(2\pi\xi)}{2\pi\xi},$$

$$\begin{aligned}
(\delta n)_4^2 &= s - \frac{2}{(2\pi)^2}\int_0^{2\pi s} (2\pi s - \xi)\left(\left(\frac{\sin\xi}{\xi}\right)^2 - \int_0^\xi \frac{\sin t}{t}\, dt \cdot \frac{d}{d\xi}\frac{\sin\xi}{\xi}\right) \\
&= s - \frac{1}{\pi^2}\int_0^{2\pi s} (2\pi s - \xi)\left(\frac{\sin\xi}{\xi}\right)^2 d\xi + \frac{1}{4\pi^2}\left(\int_0^{2\pi s} \frac{\sin t}{t}\, dt\right)^2 \\
&= \frac{1}{2\pi^2}\left(\ln(4\pi s) + \gamma + 1 + \frac{\pi^2}{8}\right) + \mathrm{O}(s^{-1}).
\end{aligned} \tag{A.38.5}$$

A.38.1 Averages of the Powers of n via n-Level Spacings. Consider a distribution of points or levels on a straight line. Let these levels be the eigenvalues of a random matrix taken from one of the three ensembles, unitary, orthogonal or symplectic. The probability $E(n, s)$ that a randomly chosen interval of length s contains exactly n levels obviously must satisfy

$$\sum_{n=0}^{\infty} E(n, s) = 1. \tag{A.38.6}$$

It is convenient to choose the length scale so that the average spacing is unity. The mean value of n is then given by

$$\sum_{n=0}^{\infty} nE(n, s) = s. \tag{A.38.7}$$

And we want to compute the mean values of powers of n. For the three ensembles studied in Chapters 6, 7 and 8, we have the following expressions of $E(n, s)$.

Unitary Ensemble, $\beta = 2$,

$$E_2(0,s) = \prod_j (1-\lambda_j) \equiv \prod_j (1+x_j)^{-1}, \tag{A.38.8}$$

$$E_2(n,s) = E_2(0,s) \cdot \sum_{(i)} x_{i_1} \cdots x_{i_n}. \tag{A.38.9}$$

Orthogonal Ensemble, $\beta = 1$,

$$E_1(0,s) = \prod_j (1-\lambda_{2j}) \equiv \prod_j (1+y_j)^{-1}, \tag{A.38.10}$$

$$E_1(2n,s) = E_1(0,s) \cdot \sum_{(i)} y_{i_1} \cdots y_{i_n} \big(1 - (b_{i_1} + \cdots + b_{i_n})\big), \tag{A.38.11}$$

$$E_1(2n-1,s) = E_1(0,s) \cdot \sum_{(i)} y_{i_1} \cdots y_{i_n} (b_{i_1} + \cdots + b_{i_n}). \tag{A.38.12}$$

Symplectic Ensemble, $\beta = 4$,

$$E_4(0,s/2) = E_1(0,s) \cdot \left(1 + \frac{1}{2} \sum_i b_i y_i\right), \tag{A.38.13}$$

$$E_4(n,s/2) = E_1(0,s) \cdot \left(\sum_{(i)} y_{i_1} \cdots y_{i_n} \left(1 - \frac{1}{2}(b_{i_1} + \cdots + b_{i_n})\right) + \frac{1}{2} \sum_{(i)} y_{i_1} \cdots y_{i_{n+1}} (b_{i_1} + \cdots + b_{i_{n+1}})\right), \tag{A.38.14}$$

where

$$x_i = \lambda_i (1-\lambda_i)^{-1}, \quad y_i = x_{2i} = \lambda_{2i}(1-\lambda_{2i})^{-1}, \tag{A.38.15}$$

$$b_i = f_{2i}(1) \int_{-1}^{1} f_{2i}(x)\,dx \Big/ \int_{-1}^{1} f_{2i}^2(x)\,dx, \tag{A.38.16}$$

λ_i and $f_i(x)$ are respectively the eigenvalues and eigenfunctions of the integral equation

$$\lambda_i f_i(x) = \int_{-1}^{1} K(x-y) f_i(y)\,dy, \tag{A.38.17}$$

with the kernel

$$K(x) = \frac{\sin(\pi s x/2)}{\pi x}. \tag{A.38.18}$$

The sums or products in Eqs. (A.38.8)–(A.38.14) are taken over all integers with the restrictions $0 \leqslant j < \infty$, $0 \leqslant i_1 < i_2 < i_3 < \cdots$.

The $f_i(x)$ are known as spheroidal functions. They are even or odd according as the index is even or odd,

$$f_i(-x) = (-1)^i f_i(x). \tag{A.38.19}$$

They are orthogonal and may be normalized,

$$\int_{-1}^{1} f_i(x) f_j(x)\, dx = \delta_{ij}. \tag{A.38.20}$$

One may then write

$$K(x-y) = \sum_i \lambda_i f_i(x) f_i(y), \tag{A.38.21}$$

so that

$$\sum_i \lambda_i = \int_{-1}^{1} K(0)\, dx = s. \tag{A.38.22}$$

We will need the following identities:

$$\begin{aligned} 1 + \sum_n \sum_{(i)} \xi_{i_1} \cdots \xi_{i_n} &\equiv 1 + \sum_i \xi_i + \sum_{i<j} \xi_i \xi_j + \sum_{i<j<k} \xi_i \xi_j \xi_k + \cdots \\ &= \prod_i (1+\xi_i), \end{aligned} \tag{A.38.23}$$

$$\begin{aligned} \sum_n \sum_{(i)} (b_{i_1} + \cdots + b_{i_n}) \xi_{i_1} \cdots \xi_{i_n} &\equiv \sum_i b_i \xi_i + \sum_{i<j} (b_i + b_j) \xi_i \xi_j + \cdots \\ &= \prod_i (1+\xi_i) \sum_j \frac{b_j \xi_j}{1+\xi_j}. \end{aligned} \tag{A.38.24}$$

One can prove these identities by recurrence. One can deduce others by replacing ξ_i by $z\xi_i$, differentiating several times with respect to z and finally setting $z = 1$. For example, from Eq. (A.38.24) we deduce

$$\begin{aligned} &\sum_n n \sum_{(i)} (b_{i_1} + \cdots + b_{i_n}) \xi_{i_1} \cdots \xi_{i_n} \\ &\quad \equiv \sum_i b_i \xi_i + 2 \sum_{i<j} (b_i + b_j) \xi_i \xi_j + \cdots \\ &\quad = \prod_i (1+\xi_i) \left(\sum_j \frac{b_j \xi_j}{(1+\xi_j)^2} + \sum_j \frac{b_j \xi_j}{1+\xi_j} \sum_k \frac{\xi_k}{1+\xi_k} \right). \end{aligned} \tag{A.38.25}$$

Setting all b_i equal to 1, we get from Eqs. (A.38.24) and (A.38.25) other identities. For later convenience we write one of them,

$$\sum_n n^2 \sum_{(i)} \xi_{i_1} \cdots \xi_{i_n} \equiv \sum_i \xi_i + 2^2 \sum_{i<j} \xi_i \xi_j + 3^2 \sum_{i<j<k} \xi_i \xi_j \xi_k + \cdots$$
$$= \prod_i (1+\xi_i) \left(\sum_j \frac{\xi_j}{(1+\xi_j)^2} + \left(\sum_j \frac{\xi_j}{1+\xi_j} \right)^2 \right). \quad \text{(A.38.26)}$$

Unitary Ensemble, $\beta = 2$

This case is the simplest. Take $\xi_i = x_i$ in Eqs. (A.38.23), (A.38.24) and (A.38.26) with all $b_i = 1$. This gives with Eq. (A.38.22), Eqs. (A.38.6) and (A.38.7) and

$$\sum_n n^2 E_2(n,s) = \sum_i \lambda_i (1-\lambda_i) + \left(\sum_i \lambda_i \right)^2, \quad \text{(A.38.27)}$$

so that the variance of n is

$$(\delta n)_2^2 = \langle n^2 \rangle_2 - \langle n \rangle^2 = \sum_i \lambda_i (1-\lambda_i). \quad \text{(A.38.28)}$$

Orthogonal Ensemble, $\beta = 1$

Take $\xi_i = y_i$ and b_i defined by Eqs. (A.38.15) and (A.38.16). Now Eqs. (A.38.11), (A.38.12), (A.38.23), (A.38.25) and (A.38.26) give in addition to (A.38.6)

$$\sum_n n E_1(n,s) = 2 \sum_i \lambda_{2i} - \sum_i b_i \lambda_{2i}, \quad \text{(A.38.29)}$$

and

$$\sum_n n^2 E_1(n,s) = \left(\sum_i (2\lambda_{2i} - b_i) \right)^2 + 4 \sum_i \lambda_{2i} (1-\lambda_{2i})(1-b_i)$$
$$+ \left(\sum_i b_i \lambda_{2i} \right) \left(1 - \sum_i b_i \lambda_{2i} \right). \quad \text{(A.38.30)}$$

From Eqs. (A.38.19) and (A.38.21) one obtains

$$\sum_i b_i \lambda_{2i} = \sum_i \lambda_i f_i(1) \int_{-1}^{1} f_i(x)\, dx = \int_{-1}^{1} K(1-x)\, dx, \quad \text{(A.38.31)}$$

and

$$\sum_i (\lambda_{2i} - \lambda_{2i+1}) = \sum_i \lambda_i \int_{-1}^{1} f_i(x) f_i(-x)\, dx = \int_{-1}^{1} K(2x)\, dx. \tag{A.38.32}$$

After some simplifications, one has therefore

$$\sum_i b_i \lambda_{2i} = \sum_i (\lambda_{2i} - \lambda_{2i+1}) = \frac{1}{\pi} \int_0^{\pi s} \frac{\sin x}{x}\, dx. \tag{A.38.33}$$

From Eqs. (A.38.29), (A.38.33) and (A.38.22) we get (A.38.7). Equations (A.38.29) and (A.38.30) give the variance of n,

$$\begin{aligned}(\delta n)_1^2 &= \langle n^2 \rangle_1 - \langle n \rangle^2 \\ &= 4 \sum_i \lambda_{2i}(1-\lambda_{2i})(1-b_i) + \left(\sum_i b_i \lambda_{2i}\right)\left(1 - \sum_i b_i \lambda_{2i}\right). \end{aligned} \tag{A.38.34}$$

Symplectic Ensemble, $\beta = 4$

Take $\xi_i = y_i$ and b_i, Eqs. (A.38.15) and (A.38.16). Equations (A.38.13), (A.38.14), (A.38.22)–(A.38.24), (A.38.26) and (A.38.33) give in addition to (A.38.6) and (A.38.7),

$$\begin{aligned}&\sum_n n^2 E_4(n, s/2) \\ &= E_1(0,s)\left(\sum_n n^2 \sum_{(i)} y_{i_1} \cdots y_{i_n} - \frac{1}{2}\sum_n (2n-1) \sum_{(i)} (b_{i_1} + \cdots + b_{i_n}) y_{i_1} \cdots y_{i_n}\right) \\ &= \sum_i \lambda_{2i}(1-\lambda_{2i}) + \left(\sum_i \lambda_{2i}\right)^2 - \sum_i b_i \lambda_{2i}(1-\lambda_{2i}) \\ &\quad - \sum_i b_i \lambda_{2i} \sum_j \lambda_{2j} + \frac{1}{2}\sum_i b_i \lambda_{2i}. \end{aligned} \tag{A.38.35}$$

The variance of n is therefore

$$\begin{aligned}(\delta n)_4^2 &= \sum_n n^2 E_4(n, s/2) - \left(\sum_i \lambda_{2i} - \frac{1}{2}\sum_i b_i \lambda_{2i}\right)^2 \\ &= \sum_i \lambda_{2i}(1-\lambda_{2i})(1-b_i) + \frac{1}{2}\sum_i b_i \lambda_{2i}\left(1 - \frac{1}{2}\sum_j b_j \lambda_{2j}\right). \end{aligned} \tag{A.38.36}$$

To see that Eqs. (A.38.28), (A.38.34) and (A.38.36) are identical to (A.38.3)–(A.38.5), one has to evaluate the various sums.

Equations (A.38.20) and (A.38.21) give

$$\sum_i \lambda_i^2 = \int_{-1}^{1} \int_{-1}^{1} K^2(x-y)\,dx\,dy = 2\int_0^1 \left(\frac{\sin(\pi s\xi)}{\pi\xi}\right)^2 (1-\xi)\,d\xi. \quad \text{(A.38.37)}$$

So that with (A.38.22),

$$\begin{aligned}\sum_i \lambda_i(1-\lambda_i) &= \frac{2s}{\pi}\int_{\pi s}^{\infty}\left(\frac{\sin x}{x}\right)^2 dx + \frac{1}{\pi^2}\int_0^{2\pi s}\frac{1-\cos x}{x}\,dx \\ &= \frac{1}{\pi^2}\left(\ln(2\pi s)+\gamma+1+\frac{\sin(2\pi s)}{2\pi s}+\mathrm{O}(s^{-2})\right). \quad \text{(A.38.38)}\end{aligned}$$

From Eqs. (A.38.22) and (A.38.33), one has

$$\begin{aligned}\sum_i \lambda_{2i} &= \frac{1}{2}\sum_i \lambda_i + \frac{1}{2}\sum_i(\lambda_{2i}-\lambda_{2i+1}) = \frac{s}{2}+\frac{1}{2\pi}\int_0^{\pi s}\frac{\sin x}{x}\,dx \\ &= \frac{s}{2}+\frac{1}{4}-\frac{1}{2\pi^2 s}\cos(\pi s)+\mathrm{O}(s^{-2}). \quad \text{(A.38.39)}\end{aligned}$$

From

$$K(x-y)+K(x+y) = 2\sum_i \lambda_{2i} f_{2i}(x) f_{2i}(y) \quad \text{(A.38.40)}$$

one gets

$$\begin{aligned}4\sum_i \lambda_{2i}^2 &= \int_{-1}^{1}\int_{-1}^{1}\left(\frac{\sin((x-y)\pi s/2)}{(x-y)\pi}+\frac{\sin((x+y)\pi s/2)}{(x+y)\pi}\right)^2 dx\,dy \\ &= 2\sum_i \lambda_i^2 + \frac{4}{\pi^2}\int_0^{\pi s} d\xi \int_0^{\pi s-\xi} d\eta\, \frac{\sin\xi}{\xi}\frac{\sin\eta}{\eta} \\ &= 2\sum_i \lambda_i^2 + 2\left(\frac{1}{\pi}\int_0^{\pi s}\frac{\sin\xi}{\xi}\,d\xi\right)^2, \quad \text{(A.38.41)}\end{aligned}$$

and

$$\begin{aligned}\sum_i b_i\lambda_{2i}^2 &= \int_{-1}^{1}\int_{-1}^{1}\frac{\sin((1-y)\pi s/2)}{(1-y)\pi}\frac{\sin((x-y)\pi s/2)}{(x-y)\pi}\,dx\,dy \\ &= \frac{1}{\pi^2}\int_0^{\pi s} d\xi \int_{\xi-\pi s}^{\xi} d\eta\,\frac{\sin\xi}{\xi}\frac{\sin\eta}{\eta}\end{aligned}$$

$$= \frac{1}{2}\left(\frac{1}{\pi}\int_0^{\pi s}\frac{\sin\xi}{\xi}\,d\xi\right)^2 + \frac{1}{2}\left(\frac{1}{\pi}\int_0^{\pi s}\frac{\sin\xi}{\xi}\,d\xi\right)^2$$

$$= \frac{1}{2}\left(\frac{1}{2} - \frac{\cos(\pi s)}{\pi^2 s}\right)^2 + \frac{1}{\pi^2}\int_0^{\pi s} d\xi \int_0^{\pi s-\xi} d\eta\,\frac{\sin\xi}{\xi}\frac{\sin\eta}{\eta} + O(s^{-2}). \tag{A.38.42}$$

From Eqs. (A.38.39), (A.38.41), (A.38.33) and (A.38.42) one gets therefore

$$4\sum_i \lambda_{2i}(1-\lambda_{2i})(1-b_i) = 4\sum_i\left(\lambda_{2i} - \lambda_{2i}^2 - b_i\lambda_{2i} + b_i\lambda_{2i}^2\right)$$

$$= \frac{2}{\pi^2}\left(\ln(2\pi s) + \gamma + 1 - \frac{\pi^2}{4} + O(s^{-2})\right). \tag{A.38.43}$$

Substituting these asymptotic expressions in (A.38.28), (A.38.34) and (A.38.36), we get finally

$$(\delta n)_2^2 = \frac{1}{\pi^2}\left(\ln(2\pi s) + \gamma + 1\right) + O(s^{-1}), \tag{A.38.44}$$

$$(\delta n)_1^2 = \frac{2}{\pi^2}\left(\ln(2\pi s) + \gamma + 1 - \frac{\pi^2}{8}\right) + O(s^{-1}), \tag{A.38.45}$$

and

$$(\delta n)_4^2 = \frac{1}{2\pi^2}\left(\ln(4\pi s) + \gamma + 1 + \frac{\pi^2}{8}\right) + O(s^{-1}), \tag{A.38.46}$$

taking care of the fact that for the symplectic ensemble E_4 was given at $s/2$ and not at s.

For comparison with the Poisson process when levels have no correlations, we note that

$$Y(\xi) = 0, \quad E_0(n,s) = \frac{s^n}{n!}e^{-s} \quad \text{and} \quad (\delta n)_0^2 = s. \tag{A.38.47}$$

A.39 Optimum Linear Statistic. Section 16.1

The general linear statistic

$$W = \sum_i f(E_i), \tag{A.39.1}$$

has the mean value

$$\langle W\rangle = D^{-1}\int_{-L}^{L} f(x)\,dx, \tag{A.39.2}$$

and the variance

$$
\begin{aligned}
V_W &= \langle W \rangle^2 - \langle W \rangle^2 \\
&= \int_{-L}^{L} \int_{-L}^{L} \big(R_2(x,y) + R_1(x)\delta(x-y) - R_1(x)R_1(y) \big) f(x) f(y)\, dx\, dy \\
&= \int_{-s/2}^{s/2} \int_{-s/2}^{s/2} f(xD) f(yD) \big(\delta(x-y) - Y(x-y) \big)\, dx\, dy, \qquad \text{(A.39.3)}
\end{aligned}
$$

where $R_1(x) = D^{-1}$ is the level density, $2L = sD$, and $Y(x)$ is the two level cluster function. Going to the Fourier transforms,

$$
f(x) = \int_{-\infty}^{\infty} e^{2\pi i k x} \varphi(k)\, dk, \qquad \text{(A.39.4)}
$$

$$
\varphi(k) = \int_{-\infty}^{\infty} e^{-2\pi i k x} f(x)\, dx, \qquad \text{(A.39.5)}
$$

$$
\delta(x) = \int_{-\infty}^{\infty} e^{-2\pi i k x}\, dk, \qquad Y(x) = \int_{-\infty}^{\infty} e^{-2\pi i k x} b(k)\, dk, \qquad \text{(A.39.6)}
$$

we have

$$
V_W = D^{-1} \int_{-\infty}^{\infty} \varphi(k)\varphi(-k)\big(1 - b(Dk)\big)\, dk, \qquad \text{(A.39.7)}
$$

where $b(k)$, the Fourier transform of $Y(x)$, is given by Eqs. (6.2.17), (7.2.46) and (8.2.9) for the three ensembles, respectively.

If $f(x)$ is a smooth function, $\varphi(k)$ is large only for values of k of the order of L^{-1}, and the whole of the integral (A.39.7) comes from values of k in this neighborhood. Therefore, the approximation

$$
b(k) = 1 - \frac{2}{\beta}|k| \qquad \text{(A.39.8)}
$$

may be used with an error of the order of $D/2L = s^{-1}$,

$$
\begin{aligned}
V_W &\cong \frac{2}{\beta} \int_{-\infty}^{\infty} \varphi(k)\varphi(-k)|k|\, dk \\
&= -\frac{2}{\beta \pi^2} \int_{-L}^{L} \int_{-L}^{L} f'(x) f'(y) \ln|x-y|\, dx\, dy. \qquad \text{(A.39.9)}
\end{aligned}
$$

Setting $f(x) = g(x/L)$, the figure of merit $\Phi_W = V_W / \langle W \rangle^2$ is

$$
\Phi_W = \frac{4}{\pi^2 s^2 \beta} \frac{\int_{-1}^{1} \int_{-1}^{1} g'(x) g'(y) \ln|x-y|\, dx\, dy}{\left(\int_{-1}^{1} g'(x)\, dx\right)^2}. \qquad \text{(A.39.10)}
$$

This is minimum if $g'(x)$ satisfies the integral equation

$$\int_{-1}^{1} g'(y) \ln |x - y| \, dy = x, \tag{A.39.11}$$

i.e., when $g(x) = (1 - x^2)^{1/2}/\pi$ (cf. Section 4.2). The optimum statistic is therefore

$$W = \sum_{i=1}^{n} \left(1 - (E_i/L)^2\right)^{1/2}, \tag{A.39.12}$$

with $\langle W \rangle = \pi L/(2D)$ and $V_W = 1/(2\beta)$.

Note that Eqs. (A.39.3) and (A.39.7) are exact, while the approximations (A.39.8), (A.39.9) and the following are valid only for smooth functions.

A.40 Mean Value of Δ. Section 16.2

The best straight line fitting an observed staircase graph is chosen by least square deviation

$$\begin{aligned} \Delta &= \min_{A,B} \frac{1}{2L} \int_{-L}^{L} \left(N(E) - \frac{A}{D} E - B \right)^2 dE \\ &= \min_{A,B} \frac{1}{s} \int_{-s/2}^{s/2} \left(n(x) - Ax - B \right)^2 dx, \end{aligned} \tag{A.40.1}$$

where D is the mean spacing, $2L = sD$, $E = xD$ and $N(E) = n(x)$. Equating to zero the partial derivatives with respect to A and B fixes the values of A, B giving this minimum

$$A = \int x n(x) \, dx \Big/ \int x^2 dx \equiv \frac{12}{s} n_1, \tag{A.40.2}$$

$$B = \int n(x) \, dx \Big/ \int dx = n_0, \tag{A.40.3}$$

with

$$n_0 = \frac{1}{s} \int n(x) \, dx, \quad n_1 = \frac{1}{s^2} \int x n(x) \, dx. \tag{A.40.4}$$

Here and in what follows, all integrals are taken from $-s/2$ to $s/2$, unless explicitly indicated otherwise. Substituting the values of A, B in (A.40.1) we get

$$\Delta = \frac{1}{s} \int n^2(x) \, dx - 12 n_1^2 - n_0^2. \tag{A.40.5}$$

Taking averages

$$\langle \Delta \rangle = \frac{1}{s} \int \langle n^2(x) \rangle dx - 12\langle n_1^2 \rangle - \langle n_0^2 \rangle. \tag{A.40.6}$$

Now

$$\langle n_1^2 \rangle = \frac{1}{s^4} \iint xy \langle n(x)n(y) \rangle dx\, dy, \tag{A.40.7}$$

$$\langle n_0^2 \rangle = \frac{1}{s^2} \iint \langle n(x)n(y) \rangle dx\, dy, \tag{A.40.8}$$

and

$$\langle n(x)n(y) \rangle = \int_{-s/2}^{x} d\xi \int_{-s/2}^{y} d\eta \{ R_1(\xi)\delta(\xi - \eta) + R_2(\xi, \eta) \}, \tag{A.40.9}$$

where $R_1(\xi)$ is the level density and $R_2(\xi, \eta)$ is the two-level correlation function. As we are measuring distances in terms of the mean spacing D,

$$R_1(\xi) = 1, \qquad R_2(\xi, \eta) = 1 - Y(\xi - \eta), \tag{A.40.10}$$

where $Y(r)$ is the two-level cluster function given by Eqs. (6.2.14), (7.2.43) or (8.2.6);

$$Y(r) = \left(\frac{\sin(\pi r)}{\pi r} \right)^2, \quad \beta = 2, \tag{A.40.11}$$

$$Y(r) = \left(\frac{\sin(\pi r)}{\pi r} \right)^2 + \frac{d}{dr} \left(\frac{\sin(\pi r)}{\pi r} \right) \int_r^{\infty} \frac{\sin(\pi t)}{\pi t} dt, \quad \beta = 1, \tag{A.40.12}$$

and

$$Y(r) = \left(\frac{\sin(\pi r)}{\pi r} \right)^2 - \frac{d}{dr} \left(\frac{\sin(2\pi r)}{2\pi r} \right) \int_0^{r} \frac{\sin(2\pi t)}{2\pi t} dt, \quad \beta = 4. \tag{A.40.13}$$

If one imposes $\xi \geqslant \eta$, then because of the symmetry one should multiply by 2, except for the delta function. Introducing the variables $u = \xi - \eta$, $v = \xi$, integrations over all variables except u are elementary. Thus for any function $f(x)$

$$\begin{aligned} \iint dx\, dy \int_{-s/2}^{x} d\xi \int_{-s/2}^{y} d\eta f(\xi - \eta) &= \iint d\xi\, d\eta \left(\frac{s}{2} - \xi \right) \left(\frac{s}{2} - \eta \right) f(\xi - \eta) \\ &= 2 \int_0^s du \int_{u-s/2}^{s/2} dv \left(\frac{s}{2} - v \right) \left(\frac{s}{2} - v + u \right) f(u) \end{aligned}$$

$$= 2\int_0^s du \int_0^{s-u} dv\, v(v+u) f(u)$$

$$= \frac{1}{3}\int_0^s du\, (s-u)^2(2s+u) f(u), \qquad \text{(A.40.14)}$$

$$\int dx \int_{-s/2}^{x} d\xi \int_{-s/2}^{x} d\eta\, f(\xi-\eta) = 2\int d\xi \int_{-s/2}^{\xi} d\eta \left(\frac{s}{2}-\xi\right) f(\xi-\eta)$$

$$= 2\int_0^s du \int_0^{s-u} dv \cdot v f(u)$$

$$= \int_0^s du\, (s-u)^2 f(u), \qquad \text{(A.40.15)}$$

and

$$\iint dx\, dy \int_{-s/2}^{x} d\xi \int_{-s/2}^{y} d\eta \cdot xy f(\xi-\eta)$$

$$= \frac{1}{4}\iint d\xi\, d\eta \left(\frac{s^2}{4}-\xi^2\right)\left(\frac{s^2}{4}-\eta^2\right) f(\xi-\eta)$$

$$= \frac{1}{2}\int_0^s du \int_0^{s-u} dv\, v(s-v)(v+u)(s-v-u) f(u)$$

$$= \frac{1}{60}\int_0^s du\, (s-u)^3(s^2+3su+u^2) f(u). \qquad \text{(A.40.16)}$$

Collecting the results we get after some simplification

$$\langle\Delta\rangle = \frac{1}{15}s^{-4}\int_0^s du\, (s-u)^3(2s^2-9su-3u^2)\left(\frac{1}{2}\delta(u)+1-Y(u)\right)$$

$$= \frac{1}{15}s^{-4}\int_0^s du\, (s-u)^3(2s^2-9su-3u^2)\left(\frac{1}{2}\delta(u)-Y(u)\right). \qquad \text{(A.40.17)}$$

For a random level sequence without correlations (Poisson process), $Y(u) = 0$, and

$$\langle\Delta\rangle = \frac{s}{15}. \qquad \text{(A.40.18)}$$

For the cases when $Y(u)$ is given by Eqs. (A.40.11)–(A.40.13), we use the asymptotic estimations,

$$\int_0^s \left(\frac{\sin(\pi u)}{\pi u}\right)^2 du = \frac{1}{2} - \int_{\pi s}^{\infty} \left(\frac{\sin x}{x}\right)^2 dx$$

$$= \frac{1}{2} - \frac{1}{2\pi s} + \mathrm{O}(s^{-2}), \qquad \text{(A.40.19)}$$

$$\int_0^s u\left(\frac{\sin(\pi u)}{\pi u}\right)^2 du = \frac{1}{2\pi^2}\int_0^{2\pi s} \frac{1-\cos x}{x}\, dx$$
$$= \frac{1}{2\pi^2}\big(\ln(2\pi s) + \gamma\big) + \mathrm{O}(s^{-1}), \tag{A.40.20}$$

and

$$\int_0^s u^j \left(\frac{\sin(\pi u)}{\pi u}\right)^2 du = \frac{1}{2\pi^2}\frac{s^{j-1}}{j-1} + \mathrm{O}(s^{j-2}), \quad j \geqslant 2. \tag{A.40.21}$$

Changing variables from πu, πt to u, t and integrating by parts, one has

$$\int_0^s \frac{d}{du}\left(\frac{\sin(\pi u)}{\pi u}\right)\int_u^\infty \frac{\sin(\pi t)}{\pi t}\, dt\, du = \frac{1}{\pi}\int_0^{\pi s}\left(\frac{\sin u}{u}\right)\int_u^\infty \frac{\sin t}{t}\, dt\, du$$
$$= -\frac{1}{2\pi^2 s} + \mathrm{O}(s^{-2}), \tag{A.40.22}$$

$$\int_0^s u \frac{d}{du}\left(\frac{\sin(\pi u)}{\pi u}\right)\int_u^\infty \frac{\sin(\pi t)}{\pi t}\, dt\, du$$
$$= \frac{1}{\pi^2}\int_0^{\pi s} u\left(\frac{\sin u}{u}\right)\int_u^\infty \frac{\sin t}{t}\, dt\, du$$
$$= \frac{1}{\pi^2}\int_0^{\pi s} u\left(\frac{\sin u}{u}\right)^2 du - \frac{1}{2\pi^2}\left(\frac{\pi}{2}\right)^2 + \int_{\pi s}^\infty du \int_u^\infty dt \left(\frac{\sin u}{u}\right)\left(\frac{\sin t}{t}\right)$$
$$= \frac{1}{2\pi^2}\big(\ln(2\pi s) + \gamma\big) - \frac{1}{8} + \mathrm{O}(s^{-1}), \tag{A.40.23}$$

$$\int_0^s u^j \frac{d}{du}\left(\frac{\sin(\pi u)}{\pi u}\right)\int_u^\infty \frac{\sin(\pi t)}{\pi t}\, dt\, du = \frac{1}{2\pi^2}\frac{s^{j-1}}{j-1} + \mathrm{O}(s^{j-2}), \quad j \geqslant 2, \tag{A.40.24}$$

and similarly for integrals containing

$$\frac{d}{du}\left(\frac{\sin(2\pi u)}{2\pi u}\right)\int_0^u \frac{\sin(2\pi t)}{2\pi t}\, dt.$$

Substituting these asymptotic estimations in Eqs. (A.40.17) we get finally

$$\langle\Delta\rangle = \frac{1}{\pi^2}\left(\ln(2\pi s) + \gamma - \frac{5}{4} - \frac{\pi^2}{8}\right) + \mathrm{O}(s^{-1}), \quad \beta = 1, \tag{A.40.25}$$

$$\langle\Delta\rangle = \frac{1}{2\pi^2}\left(\ln(2\pi s) + \gamma - \frac{5}{4}\right) + \mathrm{O}(s^{-1}), \quad \beta = 2, \tag{A.40.26}$$

$$\langle\Delta\rangle = \frac{1}{4\pi^2}\left(\ln(4\pi s) + \gamma - \frac{5}{4} + \frac{\pi^2}{8}\right) + \mathrm{O}(s^{-1}), \quad \beta = 4. \tag{A.40.27}$$

The evaluation of $\langle \Delta^2 \rangle$ involves 3- and 4-level correlations as well. They are tedious but present no difficulty in principle, specially for the unitary ensemble. The final result is that the variance of Δ, i.e., $\langle \Delta^2 \rangle - \langle \Delta \rangle^2$, is a small constant independent of s in the three cases.

A.41 Tables of Functions $\mathcal{B}_\beta(x_1, x_2)$ and $\mathcal{P}_\beta(x_1, x_2)$ for $\beta = 1$ and 2

Table A.41.1. $\mathcal{B}_1(x_1, x_2)$. It is the probability that a distance x_1 on one side and a distance x_2 on the other side of a randomly chosen eigenvalue do not contain any other eigenvalue when the matrix is chosen from the Gaussian orthogonal ensemble. $\mathcal{B}_1(x_1, x_2)$ being symmetric in x_1 and x_2, its values are given only for $x_1 \leqslant x_2$

	$\frac{\pi x_1}{2}$	0.	0.1	0.2	0.3	0.4	0.5	0.6	0.7	0.8
$\frac{\pi x_2}{2}$	$x_2 \backslash x_1$	0.	0.064	0.127	0.191	0.255	0.318	0.382	0.446	0.509
0.	0.	1.								
0.1	0.064	0.996673	0.993347							
0.2	0.127	0.986770	0.983448	0.973565						
0.3	0.191	0.970517	0.967204	0.957354	0.941211					
0.4	0.255	0.948273	0.944975	0.935182	0.919149	0.897263				
0.5	0.318	0.920509	0.917236	0.907528	0.891656	0.870019	0.843124			
0.6	0.382	0.887796	0.884559	0.874967	0.859312	0.838005	0.811560	0.780578		
0.7	0.446	0.850776	0.847584	0.838145	0.822766	0.801871	0.775987	0.745717	0.711721	
0.8	0.509	0.810140	0.807006	0.797756	0.782713	0.762317	0.737102	0.707674	0.674690	0.638836
0.9	0.573	0.766610	0.763547	0.754522	0.739878	0.720065	0.695624	0.667162	0.635330	0.600805
1.0	0.637	0.720917	0.717936	0.709174	0.694987	0.675839	0.652271	0.624889	0.594338	0.561279
1.2	0.764	0.625886	0.623106	0.614971	0.601863	0.584259	0.562699	0.537776	0.510106	0.480318
1.4	0.891	0.530374	0.527836	0.520442	0.508593	0.492763	0.473479	0.451307	0.426825	0.400610
1.6	1.019	0.438855	0.436588	0.430019	0.419547	0.405634	0.388783	0.369515	0.348361	0.325840
1.8	1.146	0.354707	0.352729	0.347024	0.337982	0.326038	0.311654	0.295305	0.277461	0.258575
2.0	1.273	0.280137	0.278449	0.273610	0.265984	0.255970	0.243983	0.230439	0.215745	0.200289
2.2	1.401	0.216245	0.214840	0.210830	0.204550	0.196351	0.186597	0.175643	0.163833	0.151486
2.4	1.528	0.163196	0.162053	0.158809	0.153758	0.147204	0.139454	0.130805	0.121539	0.111912
2.6	1.655	0.120437	0.119529	0.116966	0.112999	0.107883	0.101871	0.095204	0.088105	0.080778
2.8	1.783	0.086932	0.086228	0.084251	0.081208	0.077309	0.072755	0.067736	0.062426	0.056981
3.0	1.910	0.061384	0.060851	0.059361	0.057082	0.054179	0.050809	0.047120	0.043242	0.039291
3.2	2.037	0.042408	0.042014	0.040917	0.039250	0.037139	0.034704	0.032055	0.029288	0.026488
3.4	2.165	0.028670	0.028385	0.027596	0.026405	0.024905	0.023186	0.021328	0.019400	0.017462
3.6	2.292	0.018969	0.018768	0.018214	0.017381	0.016341	0.015155	0.013882	0.012570	0.011259
3.8	2.419	0.012285	0.012145	0.011765	0.011197	0.010491	0.009693	0.008840	0.007967	0.007101
4.0	2.546	0.007787	0.007694	0.007439	0.007060	0.006592	0.006066	0.005508	0.004940	0.004381
4.2	2.674	0.004833	0.004771	0.004604	0.004357	0.004054	0.003715	0.003358	0.002998	0.002645
4.4	2.801	0.002937	0.002897	0.002789	0.002632	0.002440	0.002227	0.002004		
4.6	2.928	0.001747	0.001722	0.001655	0.001557	0.001438				
4.8	3.056	0.0010177	0.0010023	0.0009611						
5.0	3.183	0.0005805								

(continued on next page)

Table A.41.1. ($\mathcal{B}_1(x_1, x_2)$. *Continued*)

	$\frac{\pi x_1}{2}$	0.9	1.0	1.2	1.4	1.6	1.8	2.0	2.2	2.4
$\frac{\pi x_2}{2}$	$x_2\backslash x_1$	0.573	0.637	0.764	0.891	1.019	1.146	1.273	1.401	1.528
0.9	0.573	0.564265								
1.0	0.637	0.526374	0.490265							
1.2	0.764	0.449025	0.416818	0.351814						
1.4	0.891	0.373222	0.345188	0.189074	0.235515					
1.6	1.019	0.302445	0.278636	0.231395	0.186831	0.146802				
1.8	1.146	0.239074	0.219347	0.180558	0.144411	0.112341	0.085076			
2.0	1.273	0.184426	0.168479	0.137413	0.108825	0.083782	0.062763	0.045785		
2.2	1.401	0.138895	0.126316	0.102044	0.079993	0.060927	0.045133	0.032545	0.022860	
2.4	1.528	0.102157	0.092475	0.073971	0.057380	0.043223	0.031652	0.022555	0.015650	0.010581
2.6	1.655	0.073401	0.066127	0.052361	0.040180	0.029926	0.021658	0.015247	0.010448	0.006975
2.8	1.783	0.051535	0.046200	0.036203	0.027475	0.020228	0.014465	0.010057	0.006805	
3.0	1.910	0.035365	0.031544	0.024456	0.018352	0.013353	0.009432	0.006476		
3.2	2.037	0.023724	0.021052	0.016145	0.011977	0.008610	0.006007			
3.4	2.165	0.015561	0.013736	0.010419	0.007639	0.005425				
3.6	2.292	0.009982	0.008764	0.006573	0.004762					
3.8	2.419	0.006263	0.005469	0.004055						
4.0	2.546	0.003843	0.003338							

Table A.41.2. $\mathcal{P}_1(x_1, x_2)$. It is the probability density of two consecutive spacings between the eigenvalues of a random matrix taken from the Gaussian orthogonal ensemble; $\mathcal{P}_1(x_1, x_2) = \partial^2 \mathcal{B}_1(x_1, x_2)/\partial x_1 \partial x_2$. $\mathcal{P}_1(x_1, x_2)$ being symmetric in x_1 and x_2, its values are given only for $x_1 \leqslant x_2$

	$\frac{\pi x_1}{2}$	0.1	0.2	0.3	0.4	0.5	0.6	0.7	0.8	0.9
$\frac{\pi x_2}{2}$	$x_2\backslash x_1$	0.064	0.127	0.191	0.255	0.318	0.382	0.446	0.509	0.573
0.1	0.064	0.00								
0.2	0.127	0.004	0.00849							
0.3	0.191	0.006	0.01595	0.02793						
0.4	0.255	0.010	0.02474	0.04207	0.06249					
0.5	0.318	0.016	0.03484	0.05842	0.08446	0.11249				
0.6	0.382	0.020	0.04640	0.07560	0.10768	0.14205	0.17725			
0.7	0.446	0.027	0.05811	0.09341	0.13184	0.17185	0.21195	0.25135		
0.8	0.509	0.033	0.07016	0.11175	0.15582	0.20074	0.24554	0.28898	0.32978	
0.9	0.573	0.039	0.08241	0.12966	0.17867	0.22829	0.27713	0.32374	0.36680	0.40534
1.0	0.637	0.044	0.09421	0.14642	0.20014	0.25380	0.30580	0.35468	0.39926	0.43856
1.2	0.764	0.055	0.11484	0.17578	0.23672	0.29583	0.35163	0.40277	0.44805	0.48647
1.4	0.891	0.064	0.13065	0.19708	0.26182	0.32317	0.37963	0.42983	0.47296	0.50824
1.6	1.019	0.070	0.14003	0.20867	0.27400	0.33434	0.38850	0.43541	0.47426	0.50457
1.8	1.146	0.072	0.14273	0.21029	0.27317	0.32998	0.37961	0.42124	0.45444	0.47899
2.0	1.273	0.071	0.13906	0.20287	0.26094	0.31213	0.35569	0.39106	0.41805	0.43672
2.2	1.401	0.067	0.13015	0.18804	0.23962	0.28404	0.32077	0.34957	0.37045	0.38362
2.4	1.528	0.061	0.11730	0.16800	0.21219	0.24936	0.27919	0.30164	0.31693	0.32547
2.6	1.655	0.054	0.10210	0.14500	0.18161	0.21162	0.23495	0.25177	0.26237	0.26721
2.8	1.783	0.046	0.08600	0.12113	0.15046	0.17392	0.19156	0.20360	0.21045	0.21261
3.0	1.910	0.038	0.07017	0.09808	0.12089	0.13863	0.15146	0.15972	0.16379	0.16417

(continued on next page)

Table A.41.2. ($\mathcal{P}_1(x_1, x_2)$. *Continued*)

$\frac{\pi x_2}{2}$	$\frac{\pi x_1}{2}$ / $x_2 \backslash x_1$	0.1 / 0.064	0.2 / 0.127	0.3 / 0.191	0.4 / 0.255	0.5 / 0.318	0.6 / 0.382	0.7 / 0.446	0.8 / 0.509	0.9 / 0.573
3.2	2.037	0.030	0.05557	0.07708	0.09426	0.10727	0.11629	0.12169	0.12382	0.12314
3.4	2.165	0.023	0.04274	0.05884	0.07142	0.08067	0.08680	0.09013	0.09100	0.08979
3.6	2.292	0.018	0.03195	0.04369	0.05263	0.05901	0.06301	0.06493	0.06505	0.06371
3.8	2.419	0.013	0.02325	0.03156	0.03775	0.04200	0.04451	0.04554	0.04532	0.04392
4.0	2.546	0.009	0.01647	0.02219	0.02635	0.02911	0.03068	0.03098	0.03054	0.03051
4.2	2.674	0.006	0.01135	0.01520	0.01799	0.01951	0.02032	0.02182	0.01811	
4.4	2.801	0.004	0.00771	0.00998	0.01160	0.01406	0.01099			
4.6	2.928	0.002	0.00499	0.00738	0.00622					
4.8	3.056	0.00	0.00271							
5.0	3.183									

$\frac{\pi x_2}{2}$	$\frac{\pi x_1}{2}$ / $x_2 \backslash x_1$	1.0 / 0.637	1.2 / 0.764	1.4 / 0.891	1.6 / 1.019	1.8 / 1.146	2.0 / 1.273	2.2 / 1.401	2.4 / 1.528
1.0	0.637	0.47173							
1.2	0.764	0.51757	0.55629						
1.4	0.891	0.53524	0.56405	0.56133					
1.6	1.019	0.52626	0.54434	0.53206	0.49548				
1.8	1.146	0.49501	0.50292	0.48297	0.44217	0.38796			
2.0	1.273	0.44735	0.44659	0.42166	0.37955	0.32751	0.27196		
2.2	1.401	0.38956	0.38240	0.35501	0.31429	0.26678	0.21795	0.17182	
2.4	1.528	0.32777	0.31640	0.28893	0.25164	0.21015	0.16889	0.13098	0.09848
2.6	1.655	0.26689	0.25345	0.22771	0.19514	0.16032	0.12676	0.09698	0.07311
2.8	1.783	0.21065	0.19685	0.17404	0.14674	0.11861	0.09252	0.07086	
3.0	1.910	0.16137	0.14842	0.12912	0.10713	0.08542	0.06647		
3.2	2.037	0.12009	0.10871	0.09309	0.07615	0.06023			
3.4	2.165	0.08688	0.07745	0.06534	0.05258				
3.6	2.292	0.06120	0.05366	0.04401					
3.8	2.419	0.04184	0.03506						
4.0	2.546	0.02627							

Table A.41.3. $\mathcal{B}_2(x_1, x_2)$. Same as in Table A.41.1, but when the matrix is chosen from the Gaussian unitary ensemble

$\pi x_2/4$	$\pi x_1/4$ / $x_2 \backslash x_1$	0. / 0.	0.1 / 0.127	0.2 / 0.255	0.3 / 0.382	0.4 / 0.509	0.5 / 0.637	0.6 / 0.764
0.	0.	1.000000						
0.1	0.127	0.997766	0.995532					
0.2	0.255	0.982791	0.980559	0.965609				
0.3	0.382	0.945446	0.943223	0.928374	0.891497			
0.4	0.509	0.881362	0.879168	0.864573	0.828545	0.767460		
0.5	0.637	0.792044	0.789907	0.775812	0.741338	0.683475	0.604792	
0.6	0.764	0.683930	0.681896	0.668613	0.636520	0.583368	0.512099	0.429375
0.7	0.891	0.566398	0.564515	0.552374	0.523477	0.476374	0.414260	0.343398
0.8	1.019	0.449406	0.447720	0.437009	0.411952	0.371839	0.319924	0.261827
0.9	1.146	0.341475	0.340020	0.330935	0.310078	0.277340	0.235820	0.190312
1.0	1.273	0.248436	0.247232	0.239841	0.223213	0.197652	0.165922	0.131900

(continued on next page)

Table A.41.3. ($\mathcal{B}_2(x_1, x_2)$. *Continued*)

	$\pi x_1/4$	0.	0.1	0.2	0.3	0.4	0.5	0.6
$\pi x_2/4$	$x_2 \backslash x_1$	0.	0.127	0.255	0.382	0.509	0.637	0.764
1.1	1.401	0.173072	0.172116	0.166361	0.153681	0.134609	0.111458	0.087198
1.2	1.528	0.115466	0.114741	0.110456	0.101215	0.087625	0.071506	0.055007
1.3	1.655	0.073788	0.073262	0.070212	0.063780	0.054536	0.043826	0.033125
1.4	1.783	0.045176	0.044812	0.042738	0.038463	0.032460	0.025670	0.019051
1.5	1.910	0.026505	0.026263	0.024917	0.022203	0.018482	0.014374	0.010467
1.6	2.037	0.014904	0.014752	0.013917	0.012272	0.010069	0.007697	0.005497
1.7	2.165	0.008035	0.007942	0.007448	0.006496	0.005250	0.003942	0.002759
1.8	2.292	0.004153	0.004100	0.003820	0.003293	0.002621	0.001932	0.001324
1.9	2.419	0.002059	0.002029	0.001878	0.001599	0.001253	0.000906	0.000608
2.0	2.546	0.000979	0.000963	0.000885	0.000744	0.000573	0.000407	
2.1	2.674	0.000446	0.000439	0.000400	0.000332	0.000251		
2.2	2.801	0.000195	0.000192	0.000173	0.000142			
2.3	2.928	0.000082	0.000080	0.000072				
2.4	3.056	0.000033	0.000032					
2.5	3.183	0.000013						

	$\pi x_1/4$	0.7	0.8	0.9	1.0	1.1	1.2
$\pi x_2/4$	$x_2 \backslash x_1$	0.891	1.019	1.146	1.273	1.401	1.528
0.7	0.891	0.271058					
0.8	1.019	0.203681	0.150614				
0.9	1.146	0.145725	0.105911	0.073107			
1.0	1.273	0.099308	0.070862	0.047965	0.030828		
1.1	1.401	0.064492	0.045139	0.029935	0.018830	0.011248	
1.2	1.528	0.039932	0.027394	0.017784	0.010940	0.006386	0.003540
1.3	1.655	0.023586	0.015847	0.010064	0.006051	0.003449	0.001866
1.4	1.783	0.013295	0.008743	0.005429	0.003188	0.001773	
1.5	1.910	0.007155	0.004603	0.002793	0.001601		
1.6	2.037	0.003678	0.002313	0.001371			
1.7	2.165	0.001807	0.001110				
1.8	2.292	0.00084					
1.9	2.419						

Table A.41.4. $\mathcal{P}_2(x_1, x_2)$. Same as in Table A.41.2, but when the matrix is chosen from the Gaussian unitary ensemble

	πx_1	0.1	0.2	0.3	0.4	0.5	0.6	0.7	0.8
πx_2	$x_2 \backslash x_1$	0.032	0.064	0.095	0.127	0.159	0.191	0.223	0.255
0.1	0.032	0.00							
0.2	0.064	0.001	0.0041						
0.3	0.095	0.003	0.0126	0.0354					
0.4	0.127	0.007	0.0275	0.0712	0.13458				
0.5	0.159	0.012	0.0476	0.1161	0.20865	0.30913			
0.6	0.191	0.018	0.0703	0.1629	0.27969	0.39789	0.49321		
0.7	0.223	0.024	0.0915	0.2026	0.33410	0.45773	0.54773	0.58807	
0.8	0.255	0.028	0.1072	0.2280	0.36212	0.47897	0.55414	0.57610	0.54681
0.9	0.286	0.030	0.1148	0.2351	0.36053	0.46110	0.51663	0.52042	0.47881
1.0	0.318	0.030	0.1134	0.2242	0.33246	0.41188	0.44721	0.43670	0.38987

(*continued on next page*)

Table A.41.4. ($\mathcal{P}_2(x_1, x_2)$. *Continued*)

	πx_1	0.1	0.2	0.3	0.4	0.5	0.6	0.7	0.8
πx_2	$x_2 \backslash x_1$	0.032	0.064	0.095	0.127	0.159	0.191	0.223	0.255
1.1	0.350	0.028	0.1040	0.1989	0.28587	0.34328	0.36137	0.34247	0.29664
1.2	0.382	0.024	0.0891	0.1652	0.23018	0.28806	0.27392	0.25191	0.2117
1.3	0.414	0.020	0.0716	0.1288	0.17412	0.19691	0.19532	0.17436	0.1424
1.4	0.446	0.015	0.0542	0.0946	0.12420	0.13638	0.13136	0.11387	0.0900
1.5	0.477	0.011	0.0386	0.0656	0.08367	0.08926	0.08343	0.07046	
1.6	0.509	0.008	0.0261	0.0430	0.05335	0.05519	0.0506		
1.7	0.541	0.005	0.0166	0.0267	0.03213	0.0329			
1.8	0.573	0.003	0.0101	0.0159	0.0189				
1.9	0.605	0.001	0.0058	0.008					
2.0	0.637	0.00	0.003						

	πx_1	0.9	1.0	1.1
πx_2	$x_2 \backslash x_1$	0.286	0.318	0.350
0.9	0.286	0.40681		
1.0	0.318	0.32136	0.24628	
1.1	0.350	0.2372	0.1770	0.1213
1.2	0.382	0.1647	0.1176	
1.3	0.414	0.1066		

A.42 Sums $a_{jn}^{(\pm)}$ and $a_{jn}^{(2)}$ for $n = 1$, 2 and 3, Section 20.5

To compute $a_{j1}^{(\pm)}$ or $a_{jn}^{(2)}$ for small values of n, we need sums of the form

$$\sum_{t=0}^{j} \binom{2j}{2t} = \sum_{t=0}^{j} \binom{2j}{2t+1} = 2^{2j-1}, \tag{A.42.1}$$

$$\sum_{t_1=0}^{2m} \sum_{t_2=0}^{2j-2m} (-1)^{t_1+t_2} \binom{2j}{t_1} \binom{2j-2m}{t_2} = 0, \tag{A.42.2}$$

$$\begin{aligned}
&\sum_{t_1=0}^{2m} \sum_{t_2=0}^{2j-2m} \frac{(-1)^{t_1+t_2}}{(2m-t_1+t_2+1)(2j-2m+t_1-t_2+1)} \binom{2m}{t_1} \binom{2j-2m}{t_2} \\
&= \sum_{t_1=0}^{2m} \sum_{t_2=0}^{2j-2m} (-1)^{t_1+t_2} \binom{2m}{t_1} \binom{2j-2m}{t_2} \int_0^1 \int_0^1 dx\, dy\, x^{2m-t_1+t_2} y^{2j-2m+t_1-t_2} \\
&= \int_0^1 \int_0^1 dx\, dy\, x^{2m} y^{2j-2m} \left(1 - \frac{y}{x}\right)^{2m} \left(1 - \frac{x}{y}\right)^{2j-2m} \\
&= \int_0^1 \int_0^1 dx\, dy\, (x-y)^{2j} = \sum_{r=0}^{2j} \binom{2j}{r} (-1)^r \frac{1}{(r+1)(2j-r+1)} \\
&= \frac{(2j)!}{(2j+2)!} \sum_{r=0}^{2j} \binom{2j+2}{r+1} (-1)^r = -\frac{2}{(2j+1)(2j+2)},
\end{aligned} \tag{A.42.3}$$

$$\sum_{m=0}^{j} \frac{1}{(2m+1)!(2j-2m+1)!} = \frac{1}{(2j+2)!} \sum_{m=0}^{j} \binom{2j+2}{2m+1} = \frac{2^{2j+1}}{(2j+2)!}, \tag{A.42.4}$$

$$\begin{aligned}
&\sum_{t_1=0}^{m} \sum_{t_2=0}^{j-m} \frac{1}{(2m-2t_1+2t_2+1)(2j-2m+2t_1-2t_2+1)} \binom{2m}{2t_1} \binom{2j-2m}{2t_2} \\
&= \sum_{t_1=0}^{m} \sum_{t_2=0}^{j-m} \binom{2m}{2t_1} \binom{2j-2m}{2t_2} \int_0^1 \int_0^1 dx\, dy\, x^{2m-2t_1+2t_2} y^{2j-2m+2t_1-2t_2} \\
&= \frac{1}{4} \int_0^1 \int_0^1 dx\, dy \left\{(x+y)^{2m} + (x-y)^{2m}\right\}\left\{(x+y)^{2j-2m} + (x-y)^{2j-2m}\right\} \\
&= \frac{1}{4} \int_0^1 \int_0^1 dx\, dy \left\{(x+y)^{2j} + (x-y)^{2j} + (x+y)^{2j} \left(\frac{x-y}{x+y}\right)^{2m} \right. \\
&\left. + (x-y)^{2j} \left(\frac{x+y}{x-y}\right)^{2m} \right\},
\end{aligned} \tag{A.42.5}$$

$$\begin{aligned}
&\sum_{t_1=0}^{m} \sum_{t_2=0}^{j-m} \frac{1}{(2m-2t_1+2t_2+1)(2j-2m+2t_1-2t_2+1)} \binom{2m}{2t_1+1} \binom{2j-2m}{2t_2+1} \\
&= \frac{1}{4} \int_0^1 \int_0^1 dx\, dy \left\{(x+y)^{2j} + (x-y)^{2j} - (x+y)^{2j} \left(\frac{x-y}{x+y}\right)^{2m} \right. \\
&\left. - (x-y)^{2j} \left(\frac{x+y}{x-y}\right)^{2m} \right\},
\end{aligned} \tag{A.42.6}$$

$$\begin{aligned}
\int_0^1 \int_0^1 dx\, dy \left\{(x+y)^{2j} + (x-y)^{2j}\right\} &= \frac{(2j)!.2}{(2j+2)!} \sum_{r=0}^{j} \binom{2j+2}{2r+1} \\
&= \frac{2^{2j+2}}{(2j+1)(2j+2)}.
\end{aligned} \tag{A.42.7}$$

For the triple sum in $a_{j3}^{(2)}$ one needs to evaluate the integral

$$\begin{aligned}
&\int_0^1 \int_0^1 \int_0^1 dx\, dy\, dz \frac{(x-y)^{2j+3} + (y-z)^{2j+3} + (z-x)^{2j+3}}{(x-y)(y-z)(z-x)} \\
&= 3 \int_0^1 \int_0^1 \int_0^1 dx\, dy\, dz\, (x-y)^{2j+1} \left(\frac{1}{x-z} + \frac{1}{z-y}\right)
\end{aligned}$$

$$= 6 \int_0^1 \int_0^1 \int_0^1 dx\, dy\, dz\, (x-y)^{2j+1} \frac{1}{x-z}$$
$$= \frac{6}{2j+2} \int_0^1 dx\, \big((1-x)^{2j+2} - x^{2j+2}\big)\big(\log(1-x) - \log x\big)$$
$$= \frac{6}{j+1} \int_0^1 dx\, \big(x^{2j+2} \log(1-x) - x^{2j+2} \log x\big). \qquad \text{(A.42.8)}$$

The last integral is elementary if one expands $\log(1-x)$ in powers of x.

Using the sums in the above equations one gets then the formulas for $a_{j1}^{(\pm)}$, $a_{j2}^{(\pm)}$, $a_{j2}^{(2)}$ and $a_{j3}^{(2)}$ as announced in Section 20.5.

A.43 Values of $a_{jn}^{(+)}$, $a_{jn}^{(-)}$, and $a_{jn}^{(2)}$, for Low Values of j and n

Table A.43.1 gives their rational values, and Table A.43.2 gives their numerical values. They were calculated with the program MATHEMATICA. (The rational values of $a_{j2}^{(+)}$, $a_{j3}^{(+)}$ for $j \leqslant 11$ agree with the power series of $E_1(0,s)$ kindly supplied by Dietz and Haake.) It is easier to deduce the values of $a_{jn}^{(2)}$ from those of $a_{jn}^{(\pm)}$, Eqs. (20.5.20), (20.5.21) and the relation $E_2(0,s) = E_+(0,s) \cdot E_-(0,s)$ rather than to compute them from Eq. (20.5.28).

Table A.43.1. Rational values of $a_{jn}^{(\pm)}$, $a_{jn}^{(4)}$ and $a_{jn}^{(2)}$

j	$a_{j2}^{(+)}$
2	$1/(2 \cdot 3^4 \cdot 5^2)$
3	$1/(2^2 \cdot 3^3 \cdot 5^2 \cdot 7^2)$
4	$29/(2^5 \cdot 3^5 \cdot 5^4 \cdot 7^2)$
5	$47/(2^6 \cdot 3^6 \cdot 5^3 \cdot 7^2 \cdot 11^2)$
6	$19 \cdot 487/(2^9 \cdot 3^5 \cdot 5^3 \cdot 7^4 \cdot 11^2 \cdot 13^2)$
7	$307/(2^{10} \cdot 3^6 \cdot 5^4 \cdot 7^3 \cdot 11^2 \cdot 13^2)$
8	$1797581/(2^{12} \cdot 3^{12} \cdot 5^4 \cdot 7^3 \cdot 11^2 \cdot 13^2 \cdot 17^2)$
9	$37 \cdot 121501/(2^{13} \cdot 3^{10} \cdot 5^5 \cdot 7^3 \cdot 11^2 \cdot 13^2 \cdot 17^2 \cdot 19^2)$
10	$28943983/(2^{17} \cdot 3^{11} \cdot 5^4 \cdot 7^3 \cdot 11^4 \cdot 13^2 \cdot 17^2 \cdot 19^2)$
11	$37 \cdot 349 \cdot 13309/(2^{18} \cdot 3^{13} \cdot 5^3 \cdot 7^4 \cdot 11^3 \cdot 13^2 \cdot 17^2 \cdot 19^2 \cdot 23^2)$
12	$79 \cdot 6633208213/(2^{20} \cdot 3^{12} \cdot 5^8 \cdot 7^4 \cdot 11^3 \cdot 13^4 \cdot 17^2 \cdot 19^2 \cdot 23^2)$
13	$25601 \cdot 2285167/(2^{21} \cdot 3^{15} \cdot 5^8 \cdot 7^5 \cdot 11^3 \cdot 13^3 \cdot 17^2 \cdot 19^2 \cdot 23^2)$
14	$71 \cdot 6043 \cdot 100870463/(2^{24} \cdot 3^{17} \cdot 5^{10} \cdot 7^4 \cdot 11^3 \cdot 13^3 \cdot 17^2 \cdot 19^2 \cdot 23^2 \cdot 29^2)$
15	$2427085016077/(2^{25} \cdot 3^{16} \cdot 5^9 \cdot 7^3 \cdot 11^3 \cdot 13^3 \cdot 17^2 \cdot 19^2 \cdot 23^2 \cdot 29^2 \cdot 31^2)$
16	$239 \cdot 7547 \cdot 990991181/(2^{27} \cdot 3^{17} \cdot 5^9 \cdot 7^4 \cdot 11^4 \cdot 13^3 \cdot 17^4 \cdot 19^2 \cdot 23^2 \cdot 29^2 \cdot 31^2)$
17	$1156813282236901/(2^{28} \cdot 3^{19} \cdot 5^{10} \cdot 7^6 \cdot 11^4 \cdot 13^3 \cdot 17^3 \cdot 19^2 \cdot 23^2 \cdot 29^2 \cdot 31^2)$
18	$179 \cdot 1493 \cdot 17242537150547/(2^{33} \cdot 3^{17} \cdot 5^{10} \cdot 7^6 \cdot 11^4 \cdot 13^3 \cdot 17^3 \cdot 19^4 \cdot 23^2 \cdot 29^2 \cdot 31^2 \cdot 37^2)$
19	$300929 \cdot 3250687115171/(2^{34} \cdot 3^{18} \cdot 5^{12} \cdot 7^6 \cdot 11^4 \cdot 13^4 \cdot 17^3 \cdot 19^3 \cdot 23^2 \cdot 29^2 \cdot 31^2 \cdot 37^2)$
20	$71 \cdot 109229 \cdot 1465837 \cdot 290070607/(2^{36} \cdot 3^{21} \cdot 5^{11} \cdot 7^8 \cdot 11^4 \cdot 13^4 \cdot 17^3 \cdot 19^3 \cdot 23^2 \cdot 29^2 \cdot 31^2 \cdot 37^2 \cdot 41^2)$
21	$11124669599689437865907/(2^{37} \cdot 3^{20} \cdot 5^{11} \cdot 7^7 \cdot 11^5 \cdot 13^4 \cdot 17^3 \cdot 19^3 \cdot 23^2 \cdot 29^2 \cdot 31^2 \cdot 37^2 \cdot 41^2 \cdot 43^2)$

(continued on next page)

Table A.43.1. (*Continued*)

j	$a_{j2}^{(-)}$
4	$1/(2^5 \cdot 3^4 \cdot 5^4 \cdot 7^2)$
5	$1/(2^6 \cdot 3^8 \cdot 5^3 \cdot 7^2)$
6	$251/(2^9 \cdot 3^7 \cdot 5^3 \cdot 7^4 \cdot 11^2)$
7	$1973/(2^{10} \cdot 3^8 \cdot 5^4 \cdot 7^3 \cdot 11^2 \cdot 13^2)$
8	$1777/(2^{12} \cdot 3^{11} \cdot 5^4 \cdot 7^3 \cdot 11^2 \cdot 13^2)$
9	$19441/(2^{13} \cdot 3^9 \cdot 5^6 \cdot 7^3 \cdot 11^2 \cdot 13^2 \cdot 17^2)$
10	$328996081/(2^{17} \cdot 3^{10} \cdot 5^5 \cdot 7^4 \cdot 11^4 \cdot 13^2 \cdot 17^2 \cdot 19^2)$
11	$2680697/(2^{18} \cdot 3^{12} \cdot 5^5 \cdot 7^4 \cdot 11^3 \cdot 13^2 \cdot 17^2 \cdot 19^2)$
12	$6961905647/(2^{20} \cdot 3^{11} \cdot 5^6 \cdot 7^4 \cdot 11^3 \cdot 13^4 \cdot 17^2 \cdot 19^2 \cdot 23^2)$
13	$2617 \cdot 10419041/(2^{21} \cdot 3^{14} \cdot 5^7 \cdot 7^6 \cdot 11^3 \cdot 13^3 \cdot 17^2 \cdot 19^2 \cdot 23^2)$
14	$37 \cdot 467 \cdot 2500639/(2^{24} \cdot 3^{18} \cdot 5^8 \cdot 7^5 \cdot 11^3 \cdot 13^3 \cdot 17^2 \cdot 19^2 \cdot 23^2)$
15	$14848265777/(2^{25} \cdot 3^{17} \cdot 5^7 \cdot 7^5 \cdot 11^3 \cdot 13^3 \cdot 17^2 \cdot 19^2 \cdot 23^2 \cdot 29^2)$
16	$4019 \cdot 83813 \cdot 4457293/(2^{27} \cdot 3^{18} \cdot 5^7 \cdot 7^5 \cdot 11^4 \cdot 13^3 \cdot 17^4 \cdot 19^2 \cdot 23^2 \cdot 29^2 \cdot 31^2)$
17	$1229 \cdot 5082746465497/(2^{28} \cdot 3^{22} \cdot 5^9 \cdot 7^6 \cdot 11^4 \cdot 13^3 \cdot 17^3 \cdot 19^2 \cdot 23^2 \cdot 29^2 \cdot 31^2)$
18	$71 \cdot 181 \cdot 2647 \cdot 6983 \cdot 76519/(2^{33} \cdot 3^{20} \cdot 5^9 \cdot 7^6 \cdot 11^4 \cdot 13^3 \cdot 17^3 \cdot 19^4 \cdot 23^2 \cdot 29^2 \cdot 31^2)$
19	$1056482481642539687/(2^{34} \cdot 3^{21} \cdot 5^{10} \cdot 7^6 \cdot 11^4 \cdot 13^4 \cdot 17^3 \cdot 19^3 \cdot 23^2 \cdot 29^2 \cdot 31^2 \cdot 37^2)$
20	$719 \cdot 15110472455873/(2^{36} \cdot 3^{23} \cdot 5^8 \cdot 7^8 \cdot 11^4 \cdot 13^3 \cdot 17^3 \cdot 19^3 \cdot 23^2 \cdot 29^2 \cdot 31^2 \cdot 37^2)$
21	$163 \cdot 829 \cdot 13563127538117/(2^{37} \cdot 3^{22} \cdot 5^9 \cdot 7^7 \cdot 11^6 \cdot 13^3 \cdot 17^3 \cdot 19^3 \cdot 23^2 \cdot 29^2 \cdot 31^2 \cdot 37^2 \cdot 41^2)$
j	$a_{j2}^{(4)}$
2	$2^4/(3^4 \cdot 5^2)$
3	$2^5/(3^3 \cdot 5^2 \cdot 7^2)$
4	$2^9/(3^5 \cdot 5^4 \cdot 7^2)$
5	$2^{10} \cdot 17/(3^8 \cdot 5^3 \cdot 7^2 \cdot 11^2)$
6	$2^{12} \cdot 491/(3^7 \cdot 5^3 \cdot 7^4 \cdot 11^2 \cdot 13^2)$
7	$2^{12} \cdot 37/(3^8 \cdot 5^4 \cdot 7^3 \cdot 11^2 \cdot 13^2)$
8	$2^{17} \cdot 163/(3^{12} \cdot 5^3 \cdot 7^3 \cdot 11^2 \cdot 13^2 \cdot 17^2)$
9	$2^{20} \cdot 2657/(3^{10} \cdot 5^6 \cdot 7^3 \cdot 11^2 \cdot 13^2 \cdot 17^2 \cdot 19^2)$
10	$2^{21} \cdot 15259/(3^{11} \cdot 5^5 \cdot 7^4 \cdot 11^4 \cdot 13^2 \cdot 17^2 \cdot 19^2)$
11	$2^{22} \cdot 89 \cdot 733/(3^{13} \cdot 5^5 \cdot 7^4 \cdot 11^3 \cdot 13^2 \cdot 17^2 \cdot 19^2 \cdot 23^2)$
12	$2^{26} \cdot 223 \cdot 2237/(3^{12} \cdot 5^8 \cdot 7^4 \cdot 11^3 \cdot 13^4 \cdot 17^2 \cdot 19^2 \cdot 23^2)$
13	$2^{26} \cdot 41 \cdot 79 \cdot 241/(3^{15} \cdot 5^8 \cdot 7^6 \cdot 11^3 \cdot 13^3 \cdot 17^2 \cdot 19^2 \cdot 23^2)$
14	$2^{28} \cdot 641 \cdot 337973/(3^{18} \cdot 5^{10} \cdot 7^5 \cdot 11^3 \cdot 13^3 \cdot 17^2 \cdot 19^2 \cdot 23^2 \cdot 29^2)$
15	$2^{27} \cdot 340228597/(3^{17} \cdot 5^9 \cdot 7^5 \cdot 11^3 \cdot 13^3 \cdot 17^2 \cdot 19^2 \cdot 23^2 \cdot 29^2 \cdot 31^2)$
16	$2^{33} \cdot 71 \cdot 73 \cdot 311 \cdot 347/(3^{18} \cdot 5^9 \cdot 7^5 \cdot 11^4 \cdot 13^3 \cdot 17^4 \cdot 19^2 \cdot 23^2 \cdot 29^2 \cdot 31^2)$
17	$2^{37} \cdot 109 \cdot 533737/(3^{22} \cdot 5^{10} \cdot 7^6 \cdot 11^4 \cdot 13^3 \cdot 17^3 \cdot 19^2 \cdot 23^2 \cdot 29^2 \cdot 31^2)$
18	$2^{37} \cdot 22721 \cdot 1274939/(3^{20} \cdot 5^{10} \cdot 7^6 \cdot 11^4 \cdot 13^3 \cdot 17^3 \cdot 19^4 \cdot 23^2 \cdot 29^2 \cdot 31^2 \cdot 37^2)$
19	$2^{38} \cdot 6149542129/(3^{21} \cdot 5^{12} \cdot 7^6 \cdot 11^4 \cdot 13^4 \cdot 17^3 \cdot 19^3 \cdot 23^2 \cdot 29^2 \cdot 31^2 \cdot 37^2)$
20	$2^{42} \cdot 431865262057/(3^{23} \cdot 5^{11} \cdot 7^8 \cdot 11^4 \cdot 13^4 \cdot 17^3 \cdot 19^3 \cdot 23^2 \cdot 29^2 \cdot 31^2 \cdot 37^2 \cdot 41^2)$
21	$2^{43} \cdot 180547 \cdot 88767127/(3^{22} \cdot 5^{11} \cdot 7^7 \cdot 11^6 \cdot 13^4 \cdot 17^3 \cdot 19^3 \cdot 23^2 \cdot 29^2 \cdot 31^2 \cdot 37^2 \cdot 41^2 \cdot 43^2)$
j	$a_{j3}^{(+)}$
6	$-1/(2^3 \cdot 3^9 \cdot 5^6 \cdot 7^4)$
7	$-1/(2^5 \cdot 3^8 \cdot 5^5 \cdot 7^4 \cdot 11^2)$
8	$-1/(2^2 \cdot 3^9 \cdot 5^5 \cdot 7^4 \cdot 11^2 \cdot 13^2)$
9	$-9059/(2^9 \cdot 3^{12} \cdot 5^6 \cdot 7^6 \cdot 11^2 \cdot 13^2)$

(*continued on next page*)

Table A.43.1. (*Continued*)

10	$-492083/(2^{11} \cdot 3^{11} \cdot 5^6 \cdot 7^5 \cdot 11^4 \cdot 13^2 \cdot 17^2)$
11	$-61 \cdot 5789957/(2^{13} \cdot 3^{12} \cdot 5^8 \cdot 7^5 \cdot 11^4 \cdot 13^2 \cdot 17^2 \cdot 19^2)$
12	$-2029 \cdot 1604263/(2^{13} \cdot 3^{17} \cdot 5^6 \cdot 7^5 \cdot 11^4 \cdot 13^4 \cdot 17^2 \cdot 19^2)$
13	$-43 \cdot 4241 \cdot 95773/(2^{16} \cdot 3^{14} \cdot 5^6 \cdot 7^6 \cdot 11^4 \cdot 13^4 \cdot 17^2 \cdot 19^2 \cdot 23^2)$
14	$-191 \cdot 5981412649/(2^{19} \cdot 3^{15} \cdot 5^{10} \cdot 7^6 \cdot 11^4 \cdot 13^4 \cdot 17^2 \cdot 19^2 \cdot 23^2)$
15	$-347 \cdot 26915039197/(2^{21} \cdot 3^{19} \cdot 5^{10} \cdot 7^6 \cdot 11^6 \cdot 13^4 \cdot 17^2 \cdot 19 \cdot 23^2)$
16	$-53 \cdot 55215431280941759/(2^{20} \cdot 3^{18} \cdot 5^{12} \cdot 7^8 \cdot 11^5 \cdot 13^4 \cdot 17^4 \cdot 19^2 \cdot 23^2 \cdot 29^2)$
17	$-109 \cdot 362458560948299393/(2^{24} \cdot 3^{20} \cdot 5^{11} \cdot 7^7 \cdot 11^5 \cdot 13^4 \cdot 17^4 \cdot 19^2 \cdot 23^2 \cdot 29^2 \cdot 31^2)$
18	$-47 \cdot 79 \cdot 263 \cdot 347 \cdot 173347 \cdot 18207797/(2^{26} \cdot 3^{20} \cdot 5^{11} \cdot 7^7 \cdot 11^5 \cdot 13^6 \cdot 17^4 \cdot 19^4 \cdot 23^2 \cdot 29^1 \cdot 31^2)$
19	$-53 \cdot 277 \cdot 47807 \cdot 69035017327/(2^{28} \cdot 3^{19} \cdot 5^{12} \cdot 7^7 \cdot 11^5 \cdot 13^5 \cdot 17^4 \cdot 19^4 \cdot 23^2 \cdot 29^2 \cdot 31^2)$
20	$-575236932889423567981/(2^{28} \cdot 3^{20} \cdot 5^{12} \cdot 7^7 \cdot 11^5 \cdot 13^5 \cdot 17^4 \cdot 19^4 \cdot 23^2 \cdot 29^2 \cdot 31^2 \cdot 37^2)$
j	$a_{j3}^{(-)}$
9	$1/(2^9 \cdot 3^{13} \cdot 5^6 \cdot 7^6 \cdot 11^2)$
10	$1/(2^{11} \cdot 3^{12} \cdot 5^6 \cdot 7^5 \cdot 11^2 \cdot 13^2)$
11	$571/(2^{13} \cdot 3^{13} \cdot 5^8 \cdot 7^5 \cdot 11^4 \cdot 13^2)$
12	$227/(2^{13} \cdot 3^{17} \cdot 5^4 \cdot 7^5 \cdot 11^4 \cdot 13^2 \cdot 17^2)$
13	$859 \cdot 6733/(2^{16} \cdot 3^{14} \cdot 5^6 \cdot 7^6 \cdot 11^4 \cdot 13^4 \cdot 17^2 \cdot 19^2)$
14	$22614847/(2^{19} \cdot 3^{15} \cdot 5^8 \cdot 7^6 \cdot 11^4 \cdot 13^4 \cdot 17^2 \cdot 19^2)$
15	$286173859133/(2^{21} \cdot 3^{17} \cdot 5^8 \cdot 7^6 \cdot 11^6 \cdot 13^4 \cdot 17^2 \cdot 19^2 \cdot 23^2)$
16	$31 \cdot 67619 \cdot 301463/(2^{20} \cdot 3^{16} \cdot 5^{12} \cdot 7^8 \cdot 11^5 \cdot 13^4 \cdot 17^2 \cdot 19^2 \cdot 23^2)$
17	$610429 \cdot 5180897/(2^{24} \cdot 3^{18} \cdot 5^{11} \cdot 7^7 \cdot 11^5 \cdot 13^4 \cdot 17^4 \cdot 19^2 \cdot 23^2)$
18	$183536658137955389/(2^{26} \cdot 3^{21} \cdot 5^{11} \cdot 7^7 \cdot 11^5 \cdot 13^6 \cdot 17^4 \cdot 19^2 \cdot 23^2 \cdot 29^2)$
19	$43 \cdot 2203 \cdot 53653 \cdot 22079891863/(2^{28} \cdot 3^{20} \cdot 5^{12} \cdot 7^7 \cdot 11^5 \cdot 13^5 \cdot 17^4 \cdot 19^4 \cdot 23^2 \cdot 29^2 \cdot 31^2)$
20	$7431467315016491353/(2^{28} \cdot 3^{21} \cdot 5^{12} \cdot 7^8 \cdot 11^5 \cdot 13^5 \cdot 17^4 \cdot 19^4 \cdot 23^2 \cdot 29^2 \cdot 31^2)$
j	$a_{j3}^{(4)}$
6	$-2^{11}/(3^9 \cdot 5^6 \cdot 7^4)$
7	$-2^{11}/(3^8 \cdot 5^5 \cdot 7^4 \cdot 11^2)$
8	$-2^{16}/(3^9 \cdot 5^5 \cdot 7^4 \cdot 11^2 \cdot 13^2)$
9	$-2^{18} \cdot 211/(3^{13} \cdot 5^6 \cdot 7^6 \cdot 11^2 \cdot 13^2)$
10	$-2^{20} \cdot 563/(3^{12} \cdot 5^5 \cdot 7^5 \cdot 11^4 \cdot 13^2 \cdot 17^2)$
11	$-2^{21} \cdot 976553/(3^{13} \cdot 5^8 \cdot 7^5 \cdot 11^4 \cdot 13^2 \cdot 17^2 \cdot 19^2)$
12	$-2^{26} \cdot 61 \cdot 5821/(3^{17} \cdot 5^6 \cdot 7^5 \cdot 11^4 \cdot 13^4 \cdot 17^2 \cdot 19^2)$
13	$-2^{25} \cdot 251219/(3^{14} \cdot 5^6 \cdot 7^5 \cdot 11^4 \cdot 13^4 \cdot 17^2 \cdot 19^2 \cdot 23^2)$
14	$-2^{26} \cdot 25737563/(3^{15} \cdot 5^{10} \cdot 7^6 \cdot 11^4 \cdot 13^4 \cdot 17^2 \cdot 19^2 \cdot 23^2)$
15	$-2^{25} \cdot 193 \cdot 971 \cdot 1601/(3^{19} \cdot 5^{10} \cdot 7^6 \cdot 11^6 \cdot 13^4 \cdot 17^2 \cdot 19^2 \cdot 23)$
16	$-2^{31} \cdot 24953 \cdot 94423283/(3^{18} \cdot 5^{11} \cdot 7^8 \cdot 11^5 \cdot 13^4 \cdot 17^4 \cdot 19^2 \cdot 23^2 \cdot 29^2)$
17	$-2^{33} \cdot 612511 \cdot 2569667/(3^{20} \cdot 5^{10} \cdot 7^7 \cdot 11^5 \cdot 13^4 \cdot 17^4 \cdot 19^2 \cdot 23^2 \cdot 29^2 \cdot 31^2)$
18	$-2^{35} \cdot 4231 \cdot 33105241523/(3^{21} \cdot 5^9 \cdot 7^7 \cdot 11^5 \cdot 13^6 \cdot 17^4 \cdot 19^4 \cdot 23^2 \cdot 29^2 \cdot 31^2)$
19	$-2^{36} \cdot 5233 \cdot 377431757/(3^{20} \cdot 5^{12} \cdot 7^7 \cdot 11^5 \cdot 13^5 \cdot 17^4 \cdot 19^4 \cdot 23^2 \cdot 29^2 \cdot 31^2)$
20	$-2^{40} \cdot 1187 \cdot 5441 \cdot 4398263/(3^{21} \cdot 5^{12} \cdot 7^8 \cdot 11^5 \cdot 13^5 \cdot 17^4 \cdot 19^4 \cdot 23^2 \cdot 29^2 \cdot 31^2 \cdot 37^2)$
j	$a_{j4}^{(+)}$
12	$1/(2^5 \cdot 3^{17} \cdot 5^8 \cdot 7^8 \cdot 11^4 \cdot 13^2)$
13	$1/(2^5 \cdot 3^{19} \cdot 5^{10} \cdot 7^7 \cdot 11^4 \cdot 13^2)$
14	$409/(2^8 \cdot 3^{18} \cdot 5^9 \cdot 7^7 \cdot 11^4 \cdot 13^4 \cdot 17^2)$

(*continued on next page*)

Table A.43.1. (*Continued*)

15	$41 \cdot 863/(2^4 \cdot 3^{19} \cdot 5^9 \cdot 7^7 \cdot 11^6 \cdot 13^4 \cdot 17^2 \cdot 19^2)$
16	$173 \cdot 2372659/(2^{13} \cdot 3^{23} \cdot 5^{10} \cdot 7^8 \cdot 11^6 \cdot 13^4 \cdot 17^2 \cdot 19^2)$
17	$155592071/(2^{13} \cdot 3^{21} \cdot 5^{10} \cdot 7^8 \cdot 11^6 \cdot 13^4 \cdot 17^2 \cdot 19^2 \cdot 23^2)$
18	$275911 \cdot 2239105373/(2^{16} \cdot 3^{22} \cdot 5^{14} \cdot 7^8 \cdot 11^6 \cdot 13^6 \cdot 17^4 \cdot 19^2 \cdot 23^2)$
19	$937 \cdot 408610135097/(2^{15} \cdot 3^{25} \cdot 5^{13} \cdot 7^{10} \cdot 11^6 \cdot 13^6 \cdot 17^4 \cdot 19^2 \cdot 23^2)$
20	$863 \cdot 205262318558473373/(2^{21} \cdot 3^{24} \cdot 5^{13} \cdot 7^9 \cdot 11^8 \cdot 13^6 \cdot 17^4 \cdot 19^4 \cdot 23^2 \cdot 29^2)$
j	$a_{j4}^{(-)}$
16	$1/(2^{13} \cdot 3^{23} \cdot 5^{10} \cdot 7^8 \cdot 11^6 \cdot 13^4)$
17	$1/(2^{13} \cdot 3^{21} \cdot 5^{10} \cdot 7^8 \cdot 11^6 \cdot 13^4 \cdot 17^2)$
18	$43 \cdot 79/(2^{16} \cdot 3^{22} \cdot 5^{12} \cdot 7^8 \cdot 11^6 \cdot 13^4 \cdot 17^2 \cdot 19^2)$
19	$127 \cdot 2099/(2^{15} \cdot 3^{24} \cdot 5^{11} \cdot 7^{10} \cdot 11^6 \cdot 13^6 \cdot 17^2 \cdot 19^2)$
20	$13037 \cdot 9166019/(2^{21} \cdot 3^{23} \cdot 5^{11} \cdot 7^9 \cdot 11^8 \cdot 13^6 \cdot 17^4 \cdot 19^2 \cdot 23^2)$
j	$a_{j4}^{(4)}$
12	$2^{22}/(3^{17} \cdot 5^8 \cdot 7^8 \cdot 11^4 \cdot 13^2)$
13	$2^{24}/(3^{19} \cdot 5^{10} \cdot 7^7 \cdot 11^4 \cdot 13^2)$
14	$2^{23} \cdot 409/(3^{18} \cdot 5^9 \cdot 7^7 \cdot 11^4 \cdot 13^4 \cdot 17^2)$
15	$2^{29} \cdot 41 \cdot 863/(3^{19} \cdot 5^9 \cdot 7^7 \cdot 11^6 \cdot 13^4 \cdot 17^2 \cdot 19^2)$
16	$2^{31} \cdot 267301/(3^{22} \cdot 5^{10} \cdot 7^8 \cdot 11^6 \cdot 13^4 \cdot 17^2 \cdot 19^2)$
17	$2^{31} \cdot 67 \cdot 173/(3^{20} \cdot 5^9 \cdot 7^7 \cdot 11^6 \cdot 13^4 \cdot 17^2 \cdot 19^2 \cdot 23^2)$
18	$2^{32} \cdot 1361 \cdot 42367799/(3^{21} \cdot 5^{14} \cdot 7^7 \cdot 11^6 \cdot 13^6 \cdot 17^4 \cdot 19^2 \cdot 23^2)$
19	$2^{36} \cdot 173 \cdot 2178492157/(3^{25} \cdot 5^{13} \cdot 7^{10} \cdot 11^6 \cdot 13^6 \cdot 17^4 \cdot 19^2 \cdot 23^2)$
20	$2^{37} \cdot 541 \cdot 10145957816573/(3^{24} \cdot 5^{13} \cdot 7^9 \cdot 11^8 \cdot 13^6 \cdot 17^4 \cdot 19^4 \cdot 23^2 \cdot 29^2)$
j	$a_{j3}^{(2)}$
3	$1/(2^3 \cdot 3^5 \cdot 5^2)$
4	$1/(3^3 \cdot 5^4 \cdot 7^2)$
5	$1/(2 \cdot 3^7 \cdot 5^3 \cdot 7^2)$
6	$2 \cdot 2293/(3^8 \cdot 5^5 \cdot 7^4 \cdot 11^2)$
7	$3581/(2^5 \cdot 3^7 \cdot 5^4 \cdot 7^3 \cdot 11^2 \cdot 13^2)$
8	$71/(2 \cdot 3^{10} \cdot 5^4 \cdot 7^3 \cdot 11^2 \cdot 13^2)$
9	$94789/(2^2 \cdot 3^{10} \cdot 5^5 \cdot 7^5 \cdot 11^2 \cdot 13^2 \cdot 17^2)$
10	$31 \cdot 2467/(3^9 \cdot 5^4 \cdot 7^4 \cdot 11^4 \cdot 13^2 \cdot 17^2 \cdot 19^2)$
11	$407221/(2^3 \cdot 3^{11} \cdot 5^7 \cdot 7^4 \cdot 11^3 \cdot 13^2 \cdot 17^2 \cdot 19^2)$
12	$2477 \cdot 99017/(3^{16} \cdot 5^5 \cdot 7^4 \cdot 11^3 \cdot 13^4 \cdot 17^2 \cdot 19^2 \cdot 23^2)$
13	$40956413/(2 \cdot 3^{13} \cdot 5^8 \cdot 7^6 \cdot 11^3 \cdot 13^3 \cdot 17^2 \cdot 19^2 \cdot 23^2)$
14	$2 \cdot 131 \cdot 197 \cdot 479/(3^{17} \cdot 5^{10} \cdot 7^5 \cdot 11^3 \cdot 13^3 \cdot 17^2 \cdot 19^2 \cdot 23^2)$
15	$89 \cdot 3617 \cdot 2204977/(2^7 \cdot 3^{18} \cdot 5^9 \cdot 7^5 \cdot 11^5 \cdot 13^3 \cdot 17^2 \cdot 19^2 \cdot 23^2 \cdot 29^2)$
16	$59 \cdot 557 \cdot 22682791447/(2^2 \cdot 3^{17} \cdot 5^{11} \cdot 7^7 \cdot 11^4 \cdot 13^3 \cdot 17^4 \cdot 19^2 \cdot 23^2 \cdot 29^2 \cdot 31^2)$
17	$53 \cdot 92174813/(2^3 \cdot 3^{21} \cdot 5^{10} \cdot 7^6 \cdot 11^2 \cdot 13^3 \cdot 17^3 \cdot 19^2 \cdot 23^2 \cdot 29^2 \cdot 31^2)$
18	$259219 \cdot 8279597/(2 \cdot 3^{21} \cdot 5^{10} \cdot 7^6 \cdot 11^3 \cdot 13^5 \cdot 17^3 \cdot 19^4 \cdot 23^2 \cdot 29^2 \cdot 31^2)$
19	$1877321357027/(2^4 \cdot 3^{20} \cdot 5^{12} \cdot 7^6 \cdot 11^3 \cdot 13^4 \cdot 17^3 \cdot 19^3 \cdot 23^2 \cdot 29^2 \cdot 31^2 \cdot 37^2)$
20	$3767 \cdot 1688147/(2^2 \cdot 3^{23} \cdot 5^{11} \cdot 7^7 \cdot 11^3 \cdot 13^4 \cdot 17^3 \cdot 19^3 \cdot 23^2 \cdot 29^2 \cdot 31^2 \cdot 37^2)$
j	$a_{j4}^{(2)}$
6	$1/(2^5 \cdot 3^7 \cdot 5^6 \cdot 7^2)$
7	$1/(2 \cdot 3^{10} \cdot 5^5 \cdot 7^4)$

(*continued on next page*)

Table A.43.1. (*Continued*)

8	$83/(2^4 \cdot 3^8 \cdot 5^7 \cdot 7^4 \cdot 11^2)$
9	$174931/(2^2 \cdot 3^{13} \cdot 5^6 \cdot 7^6 \cdot 11^2 \cdot 13^2)$
10	$17 \cdot 113/(2^5 \cdot 3^{14} \cdot 5^5 \cdot 7^5 \cdot 11^2 \cdot 13^2)$
11	$4585051/(2^2 \cdot 3^{13} \cdot 5^8 \cdot 7^5 \cdot 11^4 \cdot 13^2 \cdot 17^2)$
12	$257 \cdot 22926661/(2^3 \cdot 3^{17} \cdot 5^6 \cdot 7^7 \cdot 11^4 \cdot 13^2 \cdot 17^2 \cdot 19^2)$
13	$83 \cdot 1109 \cdot 6367/(2 \cdot 3^{16} \cdot 5^7 \cdot 7^6 \cdot 11^4 \cdot 13^4 \cdot 17^2 \cdot 19^2)$
14	$4157 \cdot 5665783/(2^8 \cdot 3^{15} \cdot 5^6 \cdot 7^6 \cdot 11^4 \cdot 13^4 \cdot 17^2 \cdot 19^2 \cdot 23^2)$
15	$131 \cdot 307 \cdot 397 \cdot 254929/(2^3 \cdot 3^{17} \cdot 5^{10} \cdot 7^6 \cdot 11^6 \cdot 13^4 \cdot 17^2 \cdot 19^2 \cdot 23^2)$
16	$67 \cdot 21341 \cdot 512061481/(2^6 \cdot 3^{20} \cdot 5^{12} \cdot 7^8 \cdot 11^5 \cdot 13^4 \cdot 17^2 \cdot 19^2 \cdot 23^2)$
17	$1609 \cdot 7687 \cdot 79003289/(2^4 \cdot 3^{18} \cdot 5^{11} \cdot 7^7 \cdot 11^5 \cdot 13^4 \cdot 17^4 \cdot 19^2 \cdot 23^2 \cdot 29^2)$
18	$1651095213293251654 2649/(2^7 \cdot 3^{22} \cdot 5^{13} \cdot 7^7 \cdot 11^5 \cdot 13^6 \cdot 17^4 \cdot 19^2 \cdot 23^2 \cdot 29^2 \cdot 31^2)$
19	$39863 \cdot 6145849 \cdot 41097795887/(2^4 \cdot 3^{23} \cdot 5^{12} \cdot 7^9 \cdot 11^5 \cdot 13^5 \cdot 17^4 \cdot 19^4 \cdot 23^2 \cdot 29^2 \cdot 31^2)$
20	$89 \cdot 53693 \cdot 135257537428817/(2^8 \cdot 3^{23} \cdot 5^{12} \cdot 7^8 \cdot 11^7 \cdot 13^5 \cdot 17^4 \cdot 19^4 \cdot 23^2 \cdot 29^2 \cdot 31^2)$

j	$a_{j5}^{(2)}$
10	$-1/(2^7 \cdot 3^{13} \cdot 5^9 \cdot 7^6)$
11	$-1/(2^4 \cdot 3^{16} \cdot 5^7 \cdot 7^6 \cdot 11^2)$
12	$-19/(2^5 \cdot 3^{13} \cdot 5^7 \cdot 7^8 \cdot 11^2 \cdot 13^2)$
13	$-31247/(2^3 \cdot 3^{19} \cdot 5^9 \cdot 7^7 \cdot 11^4 \cdot 13^2)$
14	$-751 \cdot 7309/(2^9 \cdot 3^{20} \cdot 5^8 \cdot 7^7 \cdot 11^4 \cdot 13^2 \cdot 17^2)$
15	$-54664654529/(2^4 \cdot 3^{19} \cdot 5^{10} \cdot 7^9 \cdot 11^4 \cdot 13^4 \cdot 17^2 \cdot 19^2)$
16	$-79 \cdot 197 \cdot 1511 \cdot 14009/(2^5 \cdot 3^{23} \cdot 5^9 \cdot 7^8 \cdot 11^6 \cdot 13^4 \cdot 17^2 \cdot 19^2)$
17	$-103 \cdot 199 \cdot 1523 \cdot 12979/(2^3 \cdot 3^{23} \cdot 5^9 \cdot 7^8 \cdot 11^6 \cdot 13^4 \cdot 17^2 \cdot 19^2 \cdot 23^2)$
18	$-2603843378303/(2^9 \cdot 3^{22} \cdot 5^{13} \cdot 7^8 \cdot 11^6 \cdot 13^4 \cdot 17 \cdot 19^2 \cdot 23^2)$
19	$-4139 \cdot 6024648954283/(2^6 \cdot 3^{24} \cdot 5^{12} \cdot 7^9 \cdot 11^6 \cdot 13^6 \cdot 17^4 \cdot 19^2 \cdot 23^2)$

j	$a_{j6}^{(2)}$
15	$1/(2^{10} \cdot 3^{22} \cdot 5^{11} \cdot 7^{10} \cdot 11^2)$
16	$1/(2^8 \cdot 3^{20} \cdot 5^{10} \cdot 7^9 \cdot 11^4 \cdot 13^2)$
17	$503/(2^{11} \cdot 3^{24} \cdot 5^{12} \cdot 7^9 \cdot 11^4 \cdot 13^2)$
18	$227839259/(2^8 \cdot 3^{25} \cdot 5^{11} \cdot 7^{11} \cdot 11^6 \cdot 13^4 \cdot 17^2)$
19	$123429487/(2^{13} \cdot 3^{26} \cdot 5^{12} \cdot 7^{10} \cdot 11^6 \cdot 13^4 \cdot 17^2 \cdot 19^2)$

Table A.43.2. Numerical Values of $a_{jn}^{(\pm)}$ and $a_{jn}^{(4)}$

n	j	$a_{jn}^{(+)}$	$a_{jn}^{(-)}$	$a_{jn}^{(4)}$
2	2	$2.469135802 \times 10^{-4}$	0.	$7.901234568 \times 10^{-3}$
2	3	$7.558578987 \times 10^{-6}$	0.	$9.674981104 \times 10^{-4}$
2	4	$1.217771059 \times 10^{-7}$	$1.259763165 \times 10^{-8}$	$6.879986563 \times 10^{-5}$
2	5	$1.359248593 \times 10^{-9}$	$3.888157915 \times 10^{-10}$	$3.580035861 \times 10^{-6}$
2	6	$1.211803157 \times 10^{-11}$	$6.172590045 \times 10^{-12}$	$1.498367723 \times 10^{-7}$
2	7	$9.381335727 \times 10^{-14}$	$6.699013894 \times 10^{-14}$	$5.269208964 \times 10^{-9}$
2	8	$6.518225927 \times 10^{-16}$	$5.586598567 \times 10^{-16}$	$1.586603556 \times 10^{-10}$
2	9	$4.064038601 \times 10^{-18}$	$3.806740744 \times 10^{-18}$	$4.126555161 \times 10^{-12}$
2	10	$2.252568149 \times 10^{-20}$	$2.194641461 \times 10^{-20}$	$9.326474527 \times 10^{-14}$
2	11	$1.103643203 \times 10^{-22}$	$1.092800194 \times 10^{-22}$	$1.842510265 \times 10^{-15}$
2	12	$4.778937767 \times 10^{-25}$	$4.761787936 \times 10^{-25}$	$3.201336318 \times 10^{-17}$

(*continued on next page*)

Table A.43.2. (*Continued*)

n	j	$a_{jn}^{(+)}$	$a_{jn}^{(-)}$	$a_{jn}^{(4)}$
2	13	$1.834877519 \times 10^{-27}$	$1.832554482 \times 10^{-27}$	$4.922343908 \times 10^{-19}$
2	14	$6.276777048 \times 10^{-30}$	$6.274051223 \times 10^{-30}$	$6.73817462 \times 10^{-21}$
2	15	$1.923014905 \times 10^{-32}$	$1.922735104 \times 10^{-32}$	$8.258685258 \times 10^{-23}$
2	16	$5.303589326 \times 10^{-35}$	$5.303335921 \times 10^{-35}$	$9.111279409 \times 10^{-25}$
2	17	$1.323126737 \times 10^{-37}$	$1.323106336 \times 10^{-37}$	$9.092387607 \times 10^{-27}$
2	18	$2.999397805 \times 10^{-40}$	$2.999383108 \times 10^{-40}$	$8.244661708 \times 10^{-29}$
2	19	$6.204092462 \times 10^{-43}$	$6.204082931 \times 10^{-43}$	$6.821466562 \times 10^{-31}$
2	20	$1.175458268 \times 10^{-45}$	$1.175457709 \times 10^{-45}$	$5.169718905 \times 10^{-33}$
2	21	$2.047237978 \times 10^{-48}$	$2.047237679 \times 10^{-48}$	$3.601538875 \times 10^{-35}$
3	6	$-1.692803446 \times 10^{-13}$	0.	$-2.773489166 \times 10^{-9}$
3	7	$-5.246291672 \times 10^{-15}$	0.	$-3.43820971 \times 10^{-10}$
3	8	$-8.278172263 \times 10^{-17}$	0.	$-2.17007319 \times 10^{-11}$
3	9	$-8.856760503 \times 10^{-19}$	$5.507570832 \times 10^{-21}$	$-9.229235435 \times 10^{-13}$
3	10	$-7.222858276 \times 10^{-21}$	$1.710931767 \times 10^{-22}$	$-2.957724656 \times 10^{-14}$
3	11	$-4.786823875 \times 10^{-23}$	$2.691300383 \times 10^{-24}$	$-7.579432532 \times 10^{-16}$
3	12	$-2.685634577 \times 10^{-25}$	$2.856597672 \times 10^{-26}$	$-1.610595831 \times 10^{-17}$
3	13	$-1.313376346 \times 10^{-27}$	$2.300736175 \times 10^{-28}$	$-2.907968618 \times 10^{-19}$
3	14	$-5.727370043 \times 10^{-30}$	$1.499365267 \times 10^{-30}$	$-4.53978556 \times 10^{-21}$
3	15	$-2.269163679 \times 10^{-32}$	$8.23379802 \times 10^{-33}$	$-6.209594468 \times 10^{-23}$
3	16	$-8.295414237 \times 10^{-35}$	$3.918343826 \times 10^{-35}$	$-7.519749707 \times 10^{-25}$
3	17	$-2.832492468 \times 10^{-37}$	$1.649247798 \times 10^{-37}$	$-8.131195461 \times 10^{-27}$
3	18	$-9.111962196 \times 10^{-40}$	$6.235369017 \times 10^{-40}$	$-7.907119122 \times 10^{-29}$
3	19	$-2.775755352 \times 10^{-42}$	$2.142975807 \times 10^{-42}$	$-6.957484681 \times 10^{-31}$
3	20	$-8.023921599 \times 10^{-45}$	$6.757695832 \times 10^{-45}$	$-5.568919818 \times 10^{-33}$
4	12	$4.34297062 \times 10^{-29}$	0.	$5.829036494 \times 10^{-21}$
4	13	$1.351146415 \times 10^{-30}$	0.	$7.253912081 \times 10^{-22}$
4	14	$2.121497117 \times 10^{-32}$	0.	$4.555880368 \times 10^{-23}$
4	15	$2.240886943 \times 10^{-34}$	0.	$1.924907227 \times 10^{-24}$
4	16	$1.790950977 \times 10^{-36}$	$4.552053044 \times 10^{-40}$	$6.155224774 \times 10^{-26}$
4	17	$1.154985597 \times 10^{-38}$	$1.417594374 \times 10^{-41}$	$1.589348444 \times 10^{-27}$
4	18	$6.259747967 \times 10^{-41}$	$2.223253965 \times 10^{-43}$	$3.453555307 \times 10^{-29}$
4	19	$2.932258013 \times 10^{-43}$	$2.34090826 \times 10^{-45}$	$6.499580679 \times 10^{-31}$
4	20	$1.211781268 \times 10^{-45}$	$1.861345908 \times 10^{-47}$	$1.082266648 \times 10^{-32}$

Table A.43.3. Numerical Values of $a_{jn}^{(2)}$

n	j	$a_{jn}^{(2)}$	n	j	$a_{jn}^{(2)}$
3	3	$2.057613169 \times 10^{-5}$	4	6	$1.866315799 \times 10^{-11}$
3	4	$1.209372638 \times 10^{-6}$	4	7	$1.128535631 \times 10^{-12}$
3	5	$3.732631599 \times 10^{-8}$	4	8	$3.48353767 \times 10^{-14}$
3	6	$7.699037954 \times 10^{-10}$	4	9	$7.297097265 \times 10^{-16}$
3	7	$1.167237811 \times 10^{-11}$	4	10	$1.168604418 \times 10^{-17}$
3	8	$1.371417205 \times 10^{-13}$	4	11	$1.531448318 \times 10^{-19}$
3	9	$1.292930747 \times 10^{-15}$	4	12	$1.716934864 \times 10^{-21}$
3	10	$1.003009124 \times 10^{-17}$	4	13	$1.697645288 \times 10^{-23}$

(*continued on next page*)

Table A.43.3. (*Continued*)

n	j	$a_{jn}^{(2)}$	n	j	$a_{jn}^{(2)}$
3	11	$6.527612749 \times 10^{-20}$	4	14	$1.511360414 \times 10^{-25}$
3	12	$3.619439586 \times 10^{-22}$	4	15	$1.227975163 \times 10^{-27}$
3	13	$1.731803061 \times 10^{-24}$	4	16	$9.183002113 \times 10^{-30}$
3	14	$7.227408932 \times 10^{-27}$	4	17	$6.353423592 \times 10^{-32}$
3	15	$2.655092098 \times 10^{-29}$	4	18	$4.080343505 \times 10^{-34}$
3	16	$8.654606305 \times 10^{-32}$	4	19	$2.438225173 \times 10^{-36}$
3	17	$2.520687841 \times 10^{-34}$	4	20	$5.659333071 \times 10^{-40}$
3	18	$6.600553579 \times 10^{-37}$	5	10	$-2.132531426 \times 10^{-20}$
3	19	$1.562527668 \times 10^{-39}$	5	11	$-1.305498271 \times 10^{-21}$
3	20	$5.600968994 \times 10^{-43}$	5	12	$-4.043718229 \times 10^{-23}$
6	15	$1.864646977 \times 10^{-32}$	5	13	$-8.443854407 \times 10^{-25}$
6	16	$1.148934026 \times 10^{-33}$	5	14	$-1.336599335 \times 10^{-26}$
6	17	$3.56736923 \times 10^{-35}$	5	15	$-1.709823129 \times 10^{-28}$
6	18	$7.440144587 \times 10^{-37}$	5	16	$-1.839827713 \times 10^{-30}$
6	19	$1.628250951 \times 10^{-41}$	5	17	$-1.710992065 \times 10^{-32}$
			5	18	$-1.401979299 \times 10^{-34}$
			5	19	$-1.026691317 \times 10^{-36}$

A.44 A Personal Recollection

In the 1970s the Kyoto school of mathematics was very active and every month or so we received a new article. These papers piled up on my desk and I said to myself; some day I will read them. That "some day" almost never came. The paper by M. Jimbo, T. Miwa, A. Mori and M. Sato on impenetrable Bose gas containing the differential equation satisfied by the spacing function was one of them and its importance to the random matrix theory was mentioned to me by Barry McCoy. In the 1980s on every occasion I met Barry he emphatically reminded me to read this particular paper and I made at least three serious attempts to do it; once on my own, a second time with the help of my colleagues in Saclay and a third time with the help of the four authors during their extended stay at Saclay. Unfortunately, the authors considered me intelligent enough and with the characteristic eastern politeness evaded to answer my stupid questions. As a result I never got beyond the first 10 or 15 pages of the 80 or so of the said paper.

In the second edition of "Random Matrices" I mentioned therefore their work saying "... We will not need these and so we do not have to copy their proofs here." And as fate had it, I copied the wrong equation. When the second edition came on the market, knowledgeable people like Barry McCoy and Craig Tracy started laughing. This was quite irritating and this time I set aside the paper and tried on my own to get a differential equation for the spacing function. This way I got not only their Eq. (21.3.4) but a few similar ones more. It was comforting later to hear Harold Widom saying "... It was fortunate that you did not understand..." or Barry McCoy "... Now I understand it."

Lesson: If you are stupid, don't get discouraged but try a different way to get the result.

A.45 About Painlevé Transcendents

About one hundred years back Painlevé studied differential equations of the form

$$\frac{d^2y}{dt^2} = R\left(\frac{dy}{dt}, y, t\right), \tag{A.45.1}$$

where R is polynomial in dy/dt, rational in y, analytic in a certain domain D of the complex variable t and (to deserve the name "function") the solution should not have a movable critical point. In other words, the location and nature of the singular points of the solutions are all fixed by the parameters appearing explicitly in the differential equation itself and do not depend on the initial conditions. Painlevé and Gambier found that any such equation, by a change of independent and/or dependent variable, can be reduced to one of the 50 or so standard forms. Out of these standard forms six were new, all others can be integrated in terms of either elementary functions, or in terms of solutions of certain linear equations, or in terms of solutions of the first order equations, or finally, in terms of solutions of the remaining 6 equations. The six remaining equations of Painlevé and Gambier which can not be so integrated are now (somewhat improperly) called Painlevé equations, and their solutions Painlevé transcendents. They have the following form.

A.45.1 Six Painlevé Equations

$$\text{(P1):} \quad \frac{d^2y}{dt^2} = 6y^2 + t, \tag{A.45.2}$$

$$\text{(P2):} \quad \frac{d^2y}{dt^2} = 2y^3 + ty + \alpha, \tag{A.45.3}$$

$$\text{(P3):} \quad \frac{d^2y}{dt^2} = \frac{1}{y}\left(\frac{dy}{dt}\right)^2 - \frac{1}{t}\frac{dy}{dt} + \frac{1}{t}(\alpha y^2 + \beta) + \gamma y^3 + \frac{\delta}{y}, \tag{A.45.4}$$

$$\text{(P3}'\text{):} \quad \frac{d^2w}{dx^2} = \frac{1}{w}\left(\frac{dw}{dx}\right)^2 - \frac{1}{x}\frac{dw}{dx} + \frac{1}{4x^2}(\alpha w^2 + \gamma w^3) + \frac{\beta}{4x} + \frac{\delta}{4w}, \tag{A.45.4$'$}$$

$$\text{(P4):} \quad \frac{d^2y}{dt^2} = \frac{1}{2y}\left(\frac{dy}{dt}\right)^2 + \frac{3}{2}y^3 + 4ty^2 + 2(t^2 - \alpha)y + \frac{\beta}{y}, \tag{A.45.5}$$

$$\text{(P5):} \quad \frac{d^2y}{dt^2} = \left(\frac{1}{2y} + \frac{1}{y-1}\right)\left(\frac{dy}{dt}\right)^2 - \frac{1}{t}\frac{dy}{dt} + \frac{(y-1)^2}{t^2}\left(\alpha y + \frac{\beta}{y}\right)$$

$$+ \gamma\frac{y}{t} + \delta\frac{y(y+1)}{y-1}, \tag{A.45.6}$$

$$(\text{P6}): \quad \frac{d^2y}{dt^2} = \frac{1}{2}\left(\frac{1}{y} + \frac{1}{y-1} + \frac{1}{y-t}\right)\left(\frac{dy}{dt}\right)^2 - \left(\frac{1}{t} + \frac{1}{t-1} + \frac{1}{y-t}\right)\frac{dy}{dt}$$
$$+ \frac{y(y-1)(y-t)}{t^2(t-1)^2}\left(\alpha + \beta\frac{t}{y^2} + \gamma\frac{t-1}{(y-1)^2} + \delta\frac{t(t-1)}{(y-t)^2}\right), \quad (\text{A.45.7})$$

where $\alpha, \beta, \gamma, \delta$ are (complex) parameters. For the third Painlevé equation some people prefer the form (P3′) rather than (P3); the two are equivalent by the change of variables $y = wx, t = x^2$.

A.45.2 Lax Pairs. The Painlevé equations can be thought of as the compatibility condition of certain pairs of linear equations called Lax pairs. The pair is of the form

$$\frac{\partial Y}{\partial x} = AY, \qquad \frac{\partial Y}{\partial t} = BY, \quad (\text{A.45.8})$$

with the corresponding compatibility condition

$$AB - BA + \frac{\partial A}{\partial t} - \frac{\partial B}{\partial x} = 0. \quad (\text{A.45.9})$$

Here and in what follows Y, A, B are 2×2 matrices; u, v, w, f, y, p, q are functions only of t; $\alpha, \beta, \gamma, \delta, \eta, \theta$ with any index are complex constants and σ_j, $j = 1, 2, 3$, are the Pauli matrices

$$\sigma_1 = \begin{bmatrix} 0 & 1 \\ 1 & 0 \end{bmatrix}, \quad \sigma_2 = \begin{bmatrix} 0 & -i \\ i & 0 \end{bmatrix}, \quad \sigma_3 = \begin{bmatrix} 1 & 0 \\ 0 & -1 \end{bmatrix}.$$

The elimination of functions other than y, usually quite a long and tedious process, gives the Painlevé equation, t being the independent variable.

Note. In fact, if the matrix A is given then one can compute B. The method is to compute Y near all its singular points, then compute $(\partial Y/\partial t)Y^{-1}$ near those same points, which gives B. The actual calculations are usually long. But I do not know how to find A (or B). The pair of matrices A and B are not unique.

P-1:

$$A = \begin{bmatrix} -v & x^2 + xy + y^2 + t/2 \\ 4(x-y) & v \end{bmatrix}, \quad (\text{A.45.10})$$

$$B = \begin{bmatrix} 0 & x/2 + y \\ 2 & 0 \end{bmatrix}. \quad (\text{A.45.11})$$

P-2:

$$A = \begin{bmatrix} x^2 + v + t/2 & x - y \\ -2v(x+y) - 2\theta & -x^2 - v - t/2 \end{bmatrix}, \tag{A.45.12}$$

$$B = \frac{1}{2}\begin{bmatrix} x + y & 1 \\ -2v & -x - y \end{bmatrix}, \tag{A.45.13}$$

$$\alpha = \frac{1}{2} - \theta. \tag{A.45.14}$$

P-3:

$$\begin{aligned} A &= \frac{t}{2}\begin{bmatrix} 1 & 0 \\ 0 & -1 \end{bmatrix} + \frac{1}{x}\begin{bmatrix} -\theta_\infty/2 & u \\ v & \theta_\infty/2 \end{bmatrix} + \frac{a}{x^2}\begin{bmatrix} s - bt/2 & -ws \\ (s-bt)/w & -s + bt/2 \end{bmatrix} \\ &= \frac{1}{2}\left(t - \frac{\theta_\infty}{x} + a\frac{2s - bt}{x^2}\right)\sigma_3 + \frac{1}{x}\begin{bmatrix} 0 & u \\ v & 0 \end{bmatrix} + \frac{a}{x^2}\begin{bmatrix} 0 & -ws \\ (s-bt)/w & 0 \end{bmatrix}, \end{aligned} \tag{A.45.15}$$

$$\begin{aligned} B &= \frac{x}{2}\begin{bmatrix} 1 & 0 \\ 0 & -1 \end{bmatrix} + \frac{1}{t}\begin{bmatrix} 0 & u \\ v & 0 \end{bmatrix} - \frac{a}{xt}\begin{bmatrix} s - bt/2 & -ws \\ (s-bt)/w & -s + bt/2 \end{bmatrix} \\ &= \frac{1}{2}\left(x - a\frac{2s - bt}{xt}\right)\sigma_3 + \frac{1}{t}\begin{bmatrix} 0 & u \\ v & 0 \end{bmatrix} - \frac{a}{xt}\begin{bmatrix} 0 & -ws \\ (s-bt)/w & 0 \end{bmatrix}; \end{aligned} \tag{A.45.16}$$

$y = -u/(ws)$ satisfies Painlevé 3 with

$$\alpha = 4b\theta_0, \quad \beta = -4a(\theta_\infty - 1), \quad \gamma = 4b^2, \quad \delta = -4a^2. \tag{A.45.17}$$

Note that the Lax pair is independent of α. How can one choose β and δ mutually independent? For example, $\beta = 0$, $\delta \neq 0$; or $\beta \neq 0$, $\delta = 0$? Similarly, for α and γ? For a detailed discussion of this point see, e.g., Lin et al. (2003).

P-4:

$$A = (x + t)\sigma_3 + \frac{1}{x}\begin{bmatrix} -v + \theta_0 & x - y/2 \\ 2x(v - \theta_0 - \theta_\infty) + 2v(v - 2\theta_0)/y & v - \theta_0 \end{bmatrix}, \tag{A.45.18}$$

$$B = (x + t)\sigma_3 + \begin{bmatrix} y/2 & 1 \\ 2(v - \theta_0 - \theta_\infty) & -y/2 \end{bmatrix}, \tag{A.45.19}$$

$$\alpha = 2\theta_\infty - 1, \qquad \beta = -\theta_0^2/2. \tag{A.45.20}$$

P-5:

$$A = \frac{a}{2} t\sigma_3 + \frac{1}{x} \begin{bmatrix} v + \theta_0/2 & -v - \theta_0 \\ v & -v - \theta_0/2 \end{bmatrix}$$

$$+ \frac{1}{x-1} \begin{bmatrix} -w & y(w - \theta_1/2) \\ -(w + \theta_1/2)/y & w \end{bmatrix}, \tag{A.45.21}$$

$$B = \frac{a}{2} x\sigma_3 + \frac{1}{t} \begin{bmatrix} f & -v - \theta_0 + y(w - \theta_1/2) \\ v - (w + \theta_1/2)/y & -f \end{bmatrix}, \tag{A.45.22}$$

$$w = v + \frac{1}{2}(\theta_0 + \theta_\infty), \tag{A.45.23}$$

$$2f = 2v + \theta_0 - y\left(w - \frac{\theta_1}{2}\right) - \frac{1}{y}\left(w + \frac{\theta_1}{2}\right), \tag{A.45.24}$$

$$\alpha = \frac{1}{8}(\theta_0 - \theta_1 + \theta_\infty)^2, \qquad \beta = -\frac{1}{8}(\theta_0 - \theta_1 - \theta_\infty)^2, \tag{A.45.25}$$

$$\gamma = a(1 - \theta_0 - \theta_1), \qquad \delta = -\frac{a^2}{2}. \tag{A.45.25'}$$

What about the cases $\gamma \neq 0$, $\delta = 0$; or $\gamma = 0$, $\delta \neq 0$ and arbitrary values of α and β, each zero or non-zero? In other words I do not know the Lax pairs for P-3 and P-5 in terms of the parameters $\alpha, \beta, \gamma, \delta$. Cf. Lin et al. (2003).

P-6:

$$A = \frac{A_0}{x} + \frac{A_1}{x-1} + \frac{A_t}{x-t}, \qquad B = -\frac{A_t}{x-t}, \tag{A.45.26}$$

$$A_j = \begin{bmatrix} -z_j & u_j(\theta_j - z_j) \\ (\theta_j + z_j)/u_j & z_j \end{bmatrix}, \quad j = 0, 1, t, \tag{A.45.27}$$

with

$$A_0 + A_1 + A_t = \begin{bmatrix} -\theta & 0 \\ 0 & \theta \end{bmatrix}. \tag{A.45.28}$$

Then

$$y = \frac{tu_0(\theta_0 - z_0)}{tu_0(\theta_0 - z_0) + (t-1)u_1(\theta_1 - z_1)} \tag{A.45.29}$$

satisfies Painlevé 6 with

$$\alpha = \frac{1}{2}(\theta - 1)^2, \quad \beta = -\frac{1}{2}\theta_0^2, \quad \gamma = \frac{1}{2}(\theta^2 + \theta_1), \quad \delta = \frac{1}{2}(1 - \theta_t^2). \tag{A.45.30}$$

A.45.3 Hamiltonian Equations. Painlevé equations can also be thought of as equations of movement resulting from certain Hamiltonians (not unique).

$$H = H(p, q), \tag{A.45.31}$$

$$\frac{dq}{dt} = \frac{\partial H}{\partial p}, \qquad \frac{dp}{dt} = -\frac{\partial H}{\partial q}; \tag{A.45.32}$$

elimination of p give the corresponding Painlevé equation for q.

H-1:

$$H = \frac{1}{2}p^2 - (2q^3 + tq),$$
$$\frac{dq}{dt} = p, \qquad \frac{dp}{dt} = 6q^2 + t.$$

H-2:

$$H = \frac{1}{2}p^2 \pm \left(q^2 + \frac{t}{2}\right)p - \left(\alpha \mp \frac{1}{2}\right)q,$$
$$\frac{dq}{dt} = p \pm \left(q^2 + \frac{t}{2}\right), \qquad \frac{dp}{dt} = \mp 2qp + \alpha \mp \frac{1}{2},$$

or

$$H = \frac{1}{2}p^2 - \frac{1}{2}(q^4 + tq^2 + 2\alpha q),$$
$$\frac{dq}{dt} = p, \qquad \frac{dp}{dt} = 2q^3 + tq + \alpha.$$

H-3:

$$tH = 2p^2q^2 - \left[2btq^2 - (2\theta_\infty - 1)q - 2at\right]p - b(\theta_0 + \theta_\infty)tq,$$
$$t\frac{dq}{dt} = 4pq^2 - 2btq^2 + (2\theta_\infty - 1)q + 2at,$$
$$t\frac{dp}{dt} = -4p^2q + (4btq - 2\theta_\infty + 1)p + (\theta_0 + \theta_\infty)bt;$$
$$\alpha = 4b\theta_0, \quad \beta = -4a(\theta_\infty - 1), \quad \gamma = 4b^2, \quad \delta = -4a^2,$$

or

$$tH = 2\eta p^2q^2 - 3pq + \eta t^2\left(aq^2 + \frac{b}{q^2}\right) + \eta t\left(cq + \frac{d}{q}\right),$$
$$t\frac{dq}{dt} = 4\eta pq^2 - 3q,$$

$$t\frac{dp}{dt} = -4\eta p^2 q + 3p - 2\eta t^2\left(aq - \frac{b}{q^3}\right) - \eta t\left(c - \frac{d}{q^2}\right),$$

$$\alpha = -4c\eta^2, \quad \beta = 4d\eta^2, \quad \gamma = -8a\eta^2, \quad \delta = 8b\eta^2.$$

One may take $\eta = 1$.

H-4:

$$H = 2\varepsilon tpq + \varepsilon(qp^2 + q^2 p) + 2kp + (\varepsilon\alpha + k + 1)q, \quad \varepsilon^2 = 1;$$

$$\frac{dq}{dt} = 2\varepsilon pq + \varepsilon q^2 + 2\varepsilon tq + 2k, \quad \frac{dp}{dt} = -\varepsilon p^2 - 2\varepsilon pq - 2\varepsilon tp - (\varepsilon\alpha + k + 1),$$

$$\beta = -2k^2;$$

or

$$H = 2qp^2 - (q^2 + 2tq + 2k)p + \frac{1}{2}(\alpha + k - 1)q,$$

$$\frac{dq}{dt} = 4pq - (q^2 + 2tq + 2k), \quad \frac{dp}{dt} = -2p^2 + p(2q + 2t) - \frac{1}{2}(\alpha + k - 1),$$

$$\beta = -2k^2;$$

or

$$H = 2\eta p^2 q - 2p - \eta\left(\frac{2\alpha_0}{q} + 2\alpha_1 q\right) - \frac{1}{8\eta}q(q + 2t)^2,$$

$$\frac{dq}{dt} = 4\eta pq - 2, \quad \frac{dp}{dt} = \eta\left(-2p^2 - \frac{2\alpha_0}{q^2} + 2\alpha_1\right) + \frac{1}{8\eta}(q + 2t)(3q + 2t),$$

$$\alpha = -4\alpha_1\eta^2, \quad \beta = -8\alpha_0\eta^2 - 2.$$

H-5:

$$\begin{aligned} tH &= q(q-1)^2 p^2 - \left[\theta_0(q-1)^2 + \theta_1 q(q-1) - \eta tq\right]p \\ &\quad + \frac{1}{4}(q-1)\left[(\theta_0 + \theta_1)^2 - \theta_\infty^2\right], \end{aligned}$$

$$t\frac{dq}{dt} = 2q(q-1)^2 p - \theta_0(q-1)^2 - \theta_1 q(q-1) + \eta tq,$$

$$\begin{aligned} t\frac{dp}{dt} &= -p^2\left[(q-1)^2 + 2q(q-1)\right] + p\left[2\theta_0(q-1) + \theta_1(2q-1) - \eta t\right] \\ &\quad - \frac{1}{4}\left[(\theta_0 + \theta_1)^2 - \theta_\infty^2\right], \end{aligned}$$

$$\alpha = \frac{1}{2}\theta_\infty^2, \quad \beta = -\frac{1}{2}\theta_0^2, \quad \gamma = \eta(\theta_1 + 1), \quad \delta = -\frac{1}{2}\eta^2,$$

or

$$\frac{t}{\eta}H = q(q-1)^2 p^2 - \frac{1}{\eta}(q-1)(2q-1)p$$
$$- \left[a_0 \frac{(q-1)^2}{q} + a_1 \frac{t^2 q}{(q-1)^2} + a_2 \frac{tq}{q-1} + a_3 q\right],$$
$$t\frac{dq}{dt} = \eta \cdot 2pq(q-1)^2 - (q-1)(2q-1),$$
$$t\frac{dp}{dt} = -\eta \cdot p^2(q-1)(3q-1) + p(4q-3)$$
$$+ \eta\left[a_0 \frac{q^2-1}{q^2} - a_1 \frac{t^2(q+1)}{(q-1)^3} - a_2 \frac{t}{(q-1)^2} + a_3\right],$$
$$\alpha = 2 + 2\eta^2(a_0 + a_3), \quad \beta = -\frac{1}{2} - 2a_0\eta^2, \quad \gamma = -2a_2\eta^2, \quad \delta = -2a_1\eta^2.$$

H-6:

$$t(t-1)H = q(q-1)(q-t)p^2 - \left[k_0(q-1)(q-t) + k_1 q(q-t) + k_t q(q-1)\right]p$$
$$+ k(q-t),$$
$$t(t-1)\frac{dq}{dt} = 2q(q-1)(q-t)p - k_0(q-1)(q-t) - k_1 q(q-t) - k_t q(q-1),$$
$$-t(t-1)\frac{dp}{dt} = p^2\left[q(q-1) + q(q-t) + (q-1)(q-t)\right]$$
$$- p\left[k_0(2q-t-1) + k_1(2q-t) + k_t(2q-1)\right] + k,$$
$$\alpha = \frac{1}{2}(k_0 + k_1 + k_t)^2 - 2k, \quad \beta = -\frac{1}{2}k_0^2,$$
$$\gamma = \frac{1}{2}k_1^2, \quad \delta = -\frac{1}{2}k_t(k_t + 2),$$

or

$$t(t-1)H = \eta q(q-1)(q-t)p^2 - (2q-1)(q-t)p$$
$$- \eta\left[\frac{\alpha_0(q-1)(q-t)}{q} + \frac{\alpha_1 q(q-t)}{q-1} + \frac{\alpha_t q(q-1)}{q-t} + \alpha_\infty(q-t)\right],$$
$$t(t-1)\frac{dq}{dt} = 2\eta q(q-1)(q-t)p - (2q-1)(q-t),$$

$$-t(t-1)\frac{dp}{dt} = \eta p^2\big[q(q-1)+q(q-t)+(q-1)(q-t)\big]-(4q-2t-1)p$$
$$+\eta\left[\frac{\alpha_0(q-1)(q-t)}{q^2}+\frac{\alpha_1 q(q-t)}{(q-1)^2}+\frac{\alpha_t q(q-1)}{(q-t)^2}\right]$$
$$-\eta\left[\frac{\alpha_0(2q-t-1)}{q}+\frac{\alpha_1(2q-t)}{q-1}+\frac{\alpha_t(2q-1)}{q-t}+\alpha_\infty\right],$$
$$\alpha = 2+2\eta^2(\alpha_0+\alpha_1+\alpha_t+\alpha_\infty), \quad \beta = -2\alpha_0\eta^2-\frac{1}{2},$$
$$\gamma = 2\alpha_1\eta^2+\frac{1}{2}, \quad \delta = -2\alpha_t\eta^2.$$

Note. If the constants are such that $H(p=0,q)=0$, then $p=0$ is a solution of the Hamiltonian equations. In this case any solution of $dq/dt = \partial H/\partial p|_{p=0}$ is a solution of the corresponding Painlevé equation. This provides us a one parameter family of solutions of the Painlevé equation for the particular values of the constants $\alpha, \beta, \gamma, \delta$; however, all the solutions are not of this form.

A.45.4 Confluences of the Hamiltonians and the Painlevé Equations. If one replaces $t \to 1+\varepsilon t$, $\gamma \to \gamma\varepsilon^{-1}-\delta\varepsilon^{-2}$, $H \to \varepsilon^{-1}H$, in H-6 and P-6, and takes the limit $\varepsilon \to 0$, one gets H-5 and P-5. This fact we will express as

P-6 → P-5:

$$q \to q, \quad p \to p, \quad t \to 1+\varepsilon t, \quad H \to H\varepsilon^{-1}, \quad \alpha \to \alpha, \quad \beta \to \beta,$$
$$\gamma \to \gamma\varepsilon^{-1}-\delta\varepsilon^{-2}, \quad \delta \to \delta\varepsilon^{-2}, \quad \varepsilon \to 0.$$

Other confluences are

P-5 → P-4:

$$q \to \varepsilon q/\sqrt{2}, \quad p \to p\sqrt{2}/\varepsilon, \quad t \to 1+\varepsilon\sqrt{2}t,$$
$$H+\frac{1}{4}\big[(\theta_0+\theta_1)^2-\theta_\infty^2\big] \to \big(\varepsilon\sqrt{2}\big)^{-1}H, \quad \alpha \to (2\varepsilon^4)^{-1}-\varepsilon^{-2}\alpha,$$
$$\beta \to \beta/4, \quad \gamma \to -\varepsilon^{-4}, \quad \delta \to -(2\varepsilon^4)^{-1}, \quad \varepsilon \to 0.$$

P-5 → P-3:

$$q \to 1+\varepsilon t q, \quad p \to p/(\varepsilon t), \quad t \to t^2, \quad H \to H/(2t)+pq/(2t^2),$$
$$\alpha \to (8\varepsilon)^{-1}(\alpha+\varepsilon^{-1}\gamma), \quad \beta \to (8\varepsilon)^{-1}(\alpha-\varepsilon^{-1}\gamma),$$
$$\gamma \to \varepsilon\beta/4, \quad \delta \to \varepsilon^2\delta/8, \quad \varepsilon \to 0.$$

P-4 → P-2:

$$q \to 2^{2/3}q/\varepsilon + \varepsilon^{-3}, \quad p \to 2^{-2/3}\varepsilon p, \quad t \to 2^{-2/3}\varepsilon t - \varepsilon^{-3},$$
$$H \to 2^{2/3}\varepsilon^{-1}H - (\alpha + 1/2)\varepsilon^{-3}, \quad \alpha \to -2\alpha - (2\varepsilon^6)^{-1},$$
$$\beta \to -(2\varepsilon^{12})^{-1}, \quad \varepsilon \to 0.$$

P-3 → P-2:

$$q \to 1 + 2\varepsilon q, \quad p \to p/(2\varepsilon), \quad t \to 1 + \varepsilon^2 t, \quad H + b(\theta_0 + \theta_1) \to \varepsilon^{-2}H,$$
$$\alpha \to -(2\varepsilon^6)^{-1} + 2\alpha\varepsilon^{-3}, \quad \beta \to (2\varepsilon^6)^{-1}, \quad \gamma \to (4\varepsilon^6)^{-1},$$
$$\delta \to -(4\varepsilon^6)^{-1}, \quad \varepsilon \to 0.$$

P-2 → P-1:

$$q \to \varepsilon q + \varepsilon^{-5}, \quad p \to p/\varepsilon, \quad t \to \varepsilon^2 t - 6\varepsilon^{-10},$$
$$H \to \varepsilon^{-2}H - 3\varepsilon^{-20}/2 - \varepsilon^{-8}t/2, \quad \alpha \to 4\varepsilon^{-15}, \quad \varepsilon \to 0.$$

Lax pair for P-6 suggested by Lin et al., said to be better.

$$A = -\frac{B_x}{t - x} + \frac{A_t}{t - y} + A_\infty,$$

$$B = \frac{B_0}{t} + \frac{B_1}{t - 1} + \frac{B_x}{t - x} + \frac{B_t}{t - y},$$

$$A_t = \frac{1}{4}\left[\frac{x^2(x-1)^2}{2y(y-1)(y-x)^2}\left(\frac{dy}{dt}\right)^2 - \frac{(y-x)^2}{2x(x-1)} + \frac{\theta_\infty^2(y(y-x) - x(y-1))}{2x(x-1)}\right.$$
$$\left. + \frac{\theta_0^2}{2y} - \frac{\theta_1^2}{2(y-1)} - \frac{(1-\theta_x^2)(y(x-1) + x(y-1))}{2(y-x)^2}\right](\sigma_1 - \sigma_2/i) - \frac{1}{2}\sigma_1\frac{dy}{dt},$$

$$A_\infty = \left[\frac{1}{8}\theta_\infty^2\sigma_3 - \sigma_1 + \left(1 - \frac{1}{8}\theta_\infty^2\right)\frac{\sigma_2}{i}\right]\frac{y - x}{x(x-1)},$$

$$B_0 = -\frac{1}{4}\left[\frac{x^3(x-1)^2}{2y^2(y-x)^2}\left(\frac{dy}{dt}\right)^2 + \frac{2x^2(x-1)}{y(y-x)}\left(\frac{dy}{dt}\right) - \frac{\theta_0^2 x}{2y^2}\right](\sigma_3 - \sigma_2/i)$$
$$+ \frac{y^2 - x^2}{2x}\sigma_3 - \left[\frac{x(x-1)}{2(y-x)}\left(\frac{dy}{dt}\right) + y\right]\sigma_1 + \frac{y^2 + x^2}{2x}(\sigma_2/i),$$

$$B_1 = \frac{1}{4}\left[\frac{x^2(x-1)^3}{2(y-1)^2(y-x)^2}\left(\frac{dy}{dt}\right)^2 + \frac{2x(x-1)^2}{(y-1)(y-x)}\left(\frac{dy}{dt}\right) - \frac{\theta_1^2(x-1)}{2(y-1)^2}\right](\sigma_3 - \sigma_2/i)$$
$$+ \frac{(2-y-x)(y-x)}{2(x-1)}\sigma_3 + \left[\frac{x(x-1)}{2(y-x)}\left(\frac{dy}{dt}\right) + y - 1\right]\sigma_1$$
$$- \frac{(y-1)^2 + (x-1)^2}{2(x-1)}(\sigma_2/i),$$

$$B_x = \frac{(y-x)^2}{2x(x-1)}(\sigma_3 + \sigma_2/i) - \frac{(1-\theta_x^2)x(x-1)}{8(y-x)^2}(\sigma_3 - \sigma_2/i) - \frac{\sigma_1}{2},$$

$$B_t = \frac{1}{4}\left[\frac{(y(y-x) - x(y-1))x^2(x-1)^2}{2y^2(y-1)^2(y-x)^2}\left(\frac{dy}{dt}\right)^2 + \frac{2x(x-1)}{y(y-1)}\frac{dy}{dt}\right.$$
$$\left. + \frac{\theta_\infty^2}{2} - \frac{\theta_0^2 x}{2y^2} + \frac{\theta_1^2(x-1)}{2(y-1)^2} + \frac{(1-\theta_x^2)x(x-1)}{2(y-x)^2}\right](\sigma_3 - \sigma_2/i) + \frac{\sigma_1}{2}.$$

A.46 Inverse Power Series Expansions of $\mathcal{S}_n(\tau)$, $\mathcal{A}_n(\tau)$, $\mathcal{B}_n(\tau)$, etc.

For $\tau \gg 1$, the coefficient functions $\mathcal{S}_n(\tau)$, $\mathcal{A}_n(\tau)$ and $\mathcal{B}_n(\tau)$ in Eqs. (21.4.1)–(21.4.3) can be expanded in powers of τ^{-1}. As explained in Section 21.4 we get

$$\mathcal{S}_0(\tau) = \frac{(1+i)}{2}\left[1 - \frac{1}{2^5\tau^2} - \frac{85}{2^{11}\tau^4} - \frac{11813}{2^{16}\tau^6} - \frac{14121997}{2^{23}\tau^8} - \frac{7374679967}{2^{28}\tau^{10}} + \cdots\right.$$
$$\left. - i\left(\frac{1}{4\tau} + \frac{5}{2^7\tau^3} + \frac{719}{2^{13}\tau^5} + \frac{145117}{2^{18}\tau^7} + \frac{228661979}{2^{25}\tau^9} + \cdots\right)\right], \qquad \text{(A.46.1)}$$

$$\mathcal{S}_1(\tau) = 1 + \frac{3}{2^4\tau} + \frac{5}{2^9\tau^2} + \frac{469}{2^{13}\tau^3} + \frac{2979}{2^{19}\tau^4} + \frac{1227029}{2^{23}\tau^5} + \frac{4411073}{2^{28}\tau^6} + \frac{4184326389}{2^{32}\tau^7}$$
$$+ \cdots + i\left(\frac{1}{8\tau} + \frac{7}{2^7\tau^2} + \frac{213}{2^{12}\tau^3} + \frac{4465}{2^{16}\tau^4} + \frac{501667}{2^{22}\tau^5} + \frac{17524881}{2^{26}\tau^6} + \cdots\right), \qquad \text{(A.46.2)}$$

$$\mathcal{S}_2(\tau) = 1 + \frac{5}{8\tau} + \frac{63}{2^7\tau^2} + \frac{771}{2^{10}\tau^3} + \frac{43063}{2^{15}\tau^4} + \frac{864555}{2^{18}\tau^5} + \frac{37620327}{2^{22}\tau^6} + \frac{1053966787}{2^{25}\tau^7}$$
$$+ \cdots + i\left(\frac{1}{2^6\tau^2} + \frac{13}{2^9\tau^3} + \frac{387}{2^{13}\tau^4} + \frac{6815}{2^{16}\tau^5} + \frac{566999}{2^{21}\tau^6} + \frac{13720419}{2^{24}\tau^7} + \cdots\right), \qquad \text{(A.46.3)}$$

$$\mathcal{S}_3(\tau) = 1 + \frac{17}{2^4\tau} + \frac{581}{2^9\tau^2} + \frac{14439}{2^{13}\tau^3} + \frac{1702083}{2^{19}\tau^4} + \frac{65467383}{2^{23}\tau^5} + \frac{5738450209}{2^{28}\tau^6} + \cdots$$
$$+ i\left(\frac{1}{2^9\tau^3} + \frac{45}{2^{13}\tau^4} + \frac{3557}{2^{18}\tau^5} + \frac{145979}{2^{22}\tau^6} + \frac{26141411}{2^{28}\tau^7} + \cdots\right), \qquad \text{(A.46.4)}$$

$$\mathcal{S}_4(\tau) = 1 + \frac{3}{2\tau} + \frac{63}{2^5\tau^2} + \frac{823}{2^8\tau^3} + \frac{24999}{2^{12}\tau^4} + \frac{117585}{2^{13}\tau^5} + \frac{1274445}{2^{15}\tau^6} + \frac{33179409}{2^{18}\tau^7} + \cdots$$
$$+ i\left(\frac{1}{2^{12}\tau^4} + \frac{1}{2^{10}\tau^5} + \frac{393}{2^{17}\tau^6} + \frac{9219}{2^{20}\tau^7} + \frac{444981}{2^{24}\tau^8} + \cdots\right), \tag{A.46.5}$$

$$\mathcal{A}_0(\tau) = \frac{\tau}{2} + \frac{1}{8\tau} + \frac{1}{2^5\tau^3} + \frac{5}{2^6\tau^5} + \frac{131}{2^8\tau^7} + \frac{6575}{2^{10}\tau^9} + \frac{1080091}{2^{13}\tau^{11}} + \frac{16483607}{2^{12}\tau^{13}} + \cdots, \tag{A.46.6}$$

$$\mathcal{A}_1(\tau) = 1 + \frac{5}{2^4\tau} + \frac{65}{2^9\tau^2} + \frac{1823}{2^{13}\tau^3} + \frac{163691}{2^{19}\tau^4} + \frac{7266843}{2^{23}\tau^5} + \frac{551317093}{2^{28}\tau^6} + \cdots, \tag{A.46.7}$$

$$\mathcal{A}_2(\tau) = 1 + \frac{5}{8\tau} + \frac{65}{2^7\tau^2} + \frac{797}{2^{10}\tau^3} + \frac{44675}{2^{15}\tau^4} + \frac{893927}{2^{18}\tau^5} + \frac{38868565}{2^{22}\tau^6} + \cdots, \tag{A.46.8}$$

$$\mathcal{A}_3(\tau) = 1 + \frac{17}{2^4\tau} + \frac{581}{2^9\tau^2} + \frac{14455}{2^{13}\tau^3} + \frac{1704963}{2^{19}\tau^4} + \frac{65581207}{2^{23}\tau^5} + \frac{5747809249}{2^{28}\tau^6} + \cdots, \tag{A.46.9}$$

$$\mathcal{A}_4(\tau) = 1 + \frac{3}{2\tau} + \frac{63}{2^5\tau^2} + \frac{823}{2^8\tau^3} + \frac{3125}{2^9\tau^4} + \frac{117593}{2^{13}\tau^5} + \frac{5098173}{2^{17}\tau^6} + \cdots, \tag{A.46.10}$$

$$\mathcal{B}_0(\tau) = \frac{1}{2} - \frac{1}{2^4\tau^2} - \frac{13}{2^8\tau^4} - \frac{413}{2^{11}\tau^6} - \frac{119197}{2^{16}\tau^8} - \frac{15278735}{2^{19}\tau^{10}} - \frac{6115520681}{2^{23}\tau^{12}} - \cdots, \tag{A.46.11}$$

$$\mathcal{B}_1(\tau) = 1 + \frac{1}{2^4\tau} + \frac{9}{2^9\tau^2} + \frac{587}{2^{13}\tau^3} + \frac{13915}{2^{19}\tau^4} + \frac{1502351}{2^{23}\tau^5} + \frac{21470173}{2^{28}\tau^6} + \cdots, \tag{A.46.12}$$

$$\mathcal{B}_2(\tau) = 1 + \frac{5}{8\tau} + \frac{61}{2^7\tau^2} + \frac{745}{2^{10}\tau^3} + \frac{41579}{2^{15}\tau^4} + \frac{839407}{2^{18}\tau^5} + \frac{36598265}{2^{22}\tau^6} + \cdots, \tag{A.46.13}$$

$$\mathcal{B}_3(\tau) = 1 + \frac{17}{2^4\tau} + \frac{581}{2^9\tau^2} + \frac{14423}{2^{13}\tau^3} + \frac{1699203}{2^{19}\tau^4} + \frac{65353559}{2^{23}\tau^5} + \frac{5729123937}{2^{28}\tau^6} + \cdots, \tag{A.46.14}$$

$$\mathcal{B}_4(\tau) = 1 + \frac{3}{2\tau} + \frac{63}{2^5\tau^2} + \frac{823}{2^8\tau^3} + \frac{12499}{2^{11}\tau^4} + \frac{117577}{2^{13}\tau^5} + \frac{5097387}{2^{17}\tau^6} + \cdots. \tag{A.46.15}$$

Similarly, let us write $\log \mathcal{F}_\pm(\zeta, \tau)$ as in Eq. (21.4.4),

$$\log \mathcal{F}_\pm(\zeta, \tau) = -\sum_{n=0}^{\infty}\left[\alpha\left(\zeta - \frac{2}{\pi}\right)\frac{e^{2\tau}}{\sqrt{\tau}}\right]^n \mathcal{F}_{\pm n}(\tau). \tag{21.4.4}$$

Then first, $\mathcal{F}_{\pm 0}(\tau)$ coincide with the right-hand side of Eq. (21.4.6), except that the constant terms remain unknown. We have listed the coefficients a_n up to $n = 25$ in Appendix A.47 below.

The $\mathcal{F}_{\pm n}(\tau)$ for $n = 1, 2, 3, \ldots$ have the expansions

$$\mathcal{F}_{+1}(\tau) = 1 + \frac{7}{16\tau} + \frac{205}{2^9\tau^2} + \frac{5305}{2^{13}\tau^3} + \frac{682963}{2^{19}\tau^4} + \frac{28971265}{2^{23}\tau^5} + \frac{2835864953}{2^{28}\tau^6} + \cdots, \tag{A.46.16}$$

$$\mathcal{F}_{+2}(\tau) = \frac{1}{2} + \frac{7}{16\tau} + \frac{119}{2^8\tau^2} + \frac{1485}{2^{11}\tau^3} + \frac{90647}{2^{16}\tau^4} + \frac{1773137}{2^{19}\tau^5} + \frac{80288703}{2^{23}\tau^6} + \cdots, \tag{A.46.17}$$

$$\mathcal{F}_{+3}(\tau) = \frac{1}{3} + \frac{7}{16\tau} + \frac{287}{2^9\tau^2} + \frac{22475}{3 \times 2^{13}\tau^3} + \frac{926961}{2^{19}\tau^4} + \frac{35417889}{2^{23}\tau^5} + \frac{9421927049}{3 \times 2^{28}\tau^6} + \cdots, \tag{A.46.18}$$

$$\mathcal{F}_{+4}(\tau) = \frac{1}{4} + \frac{7}{16\tau} + \frac{21}{2^5\tau^2} + \frac{1159}{2^{10}\tau^3} + \frac{36589}{2^{14}\tau^4} + \frac{344937}{2^{16}\tau^5} + \frac{7512339}{2^{17}\tau^6} + \cdots, \tag{A.46.19}$$

$$\mathcal{F}_{-1}(\tau) = \frac{1}{8\tau}\left(1 + \frac{19}{16\tau} + \frac{1069}{2^9\tau^2} + \frac{39293}{2^{13}\tau^3} + \frac{7099315}{2^{19}\tau^4} + \frac{378600725}{2^{23}\tau^5} + \cdots\right), \tag{A.46.20}$$

$$\mathcal{F}_{-2}(\tau) = \frac{1}{(8\tau)^2}\left(\frac{1}{2} + \frac{19}{16\tau} + \frac{691}{2^8\tau^2} + \frac{13725}{2^{11}\tau^3} + \frac{1226375}{2^{16}\tau^4} + \frac{30927109}{2^{19}\tau^5} + \cdots\right), \tag{A.46.21}$$

$$\mathcal{F}_{-3}(\tau) = \frac{1}{(8\tau)^3}\left(\frac{1}{3} + \frac{19}{16\tau} + \frac{1743}{2^9\tau^2} + \frac{236615}{3 \times 2^{13}\tau^3} + \frac{15023537}{2^{19}\tau^4} + \frac{773134357}{2^{23}\tau^5} + \cdots\right), \tag{A.46.22}$$

$$\mathcal{F}_{-4}(\tau) = \frac{1}{(8\tau)^4}\left(\frac{1}{4} + \frac{19}{16\tau} + \frac{263}{2^6\tau^2} + \frac{13427}{2^{10}\tau^3} + \frac{686667}{2^{14}\tau^4} + \frac{9145995}{2^{16}\tau^5} + \cdots\right). \tag{A.46.23}$$

For $n \geqslant 1$ we write

$$\frac{E_2(n,t)}{E_2(0,t)} = \left(\frac{e^{2\tau}}{4\sqrt{\pi\tau}}\right)^n \mathcal{E}_n(\tau), \tag{A.46.24}$$

$$\frac{E_\pm(n,t)}{E_\pm(0,t)} = \left(\frac{e^{2\tau}}{4\sqrt{\pi\tau}}\right)^n \mathcal{E}_{\pm n}(\tau). \tag{A.46.25}$$

Then

$$\mathcal{E}_1(\tau) = 1 + \frac{9}{2^4\tau} + \frac{281}{2^9\tau^2} + \frac{7443}{2^{13}\tau^3} + \frac{997307}{2^{19}\tau^4} + \frac{43169895}{2^{23}\tau^5} + \frac{4350267853}{2^{28}\tau^6} + \cdots, \tag{A.46.26}$$

$$\mathcal{E}_2(\tau) = \frac{1}{8\tau} + \frac{15}{2^6\tau^2} + \frac{485}{2^{10}\tau^3} + \frac{9327}{2^{13}\tau^4} + \frac{846251}{2^{18}\tau^5} + \frac{22426869}{2^{21}\tau^6} + \cdots, \quad \text{(A.46.27)}$$

$$\mathcal{E}_3(\tau) = \frac{1}{2^8\tau^3} + \frac{75}{2^{12}\tau^4} + \frac{8865}{2^{17}\tau^5} + \frac{515409}{2^{21}\tau^6} + \frac{126126891}{2^{27}\tau^7} + \frac{8338717077}{2^{31}\tau^8} + \cdots, \quad \text{(A.46.28)}$$

$$\mathcal{E}_4(\tau) = \frac{3}{2^{16}\tau^6} + \frac{117}{2^{18}\tau^7} + \frac{6063}{2^{21}\tau^8} + \frac{271683}{2^{24}\tau^9} + \frac{11613831}{2^{27}\tau^{10}} + \frac{248599305}{2^{29}\tau^{11}} + \cdots, \quad \text{(A.46.29)}$$

$$\mathcal{E}_{+1}(\tau) = 1 + \frac{7}{2^4\tau} + \frac{205}{2^9\tau^2} + \frac{5305}{2^{13}\tau^3} + \frac{682963}{2^{19}\tau^4} + \frac{28971265}{2^{23}\tau^5} + \frac{2835864953}{2^{28}\tau^6} + \cdots, \quad \text{(A.46.30)}$$

$$\mathcal{E}_{+2}(\tau) = \frac{1}{2^5\tau^2} + \frac{25}{2^8\tau^3} + \frac{1159}{2^{12}\tau^4} + \frac{29519}{2^{15}\tau^5} + \frac{3392647}{2^{20}\tau^6} + \frac{109966679}{2^{23}\tau^7} + \cdots, \quad \text{(A.46.31)}$$

$$\mathcal{E}_{+3}(\tau) = \frac{3}{2^{14}\tau^6} + \frac{531}{2^{18}\tau^7} + \frac{126207}{2^{23}\tau^8} + \frac{13057605}{2^{27}\tau^9} + \frac{5175417897}{2^{33}\tau^{10}} + \cdots, \quad \text{(A.46.32)}$$

$$\mathcal{E}_{+4}(\tau) = \frac{135}{2^{28}\tau^{12}} + \frac{14715}{2^{30}\tau^{13}} + \frac{473985}{2^{31}\tau^{14}} + \frac{190732185}{2^{36}\tau^{15}} + \cdots, \quad \text{(A.46.33)}$$

$$\mathcal{E}_{-1}(\tau) = \frac{1}{8\tau} + \frac{19}{2^7\tau^2} + \frac{1069}{2^{12}\tau^3} + \frac{39293}{2^{16}\tau^4} + \frac{7099315}{2^{22}\tau^5} + \frac{378600725}{2^{26}\tau^6} + \cdots, \quad \text{(A.46.34)}$$

$$\mathcal{E}_{-2}(\tau) = \frac{3}{2^{11}\tau^4} + \frac{147}{2^{14}\tau^5} + \frac{11073}{2^{18}\tau^6} + \frac{401025}{2^{21}\tau^7} + \frac{60024117}{2^{26}\tau^8} + \cdots, \quad \text{(A.46.35)}$$

$$\mathcal{E}_{-3}(\tau) = \frac{45}{2^{23}\tau^9} + \frac{12825}{2^{27}\tau^{10}} + \frac{4558545}{2^{32}\tau^{11}} + \frac{662542335}{2^{36}\tau^{12}} + \cdots, \quad \text{(A.46.36)}$$

$$\mathcal{E}_{-4}(\tau) = \frac{4614911831500349815001 1}{2^{49}\tau^{16}} + \cdots. \quad \text{(A.46.37)}$$

Formulae similar to Eq. (21.4.21) can be written for the coefficient of τ^{-k} in $\mathcal{F}_{\pm j}(\tau)$, $\mathcal{E}_{\pm j}(\tau)$ and $\mathcal{E}_j(\tau)$.

A.47 Table of Values of a_n in Eq. (21.4.6) for Small Values of n

The a_n's are rational numbers; we list below their numerator and denominator as well as their approximate numerical values.

One can compute the a_n's as rational numbers easily up to, say, $n = 100$. The numerators increase fast, and often contain large prime factors, while the denominators are always powers of 2 multiplied eventually by one or two prime factors of n. One can also compute the a_n's as floating point numbers as easily up to, say, $n = 200$; and study their large n behaviour, see the remarks in Appendix A.48 below.

n	Numerator	Denominator	Approximate numerical value
3	3	2^8	0.01171875
4	0		0.
5	45	2^{11}	0.02197265625
6	63	2^{12}	0.01538085938
7	7875	2^{16}	0.1201629639
8	25875	2^{17}	0.1974105835
9	733087	2^{19}	1.398252487
10	3820185	2^{20}	3.643212318
11	231485175	2^{23}	27.59518325
12	1655539131	2^{24}	98.67782182
13	55154065725	2^{26}	821.8596239
14	502263708975	2^{27}	3742.156245
15	737940519480303	5×2^{32}	34363.03323
16	1634013966582675	2^{33}	190224.2618
17	65920025157673275	2^{35}	1918525.236
18	859831419329548021	2^{36}	12512193.92
19	75777172758541734525	2^{39}	137837874.3
20	1138241713301308940625	2^{40}	1035224808.
21	381310670655387556644693	7×2^{42}	$1.238571552 \times 10^{10}$
22	926158171297749761450025	2^{43}	$1.052919937 \times 10^{11}$
23	191511646102089743313342375	2^{47}	$1.36077209 \times 10^{12}$
24	3636119003643910646472914703	2^{48}	$1.291808972 \times 10^{13}$
25	202042840811663082772877338251	2^{50}	$1.794500911 \times 10^{14}$

A.48 Some Remarks About the Numerical Computations of the Last Two Appendices

As compared with other methods, that described in Section 21.4 seems to provide more easily successive terms in large t asymptotic expansions. We just need algebraic calculations which can be performed routinely by software systems of symbolic programming like MATHEMATICA or AMP. This enables us to investigate numerically for large n, the behaviour of the nth coefficient of various asymptotic expansions in t. As typical examples, let us write the asymptotic expansions (A.46.1), (A.46.6) and (A.46.11) as follows

$$\mathcal{S}_0(\tau) = \frac{1+i}{2} \sum_{n=0}^{\infty} (\rho_n + i\sigma_n) \frac{1}{\tau^n}, \tag{A.48.1}$$

$$\mathcal{A}_0(\tau) = \frac{\tau}{2} + \sum_{n=0}^{\infty} \frac{\alpha_{2n+1}}{\tau^{2n+1}}, \tag{A.48.2}$$

$$\mathcal{B}_0(\tau) = \sum_{n=0}^{\infty} \frac{\beta_{2n}}{\tau^{2n}}. \tag{A.48.3}$$

Then, for large n

$$\rho_n = -\left(\frac{n}{2e}\right)^n \frac{1}{n} \exp\left\{-c + \frac{1}{3n} - \frac{1}{4n^2} + \mathrm{O}\left(\frac{1}{n^3}\right)\right\}, \tag{A.48.4}$$

$$\sigma_n = -\left(\frac{n}{2e}\right)^n \frac{1}{n} \exp\left\{-c - \frac{1}{6n} - \frac{3}{4n^2} + \mathrm{O}\left(\frac{1}{n^3}\right)\right\}, \tag{A.48.5}$$

$$\alpha_n = \left(\frac{n}{2e}\right)^n \frac{1}{n} \exp\left\{-c - \frac{1}{6n} - \frac{1}{2n^2} + \mathrm{O}\left(\frac{1}{n^3}\right)\right\}, \tag{A.48.6}$$

$$\beta_n = -\left(\frac{n}{2e}\right)^n \frac{1}{n} \exp\left\{-c + \frac{1}{3n} + \mathrm{O}\left(\frac{1}{n^3}\right)\right\}, \tag{A.48.7}$$

where the constant c is

$$c = 0.451582705289\ldots. \tag{A.48.8}$$

It is to be noticed that the same dominant factor $(n/2e)^n(1/n)$ appears everywhere.

The coefficients a_n of Eq. (21.4.6) have two similar large n expansions, one for even n and another one for odd n

$$a_n = \left(\frac{n}{2e}\right)^n \frac{1}{n} \exp\left\{\sum_{k=0}^{\infty} \frac{\kappa_k^{\pm}}{\tau^k}\right\}. \tag{A.48.9}$$

The constant term in the exponential does not depend on the parity of n

$$\kappa_0^{\pm} = -2.144729885849\ldots \tag{A.48.10}$$

while

$$\begin{array}{llll} \kappa_1^{+} = -2.763930691\ldots, & \kappa_2^{+} = -4.994470\ldots, & \ldots & (n \text{ even}), \\ \kappa_1^{-} = +0.930597358\ldots, & \kappa_2^{-} = -2.394586\ldots, & \ldots & (n \text{ odd}). \end{array} \tag{A.48.11}$$

It is not clear numerically whether these $\kappa_k^{\pm}$ are rational.

A.49 Convolution of Two Gaussian Kernels

For $V_j(x) = a_j x^2$, $j = 1, \ldots, p$, setting

$$W_{a,b,c}(x, y) := \exp\left(-\frac{1}{2}ax^2 - \frac{1}{2}by^2 + cxy\right), \tag{A.49.1}$$

one gets according to Eq. (23.1.11) the multiplication law

$$(W_{a,b,c} * W_{a',b',c'})(x,y) = \left(\frac{2\pi}{b+a'}\right)^{1/2} W_{a'',b'',c''}(x,y), \tag{A.49.2}$$

where

$$a'' = a - \frac{c^2}{b+a'}, \quad b'' = b' - \frac{c'^2}{b+a'}, \quad c'' = \frac{cc'}{b+a'}. \tag{A.49.3}$$

For $w_k(x,y) = W_{a_k,a_{k+1},c_k}(x,y)$ a repeated use of the above multiplication law yields

$$W(x,y) := (w_1 * w_2 * \cdots * w_{p-1})(x,y) = d \times W_{a,b,c}(x,y), \tag{A.49.4}$$

where a, b, c and d are constants depending on the parameters $a_1, \ldots, a_p$ and $c_1, \ldots, c_{p-1}$.

The orthogonality relation (23.1.12) of the polynomials $P_j(x)$ and $Q_j(x)$ takes the form

$$\int P_j(x) W(x,y) Q_k(y)\, dx\, dy = h_j \delta_{jk}, \tag{A.49.5}$$

namely the same relation as in the two matrix case with the weight $W(x,y)$, an exponential of a quadratic form in x and y. It follows that $P_j(x)$ and $Q_j(x)$ are Hermite polynomials of x times a constant

$$P_j(x) = H_j(\alpha x), \qquad \alpha := \left(\frac{ab-c^2}{2b}\right)^{1/2}; \tag{A.49.6}$$

$$Q_j(x) = H_j(\beta x), \qquad \beta := \left(\frac{ab-c^2}{2a}\right)^{1/2}; \tag{A.49.7}$$

$$h_j = \frac{2\pi}{(ab-c^2)^{1/2}} \left(\frac{c}{\sqrt{ab}}\right)^j 2^j j! d. \tag{A.49.8}$$

The eigenvalue density of the matrix A_1, for example, ignoring the eigenvalues of other matrices, is from Eq. (23.1.17)

$$\begin{aligned} R_1(x) = K_{11}(x,x) &= \sum_{j=0}^{n-1} \frac{1}{h_j} P_j(x) \int W(x,y) Q_j(y)\, dy \\ &= d\left(\frac{2\pi}{b}\right)^{1/2} e^{-\alpha^2 x^2} \sum_{j=0}^{n-1} \frac{1}{h_j} \left(\frac{c}{\sqrt{ab}}\right)^j H_j^2(\alpha x) \\ &= \frac{\alpha}{\sqrt{\pi}} e^{-\alpha^2 x^2} \sum_{j=0}^{n-1} \frac{H_j^2(\alpha x)}{2^j j!}, \end{aligned} \tag{A.49.9}$$

which in the large n limit is a semi-circle of radius $\sqrt{2n}/\alpha$. Thus in this particular case of coupled matrices one recovers Wigner's "semi-circle law" for the eigenvalues of a single matrix.

A.50 Method of the Change of Variables. Wick's Theorem

Important information can sometimes be obtained by a simple change of variables. For example, let us change variables from A to $A + \varepsilon A^k$, i.e., instead of the n^2 variables $A_{p,p}$, $\operatorname{Re} A_{p,q}$, $\operatorname{Im} A_{p,q}$, $1 \leqslant p < q \leqslant n$, of the complex Hermitian $n \times n$ matrix A we take $A_{p,q} + \varepsilon (A^k)_{p,q}$ as new variables. The Jacobian matrix is

$$\frac{\partial (A + \varepsilon A^k)_{j,\ell}}{\partial A_{p,q}} = \delta_{j,p}\delta_{\ell,q} + \varepsilon \sum_{s=1}^{k} (A^{s-1})_{j,p} (A^{k-s})_{q,\ell}, \tag{A.50.1}$$

so that for small ε one has

$$\det\left[\frac{\partial (A + \varepsilon A^k)_{j,\ell}}{\partial A_{p,q}}\right] = 1 + \varepsilon \sum_{s=1}^{k} \operatorname{tr} A^{s-1} \operatorname{tr} A^{k-s} + \mathrm{O}(\varepsilon^2). \tag{A.50.2}$$

This last step is valid only when A is complex Hermitian. When A is real symmetric, one should take the sums with $j \leqslant \ell$, $p \leqslant q$, and the final sum over j and q is not independent. When A is complex Hermitian, elements $j < \ell$ have real and imaginary parts and so both $j < \ell$ and $j > \ell$ can be considered appearing once each; similarly, for $p < q$ and $p > q$. When A is quaternion self-dual, it does not work either since there are too many terms in the sum.

From

$$\int e^{-\operatorname{tr} A^2} \operatorname{tr} A^\ell \, dA = \int e^{-\operatorname{tr}(A+\varepsilon A^k)^2} \operatorname{tr}(A + \varepsilon A^k)^\ell \, d(A + \varepsilon A^k), \tag{A.50.3}$$

keeping only the terms linear in ε, one has

$$2\langle \operatorname{tr} A^{k+1} \operatorname{tr} A^\ell \rangle = \ell \langle \operatorname{tr} A^{\ell+k-1} \rangle + \sum_{s=1}^{k} \langle \operatorname{tr} A^{s-1} \operatorname{tr} A^{k-s} \operatorname{tr} A^\ell \rangle, \tag{A.50.4}$$

where we have used the notation

$$\langle f(A) \rangle := \int e^{-\operatorname{tr} A^2} f(A) \, dA \div \int e^{-\operatorname{tr} A^2} \, dA, \tag{A.50.5}$$

or from

$$\int e^{-\operatorname{tr} A^2} \operatorname{tr} A^\ell \operatorname{tr} A^m \, dA$$

$$= \int e^{-\operatorname{tr}(A+\varepsilon A^k)^2} \operatorname{tr}(A + \varepsilon A^k)^\ell \operatorname{tr}(A + \varepsilon A^k)^m \, d(A + \varepsilon A^k), \tag{A.50.6}$$

one has

$$2\langle \operatorname{tr} A^{k+1} \operatorname{tr} A^{\ell} \operatorname{tr} A^{m}\rangle = \ell\langle \operatorname{tr} A^{\ell+k-1}\rangle + m\langle \operatorname{tr} A^{m+k-1}\rangle + \sum_{s=1}^{k} \langle \operatorname{tr} A^{s-1} \operatorname{tr} A^{k-s} \operatorname{tr} A^{\ell} \operatorname{tr} A^{m}\rangle. \tag{A.50.7}$$

Equation (A.50.4) (or (A.50.7)), sometimes called "loop equation", for various values of k and ℓ gives for example,

$$\langle \operatorname{tr} A \operatorname{tr} A^{\ell}\rangle = \frac{\ell}{2}\langle \operatorname{tr} A^{\ell-1}\rangle, \tag{A.50.8}$$

$$\langle \operatorname{tr} A^2 \operatorname{tr} A^{\ell}\rangle = \left(\frac{\ell}{2} + \frac{n^2}{2}\right)\langle \operatorname{tr} A^{\ell}\rangle, \tag{A.50.9}$$

$$\langle \operatorname{tr} A^{k+1}\rangle = \frac{1}{2}\sum_{s=1}^{k}\langle \operatorname{tr} A^{s-1} \operatorname{tr} A^{k-s}\rangle, \tag{A.50.10}$$

$$\langle \operatorname{tr} A \operatorname{tr} A^{k+1}\rangle = \frac{1}{2}\langle \operatorname{tr} A^{k}\rangle + \frac{1}{2}\sum_{s=1}^{k}\langle \operatorname{tr} A \operatorname{tr} A^{s-1} \operatorname{tr} A^{k-s}\rangle = \frac{k+1}{2}\langle \operatorname{tr} A^{k}\rangle, \tag{A.50.11}$$

$$\langle \operatorname{tr} A^2 \operatorname{tr} A^{k+1}\rangle = \langle \operatorname{tr} A^{k+1}\rangle + \frac{1}{2}\sum_{s=1}^{k}\langle \operatorname{tr} A^2 \operatorname{tr} A^{s-1} \operatorname{tr} A^{k-s}\rangle = \frac{n^2+k+1}{2}\langle \operatorname{tr} A^{k+1}\rangle. \tag{A.50.12}$$

Thus

$$\langle \operatorname{tr} A^2\rangle = \frac{n^2}{2}, \qquad \langle (\operatorname{tr} A)^2\rangle = \frac{n}{2}, \tag{A.50.13}$$

$$\langle \operatorname{tr} A^4\rangle = \frac{1}{2}\langle [2n \operatorname{tr} A^2 + (\operatorname{tr} A)^2]\rangle = \frac{n}{4}(2n^2-1), \tag{A.50.14}$$

$$\langle \operatorname{tr} A \operatorname{tr} A^3\rangle = \frac{1}{2}\langle \operatorname{tr} A^2\rangle + \frac{n}{2}\langle (\operatorname{tr} A)^2\rangle = \frac{3}{4}n^2, \tag{A.50.15}$$

etc.

One may think of other changes of variables giving other relations among the mean values of products of traces of various powers of A.

For Gaussian matrix elements another useful trick is to apply "Wick's theorem" which expresses the mean value of a product of Gaussian variables as a sum of products of mean values taken two factors at a time in all possible ways. If A is an $n \times n$ matrix with complex numbers as elements, $A^\dagger$ its Hermitian conjugate, has the probability density $\exp(-\operatorname{tr} AA^\dagger)$ Gaussian; then one has

$$\langle a_{ij}a_{pq}^*\rangle = \delta_{ip}\delta_{jq}\left(\frac{1}{2}+\frac{1}{2}\right) = \delta_{ip}\delta_{jq},$$

(the two $1/2$'s coming from real and imaginary parts). And one has for example,

$$\begin{aligned}\left\langle \operatorname{tr} A^{k+1} \operatorname{tr} A^{\dagger k+1}\right\rangle &= \sum_{p,q,r,s} \langle (A^k)_{pq}(A^{\dagger k})_{rs}\rangle\langle A_{qp}A_{sr}^\dagger\rangle \\ &= (k+1)\sum_{p,q,r,s} \langle (A^k)_{pq}(A^{\dagger k})_{rs}\rangle\delta_{qs}\delta_{pr} \\ &= (k+1)\left\langle \operatorname{tr} A^k A^{\dagger k}\right\rangle, \end{aligned} \tag{A.50.16}$$

the factor $(k+1)$ in the above comes from the fact that any one of A^{k+1} could have been taken to be paired with a $A^\dagger$.

A.51 Some Remarks About the Integral $I(k,n)$, Eq. (25.2.12)

Consider in general the integral

$$I(k,n,\gamma) := \frac{1}{n!}(2\pi)^{-n}\int_0^{2\pi} d\theta_1 \cdots \int_0^{2\pi} d\theta_n \left|\sum_{j=1}^n e^{i\theta_j}\right|^{2k} \left|\Delta(\theta)\right|^{2\gamma}, \tag{A.51.1}$$

with

$$\Delta(\theta) := \prod_{1\leqslant p<q\leqslant n} \left(e^{i\theta_p} - e^{i\theta_q}\right). \tag{A.51.2}$$

We have come across $I(0,n,\gamma)$ as the normalization constant in Chapter 12. Its value was computed in at least two ways; from the general theorem of Dyson–Good, Section 12.1, or from Selberg integral, Section 17.7. For the theory of random permutations one needs $I(k,n) \equiv I(k,n,1)$ specially for k and n large, since $I(k,n)$ is the number of permutations of $(1,2,\ldots,k)$ that have no increasing subsequence of length greater than n. Set $L(k,n) = I(k,n) - I(k,n-1)$ so that $L(k,n)$ is the number of permutations of

$(1,2,\ldots,k)$ having the length of the longest increasing subsequence exactly equal to n. From these considerations one has for example (see also Rogers, 1978)

$$I(k,n) = k! \quad \text{for } n \geqslant k, \tag{A.51.3}$$

$$I(k-1,k) = k! - 1, \tag{A.51.4}$$

$$I(k,1) = 1, \quad I(k,2) = \frac{(2k)!}{k!(k+1)!}, \tag{A.51.5}$$

$$I(k,3) = 2\sum_{j=0}^{k} \binom{2j}{j}\binom{k}{j}^2 \frac{3j^2 \cdot 2j - 2jk - k - 1}{(j+1)^2(j+2)(k-j+1)}, \tag{A.51.6}$$

$$L(k,k) = 1, \quad L(k,k-1) = (k-1)^2, \tag{A.51.7}$$

$$L(k,k-2) = \binom{k-1}{2} + \left[\binom{k-1}{2} - 1\right]\left[2\binom{k-1}{2} - 1\right], \tag{A.51.8}$$

$$L(k,k-3) = \text{a polynomial of order 6 in } k. \tag{A.51.9}$$

Equation (A.51.6) gives for example the values of $I(k,3)$ for $k = 0, 1, 2, \ldots, 10$ respectively as 1, 1, 2, 6, 23, 103, 513 = 19.27, 2761 = 11.251, 15767, 94359 = 3.71.443, 293295 = 15.19553.

The method of Dyson–Good (or of Selberg) can be used to compute $I(k,n,\gamma)$ for small k or for small n with some effort. As in Section 12.1 observe that

$$\begin{aligned} |\Delta(\theta)|^2 &= \prod_{1\leqslant p<q\leqslant n} \left(e^{i\theta_p} - e^{i\theta_q}\right)\left(e^{-i\theta_p} - e^{-i\theta_q}\right) \\ &= \prod_{1\leqslant p<q\leqslant n} \left(1 - e^{i(\theta_p-\theta_q)}\right)\left(1 - e^{i(\theta_q-\theta_p)}\right) \\ &= \prod_{p\neq q=1}^{n} \left(1 - e^{i(\theta_p-\theta_q)}\right), \end{aligned} \tag{A.51.10}$$

and for any integer m

$$\int_0^{2\pi} \frac{d\theta}{2\pi} e^{im\theta} = \delta_{m0}. \tag{A.51.11}$$

So that if we expand

$$\prod_{1\leqslant p\neq q\leqslant n} \left(1 - \frac{x_p}{x_q}\right)^{\gamma} \tag{A.51.12}$$

in positive and negative powers of $x_1, \ldots, x_n$, then the constant term $C(0)$ will be equal to $I(0, n, \gamma)$.

Similarly, $I(k, n, \gamma)$ is the constant term in the expansion of

$$(x_1 + \cdots + x_n)^k \left(\frac{1}{x_j} + \cdots + \frac{1}{x_n} \right)^k \prod_{1 \leqslant p \neq q \leqslant n} \left(1 - \frac{x_p}{x_q} \right)^\gamma \tag{A.51.13}$$

in positive and negative powers of all the variables. Thus we get (after long computations) for example,

$$I(0, n, \gamma) = \frac{(n\gamma)!}{n!(\gamma!)^n}, \tag{A.51.14}$$

$$I(1, n, \gamma) = \frac{(n\gamma)!}{n!(\gamma!)^n} \frac{n}{2(n-1)\gamma + 2}, \tag{A.51.15}$$

$$I(k, 1, \gamma) = 1, \quad I(k, 2, \gamma) = \frac{(2k)!(2\gamma)!}{2k!\gamma!(k+\gamma)!}. \tag{A.51.16}$$

Odlyzko et al. have computed $L(k, n)$ for $k \leqslant 120$. There seems to be no general formula.

A.52 Meijer G-functions for Small and Large Values of the Variable

A.52.1 Computation of the G-functions Near the Origin. The Meijer G-functions have been extensively studied and we give in this and the next sections their behaviour for small and large values of the variable. For convenience details will be given only for the functions appearing in the case of Gaussian unitary ensemble.

The G-functions have convergent series expansions which are convenient for their numerical evaluation. By definition (cf. Luke (1969), Chapter 5; Bateman (1953a), Sections 5.3–5.6; Gradshteyn and Rizhik (1965), Section 9.3)

$$G_{0,n}^{n,0}(y^2 \mid b_1, \ldots, b_n) = \frac{1}{2i\pi} \int_{\mathcal{L}} y^{2s} \prod_{j=1}^{n} \Gamma(b_j - s)\, ds. \tag{A.52.1}$$

The contour $\mathcal{L}$ goes from $-i\infty$ to $+i\infty$ so that all poles of $\Gamma(b_j - s)$ lie to the right of the path. It can be closed in the right half complex s-plane, so that the G-function is the negative of the sum of residues at its poles which from Eqs. (26.2.10) and (26.2.15) all lie on the non-negative real axis.

When $n = 1$, $G_{0,1}^{1,0}(y^2|0)$ has simple poles at $j = 0, 1, \ldots$ with the residue $(-1)^{j+1} y^{2j}/j!$ and

$$G_{0,1}^{1,0}(y^2 \mid 0) = e^{-y^2}. \tag{A.52.2}$$

For $n = 2$, one can either calculate the residues at the poles or consult the literature (cf. Luke (1969), Section 6.5 (8); Bateman (1953a), Section 5.6 (4)) to find that

$$G_{0,2}^{2,0}(y^2 \mid b_1, b_2) = 2|y|^{b_1+b_2} K_{|b_2-b_1|}(2|y|), \tag{A.52.3}$$

with K_ν the modified Bessel function (Bateman (1953b), Section 7.2.2; Abramowitz and Stegun (1965), Section 9.6). Thus

$$G_{0,2}^{2,0}(y^2 \mid 0, 1) = 2|y| K_1(2|y|), \tag{A.52.4}$$

$$G_{0,2}^{2,0}\left(y^2 \,\middle|\, \frac{1}{2}, \frac{1}{2}\right) = 2|y| K_0(2|y|), \tag{A.52.5}$$

$$G_{0,2}^{2,0}\left(y^2 \,\middle|\, \frac{1}{2}, \frac{3}{2}\right) = 2|y| K_1(2|y|), \tag{A.52.6}$$

$$G_{0,2}^{2,0}\left(y^2 \mid 0, \frac{1}{4}\right) = 2|y|^{1/4} K_{1/4}(2|y|). \tag{A.52.7}$$

The case $n > 2$, is difficult to find in the literature.

For $n \geqslant 2$, in the case of the $\{b_j^+(2)\}$, $s = 0$ is a simple pole with residue $-\prod_{j=1}^n (b_j^+(2))!$; $s = 1$ is a pole of order 2 or 3 according as $n = 2$ or $n \geqslant 3$; $s = 2$ is a pole of order $\min(n, 5)$; $s = 3$ is a pole of order $\min(n, 7)$; and so on. It is straightforward to calculate the residue at $s = j$, by writing

$$\Gamma(k - s) = \left[(k - s)(k + 1 - s) \cdots (j - s)\right]^{-1} \Gamma(j + 1 - s), \quad 0 \leqslant k \leqslant j. \tag{A.52.8}$$

The result is a series

$$G_{0,n}^{n,0}(y^2 \mid b_1^+(2), \ldots, b_n^+(2))$$
$$= 1 + \sum_{j=1}^{\infty} c_+(n, j, \ln|y|) y^{2j} \Big/ \prod_{k=1}^{\min(2j,n)} (j - b_k^+(2))!, \tag{A.52.9}$$

where $c_+(n, j, \ln|y|)$ is a polynomial of order at most $n - 1$ in $\ln|y|$. Similarly, in the case of $\{b_j^-(2)\}$, $s = j_1 + 1/2$ is a pole of order $\min(n, 2j + 2)$, calculation of the residue is again straightforward, and

$$G_{0,n}^{n,0}(y^2 \mid b_1^-(2), \ldots, b_n^-(2))$$
$$= \sum_{j=0}^{\infty} c_-(n, j, \ln|y|)|y|^{2j+1} \Big/ \prod_{k=1}^{\min(2j,n)} (j + 1/2 - b_k^-(2))!, \tag{A.52.10}$$

where $c_-(n, j, \ln|y|)$ is again a polynomial of order at most $n-1$ in $\ln|y|$. For large j, due to the presence of n factorials in the denominator, the convergence of the series (A.52.9) or (A.52.10) is better, larger is n.

A.52.2 G-functions for Large Values of the Variable. When $|x| \to \infty$, $|\arg x| \leqslant (n+1)\pi - \delta$, $\delta > 0$, we have the asymptotic expansion (Luke (1969), 5.7, Theorem 5, (12)–(15))

$$G^{n,0}_{0,n}(x \mid b_1, \ldots, b_n)$$
$$= (2\pi)^{(n-1)/2} n^{-1/2} \exp(-nx^{1/n}) x^{\theta_n} \left[1 + c_n x^{-1/n} + O(x^{-2/n})\right], \qquad \text{(A.52.11)}$$

where

$$\theta_n = \frac{1}{n}\left(\sum_{j=1}^{n} b_j - \frac{n-1}{2}\right), \qquad \text{(A.52.12)}$$

$$c_n = \frac{1}{2}\sum_{j=1}^{n}(b_j)^2 - \frac{1}{2n}\left(\sum_{j=1}^{n} b_j\right)^2 - \frac{n^2-1}{24n}. \qquad \text{(A.52.13)}$$

This gives the asymptotic behaviour of $g_n(\beta, y)$ for $\beta = 2$, any n and for $\beta = 1$, n odd. For example, from the expressions of the $b_j^+(2)$, Eq. (26.2.10), one has for n odd,

$$\theta_n^+(2) = \frac{(n-1)^2}{4n}, \qquad \text{(A.52.14)}$$

$$c_n^+(2) = \frac{(n^2-1)^2}{96n}. \qquad \text{(A.52.15)}$$

Similarly, for n even, Eqs. (26.2.10) and (26.2.15) give

$$\theta_n^{\pm}(2) = \frac{n^2-2n+2}{4n}, \qquad \text{(A.52.16)}$$

$$c_n^+(2) + c_n^-(2) = \frac{n^4-2n^2+4}{48n}, \qquad \text{(A.52.17)}$$

$$c_n^+(2) - c_n^-(2) = \frac{n}{8}. \qquad \text{(A.52.18)}$$

A.53 About Binary Quadratic Forms

Consider an integral primitive positive definite quadratic form $Q(x, y) = ax^2 + bxy + cy^2$;

integral: a, b, c, integers,

primitive: a, b, c have no common factor other than 1,

positive definite: $a > 0$, $c > 0$, the "discriminant" $d := b^2 - 4ac < 0$.

When x and y run through all integer values, the ensemble of integers covered by the integral positive definite quadratic (IPDQ) form $Q(x, y)$ is the same as that covered by the (IPDQ) form $Q'(x, y) := Q(\alpha x + \beta y, \gamma x + \delta y)$, if α, β, γ, δ are integers and $\alpha\delta - \beta\gamma = \pm 1$. The two IPDQ forms Q and Q' have the same discriminant d, they are said to be equivalent. Q' is the transform of Q by the matrix $A := \left(\begin{smallmatrix}\alpha & \beta \\ \gamma & \delta\end{smallmatrix}\right)$, $Q' = AQ$.

For $A = \left(\begin{smallmatrix}1 & k \\ 0 & 1\end{smallmatrix}\right)$, $Q'(x, y) = AQ(x, y) = a'x^2 + b'xy + c'y^2$, with $a' = a$, $b' = b + 2ak$, $c' = ak^2 + bk + c$, so that we can always choose the integer k so that $|b'| \leqslant a'$. Also we may interchange a and c or change the sign of b and get an equivalent Q. Thus among the equivalent IPDQ-forms we can always choose one with $0 \leqslant b \leqslant a \leqslant c$. Then from $-d = 4ac - b^2$ we get $-d \geqslant 4a^2 - a^2 = 3a^2$, or $0 \leqslant b \leqslant a \leqslant \sqrt{|d|/3}$. Thus for a given discriminant d the possible choices for the integers a and b are finite. Also for given a, b and d, c is unique. Thus the number of inequivalent IPDQ-forms for a given discriminant is finite. Among them some forms may not be primitive. The number of inequivalent primitive IPDQ-forms for a given discriminant d is called the class function and is denoted by $h(d)$.

Examples. $Q_1(x, y) = x^2 + 3y^2$ and $Q_2(x, y) = 2x^2 + 2xy + 2y^2$ both have the same discriminant $d = -12$. They are NOT equivalent. No other Q can be written with $|b| \leqslant a \leqslant \sqrt{-d/3} = 2$ and $d = -12$. However Q_2 is not primitive. So $h(-12) = 1$.

$Q_3(x, y) = x^2 + 5y^2$, $Q_4(x, y) = 2x^2 + 2xy + 3y^2$ both are primitive and have the same discriminant $d = -20$. They are NOT equivalent. No other Q can be written with $|b| \leqslant a \leqslant \sqrt{-d/3}$ and $d = -20$. So $h(-20) = 2$. $h(-3) = h(-4) = 1$ with $Q_5(x, y) = x^2 + xy + y^2$ and $Q_6(x, y) = x^2 + y^2$ as the only non equivalent IPDQ forms, respectively.

To count the inequivalent IPDQ forms we can always restrict the coefficients to $0 \leqslant b \leqslant a \leqslant \sqrt{-d/3}$. Also from $b^2 - d = 4ac$ we see that with a, b, c integers, $-d$ can not be of the form $4k + 1$ or $4k + 2$, and b should have the same parity as d. With these simple facts in mind one can with some effort prepare a table of the $h(d)$ and the primitive integral positive definite quadratic forms (PIPDQFs) for small $|d|$. We give here such a table for $-d \leqslant 100$; $Q(x, y) = ax^2 + bxy + cy^2$ is denoted as (a, b, c).

We note here a curiosity not widely known outside the number theory experts. Let us look for integer solutions of $b^2 - d = 4ac$, $-d = 4n - 1$, $0 \leqslant b \leqslant a \leqslant \sqrt{-d/3}$, $a \leqslant c$. Then b has to be odd, say $b = 2j + 1$, and $(2j + 1)^2 + 4n - 1 = 4ac$, or $j(j + 1) + n = ac$. So if $j(j + 1) + n$ is prime for $0 \leqslant j \leqslant \sqrt{n/3}$, then the only integer solution looked for will be $a = b = 1$, $c = n$ with the unique PIPDQF $x^2 + xy + ny^2$. Hence $h(1 - 4n) = 1$. A theorem of number theory then says that $j(j + 1) + n$ is prime not only for $0 \leqslant j \leqslant \sqrt{n/3}$ that we required, but for $0 \leqslant j < n - 1$. Such is the case for $n = 1$, 2, 3, 5, 11, 17 and 41. For example, $j(j + 1) + 41$ gives a prime integer for $0 \leqslant j \leqslant 39$. The largest number known with this property is 41.

Table A.53.1. Inequivalent primitive integral positive definite quadratic forms for some small values of the discriminant

$-d$	$h(d)$	*PIPDQF*	$-d$	$h(d)$	*PIPDQF*
3	1	(1, 1, 1)	4	1	(1, 0, 1)
7	1	(1, 1, 2)	8	1	(1, 0, 2)
11	1	(1, 1, 3)	12	1	(1, 0, 3)
15	2	(1, 1, 4), (2, 1, 2)	16	1	(1, 0, 4)
19	1	(1, 1, 5)	20	2	(1, 0, 5), (2, 2, 3)
23	2	(1, 1, 6), (2, 1, 3)	24	2	(1, 0, 6), (2, 0, 3)
27	1	(1, 1, 7)	28	1	(1, 0, 7)
31	2	(1, 1, 8), (2, 1, 4)	32	2	(1, 0, 8), (3, 2, 3)
35	2	(1, 1, 9), (3, 1, 3)	36	2	(1, 0, 9), (2, 2, 5)
39	3	(1, 1, 10), (2, 1, 5), (3, 3, 4)	40	2	(1, 0, 10), (2, 0, 5)
43	1	(1, 1, 11)	44	3	(1, 0, 11), (2, 2, 6), (3, 2, 4)
47	3	(1, 1, 12), (2, 1, 6), (3, 1, 4)	48	2	(1, 0, 12), (3, 0, 4)
51	2	(1, 1, 13), (3, 3, 5)	52	2	(1, 0, 13), (2, 2, 7)
55	3	(1, 1, 14), (2, 1, 7), (4, 3, 4)	56	3	(1, 0, 14), (2, 0, 7), (3, 2, 5)
59	2	(1, 1, 15), (3, 1, 5)	60	2	(1, 0, 15), (3, 0, 5)
63	3	(1, 1, 16), (2, 1, 8), (4, 1, 4)	64	2	(1, 0, 16), (4, 4, 5)
67	1	(1, 1, 17)	68	3	(1, 0, 17), (2, 2, 9), (3, 2, 6)
71	4	(1, 1, 18), (2, 1, 9), (3, 1, 6), (4, 3, 5)	72	2	(1, 0, 18), (2, 0, 9)
75	2	(1, 1, 19), (3, 3, 7)	76	2	(1, 0, 19), (3, 3, 7)
79	3	(1, 1, 20), (2, 1, 10), (4, 1, 5)	80	3	(1, 0, 20), (4, 0, 5), (3, 2, 7)
83	2	(1, 1, 21), (3, 1, 7)			
84	4	(1, 0, 21), (3, 0, 7), (2, 2, 11), (5, 4, 5)			
87	4	(1, 1, 22), (2, 1, 11), (3, 3, 8), (4, 3, 6)	88	2	(1, 0, 22), (2, 0, 11)
91	2	(1, 1, 23), (5, 3, 5)	92	2	(1, 0, 23), (3, 2, 8)
95	5	(1, 1, 24), (2, 1, 12), (3, 1, 8), (4, 1, 6), (5, 5, 6)			
96	4	(1, 0, 24), (3, 0, 8), (5, 2, 5), (4, 4, 7)			
99	2	(1, 1, 25), (5, 1, 5)	100	2	(1, 0, 25), (2, 2, 13)
103	3	(1, 1, 26), (2, 1, 13), (4, 3, 7)			
104	4	(1, 0, 26), (2, 0, 13), (3, 2, 9), (5, 4, 6)			
107	2	(1, 1, 27), (3, 1, 9)	108	2	(1, 0, 27), (4, 2, 7)

The values of $h(-d)$ listed in Table A.53.1 often differ from those of the number theory people. One reason is that they distinguish between transformations with $\alpha\delta - \beta\gamma = +1$ and $\alpha\delta - \beta\gamma = -1$, which we do not. For example, for $d = -23$, they count the PIPDQF's $(2, 1, 3)$ and $(2, -1, 3)$ as distinct. Also there are other differences. For example, for $-d = 12, 16, 27$ and 28 we list $h(d) = 1$, while the number theory people will list $h(d) = 2$. (See e.g. Cohen in Waldschmidt et al. (1990), pp. 226–227). The reason of these differences cannot be explained here by an incompetent author.

NOTES

Our knowledge of random matrices is most extensive for the following ensembles.

Gaussian unitary ensemble or GUE. This is the ensemble of $N \times N$ Hermitian matrices with the joint probability density proportional to $\exp(-\operatorname{tr} H^2)$; i.e., apart from the Hermitian character the real and imaginary parts of every matrix element is a Gaussian random variable with a common variance. This ensemble is invariant under unitary transformations and is appropriate to describe systems without time reversal symmetry. The joint probability density of the eigenvalues is proportional to

$$|\Delta(x)|^2 \exp\left(-\sum_{i=1}^{N} x_i^2\right) \tag{N.1}$$

where

$$\Delta(x) = \prod_{1 \leqslant i < j \leqslant N} (x_j - x_i). \tag{N.2}$$

The n-point correlation and cluster functions for any finite n (and finite or infinite N) are known, as is the probability $E_2(r, s)$ of having exactly r eigenvalues in a randomly chosen interval of length s. (See Chapter 6.)

Gaussian orthogonal ensemble or GOE. This is the ensemble of $N \times N$ real symmetric matrices with the joint probability density proportional to $\exp(-\operatorname{tr} H^2/2)$; i.e., apart from the symmetry every matrix element is a Gaussian random variable with the same variance. This ensemble is invariant under orthogonal transformations and is appropriate to describe systems with time reversal and rotational symmetry; i.e., most of the physical systems found in nature. The joint probability density of the eigenvalues is proportional to

$$|\Delta(x)| \exp\left(-\sum_{i=1}^{N} x_i^2/2\right) \tag{N.3}$$

with $\Delta(x)$ given by Eq. (N.2). The n-point correlation and cluster functions for any finite n (and finite or infinite N) as well as the probability $E_1(r,s)$ of having exactly r eigenvalues in a randomly chosen interval of length s are again explicitly known. (See Chapter 7.)

Gaussian symplectic ensemble or GSE. This is the ensemble of $N \times N$ quaternion self-dual matrices with the joint probability density proportional to $\exp(-2\,\mathrm{tr}\,H^2)$. This ensemble is invariant under symplectic transformations and is appropriate to describe systems with time reversal symmetry, half odd integer spin and no rotational symmetry. Such systems are rare in nature. The joint probability density of the eigenvalues is proportional to

$$|\Delta(x)|^4 \exp\left(-2\sum_{i=1}^{N} x_i^2\right) \tag{N.4}$$

with $\Delta(x)$ given by Eq. (N.2). The n-point correlation and cluster functions for any finite n (and finite or infinite N) as well as the probability $E_4(r,s)$ of having exactly r eigenvalues in a randomly chosen interval of length s are again known. (See Chapters 8 and 11.)

These three ensembles are characterized by a parameter β taking the values 2 (for unitary), 1 (for orthogonal) and 4 (for symplectic) respectively. Analytical evaluation of integrals of $|\Delta(x)|^\beta \prod_{i=1}^{N} w(x_i)$ depends on the use of polynomials which are orthogonal when $\beta = 2$ and skew-orthogonal of the quaternion or real type when $\beta = 4$ or $\beta = 1$ respectively, for the weight function $w(x)$. (See Chapter 5.)

Circular orthogonal, circular symplectic and circular unitary ensembles (COE, CSE and CUE respectively). These are the ensembles of $N \times N$ random unitary matrices, which are in addition symmetric (for orthogonal), self-dual (for symplectic), or having no other restriction (for unitary). The eigenvalues of these matrices are of the form $\exp(i\theta)$, θ real, and their joint probability density is proportional to

$$\prod_{1 \leqslant j < k \leqslant N} \left|\exp(i\theta_k) - \exp(i\theta_j)\right|^\beta \tag{N.5}$$

β taking again the values 1, 2 and 4 for COE, CUE and CSE respectively. Analytical evaluation of integrals containing the expression (N.5) as integrand is similar to the case of Gaussian ensembles. All local statistical properties, i.e., those extending to a finite number of eigenvalues, are in the limit of infinite matrices identical to those of the corresponding Gaussian ensembles. (See Chapter 11.)

The probability $E_\beta(r, s)$, $\beta = 1, 2$ or 4, of having exactly r eigenvalues in a randomly chosen interval s is expressed as linear combinations of partial derivatives with respect to z at $z = 1$ of two Fredholm determinants (or infinite products) $F_\pm(z, s)$. These Fredholm determinants have several intricate and deep relations between them, and they satisfy certain second order non-linear differential equations, the so called Painlevé equations. Their power series expansions for small s and asymptotic expansions for large s are known (see Chapters 20 and 21). Extensive numerical tables of $E_\beta(r, s)$ and their few derivatives useful for applications have been computed. (See Appendices A.13, A.14.)

Ensembles of Hermitian matrices, where apart from the Hermitian character, the real parts of the matrix elements are Gaussian random variables with a common variance, so also the imaginary parts, but the common variance of the real parts is not equal to the common variance of the imaginary parts. Here again the n-point correlation and cluster functions for any finite n (and finite or infinite N) are known. (See Chapter 14.)

One knows a little less about ensembles of complex matrices or of real matrices with no further restriction. If one takes the (real and imaginary parts of the) matrix elements as Gaussian random variables, then one can say some thing about the n-point correlation or cluster functions for matrices with complex elements and estimate the number of real eigenvalues for matrices with real elements. (See Chapter 15.)

For Hermitian matrices coupled in a linear chain one can express the n-point correlation function as a single determinant. (See Chapter 23.)

In the last few decades a huge amount of experimental as well as numerical data have been collected and analyzed for their statistical properties concerning various systems such as nuclear excitation energies, atomic energies, possible energies of a particle free to move on billiard tables of odd shapes, resonance frequencies of electro-magnetic cavities of odd shapes (classically chaotic systems), characteristic ultrasonic frequencies of structural materials like the aluminium beams, distribution of trees in Scandinavian forests, imaginary parts of the zeros of the Riemann zeta function on the line $\operatorname{Re} z = 1/2$, longest increasing subsequences in random permutations, and so on. Their agreement with the theoretical predictions of various random matrix models, specially the GOE, CUE and GUE, is quite convincing. (See Chapters 1 and 16.) But we do not yet understand why this should be so. Why the zeros of the Riemann zeta function on the line $\operatorname{Re} z = 1/2$ should behave as the eigenvalues of a matrix from the GUE (or CUE); more generally, why the Riemann zeta function $\zeta(1/2 + it)$ should resemble statistically to the characteristic function $\det[U - e^{i\theta} I]$ of a random matrix U from the CUE, I being the unit matrix while t is proportional to θ (proper scaling). Or why the possible energies of a classically chaotic system should behave as the eigenvalues of a matrix from the GOE, is not clear.

With the availability of large computers people have generated matrices with random elements. These matrices were either real symmetric or complex Hermitian. The probability density of the matrix elements, or of their real and imaginary parts, was not necessarily Gaussian. The statistical properties of a few eigenvalues of such matrices were found to be quite universal. Apparently they do not depend on the probability density of the individual matrix elements. And we do not quite understand why this should be so.

Classically, chaotic systems fall into various categories according to their chaoticity; but we cannot yet make this finer distinction by looking at their quantum mechanical energy spectrum alone. There have been attempts to see whether the eigenvectors or the wave functions of such systems carry any information about their chaotic nature; but these attempts are not very successful yet.

A generalization of the Euler gamma integral due to Selberg may be, and was, ignored at first sight. But it has deep consequences and unsuspected relations with other branches of mathematics, such as the theory of random matrices and finite groups generated by reflections (see Chapter 17), apart from the number theory for which it was originally devoted. Similarly, an integral over the unitary group, seemingly innocent, is quite unobvious and deep (see Appendix A.5). A generalization of this later integral for other Lie algebras is known from Harish-Chandra (1957). It reads as follows.

Let G be a compact simple Lie group, L its Lie algebra of order N and rank n, W the Weyl (or Coxeter) group of L, R_+ the set of its positive roots, and $m_i = d_i - 1$ its Coxeter indices. Also for X and Y elements of L let (X, Y) be a bi-linear form invariant under G, i.e., $(X_1 + X_2, Y) = (X_1, Y) + (X_2, Y)$, $(X, Y_1 + Y_2) = (X, Y_1) + (X, Y_2)$, and $(gX, gY) = (X, Y)$ for $g \in G$. Then

$$\int_{g\in G} \exp(c(X, gYg^{-1}))\, dg$$
$$= \text{const} \cdot \sum_{w\in W} \varepsilon_w \exp(c(X, wY)) \div \prod_{\alpha\in R_+} (\alpha, X)(\alpha, Y) \qquad \text{(N.6)}$$

where ε_w is the parity of w and the

$$\text{const} = c^{-(N-n)/2} \prod_{\alpha\in R_+} \frac{|\alpha|^2}{2} \prod_{i=1}^{n} m_i!. \qquad \text{(N.7)}$$

But this generalization is not good enough for our needs. For example, one knows the integral $\int \exp(c\,\text{tr}(A - QBQ^{-1})^2)\, dQ$ over the group of all real orthogonal matrices Q, when A and B are real anti-symmetric matrices, but not when A and B are real symmetric matrices.

Chapter 1

The experimental information about slow neutron resonances in various nuclei were collected in the 1960s and 1970s mainly by a few groups of workers, the most extensive being that of the Columbia university group of Camarda et al. A good analysis of all this data was done by French et al. (1985) and O. Bohigas et al. (1985). For nuclear level densities see Bethe (1937), Lang and Lecouteur (1954) and Cameron (1956). For Hardy–Ramanujan formula see Andrews (1976). The level spacing law (1.5.1) was proposed by E.P. Wigner and appeared in print in Canad. Math. Congr. Proc., University of Toronto Press, Toronto (1957). This and other important papers on the subject before 1965 together with a detailed introductory article by Porter himself can be found in Porter (1965). The possibility of choosing arbitrary global properties with disregard to the local ones was suggested by Balian (1968). For some interesting material about quantum chaos see the papers in "Quantum chaos and statistical nuclear physics", Proc. 2nd Internat. Conf. on Quantum Chaos and 4th Internat. Colloquium on Statistical Nuclear Physics, Cuernavaca, Mexico, 1986, edited by Seligman and Nishioka, Springer-Verlag, 1986 as well as "Chaos and quantum physics" Les Houches 52, edited by M.-J. Giannoni, A. Voros and J. Zinn-Justin, North-Holland (1991). A good review article about quantum chaos is Eckhardt (1988). About small metallic particles see the review article by R. Kubo in "Polarization, matière et rayonnement" Press. Univer. de France, Paris, 1970. For a critical discussion of various reaction width distributions, strength functions, nuclear reaction theories, experimental situation about small metallic particles and many other interesting topics see the thorough and exhaustive review article by Brody et al. (1981). About the zeros of the zeta functions see Montgomery (1973, 1975), Titchmarsh (1951), Davenport and Heilbronn (1936), Potter and Titchmarsh (1935), Odlyzko (1987, 1989) and Cipra (1988). About the distributions of $\log|\zeta(1/2+it)|$ and the phase of $\zeta(1/2+it)$ for large t compared to the corresponding quantities for matrices in the CUE see Keating and Snaith (2000a) and references given therein. More information about these and other related topics can be found in the review articles by Guhr et al. (1998), di Francesco et al. (1995) and in the special issue of J. Phys. A 36 (2003).

Chapter 2

Section 2.2 is based on Wigner (1959), Sections 2.3 to 2.5 are largely based on Dyson (1962a-I), and Section 2.6 on Porter and Rosenzweig (1960a). All these papers are reproduced in Porter (1965).

Chapter 3

Sections 3.1 to 3.3 are largely based on a paper of Wigner (1965a). Section 3.6 is based on Balian (1968).

Chapter 4

Section 4.2 is based on Wigner (1957a).

Chapter 5

Expressing Pfaffians as determinants of self-dual quaternion matrices and their use for correlation and cluster functions was discovered by Dyson (1970) for the circular ensembles and were adapted to the Gaussian case by Mehta (1971). Skew-orthogonal polynomials came formally into existence in 1977 in a book "Elements of Matrix Theory" by Mehta. The method of integration on alternate variables is essentially due to a remark of Gaudin during a private conversation. Sections 5.4 and 5.5 are based on Mehta and Mahoux (1991). Bi-orthogonal polynomials were first used by Mehta (1981). The integral representation of the skew-orthogonal polynomials is due to Eynard (2001) rediscovered independently by Ghosh (2002). The fact that for a class of weights the zeros of the bi-orthogonal polynomials are real and simple is due to Ercolani and McLaughlin (2001). For general weights the zeros may be complex as indicated by Deligne in a private conversation. The final remark in Section 5.14 relating the three scalar products was made by Balian in a private conversation.

Chapter 6

Section 6.3 is based on Kahn (1963), Section 6.4 on Mehta and des Cloizeaux (1972) and the remarks in Section 6.5 result from conversations with various colleagues.

Chapter 7

As we said above, the method of integration on alternate variables is essentially due to a remark of Gaudin. Sections 7.4 and 7.5 are taken from Mehta and des Cloizeaux (1972), while Section 7.6 is from Mehta (1960). Effetof (1982) used anti-commuting or super-variables to rederive the two level correlation function for the three Gaussian ensembles, orthogonal, unitary and symplectic.

Chapter 9

The papers of Uhlenbeck, Ornstein and Wang and other papers on the mathematics of Brownian motion can be looked up in the collection of Nelson Wax (1954). This chapter owes much to Dyson (1962a).

Chapter 10

This chapter is largely based on Dyson (1962a, I).

Chapter 11

Sections 11.1, 11.3 and 11.5 are based on Dyson (1970), Sections 11.2 and 11.4 on Dyson (1962a, III), Section 11.6 is based on Mehta and Dyson (1963) and Section 11.8 on Dyson (1962a).

Chapter 12

Sections 12.1–12.4 are based on Dyson (1962a, I and II).

Chapter 13

This chapter is based on Mehta and Rosenzweig (1968).

Chapter 14

This chapter resumes the three papers of Pandey and Mehta (1983).

Chapter 15

Section 15.1 is based on Ginibre (1965). Jancovici (1981) considered the case when the power in Eq. (15.1.10) is $2+\varepsilon$ instead of 2, and computed the first term in the expansion of the two-point function $R_2(z_1, z_2)$ in powers of ε. Section 15.2 is based on Ginibre (1965) and Mehta and Srivastava (1966). Section 15.3 is based on Edelman et al. (1994) and Section 15.4 on Cicuta and Mehta (2000).

Chapter 16

As this chapter is just for illustrative purposes, we did not make any effort to update it, though several papers dealing with the comparison of various systems with the predictions of random matrix theory have appeared in the last decade.

Chapter 17

As indicated in the text, Sections 17.2 and 17.3 follow Selberg (1944) and Aomoto (1987a) respectively. Not knowing Selberg's work, equality (17.6.7) has harassed many people for years; even today no argument is known which proves only that without passing through Eq. (17.1.3) or Eq. (17.6.5). Section 17.9 is based on Macdonald (1982) and Section 17.10 follows Askey and Richards (1989). A little generalization of Aomoto's integral is taken from Andrews et al.'s book "Special Functions" (1993).

All these integrals and constant term identities have their several so called q-extensions, either proved or conjectured. See Macdonald (1982) or Morris (1982) and references therein.

Chapter 18

Section 18.1 is based on des Cloizeaux and Mehta (1973), Section 18.2 on Widom (1971) and as noted in the text Sections 18.3–18.6 are copied from Dyson (1976).

Chapter 19

This chapter is based on the thesis of Ghosh (2002); see also Nagao and Wadati (1991a, 1992) and Pandey and Ghosh (2001).

Chapter 20

Sections 20.1–20.4 are based on Mehta and Pandey (1997); Section 20.5 is based on the thesis of Dietz (1991).

Chapter 21

Section 21.1 is based on Mehta (1992a); Sections 21.2 and 21.3 on Mahoux and Mehta (1993).

Chapter 22

This chapter is largely based on Mehta and Normand (2001). Section 22.4 is based on Fyodorov and Strahov (2003).

Chapter 23

This chapter is based on Eynard and Mehta (1998) and Mahoux et al. (1998).

Chapter 24

Section 24.1 resumes the thesis of Bronk (1964a); see also his two articles in *J. Math. Phys.* **5** (1964) and **6** (1965). Sections 24.2 and 24.3 are entirely based on Tracy and Widom (1994a).

Chapter 25

This chapter is largely based on unpublished notes of Odlyzko, Poonen, Wilf and Widom (1993).

Chapter 26

This chapter is largely based on Mehta and Normand (1998) and Le Caër and Delannay (2003).

Chapter 27

Section 27.1 is based on Rosenzweig (1963) and Section 27.2 on Le Caër and Delannay (2003).

Appendices

A.1. Almost all of the numerical computations corroborating Conjectures 1.2.1 and 1.2.2 were carried out in the early days of the random matrix hypothesis when very little was known analytically. Such numerical studies are lacking for the case when the matrix element densities do not have all their moments finite. For example, when

$$P(H) = \prod_{i \leqslant j} \frac{a}{\pi} (H_{ij}^2 + a^2)^{-1}$$

even the second moment of H_{ij} is infinite, and we do not know whether the level density is the "semi circle" or whether the spacing probability density resembles the " Wigner surmise".

A.5. The integral in Eq. (14.3.1) or (A.5.1) is over the group of unitary matrices U. It is equivalent to Eq. (N.6) for the compact Lie algebra A_n. The same integral over other compact Lie groups, in particular orthogonal and symplectic, is of great interest. D. Altshuler and C. Itzykson have rederived a formula due to Harish-Chandra, Eq. (N.6). But as we said earlier, this is not good enough for an analytic treatment of, say, a fixed diagonal real matrix perturbed by a real symmetric random matrix.

A.10. That the identity

$$\int_{-\infty}^{\infty} K(x, z) K(z, y)\, dz = K(x, z)$$

holds for

$$K(x, y) \equiv K_n(x, y) = \sum_{j=0}^{n-1} \varphi_j(x) \varphi_j(y)$$

is almost obvious from the orthonormality of the oscillator functions $\varphi_j(x)$. But that in the particular limits it gives rise to the same identity for $K(x, y) = \sin \pi (x - y) / \pi (x - y)$ and $K(x, y) = [Ai(x) Ai'(y) - Ai'(x) Ao(y)]/(x - y)$ will come as a surprise to many.

A.13. The power series expansions of $E_\beta(0, s)$ for $\beta = 1, 2$ and 4 up to s^{40} were first computed by Dietz in her thesis.

A.14–A.15. The tables of $\lambda_j(s)$ and $E_\beta(n, s)$ for $\beta = 1$ and 2 first appeared in Mehta and des Cloizeaux (1972).

A.16. This appendix was kindly written by G. Mahoux.

A.36. This appendix unfortunately does not give a simple and direct proof of a nice looking integral, Eq. (A.36.6). If the form of the result on the right-hand side is assumed, then one can find the two constants by considering the cases $\lambda = 0$ and $\lambda = \infty$. Appendix A.36 is based on Edelman et al. (1994) while Appendix A.37 is largely based on Edelman (1997).

A.45. The standard form of the six Painlevé equations can be found in any good book on differential equations. The "Les leçons de Stockholm" (1896) by Painlevé himself are very instructive even today. The Lax pairs and Hamiltonian equations have been largely copied from the extensive works of the Kyoto school of mathematics (Jimbo, Miwa, Okamoto, 1970s and 1980s). They have some defects as they are not general enough to deal with arbitrary values of the parameters α, β, γ and δ. This question is discussed by Lin et al. (2003).

A.50. To our knowledge this idea of changing variables to compute various averages was first used by Eynard.

A.51. The remarks of this appendix were in response to a question of Odlyzko.

REFERENCES

Ablowitz M. and Clarkson P.A. (1991). *Solitons, Non Linear Evolution Equations and Inverse Scattering*, Cambridge Univ. Press, Cambridge, UK.

Ablowitz M. and Segur H. (1977). Exact linearization of a Painlevé transcendent, *Phys. Rev. Lett.* **38**, 1103–1106.

Abramowitz M. and Stegun I.A. (1965). *Handbook of Mathematical Functions*, Dover, New York.

Aitchison J. and Brown J.A.C. (1957). *The Lognormal Distribution*, Cambridge Univ. Press, Cambridge, UK.

Aldous D. and Diaconis P. (1995). Hammersley's interacting particle process and longest increasing subsequences, *Probab. Theory Related Fields* **103**, 199–213.

Al'tshuler B.L. and Shklovskii B.I. (1986). Repulsion of energy levels and conductivity of small metal samples, *Sov. Phys. JETP* **64**, 127–135.

Anderson T.W. (1946). The noncentral Wishart distribution and certain problems of multivariate statistics, *Ann. Math. Statist.* **17**, 409–431.

Anderson T.W. (1948). The asymptotic distributions of the roots of certain determinantal equations, *J. Roy. Statist. Soc.* **10**, 132ff.

Anderson G.W. (1991). A short proof of Selberg's generalized beta formula, *Forum Math.* **3**, 415–417.

Andrews G.E. (1976). The theory of partitions, *Encyclopedia of Mathematics and Its Applications*, vol. 2, Addison–Wesley, Reading, MA.

Andrews G.E. (1980). Notes on the Dyson conjecture, *SIAM J. Math. Anal.* **11**, 787–792.

Andrews G.E., Askey R.A. and Roy R. (1993). *Special Functions*, Cambridge Univ. Press, Cambridge, UK, Chapter 8.

Aomoto K. (1987a). Jacobi polynomials associated with Selberg integrals, *SIAM J. Math. Anal.* **18**, 545–549.

Aomoto K. (1987b). The complex Selberg integral, *Quart. J. Math. Anal.* **38**, 1–15.

Aomoto K. (1988a). Scaling limit formula for 2-point correlation function of random matrices, *Adv. Stud. Pure Math.* **16**, 1–15.

Aomoto K. (1988b). Correlation functions of the Selberg integral, in: *Ramanujan Revisited*, Academic Press, San Diego, pp. 591–605.

Aomoto K. (1989). On the complex Selberg integral, in: *Proc. Sympos. Pure Math.*, vol. 49, pp. 279–281.

Arnold V.I. and Avez A. (1967). *Problèmes ergodiques de la mécanique classique*, Gauthiers–Villars, Paris; English translation, *Ergodic Problems in Classical Mechanics*, Benjamin, New York, 1968.

Artin E. (1966). *Geometric Algebra*, Interscience, New York, Chapter 4.

Askey R.A. (1980). Some basic hypergeometric extensions of integrals of Selberg and Andrews, *SIAM J. Math. Anal.* **11**, 938–951.

Askey R.A. and Ismail M. (1984). Recurrence relations, continued fractions, and orthogonal polynomials, *Mem. Amer. Math. Soc.*, vol. 49, no. 300, Amer. Math. Soc., Providence, RI, pp. 1–108.

Askey R.A. and Regev A. (1984). Maximal degrees for Young diagrams in a strip, *European J. Combin.* **5**, 189–191.

Askey R.A. and Richards D. (1989). Selberg's second beta integral and an integral of Mehta, in: Anderson T.W., Athreya K.B. and Inglehart D.L. (Eds.), *Probability, Statistics and Mathematics: Papers in Honor of S. Karlin*, Academic Press, New York, pp. 27–39.

Auluck F.C. and Kothari D.S. (1946). Statistical mechanics and the partitions of numbers, *Proc. Camb. Phil. Soc.* **42**, 272–277.

Baer R.M. and Brock P. (1968). Natural sorting over permutation spaces, *Math. Comp.* **22**, 385–410.

Baik J., Deift P. and Strahov E. (2003). Products and ratios of characteristic polynomials of random Hermitian matrices, math-ph/0304016.

Baker T.H. and Forrester P.J. (1997). The Calogero–Sutherland model and generalized classical polynomials, *Comm. Math. Phys.* **188**, 175–216, around Eqs. (5.30)–(5.33); Finite N fluctuation formulas for random matrices, *J. Stat. Phys.* **88** (1997) 1371–1386.

Balazs N.L. and Voros A. (1986). Chaos on the pseudosphere, *Phys. Reports* **143**, 109–240.

Balazs N.L. and Voros A. (1989). The quantized baker's transformation, *Ann. Phys.* **190**, 1–31.

Balian R. (1968). Random matrices and information theory, *Nuovo Cimento B* **57**, 183–193.

Barbasch D. and Vogan D. (1982). Primitive ideals and orbital integrals in complex classical groups, *Math. Ann.* **259**, 153–199.

Barnes E.W. (1900). The theory of the G-function, *Quart. J. Pure and Appl. Math.* **31**, 264–314.

Barouch E., McCoy B.M. and Wu T.T. (1973). Zero field susceptibility of the two dimensional Ising model near T_c, *Phys. Rev. Lett.* **31**, 1409–1411.

Basor E., Tracy C.A. and Widom H. (1992). Asymptotics of level spacing distribution for random matrices, *Phys. Rev. Lett.* **69**, 5–8.

Basor E. and Widom H. (1983). Toeplitz and Wiener–Hoff determinants with piecewise continuous symbols, *J. Funct. Anal.* **50**, 387–413.

Bassom A.P., Clarkson P.A., Hicks A.C. and McLeod J.B. (1992). Integral equations and exact solutions for the fourth Painlevé transcendent, *Proc. Royal Soc. London A* **437**, 1–24.

Bateman H. (1953a). *Higher Transcendental Functions*, vol. 1, Erdéyli A. et al. (Eds.), McGraw Hill, New York.

Bateman H. (1953b). *Higher Transcendental Functions*, vol. 2, Bateman manuscript project, Erdéyli A. et al. (Eds.), MacGraw Hill, New York, Sections 10.10 and 10.13, Eqs. (30)–(31).

Bateman H. (1954). *Integral Transforms*, vol. 1, McGraw Hill, New York, Section 6.9, Eq. (14).

Bauldry W.C. (1990). Estimates of asymmetric Freud polynomials, *J. Approx. Theory* **63**, 225–237.

Beenakker C.W.J. (1997). Random matrix theory of quantum transport, *Rev. Mod. Phys.* **69**, 731–808.

Berry M.V. (1985). Semi-classical theory of spectral rigidity, *Proc. Royal Soc. London A* **400**, 229–251.

Berry M.V. (1988). Semi-classical formula for the number variance of the Riemann zeros, *Nonlinearity* **1**, 399–407.

Bessis D. (1979). A new method in the combinatoric of the topological expansion, *Comm. Math. Phys.* **69**, 147–163.

Bethe H.A. (1937). Nuclear physics: nuclear dynamics, theoretical, *Rev. Mod. Phys.* **9**, 69–244.

Bleher P. and Its A. (1999). Semi-classical asymptotics of orthogonal polynomials, Riemann–Hilbert problem and universality in the matrix model, *Ann. Math.* **150**, 185–266.

Bogomolny E.B. and Keating J.P. (1995). Random matrix theory and Riemann zeros I: three and four point correlations, *Nonlinearity* **8**, 1115–1131; II: n-point correlations, ibid **9** (1996) 911–935.

Bohigas O. (1991). Random matrix theories and chaotic dynamics, in: Giannoni M.-J., Voros A. and Zinn-Justin J. (Eds.), *Chaos and Quantum Physics* (Les Houches summer school, 1989), North-Holland, Asmterdam.

Bohigas O. and Giannoni M.J. (1984). Chaotic motion and random matrix theories, in: Dehesa J.S. et al. (Eds.), *Mathematical and Computational Methods in Nuclear Physics* (Proceedings, Granada, Spain, 1983), *Lecture Notes in Phys.*, vol. 209, Springer-Verlag, Berlin, pp. 1–99.

Bohigas O., Giannoni M.J. and Schmit C. (1984a). Characterization of chaotic quantum spectra and universality of level fluctuation laws, *Phys. Rev. Lett.* **52**, 1–4.

Bohigas O., Giannoni M.J. and Schmit C. (1984b). Spectral fluctuations and chaotic motion, in: Bonche P. et al. (Eds.), *Heavy Ion Collisions*, Plenum Press, New York, pp. 145–163.

Bohigas O., Haq R.U. and Pandey A. (1983). Fluctuation properties of nuclear energy levels and widths: comparison of theory with experiment, in: Böckhoff K.H. (Ed.), *Nuclear Data for Science and Technology*, Brussels, pp. 809–814.

Bohigas O., Haq R.U. and Pandey A. (1985). Higher order correlations in spectra of complex systems, *Phys. Rev. Lett.* **54**, 1645–1648 and references therein.

Bohigas O., Tomsovic S. and Ullmo D. (1993). Manifestation of classical phase space structures in quantum mechanics, *Phys. Reports* **223**, 93–133.

Bohr A. (1956). On the theory of nuclear fission, in: *Proc. 1st Internat. Conf. Peaceful Uses of Atomic Energy, Geneva, 1955*, vol. 2, Columbia Univ. Press, New York, pp. 151–154.

Bohr A. and Mottelson B.R. (1975). *Nuclear Structure*, 2 vols., Benjamin, New York.

Bollbàs B. and Brightwell G. (1992). The height of a random partial order: concentration of measure, *Ann. Appl. Probab.* **2**, 1009–1018.

Bollbàs B. and Winkler P.M. (1988). The longest chain among random points in Euclidean space, *Proc. Amer. Math. Soc.* **103**, 347–353.

Bonan S.S. and Clark D.S. (1990). Estimates of the Hermite and Freud polynomials, *J. Approx. Theory* **63**, 210–224.

Boutroux P. (1913). Recherches sur les transcendents de Monsieur Painlevé et l'étude asymptotique des équations différentielles du second ordre, *Ann. Ecole Normale Supér.* **30**, 253–375; **31** (1914) 99–159.

Bouttier J., di Francesco P. and Guitter E. (2002). Census of planar maps: from the one-matrix model solution to a combinatorial proof, *Nucl. Phys. B* **645** [PM], 477–499.

Bouwcamp C.J. (1947). On spheroidal functions of order zero, *J. Math. Phys.* **26**, 79–92.

Bowick M.J. and Brézin E. (1991). Universal scaling of the tail of the density of eigenvalues in random matrix models, *Phys. Lett. B* **268**, 21–28.

Brendt R.P. (1979). On the zeros of the Riemann zeta function in the critical strip, *Math. Comp.* **33**, 1361–1372.

Brézin E. (1992). Large N limit and discretized two-dimensional quantum gravity, in: Gross D.J., Piran T. and Weinberg S. (Eds.), *Two-Dimensional Quantum Gravity and Random Surfaces*, World Scientific, Singapore, pp. 1–40.

Brézin E., Douglas M.R., Kazakov V. and Shenker S. (1992). The Ising model coupled to 2D gravity, a non perturbative analysis, *Phys. Lett. B* **237**, 43–51.

Brézin E. and Hikami S. (2000). Characteristic polynomials of random matrices, *Comm. Math. Phys.* **214**, 111–135.

Brézin E., Itzykson C., Parisi G. and Zuber J.B. (1978). Planar diagrams, *Comm. Math. Phys.* **59**, 35–51.

Brézin E. and Kazakov V.A. (1990). Exactly solvable field theories of closed strings, *Phys. Lett. B* **236**, 144–150.

Brézin E. and Neuberger H. (1991). Multicritical points of unoriented random surfaces, *Nucl. Phys. B* **350**, 513–553; Large N scaling limits of symmetric matrix models as systems of fluctuating unoriented surfaces, *Phys. Rev. Lett.* **65** (1990) 2098–2101.

Brody T.A. (1973). A statistical measure for the repulsion of energy levels, *Nuovo Cimento Lett.* **7** 482–489.

Brody T.A., Flores J., French J.B., Mello P.A., Pandey A. and Wong S.S.M. (1981). Random matrix physics: spectrum and strength fluctuations, *Rev. Mod. Phys.* **53**, 385–479.

Bronk B.V. (1964a). *Topics in the Theory of Random Matrices*, Thesis, Princeton University (unpublished).

Bronk B.V. (1964b). Accuracy of the semicircle approximation for the density of eigenvalues of random matrices, *J. Math. Phys.* **5**, 215–220.

Bronk B.V. (1965). Exponential ensemble for random matrices, *J. Math. Phys.* **6**, 228–237.

Bureau F.J. (1964). Differential equations with fixed critical points, *Ann. Mat. Pura Appl. (4)* **64**, 233–364.

Bureau F.J. (1972). Equations différentielles du second ordre en Y et du second degré en Y'' dont l'intégrale générale est à points critiques fixes, *Ann. Mat. Pura Appl. (4)* **91**, 163–281.

Caillol J.M. (1981). Exact results for a two-dimensional one component plasma on a sphere, *J. Phys. Lett.* **42**, L245–L247.

Camarda H.S. (1992). Statistical behaviour of eigenvalues of real symmetric and complex Hermitian band matrices: comparison with random matrix theory, *Phys. Rev. A* **45**, 579–582.

Camarda H.S., Desjardins J.S., Garg J.B., Hacken G., Havens Jr. W.W., Liou H.I., Peterson J.S., Rahn F., Rainwater J., Rosen J.L., Singh U.N., Slagowitz M. and Weinchank S. (1973). (See the series of papers on Neutron resonance spectroscopy I–XIV in the Phys. Rev. of the 1960s and 1970s.) To find references, see for example, *Phys. Rev. C* **8**, 1833–1836, or *Phys. Rev. C* **11** (1975) 1117–1121, or Neutron resonance spectroscopy, *Phys. Rev. C* **11** (1975) 1117–1121.

Camarda H.S. and Georgopulos P.D. (1983). Statistical behaviour of atomic energy levels: agreement with random matrix theory, *Phys. Rev. Lett.* **50**, 492–495.

Cameron A.G.W. (1956). Nuclear level spacings, *Canad. J. Phys.* **36**, 1040–1057.

Cassels J.W.S. (1961). Footnote to a note of Davenport and Heilbronn, *J. London Math. Soc.* **36**, 177–184.

Chadan K. and Sabatier P.C. (1977). *Inverse Scattering Problems in Quantum Scattering Theory*, Springer-Verlag, Berlin/New York, Eqs. (III.5.7) and (IV.1.10).

Chaddha S., Mahoux G. and Mehta M.L. (1981). A method of integration over matrix variables II, *J. Phys. A* **14**, 579–586.

Chandrasekhar S. (1943). Stochastic problems in physics and astronomy, *Rev. Mod. Phys.* **15**, 1–89.

Chen Y. and Manning S.M. (1994). Distribution of linear statistics in random matrix models, *J. Phys. Cond. Mat.* **6**, 3039–3044.

Chevalley C. (1946). *Theory of Lie Groups*, Princeton Univ. Press, Princeton, NJ, pp. 16–24.

Chihara T.S. (1979). *An Introduction to Orthogonal Polynomials*, Gordon and Breach, New York.

Cicuta G.M. and Mehta M.L. (2000). Probability density of determinants of random matrices, *J. Phys. A* **33**, 8029–8035.

Cipra B.A. (1988/89). Zeroing in on the zeta function, *Science*, 11 March 1988, pp. 1241–1242; Zeta zero update, *Science*, 3 March 1989, p. 1143.

Clarkson P.A. and McLeod J.B. (1992). Integral equations and connection formulae for the Painlevé equations, in: Levy D. and Winternitz P. (Eds.), *Painlevé Transcendents, Their Asymptotics and Physical Applications*, Plenum Press, New York, pp. 1–31.

Cohen P.S. and Regev A. (1982). Asymptotic estimates of some S_n characters and the identities of the 2×2 matrices, *Comm. Algebra* **10**, 71–85.

Cohen P.S. and Regev A. (1988a). On maximal degrees for Young diagrams, *European J. Combin.* **9**, 607–610.

Cohen P.S. and Regev A. (1988b). Asymptotics of combinatorial sums and the central limit theorem, *SIAM J. Math. Anal.* **19**, 301–305.

Conrey J.B. (2001). L-functions and random matrices, in: Enquist B. and Schmid W. (Eds.), *Mathematics Unlimited: 2001 and Beyond*, Springer-Verlag, Berlin, pp. 331–352.

Conte R. and Musette M. (2001). First degree bi-rational transformations of the Painlevé equations and their contiguity relations, *J. Phys. A* **34**, 10507–10522.

Cosgrove C.M. and Scoufis G. (1993). Painlevé classification of a class of differential equations of the second order and second degree, *Stud. Appl. Math.* **88**, 25–87.

Cristofori F., Sona P.G. and Tonolini F. (1966). The statistics of the eigenvalues of random matrices, *Nucl. Phys.* **78**, 553–556.

Davenport H. and Heilbronn H. (1936). On the zeros of certain Dirichlet series I, II, *J. London Math. Soc.* **11**, 181–185 and 307–312.

David F. (1993). Non-perturbative effects in matrix models and vacua of two-dimensional gravity, *Phys. Lett. B* **302**, 403–410.

David F. (1995). Simplicial quantum gravity and random lattices, in: Julia B. and Zinn-Justin J. (Eds.), *Gravitation and Quantization* (Les Houches session 56), Elsevier, Amsterdam, pp. 679–746.

de Bruijn N.G. (1955). On some multiple integrals involving determinants, *J. Indian Math. Soc.* **19**, 133–151.

Deift P. (2000). *Orthogonal Polynomials and Random Matrices: A Riemann–Hilbert Approach*, Amer. Math. Soc., Providence, RI.

Deift P., Its A.R. and Zhou X. (1997). A Riemann–Hilbert approach to asymptotic problems arising in the theory of random matrix models and also in theory of integrable statistical mechanics, *Ann. Math. (2)* **146**, 149–235.

Deift P., McLaughlin K.T.-R., Kriecherbauer T., Venakides S. and Zhou X. (1999a). Uniform asymptotics for polynomials orthogonal with respect to varying exponential weights and applications to universality questions in random matrix theory, *Comm. Pure Appl. Math.* **52**, 1335–1425.

Deift P., McLaughlin K.T.-R., Kriecherbauer T., Venakides S. and Zhou X. (1999b). Strong asymptotics of orthogonal polynomials with respect to exponential weights, *Comm. Pure Appl. Math.* **52**, 1491–1552.

Delande D. and Gay J.C. (1986). Quantum chaos and statistical properties of energy levels: numerical study of the hydrogen atom in a magnetic field, *Phys. Rev. Lett.* **57**, 2006–2009; *J. Phys. B* **17** (1984) L335ff; ibid **19** (1986) L173ff.

Delannay R. and Le Caër G. (2000). Distribution of the determinant of a random real symmetric matrix from the Gaussian orthogonal ensemble, *Phys. Rev. E* **62**, 1526–1536.

Derrida B. and Vannimenus J. (1982). A transfer matrix approach to random resistor networks, *J. Phys. A* **15**, L557–L564.

des Cloizeaux J. and Mehta M.L. (1972). Some asymptotic expressions for prolate spheroidal functions and for the eigenvalues of differential and integral equations of which they are solutions, *J. Math. Phys.* **13**, 1745–1754.

des Cloizeaux J. and Mehta M.L. (1973). Asymptotic behaviour of spacing distributions for the eigenvalues of random matrices, *J. Math. Phys.* **14**, 1648–1650.

Diaconis P. and Shahshahani M. (1994). On the eigenvalues of random matrices, *J. Appl. Probab.* **31**, 49–61.

Dietz B. (1991). *Zufallsmatrixtheorie und Gleichgewichtsstatistik für Quasienergien klassisch chaotischer Systeme*, Thesis, Essen, Germany (unpublished).

Dietz B. and Haake F. (1990). Taylor and Padé analysis of the level spacing distributions of random matrix ensembles, *Z. Phys. B* **80**, 153–158.

di Francesco P., Ginsparg P. and Zinn-Justin J. (1995). 2-D gravity and random matrices, *Phys. Rep.* **254**, 1–133.

Dieudonné J. (1955). *La géometrie des groupes classiques*, in: *Ergeb. Math.*, vol. 5, Springer-Verlag, Berlin.

Dijkgraaf R. and Witten E. (1990). Mean field theory, topological field theory, and multimatrix models, *Nucl. Phys. B* **342**, 486–522.

Disteler J. (1990). 2D quantum gravity, topological field theory and multicritical matrix models, *Nucl. Phys. B* **342**, 523–538.

Douglas M.R. (1990). Strings in less than one dimension and the generalized KdV hierarchies, *Phys. Lett. B* **238**, 176–180.

Douglas M.R. and Shenker S.H. (1990). Strings in less than one dimension, *Nucl. Phys. B* **335**, 635–654.

Dyson F.J. (1953). The dynamics of a disordered linear chain, *Phys. Rev.* **92**, 1331–1338.

Dyson F.J. (1962a). Statistical theory of energy levels of complex systems I, II and III, *J. Math. Phys.* **3**, 140–156, 157–165, 166–175.

Dyson F.J. (1962b). A Brownian motion model for the eigenvalues of a random matrix, *J. Math. Phys.* **3**, 1191–1198.

Dyson F.J. (1962c). The three fold way. Algebraic structure of symmetry groups and ensembles in quantum mechanics, *J. Math. Phys.* **3**, 1199–1215.

Dyson F.J. (1970). Correlations between the eigenvalues of a random matrix, *Comm. Math. Phys.* **19**, 235–250.

Dyson F.J. (1972a). Quaternion determinants, *Helv. Phys. Acta* **45**, 289–302.

Dyson F.J. (1972b). A class of matrix ensembles, *J. Math. Phys.* **13**, 90–97.

Dyson F.J. (1976). Fredholm determinants and inverse scattering problems, *Comm. Math. Phys.* **47**, 171–183.

Dyson F.J. (1995). The Coulomb fluid and the fifth Painlevé transcendent, in: Liu C.S. and Yau S.T. (Eds.), *Chen Ning Yang, a Great Physicist of the 20th Century*, International Press, Cambridge, MA, pp. 131–146.

Dyson F.J. and Mehta M.L. (1963). Statistical theory of energy levels of complex systems IV, *J. Math. Phys.* **4**, 701–712.

Ebeling K.J. (1984). Statistical properties of random wave fields, in: Masson W.P. (Ed.), *Physical Acoustics*, vol. 17, Academic Press, New York, pp. 233–310.

Eckhardt B. (1988). Quantum mechanics of classically non-integrable systems, *Phys. Reports* **163**, 205–297.

Edelman A. (1989). Eigenvalues and condition numbers of random matrices, Ph.D. Thesis, Mass. Inst. Tech. (unpublished).

Edelman A. (1997). The probability that a random real Gaussian matrix has k real eigenvalues, related distributions, and the circular law, *J. Multivariate Anal.* **60**, 203–232.

Edelman A., Kostlan E. and Shub M. (1994). How many eigenvalues of a random matrix are real?, *J. Amer. Math. Soc.* **7**, 247–267.

Effetof K.B. (1982). Statistics of the levels in small metallic particles, *Zh. Eksp. Teor. Fiz.* **83**, 833–847; English translation, *Sov. Phys. JETP* **56** (1982) 467–475.

Effetof K.B. (1997). *Supersymmetry in Disorder and Chaos*, Cambridge Univ. Press, Cambridge, UK.

Engleman R. (1958). The eigenvalues of a randomly distributed matrix, *Nuovo Cimento* **10**, 615–621.

Ercolani N.M. and McLaughlin K.T.-R. (2001). Asymptotic and integrable structures for bi-orthogonal polynomials associated to a random two matrix model, *Physica D* **152–153**, 232–268.

Erdelyi A. (1960). Asymptotic forms for Laguerre polynomials, *J. Indian Math. Soc., Golden Jubilee Commemoration* **1907–1908**, 235–250.

Erdös P. and Szekeres G. (1938). A combinatorial problem in geometry, *Compositio Math.* **2**, 463–470.

Eynard B. (1997). Eigenvalue distribution of large random matrices, from one matrix to several coupled matrices, *Nucl. Phys. B* **506**, 633–664.

Eynard B. (1998). Correlation functions of eigenvalues of multi-matrix models, and the limit of a time dependent matrix, *J. Phys. A* **31**, 8081–8102.

Eynard B. (2001). Asymptotics of skew-orthogonal polynomials, *J. Phys. A* **34**, 7591–7605.

Eynard B. and Mehta M.L. (1998). Matrices coupled in a chain: I. Eigenvalue correlations, *J. Phys. A* **31**, 4449–4456.

Faddeev L.D. (1959). The inverse problem in quantum theory of scattering, *Uspekhi Mat. Nauk* **14** (4) 57–119; English translation in *J. Math. Phys.* **4**, (1963) 72–104.

Flaschka H. and Newell A.C. (1980). Monodromy and spectrum preserving deformations I, *Comm. Math. Phys.* **76**, 65–116.

Fokas A.S., Its A.R. and Kitaev A.V. (1990). An isomonodromy approach to the theory of 2-dimensional quantum gravity, *Uspekhi Mat. Nauk* **45** (6), 135–136 (in Russian); English translation in *Russian Math. Surveys* **45** (6) (1990) 155–157.

Fokas A.S., Its A.R. and Kitaev A.V. (1991). Discrete Painlevé equations and their appearance in quantum gravity, *Comm. Math. Phys.* **142**, 313–344.

Forrester P.J. (1993). The spectrum edge of random matrix ensembles, *Nucl. Phys. B* **402**, 709–728.

Forrester P.J., Snaith N.C. and Verbaarschot J.J.M. (Eds.) (2003). *J. Phys. A* **36**, 2859–3645 (special issue on random matrices).

Forrester P.J. and Witte N.S. (2001). Application of the τ-function theory of Painlevé equations to random matrices: PIV, PII and GUE, *Comm. Math. Phys.* **219**, 357–398, Eq. (4.43) and the paragraph following Eq. (4.44).

Fox D. and Kahn P.B. (1964). Higher order spacing distributions for a class of matrix ensembles, *Phys. Rev. B* **134**, 1151–1155.

French J.B. and Kota V.K.B. (1983). Nuclear level densities and partition functions with interactions, *Phys. Rev. Lett.* **51**, 2183–2186.

French J.B., Kota V.K.B., Pandey A. and Tomosovic S. (1985). Bounds on time reversal non-invariance in the nuclear Hamiltonian, *Phys. Rev. Lett.* **54**, 2313–2316.

French J.B., Kota V.K.B., Pandey A. and Tomosovic S. (1988). Statistical properties of many particle spectra, V and VI, *Ann. Phys.* **181**, 198–234 and 235–260. For earlier papers in this series, see references therein.

Frieze A. (1991). On the length of the longest monotone subsequence in a random permutation, *Ann. Appl. Probab.* **1**, 301–305.

Fröhlich H. (1937). Die spezifische Wärme der Electronen kleiner Metallteilchen bei tiefen Temperturen, *Physica* **4**, 406–412.

Fuchs W.H.J. (1964). On the eigenvalues of an integral equation arising in the theory of band limited signals, *J. Math. Anal. Appl.* **9**, 317–330.

Fukazawa K., Hamada K. and Sato H. (1990). Phase diagrams of 3 matrix model, *Mod. Phys. Lett. A* **5**, 2431–2438.

Fyodorov Y.V. and Strahov E. (2003). An exact formula for general spectral correlation function of random Hermitian matrices, *J. Phys. A* **36**, 3203–3213.

Gallagar P.X. and Muller J.H. (1978). Primes and zeros in short intervals, *J. Reine Angew. Math. (Crelle)* **303–304**, 205–220.

Gambier B. (1906a). Sur les équations différentielles du second ordre et du premier degré dont l'intégrale générale est à points critiques fixes, *C. R. Acad. Sci. Paris Sér. I Math.* **143**, 741–743; **144** (1907) 827–830; 962–964; Thesis, Paris (1909); *Acta Math.* **33** (1910) 1–55.

Gambier B. (1906b). Sur les équations différentielles du second ordre et du premier degré dont l'intégrale générale est uniforme, *C. R. Acad. Sci. Paris Sér. I Math.* **142**, 266–269; 1403–1406; 1497–1500.

Garg J.B., Rainwater J., Peterson J.S. and Havens Jr. W.W. (1964). Neutron resonance spectroscopy III, Th^{232} and U^{238}, *Phys. Rev.* **134**, B985–B1009.

Garvan F.G. (1989). Some McDonald–Mehta integrals by brute force, in: Stanton D. (Ed.), *q-Series and Partitions, IMA Volumes in Mathematics and Its Applications*, vol. 18, Springer-Verlag, Berlin, pp. 77–98.

Gaudin M. (1961). Sur la loi limite de l'espacement des valeurs propres d'une matrice aléatoire, *Nucl. Phys.* **25**, 447–458.

Gaudin M. (1996). *Modèles exactement résolus*, Les Editions de Physique, France.

Gel'fand I.M. and Levitan B.M. (1951). On the determination of a differential equation by its spectral function, *Izv. Akad. Nauk SSSR Ser. Mat.* **15**, 309–360; English translation in *Amer. Math. Soc. Transl. Ser.* 2 **1** (1955) 253–304.

Gessel I.M. (1990). Symmetric functions and P-recursiveness, *J. Combin. Theory A* **53**, 257–285.

Ghosh S. (2002). *Non Gaussian ensembles of random matrices*, Thesis, Jawaharlal Nehru University, New Delhi (unpublished).

Ghosh S. and Pandey A. (2002). Skew orthogonal polynomials and random matrix ensembles, *Phys. Rev. E* **65**, 046221-1-21.

Ghosh S., Pandey A., Puri S. and Saha R. (2003). Non-Gaussian random matrix ensembles with banded spectra, *Phys. Rev. E* **67**, 025201 (R) 1–4.

Giannoni M.-J., Voros A. and Zinn-Justin J. (Eds.) (1991). *Chaos and Quantum Physics* (Les Houches summer school 52) North-Holland, Amsterdam.

Ginibre J. (1965). Statistical ensembles of complex, quaternion and real matrices, *J. Math. Phys.* **6**, 440–449.

Ginsparg P. and Zinn-Justin J. (1990). 2D gravity + 1D matter, *Phys. Lett. B* **240**, 333–340.

Girko V.L. (1990). *Theory of Random Determinants*, in: *Math. Appl. Soviet Ser.*, vol. 45, Kluwer Academic, Dordrecht.

Good I.J. (1970). Short proof of a conjecture by Dyson, *J. Math. Phys.* **11**, 1884.

Gorkov L.P. and Eliashberg G.M. (1965). Minute metallic particles in an electromagnetic field, *Zh. Eksp. i Teor. Fiz.* **48**, 1407–1418; (*Sov. Phys. JETP* **21** (1965) 940–947).

Goursat E. (1956). *Cours d'analyse mathématique*, vol. 3, Gauthier–Villars, Paris, pp. 389, 454.

Grabiner D.J. and Magyar P. (1993). Random walks in Weyl chambers and the decomposition of tensor powers, *Algebraic Combin.* **2**, 239–260.

Gradshteyn I.S. and I.M. Rizhik I.M. (1965). *Tables of Integrals, Series and Products*, Academic Press, New York.

Grobe R. and Haake F. (1988). Quantum distinction of regular and chaotic dissipative motions, *Phys. Rev. Lett.* **61**, 1899–1902.

Grobe R. and Haake F. (1989). Universality of cubic level repulsion for dissipative quantum chaos, *Phys. Rev. Lett.* **62**, 2803–2806.

Gromak V.I. and Lukashevich N.A. (1990). *Analytical Properties of Solutions of Painlevé Equations* (in Russian), University Press, Minsk, and references therein.

Gross D.J. and Klebanov I. (1990). One dimensional string theory on a circle, *Nucl. Phys. B* **344**, 475–498; Fermionic string field theory of $c = 1$ two dimensional quantum gravity, *Nucl. Phys. B* **352** (1991) 671–688.

Gross D.J. and Migdal A.A. (1990). Non perturbative two dimensional quantum gravity, *Phys. Rev. Lett.* **64**, 127–130; Non perturbative solution of the Ising model on a random surface, *Phys. Rev. Lett.* **64** (1990) 717–720; A non perturbative treatment of two dimensional quantum gravity, *Nucl. Phys. B* **340** (1990) 333–363.

Gross D.J. and Taylor W. (1993a). Two dimensional QCD is a string theory, *Nucl. Phys. B* **400**, 181–208.

Gross D.J. and Taylor W. (1993b). Twists and Wilson loops in the string theory of two dimensional QCD, *Nucl. Phys. B* **403**, 395–452.

Gross D.J. and Witten E. (1980). Possible third order phase transition in the large N lattice gauge theory, *Phys. Rev. D* **21**, 446–453.

Guhr T., Müler-Groeling A. and Weidenmüller H.A. (1998). Random matrix theories in quantum physics: common concepts, *Phys. Rep.* **299**, 189–425 (with some 800 references).

Habsieger L. (1987). Conjecture de Macdonald et q-integral de Selberg–Askey, Thèse, Université Louis Pasteur, Strasbourg (unpublished).

Hammersley J.M. (1972). A few seedlings of research, in: *Proc. 6th Berkeley Symp. Math. Stat. Probab.*, Univ. of California Press, Berkeley, pp. 345–394.

Hannay J.H. and Ozorio de Almeida A.M. (1984). Periodic orbits and a correlation function for the semi-classical density of states, *J. Phys. A* **17**, 3429–3440.

Haq R.U., Pandey A. and Bohigas O. (1982). Fluctuation properties of nuclear energy levels: do theory and experiment agree?, *Phys. Rev. Lett.* **48**, 1086–1089.

Hardy G.H., Littlewood J.E. and Polya G. (1964). *Inequalities*, Cambridge Univ. Press, Cambridge, UK, (1.3.3).

Hardy G.H. and Ramanujan S. (1918). Asymptotic formulae in combinatory analysis, *Proc. London Math. Soc.* **17**, 75–115.

Harer J. and Zagier D. (1986). The Euler characteristic of the moduli space of curves, *Invent. Math.* **85**, 457–485.

Harish-Chandra (1957). Differential operators on a semi-simple Lie algebra, *Amer. J. Math.* **79**, 87–120.

Harnad J., Tracy C.A. and Widom H. (1993). Hamiltonian structure of equations appearing in random matrices, in: Osborn H. (Ed.), *Low Dimensional Topology and Quantum Field Theory*, Plenum Press, New York, pp. 231–245.

Hartwig R.E. and Fisher M.E. (1969). Asymptotic behavior of Toeplitz matrices and determinants, *Arch. Rational Mech. Anal.* **32**, 190–225.

Harvey J.A. and Hughes D.J. (1958). Spacings of nuclear energy levels, *Phys. Rev.* **109**, 471–479.

Hastings S.P. and McLeod J.B. (1980). A boundary value problem associated with the second Painlevé transcendent and the Korteweg–de Vries equation, *Arch. Rational Mech. Anal.* **73**, 31–51.

Heijhal D.A. (1976). The Selberg trace formula and the Riemann zeta function, *Duke Math. J.* **43**, 441–482.

Hermann H., Derrida B. and Vannimenus J. (1984). Superconductivity exponents in two- and three-dimensional percolation, *Phys. Rev. B* **30**, 4080–4082.

Hsu P.L. (1939). On the distribution of roots of certain determinantal equations, *Ann. Eugenics* **9**, 250–258.

Hua L.K. (1963). *Harmonic Analysis of Functions of Several Complex Variables in the Classical Domains*, Amer. Math. Soc., Providence, RI, Chapter 3.4 (original in Chinese, translated first in Russian and then in English).

Hughes C.P., Keating J.P. and O'Connell N. (2000). Random matrix theory and the derivative of the Riemann zeta function, *Proc. Royal Soc. London A Math.* **456**, 2611–2627.

Ince E.L. (1956), *Ordinary Differential Equations*, Dover, New York.

Its A.R., Isergin A.G., Korepin V.E. and Slavnov N.A. (1990). Differential equations for quantum correlation functions, *Internat. J. Mod. Phys. B* **4**, 1003–1037.

Its A.R. and Novokshenov V.Yu. (1986). *The Isomonodromic Deformation Methods in the Theory of Painlevé Equations*, in: *Lecture Notes in Math.*, vol. 1191, Springer-Verlag, Berlin/New York.

Itzykson C. and Zuber J.B. (1980). The planar approximation II, *J. Math. Phys.* **21**, 411–421.

Itzykson C. and Zuber J.B. (1990). Matrix integration and combinatorics of modular groups, *Comm. Math. Phys.* **134**, 197–207.

Iwasaki K., Kimura K., Shimomura S. and Yoshida M. (1991). *From Gauss to Painlevé: A Modern Theory of Special Functions*, Vieweg, Braunschweig.

James A.T. (1964). Distribution of matrix variates and latent roots derived from normal samples, *Ann. Math. Statist.* **35**, 475–501.

Jancovici B. (1981). Exact results for the two-dimensional one-component plasma, *Phys. Rev. Lett.* **46**, 386–388.

Jancovici B. and Forrester P.J. (1994). Derivation of an asymptotic expression in Binnakker's general fluctuation formula for random matrix correlations near an edge, *Phys. Rev. B* **50**, 14599–14600.

Jimbo M. (1982). Monodromy problem and the boundary condition for some Painlevé equations, *Publ. Res. Inst. Math. Sci.* **18**, 1137–1161.

Jimbo M. and Miwa T. (1981). Monodromy preserving deformations of linear ordinary differential equations with rational coefficients, *Physica D* **2**, 407–448.

Jimbo M., Miwa T., Mori Y. and Sato M. (1979). Density matrix of impenetrable Bose gas and the fifth Painlevé transcendent, *Japan Acad. Ser. A Math. Sci.* **55**, 317–322; *Physica D* **1** (1980) 80–158.

Jimbo M., Miwa T. and Sato M. (1981). Holonomic quantum fields IV, *Publ. Res. Inst. Math. Sci.* **17**, 137–151.

Johansson K. (1988). On Szegö's asymptotic formula for Toeplitz determinants and generalizations, *Bull. Soc. Math. (2)* **112**, 257–304.

Johansson K. (1998). The longest increasing subsequence in a random permutation and a unitary matrix model, *Math. Res. Lett.* **5**, 63–82.

Joshi N. (1987). *The connection problem for the first and second Painlevé transcendents*, Thesis, Princeton, NJ (unpublished).

Joshi N. and Kruskal M.D. (1988). An asymptotic approach to the connection problem for the first and second Painlevé equations, *Phys. Lett. A* **130**, 129–137; The connection problem for Painlevé transcendents, *Physica D* **18** (1986) 215–216; The Painlevé connection problem: an asymptotic approach, *Stud. Appl. Math.* **86** (1992) 315–376; A direct proof that the six Painlevé equations have no movable singularities except poles, *Stud. Appl. Math.* **93** (1994) 187–207.

Jost R. (1947). Über die falschen Nullstellen der Eigenwerte der S-Matrix, *Helv. Phys. Acta* **20**, 256–266.

Kac M. (1954). Toeplitz matrices, translation kernels and a related problem in probability theory, *Duke Math. J.* **21**, 501–509.

Kahn P.B. (1963). Energy level spacing distributions, *Nucl. Phys.* **41**, 159–166.

Kamien R.D., Politzer H.D. and Wise M.B. (1988). Universality of random matrix predictions for the statistics of energy levels, *Phys. Rev. Lett.* **60**, 1995–1998.

Karliner M. and Migdal S. (1990). Nonperturbative 2D quantum gravity via supersymmetric string, *Mod. Phys. Lett. A* **5**, 2565–2572.

Karliner M., Migdal A. and Rusakov B. (1993). Ground state of 2D quantum gravity and spectral density of random matrices, *Nucl. Phys. B* **399**, 514–526.

Katz N.M. and Sarnak P. (1999). *Random Matrices, Frobenius Eigenvalues and Monodromy*, Amer. Math. Soc., Providence, RI.

Keating J. and Berry M. (1999). The Riemann zeros and eigenvalue asymptotics, *SIAM Rev.* **41**, 236–266.

Keating J. and Snaith N.C. (2000a). Random matrix theory and $\zeta(1/2+it)$, *Comm. Math. Phys.* **214**, 57–89.

Keating J. and Snaith N.C. (2000b). Random matrix theory and L-functions at $s = 1/2$, *Comm. Math. Phys.* **214**, 91–110.

Kerov S.V. and Vershik A.M. (1986). The characters of the infinite symmetric group and probability properties of the Robinson–Schensted–Knuth algorithm, *SIAM J. Alg. Discr. Math.* **7**, 116–124.

Khinchin A.I. (1957). *Mathematical Foundations of Information Theory*, Dover, New York.

Kingman J.F.C. (1973). Subadditive ergodic theory, *Ann. Probab.* **1**, 883–899.

Kingman J.F.C. (1990). Some random collections of finite subsets, in: Grimmett G.R. and Welsh D.J.A. (Eds.), *Disorder in Physical Systems*, Oxford Univ. Press, London, pp. 241–247.

Kirilov A.N., Lascoux A., Leclerc B. and Thibon J.Y. (1994). Séries génératrices pour les tableaux de dominos, *C. R. Acad. Sci. Paris Sér. I Math.* **318** (5), 395–400.

Kisslinger L.S. and Sorenson R.A. (1960). Pairing plus long range force for single closed shell nuclei, *Kgl. Danske Videnskab. Selskab. Mat.-Fys. Medd.* **32** (9).

Klarkson P.A. and McLeod J.B. (1992). Integral equations and connection formulae for the Painlevé equations, in: Levi D. and Winternitz P. (Eds.), *Painlevé Transcendents, Their Asymptotics and Physical Applications*, Plenum Press, New York, pp. 1–31.

Knuth D. (1973). *The Art of Computer Programming*, vol. 3, *Sorting and Searching*, second ed., Addison–Wesley, Reading, MA.

Korepin V.E., Bogolubov N.M. and Izergin A.G. (1993). *Quantum Inverse Scattering Method and Correlation Functions*, Cambridge Univ. Press, Cambridge, UK.

Kramer H.A. (1930). *Proc. Acad. Sci. Amsterdam* **33**, 959.

Krattenthaler C., Advanced determinant calculus, Preprint, http://radon.mat.univie.ac.at/People/kratt and private communication.

Kubo R. (1969). Electrons in small metallic particles, in: *Polarization, matière et rayonnement*, volume jubiliaire en l'honneur d'Alfred Kastler, Société Fançaise de Physique, Presses Univ. de France, pp. 325–339.

Kullback S. (1934). An application of characteristic functions to the distribution problem of statistics, *Ann. Math. Stat.* **5**, 263–307.

Landau L. and Smorodinski Ya. (1955). *Lektsii po teori atomnogo yadra*, Gos. Izd. Tex.-Teoreticheskoi Lit., Moscow, pp. 92–93.

Lang J.M.B. and Lecouteur K.J. (1954). Statistics of nuclear levels, *Proc. Phys. Soc. London A* **67**, 586–600.

Lebowitz J. and Martin Ph.A. (1984). On potential and field fluctuations in classical charged systems, *J. Stat. Phys.* **34**, 287–311.

Le Caër G. (1989). Do Swedish pines diagonalize complex random matrices, unpublished notes.

Le Caër G. and Delannay R. (1993). The administrative divisions of mainland France as 2-D random structures, *J. Phys. I (Phys. Statist.)* **3**, 1777–1800.

Le Caër G. and Delannay R. (2003). The distributions of the determinant of fixed trace ensembles of real symmetric and of Hermitian random matrices, *J. Phys. A* **36**, 9885–9898.

Le Caër G. and Ho J.S. (1990). Voronoi tessellations generated from eigenvalues of complex random matrices, *J. Phys. A* **23**, 3279–3295.

Leff H.S. (1963). *Statistical theory of energy level spacing distributions for complex spectra*, Thesis, State Univ. Iowa, SUI-63-23 (unpublished).

Leff H.S. (1964a). Systematic characterization of m-th order energy level spacing distributions, *J. Math. Phys.* **5**, 756–762.

Leff H.S. (1964b). Class of ensembles in the statistical theory of energy level spectra, *J. Math. Phys.* **5**, 763–768.

Levin E. and Lubinski D.S. (2001). *Orthogonal Polynomials for Exponential Weights*, in: *CMS Books in Mathematics*, vol. 4, Springer-Verlag, New York.

Levinson N. (1949). On the uniqueness of the potential in a Schrödinger equation for a given asymptotic phase, *Kgl. Danske Vidensk. Selsk. Mat.-Fys. Medd.* **25** (9), 1–29.

Lin R., Conte R. and Musette M. (2003). On the Lax Pair of the continuous and discrete sixth Painlevé equations, *J. Nonlinear Math. Phys.* **10** (Suppl. 2), 107–118.

Liou H.I., Camarda H.S. and Rahn F. (1972a). Applications of statistical tests for single level populations to neutron resonance spectroscopy data, *Phys. Rev. C* **5**, 1002–1015.

Liou H.I., Camarda H.S., Wynchank S., Slagowitz M., Hacken G., Rahn F. and Rainwater J. (1972b). Neutron resonance spectroscopy VIII. The separated isotopes of Erbium: Evidence for Dyson's theory concerning level spacings, *Phys. Rev. C* **5**, 974–1001.

Logan B.F. and Shepp L.A. (1977). A variational problem for random Young tableaux, *Adv. Math.* **26**, 206–222.

Lubinsky D.S. (1990). The approximate approach to orthogonal polynomials for weights on $(-\infty, +\infty)$, in: Nevai P. (Ed.), *Orthogonal Polynomials: Theory and Practice, NATO ASI Ser. C: Math. Phys. Sci.*, vol. 294, Kluwer Academic, Dordrecht.

Luke Y.L. (1969). *The Special Functions and Their Applications*, vol. 1, Academic Press, New York, Chapter 5; 6.5 (8).

Macdonald I.G. (1979). *Symmetric Functions and Hall Polynomials*, Clarendon Press, Oxford.

Macdonald I.G. (1982). Some conjectures for root systems, *SIAM J. Math. Anal.* **13**, 988–1004.

Mahoux G. and Mehta M.L. (1991). A method of integration over matrix variables IV, *J. Phys. I France* **1**, 1093–1108.

Mahoux G. and Mehta M.L. (1993). Level spacing functions and non linear differential equations, *J. Phys. I France* **3**, 697–715; 1507–1508.

Mahoux G., Mehta M.L. and Normand J.M. (1998). Matrices coupled in a chain: II. Spacing functions, *J. Phys. A* **31**, 4457–4464.

Mailly D., Sanquer M., Pichard J.L. and Pari P. (1989). Reduction of quantum noise in a Ga–Al–As/Ga–As heterojunction by a magnetic field: an orthogonal to unitary Wigner statistics transition, *EuroPhys. Lett.* **8**, 471–476.

Mallows C.L. (1973). Patience sorting, *Bull. Inst. Math. Appl.* **9**, 216–224.

Maradudin A.A., Mazur P., Montroll E.W. and Weiss G.H. (1958). Remarks on the vibrations of diatomic lattices, *Rev. Mod. Phys.* **30**, 175–196, Section IV, 186–195.

Marchenko V.A. (1950). Concerning the theory of a differential operator of the second order, *Dokl. Akad. Nauk SSSR* **72**, 457–460; The construction of the potential energy from the phases of the scattered waves, ibid **104** (1955) 695–698.

Marshakov A., Mironov A. and Morozov A. (1991). Generalized matrix models as conformal field theories: discrete case, *Phys. Lett. B* **265**, 99–107.

Mathai A.M. (1993). *A Handbook of Generalized Special Functions for Statistical and Physical Sciences*, Clarendon Press, Oxford.

Mayer M.G. and Jensen J.H.D. (1955). *Elementary Theory of Nuclear Shell Structure*, Wiley, New York.

McCoy B.M. and Tang S. (1986). Connection formula for Painlevé functions, *Physica D* **18**, 190–196; Connection formula for Painlevé V functions, *Physica D* **19** (1986) 42–72; Connection formula for Painlevé V functions II, The δ-function Bose gas problem, *Physica D* **20** (1986) 187–216.

McCoy B.M., Tracy C.A. and Wu T.T. (1977a). Painlevé functions of the third kind, *J. Math. Phys.* **18**, 1058–1092.

McCoy B.M., Tracy C.A. and Wu T.T. (1977b). Connection between the KdV equation and the two dimensional Ising model, *Phys. Lett. A* **61**, 283–284.

McCoy B.M. and Wu T.T. (1973). *The Two Dimensional Ising Model*, Harvard Univ. Press, Cambridge, MA.

McKay B.D. (1990). The asymptotic numbers of regular tournaments, Eulerian digraphs and Eulerian oriented graphs, *Combinatorica* **10**, 367–377.

McKay B.D. and Wormald N.C. (1990). Asymptotic enumeration by degree sequence of graphs of high degree, *European J. Combin.* **11**, 565–580.

Mehta M.L. (1960). On the statistical properties of the level spacings in nuclear spectra, *Nucl. Phys.* **18**, 395–419.

Mehta M.L. (1971). A note on correlations between eigenvalues of a random matrix, *Comm. Math. Phys.* **20**, 245–250.

Mehta M.L. (1974). Determinants of quaternion matrices, *J. Math. Phys. Sci.* **8**, 559–570.

Mehta M.L. (1976). A note on certain multiple integrals, *J. Math. Phys.* **17**, 2198–2202.

Mehta M.L. (1977). *Elements of Matrix Theory*, Hindustan Publishing Corporation, Delhi, India.

Mehta M.L. (1981). A method of integration over matrix variables, *Comm. Math. Phys.* **79**, 327–340.

Mehta M.L. (1986). Random matrices in nuclear physics and number theory, *Contemp. Math.* **50**, 295–309.

Mehta M.L. (1989MT). *Matrix Theory*, Les Editions de Physique, 91944 Les Ulis Cedex, France.
Mehta M.L. (1992a). Power series for level spacing functions of random matrix ensembles, *Z. Phys. B* **86**, 285–290.
Mehta M.L. (1992b). A non-linear differential equation and a Fredholm determinant, *J. Phys. I France* **2**, 1721–1729.
Mehta M.L. (1997). Random matrices and matrix models: JNU lectures, *Pramana* **48**, 7–48.
Mehta M.L. (2002). Zeros of some bi-orthogonal polynomials, *J. Phys. A* **35**, 517–525.
Mehta M.L. and des Cloizeaux J. (1972). The probabilities for several consecutive eigenvalues of a random matrix, *Indian J. Pure Appl. Math.* **3**, 329–351.
Mehta M.L. and Dyson F.J. (1963). Statistical theory of energy levels of complex systems V, *J. Math. Phys.* **4**, 713–719.
Mehta M.L. and Gaudin M. (1960). On the density of eigenvalues of a random matrix, *Nucl. Phys.* **18**, 420–427.
Mehta M.L. and Mahoux G. (1991). A method of integration over matrix variables III, *Indian J. Pure Appl. Math.* **22**, 531–546.
Mehta M.L. and Mahoux G. (1993). Level spacing functions and non linear differential equations, *J. Phys. I France* **3**, 697–715.
Mehta M.L. and Mehta G.C. (1975). Discrete Coulomb gas in one dimension: correlation functions, *J. Math. Phys.* **16**, 1256–1258.
Mehta M.L. and Normand J.M. (1998). Probability density of the determinant of a random Hermitian matrix, *J. Phys. A* **31**, 5377–5391.
Mehta M.L. and Normand J.M. (2001). Moments of the characteristic polynomial in the three ensembles of random matrices, *J. Phys. A* **34**, 4627–4639.
Mehta M.L. and Pandey A. (1983a). Spacing distributions for some Gaussian ensembles of Hermitian matrices, *J. Phys. A* **16**, L601–606.
Mehta M.L. and Pandey A. (1983b). On some Gaussian ensembles of Hermitian matrices, *J. Phys. A* **16**, 2655–2684.
Mehta M.L. and Pandey A. (1997). About the spacing functions of the three matrix ensembles, *J. Phys. A* **30**, 1243–1251.
Mehta M.L. and Rosenzweig N. (1968). Distribution laws for the roots of a random anti-symmetric Hermitian matrix, *Nucl. Phys. A* **109**, 449–456.
Mehta M.L. and Shukla P. (1994). Two coupled matrices: eigenvalue correlations and spacing functions, *J. Phys. A* **27**, 7793–7803.
Mehta M.L. and Srivastava P.K. (1966). Correlation functions for eigenvalues of real quaternion matrices, *J. Math. Phys.* **7**, 341–344.
Mehta M.L. and Wang R. (2000). Calculation of a certain determinant, *Comm. Math. Phys.* **214**, 227–232.
Mello P.A. (1988). Macroscopic approach to universal conductance fluctuations in disordered metals, *Phys. Rev. Lett.* **11**, 1089–1092.

Mello P.A., Akkermans E. and Shapiro B. (1988a). Macroscopic approach to correlations in the electronic transmission and reflection from disordered conductors, *Phys. Rev. Lett.* **61**, 459–462.

Mello P.A., Pereyra P. and Kumar N. (1988b). Macroscopic approach to multichannel disordered conductors, *Ann. Phys.* **181**, 290–317.

Mello P.A. and Pichard J.L. (1989). Maximum entropy approaches to quantum electronic transport, *Phys. Rev. B* **40**, 5276–5278.

Mezzadri F. (2003). Random matrix theory and the zeros of $\zeta'(s)$, *J. Phys. A* **36**, 2945–2962.

Mhaskar H.N. (1990). Bounds for certain Freud-type orthogonal polynomials, *J. Approx. Theory* **63**, 238–254.

Mon K.K. and French J.B. (1975). Statistical properties of many particle spectra, *Ann. Phys.* **95**, 90–111.

Monahan J.E. and Rosenzweig N. (1972). Analysis of the distributions of the spacings between nuclear energy levels II, *Phys. Rev. C* **5**, 1078–1083.

Montgomery H.L. (1973). The pair correlation of zeros of the zeta function, in: *Proc. Sympos. Pure Math.*, vol. 24, Amer. Math. Soc., Providence, RI, pp. 181–193; Distribution of the zeros of the Riemann zeta function, in: *Proc. Internat. Congr. Mathematicians*, vol. 1, Vancouver, BC (1974), pp. 379–381; Canad. Math. Congr., Montreal, Quebeque (1975).

Moore C.E. (1949). *Atomic energy levels*, NBS circular 467, Washington, DC, I, II (1952), III (1958).

Moore E.H. (1935). *General Analysis* I, in: *Mem. Amer. Math. Soc.*, vol. 1, Philadelphia.

Moore G. (1990). Matrix models of 2D gravity and isomonodromic deformations, *Prog. Theor. Phys.* **102** (Suppl.), 255–285.

Morris W.G. (1982). *Constant Term Identities for Finite and Affine Root Systems: Conjectures and Theorems*, Thesis, Madison, WI (unpublished).

Morse P.M. and Feshbach H. (1953). *Methods of Mathematical Physics*, McGraw Hill, New York, Chapter 2.4.

Moser J. (1980). Geometry of quadrics and spectral theory, in: *Chern Symposium*, Springer-Verlag, Berlin, Heidelberg, New York, pp. 147–188.

Muirhead R.J. (1982). *Aspects of Multivariate Statistical Theory*, Wiley, New York.

Mushkelishvili N.I. (1953). *Singular Integral Equations*, Groningen, Netherlands.

Muttalib K.A., Pichard J.L. and Stone A.D. (1987). Random matrix theory and universal statistics for disordered quantum conductors, *Phys. Rev. Lett.* **59**, 2475–2478.

Myers R.C. and Perival V. (1990). Exact solution of critical self-dual unitary matrix models, *Phys. Rev. Lett.* **65**, 1088–1091.

Nagao T. and Forrester P.J. (1995). Asymptotic correlations at the spectrum edge of random matrices, *Nucl. Phys. B* **435**, (FS) 401–420.

Nagao T. and Selvin K. (1993). Laguere ensembles of random matrices: nonuniversal correlation functions, *J. Math. Phys.* **34**, 2317–2330.

Nagao T. and Wadati M. (1991a). Correlation functions of random matrix ensembles related to classical orthogonal polynomials, *J. Phys. Soc. Japan* **60**, 3298–3322.

Nagao T. and Wadati M. (1991b). Thermodynamics of particle systems related to random matrices, *J. Phys. Soc. Japan* **60**, 1943–1951.

Nagao T. and Wadati M. (1992). Correlation functions of random matrix ensembles related to classical orthogonal polynomials II, *J. Phys. Soc. Japan* **61**, 78–88.

Nagao T. and Wadati M. (1993). Eigenvalue distribution of random matrices at the spectrum edge, *J. Phys. Soc. Japan* **62**, 3845–3856.

Neuberger H. (1990). Scaling regime at the large N phase transition of two dimensional pure gauge theories, *Nucl. Phys. B* **340**, 703–720.

Neuberger H. (1991). Regularized string and flow equations, *Nucl. Phys. B* **352**, 689–722.

Nevai P. (1979). *Orthogonal Polynomials*, in: *Mem. Amer. Math. Soc.*, Amer. Math. Soc., Providence, RI.

Nyquist H., Rice S.O. and Riordan J. (1954). The distribution of random determinants, *Quart. Appl. Math.* **12**, 97–104.

Odlyzko A.M. (1992). Explicit Tauberian estimates for functions with positive coefficients, *J. Comput. Appl. Math.* **41**, 187–197.

Odlyzko A.M. (1995). Asymptotic enumeration methods, in: Graham R.L., Gröthschel M. and Lovàsz L. (Eds.), *Handbook of Combinatorics*, vol. 2, North-Holland, Amsterdam, pp. 1063–1229.

Odlyzko A.M. (1987). On the distribution of spacings between zeros of the zeta function, *Math. Comput.* **48**, 273–308.

Odlyzko A.M. (1989). The 10^{20}-th zero of the Riemann zeta function and 70 million of its neighbours, AT & T Bell lab., Preprint.

Odlyzko A.M. (2001). The 10^{22}-nd zero of the Riemann zeta function, in: van Frankenhuysen M. and Lapidus M.L. (Eds.), *Dynamical, Spectral and Arithmetic Zeta Functions*, in: *Amer. Math. Soc. Contemp. Math. Ser.*, vol. 290, Amer. Math. Soc., Providence, RI, pp. 139–144.

Odlyzko A.M., Poonen B., Widom H. and Wilf H.S. (1993). On the distribution of longest increasing subsequences in random permutations (unpublished notes).

Odlyzko A.M. and Rains E.M. (2000). On longest increasing subsequences in random permutations, in: Grinberg E.L., Berhanu S., Knopp M., Mendoza G. and Quinto E.T., (Eds.), *Analysis, Geometry, Number theory: The mathematics of Leon Ehrenpreis*, in: *Contemp. Math. Ser.*, vol. 251, Amer. Math. Soc., Providence, RI, pp. 439–451.

Odlyzko A.M. and Schönhage A. (1988). Fast algorithms for multiple evaluations of the Riemann zeta function, *Trans. Amer. Math. Soc.* **309**, 797–809.

Okamoto K. (1986). Studies on the Painlevé equations III: second and fourth Painlevé equations P_{II} and P_{IV}, *Math. Ann.* **275**, 221–255.

Okamoto K. (1987a). Studies on the Painlevé equations II: fifth Painlevé equation P_V, *Japan J. Math.* **13**, 47–76.

Okamoto K. (1987b). Studies on the Painlevé equations IV: third Painlevé equation P_{III}, *Funcial. Ekvac.* **30**, 305–332.

Olson W.H. and Uppulury V.R.R. (1972). In: Le Cam L.M., Neyman J., and Scott E.L. (Eds.), *Probability Theory*, Proc. of the Sixth Berkeley Symposium on Mathematical Statistics and Probability, vol. 3, Univ. of California Press, Berkeley, p. 615.

Opdam E.M. (1989). Some applications of hyper-geometric shift operators, *Invent. Math.* **98**, 1–18, Section 6.5.

Opdam E.M. (1993). Dunkl operators, Bessel functions and the discriminant of a finite Coxeter group, *Comp. Math.* **85**, 333–373.

Painlevé P. (1897). *Leçons sur la théorie analytique des équations différentielles (leçons de Stockholm)*, Herman, Paris (1897); Reprinted in *Oeuvres de Paul Painlevé, 3 vols.*, vol. 1, Editions du CNRS, Paris (1973), (1974) and (1976).

Painlevé P. (1902). Sur les équations différentielles du second ordre et d'ordre supérieur dont l'intégral générale est uniforme, *Acta Math.* **25**, 1–85.

Painlevé P. (1906). Sur les équations différentielles du second ordre à points critiques fixes, *C. R. Acad. Sci. Paris* **143**, 1111–1117.

Pandey A. (1979). Statistical properties of many particle spectra III. Ergodic behaviour in random matrix ensembles, *Ann. Phys.* **119**, 170–191.

Pandey A. (1981). Statistical properties of many particle spectra IV. New ensembles by Stieltjes transform method, *Adv. Phys.* **134**, 110–127.

Pandey A. and Ghosh S. (2001). Skew orthogonal polynomials and universality of energy level correlations, *Phys. Rev. Lett.* **87** 024102-1-4.

Pandey A. and Mehta M.L. (1983). Gaussian ensembles of random Hermitian matrices intermediate between orthogonal and unitary ones, *Comm. Math. Phys.* **87**, 449–468.

Pastur L.A. (1972). On the spectrum of random matrices, *Theor. Math. Phys.* **10**, 67–74 (Russian original); English translation: *Teor. i Math. Fisika* **10** (1972) 102–112.

Pastur L. (1992). On the universality of the level spacing distribution for some ensembles of random matrices, *Lett. Math. Phys.* **25**, 259–265.

Pastur L. and Scherbina M. (1997). Universality of the local eigenvalue statistics for a class of unitary invariant random matrix ensembles, *J. Stat. Phys.* **86**, 109–147.

Pechukas P. (1983). Distribution of energy eigenvalues in the irregular spectrum, *Phys. Rev. Lett.* **51**, 943–946.

Penner R.C. (1986). The moduli space of a punctured surface and perturbation series, *Bull. Amer. Math. Soc.* **15**, 73–77.

Penner R.C. (1988). Perturbation series and the moduli space of Riemann surfaces, *J. Differential Geom.* **27**, 35–53.

Perival V. and Sheviz D. (1990a). Unitary matrix models as exactly solvable string theories, *Phys. Rev. Lett.* **64**, 1326–1329.

Perival V. and Sheviz D. (1990b). Exactly solvable unitary matrix models: multicritical potentials and correlations, *Nucl. Phys. B* **344**, 731–746.

Pipel S. (1990). Descending subsequences of random permutations, *J. Combin. Theory A* **53**, 96–116.

Politzer H.D. (1989). Random matrix description of the distribution of mesoscopic conductance, *Phys. Rev. B* **40**, 11917–11919.

Porter C.E. (Ed.) (1965). *Statistical Theories of Spectra: Fluctuations*, Academic Press, New York.

Porter C.E. and Rosenzweig N. (1960a). Statistical properties of atomic and nuclear spectra, *Ann. Acad. Sci. Fennicae, Ser. A, VI Physica* **44**, 1–66.

Porter C.E. and Rosenzweig N. (1960b). "Repulsion of energy levels" in complex atomic spectra, *Phys. Rev.* **120**, 1698–1714.

Porter C.E. and Thomas R.G. (1956). Fluctuations of nuclear reaction widths, *Phys. Rev.* **104**, 483–491.

Potter H.S.A. and Titchmarsh E.C. (1935). The zeros of Epstein zeta functions, *Proc. London Math. Soc. (2)* **39**, 372–384.

Rahn F., Camarda H.S., Hacken G., Havens Jr. W.W., Liou H.I., Rainwater J., Slagowitz M. and Wynchank S. (1972). Neutron resonance spectroscopy, X, *Phys. Rev. C* **6**, 1854–1869.

Rains E.M. (1995). *Topics in probability on compact Lie groups*, Ph.D. Thesis, Harvard Univ., Cambridge, MA (unpublished).

Rains E.M. (1998). Increasing subsequences and the classical groups, *Electron J. Combin.* **5**, R 12.

Regev A. (1981). Asymptotic values for degrees associated with strips of Young diagrams, *Adv. Math.* **41**, 115–136.

Riemann B. (1876). *Gesamelte Werke*, Teubner, Leipzig, reprinted by Dover, New York, 1973.

Robin L. (1959). *Fonctions sphériques de Legendre et fonctions sphéroidales*, vol. 3, Gauthier–Villars, Paris, p. 250, formula 255.

Rogers D.G. (1978). Ascending sequences in permutations, *Discrete Math.* **22**, 35–40.

Rosen J.L. (1959). *Neutron resonances in* U^{238}, Thesis, Columbia Univ., New York (unpublished).

Rosen J.L., Desjardins J.S., Rainwater J. and Havens Jr. W.W. (1960). Slow neutron resonance spectroscopy, *Phys. Rev.* **118**, 687–697.

Rosenzweig N. (1963). In: *Statistical Physics*, vol. 3, Brandeis Summer Institute, Benjamin, New York.

Rosenzweig N., Monahan J.E. and Mehta M.L. (1968). Perturbation of statistical properties of nuclear states and transitions by interactions that are odd under time reversal, *Nucl. Phys. A* **109**, 437–448.

Rubinstein M. (2001). Low lying zeros of L-functions and random matrix theory, *Duke Math. J.* **109**, 147–181.

Rudnick Z. and Sarnak P. (1996). Zeros of principle L-functions and random matrix theory, *Duke Math. J.* **81**, 269–322.

Sagan B. (1991). *The Symmetric Group: Representations, Combinatorial Algorithms and Symmetric Functions*, Wadsworth and Brooks/Cole, Pacific Grove, CA.

Sato M., Miwa T. and Jimbo M. (1979). Holonomic quantum fields, III, IV, *Publ. Res. Inst. Math. Sci.* **15**, 577–629, **15** (1979) 871–972.

Schensted C. (1961). Longest increasing and decreasing subsequences, *Canad. J. Math.* **13**, 179–191.

Schutzenberger M.P. (1963). Quelques remarques sur une construction de Schensted, *Math. Scand.* **12**, 117–128.

Scott J.M.C. (1954). Neutron widths and the density of the nuclear levels, *Phil. Mag.* **45**, 1322–1331.

Selberg A. (1944). Bemerkninger om et multiplet integral, *Norsk Matematisk Tidsskrift* **26**, 71–78.

Seligman T.H. and Nishioka H. (Eds.) (1986). *Quantum Chaos and Statistical Nuclear Physics*, Proc. 2nd Internat. Conf. on Quantum Chaos and 4th Internat. Colloq. on Statistical Nuclear Physics, Cuernavaca, Mexico, *Lecture Notes in Phys.*, vol. 263, Springer-Verlag, Berlin/New York.

Seligman T.H. and Verbaarschote J.J.M. (1985a). Quantum spectra of classically chaotic systems without time reversal invariance, *Phys. Lett. A* **108**, 183–187.

Seligman T.H. and Verbaarschote J.J.M. (1985b). Fluctuations of quantum spectra and their semi-classical limit in the transition between order and chaos, *J. Phys. A* **18**, 2227–2234.

Seligman T.H., Verbaarschote J.J.M. and Zirnbauer M.R. (1985). Spectral fluctuation properties of Hamiltonian systems: the transition between order and chaos, *J. Phys. A* **18**, 2751–2770.

Shannon C.E. (1948). The mathematical theory of communication, *Bell Syst. Tech. J.* **27**, 379–423, 623–656.

Shannon C.E. and Weaver W. (1962). *The Mathematical Theory of Communication*, Univ. of Illinois Press, Champaign. (This is a reprint of the Bell Syst. Tech. J. paper of Shannon cited above.)

Shiroishi M., Nagao T. and Wadati M. (1991). Level spacing distributions of random matrix ensembles, *J. Phys. Soc. Japan* **62**, 2248–2259.

Shohat J.A. and Tamarkin J.D. (1943). *The Problem of Moments*, Amer. Math. Soc., Providence, RI, p. 8.

Sieber M. and Steiner F. (1990). Classical and quantum mechanics of a strongly chaotic billiard system, *Physica D* **44**, 248–266.

Slepian D. (1965). Some asymptotic expansions for prolate spheroidal functions, *J. Math. Phys.* **44**, 99–140.

Slepian D. and Pollak H.O. (1961). Prolate spheroidal wave functions, Fourier analysis and uncertainty I, *Bell Systems Tech. J.* **40**, 43–64.

Smythe W.R. (1950). *Static and Dynamic Electricity*, McGraw Hill, New York, p. 104, problems 29 and 30.

Sommers H.J., Crisanti A., Sompolinski H. and Stein Y. (1988). The spectrum of large random asymmetric matrices, *Phys. Rev. Lett.* **60**, 1895–1898.

Soundararajan K. (1998). The horizontal distribution of zeros of $\zeta'(s)$, *Duke Math. J.* **91**, 33–59.

Springer M.D. (1979). *The Algebra of Random Variables*, Wiley, New York.

Staton D.W. and White D.E. (1985). A Schensted algorithm for the hook tableaux, *J. Combin. Theory A* **40**, 211–247.

Stieltjes T.J. (1914). Sur quelques théorèmes d'algèbre, in: *Oeuvres Complètes*, vol. 1, Noordhoff, Groningen, p. 440.

Stratton J.A., Morse P.M., Chu L.J., Little J.D.C. and Corbato F.J. (1956). *Spheroidal Wave Functions*, MIT Press, Cambridge, MA.

Sulemanov B.L. (1987). The relation between asymptotic properties of the second Painlevé equation in different directions towards infinity, *Differential Equations* **23**, 569–576.

Szegö G. (1939). *Orthogonal Polynomials*, Amer. Math. Soc., New York (1959), (1966), (1975), Sections 6.7 and 6.7.1, pp. 139 and 142.

't Hooft G. (1974). A planer diagram theory for strong interactions, *Nucl. Phys. B* **72**, 461–473.

Titchmarsh E.C. (1939). *Theory of Functions*, Oxford Univ. Press, London and New York, p. 186.

Titchmarsh E.C. (1951). *The Theory of the Riemann Zeta Function*, Clarendon Press, Oxford, Chapter 10; second ed. (1986), p. 212, Theorem 9.3.

Tracy C.A. and McCoy B.M. (1973). Neutron scattering and correlation functions of the Ising model near T_c, *Phys. Rev. Lett.* **31**, 1500–1504.

Tracy C.A. and Widom H. (1993a). Introduction to random matrices, in: Helminck G.F. (Ed.), *Geometric and Quantum Aspects of Integrable Systems, Lecture Notes in Phys.*, vol. 424, Springer, Berlin, Heidelberg, New York, pp. 407–424.

Tracy C.A. and Widom H. (1993b). Level spacing distributions and the Airy kernel, *Phys. Lett. B* **305**, 115–118.

Tracy C.A. and Widom H. (1994a). Level spacing distributions and the Bessel kernel, *Comm. Math. Phys.* **161**, 289–309.

Tracy C.A. and Widom H. (1994b). Level spacing distributions and the Airy kernel, *Comm. Math. Phys.* **159**, 151–174.

Tracy C.A. and Widom H. (1994c). Fredholm determinants, differential equations, and matrix models, *Comm. Math. Phys.* **163**, 33–72.

Tracy C.A. and Widom H. (1996). On orthogonal and symplectic matrix ensembles, *Comm. Math. Phys.* **177**, 727–754.

Tracy C.A. and Widom H. (1998). Correlation functions, cluster functions, and spacing distributions for random matrices, *J. Stat. Phys.* **92**, 809–835.

Tricomi F.G. (1949). Sul comportamento asintotico dei polinomi de Laguerre, *Ann. Mat. Pura Appl.* **28**, 263–289.

Tutte W. (1962). A Census of planar triangulations, *Canad. J. Math.* **14**, 21–38; A Census of Hamiltonian polygons, *Canad. J. Math.* **14** (1962) 402–417; A Census of slicings, *Canad. J. Math.* **14** (1962) 708–722; A Census of Planar Maps, *Canad. J. Math.* **15** (1963) 249–271.

Uhlenbeck G.E. and Ornstein L.S. (1930). On the theory of the Brownian motion, *Phys. Rev.* **36**, 823–841.

Ulam S.M. (1961). Monte Carlo calculations in problems of mathematical physics, in: Beckenbach E.F. (Ed.), *Modern Mathematics for the Engineer*, McGraw Hill, New York.

Ullah N. (1964). Invariance hypothesis and higher order correlations of Hamiltonian matrix elements, *Nucl. Phys.* **58**, 65–71.

Ullah N. (1966). Asymptotic solution for the Brownian motion of the eigenvalues of a random matrix, *Nucl. Phys.* **78**, 557–560.

Ullah N. (1986). Ensemble average of an arbitrary number of pairs of different eigenvalues using Grassmann integration, *Comm. Math. Phys.* **104**, 693–695.

Ullah N. and Porter C.E. (1963). Invariance hypothesis and Hamiltonian matrix elements correlations, *Phys. Lett.* **6**, 301–302.

Van Buren A.L., *A Fortran computer program for calculating the linear prolate functions*, Report 7994, Naval Research Lab., Washington, DC, May 1976.

Vershik A.M. and Kerov C.V. (1977). Asymptotics of the Plancheral measure of the symmetric group and a limiting form for Young tableaux, *Dokl. Akad. Nauk USSR* **233**, 1024–1027.

Vo-Dai T. and Derome J.R. (1975). Correlations between eigenvalues of random matrices, *Nuovo Cimento B* **30**, 239–253.

Waldschmidt M., Moussa P., Luck J.-M. and Itzykson C. (1990). *From Number Theory to Physics*, Springer-Verlag, Berlin.

Wang M.C. and Uhlenbeck G.E. (1945). On the theory of the Brownian motion II, *Rev. Mod. Phys.* **17**, 323–342.

Wax N. (Ed.) (1954). *Selected Topics in Noise and Stochastic Processes*, Dover, New York.

Weaver R.L. (1989). Spectral statistics in elastodynamics, *J. Acoust. Soc. Amer.* **85**, 1005–1013.

Weyl H. (1946). *Classical Groups*, Princeton Univ. Press, Princeton, NJ.

White D.E. (1983). A bijection proving orthogonality of the characters of S_n, *Adv. Math.* **50**, 160–186.

Widder D.V. (1971). *Transform Theory*, Academic Press, New York, 5.7, Corollary 7.3a.

Widom H. (1964). Asymptotic behaviour of the eigenvalues of certain integral equations II, *Arch. Rational Mech. Anal.* **17**, 215–229.

Widom H. (1971). Strong Szegö limit theorem on circular arcs, *Indiana Univ. Math. J.* **21** 277–283.

Widom H. (1973). Toeplitz determinants with singular generating functions, *Amer. J. Math.* **95**, 333–383.

Widom H. (1994). The asymptotics of a continuous analogue of orthogonal polynomials, *J. Approx. Theory* **76**, 51–64.

Wigner E.P. (1951). On the statistical distribution of the widths and spacings of nuclear resonance levels, *Proc. Cambridge Phil. Soc.* **47**, 790–798.

Wigner E.P. (1955). Characteristic vectors of bordered matrices with infinite dimensions I and II, *Ann. Math.* **62**, 548–564; **65** (1957) 203–207.

Wigner E.P. (1957a). Results and theory of resonance absorption, in: *Gatlinberg Conf. on Neutron Phys. by Time of Flight, 1956*, Oak Ridge Natl. Lab. Rept. ORNL-2309, pp. 59–70.

Wigner E.P. (1957b). Statistical properties of real symmetric matrices with many dimensions, in: *Canadian Mathematical Congress Proceedings*, Univ. of Toronto Press, Toronto, Canada, pp. 174–184. Reproduced in: Porter C.E. (Ed.), *Statistical Theories of Spectra: Fluctuations*, Academic Press, New York (1965).

Wigner E.P. (1959). *Group Theory*, Academic Press, New York, Chapter 26.

Wigner E.P. (1965a). Distribution laws for the roots of a random Hermitian matrix, in: Porter C.E. (Ed.), *Statistical Theories of Spectra: Fluctuations*, Academic Press, New York, pp. 446–461.

Wigner E.P. (1965b). Statistical properties of real symmetric matrices with many dimensions, in: Porter C.E. (Ed.), *Statistical Theories of Spectra: Fluctuations*, Academic Press, New York, pp. 188–198.

Wigner E.P. (1968). On the distribution of the roots of certain symmetric matrices, *Ann. Math.* **67**, 325–327.

Wilf H.S. (1962). *Mathematics for the Physical Sciences*, Wiley, New York.

Wilson K.G. (1962). Proof of a conjecture by Dyson, *J. Math. Phys.* **3**, 1040–1043.

Wishart J. (1928). The generalized product moment distribution in samples from a normal multivariate population, *Biometrika A* **20**, 32–43.

Wu T.T., McCoy B.M., Tracy C.A. and Barouch E. (1976). Spin–spin correlation functions of the two dimensional Ising model: Exact theory in the scaling region, *Phys. Rev. B* **13**, 316–374.

Xu Y. (2002). Bi-orthogonal polynomials and positive weight functions, *J. Phys. A* **35**, 4499–4510.

Zano N. and Pichard J.L. (1988). Random matrix theory and universal statistics for disordered quantum conductors with spin dependent hopping, *J. Phys.* **49**, 907–920.

AUTHOR INDEX

U

V

W

Z

SUBJECT INDEX

V

W

Y